Autodesk 3ds Max 2022 for Beginners

A Tutorial Approach

(22nd Edition)

CADCIM Technologies
525 St. Andrews Drive
Schererville, IN 46375, USA
(www.cadcim.com)

Contributing Authors

Sham Tickoo
Professor
Purdue University Northwest
Hammond, Indiana, USA

Mable Thomas
CADCIM Technologies
USA

CADCIM Technologies

Autodesk 3ds Max 2022 for Beginners: A Tutorial Approach, 22nd Edition
Sham Tickoo

CADCIM Technologies
525 St Andrews Drive
Schererville, Indiana 46375, USA
www.cadcim.com

ISBN 978-1-64057-134-1

Copy Editor
Anju Jethwani

Technical Editor
Arti Deshpande

www.cadcim.com

DEDICATION

To teachers, who make it possible to disseminate knowledge to enlighten the young and curious minds of our future generations

To students, who are dedicated to learning new technologies and making the world a better place to live in

THANKS

To employees of CADCIM Technologies and Tickoo Institute of Emerging Technologies (TIET) for their valuable help

Online Training Program Offered by CADCIM Technologies

CADCIM Technologies provides effective and affordable virtual online training on various software packages including Computer Aided Design, Manufacturing and Engineering (CAD/CAM/CAE), computer programming languages, animation, architecture, and GIS. The training is delivered 'live' via Internet at any time, any place, and at any pace to individuals as well as the students of colleges, universities, and CAD/CAM/CAE training centers. The main features of this program are:

Training for Students and Companies in a Classroom Setting

Highly experienced instructors and qualified engineers at CADCIM Technologies conduct the classes under the guidance of Prof. Sham Tickoo of Purdue University Northwest, USA. This team has authored several textbooks that are rated "one of the best" in their categories and are used in various colleges, universities, and training centers in North America, Europe, and in other parts of the world.

Training for Individuals

CADCIM Technologies with its cost effective and time saving initiative strives to deliver the training in the comfort of your home or work place, thereby relieving you from the hassles of traveling to training centers.

Training Offered on Software Packages

CADCIM provides basic and advanced training on the following software packages:

***CAD/CAM/CAE**: CATIA, Pro/ENGINEER Wildfire, PTC Creo Parametric, Creo Direct, SOLIDWORKS, Autodesk Inventor, Solid Edge, NX, AutoCAD, AutoCAD LT, AutoCAD Plant 3D, Customizing AutoCAD, EdgeCAM, and ANSYS*

***Architecture and GIS**: Autodesk Revit (Architecture, Structure, MEP), AutoCAD MAP 3D, AutoCAD Civil 3D, Navisworks, Primavera, and Bentley STAAD Pro*

***Animation and Styling**: Autodesk 3ds Max, Autodesk Maya, Autodesk Alias, The Foundry NukeX, MAXON CINEMA 4D, and Adobe Premiere*

***Computer Programming**: C++, VB.NET, Oracle, AJAX, and Java*

For more information, please visit the following link: ***https://www.cadcim.com***

Note

If you are a faculty member, you can register by clicking on the following link to access the teaching resources: ***https://cadcim.com/Registration.aspx***. The student resources are available at ***www.cadcim.com***. We also provide **Live Virtual Online Training** on various software packages. For more information, write us at ***sales@cadcim.com***.

Table of Contents

Preface

Autodesk 3ds Max 2022

Welcome to the world of Autodesk 3ds Max, a 3D modeling, animation, and rendering software package developed by Autodesk Inc. It is widely used by architects, game developers, design visualization specialists, and visual effects artists. A wide range of modeling and texturing tools make it an ideal platform for 3D modelers and animators. The intuitive user interface and workflow tools of Autodesk 3ds Max have made the job of design visualization specialists easier.

Autodesk 3ds Max 2022 for Beginners: A Tutorial Approach is a tutorial-based textbook that introduces the readers to the features of 3ds Max 2022 such as modeling, texturing, lighting, Animation, and Arnold rendering in an effective and simple manner. In this edition, the readers will be able to learn about the Smart Extrude concept introduced in 3ds Max 2022. This textbook will help readers unleash their creativity and help them create simple 3D models and animations. The textbook will help the learners transform their imagination into reality with ease.

The main features of this textbook are as follows:

- **Tutorial Approach**
  The author has adopted the tutorial point-of-view and the learn-by-doing approach throughout the textbook. This approach helps the users create 3D models and animations in the tutorials with ease.

- **Projects based on Real-World Models**
  The author has used 5 projects based on real-world models that allow the users to apply the skills learned in the text. In addition, there are 21 exercises that can be used by the readers to assess their knowledge.

- **Notes and Tips**
  Additional information related to various topics is provided to the users in the form of notes and tips.

- **Learning Objectives**
  The first page of every chapter summarizes the topics that are covered in the chapter. This helps the users to easily refer to a topic.

- **Self-Evaluation Test, Review Questions, and Exercises**
  Every chapter ends with a Self-Evaluation Test so that the users can assess their knowledge of the chapter. The answers to Self-Evaluation Test are given at the end of the chapter. Also, Review Questions and Exercises are given at the end of the chapters and they can be used by the instructors as test questions and exercises.

Symbols Used in the Textbook

Note
The author has provided additional information to the users about the topic being discussed in the form of notes.

Tip
Special information and techniques are provided in the form of tips that helps in increasing the efficiency of the users.

The author has provided this symbol next to the new tutorials added in this edition of the textbook.

The author has provided this symbol next to the topics that are enhanced.

Formatting Conventions Used in the Textbook

Refer to the following list for the formatting conventions used in this textbook.

• Names of tools, buttons, options, and menus are written in boldface.	Example: The **Select and Move** tool, the **Geometry** button, the **Multiply** option, the **Create** menu, and so on.
• Names of dialog boxes, drop-down lists, windows, radio buttons, spinners, areas, and check boxes are written in boldface.	Example: The **Save** dialog box, the **View** drop-down list, the **Frames** radio button, the **Material Editor** dialog box, the **Track View - Dope Sheet** window, the **Size** spinner, the **Mapping** area, the **Save File** check box, and so on.
• Values entered in spinners are written in boldface.	Example: Set the value **0.5** in the **Amount** spinner.
• The path used for accessing a tool from the menu bar is written in boldface.	Example: Choose **Tools > Align > Normal Align** from the menu bar.
• Names of the files are italicized	Example: *c05_tut1_start.max*

Naming Conventions Used in the Textbook

The naming conventions used in this textbook are as follows:

Tool

If on selecting an item either from the **Command Panel** or from the **Main Toolbar**, a command is invoked to create/edit an object or perform some action, then that item is termed as **Tool**. For example: **Select and Rotate** tool, **Render Setup** tool, **Align** tool, **Mirror** tool, and so on.

Flyout

If on invoking a tool, a menu is displayed with options that have similar functions, then that menu is called a flyout.

Quad Menu

The quad menus provide quick access to the commonly used commands that are related to the current selection of an object. When you right-click on an object, a quad menu is displayed, as shown in Figure 1. Some of the options in the quad menus have an arrow on their right side. If you move the cursor on such options, a cascading menu will be displayed showing some more options related to the selected option, refer to Figure 1.

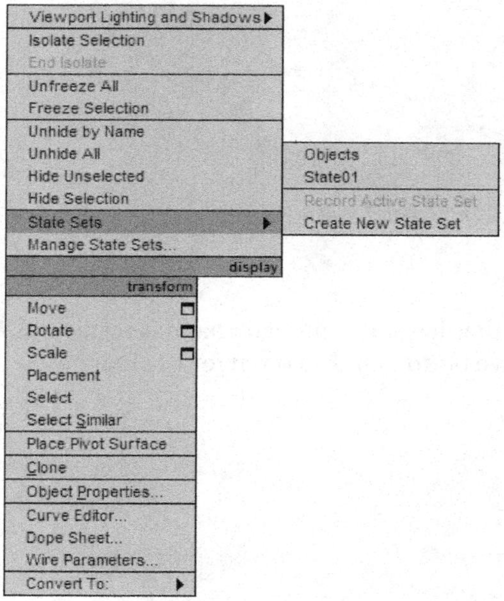

***Figure 1** Quad menu displayed on right-clicking in the viewport*

Dialog Box

In a dialog box, different terms are used to refer to its components. Refer to Figure 2 for the terminologies used.

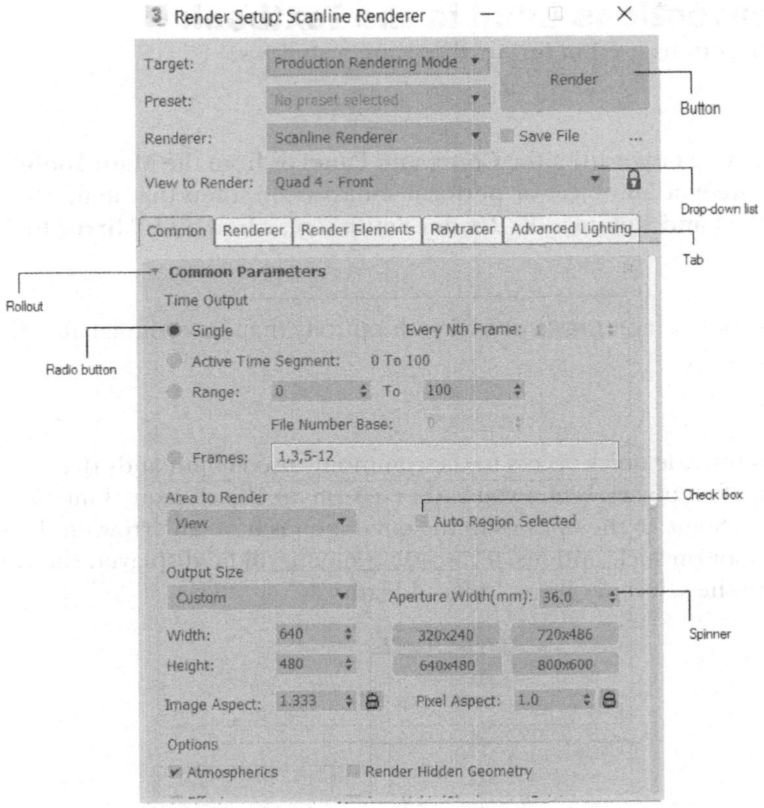

Figure 2 Different terms used for the options in a dialog box

Button

The item in a dialog box that has a rectangular shape is termed as **Button**. For example, **OK** button, **Cancel** button, **Save** button, and so on, refer to Figure 3.

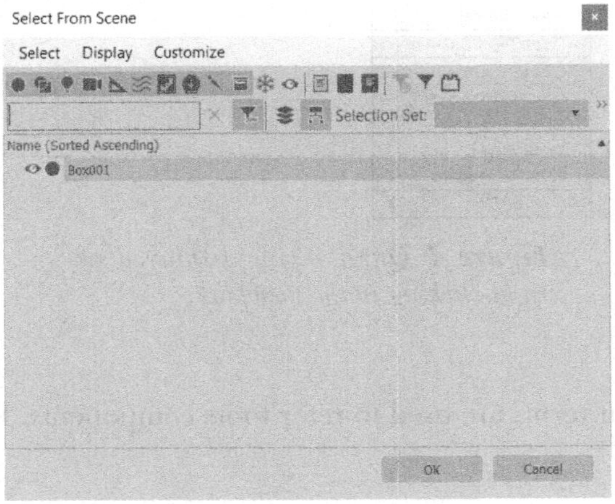

Figure 3 Buttons in the dialog box

Scene Explorer

The **Scene Explorer** is used to view, select, filter, and sort objects. It is also used to rename, delete, group, freeze, and hide the objects. The **Scene Explorer** is docked by default at the left of the interface in the default workspace, refer to Figure 4.

Figure 4** The **Scene Explorer

Drop-down List

A drop-down list is the one in which a set of options are grouped together. You can set parameters using these options. You can identify a drop-down list with a down arrow on it. For example, a drop-down list displayed in the **Output Size** area, refer to Figure 5.

Options

Options are the items that are available in shortcut menus, drop-down lists, dialog boxes, and so on, refer to Figure 6.

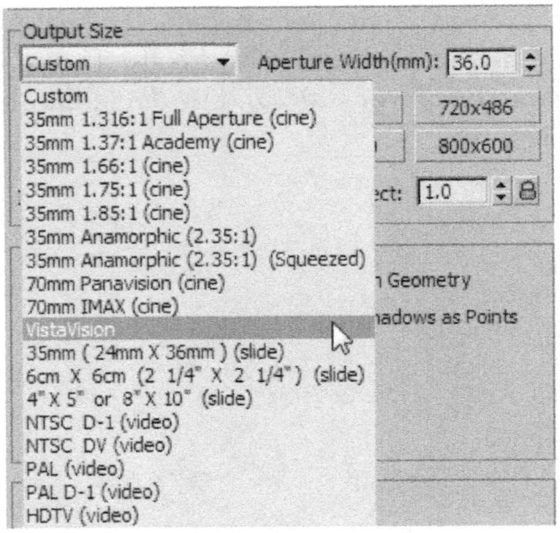

Figure 5 Selecting an option from the drop-down list

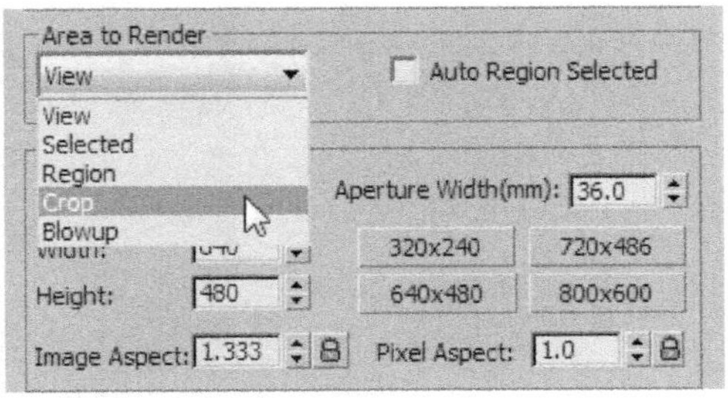

Figure 6 Options displayed in the drop-down list

Window

A window consists of various components such as tools, buttons, main menu, and so on. Different types of windows are available in Autodesk 3ds Max. Figure 7 shows the **Track View - Dope Sheet** window.

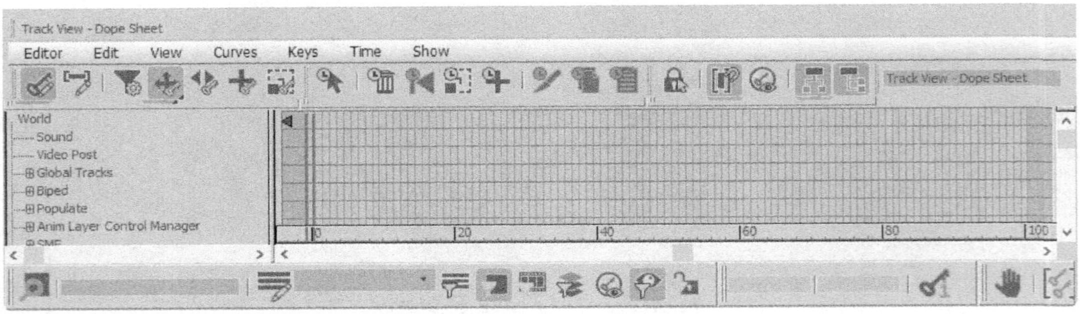

*Figure 7 The **Track View - Dope Sheet** window*

Free Companion Website

It has been our constant endeavor to provide you the best textbooks and services at affordable price. In this endeavor, we have come out with a Free Companion website that will facilitate the process of teaching and learning of Autodesk 3ds Max. If you purchase this textbook, you will get access to the files on the Companion website.

To access the files, you need to visit the **Resources** section of the CADCIM website:

Faculty Resources

- **Technical Support**
  You can get online technical support by contacting *techsupport@cadcim.com*.

- **Instructor Guide**
  Solutions to all review questions and exercises in the textbook are provided in this guide to help the faculty members test the skills of the students.

- **3ds Max Files**
  The 3ds Max file used in tutorials and exercises are available for free download.

- **Rendered Images and Media Files**
  The media files used in the tutorials and rendered images of all tutorials are provided in the CADCIM website. You can use these images to compare them with your rendered images.

- **Additional Resources**
  You can access additional learning resources by visiting: *https://3dsmaxexperts.blogspot.com*

- **Colored Images**
  You can download the PDF file containing colored images of the screenshots used in this textbook from CADCIM website.

Student Resources

- **Technical Support**
  You can get online technical support by contacting *techsupport@cadcim.com.*

- **3ds Max Files**
  The 3ds Max files used in tutorials are available for free download.

- **Rendered Images and Media Files**
  The media files used in the tutorials and rendered images of all tutorials are provided in the CADCIM website. You can use these images to compare them with your rendered images.

- **Additional Resources**
  You can access additional learning resources by visiting: *https://3dsmaxexperts.blogspot.com*

- **Colored Images**
  You can download the PDF file containing colored images of the screenshots used in this textbook from CADCIM website.

- **Additional Project**
  Apart from tutorials, one student project has also been added in this textbook for the students to practice the tools learned.

If you face any problem in accessing these files, please contact the publisher at *sales@cadcim.com* or the author at *stickoo@pnw.edu* or *tickoo525@gmail.com*.

Video Courses

CADCIM offers video courses in CAD, CAE Simulation, BIM, Civil/GIS, and Animation domains on various e-Learning/Video platforms. To enroll for the video courses, please visit the CADCIM website using the following link: https://www.cadcim.com/video-courses

Stay Connected

You can now stay connected with us through Facebook and Twitter to get the latest information about our textbooks, videos, and teaching/learning resources. To stay informed of such updates, follow us on Facebook ***(www.facebook.com/cadcim)*** and Twitter (***@cadcimtech***). You can also subscribe to our YouTube channel ***(www.youtube.com/cadcimtech)*** to get the information about our latest video tutorials.

1

Introduction to Autodesk 3ds Max 2022

Learning Objectives

After completing this chapter, you will be able to:

- *Understand the Autodesk 3ds Max interface components*
- *Use controls for creating or modifying objects*
- *Use and customize hotkeys in Autodesk 3ds Max*
- *Customize the colors of the scene elements*

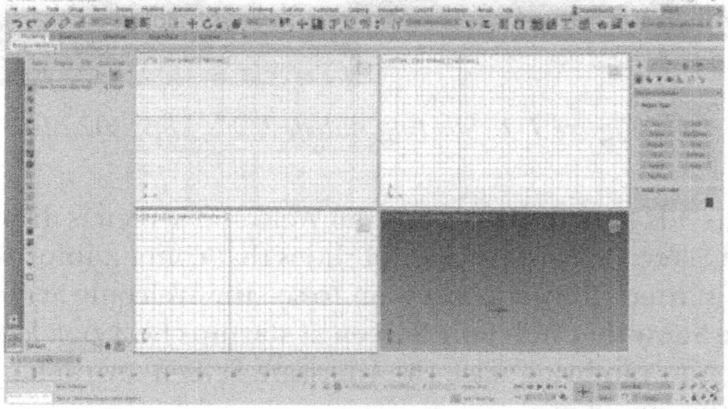

INTRODUCTION TO Autodesk 3ds Max 2022

Welcome to the world of Autodesk 3ds Max, an advanced application that is used to create still or animated 3D models and objects. With the help of this application, you can create realistic scenes by modifying objects, applying maps and materials to a scene, assigning environment to a scene, adding lights and cameras, and so on. Before working with Autodesk 3ds Max, you should have the basic knowledge of various tools and commands available in this software. In this chapter, you will learn the basic features of Autodesk 3ds Max.

GETTING STARTED WITH Autodesk 3ds Max

First, you need to install Autodesk 3ds Max 2022 on your computer. On installing the software, the **3ds Max 2022** shortcut icon will be created automatically on the desktop. Double-click on this icon to start Autodesk 3ds Max. Alternatively, you can start Autodesk 3ds Max from the Start menu. To do so, click the **Start** button from the taskbar to display the Start menu and then choose **Autodesk > 3ds Max 2022**, refer to Figure 1-1.

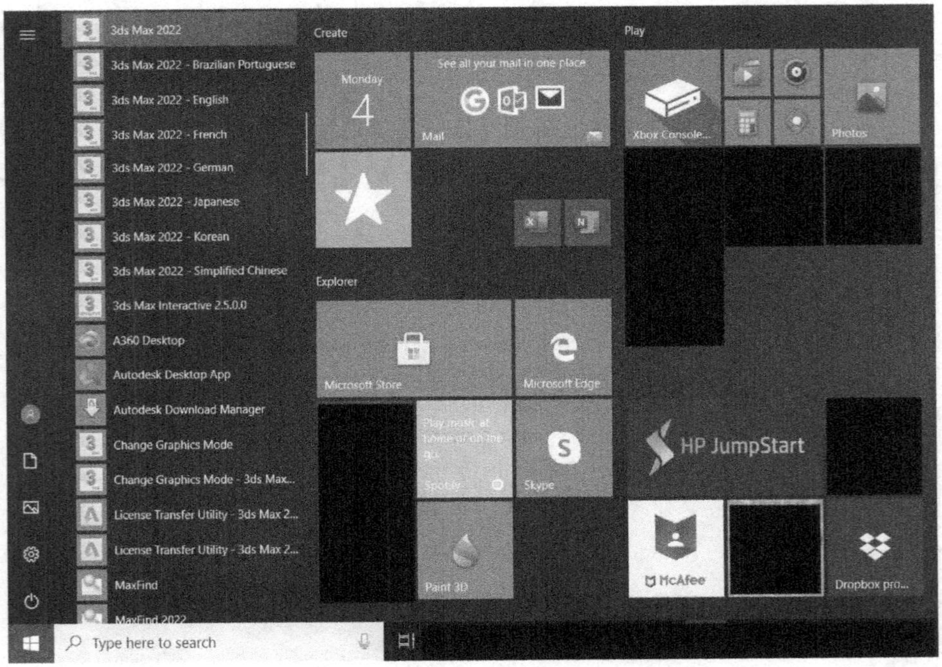

Figure 1-1 *Starting Autodesk 3ds Max 2022 from the taskbar*

When you start 3ds Max the first time, the Welcome Screen is displayed, as shown in Figure 1-2. The Welcome Screen consists of a set of slides that contain information for new users to inspire and get them started. If you do not want to see the Welcome Screen the next time you start 3ds Max, clear the **Show this Welcome Screen at startup** check box located in the bottom-left corner of the screen. You can bring back the Welcome Screen anytime by choosing **Help > Welcome Screen** from the menubar.

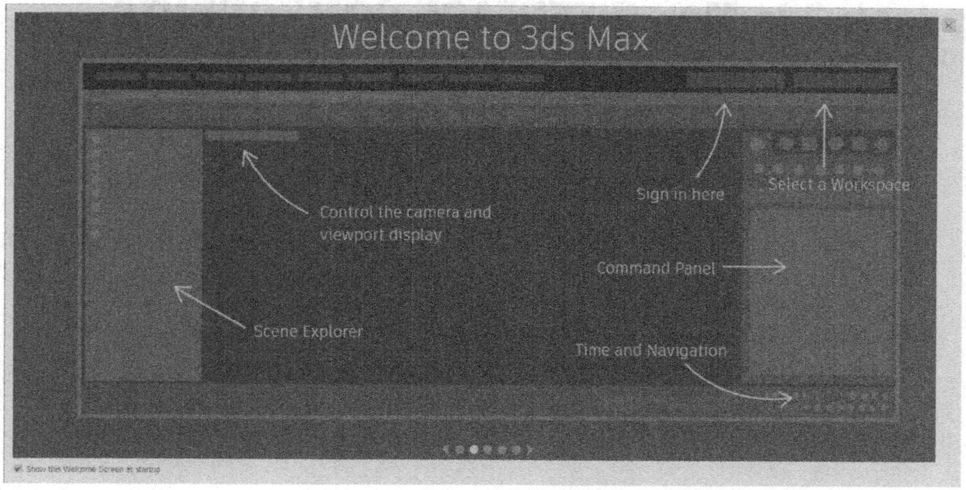

Figure 1-2 The Welcome Screen

STARTING A NEW FILE IN Autodesk 3ds Max

To start a new file in Autodesk 3ds Max, choose the **File > New** from the menu bar; a new file will be displayed in the 3ds Max interface. The new file will clear all the contents of the current file. Alternatively, press the CTRL+N keys; the **New Scene** dialog box will be displayed, as shown in Figure 1-3. By default, the **New All** radio button is selected in this dialog box. Choose the **OK** button; a new file will be displayed.

You can also reuse the objects from the current scene in the new scene. Select the **Keep Objects** radio button in the **New Scene** dialog box to keep only the objects from the current scene for the new file. However, on selecting this radio button, all the animation keys and links between the objects will be cleared. To keep the objects and the links between them, select the **Keep Objects and Hierarchy** radio button. However, in this case, the animation keys will be deleted.

Before starting a new scene in Autodesk 3ds Max, it is recommended to reset Autodesk 3ds Max and start afresh. By doing so, you will be able to reset all settings for the new scene. To reset Autodesk 3ds Max, choose **Reset** from the **File** menu; the **3ds Max** message box will be displayed, as shown in Figure 1-4. The message box will ask if you really want to reset 3ds Max. Choose the **Yes** button; the 3ds Max will be reset.

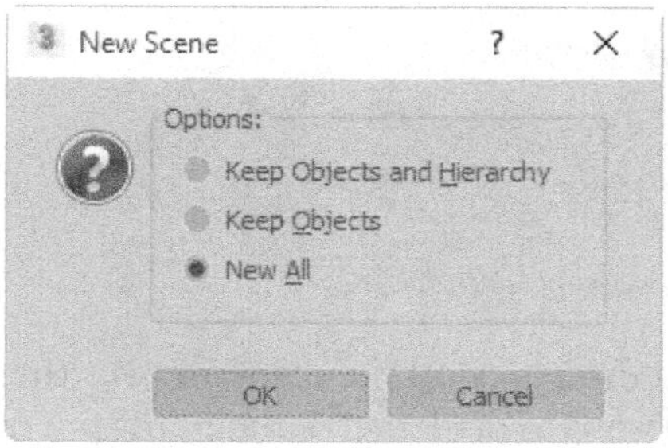

*Figure 1-3 The **New Scene** dialog box*

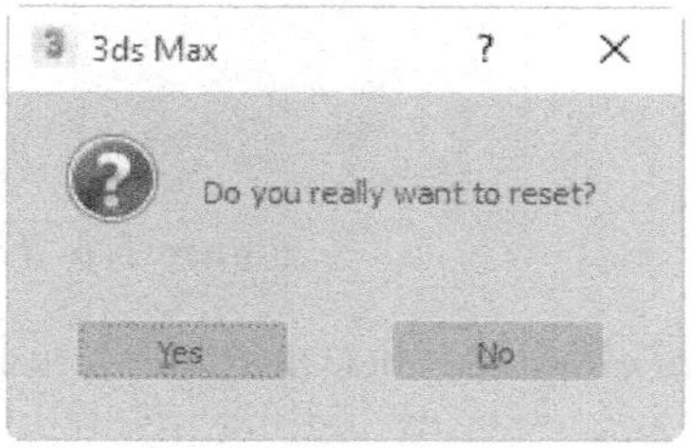

*Figure 1-4 The **3ds Max** message box*

Autodesk 3ds Max INTERFACE COMPONENTS

The 3ds Max interface consists of different components, as shown in Figure 1-5.

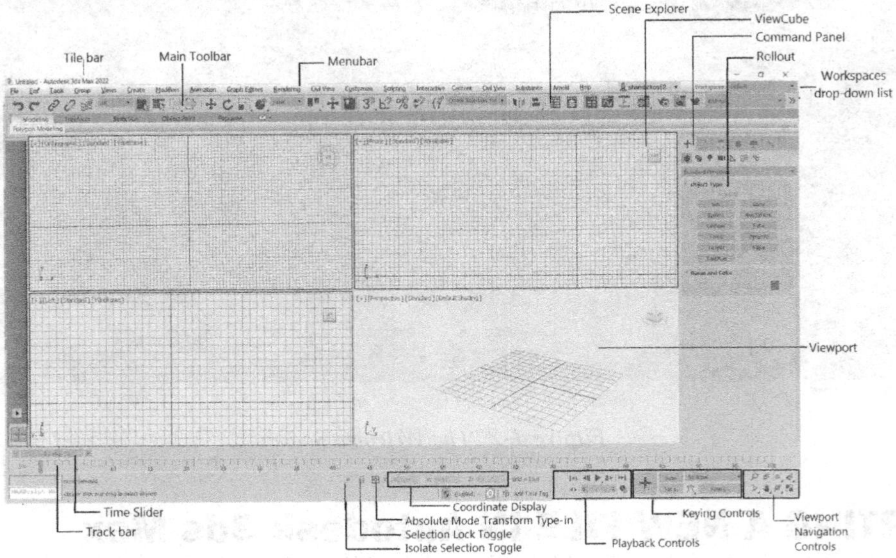

Figure 1-5 Different screen components of Autodesk 3ds Max interface

You can customize the interface in a variety of ways by adding toolbars, moving toolbars and Command Panel, and so on. The interface in 3ds Max is high DPI aware that ensures that the interface scales correctly to the latest high DPI displays.

You can easily customize the workspace by floating and docking elements of a scene such as panels, windows, menu, and so on. You can dock or float any element that has the handle. A handle is denoted by double dotted line. A handle can be on the top or left of the element, refer to Figure 1-6.

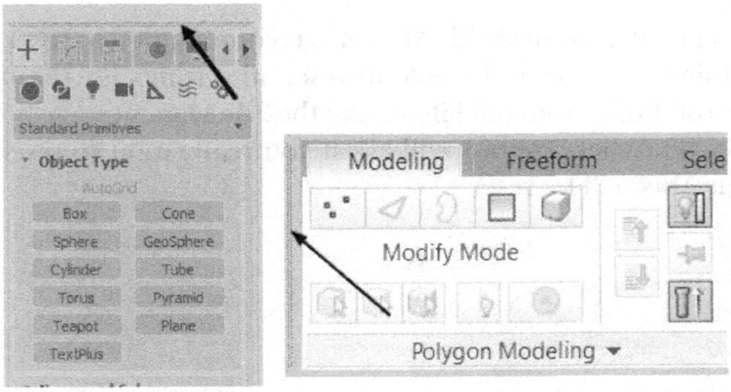

Figure 1-6 The handles marked with arrows

To float an element, click-drag the handle. As you drag around the element over the interface, valid docking areas are highlighted in blue. If you want to dock the element, drop it on the blue highlighted area. Keep in mind that the toolbars can only be docked on the outer edge of the interface. When you move around the elements over the interface, not all elements will resize automatically. Sometimes, manual adjustment might be required.

The 3ds Max interface components are discussed next.

Menu Bar

The menu bar is located just below the title bar, refer to Figure 1-5 and contains various pull-down menus. Some of the pull-down menus are standard window menus such as **File**, **Edit**, **Help**, and so on while others are 3ds Max pull-down menus such as **Create**, **Modifiers**, **Animation**, **Graph Editors**, **Rendering**, **Customize**, and so on. The title of each pull-down menu indicates the purpose of commands in the menu. When you choose one of the menu titles, Autodesk 3ds Max displays the corresponding pull-down menu. Each menu consists of a collection of commands. In a pull-down menu, the dots after a command indicate that a dialog box will be displayed on choosing that command. An arrow next to a command indicates that a cascading menu will be displayed on placing the cursor on that command. For some of the commands in the pull-down menus, the keyboard shortcuts are displayed on their right side, as shown in Figure 1-7.

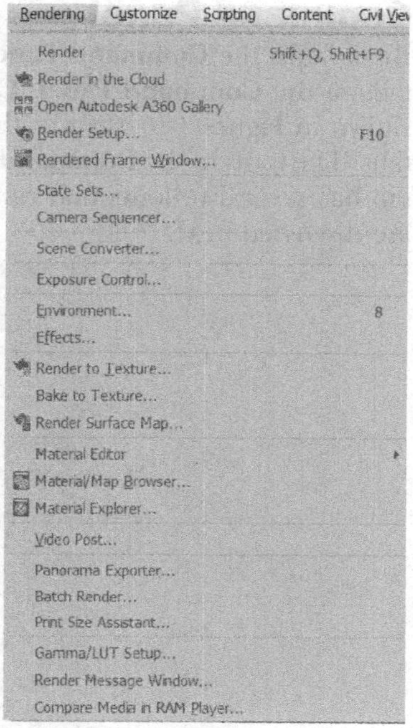

***Figure 1-7** The keyboard shortcuts in the **Rendering** pull-down menu*

Workspaces

The workspace includes toolbars, menus, the ribbon, hotkeys, quad menus, and viewport layout presets. You can switch between different workspaces by selecting the required option from the **Workspaces** drop-down list located on the top-right corner of the interface, refer to Figure 1-5. To create a new workspace, you need to change the interface setup as required and then choose the **Manage Workspaces** option from the **Workspaces** drop-down list; the **Manage Workspaces** dialog box will be displayed. In this dialog box, choose the **Save as New Workspace** button; the **Create New Workspace** dialog box will be displayed. Enter the name for the workspace in the **Name** text box in the **New Workspace** area and then choose the **OK** button to close the dialog box. Next, close the **Manage Workspaces** dialog box. The newly created workspace will be active now.

Toolbars

In Autodesk 3ds Max, various commands can be invoked by using the buttons or tools in the toolbars. By default, only the **Main Toolbar** will be displayed on Autodesk 3ds Max screen. However, you can display other toolbars such as **Snaps**, **Axis Constraints**, **Extras**, **MassFX Toolbar**, and so on in the 3ds Max interface. Also, you can move, resize, and undock them based on your requirements. To display these toolbars, right-click in the blank area on the **Main Toolbar**; a shortcut menu will be displayed with the names of all toolbars, as shown in Figure 1-8. Next, choose the required toolbar; the chosen toolbar will be displayed on the screen. Also, you can hide any of the displayed toolbars by choosing its label from the shortcut menu.

The **Main Toolbar** provides quick access to many tools and dialog boxes such as **Select and Link**, **Unlink Selection**, **Select Object**, **Material Editor**, and so on. This toolbar is docked just below the menu bar. You will learn more about the tools available in various toolbars in the later chapters.

Command Panel

By default, the **Command Panel** is docked on the right in the 3ds Max screen. There are six tabs in the **Command Panel**: **Create**, **Modify**, **Hierarchy**, **Motion**, **Display**, and **Utilities**, as shown in Figure 1-9. Most of the 3ds Max modeling and animation tools are placed in these tabs. The tools in the **Command Panel** are used to create, modify, and animate the objects. Each tab has several rollouts that can be expanded or collapsed. The tabs in the **Command Panel** are discussed next.

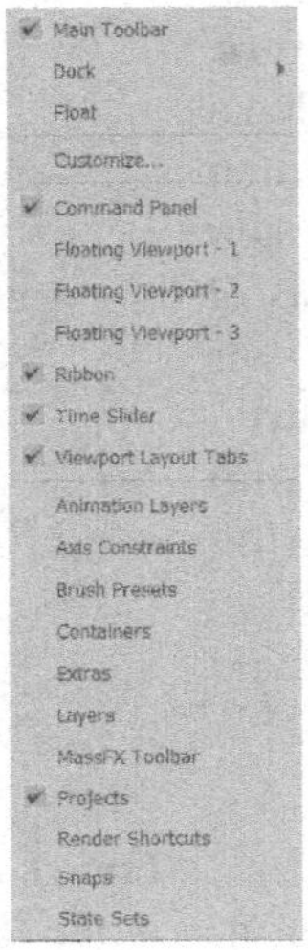

Figure 1-8 The shortcut menu displayed to view the hidden toolbars

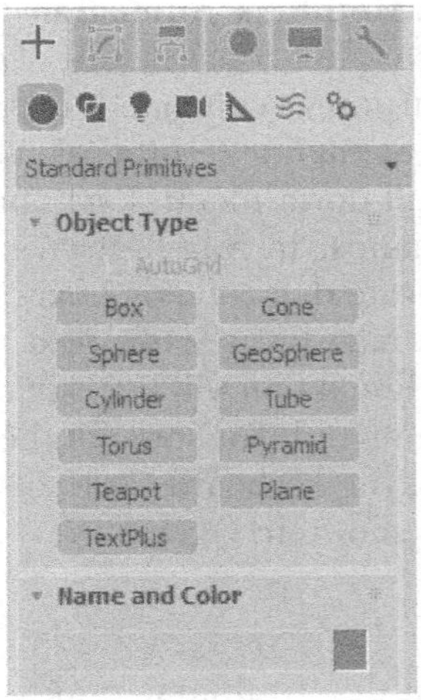

*Figure 1-9 The **Command Panel***

The **Create** tab is chosen by default. The tools in the **Create** tab are used to create objects, cameras, lights, and so on.

The **Modify** tab is used to modify the selected objects by modifying their parameters, applying various modifiers, and editing the mesh as well as polygonal and patch objects.

The **Hierarchy** tab is used to control the links in the hierarchy, joints, and inverse kinematics.

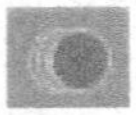
The **Motion** tab is used to control the animation controllers and trajectories.

The **Display** tab is used to hide and unhide the objects in the viewports.

The **Utilities** tab is used to access various utility programs.

Scene Explorer

The Scene Explorer is used to view, select, filter, and sort objects. It is also used to rename, delete, group, freeze, and hide objects. By default, the Scene Explorer is docked on the left in the default workspace, refer to Figure 1-5. It is discussed in detail in Chapter 2.

Viewports

When you start Autodesk 3ds Max, the default interface screen appears. This interface consists of four equal sized viewports surrounded by tools and commands, refer to Figure 1-5. These viewports are labeled as Top, Front, Left, and Perspective. The viewports in Autodesk 3ds Max are used to create 3D scenes. Also, they enable you to view a scene from different angles. When you create an object in the viewport, the Top, Front, and Left viewports will display the top, front, and left orthographic views of the object, respectively.

You can loop between viewports to make a particular viewport active by using the WINDOWS+SHIFT keys. The active viewport in 3ds Max is highlighted with a yellow border. Only one viewport can remain active at a time. All commands and actions in 3ds Max are performed in the active viewport. You can switch between the viewports by using the WINDOWS + SHIFT keys. However, if only one viewport is maximized, then on repeatedly pressing the WINDOWS + SHIFT keys, a window with available viewports will be displayed, refer to Figure 1-10. When the WINDOWS + SHIFT keys are released, the window will disappear and the viewport you have chosen will become active.

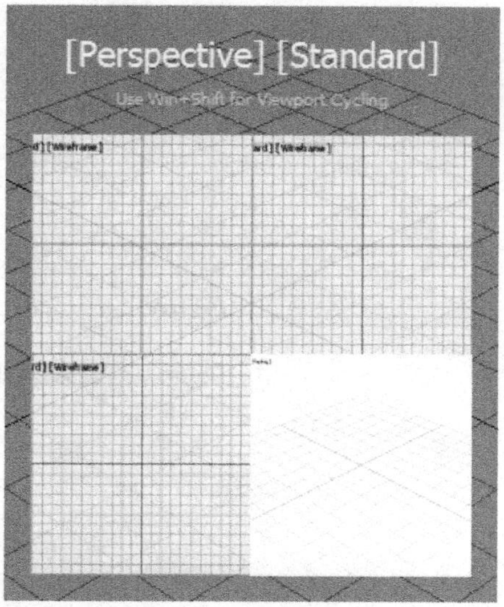

Figure 1-10 *Selecting a viewport to make it active*

You can modify the size of the viewports by dragging the intersection of the viewports on the splitter bars. To restore the original layout, right-click on the intersection of the dividing lines; a shortcut menu will be displayed, as shown in Figure 1-11. Choose the **Reset Layout** option from the shortcut menu; the viewports will be restored to their default size.

On the bottom left corner of each viewport, there is a world-space tripod, as shown in Figure 1-12. The world-space tripod has three axes, X, Y, and Z, which are displayed in red, green, and blue colors, respectively. The tripod always refers to the world coordinate system, regardless of the local coordinate system. ViewCube is placed at the top right corner of the viewport, refer to Figure 1-12. The ViewCube provides visual feedback of the current orientation of the viewport.

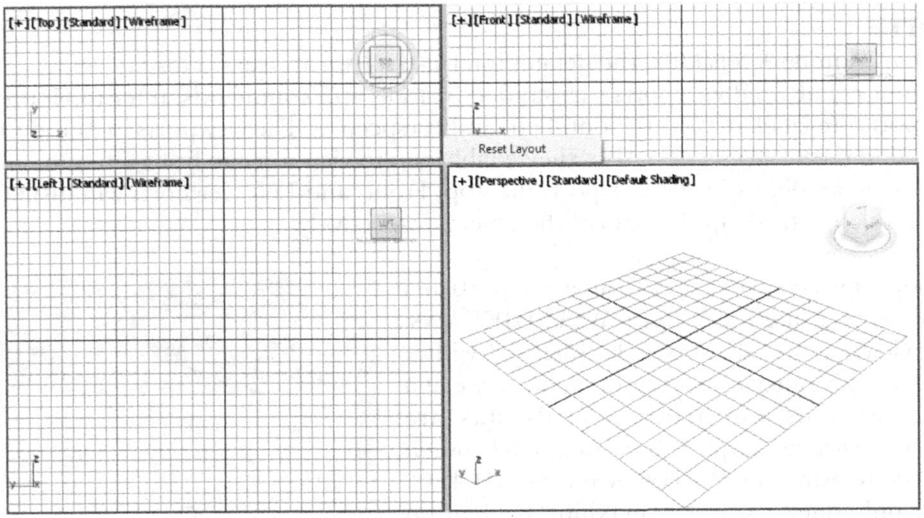

*Figure 1-11 The **Reset Layout** option in the shortcut menu*

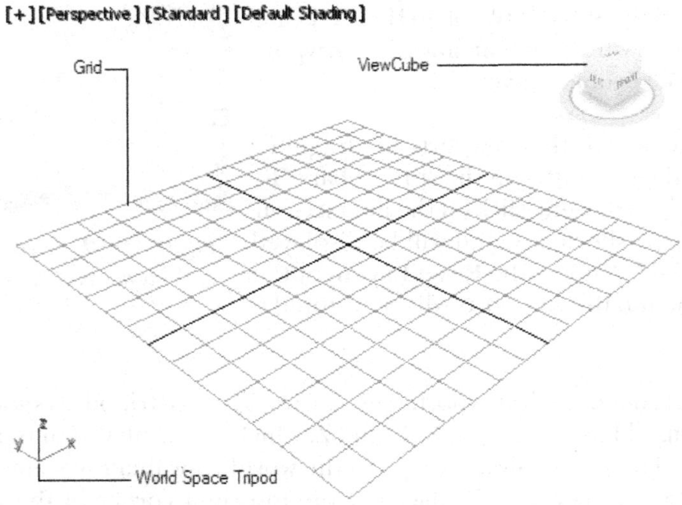

Figure 1-12 The world space tripod, grid, and ViewCube in the Perspective viewport

Note

The ViewCube will not be visible in the camera and light viewports.

It is important to note that the Local coordinate system defines local position of an object in a scene whereas the World coordinate system uses fixed axes to define the position of all the objects in the world space. Each viewport has a grid placed in it, refer to Figure 1-12. It is like a graph paper in which all the lines intersect each other at right angles. You can modify the spacing in the grids. The grids in all viewports act as an aid to visualize the spacing and distance while creating objects. Also, they are used as a construction plane to create and align the objects. You can also use the grids as a reference system while using the snap tools to align the objects. You can also hide the grid in the viewport. To do so, press the G key; the grid will disappear from the viewport. To make the grid visible, press G again.

At the top left corner of each viewport, there are four viewport labels: General viewport label, Point of view (POV) viewport label, Settings viewport label, and Shading viewport label, refer to Figure 1-13. When you click on any of the viewport labels, the corresponding flyout will be displayed, as shown in Figure 1-14. The options in these shortcut menus are used to modify various aspects of the active viewport.

When you click on the General viewport label, general options related to viewport are displayed in a flyout. In this flyout, the **Float Viewport** option is added as a new option in the list. When you choose the **Float Viewport** option, a cascading menu will be displayed, as shown in Figure 1-14 (a). The options in the cascading menu are used to make the current viewport a floating viewport. These options do not remove the current viewport but simply make a copy of the current viewport and make it floating. The floating viewports can be moved as needed and still have the same viewport labels, refer to Figure 1-14 (b).

You can configure the active viewport by using the options in the General viewport label menu. Choose the **Configure Viewports** option from this menu; the **Viewport Configuration** dialog box will be displayed. Various options in this dialog box can be used to configure the viewports. You already know that four equally sized viewports are displayed on the screen. However, you can change the viewport configuration based on your requirement. To change the basic configuration of the viewports, choose the **Layout** tab of the **Viewport Configuration** dialog box, refer to Figure 1-15. In the **Layout** tab, you can specify the division method of the viewports. There are 14 types of configurations displayed at the top in the tab. Select the required configuration and then choose the **OK** button; the viewports will be displayed according to the configuration that you have selected in the **Viewport Configuration** dialog box.

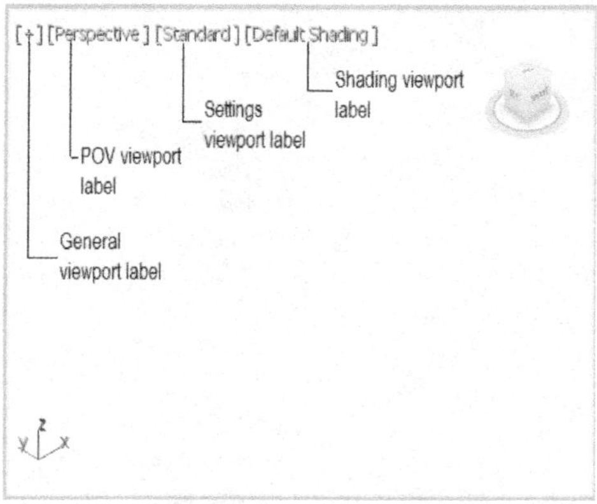

Figure 1-13 The viewport labels in the Front viewport

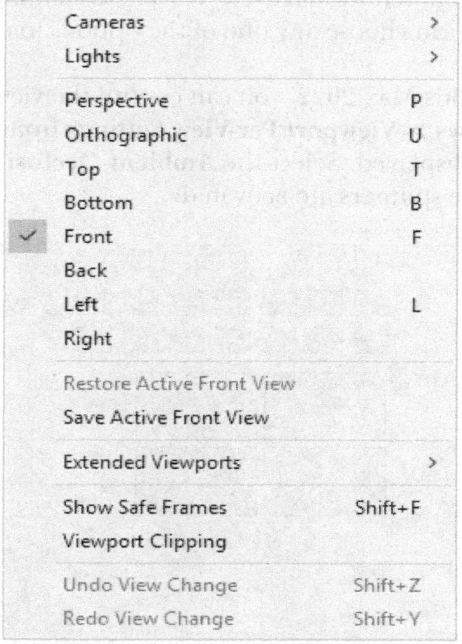

Figure 1-14 The flyout displayed on clicking the POV viewport label

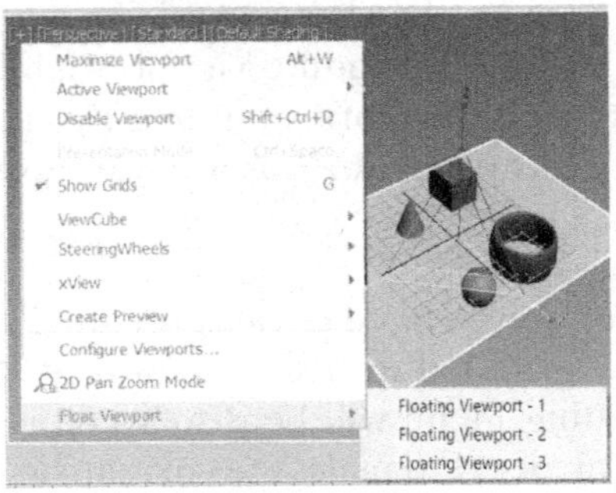

Figure 1-14 (a) *The cascading manu displayed on choosing the* ***Float Viewport*** *option*

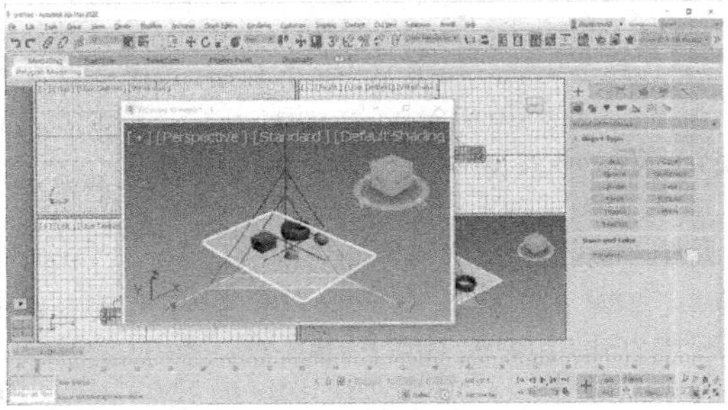

Figure 1-14 (b) *The Floating viewport*

Note

The viewport configuration specifies how the viewports will be arranged on the screen.

You can change the default viewport to any other viewport type available such as Bottom, Right, and so on, by using the options in the POV viewport label menu. To do so, click on the POV viewport label; a flyout will be displayed. Choose the viewport that you want to display. Using the Settings viewport label, you can change the display quality in viewports as well as lighting, shadows, and material settings. Different types of shading types that can be defined using the Settings Viewport label are: **Default shading**, **Facets**, **Flat color**, **Bounding Box**, **Clay**, and so on. However, some other shading types are available in the cascading menu of the **Stylized** option, refer to Figure 1-16. These shading types are **Graphite**, **Color Pencil**, **Ink**, and so on. You can choose any one of the options to change the shading.

In 3ds Max 2022, you can control the viewport ambient occlusion sampling. To do so, choose **Views > Viewport Per-View Settings** from the menubar, the **Viewport settings** dialog box will be displayed. Select the **Ambient Occlusion** check box, the **Samples**, **Radius**, and **Intensity/Fade** spinners are activated.

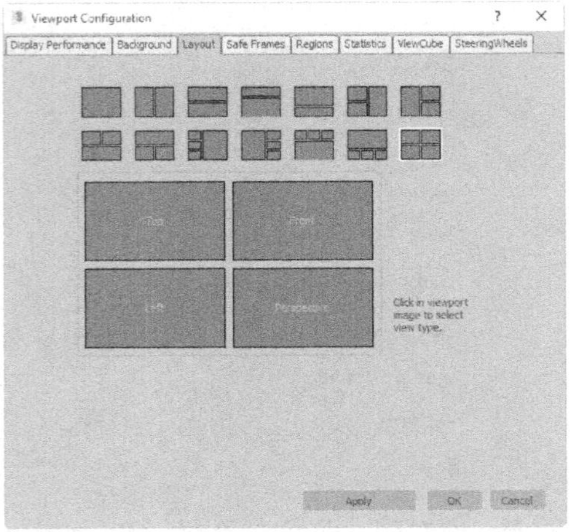

Figure 1-15 *The* ***Layout*** *tab of the* ***Viewport Configuration*** *dialog box*

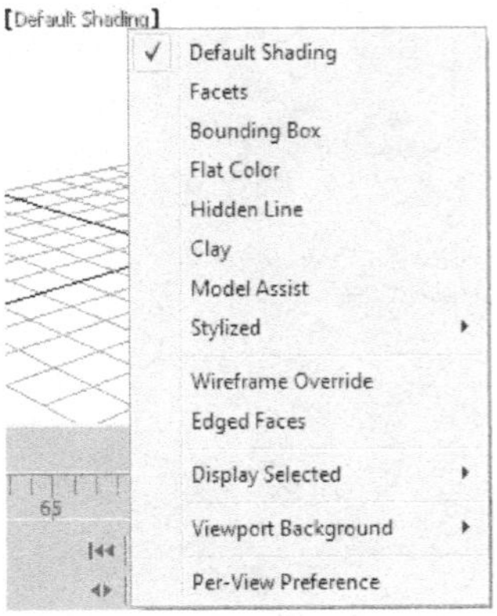

Figure 1-16 *The flyout displayed on clicking the Shading viewport label*

Viewport Navigation Controls

There are various tools available at the bottom right corner of the Autodesk 3ds Max screen, as shown in Figure 1-17. These tools are known as viewport navigation controls and they are used to control the display and navigation of the viewport. The tools displayed in the viewport navigation controls depend on the viewport selected. For example, if the Camera viewport is selected, its corresponding tools will be displayed in the viewport navigation control. These tools are discussed in detail in the later chapters.

Figure 1-17 *The viewport navigation controls*

Viewport Layout Tab Bar

The Viewport Layout tab bar enables you to store multiple viewport setups in a single scene. You can switch between different viewport setups with a click. To display the viewport layout tab bar if it is not displayed, right-click in the blank area on the **Main Toolbar**; a shortcut menu will be displayed with the names of all toolbars. Next, choose the **Viewport Layout Tabs** option; the Viewport Layout tab bar will be displayed on the screen. By default, there is a single tab at the bottom of the bar that represents the startup layout. To add more layout tabs to the bar, click on the arrow button on the bar; the **Standard Viewport Layouts** flyout will be displayed. Next, choose the required option from the flyout; the chosen layout tab will be added to the bar. To remove a tab from the bar, right-click on the tab; a shortcut menu will be displayed. Next, choose **Delete Tab** from the shortcut menu.

Animation Playback Controls

Figure 1-18 *The animation playback controls*

The tools in the animation playback controls are displayed on the left side of the viewport navigation controls, refer to Figure 1-18. These tools are used to control the animation in the active viewport. Also, you can set the total number of frames, animation length, and other settings of the animation using these tools.

Animation Keying Controls

The tools in the animation keying controls are displayed on the left side of the animation playback controls, refer to Figure 1-19. These tools are used to enter or exit different animation modes.

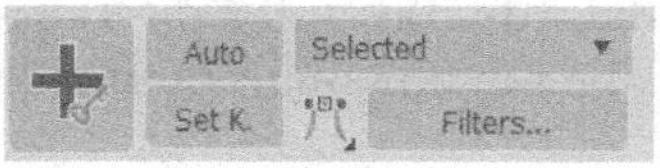

Figure 1-19 *The animation keying controls*

Track Bar

The track bar lies between the time slider and the status bar, refer to Figure 1-20. It displays a timeline along with the frame numbers.

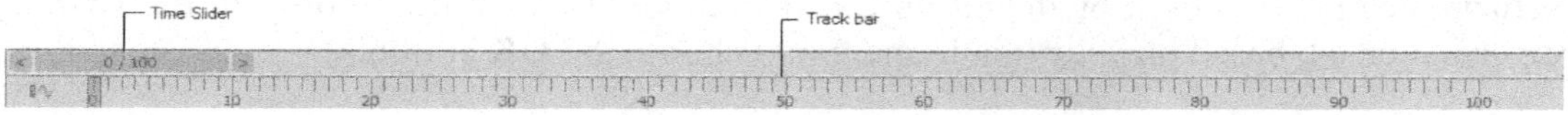

Figure 1-20 *The track bar and the time slider*

Time Slider

The time slider displays the current frame and the total number of frames in the current time segment, refer to Figure 1-20. You can view the animation at each frame by dragging the time slider. The time segment is the total range of frames that you can access using the time slider. By default, it ranges from 0 to 100. You can set the range using the **Time Configuration** dialog box, about which you will learn in the later chapters.

Status Bar

There are various tools in the status bar that provide information about the scene and the active command, as shown in Figure 1-21. The prompt line, which is located at the bottom of the screen, displays information about the active command or tool. On top of the status bar, a text box known as the status line is available. This status line displays the number of currently selected objects (current selection set). The **Selection Lock Toggle** tool available on the right side of the status bar is used to lock the selection set. The Coordinate display/transform type-in area displays the X, Y, and Z coordinates of the cursor or the currently selected object. The Coordinate display/transform type-in area can also be used to enter transform values while moving, scaling, or rotating the selected object(s).

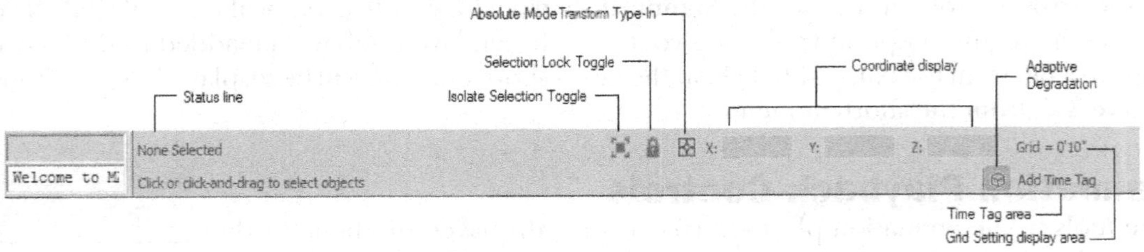

Figure 1-21 The status bar

The **Grid** setting display area is placed on the right of the Coordinate display area. It displays the size of the grid. The **Time tag** area located below the **Grid** setting display area is used to assign the text labels at any point of time in your animation. Click on the **Time tag** area; a flyout with the **Add Tag** and **Edit Tag** options will be displayed. Use these options to add or edit the text labels at any point of time in your animation.

The **Adaptive Degradation** button placed on the right of the prompt line is used to improve the viewport performance in a complex scene by decreasing the visual fidelity of some of the objects temporarily. This results in smoother viewport motions and object transformations in such scenes. It also improves viewport quality incrementally, depending on the availability of processing time. To activate this feature, right-click on the **Adaptive Degradation** button; the **Viewport Configuration** dialog box will be displayed, as shown in Figure 1-22. The **Display Performance** tab is chosen by default in this dialog box. Change the settings in the **Display Performance** tab based on your requirement and choose the **OK** button.

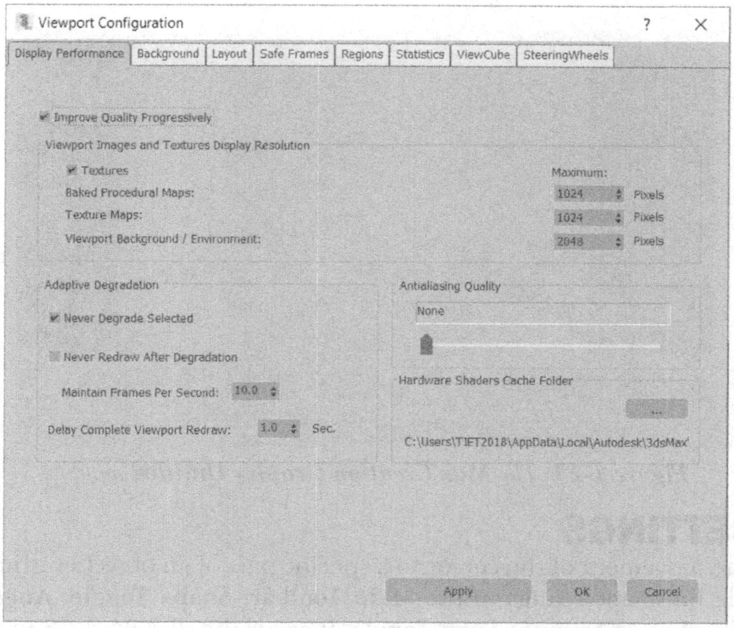

*Figure 1-22 The **Viewport Configuration** dialog box*

MAX CREATION GRAPH

Max Creation Graph uses node based Max Creation Graph Editor. Using this editor, you can create customized tools which are used to create various geometrical shapes, create modifiers, specify render settings, create utilities, and so on. To open Max Creation Graph Editor, choose **Scripting > Max Creation Graph Editor** from the menu bar; the **Max Creation Graph-Untitled** window will be displayed, as shown in Figure 1-23. This window has a menu bar at the top and four panels: **Operator Depot**, **Node Properties**, **View Navigator**, **Message Log**, and active graph view. All these panels in the **Max Creation Graph-Untitled** window can be resized, undocked, and closed.

The **Operator Depot** panel consists of a number of nodes and operators under various categories. These categories are **3ds Max**, **Array**, **Bitmap**, and so on. Click on the arrow at the left of the category name to expand it. The **Node Properties** panel provides detailed information of selected operator/node.

The view panel is used to connect nodes and operators. You need to drag the nodes and operators from the **Operator Depot** panel to the active view to create a graph. These nodes and operators has input and/or output connector(s) to create a network of nodes and operators. This network is then evaluated using the commands in the menu bar to create customized tools. The **Message Log** panel displays messages for the network created in the **Main Graph Window** panel once the networks are evaluated. On creating the network of nodes and operators, you need to evaluate it and save it at the default location with desired name to use it as a customized tool in 3ds Max interface.

***Figure 1-23** The **Max Creation Graph - Untitled** window*

SNAPS SETTINGS

Snaps restrict the movement of the cursor to a specific part of an object or grid. There are four buttons available for snap settings in the **Main Toolbar**: **Snaps Toggle**, **Angle Snap Toggle**, **Percent Snap Toggle**, and **Spinner Snap Toggle**. If you right-click on the **Snaps Toggle**, **Angle Snap Toggle**, or **Percent Snap Toggle** button, the **Grid and Snap Settings** dialog box will be displayed, as shown in Figure 1-24. In this dialog box, you can select different parts of the objects or grid where the cursor will snap to. You can turn the snap command on and off by pressing the S key or by choosing the **Snaps Toggle** tool. If you choose and hold the **Snaps Toggle** tool, a flyout will be displayed. This flyout contains the **2D Snap**, **2.5 Snap**, and **3D Snap** tools, which can be chosen to snap the cursor.

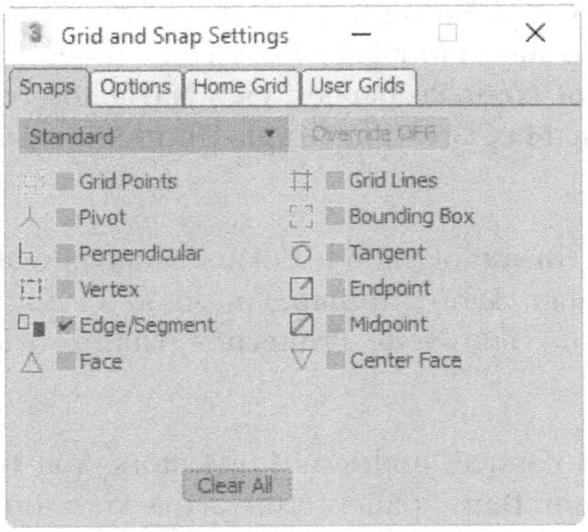

***Figure 1-24** The **Grid and Snap Settings** dialog box*

Snaps Toggle

Main Toolbar: Snaps Toggle
Menu bar: Tools > Grids and Snaps > Snaps Toggle
Keyboard: S

*Figure 1-25 The **Snaps Toggle** flyout*

The **Snaps Toggle** tool is used to snap the objects on the grid. On invoking the **Snaps Toggle** tool, a flyout will be displayed, as shown in Figure 1-25. This flyout has three tools which are discussed next.

2D Snap

If you choose the **2D Snap** tool from the **Snaps Toggle** flyout, then the cursor snaps to the active grid in two dimensions, X and Y. The Z-axis is not taken into consideration.

2.5D Snap

If you choose the **2.5D Snap** tool from the **Snaps Toggle** flyout, then the cursor snaps to the vertices and edges of the objects projected on the active grid.

3D Snap

If you choose the **3D Snap** tool from the **Snaps Toggle** flyout, then the cursor snaps to any object in 3D space using the **3D Snap** tool.

On moving the gizmo or snap handle, this axis center will act as the start snap point. This feature also helps in increasing the accuracy of snaps.

Angle Snap Toggle

Main Toolbar: Angle Snap Toggle
Menu bar: Tools > Grids and Snaps > Angle Snap Toggle
Keyboard: A

The **Angle Snap Toggle** tool enables you to rotate an object in angular increments. The increment value is specified in the **Angle** spinner of the **Grid and Snap Settings** dialog box. By default, the value in the **Angle** spinner is set to 5.0.

Percent Snap Toggle

Main Toolbar: Percent Snap Toggle
Menu bar: Tools > Grids and Snaps > Percent Snap Toggle
Keyboard: CTRL+SHIFT+P

The **Percent Snap Toggle** tool enables you to scale an object in percent increments. The increment value can be specified in the **Percent** spinner of the **Grid and Snap Settings** dialog box. By default, the value set in the **Percent** spinner is set to 10.0.

Spinner Snap Toggle

Main Toolbar: Spinner Snap Toggle

The **Spinner Snap Toggle** tool is used to set the single increment or decrement value for all the spinners in Autodesk 3ds Max. By default, the increment or decrement value is set to 1. To set the increment value, right-click on the **Spinner Snap Toggle** tool; the **Preference Settings** dialog box will be displayed. In this dialog box, choose the **General** tab, if it is not already chosen. Now, in the **Spinners** area, set a value in the **Snap** spinner, refer to Figure 1-26.

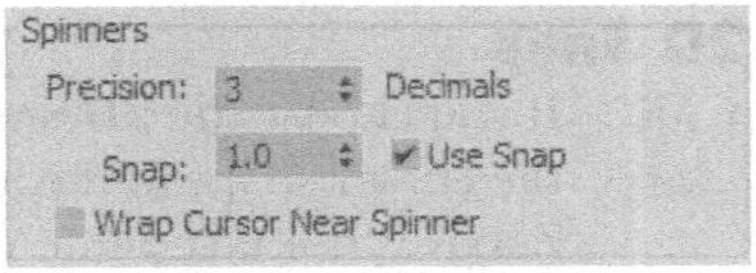

***Figure 1-26** The **Spinners** area in the **Preference Settings** dialog box*

Also, select the **Use Snap** check box and then choose the **OK** button; the **Spinner Snap Toggle** tool in the **Main Toolbar** is chosen. Now, when you use any spinner in 3ds Max, the value will increase or decrease according to the value that you have specified in the **Preference Settings** dialog box.

UNITS SETUP

The units setup in 3ds Max is used to specify the units that help in measuring the geometry in a scene. You can change the settings for units by using the **Customize** pull-down menu. To do so, choose **Customize > Units Setup** from the menu bar; the **Units Setup** dialog box will be displayed, as shown in Figure 1-27. By default, the **Generic Units** radio button is selected in the **Display Unit Scale** area in this dialog box. You can select any other radio button as per the requirement from the **Display Unit Scale** area of this dialog box and then choose the **OK** button; the limits in all the spinners in 3ds Max will be modified accordingly.

Setting Grid Spacing

To set the spacing between the visible grids in the viewports, choose **Tools > Grids and Snaps > Grid and Snap Settings** from the menu bar; the **Grid and Snap Settings** dialog box will be displayed. Choose the **Home Grid** tab in this dialog box, as shown in Figure 1-28. In the **Grid Dimensions** area, set the value in the **Grid Spacing** spinner to specify the size of the smallest square of the grid. The value in the spinners will be measured in the units that you specify in the **Units Setup** dialog box. Set the value in the **Major Lines every Nth Grid Line** spinner to specify the number of squares between the major lines in the grid. Set the value in the **Perspective View Grid Extent** spinner to specify the size of the home grid in the Perspective viewport. Note that the default grid displayed in the viewports on starting 3ds Max is known as the home grid.

In the **Dynamic Update** area, the **Active Viewport** radio button is selected by default. It is used to update the active viewport according to the new values of the **Grid and Snap Settings** dialog box. Select the **All Viewports** radio button to update all viewports simultaneously according to the new values that you set in the spinners of the **Grid and Snap Settings** dialog box.

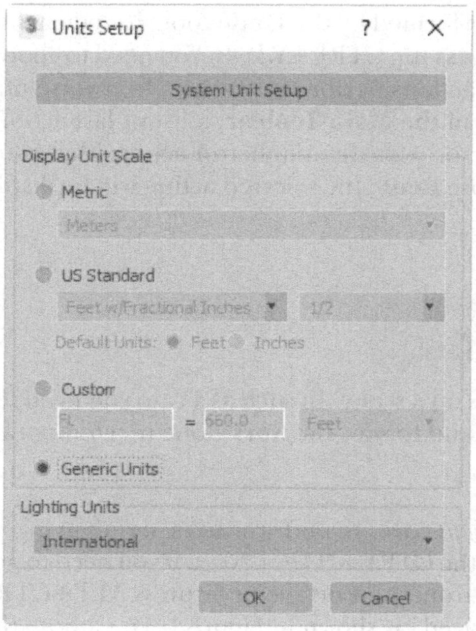

*Figure 1-27 The **Units Setup** dialog box*

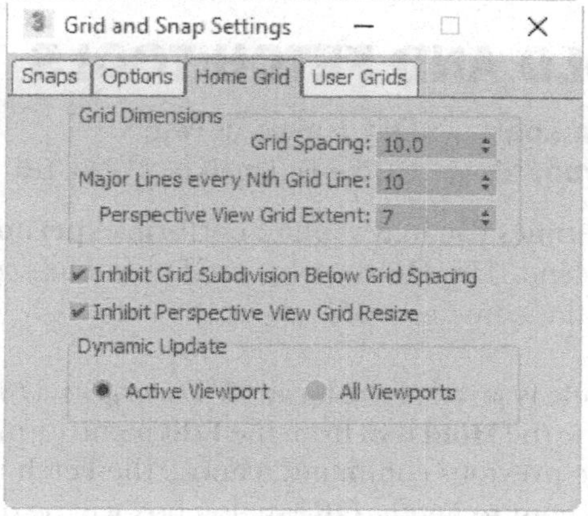

*Figure 1-28 The **Home Grid** tab chosen in the **Grid and Snap Settings** dialog box*

UNDO AND REDO TOOLS

Menu bar:	Edit > Undo or Redo
Keyboard:	CTRL+Z (Undo) or CTRL+Y (Redo)

The **Undo** tool is used to revert the last actions performed while creating or modifying a model in Autodesk 3ds Max. To undo an action, choose the **Undo** tool from the **Main Toolbar** or press the CTRL+Z keys. You need to choose the **Undo** tool repeatedly till all the previously performed actions are reversed. To reverse a number of actions at a time, right-click on the **Undo** tool in the **Main Toolbar**; a shortcut menu will be displayed, refer to Figure 1-29. Move the cursor over the number of actions that you want to reverse; the actions will be selected and then choose **Undo**. By default, you can reverse your actions up to 20 times. If you want to change this number, choose **Customize > Preferences** from the menu bar; the **Preference Settings** dialog box will be displayed. By default, the **General** tab is chosen in this dialog box. In the **Scene Undo** area, set the new value in the **Levels** spinner, as shown in Figure 1-30.

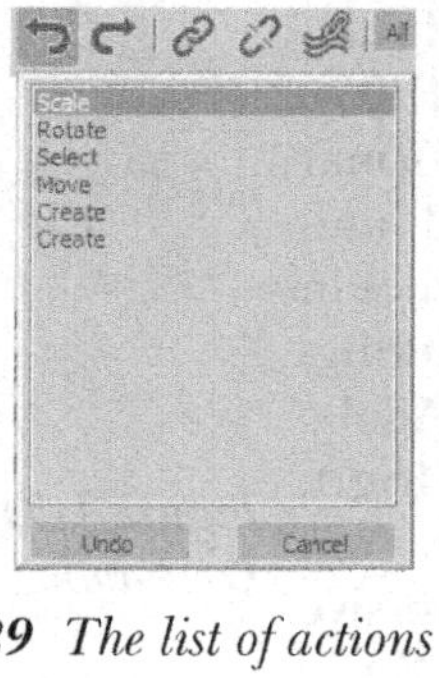

Figure 1-29 The list of actions displayed

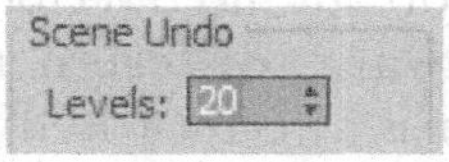

*Figure 1-30 The **Scene Undo** area in the **Preference Settings** dialog box*

The **Redo** tool is used to revert the last actions performed by the **Undo** tool. To redo an action, choose the **Redo** tool from the **Main Toolbar** or press the CTRL+Y keys. You need to choose the **Redo** tool repeatedly till you want to reverse the actions performed earlier. To reverse number of actions at a time, right-click on the **Redo** tool in the **Main Toolbar**; a list of last actions will be displayed in the shortcut menu. Move the cursor over the number of actions that you want to reverse; the actions will be selected. Next, choose **Redo**; the selected action will be displayed.

HOLD AND FETCH TOOLS

Menu bar:	Edit > Hold or Fetch
Keyboard:	CTRL+H (Hold) or ALT+CTRL+F (Fetch)

Sometimes you may want to perform experiments on a scene. In such as a case you need to hold the scene. The **Hold** tool is used to hold a scene and to save the work done in a temporary file with the name *maxhold.mx*.

The file is saved at the location *\Documents\3dsmax\autoback*. To perform an experiment in a scene, choose the **Hold** tool from the **Edit** menu or press the CTRL+H keys. Next, if you need to go back to the previous command, choose the **Fetch** tool from the **Edit** menu or press ALT+CTRL+F; the **About to Fetch. OK?** dialog box will be displayed, as shown in Figure 1-31. Choose the **Yes** button; the scene with the previous command will be displayed. In this way, you can go back to a series of commands using the **Hold** tool.

Note

*When you use the **Fetch** tool in a scene, the history of the actions performed so far will be deleted. As a result, you cannot undo or redo the actions performed before invoking this tool.*

IMPORTING FILES

Enhanced

In 3ds Max, you can import files from different formats. The **Import** command in the **File** menu is used to import a 3D geometry from other softwares for photorealistic rendering and perfect animation. To invoke this command, choose **File > Import** from the menubar; a cascading menu will be displayed. The options in the cascading menu are: **Import**, **Merge**, **Replace**, **Link Revit**, **Link FBX**, and **Link AutoCAD**. These options are discussed next.

Import

The **Import** option is used to load files that are in different file formats. You can import FBX, DWG, DXF, OBJ, IGES (Initial Graphics Exchange Specification), CATIA V4 (MODEL, DLV4, DLV3, DLV, EXP, SESSION, MDL), SolidWorks file format (SLDPRT and SLDASM), and Autodesk Inventor (IPT and IAM) file formats. You can also import Pro/ENGINEER files in 3ds Max to enhance and texture your model for final rendering. 3ds Max can import advanced vector drawing files from Adobe Illustrator software to design the 3D company logo. Adobe Illustrator is a vector-based graphics application. You can also import Virtual Reality file formats to create immersive

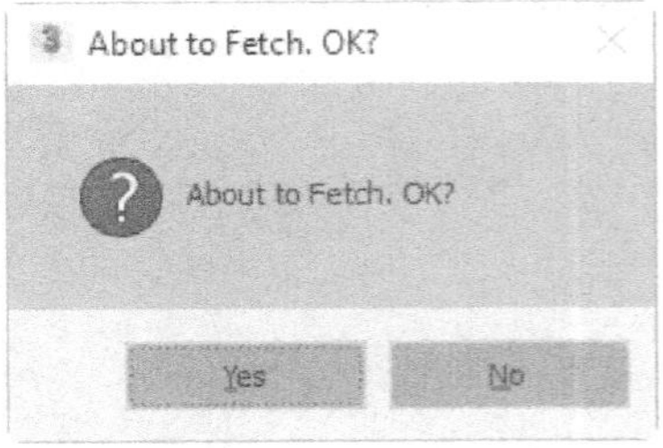

*Figure 1-31 The **About to Fetch. OK?** dialog box*

and interactive architectural visualizations. You can import and export IGES (Initial Graphics Exchange Specification) files from 3ds Max. The IGES file format is used for 3D printing. In this version of 3ds max, the new importer allows you to import any version of SketchUp files, whereas using the legacy importer, you can import files only up to SketchUp 2014.

Merge

The **Merge** option is used to combine two files in such a way that the resulting file has all the objects from both the files. You can merge a file into your existing file. To do so, choose **File > Import >Merge** from the menubar.

Replace

The **Replace** option is used to replace the objects in the current 3ds Max scene with objects from an external file. To do so, choose **File > Import > Replace** from the menubar.

Link Revit, Link FBX, and Link AutoCAD

You can link 3ds max file with the files in other Autodesk software. The **Link Revit**, **Link FBX**, and **Link AutoCAD** options are used to maintain a live link between two Autodesk softwares. When you are using two or more Autodesk products, you can link the files using the File Link Manager. To do so, choose **File > Import > Link Revit/Link FBX/Link AutoCAD** from the menubar. You can also choose **Utilities > More > File Link Manager** from the **Command Panel**.

EXPORTING FILES

In 3ds Max, you can export files to different file formats. To do so, choose **File > Export** from the menubar; a cascading menu will be displayed. The options in the cascading menu are: **Export**, **Export Selected**, **Publish to DWF**, and **Game Exporter**. The **Export** option is used to export the entire file. The **Export Selected** option is used to export selected 3D objects from the scene to different file formats. By default, the **Export Selected** option is hidden. When you select a 3D object in the scene then only this option will be enabled. The **Publish to DWG** option is used to export all the AutoCAD drawing files. The **Game Exporter** option is used to easily export models and animation clips in the game engine.

HOT KEYS

Enhanced

In 3ds Max, you can use the hot keys to choose some of the commonly used tools and commands. These keys are known as the hot keys. You can work faster and more efficiently using the hot keys. The major hot keys and their functions are listed next.

Main Toolbar

The hot keys that can be used to invoke the tools available in the **Main Toolbar** are given next:

H	Invokes the **Select From Scene** dialog box
S	Invokes the **Snaps Toggle** tool
A	Invokes the **Angle Snap Toggle** tool

CTRL+SHIFT+P	Invokes the **Percent Snap Toggle** tool
M	Invokes the **Material Editor** dialog box
SHIFT+Q	Invokes the **Render Production** tool

Viewport Navigation Controls

The hot keys that can be used to invoke the tools available in the viewport navigation controls are given next:

ALT+CTRL+Z	Performs the action of the **Zoom Extents** tool
ALT+W	Invokes the **Maximize Viewport Toggle** tool
ALT+Z	Invokes the **Zoom** tool
CTRL+W	Invokes the **Zoom Region** tool
CTRL+P	Invokes the **Pan View** tool
Scroll the middle mouse button	Zooms in or out the active viewport
CTRL+R	Invokes the **Orbit** tool
SHIFT+Z	Used to undo the **Zoom** or **Pan** command actions
ALT+ press and hold the middle mouse button and move the mouse	Performs the actions of the **Orbit** tool

The following hot keys are used to change the POV viewport labels:

V	Invokes the viewport quad menu
T	Invokes the Top viewport
F	Invokes the Front viewport
L	Invokes the Left viewport
P	Invokes the Perspective viewport
B	Invokes the Bottom viewport
U	Invokes the Orthographic viewport

Material Editor Hot Keys

The hot keys that can be used to invoke the tools available in the **Material Editor** dialog box are given next:

A	Assign Material to Selection
O	Get Material
F9	Make Preview

Animation Playback Controls

The hot keys that can be used to invoke the tools available in the animation playback controls are given next:

N	Invokes the **Auto Key** tool
Home	Go to start frame
End	Go to end frame
/ (backslash)	Plays animation
ESC	Stop the animation
, (comma)	Go to previous frame
. (period)	Go to next frame

Customizing the Hot Keys

In 3ds Max, you can create your own keyboard shortcuts. To do so, choose **Customize > Customize User Interface** from the menu bar; the **Customize User Interface** dialog box will be displayed, as shown in Figure 1-32. In this dialog box, the **Keyboard** tab is chosen by default. Next, select a command from the **Group** and **Category** drop-down lists each; the list of corresponding actions will be displayed just below the **Category** drop-down list. Now, select one of the actions from the list and then enter the key that you want to assign to the selected action in the **Hotkey** text box. Next, choose the **Assign** button; the key is assigned to the selected action.

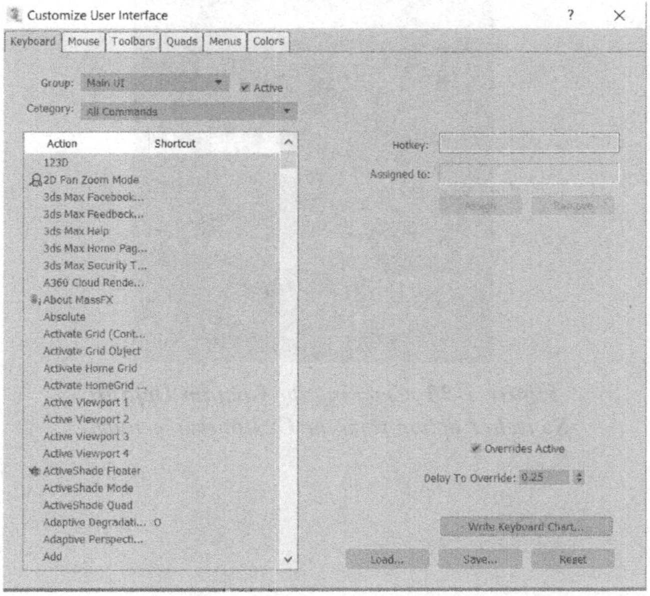

*Figure 1-32 The **Customize User Interface** dialog box*

CUSTOMIZING THE COLORS OF THE USER INTERFACE

3ds Max allows you to modify the colors of the interface. You can modify the colors for almost every element in the interface. To modify the colors, choose **Customize > Customize User Interface** from the menu bar; the **Customize User Interface** dialog box will be displayed. Choose the **Colors** tab from this dialog box, refer to Figure 1-32. Next, select the category of the interface element from the **Elements** drop-down list; the list of corresponding elements will be displayed just below the **Elements** drop-down list. Now, select one of the elements from the list and click on the **Color** swatch on the right of the **Elements** drop-down list; the **Color Selector** dialog box will be displayed. Select a new color and choose the **OK** button to close the **Color Selector** dialog box.

You can switch the color of interface from dark to light. To modify the interface color, choose **Customize > Custom Defaults Switcher** from the menu bar, as shown in Figure 1-33; the **Custom UL and Defaults Switcher** dialog box will be displayed, as shown in Figure 1-34. Make sure the **Max** option is selected in the **Default Setting** area. Next, choose the **ame-light** option from the list in the **User Interface schemes** area, refer to Figure 1-35. Now, choose the **Set** button; the **Custom Scheme** message box is displayed with the message **Loading Custom Scheme**, refer to Figure 1-35. You will notice that now the interface has changed to a light color, as shown in Figure 1-36.

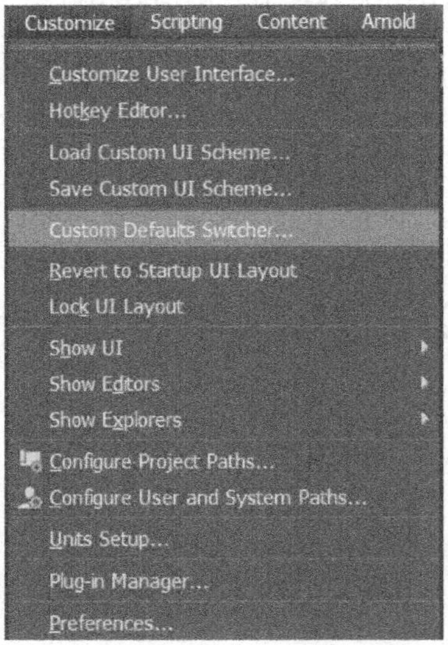

Figure 1-33 *Choosing the **Custom Default Switcher** option from the **Customize** menubar*

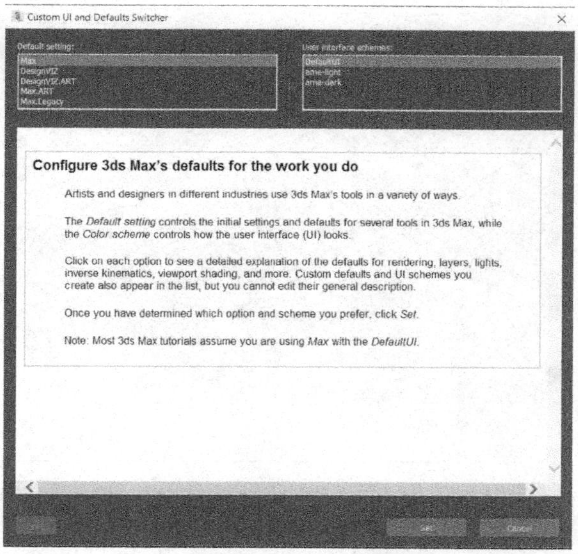

Figure 1-34 *The* ***Custom UI and Default Switcher*** *dialog box*

Figure 1-35 *The* ***Custom Scheme*** *message box*

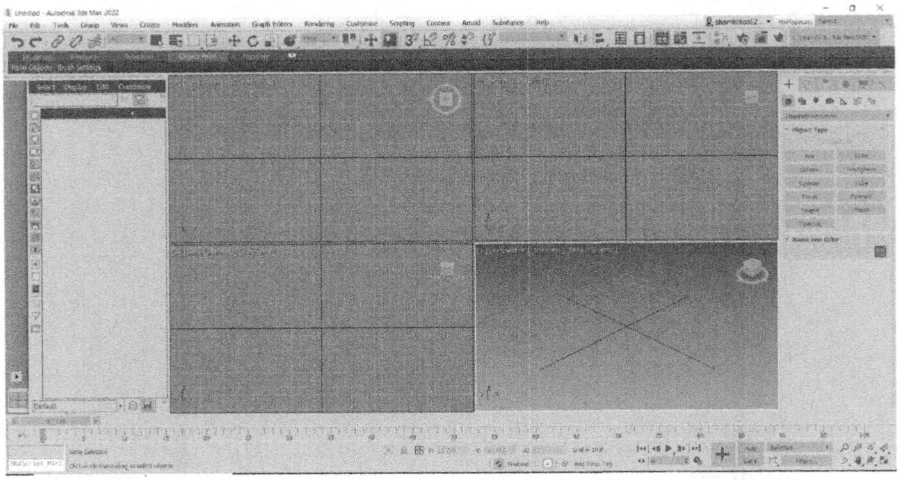

Figure 1-36 *The Interface changed into light color*

In case you want to reset the new color to the default one, choose the **Reset** button located next to the **Color** color swatch. You can also reset all the changes you made to colors. To do so, choose the **Reset** button at the bottom of the **Customize User Interface** dialog box; the **Revert Color File** message box will be displayed, as shown in Figure 1-37. Choose the **Yes** button; the default colors will be displayed in the color swatches.

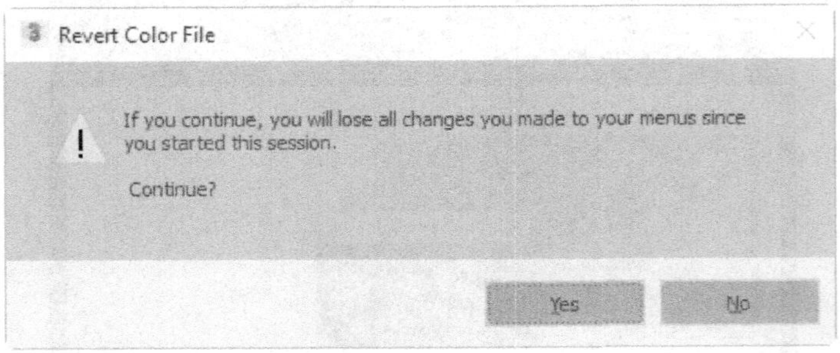

Figure 1-37 *The* ***Revert Color File*** *message box*

HELP PREFERENCES

When you choose **Help > Autodesk 3ds Max Help** from the menu bar, the 3ds Max help will be displayed. If you want to open help from a local drive, you need to download it from *http://www.autodesk.com/3dsmax-helpdownload-enu* and then install it on your system. Next, choose **Customize > Preferences** from the menu bar; the **Preference Settings** dialog box will be displayed. Choose the **Help** tab in this dialog box, refer to Figure 1-38. Select the **Local Computer/ Network** radio button from the **Help Location** area and then choose the **Browse** button; the **Browse For Folder** dialog box will be displayed. Select the path where you have saved the file and then choose the **OK** button to exit the dialog box. Next, choose the **OK** button in the **Preference Settings** dialog box; the dialog box will be closed. Now, you can access the help from your local drive.

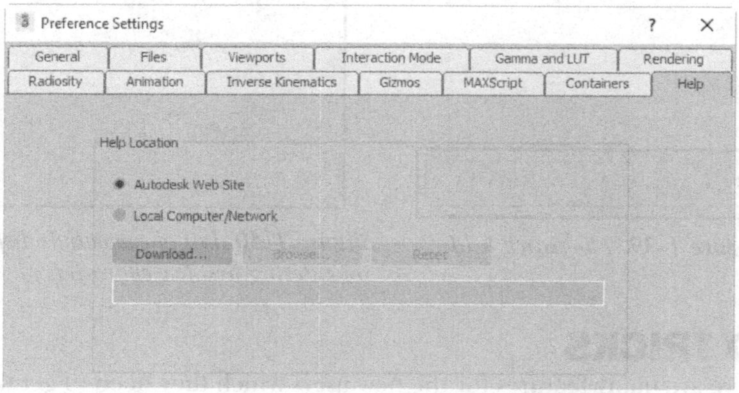

*Figure 1-38 The **Help** tab chosen in the **Preference Settings** dialog box*

If you choose **Help > Search 3ds Max Commands** from the menu bar or press X, a search field will be displayed, refer to Figure 1-39. Next, enter the initial characters of a command; a list of command names that contain the specified characters will be displayed, refer to Figure 1-40. Now, you can execute the desired command by choosing it from the list displayed.

Note

*For the printing purpose, this textbook will follow the white background. However, for better understanding and clear visualization, this textbook will follow other color schemes as well at various places. In addition, the **Default Shading** shading type has been used throughout the textbook in all screen captures. Moreover, at some places in figures, grids have been hidden for better understanding and visualization.*

Figure 1-39 The search field

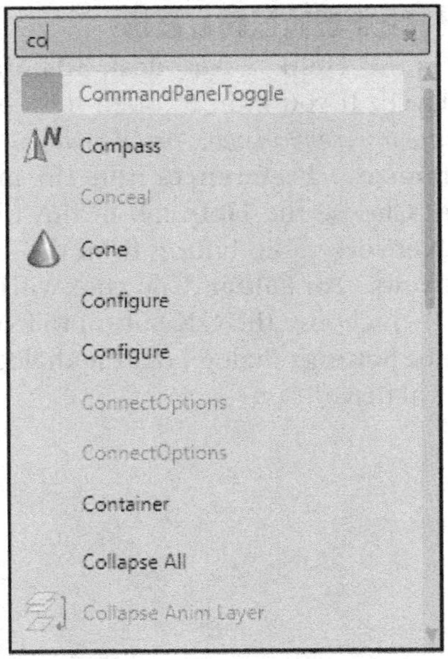

Figure 1-40 List of commands displayed on entering first few characters

TIPS AND TRICKS

In 3ds Max, there are many features for the new users which they need to get familiar with. We have provided information of some of the features in the form of tips and tricks below. This will help users to work efficiently.

- To avoid crashing of system while working on complicated scenes, you can disable autobackup (save often) option. To enable or disable autobackup, choose **Customize > Preferences** in the menubar; the **Preferences Settings** dialog box will be displayed. In this dialog box, choose the **Files** tab. In the **Auto Backup** area, select or clear the **Enable** check box. You can save incrementally to keep a record of progress.

- To make the selected object transparent, press ALT + X (toggle key).

- To speed-up the workflow, choose the **Bounding Box** option from the **Shading viewport label** flyout.

- Press 7 to display the total polygons and vertex count in the viewport.

- To increase scene undo levels, you can change the value in the **Levels** spinner of the **General** tab in the **Preference Settings** dialog box.

- Right-click on the Snap icon in the **Main Toolbar** to access the snap settings.

- Press CTRL + C to insert camera in the Perspective view.

- Right-click on the **Select**, **Rotate**, and **Scale** icons in the **Main Toolbar** to input values for transformation of an object.
- While creating a spline, press BACKSPACE to remove the last created vertex.
- Hold the CTRL key to pan quickly in the viewport.
- Press I to get the cursor position at the center of the viewport.
- Right-click on the **Modifiers List** drop-down list in the **Command Panel** to access new list of hidden modifiers.
- Use CTRL + L to deactivate active lights in viewports.
- If you have problem with the display of your textures, it is probably because some of the faces of your model are reversed. To ensure that the polygons have the proper orientation, use the **Normal** modifier.
- To create the project folders, navigate to **File > Manage > Set Project Folder** in the menubar.
- If the transform gizmo disappears, press X to bring it back.
- Press ALT + W to maximize or minimize a viewport.
- Use G to turn the Grid off and on.
- To lock any object, select it and then press SPACEBAR.
- To select an edge loop, pick one of the edges from it, hold SHIFT, and then select the next edge.
- Press the ALT + Q keys to isolate a selected object.

Self-Evaluation Test

Answer the following questions and then compare them to those given at the end of this chapter:

1. Which of the following buttons is used to improve the performance of a viewport in a complex scene by temporarily decreasing the visual fidelity of some of the objects?

   (a) **Adaptive Degradation** (b) **Time Configuration**
   (c) **Auto Key** (d) All of these

2. Which of the following tabs should be chosen to control the animation controllers?

   (a) **Motion** (b) **Display**
   (c) **Hierarchy** (d) None of these

3. Which of the following dialog boxes is used to set the spacing in the grids displayed in the viewports?

   (a) **Customize User Interface** (b) **Grid and Snap Settings**
   (c) **Units Setup** (d) All of these

4. The __________ provides visual feedback of the current orientation of the viewport.

5. The __________ is located at the bottom of the screen and displays the information about the active command or tool.

6. The __________ is used to float an element of the interface.

7. Most of the 3ds Max modeling and animation tools can be chosen from the tabs located in the **Command Panel**. (T/F)

8. The default interface of 3ds Max consists of three equal sized viewports surrounded by tools and commands. (T/F)

9. In 3ds Max, you can modify the colors of almost every element in the interface. (T/F)

10. The tools in the animation playback controls are used to control the display of a viewport. (T/F)

Review Questions

Answer the following questions:

1. Which of the following combinations of keys is used to choose the **Pan** tool?

   (a) CTRL+P (b) SHIFT+Z
   (c) CTRL+A (d) CTRL+W

2. The __________ is located between the time slider and the status bar.

3. The __________ tool located on the right of the status bar is used to lock a selection set.

4. The Viewport Layout Tabs bar enables you to store multiple viewport setups in a single scene. (T/F)

5. Snapping restricts the movement of the cursor to a specific part of an object or grid. (T/F)

6. The options in the General viewport label menu are used for defining the type of shadings displayed in the viewport. (T/F)

Answers to Self-Evaluation Test

1. a, **2.** a, **3.** b, **4.** ViewCube, **5.** prompt line, **6.** handle, **7.** T, **8.** F, **9.** T, **10.** F

Chapter 2

Primitive Objects-I

Learning Objectives

After completing this chapter, you will be able to:
- *Create primitive objects*
- *Modify primitive objects*
- *Use the Mirror and Align tools*

INTRODUCTION

In this chapter, you will learn to create primitive objects and modify them using modifiers such as **Edit Mesh** and **Edit Poly**. In addition, you will learn to use the **Mirror** and **Align** tools.

TUTORIALS

Tutorial 1

In this tutorial, you will create the model of a temple, as shown in Figure 2-1, using various primitive objects. **(Expected time: 20 min)**

Figure 2-1 *The model of a temple*

The following steps are required to complete this tutorial:

a. Create the project folder.
b. Create bottom part of the temple.
c. Create pillars.
d. Create the roof.
e. Modify the roof objects.
f. Add details to the dome.
g. Change the background color of the scene.
h. Save and render the scene.

Creating the Project Folder

Before starting a new scene, it is recommended that you create the project folder. Creating a project folder helps you keep all files of a project in an organized manner. Open File Explorer and browse to the *Documents* folder. In this folder, create a new folder with the name *3dsmax 2022*. The *3dsmax 2022* folder will be the main folder and it will contain all project folders that you will create while working on the tutorials of this textbook. Next, you will create first project folder for Tutorial 1 of this chapter. To do so, you need to follow the steps given next.

1. Start 3ds Max 2022.

2. If 3ds Max is already running, you need to reset it. To do so, choose the **Reset** option from the **File** menu; the **3ds Max** message box is displayed. Choose the **Yes** button from the message box; a new screen is displayed with default settings.

Note
*The **Reset** option is used to restore 3ds Max settings to its startup defaults.*

3. Create a new folder with the name *c02_tut1* at *\Documents\3dsmax2022*. Next, choose **File > Project > Create Default** from the menubar; the **Choose a folder** dialog box is displayed.

4. Navigate to *\Documents\3dsmax2022*. Next, select the **c02_tut1** folder and then choose the **Select Folder** button to close the **Choose a folder** dialog box; the *c02_tut1* folder is created.

5. Choose **Save** from the **File** menu; the **Save File As** dialog box is displayed.

Note
*1. The scenes created in 3ds Max are saved with the .max extension. As the project folder is already created, the **scenes** folder is displayed in the **Save in** drop-down list of the **Save File As** dialog box.*

*2. You can also save a scene to previous versions (2017, 2018, 2019, and 2020) of 3ds Max. To do so, select the desired option from the **Save as type** drop-down list in the **Save File As** dialog box.*

*3. After setting the project folder when you open or save a scene, the **scenes** folder of this project is selected by default in the **Save File As** and **Open File** dialog boxes.*

6. Enter **c02tut1** in the **File name** text box and then choose the **Save** button to close the dialog box.

When you start 3ds Max, the last project that you worked on will be active and an empty scene will be displayed with the name **Untitled**. It is recommended that you frequently save the files while you are working on them by pressing the CTRL+S keys.

Autodesk 3ds Max creates a backup of the scene and saves changes periodically. In case of system failure, you can open the auto backup scene file and continue working on it. If you have not created the project folder, the default path for windows to save the auto backup file is as follows: *C:\users\<username>\My Documents\3dsMax\autoback*. However, if you have created the project folder, the backup file will be saved in the *autoback* subfolder of the project folder. When 3ds Max is auto saving a file, the information, **Autosave in progress... (Press ESC to cancel)** is displayed in the prompt line at the bottom of the interface. If the size of the file is too large and consuming long time to save, you can press the ESC key to interrupt the saving process. You can also set the number of autoback files, specify the names for these files, and set the backup time interval. To do so, choose **Customize > Preferences** from the menu bar; the **Preferences Settings** dialog box will be displayed. Next, choose the **Files** tab and then set the options as required in the **Auto Backup** area of the dialog box.

Creating Bottom Part of the Temple

In this section, you will create the bottom part of the temple using the **Box** tool.

1. Choose **Create > Geometry** in the **Command Panel**, if it is not already chosen; different tools and rollouts pertaining to the **Create** command are displayed in the **Command Panel**, refer to Figure 2-2. 

   When you start 3ds Max, the **Geometry** button is chosen by default.

2. Invoke the **Box** tool from the **Object Type** rollout; the rollouts corresponding to the **Box** tool such as **Name and Color**, **Creation Method**, **Keyboard Entry**, and **Parameters** are displayed.

   The arrow sign on the left side of the rollout is used to expand or collapse the rollout, refer to Figure 2-2.

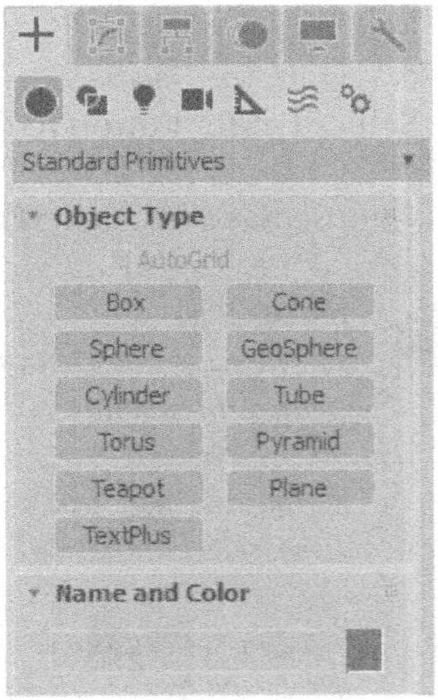

Figure 2-2** The tools and rollouts in the **Command Panel

3. Expand the **Keyboard Entry** rollout by clicking on its title bar and set the values as given next.

   Length: **360**　　　Width: **360**
   Height: **20**

   Choose the **Create** button; the box is created in all viewports with the name *Box001*. Enter **Bottom** in the **Name and Color** rollout; the box is renamed as *Bottom*.

4. Invoke the **Zoom Extents All** tool from the viewport navigation controls; *Bottom* is zoomed to its extents, refer to Figure 2-3. 

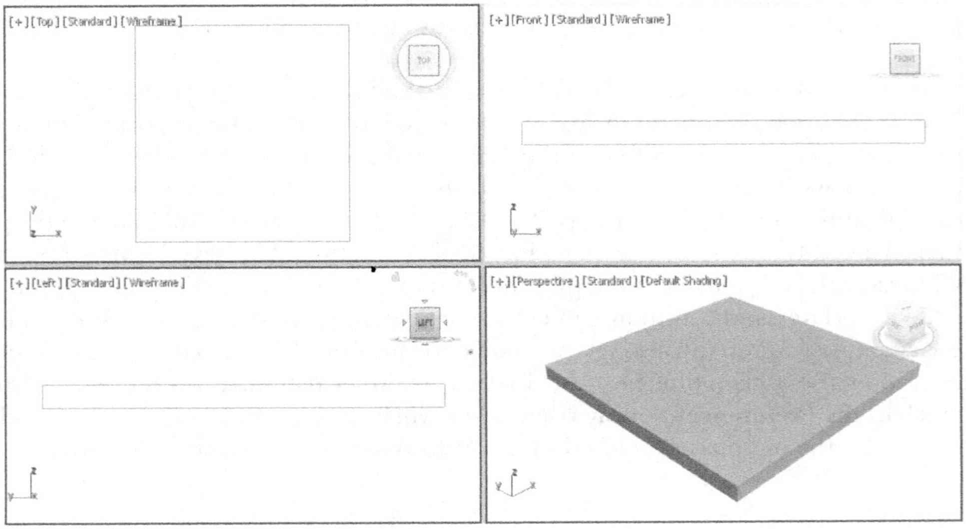

***Figure 2-3** Bottom zoomed to its extents*

5. Activate the Top viewport. Make sure the **Box** tool is invoked.

6. In the **Keyboard Entry** rollout, set the values as given next.

   Length: **60** Width: **60** Height: **30**

   Choose the **Create** button; a box is created at the center of *Bottom*. Enter **Base01** in the **Name and Color** rollout; the box is renamed as *Base01*. Next, right-click to exit the **Box** tool.

7. Make sure the Top, Front, and Left viewports are in the wireframe mode, refer to Figure 2-4.

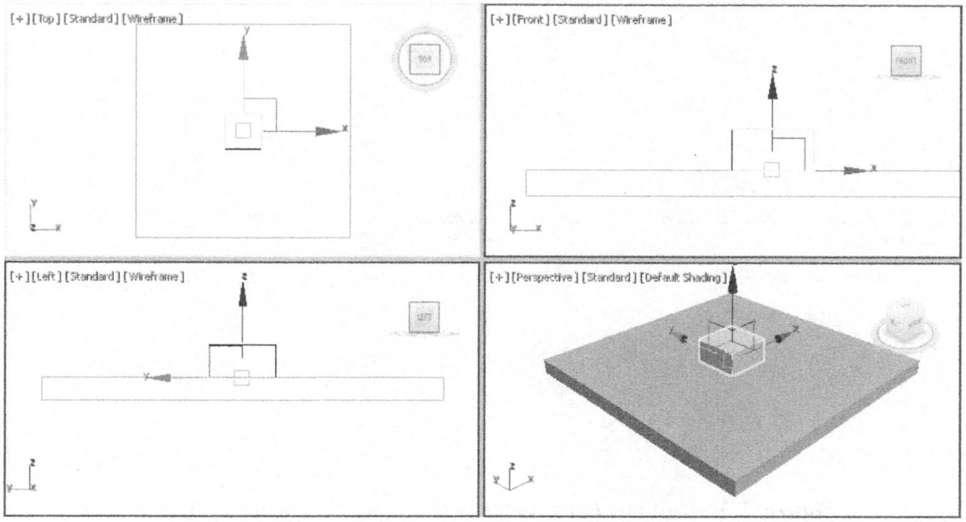

Figure 2-4 *Bottom and Base01 displayed in viewports*

Tip: *You can press ALT+W to toggle between the active viewport display and four viewport display. You can use the G key to toggle the display of the grid in the viewports.*

8. Activate the Front viewport by clicking the MMB button. Next, invoke the **Zoom** tool from the viewport navigation controls and zoom in the small box in the Front viewport. 

Note

You can switch between the viewports to make a particular viewport active by using the WINDOWS +SHIFT keys. The active viewport in 3ds Max is marked with a yellow highlighted border. Only one viewport can remain active at a time. If all viewports are visible, press the WINDOWS+SHIFT keys to make the subsequent viewports active. However, if only one viewport is maximized, then on pressing the WINDOWS+SHIFT keys repeatedly, a window with available viewports will be displayed. When the WINDOWS+SHIFT keys are released, the window will disappear and the selected viewport will become active. You can use this process to activate a viewport whenever you need to do so.

9. Invoke the **Select and Move** tool by pressing the W key. In the Front viewport, select *Base01* and move it up along the Y-axis such that its bottom touches the top of the object *Bottom*. Now, release the left mouse button. Next, select the *Bottom* and invoke the **Zoom Extents** tool to zoom *Bottom*.

10. Activate the Top viewport and invoke the **Zoom Extents** tool from the viewport navigation controls area. Next, invoke the **Maximize Viewport Toggle** tool from the viewport navigation controls to maximize the viewport.

11. Move *Base01* to the lower-left corner of *Bottom*, as shown in Figure 2-5.

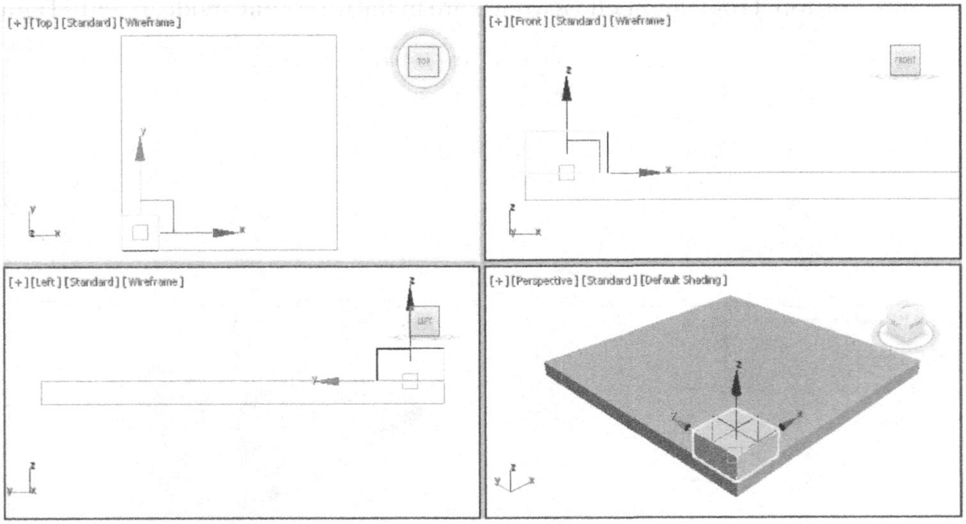

Figure 2-5 *Alignment of Base01 to the lower-left corner of Bottom*

Note
*The **Axis Constraints** toolbar is not displayed by default. To display this toolbar, right-click on the empty area of the **Main Toolbar** and then choose the **Axis Constraints** option from the shortcut menu displayed.*

12. Press and hold the left mouse button along with the SHIFT key, and then drag *Base01* to the upper-left corner of *Bottom* using the Y axis handle of the gizmo. Next, release the left mouse button; the **Clone Options** dialog box is displayed. In this dialog box, make sure the **Copy** radio button is selected and then enter **Base02** in the **Name** text box, as shown in Figure 2-6. Next, choose the **OK** button to close the dialog box; *Base02* is created in the viewport.

Note
When multiple objects are selected and moved with the SHIFT key pressed, a copy of each selected object is created.

13. Make sure the **Select and Move** tool is invoked and *Base02* is selected. Next, press and hold the CTRL key and select *Base01*. Now, release the CTRL key.

The CTRL key enables you to select multiple objects.

14. Press and hold the SHIFT key and drag both the objects to the right side of *Bottom* using the X axis handle of the gizmo. Now, release the left mouse button; the **Clone Options** dialog box is displayed. In this dialog box, make sure the **Copy** radio button is selected and then enter **Base03** in the **Name** text box. Next, choose the **OK** button; a copy each of *Base01* and *Base02* is created. One of the copies is already renamed as *Base03*. Rename the other copy as **Base04**.

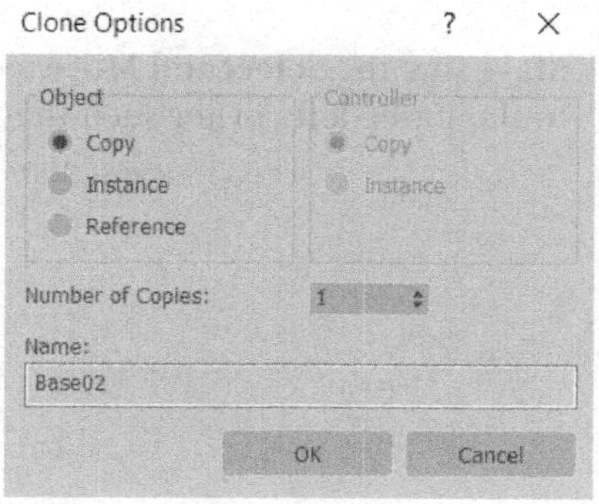

***Figure 2-6** The **Clone Options** dialog box*

15. Invoke the **Maximize Viewport Toggle** tool from the viewport navigation controls to restore the previous viewport configuration or press ALT+W. Next, invoke the **Zoom Extents All** tool. Adjust the Perspective viewport. Figure 2-7 shows the four bases of the temple.

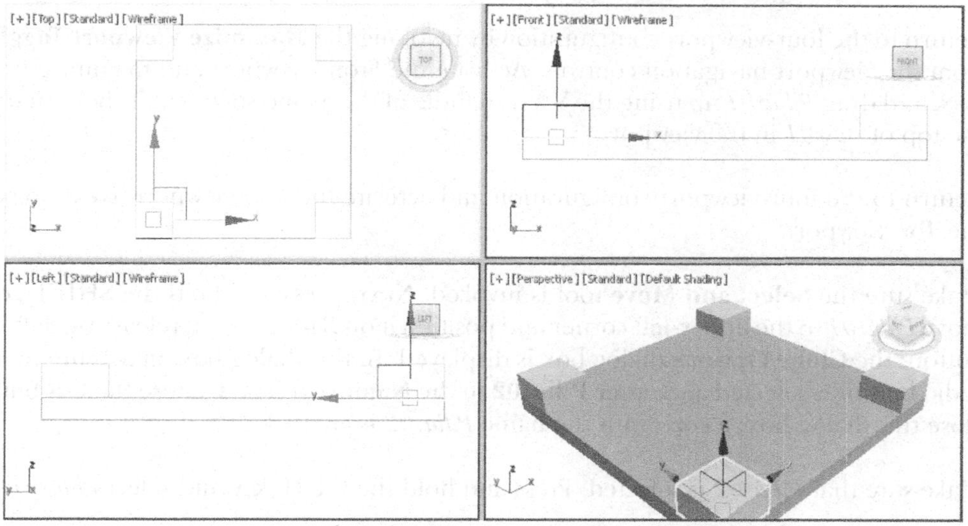

***Figure 2-7** Four bases of the temple*

Creating Pillars

In this section, you will create pillars for the temple by using the **Cylinder** tool.

1. Maximize the Top viewport using the **Maximize Viewport Toggle** tool from the viewport navigation controls. Next, choose **Create > Geometry** in the **Command Panel** and then invoke the **Cylinder** tool.

2. Expand the **Keyboard Entry** rollout and set **30** in the **Radius** spinner and **250** in the **Height** spinner. Next, choose the **Create** button from the **Keyboard Entry** rollout; the cylinder is created at the center of the object *Bottom*. Enter **Pillar01** in the **Name and Color** rollout. This cylinder will act as a pillar in the model of the temple.

3. Press the ESC key to exit the **Cylinder** tool.

4. Make sure the **Select and Move** tool is invoked from the **Main Toolbar** and then drag *Pillar01* to the lower-left corner such that it is positioned on *Base01*, as shown in Figure 2-8.

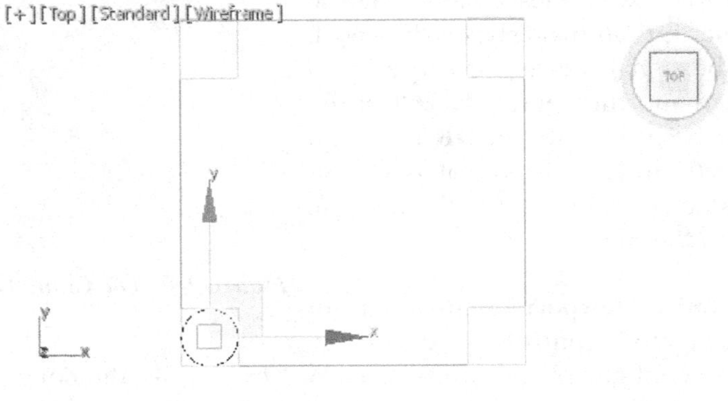

Figure 2-8 *Aligning Pillar01 to Base01 in the Top viewport*

5. Return to the four-viewport configuration by invoking the **Maximize Viewport Toggle** tool from the viewport navigation controls. Activate the Front viewport and maximize it. Next, click and drag *Pillar01* up using the Y axis handle of the gizmo such that its bottom touches the top of *Base01* in the viewport.

6. Return to the four-viewport configuration and activate the Top viewport. Next, maximize the Top viewport.

7. Make sure the **Select and Move** tool is invoked. Next, press and hold the SHIFT key and move *Pillar01* to the upper-left corner and position it on *Base02*. Next, release the left mouse button; the **Clone Options** dialog box is displayed. In this dialog box, make sure the **Copy** radio button is selected and enter **Pillar02** in the **Name** text box. Choose the **OK** button to close this dialog box; a copy with the name *Pillar02* is created.

8. Make sure that *Pillar02* is selected. Press and hold the CTRL key and select *Pillar01*.

9. Press and hold the left mouse button along with the SHIFT key and move both objects to the right using the X axis handle of the gizmo such that they are centered on the remaining two bases. Next, release the left mouse button; the **Clone Options** dialog box is displayed. In this dialog box, make sure the **Copy** radio button is selected and then enter **Pillar03** in the **Name** text box. Choose the **OK** button; two pillars are created. Name the other copy as **Pillar04**.

10. Return to the four-viewport configuration and invoke the **Zoom Extents All** tool from the viewport navigation controls, refer to Figure 2-9.

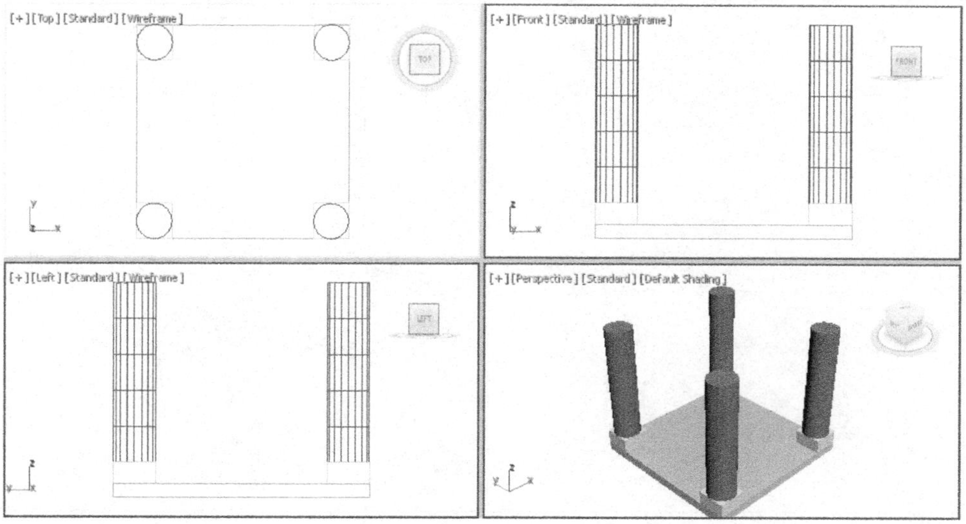

Figure 2-9 *The pillars created for the temple*

Creating the Roof

In this section, you will create the roof of the temple by using the **Sphere** tool.

1. Select *Bottom* and maximize the Front viewport. Press and hold the left mouse button along with the SHIFT key and drag *Bottom* upward. Next, release the left mouse button; the **Clone Options** dialog box is displayed. In this dialog box, make sure the **Copy** radio button is selected and then enter **Top** in the **Name** text box. Now, choose the **OK** button; a copy of *Bottom* with the name *Top* is created. Now, align it on top of the pillars using the **Select and Move** tool.

2. Return to the four-viewport configuration and invoke the **Zoom Extents All** tool to view the objects properly, refer to Figure 2-10.

3. Activate the Top viewport and choose **Create > Geometry** in the **Command Panel**. Next, invoke the **Sphere** tool.

4. In the **Keyboard Entry** rollout, set **170** in the **Radius** spinner. In the **Parameters** rollout, set **0.5** in the **Hemisphere** spinner.

   A hemisphere of 0.5 value will create one-half of a sphere.

5. Choose the **Create** button in the **Keyboard Entry** rollout. Enter the name **Dome** in the **Name and Color** rollout; the sphere is renamed as *Dome*.

6. Maximize the Front viewport. Next, invoke the **Select and Move** tool from the **Main Toolbar** and move *Dome* up such that its base is on the top surface of *Top*.

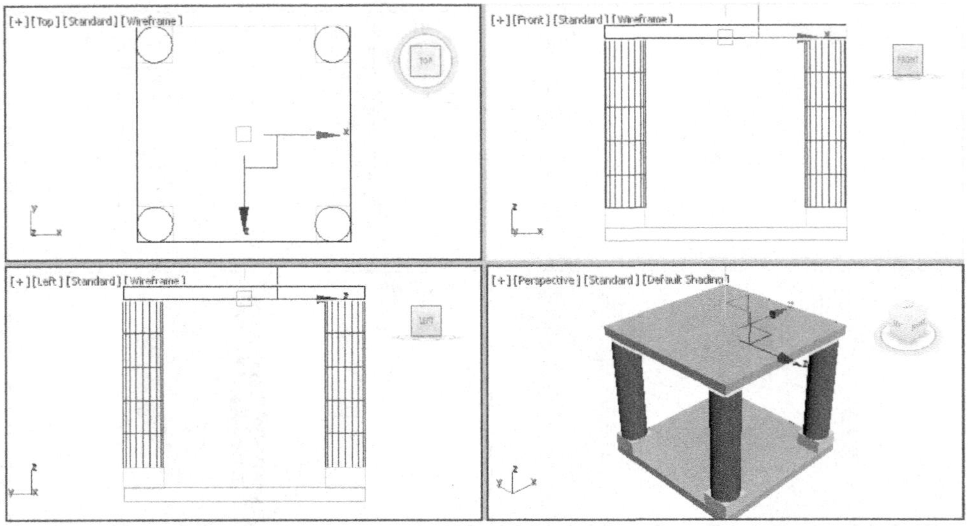

Figure 2-10 Bottom copied to create top of temple

7. Invoke the **Zoom Extents All** tool. Make sure the **Geometry** button is chosen in the **Create** tab of the **Command Panel**. Activate the Top viewport. Next, invoke the **Sphere** tool.

8. In the **Keyboard Entry** rollout, set **110** in the **Radius** spinner. In the **Parameters** rollout, set **26** in the **Segments** spinner. Next, choose the **Create** button in the **Keyboard Entry** rollout; a sphere is created with the name *Sphere001*.

9. Make sure *Sphere001* is selected and then activate the Front viewport. Next, invoke the **Align** tool from the **Main Toolbar**; the cursor changes to reflect the active command.

The **Align** tool can also be used to easily and quickly move the object.

10. Now, select *Dome*; the **Align Selection (Dome)** dialog box is displayed. Set the values as shown in Figure 2-11. Now, choose the **OK** button; *Sphere001* is now vertically aligned with the flat surface of *Dome*, as shown in Figure 2-12.

Modifying Roof Objects

In this section, you will modify roof objects using the **Boolean** tool and **Edit Mesh** modifier.

1. Maximize the Front viewport. Select *Top* from the **Scene Explorer**, as shown in Figure 2-13.

2. Select **Compound Objects** from the drop-down list below the **Geometry** button. Next, invoke the **Boolean** tool from the **Object Type** rollout; the rollouts corresponding to the **Boolean** tool are displayed. Note that Top is displayed in the **Operands** list of the **Boolean Parameters** rollout. Choose the **Add Operands** button and then select *Sphere001* in a viewport; *Sphere001* is also displayed in the **Operands** list.

3. Make sure the **Sphere001** option is selected in the **Operands** list and then choose the **Subtract** button in the **Operand Parameters** rollout; *Sphere001* is subtracted from the *Top*, thus creating a hole.

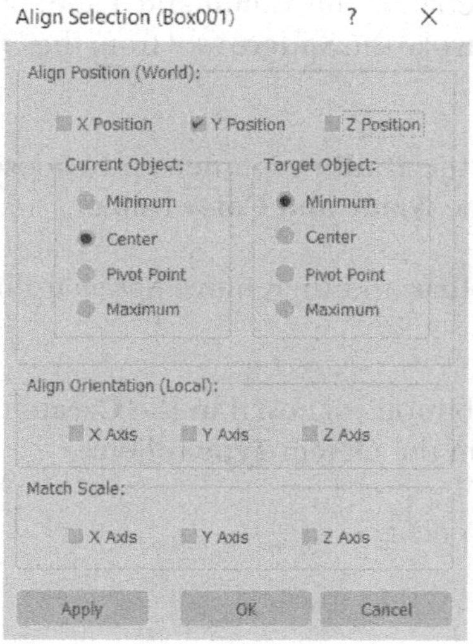

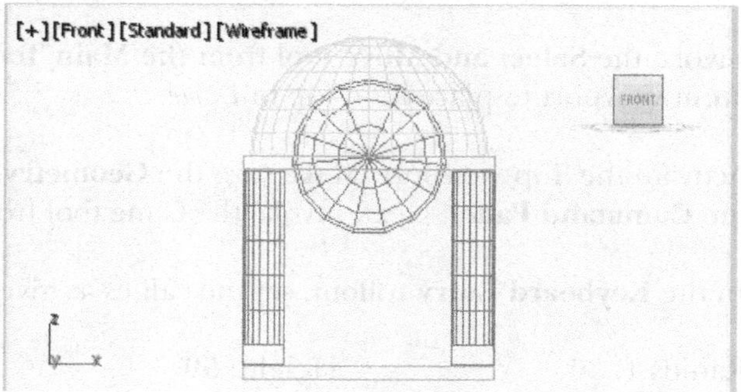

***Figure 2-11** The **Align Selection (Dome)** dialog box*

***Figure 2-12** Alignment of Sphere001 with Dome*

After performing the Boolean operation, invoke the **Select Object** tool from the **Main Toolbar** to prevent accidental selection of a different operand.

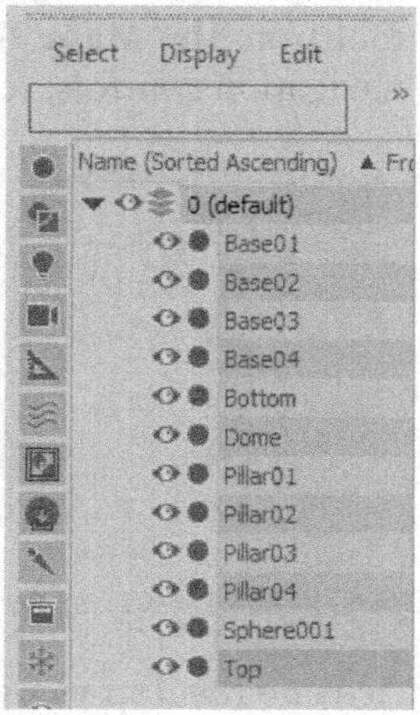

Figure 2-13** Selecting Top from the **Scene Explorer

4. Select *Dome* in the Front viewport.

5. Choose the **Modify** tab in the **Command Panel**. From the **Modifier List** drop-down list, select the **Edit Mesh** option in the **OBJECT-SPACE MODIFIER** section. Next, choose the **Vertex** button in the **Selection** rollout;  vertices of *Dome* are displayed, as shown in Figure 2-14.

   By choosing the **Vertex** button, you have entered the **Vertex** sub-object mode. The vertices which are not selected will be displayed in blue whereas the selected vertices will be displayed in red.

6. Invoke the **Select and Move** tool and then select the top single vertex of the dome and drag it down to about -3 units, using the coordinate display in the status bar. Release the left mouse button.

7. Choose the **Vertex** button from the **Selection** rollout to exit the sub-object mode. Next, return to the four-viewport configuration.

Adding Details to the Dome

In this section, you will use the **Sphere** and **Cone** tools to add more details to *Dome*.

1. Activate the Top viewport. Choose **Create > Geometry** in the **Command Panel**. Select **Standard Primitives** from the drop-down list. Then, invoke the **Sphere** tool from the **Object Type** rollout.

2. In the **Keyboard Entry** rollout, set the value **30** in the **Radius** spinner and choose the **Create** button; a sphere is created. Enter **Sphere** in the **Name and Color** rollout.

3. Invoke the **Select and Move** tool from the **Main Toolbar** and then move *Sphere* up in the Front viewport to place it on top of *Dome*.

4. Activate the Top viewport. Make sure the **Geometry** button is chosen in the **Create** tab of the **Command Panel**. Next, invoke the **Cone** tool from the **Object Type** rollout.

5. In the **Keyboard Entry** rollout, set the values as given next.

   Radius 1: **30** Height: **50**

   Choose the **Create** button; a cone is created. Enter **Cone** in the **Name and Color** rollout.

6. Activate the Front viewport and move *Cone* on the top of *Sphere* by using the **Select and Move** tool from the **Main Toolbar**. Next, invoke the **Zoom Extents All** tool from the viewport navigation controls, refer to Figure 2-15.

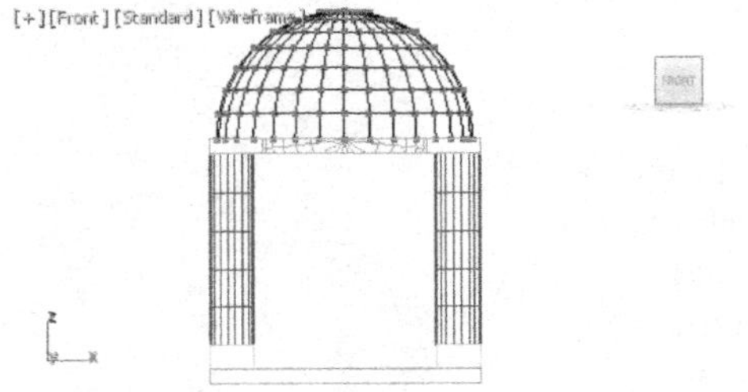

Figure 2-14 *Vertices of the Dome displayed*

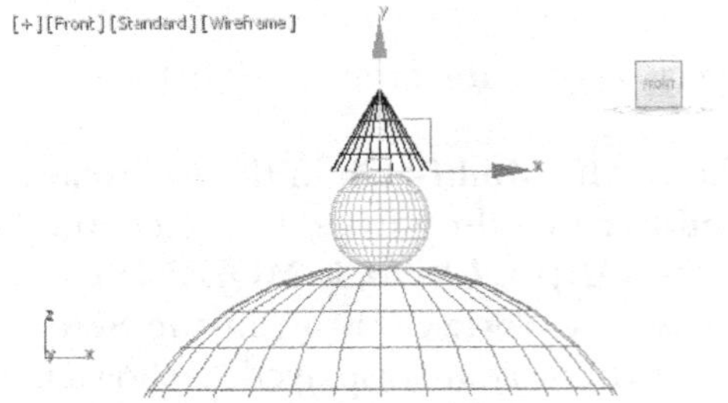

Figure 2-15 *The details added to the top of Dome*

7. Activate the Top viewport. Make sure the **Geometry** button is chosen in the **Create** tab of the **Command Panel**. Next, invoke the **Cone** tool in the **Object Type** rollout.

8. In the **Keyboard Entry** rollout, set the values as given next.

   Radius 1: **40** Radius 2: **10** Height: **70**

   Choose the **Create** button. Enter **Bell** in the **Name and Color** rollout.

9. Activate the Front viewport and invoke the **Select and Move** tool from the **Main Toolbar**. Move *Bell* such that half of it lies above *Top* and the other half of *Bell* lies below it, refer to Figure 2-16. Also, assign same color to all the objects of temple.

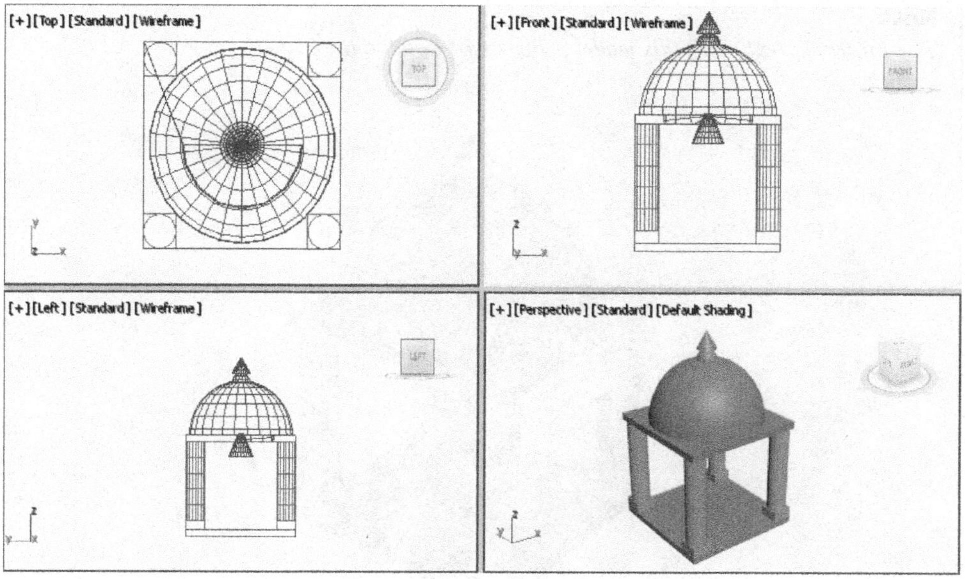

Figure 2-16 *Adjustment of Bell at the center of Dome*

Changing the Background Color of the Scene

In this section, you will change the background color of the scene.

1. Choose **Rendering > Environment** from the menu bar or press 8; the **Environment and Effects** dialog box is displayed with the **Environment** tab chosen.

2. In the **Background** area of the **Common Parameters** rollout, choose the color swatch corresponding to the **Color** parameter; the **Color Selector: Background Color** dialog box is displayed. In this dialog box, select the white color and choose the **OK** button. Next, close the **Environment and Effects** dialog box.

Saving and Rendering the Scene

In this section, you will save the scene that you have created and then render it. You can also view the final rendered image of this model by downloading the *c02_3dsmax_2022_rndr.zip* file from *www.cadcim.com*. The path of the file is as follows: *Textbooks > Animation and Visual Effects > Autodesk 3ds Max > Autodesk 3ds Max 2022 for Beginners: A Tutorial Approach*

1. Choose **Save** from the **File** menu.

2. Activate the Perspective viewport. Next, invoke the **Render Production** tool from the **Main Toolbar**; the rendered image is displayed, refer to Figure 2-17.

Tip

You can press SHIFT+Q or F9 to render the active viewport.

Note
The process of texturing this model is discussed in Chapter 11.

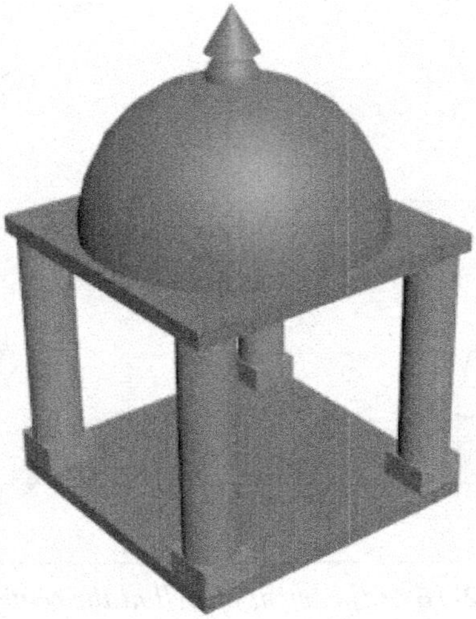

Figure 2-17 *The rendered image*

Tutorial 2

In this tutorial, you will create the model of the table and the benches, as shown in Figure 2-18. **(Expected time: 30 min)**

Figure 2-18 *The table and benches*

The following steps are required to complete this tutorial:

a. Create the project folder.
b. Create the top of the table.
c. Create the legs of the table.
d. Create the base of the table.
e. Create the rivets of the table.
f. Group the objects and smoothen the edges of table.

g. Create the benches.
h. Save and render the scene.

Creating the Project Folder

Create the project folder with the name *c02_tut2* in the *3dsmax2022* folder, as discussed in Tutorial 1.

Creating the Top of the Table

In this section, you will create the top of the table by using the **Box** tool.

1. Choose **Create > Geometry** in the **Command Panel**; **Standard Primitives** is displayed in the drop-down list located below it. Next, invoke the **Box** tool from the **Object Type** rollout; various rollouts are displayed in the **Command Panel**.

2. Expand the **Keyboard Entry** rollout and set the values as given next.

   Length: **100** Width: **70** Height: **1.5**

3. Choose the **Create** button from the **Keyboard Entry** rollout; a box is created in all viewports, refer to Figure 2-19. In the **Name and Color** rollout, enter **Top** and press the ENTER key; the box is named as *Top*.

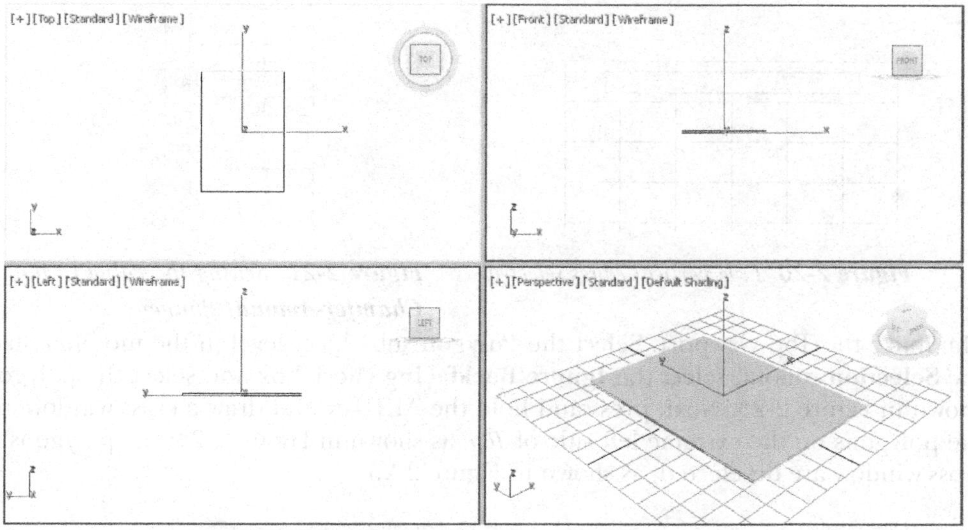

Figure 2-19 *The box created in all viewports*

4. Choose the **Modify** tab in the **Command Panel**. In the **Parameters** rollout, set the values as given next.

   Length Segs: **12** Width Segs: **6** Height Segs: **2**

   Next, you need to convert *Top* to editable poly to modify it at the sub-object level.

5. Make sure the **Modify** tab is chosen and then select the **Edit Poly** option from the **OBJECT-SPACE MODIFIERS** section of the **Modifier List** drop-down list; the **Edit Poly** modifier is displayed in the modifier stack. In the modifier stack, click on the arrow sign on the left of the **Editable Poly** to view all the sub-object levels.

6. Make sure the Top viewport is activated and press the W key to invoke the **Select and Move** tool. Next, select the **Edge** sub-object level in the modifier stack. Select five vertical edges of *Top* using the CTRL key, as shown in Figure 2-20. In the **Selection** rollout, choose the **Loop** button; five edge loops are selected.

7. In the **Edit Edges** rollout, choose the **Settings** button on the right of the **Chamfer** button; the **Chamfer** caddy control is displayed in the viewport. Set the value **0.2** in the **Chamfer-Amount** spinner, as shown in Figure 2-21, and choose the **OK** button; the selected edges are chamfered, as shown in Figure 2-22.

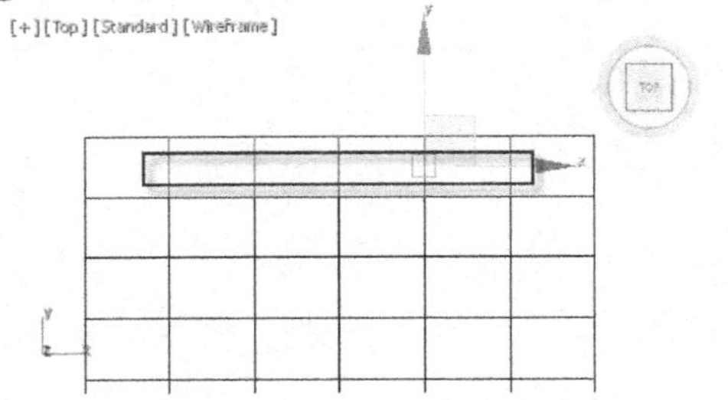

Figure 2-20 *Five vertical edges selected*

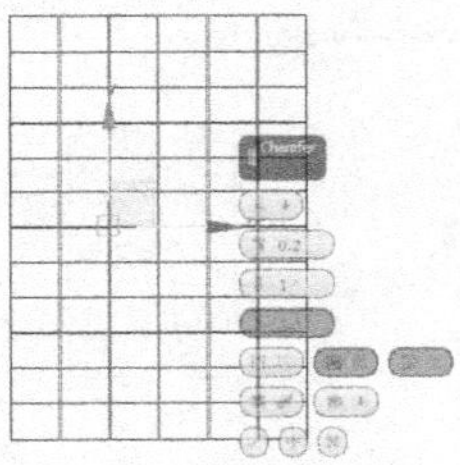

Figure 2-21 *Setting the value in the **Chamfer-Amount** spinner*

8. Maximize the Top viewport. Select the **Polygon** sub-object level in the modifier stack. In the **Selection** rollout, select the **Ignore Backfacing** check box and select the polygons, as shown in Figure 2-23. Next, press and hold the ALT key and draw a cross window around the polygons on the extreme left side of *Top*, as shown in Figure 2-24; the polygons in the cross window are deselected, as shown in Figure 2-25.

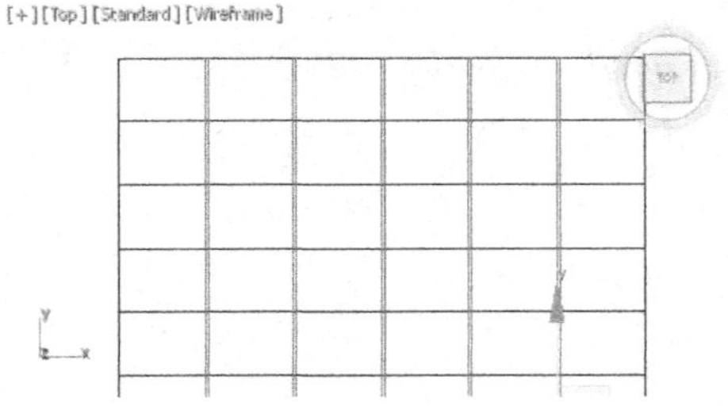

Figure 2-22 *The edges chamfered*

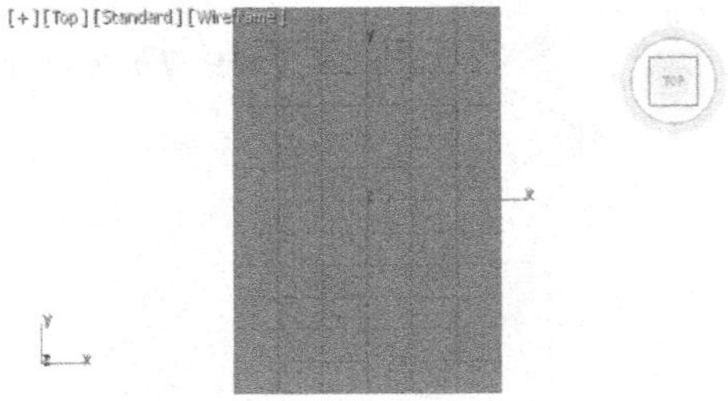

Figure 2-23 *The polygons selected*

The **Ignore Backfacing** check box is selected to avoid selection of polygons from the opposite side of the object. You will notice that the polygons on the bottom of *Top* are not selected.

9. Similarly, press the ALT key and deselect the polygons, refer to Figure 2-26. Notice that only the polygons between the chamfered edges remain selected. Now, delete these polygons.

Creating the Legs of the Table

In this section, you will create the legs of the table by extruding the polygons of *Top*.

1. Activate the Bottom viewport and select the **Edge** sub-object level in the modifier stack. Select two horizontal edges from the viewport by using the CTRL key, as shown in Figure 2-27. Next, in the **Selection** rollout, choose the **Loop** button; two edge loops are selected.

Figure 2-24 *Drawing a cross window around the polygons*

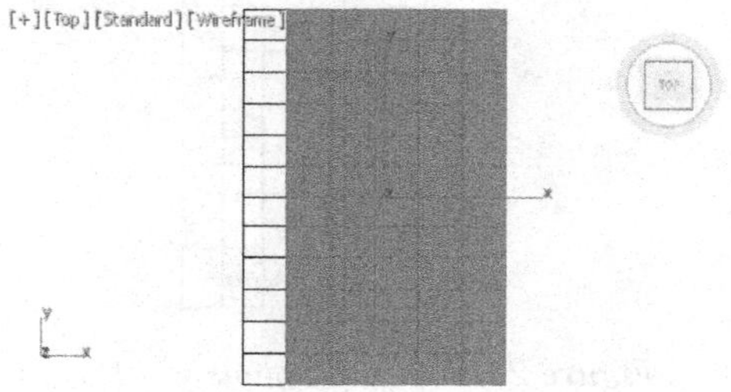

Figure 2-25 *The polygons deselected*

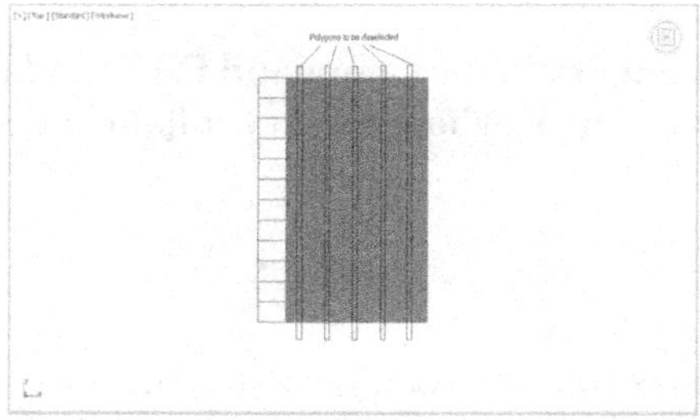

Figure 2-26 *The polygons to be deselected*

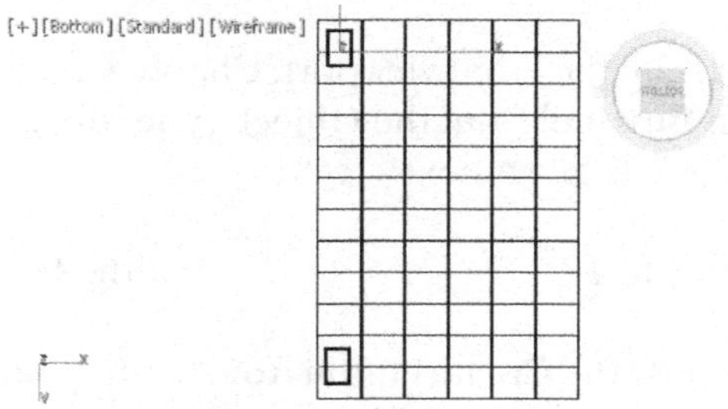

Figure 2-27 *Two horizontal edges selected in the Bottom viewport*

Note

*To activate the Bottom viewport, click on the POV viewport label in the Top viewport; a flyout is displayed. Choose the **Bottom** option; the Top viewport is switched to the Bottom viewport. Similarly, you can switch to any viewport by clicking on the POV viewport label.*

2. In the **Edit Edges** rollout, choose the **Settings** button on the right of the **Chamfer** button; the **Chamfer** caddy control is displayed in the viewport. Set the value **4** in the **Chamfer-Amount** spinner and **2** in the **Chamfer-Segments** spinner. Next, choose the **OK** button; the selected edges are chamfered.

3. Select the **Polygon** sub-object level from the modifier stack and select the polygons in the Bottom viewport by using the CTRL key, as shown in Figure 2-28. In the **Edit Polygons**

rollout, choose the **Settings** button on the right of the **Extrude** button; the **Extrude Polygons** caddy control is displayed in the viewport. Set **7** in the **Extrude Polygons-Height** spinner, and choose the **OK** button; the selected polygons are extruded.

4. Make sure the Bottom viewport is activated. Again, select the polygons by using the CTRL key, as shown in Figure 2-29. In the **Edit Polygons** rollout, choose the **Settings** button on the right of the **Extrude** button; the **Extrude Polygons** caddy control is displayed in the viewport. Set **50** in the **Extrude Polygons-Height** spinner, and choose the **OK** button; the selected polygons are extruded, refer to Figure 2-30.

   Next, you need to add edges to the extruded part of legs to add some details to them.

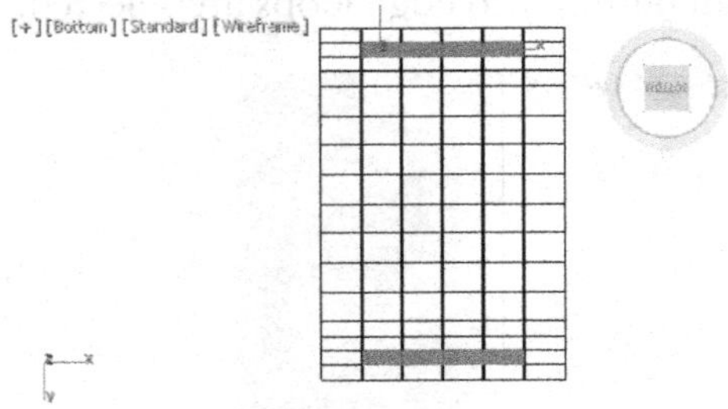

Figure 2-28 *The polygons selected in the Bottom viewport*

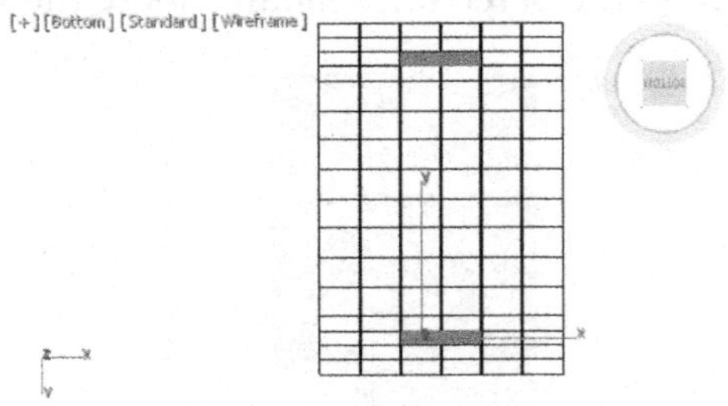

Figure 2-29 *The polygons selected*

Creating the Base of the Table

In this section, you will create the base of the table using the **Box** tool and the **Edit Poly** modifier.

1. Activate the Front viewport. Choose **Create > Geometry** in the **Command Panel** and invoke the **Box** tool from the **Object Type** rollout. Expand the **Keyboard Entry** rollout and set the values as given next.

   Length: **12** Width: **45** Height: **5**

2. Choose the **Create** button from the **Keyboard Entry** rollout; a box is created in all viewports. In the **Name and Color** rollout, enter **base1** and press ENTER; the box is named as *base1*. Now, invoke the **Select and Move** tool from the **Main Toolbar** and move *base1* at the bottom of the leg.

3. Choose the **Modify** tab in the **Command Panel**. Next, in the **Parameters** rollout, set the values as given next.

   Length Segs: **4** Width Segs: **12** Height Segs: **2**

4. Make sure the **Modify** tab is chosen and then select the **Edit Poly** modifier from the **OBJECT-SPACE MODIFIERS** section of the **Modifier List** drop-down list; the **Edit Poly** modifier is displayed in the modifier stack. In the modifier stack, click on the arrow on the left of the **Editable Poly** to view all sub-object levels.

5. Select the **Vertex** sub-object level. Make sure the Front viewport is activated. Next, select the vertices using the CTRL key, as shown in Figure 2-31 and move them slightly downward. Similarly, move other vertices of *base1* to get a shape, as shown in Figure 2-32. Next, select the **Vertex** sub-object level again in the modifier stack to deactivate it.

6. Activate the Bottom viewport and then switch to the Top viewport, as discussed earlier. Next, align *base1* in all viewports, as shown in Figure 2-33.

   Next, you need to create a copy of *base1* which will act as a base for the other leg.

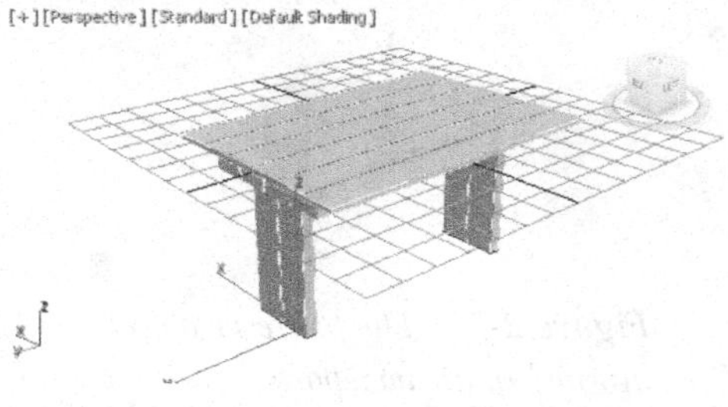

Figure 2-30 *The polygons extruded*

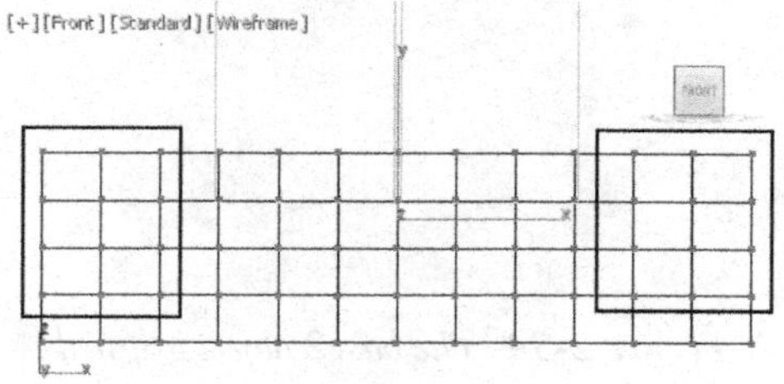

Figure 2-31 *The vertices to be selected*

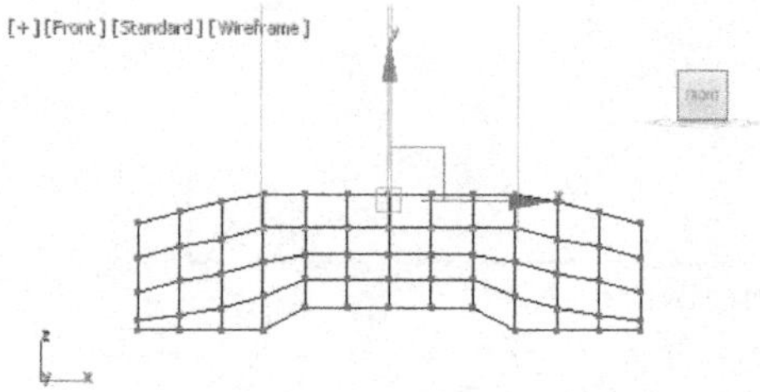

Figure 2-32 *The shape of base1*

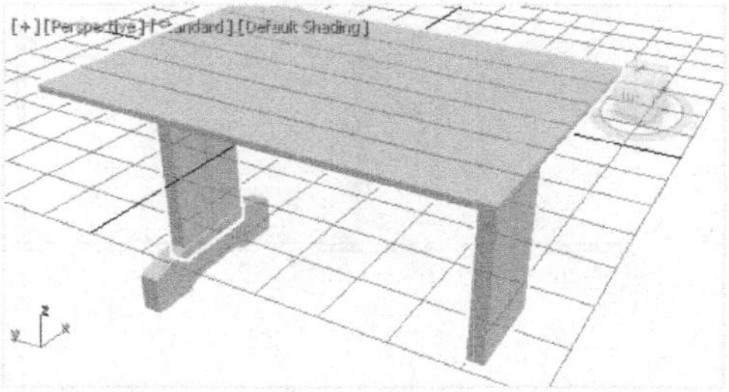

Figure 2-33 *The base1 object aligned in all viewports*

7. Activate the Left viewport and make sure *base1* is selected. Press and hold the left mouse button along with the SHIFT key and then move the cursor toward the other leg. Next, release the left mouse button; the **Clone Options** dialog box is displayed. In this dialog box, make sure the **Copy** radio button is selected in the **Object** area and 1 is displayed in the **Number of Copies** spinner. Now, enter **base2** in the **Name** text box and choose the **OK** button; a copy of *base1* with the name *base2* is created.

8. Align *base2* in all the viewports to position it at the bottom of the other leg, as shown in Figure 2-34.

   Next, you need to create the footrests for *table*.

9. Make sure the Left viewport is activated. Choose **Create > Geometry** in the **Command Panel** and invoke the **Box** tool from the **Object Type** rollout. Expand the **Keyboard Entry** rollout and set the values as given next.

   Length: **5** Width: **80** Height: **2**

10. Choose the **Create** button from the **Keyboard Entry** rollout; a box is created in all viewports. Enter **footrest1** in the **Name and Color** rollout, and press the ENTER key; the box is named as *footrest1*. Invoke the **Select and Move** tool from the **Main Toolbar** and align *footrest1* in all viewports, as shown in Figure 2-35.

11. Create the copy of *footrest1* in the Front viewport as done in Step 7 and then align it in all viewports, as shown in Figure 2-36.

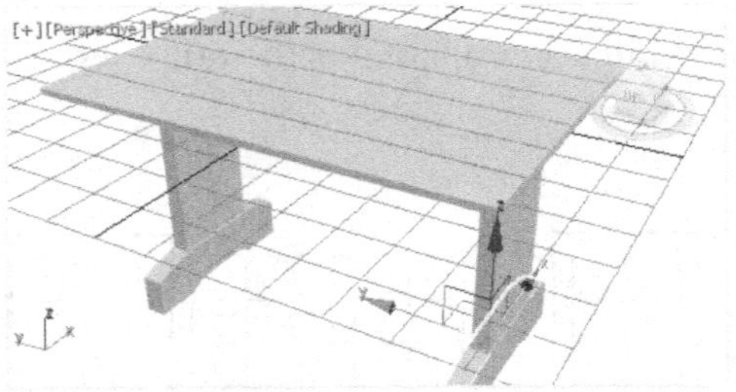

***Figure 2-34** The base2 object aligned in all viewports*

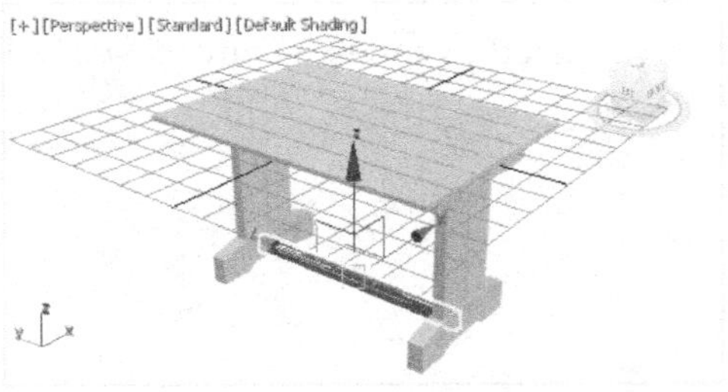

***Figure 2-35** The footrest1 object aligned in all viewports*

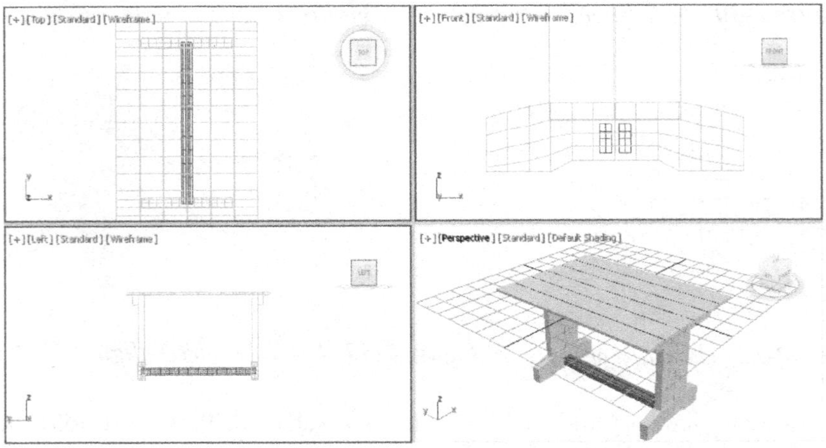

***Figure 2-36** The copy of footrest1 object aligned in all viewports*

Creating the Rivets of the Table

In this section, you will create a sphere and a cylinder to create the cap and the body of rivet, respectively.

1. Activate the Top viewport and choose **Create > Geometry** in the **Command Panel**; the **Standard Primitives** option is displayed in the drop-down list. Next, invoke the **Sphere** tool from the **Object Type** rollout; various rollouts are displayed in the **Command Panel**.

2. Expand the **Keyboard Entry** rollout and set the value **0.65** in the **Radius** spinner and choose the **Create** button; a sphere is created in all the viewports.

3. In the **Parameters** rollout, make sure that the **Smooth** check box and the **Chop** radio button are selected. Set the value **0.5** in the **Hemisphere** spinner.

4. Enter **cap** in the **Name and Color** rollout and use the color swatch in this rollout to change the color to black.

   Next, you need to create a cylinder for the body of the rivet.

5. Choose the **Cylinder** tool from **Create > Geometry > Standard Primitives > Object Type** rollout in the **Command Panel**.

6. In the **Keyboard Entry** rollout, set the values as given next.

   Radius: **0.241** Height: **3.5**

7. Choose the **Create** button from the **Keyboard Entry** rollout; a cylinder is created.

8. Enter **body** in the **Name and Color** rollout and use the color swatch in this rollout to change the color to black. Next, right-click to exit the **Cylinder** tool.

9. Activate the Left viewport and invoke the **Name** tool from the **Main Toolbar**; the **Select From Scene** dialog box is displayed. In this dialog box, select *cap* and *body* simultaneously by holding the CTRL key and then choose the **OK** button; the *cap* and the *body* are selected in the viewport.

Note

*You can also select cap and body from the **Scene Explorer** located at the left of the interface.*

10. Invoke the **Zoom Extents All Selected** tool from the **Main Toolbar** to magnify the selected objects in all viewports and click on an empty area of the screen to deselect the objects.

11. Align *cap* and *body* of the rivet in the Left viewport using the **Select and Move** tool, as shown in Figure 2-37.

Note

*You can use the **Zoom Region** and **Pan View** tools from the viewport navigation controls to align the cap and body of the rivet.*

Next, you need to group *body* and *cap* to create the rivet.

12. Select *cap* and *body* of the rivet in the **Scene Explorer**. Choose **Group > Group** from the menu bar; the **Group** dialog box is displayed. In this dialog box, enter **rivet001** in the **Group Name** text box and then choose the **OK** button; the **rivet001** group is created.

    Next, you need to align *rivet001*.

13. In the Left viewport, make sure *rivet001* is selected. Next, right-click on the **Select and Rotate** tool in the **Main Toolbar**; the **Rotate Transform Type-In** dialog box is displayed. Also, a circular gizmo along with the X, Y, and Z axes is displayed.

14. In the **Offset:Screen** area of the **Rotate Transform Type-In** dialog box, set **-90** in the **Z** spinner and press the ENTER key; *rivet001* gets rotated, as shown in Figure 2-38. Now, close the **Rotate Transform Type-In** dialog box.

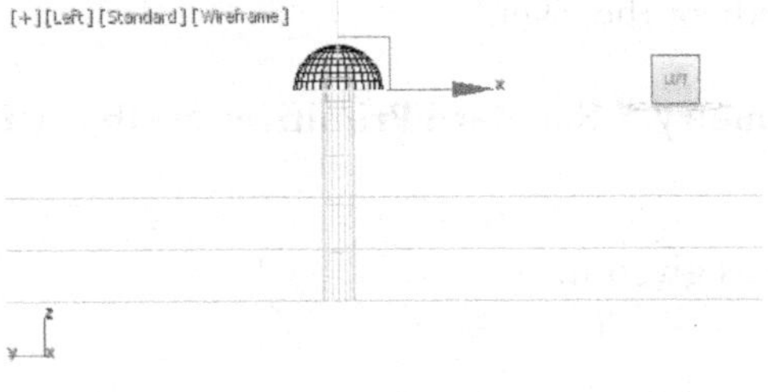

Figure 2-37 *The cap and body objects aligned in all viewports*

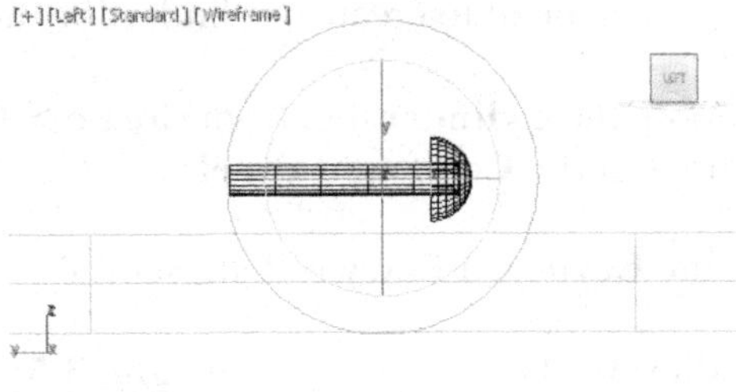

Figure 2-38 *The rivet001 object rotated*

15. Activate the Perspective viewport. Invoke the **Select and Place** tool from the **Main Toolbar**. Next, drag *rivet001* and place it on the *base2*. Now, align *rivet001* in all viewports, as shown in Figure 2-39.

The **Select and Place** tool is used to accurately position an object on the other object.

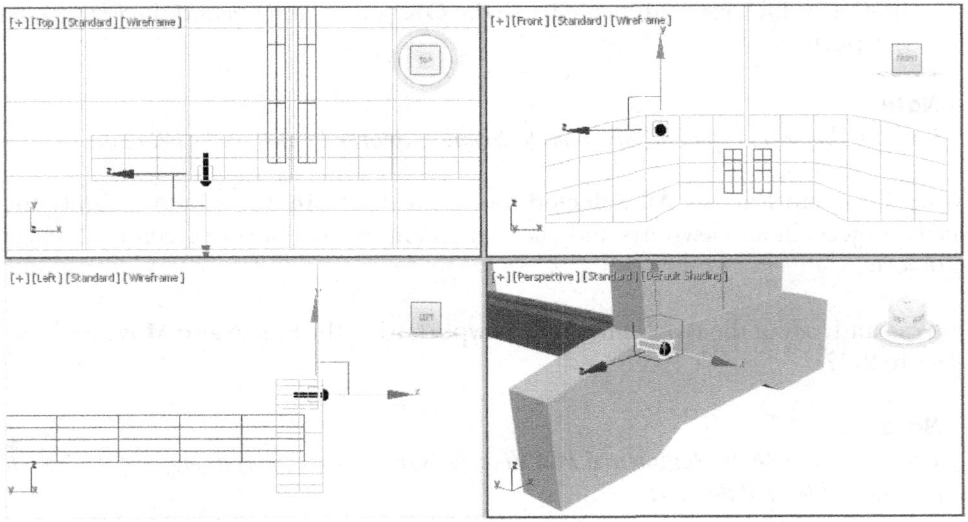

Figure 2-39 *The rivet001 object aligned in all viewports*

16. Activate the Front viewport. Create 3 more copies of *rivet001* as done earlier and align them in all viewports, as shown in Figure 2-40.

Next, you will use the **Mirror** tool to create a copy of these rivets.

17. Activate the Left viewport and select all rivets. Invoke the **Mirror** tool from the **Main Toolbar**; the **Mirror: Screen Coordinates** dialog box is displayed. In this dialog box, make sure the **X** radio button is selected in the **Mirror Axis** area and set the value **-83** in the **Offset** spinner. Next, select the **Copy** radio button in the **Clone Selection** area and choose the **OK** button; the copy of rivets is created, as shown in Figure 2-41.

Note

You may need to change the value in the ***Offset*** *spinner as per your requirement.*

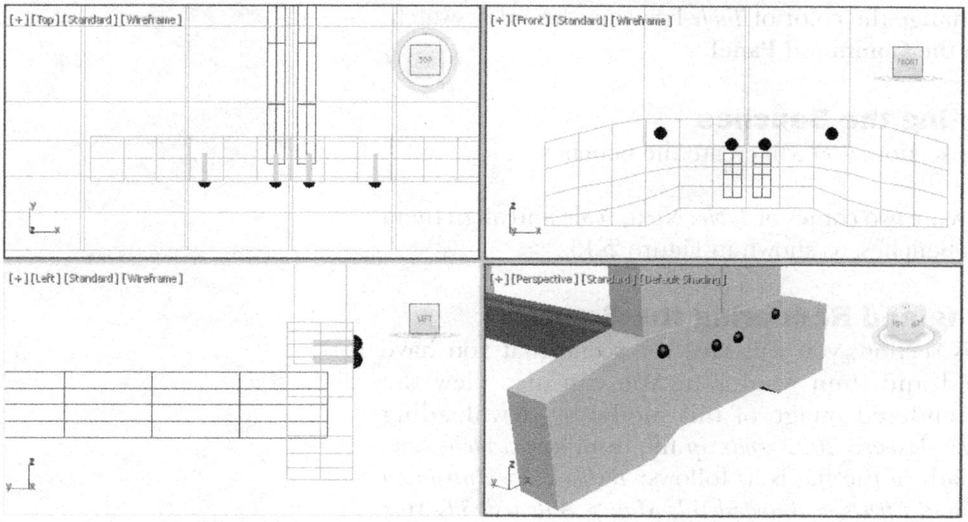

Figure 2-40 *The copies of rivet001 aligned in all viewports*

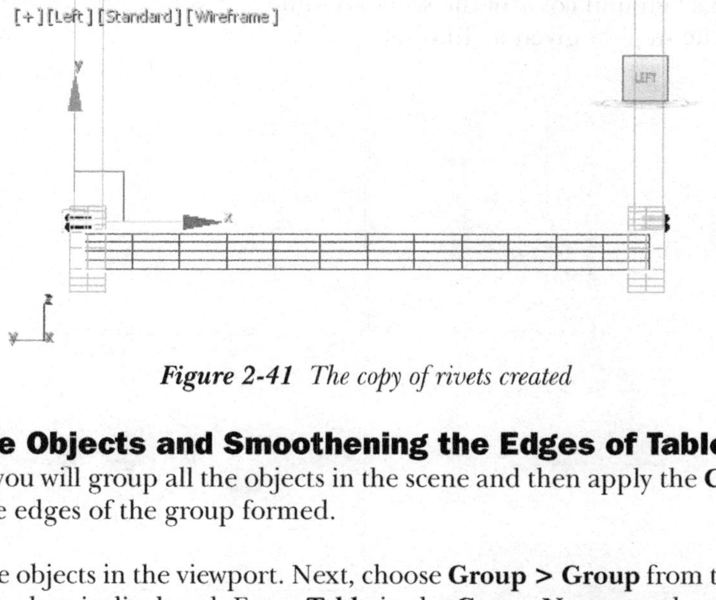

Figure 2-41 *The copy of rivets created*

Grouping the Objects and Smoothening the Edges of Table

In this section, you will group all the objects in the scene and then apply the **Chamfer** modifier to smoothen the edges of the group formed.

1. Select all the objects in the viewport. Next, choose **Group > Group** from the menu bar; the **Group** dialog box is displayed. Enter **Table** in the **Group Name** text box and then choose **OK** to close the dialog box.

2. Make sure *Table* is selected. Next, click on the **Modifier List** drop-down list in the **Command Panel** and then select **Chamfer** from it; the **Chamfer** modifier is applied to *Table* and is displayed in the modifier stack. Also, the **Parameters** rollout is displayed in the Modify panel.

3. In the **Parameters** rollout, set the parameters, as shown in Figure 2-42. You will notice all the edges of *Table* are smoothened.

4. Change the color of *Table* by using the color swatch in the **Command Panel**.

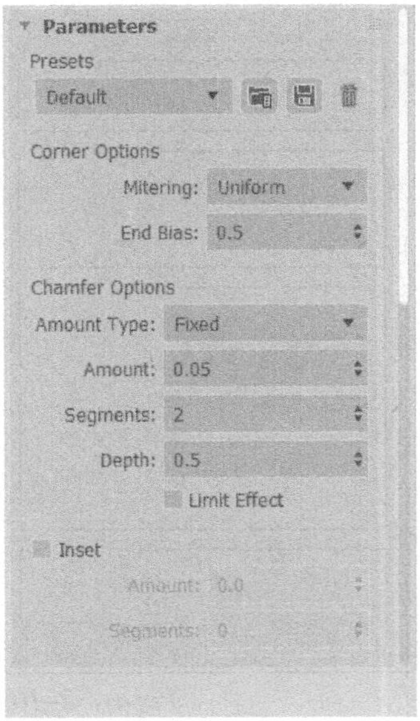

***Figure 2-42** The **Parameters** rollout*

Creating the Benches

In this section, you will create the benches.

1. Create two copies of *Table*. Next, scale and align them as benches, as shown in Figure 2-43.

Saving and Rendering the Scene

In this section, you will save the scene that you have created and then render it. You can also view the final rendered image of this model by downloading the *c02_3dsmax_2022_rndr.zip* file from *www.cadcim.com*. The path of the file is as follows: *Textbooks > Animation and Visual Effects > Autodesk 3ds Max > Autodesk 3ds Max 2022 for Beginners: A Tutorial Approach*

1. Change the background color of the scene to white by following the steps as given in Tutorial 1.

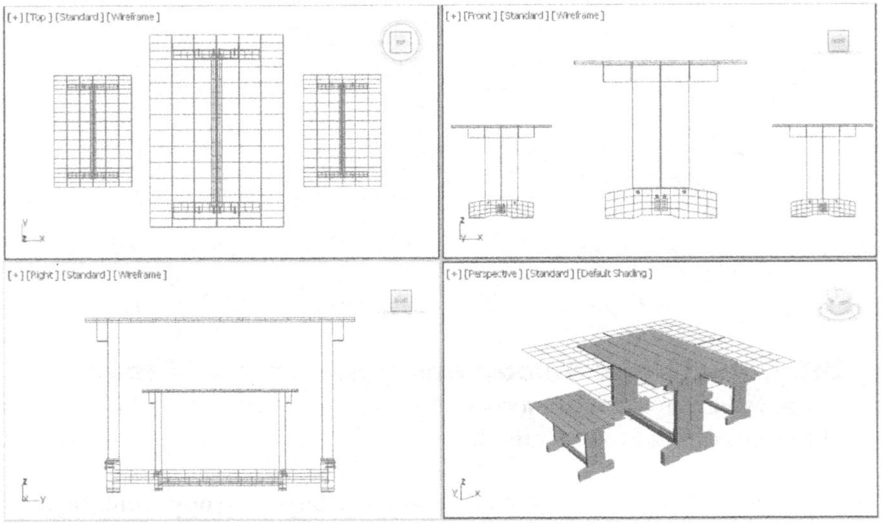

***Figure 2-43** The table and benches aligned*

2. Choose **Save** from the **File** menu.

3. Activate the Perspective viewport. Next, invoke the **Render Production** tool from the **Main Toolbar**; the rendered image is displayed, refer to Figure 2-44.

Figure 2-44 The rendered image

Self-Evaluation Test

Answer the following questions and then compare them to those given at the end of this chapter:

1. Which of the following shortcut keys is used to toggle between the active viewport and four viewport display?

   (a) ALT+W (b) ALT+S
   (c) SHIFT+Q (d) None of these

2. Which of the following tools is used to accurately position an object on the other object?

   (a) **Select and Scale** (b) **Select and Place**
   (c) **Select and Rotate** (d) **Select and Squash**

3. You need to press and hold the _________ key while moving the object using the **Select and Move** tool to create a copy of an object.

4. You need to set the value in the _________ spinner of the **Parameters** rollout to create a hemisphere.

5. You need to choose the **Geometry** button in the **Create** tab of the **Command Panel** and then select **Compound Objects** from the drop-down list to apply boolean operations on the objects. (T/F)

6. The **Ignore Backfacing** check box in the **Selection** rollout is used to avoid selection of polygons from the opposite side of the object. (T/F)

Review Questions

Answer the following questions:

1. Which of the following shortcut keys is used to render the active viewport?

   (a) ALT+Q (b) SHIFT+Q
   (c) CTRL+S (d) None of these

2. Which of the following options provides the additional functionality to select, rename, delete, freeze, and hide the objects in the viewports?

   (a) **Select Object** (b) **Select and Manipulate**
   (c) **Scene Explorer** (d) **Schematic View**

3. The __________ tool available in the **Main Toolbar** is used to align the current object with the target object.

4. The __________ dialog box is used to create the copies of the selected object(s).

5. The __________ key is used to toggle the display of grids in the viewport.

6. The **Edit Mesh** modifier is used to modify the object at sub-object level. (T/F)

EXERCISE

The rendered output of the model used in the following exercise can be accessed by downloading the *c02_3dsmax_2022_exr.zip* from *www.cadcim.com*. The path of the file is as follows: *Textbooks > Animation and Visual Effects > Autodesk 3ds Max > Autodesk 3ds Max 2022 for Beginners: A Tutorial Approach*

Exercise 1

Create the model of chairs and center table, as shown in Figure 2-45.

Figure 2-45 *The model of chairs and center table*

Answers to Self-Evaluation Test

1. a, **2.** b, **3.** SHIFT, **4. Hemisphere**, **5.** T, **6.** T

Chapter 3

Primitive Objects-II

Learning Objectives

After completing this chapter, you will be able to:

- *Create primitive objects*
- *Modify primitive objects*
- *Use the Boolean and Array tools*

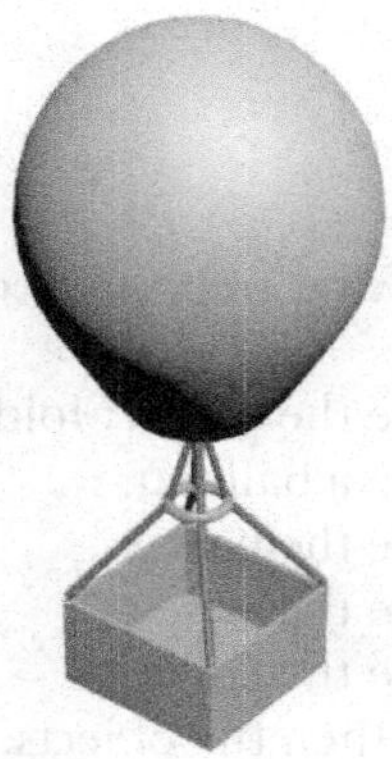

INTRODUCTION

In this chapter, you will create primitive and extended primitive objects. You will also modify them using the **Edit Poly** and **Edit Mesh** modifiers. In addition to this, you will learn to use the **Boolean** and **Array** tools.

TUTORIALS

Tutorial 1

In this tutorial, you will create the model of a hot air balloon consisting of a balloon, a basket, ropes, and a collar, refer to Figure 3-1. You will create these components using primitive objects. **(Expected time: 25 min)**

Figure 3-1 *The model of hot air balloon*

The following steps are required to complete this tutorial:

a. Create the project folder.
b. Create a balloon.
c. Create the basket.
d. Create the ropes.
e. Create the collar.
f. Smoothen the objects.
g. Save and render the scene.

Creating the Project Folder

Create the project folder with the name *c03_tut1* in the *3dsmax2022* folder as discussed in Tutorial 1 of Chapter 2.

Creating a Balloon

In this section, you will create a balloon using the **Sphere** tool.

1. Activate the Top viewport and choose **Create > Geometry** in the **Command Panel**. Now, invoke the **Sphere** tool from the **Object Type** rollout.

2. In the **Keyboard Entry** rollout, set **150** in the **Radius** spinner. In the **Parameters** rollout, set **16** in the **Segments** spinner. Next, choose the **Create** button in the **Keyboard Entry** rollout; a sphere is created. Enter **Balloon** in the **Name and Color** rollout. Right-click in the viewport to exit the tool.

3. Activate the Perspective viewport. Next, invoke the **Zoom Extents All** tool from viewport navigation controls to zoom out the object in the Perspective viewport, refer to Figure 3-2.

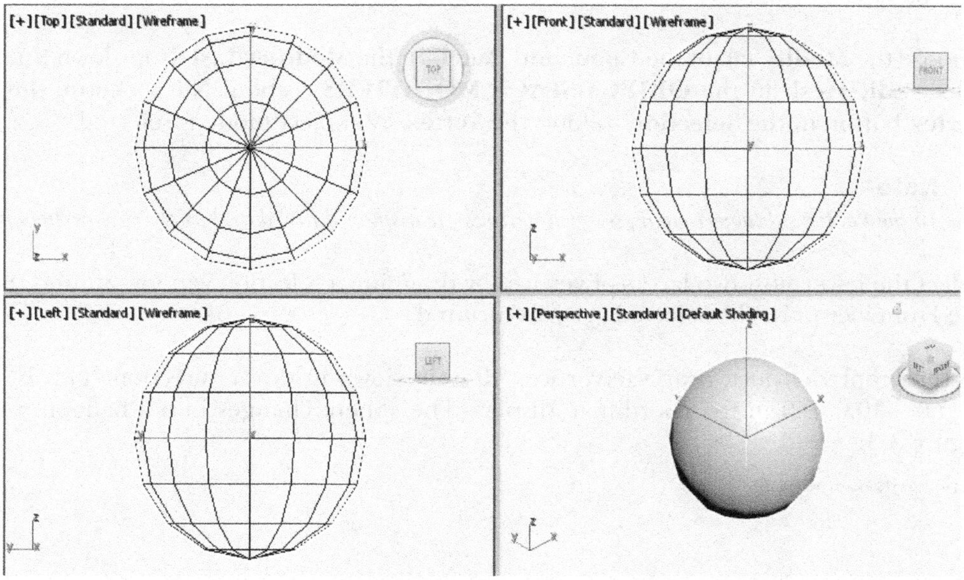

***Figure 3-2** The sphere created and zoomed in all four viewports*

4. Activate the Top viewport and click on the **Wireframe** Shading viewport label in the Top viewport to display a flyout. Choose **Default Shading** from the flyout to display the mesh in the Top viewport.

5. Invoke the **Zoom All** tool from the viewport navigation control. Next, hold the left mouse button and move the cursor downward in any viewport to zoom out to display *Balloon*. Right-click to exit the tool.

The **Zoom All** tool simultaneously zooms in/out all viewports equally.

6. Activate the Front viewport and invoke the **2D Snap** tool from the **Snaps Toggle** flyout to turn on snaps. Right-click on this tool to open the **Grid and Snap Settings** dialog box. In the **Snaps** tab of this dialog box, make sure the **Grid Points** check box is selected and all other check boxes are cleared.

7. Close the **Grid and Snap Settings** dialog box.

8. Choose the **Transform Gizmo Y Constraint** button from the **Axis Constraints** toolbar.

Note

*The **Axis Constraints** tools do not normally work with snaps on. To overcome this problem, right-click on the **Snaps Toggle** tool to display the **Grid and Snap Settings** dialog box. Choose the **Options** tab and make sure the **Enable Axis Constraints** check box is selected in the **Translation** area.*

9. Maximize the Front viewport and zoom it, if needed, to display the subgrid. Invoke the **Select and Move** tool from the **Main Toolbar**. Select *Balloon*, if it is not already selected.

10. Choose the **Modify** tab in the **Command Panel**. In the **Modifier List** drop-down list, select **Edit Mesh** in the **OBJECT-SPACE MODIFIERS** section. Next, choose the **Vertex** button in the **Selection** rollout; the **Vertex** sub-object mode is selected.

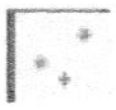

Note

*In the **Vertex** sub-object mode, selected vertices are displayed in red and unselected vertices in blue.*

11. Select the lowermost two layers of vertices by dragging a selection window around them in the Front viewport; the selected vertices turn red.

12. In the Front viewport, drag the vertices 40 units down in the -Y direction. This is shown as 0.0, –40.0, 0.0 in the coordinate display. The sphere changes into a balloon, refer to Figure 3-3.

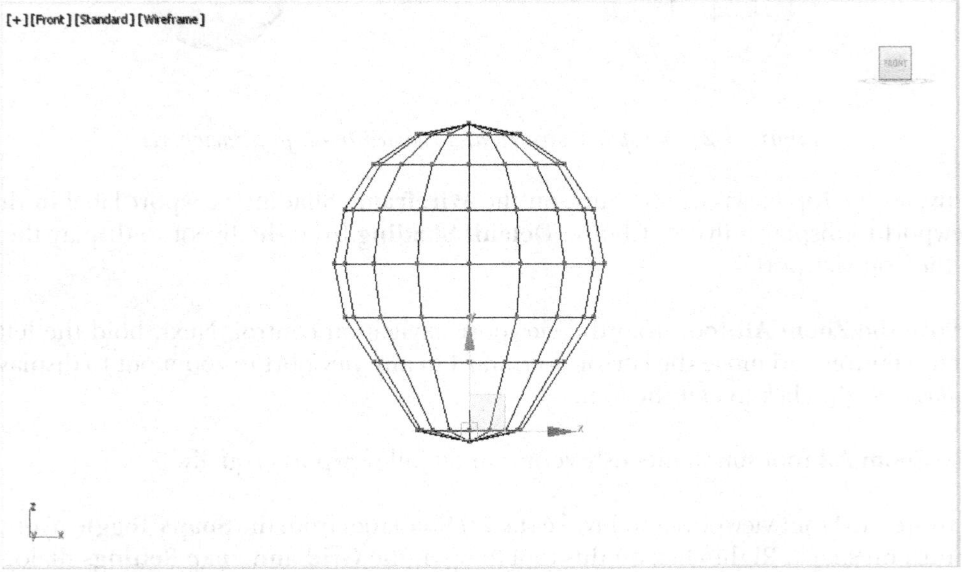

Figure 3-3 *The shape of the sphere changes into balloon shape*

Alternatively, right-click on the **Select and Move** tool from the **Main Toolbar**; the **Move Transform Type-In** dialog box is displayed. Next, set the value **–40** in the **Y** spinner of the **Offset: Screen** area.

13. Choose the **Vertex** button again in the **Selection** rollout to exit the sub-object mode. Return to four viewport configuration.

Creating the Basket

In this section, you will create a basket using the **Box** and **Boolean** tools.

1. Activate the Top viewport. Choose **Create > Geometry** in the **Command Panel**. Next, invoke the **Box** tool.

2. Expand the **Keyboard Entry** rollout and set the values as given next:

   Length: **160** Width: **160** Height: **80**

   Choose the **Create** button in the **Keyboard Entry** rollout. Enter **Basket01** in the **Name and Color** rollout; the box is named as *Basket01*.

3. Now, again in the **Keyboard Entry** rollout, set the values as given next:

   Length: **150** Width: **150** Height: **100**

   Next, choose the **Create** button and then enter **Basket02** in the **Name and Color** rollout; *Basket02* is created and it protrudes through the top of *Basket01*. Right-click to exit the **Box** tool.

4. Press and hold the CTRL key and select *Basket01* and *Basket02* in the **Scene Explorer**, as shown in Figure 3-4.

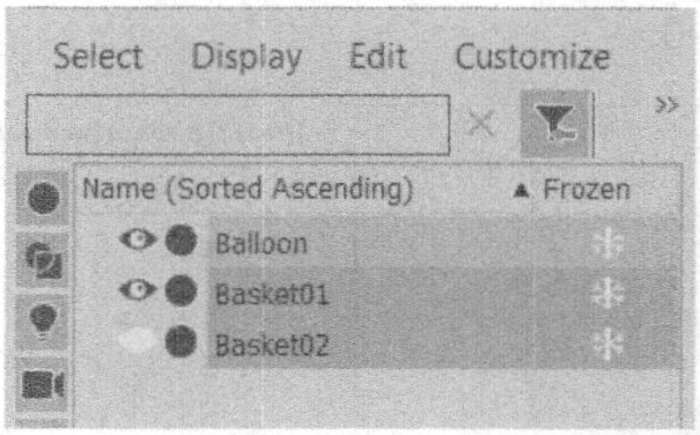

Figure 3-4 *Selecting Basket01 and Basket02 from the* ***Scene Explorer***

5. Activate the Front viewport and then invoke the **Maximize Viewport Toggle** tool from the viewport navigation controls. Next, invoke the **Select and Move** tool from the **Main Toolbar**.

6. You need to move both baskets down vertically by 430 units. To do so, right-click on the **Select and Move** tool from the **Main Toolbar**; the **Move Transform Type-In** dialog box is displayed. Next, set **–430** in the **Y** spinner in the **Offset: Screen** area and press the ENTER key, refer to Figure 3-5.

7. Close the **Move Transform Type-In** dialog box.

8. Select *Basket02* in the **Scene Explorer**; *Basket01* is deselected, while *Basket02* remains selected.

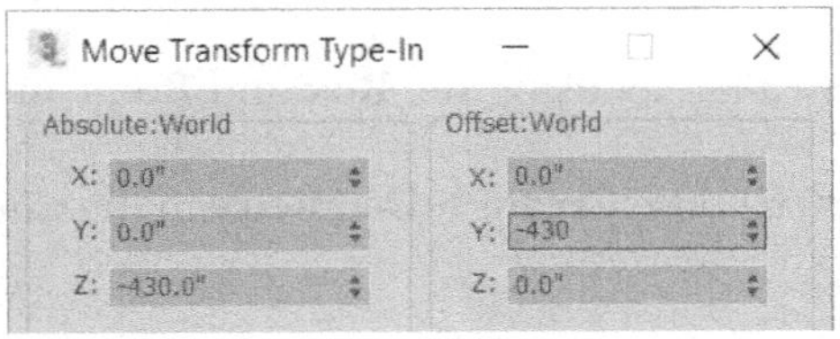

***Figure 3-5** The **Move Transform Type-In** dialog box*

9. Right-click on the **Select and Move** tool; the **Move Transform Type-In** dialog box is displayed. Next, set **10** in the **Y** spinner in the **Offset:Screen** area and press the ENTER key; *Basket02* is moved up by 10 units in the Front viewport. Next, close this dialog box.

10. Select *Basket01* in the **Scene Explorer**. Make sure the **Geometry** button is chosen in the **Create** tab of the **Command Panel**. Next, select **Compound Objects** in the drop-down list below the **Geometry** button.

11. In the **Object Type** rollout, invoke the **Boolean** tool. Choose the **Add Operands** button in the **Boolean Parameters** rollout and select *Basket02* from the Scene Explorer. Next, choose the **Subtract** button from the **Operand Parameters** rollout; *Basket02* is subtracted from *Basket01* thus creating a hollow basket. Invoke the **Select Object** tool from the **Main Toolbar** to complete the operation.

12. Invoke the **Maximize Viewport Toggle** tool from the viewport navigation controls. Next, invoke the **Zoom Extents All** tool from the viewport navigation controls to view all objects properly, refer to Figure 3-6.

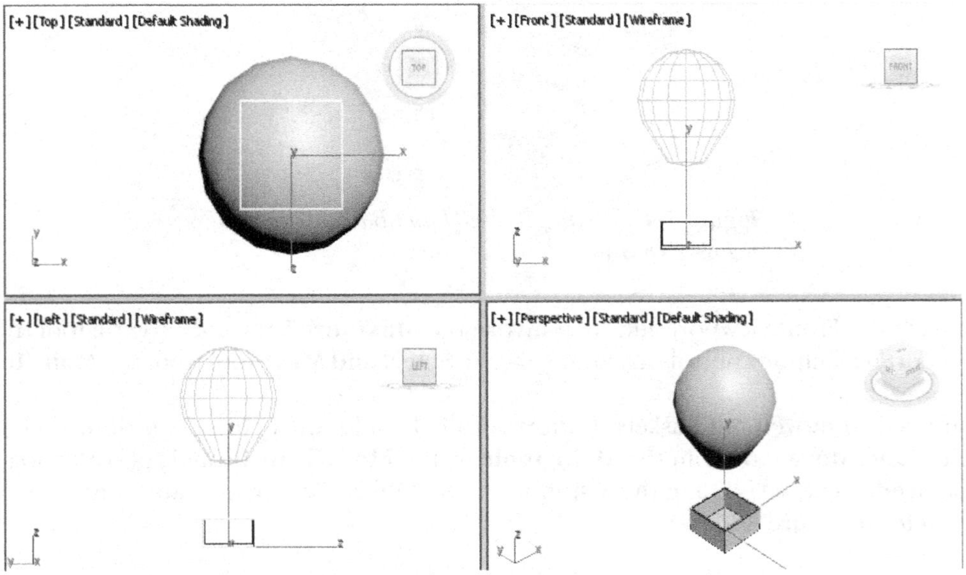

***Figure 3-6** Basket01 after performing the **Boolean** operation*

Creating the Ropes

In this section, you will create ropes using the **Cylinder** and **Array** tools.

1. Activate the Top viewport. Click on the **Default Shading** viewport label on the top-left side of the viewport and then choose **Wireframe Override** from the flyout displayed.

2. Make sure the **Geometry** button is chosen in the **Create** tab of the **Command Panel**. Next, select **Standard Primitives** in the drop-down list below the **Geometry** button. Invoke the **Cylinder** tool from the **Object Type** rollout.

3. In the **Parameters** rollout, set **5** in the **Height Segments** spinner. In the **Keyboard Entry** rollout, set **4** in the **Radius** spinner and **150** in the **Height** spinner. Next, choose the **Create** button. Enter **Rope01** in the **Name and Color** rollout; *Rope01* is created.

4. Invoke the **Select and Rotate** tool from the **Main Toolbar** and choose the **Transform Gizmo X Constraint** button from the **Axis Constraints** toolbar. Turn the snap off. Invoke the **Angle Snap Toggle** tool from the **Main Toolbar** to rotate *Rope01* in increments.

The **Angle Snap Toggle** tool in the **Main Toolbar** is used to rotate the object in the increments you set. By default, the rotations take place in 5-degree increments. Right-click on the **Angle Snap Toggle** tool; the **Grid and Snap Settings** dialog box is displayed. Set **30** in the **Angle** spinner in the **General** area of the **Options** tab, as shown in Figure 3-7. Close the **Grid and Snap Settings** dialog box.

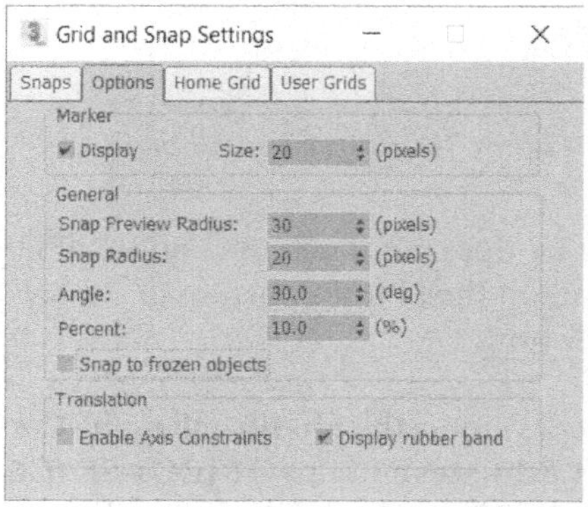

Figure 3-7 *The* ***Grid and Snap Settings*** *dialog box*

5. In the Top viewport, move the cursor over the X-axis and rotate *Rope01* through an angle of 30 degrees or enter **30** in the **X** spinner of Status Bar; the rotation value is displayed in the coordinate display below the viewports.

Be careful that you do not override the axis constraint by accidentally selecting an unwanted portion of the transform gizmo.

6. Choose the **Transform Gizmo Y Constraint** button from the **Axis Constraints** toolbar. Right-click on the **Angle Snap Toggle** tool; the **Grid and Snap Settings** dialog box is

displayed. Set **35** in the **Angle** spinner in the **General** area of the **Options** tab. Rotate *Rope01* through an angle of 35 degrees and then close this dialog box or enter **35** in the **Y** spinner of Status Bar. Turn the **Angle Snap Toggle** off.

7. Activate the Front viewport. Invoke the **Select and Move** tool from the **Main Toolbar** and also choose the **Transform Gizmo XY Plane Constraint** button from the **Axis Constraints** toolbar. Next, move *Rope01* so that its lower end touches the upper-left corner or the basket. Activate the Left viewport and move *Rope01* such that its lower end touches the upper-left corner of the basket, as shown in Figure 3-8.

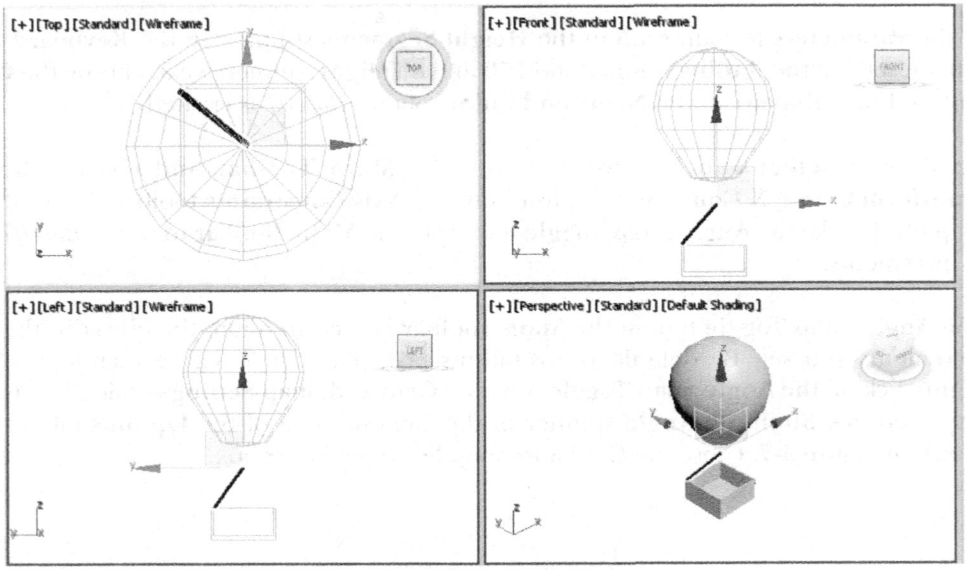

Figure 3-8 *Rope01 in all viewports*

8. Notice that the top of *Rope01* does not touch the bottom of *Balloon*. Invoke the **Zoom Region** tool. Next, zoom in on the bottom portion of *Balloon* and the upper portion of *Rope01* in the Front viewport.

9. Make sure *Rope01* is selected. Choose the **Modify** tab in the **Command Panel**. In the **Modifier List** drop-down list, select **Edit Mesh** in the **OBJECT-SPACE MODIFIERS** section. In the **Selection** rollout, choose the **Vertex** button to enter the sub-object mode and then select the upper vertex of *Rope01*; the vertex at the top turns red.

10. Choose the **Transform Gizmo Y Constraint** button from the **Axis Constraints** toolbar. Drag the selected vertices to touch the lower point of the balloon, as shown in Figure 3-9.

11. Choose the **Vertex** button in the **Selection** rollout again to exit the sub-object mode. Invoke the **Zoom Extents** tool from the viewport navigation controls.

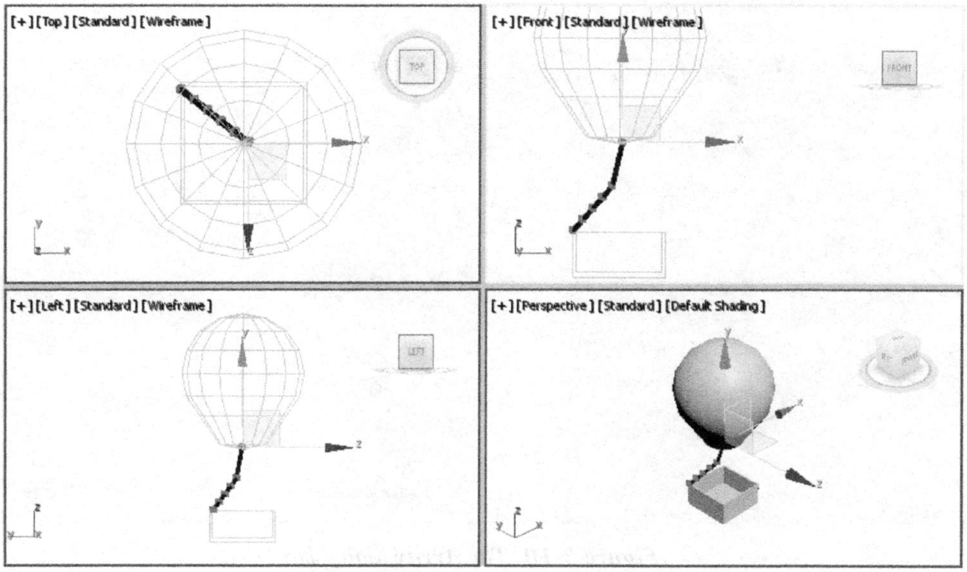

Figure 3-9 *The end vertices of the Rope01 moved to touch the bottom of the balloon*

Note

The ***View*** *option is selected in the* ***Reference Coordinate System*** *drop-down list by default. If you want to rotate objects about the world origin, you need to select the* ***World*** *option in this drop-down list, provided the viewport is not centered at the origin. After rotating the objects, you may select the* ***View*** *option in the* ***Reference Coordinate System*** *drop-down list to avoid confusion.*

12. Activate the Top viewport. Next, invoke the **Use Transform Coordinate Center** tool from the **Use Pivot Point Center** flyout from the **Main Toolbar**. Select **World** in the **Reference Coordinate System** drop-down list. Next, choose **Tools > Array** from the menu bar; the **Array** dialog box is displayed, as shown in Figure 3-10.

13. Choose the **Reset All Parameters** button available at the bottom of the **Array** dialog box and set **90** in the **Incremental Z** spinner in the **Rotate** row. In the **Type of Object** area, select the **Copy** radio button. In the **Array Dimensions** area, select the **1D** radio button and set **4** in the **1D Count** spinner. Choose the **OK** button; ropes are placed at the four corners of the basket and extended to the bottom of *Balloon,* as shown in Figure 3-11.

14. Select **View** in the **Reference Coordinate System** drop-down list and then choose the **Use Pivot Point Center** tool from the **Use Pivot Point Center** flyout from the **Main Toolbar**. Rename the three newly created ropes as **Rope02**, **Rope03**, and **Rope04**.

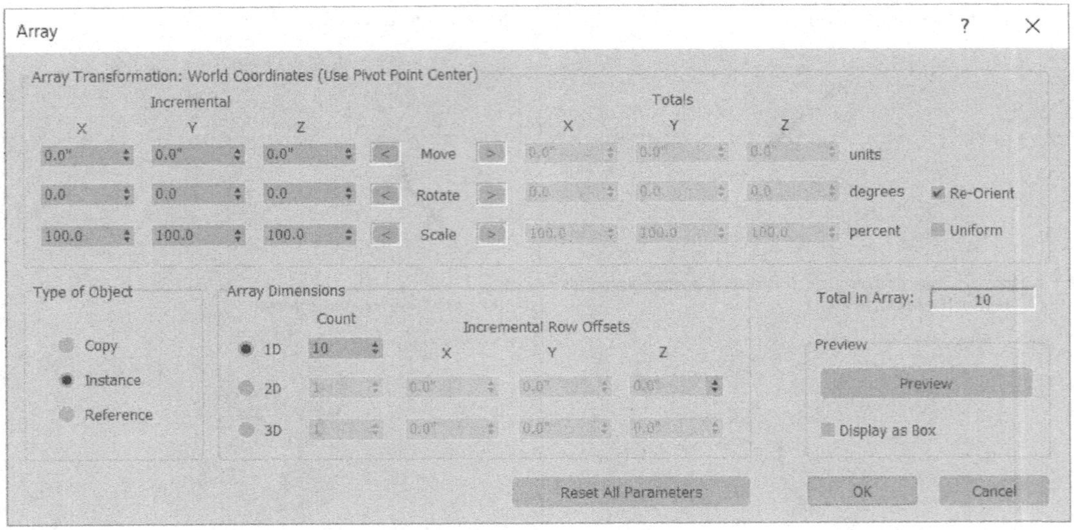

*Figure 3-10 The **Array** dialog box*

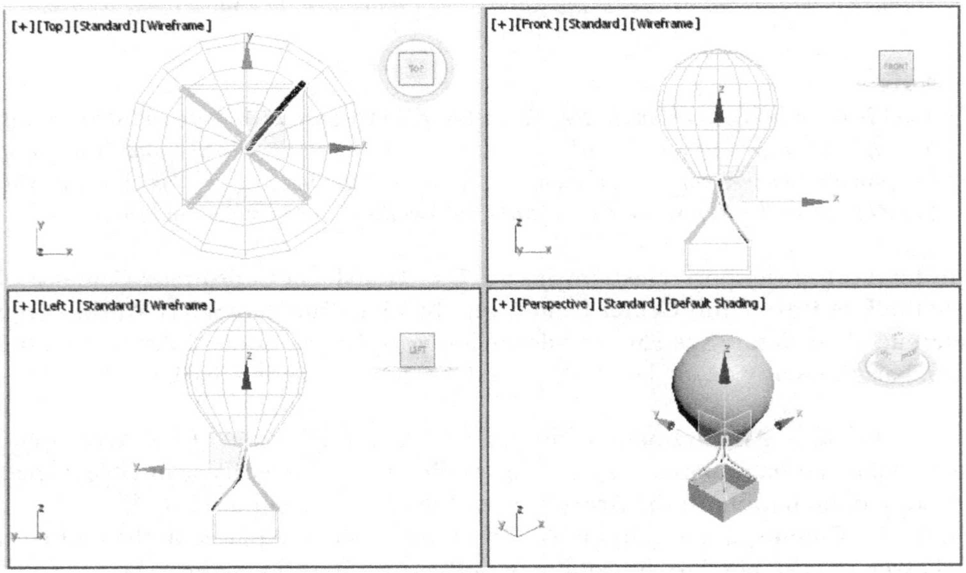

Figure 3-11 *The four ropes placed at the four corners of basket*

Creating the Collar

Next, you will create a collar and place it at the "kink" in the ropes.

1. Make sure the Top viewport is activated and then choose **Create > Geometry** in the **Command Panel**. Next, invoke the **Torus** tool from the **Object Type** rollout.

2. In the **Keyboard Entry** rollout, set the value **25** in the **Major Radius** spinner and **5** in the **Minor Radius** spinner. Choose the **Create** button and enter **Collar** in the **Name and Color** rollout.

3. Invoke the **Select and Move** tool from the **Main Toolbar**. In the Front viewport, move *Collar* down so that the "kinks" in the ropes are inside the collar, as shown in Figure 3-12 and then zoom the view properly as needed.

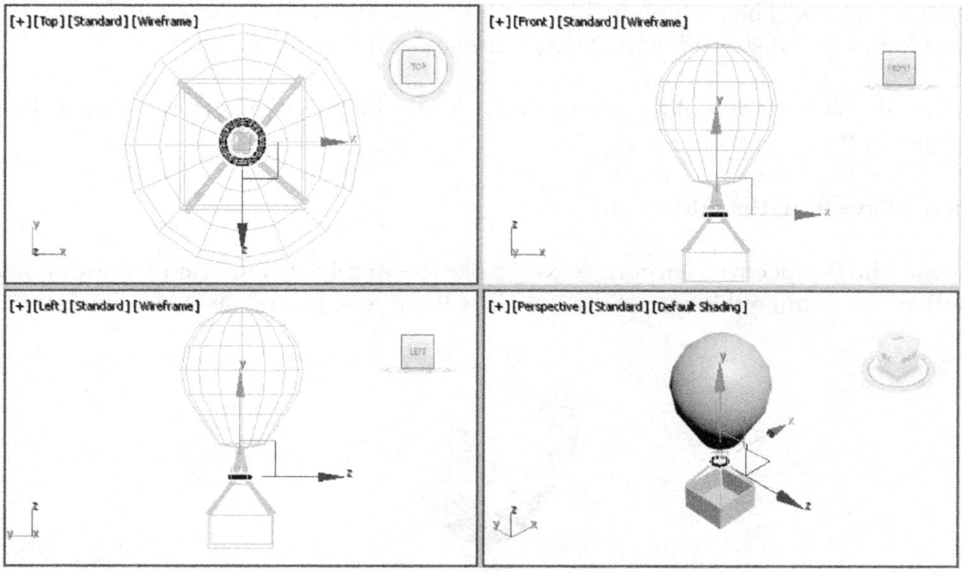

Figure 3-12 Alignment of collar with the ropes

4. Invoke the **Zoom Extents All** tool from the viewport navigation controls to view all objects properly.

Note
The process of texturing this model has been discussed in Chapter 11.

Smoothening the Objects

In this section, you will smoothen *Balloon* and *Basket* using the **MeshSmooth** modifier and the **Chamfer** modifier, respectively.

1. Select *Balloon* in the viewport. Choose the **Modify** tab in the **Command Panel**. In the **Modifier List** drop-down list, select **MeshSmooth** from the **OBJECT-SPACE MODIFIERS** section; the **MeshSmooth** modifier is applied to *Balloon*. You will notice that *Balloon* is smoothened.

2. Select *Basket* in the viewport. Choose the **Modify** tab in the **Command Panel**. In the **Modifier List** drop-down list, select **Chamfer** from the **OBJECT-SPACE MODIFIERS** section; the **Chamfer** modifier is applied to *Basket* and the **Parameters** rollout is displayed in the **Command Panel**.

3. In the **Chamfer Options** area, set **0.25** in the **Amount** spinner and **2** in the **Segments** spinner. You will notice that edges of *Basket* are smoothened.

Saving and Rendering the Scene

In this section, you will save the scene that you have created and then render it. You can also view the final rendered image of this model by downloading the *c03_3dsmax_2022_rndr.zip* file from *www.cadcim.com*. The path of the file is as follows: *Textbooks > Animation and Visual Effects > Autodesk 3ds Max > Autodesk 3ds Max 2022 for Beginners: A Tutorial Approach.*

1. Change the background color of the scene to white by following the steps as given in Tutorial 1 of Chapter 2.

2. Choose **Save** from the **File** menu.

3. Activate the Perspective viewport. Next, invoke the **Render Production** tool from the **Main Toolbar**; the rendered image of the hot air balloon is displayed, as shown in Figure 3-13.

Figure 3-13 *The rendered image of the hot air balloon*

Tutorial 2

In this tutorial, you will create the model of a shed, as shown in Figure 3-14.

(Expected time: 60 min)

Figure 3-14 *The model of a shed*

The following steps are required to complete this tutorial:

a. Create the project folder.
b. Create the sides of the shed.
c. Create supports for the side.
d. Create the top of the shed.
e. Save and render the scene.

Creating the Project Folder

Create the project folder with the name *c03_tut2* in the *3dsmax 2022* folder as discussed in Tutorial 1 of Chapter 2.

Creating the Sides of the Shed

In this section, you will create one side of the shed and then copy it using the **Mirror** tool to create other sides.

1. Activate the Top viewport and then maximize it. Choose **Create > Geometry** in the **Command Panel**; **Standard Primitives** is displayed in the drop-down list. Next, invoke the **Box** tool from the **Object Type** rollout; various rollouts are displayed in the **Command Panel**.

2. Expand the **Keyboard Entry** rollout and set the values as given next:

   Length: **3** Width: **50** Height: **19**

Note

*To view segments of objects in the Perspective viewport, you need to switch from **Default Shading** view to **Default Shading+Edged Faces** view. To do so, click on the **Default Shading** viewport label in the Perspective viewport; a flyout is displayed. Choose the **Edged Faces** option from it.*

3. Choose the **Create** button from the **Keyboard Entry** rollout; a box is created. In the **Name and Color** rollout, enter **side1** and press ENTER; the box is named as *side1*.

4. Choose the **Modify** tab in the **Command Panel**. Next, in the **Parameters** rollout, set the values as given next:

   Length Segs: **1** Width Segs: **18** Height Segs: **1**

   On setting these values, the modified *side1* is displayed in all viewports, as shown in Figure 3-15.

   Next, you need to convert *side1* to editable poly to modify it at the sub-object level.

5. Make sure the **Modify** tab is chosen and then select the **Edit Poly** modifier from the **OBJECT-SPACE MODIFIERS** section in the **Modifier List** drop-down list; the **Edit Poly** modifier is displayed in the modifier stack. In the modifier stack, click on the arrow on the left of the **Editable Poly** to view all sub-object levels.

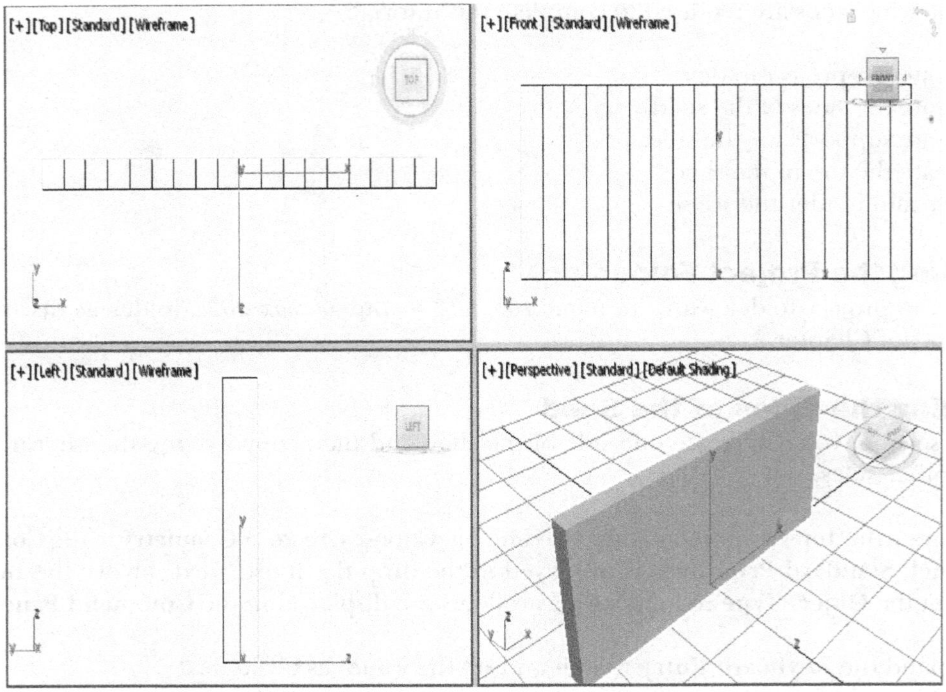

Figure 3-15 The side 1 displayed in all viewports

6. Activate the Front viewport. Select the **Polygon** sub-object level. Next, select the polygon of *side1*, as shown in Figure 3-16.

7. In the **Edit Polygons** rollout, choose the **Settings** button on the right of the **Extrude** button; the **Extrude Polygons** caddy control is displayed in the viewport, refer to Figure 3-17. Set the value **47** in the **Extrude Polygons-Height** spinner, and choose the **OK** button to close this caddy control; the selected polygon is extruded, as shown in Figure 3-18.

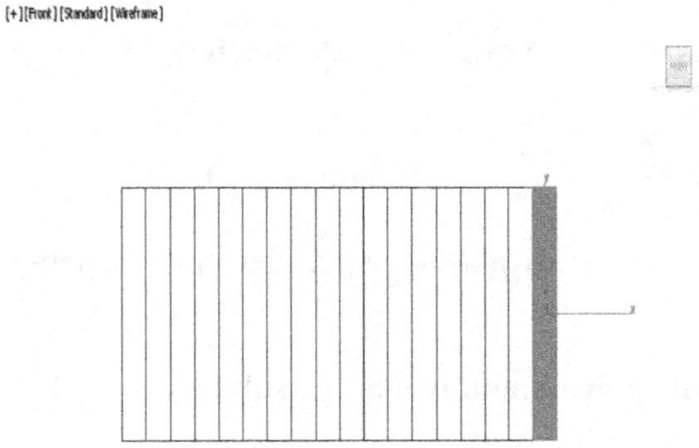

Figure 3-16 The polygon selected

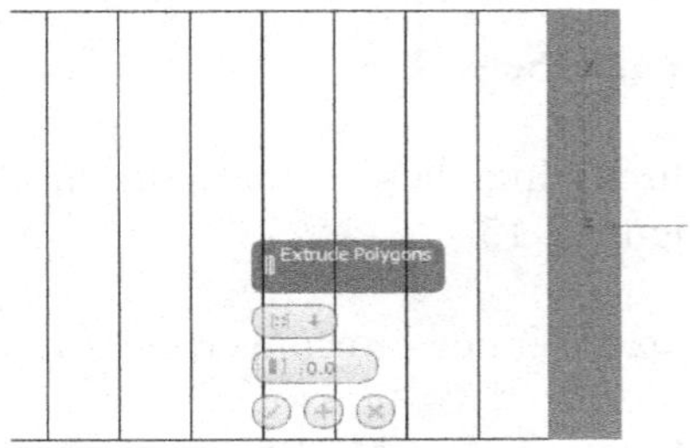

*Figure 3-17 The **Extrude Polygons** caddy control*

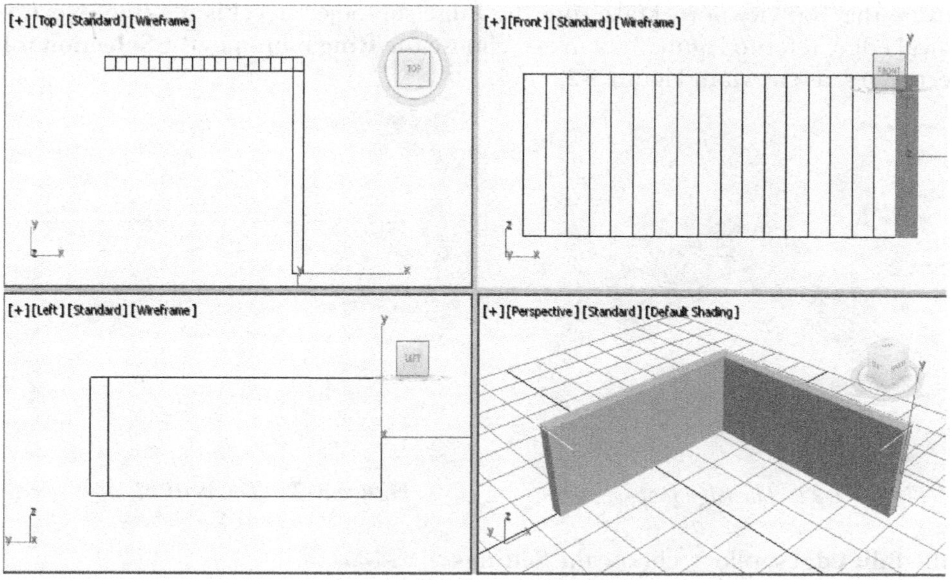

Figure 3-18 The selected polygon extruded

8. Activate the Perspective viewport and select the **Edge** sub-object level. Next, select the edge, as shown in Figure 3-19.

9. In the **Selection** rollout, choose the **Ring** button; the ring of edges is selected, as shown in Figure 3-20.

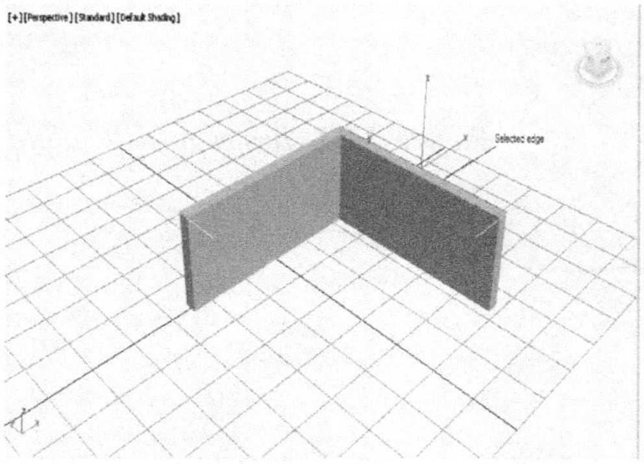

Figure 3-19 The edge selected

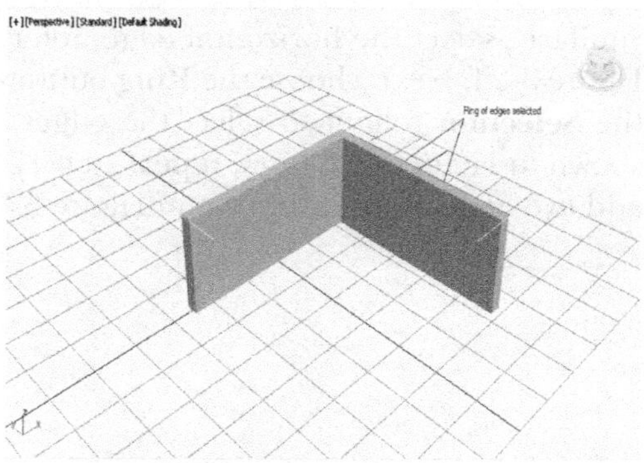

Figure 3-20 The ring of edges selected

10. In the **Edit Edges** rollout, choose the **Settings** button on the right of the **Connect** button; the **Connect Edges** caddy control is displayed in the viewport. Set the value **16** in the **Connect Edges Segments** spinner and choose the **OK** button to close this caddy control; edge loops are added, as shown in Figure 3-21.

11. Activate the Top viewport. Make sure the **Edge** sub-object level is selected. Next, select a vertical edge, refer to Figure 3-22. Next, choose the **Ring** button in the **Selection** rollout to select edges, as shown in Figure 3-22.

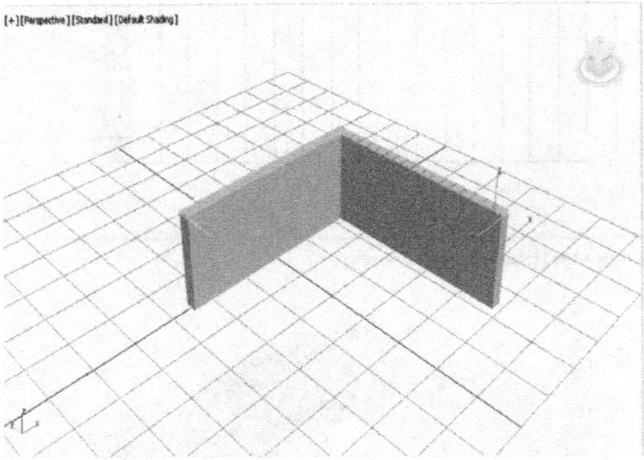

Figure 3-21 *The edge loops added*

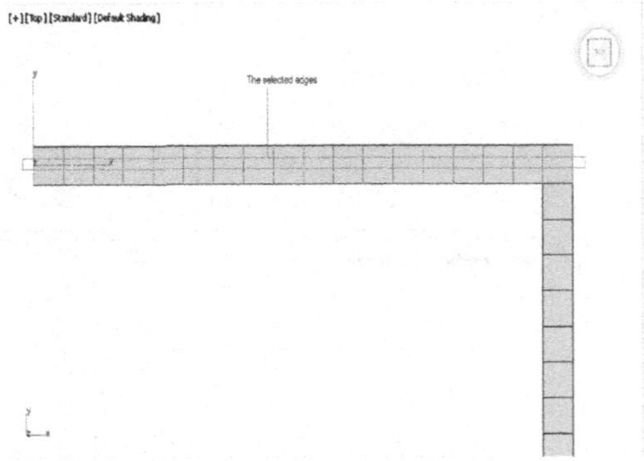

Figure 3-22 *The vertical edges selected*

12. In the **Edit Edges** rollout, choose the **Settings** button on the right of the **Connect** button; the **Connect Edges** caddy control is displayed in the viewport. Set the value **2** in the **Connect Edges Segments** spinner and **15** in the **Connect Edges Pinch** spinner. Next, choose the **OK** button to close this caddy control; two edge loops are added, as shown in Figure 3-23.

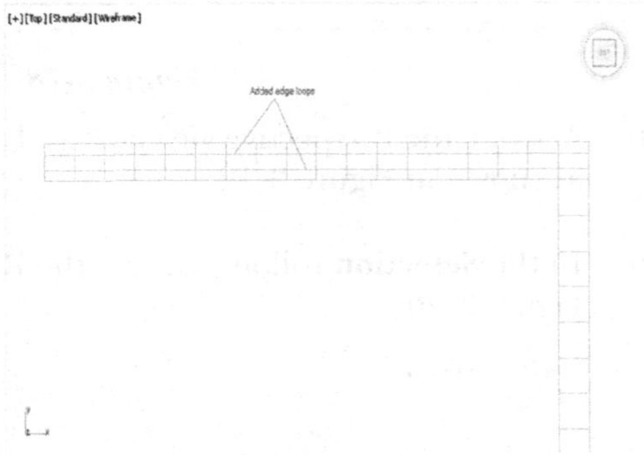

Figure 3-23 *Two edge loops added*

13. Similarly, select the horizontal edge, refer to Figure 3-24. Next, choose the **Ring** button in the **Selection** rollout to select the edges, as shown in Figure 3-24. Now, repeat step 12 to add two edge loops, as shown in Figure 3-25.

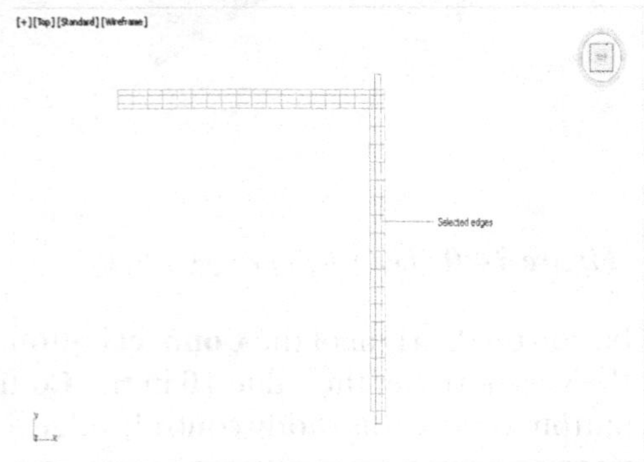

Figure 3-24 *The horizontal edges selected*

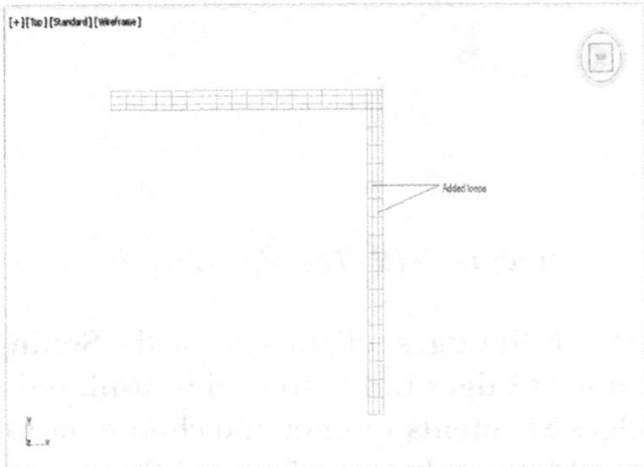

Figure 3-25 *Two edge loops added*

14. Make sure the Top viewport is activated. Select the **Polygon** sub-object level in the **Selection** rollout. Next, select the **Ignore Backfacing** check box in the **Selection** rollout and then select the polygons, as shown in Figure 3-26.

15. In the **Edit Polygons** rollout, choose the **Settings** button on the right of the **Extrude** button; the **Extrude Polygons** caddy control is displayed in the viewport. Set the value **125** in the **Extrude Polygons-Height** spinner. Next, choose the **OK** button to close this caddy control; the polygons are extruded. Figure 3-27 shows the selected polygons extruded in the Perspective viewport.

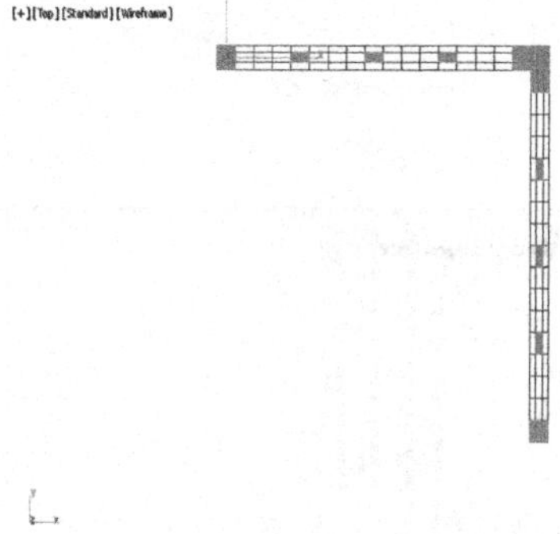

Figure 3-26 *The polygons selected*

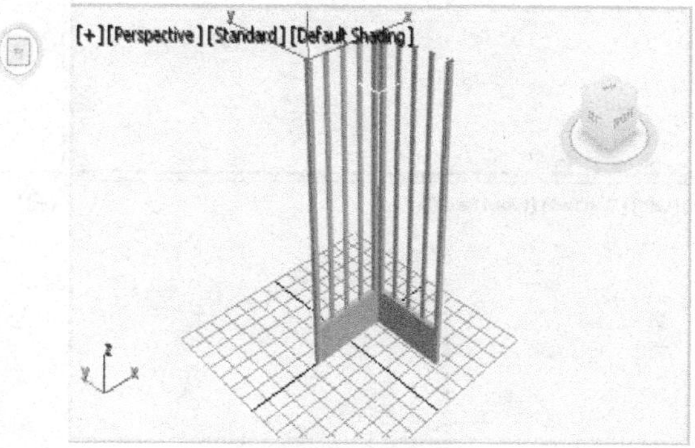

Figure 3-27 *The selected polygons extruded*

Next, you will create the horizontal strips of *side1*.

16. Make sure the Top viewport is activated. Choose **Create > Geometry** in the **Command Panel**; the **Standard Primitives** option is displayed in the drop-down list. Next, invoke the **Box** tool from the **Object Type** rollout; various rollouts are displayed in the **Command Panel**.

17. Expand the **Keyboard Entry** rollout and set the values as given next:

X: **0**	Y: **0**	Z: **27**
Length: **0.65**	Width: **47**	Height: **2.7**

18. Choose the **Create** button from the **Keyboard Entry** rollout; a box is created. In the **Name and Color** rollout, enter **hozt strip1** and press ENTER; the box is named as *hozt strip1*. Figure 3-28 shows *hozt strip1* in all viewports.

19. Activate the Front viewport and then invoke the **Select and Move** tool from the **Main Toolbar**. Make sure *hozt strip1* is selected. Next, press and hold the SHIFT key and drag the cursor in the Y direction till the value in the **Y** spinner of the coordinate display is changed to 12.7. Now, release the left mouse button; the **Clone Options** dialog box is displayed. In this dialog box, select the **Copy** radio button. Also, set the value in the **Number of Copies** spinner to **8** and then choose the **OK** button; the copies of *hozt strip1* are created. Figure 3-29 shows the copies of *hozt strip1* in the Perspective viewport.

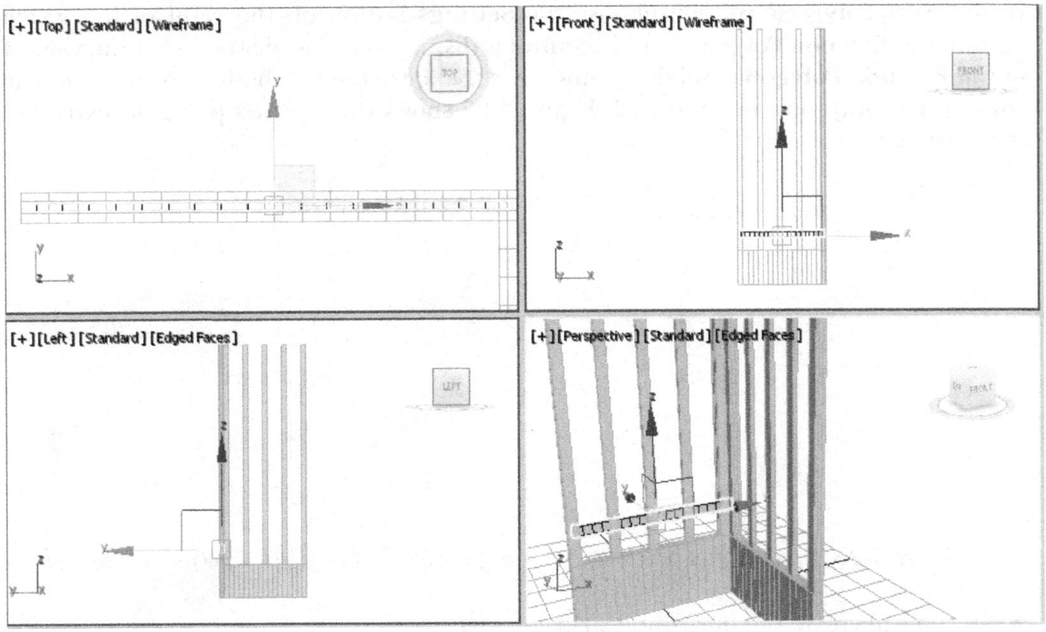

Figure 3-28 The hozt strip1 aligned

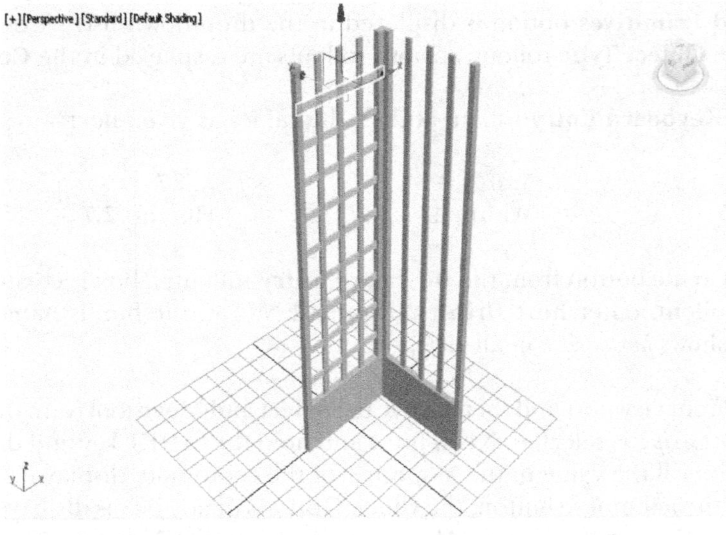

Figure 3-29 The copies of hozt strip1 created

20. Similarly, create horizontal strips for the other portion of *side1*, as shown in Figure 3-30.

    Next, you need to group all objects created in the scene.

21. Select all objects in the scene and then choose **Group > Group** from the menu bar; the **Group** dialog box is displayed. In this dialog box, enter **side01** in the **Group name** text box and choose the **OK** button to close the dialog box; *side01* group is created.

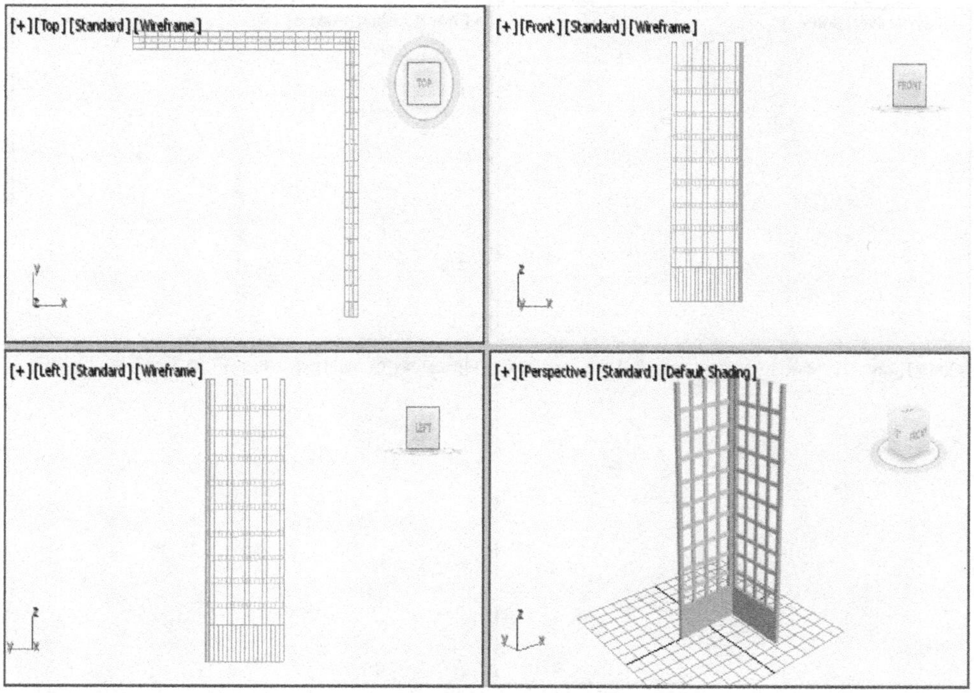

Figure 3-30 The copies of the horizontal strips created

Creating Supports for the Side

In this section, you will create supports for the side by using the **Box** tool.

1. Activate the Top viewport. Choose **Create > Geometry** in the **Command Panel**; the **Standard Primitives** option is displayed in the drop-down list. Next, invoke the **Box** tool from the **Object Type** rollout; various rollouts are displayed in the **Command Panel**.

2. Expand the **Keyboard Entry** rollout and set the values as given next:

X: **0**	Y:**0**	Z: **0**
Length: **49.8**	Width: **2.3**	Height: **17.5**

3. In the **Parameters** rollout, set the value in the **Height Segs** spinner to **6** and **Width Segs** spinner to **1**.

4. Choose the **Create** button from the **Keyboard Entry** rollout; a box is created. Enter **support1** in the **Name and Color** rollout; *support1* is created.

5. Switch the Left viewport to Right [Default Shading] viewport, as discussed earlier. Next, invoke the **Select and Move** tool and align *support1* in all viewports, as shown in Figure 3-31.

Note

For alignment of objects in the scene, you may need to switch from one viewport to another. To do so, click on the POV viewport label in a viewport; a flyout is displayed. Choose the desired option to switch to that viewport.

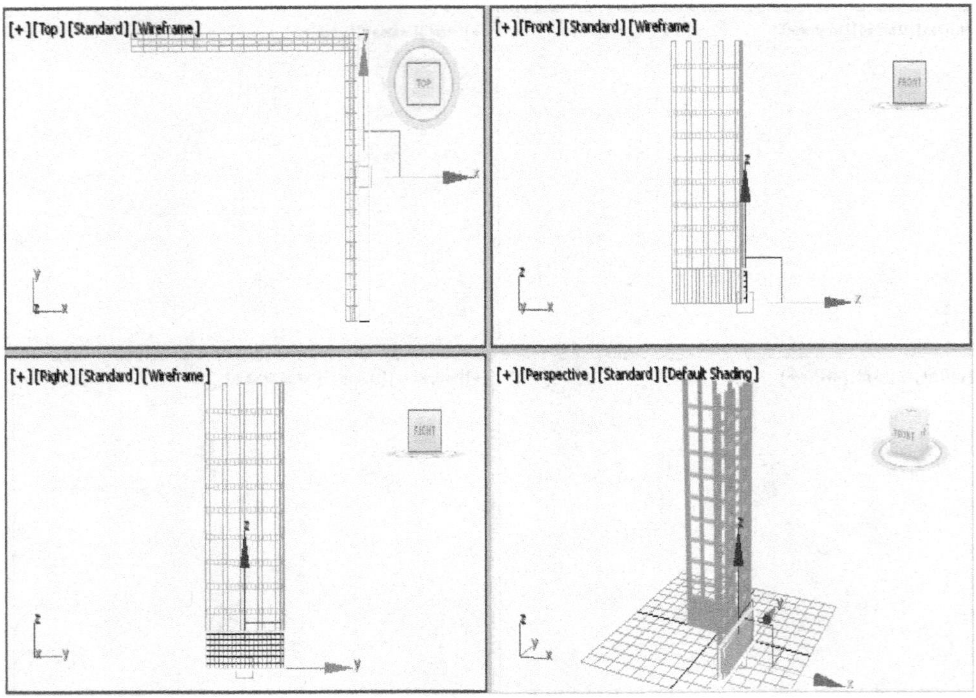

Figure 3-31 The support1 aligned in all viewports

6. Choose the **Modify** tab. Select the **Edit Poly** modifier from the **OBJECT-SPACE MODIFIERS** section in the **Modifier List** drop-down list; the **Edit Poly** modifier is displayed in the modifier stack. In the modifier stack, click on the arrow on the left of the **Editable Poly** to view all sub-object levels.

7. Activate the Perspective viewport. Select the **Edge** sub-object level from the **Selection** rollout. To display the edges of *support1*, you need to switch to the **Edge Faces** mode in the Perspective viewport. Next, select two edges using the CTRL key, as shown in Figure 3-32 and then choose the **Loop** button from the **Selection** rollout; two edge loops are selected.

8. In the **Edit Edges** rollout, choose the **Settings** button on the right of the **Chamfer** button; the **Chamfer** caddy control is displayed in the viewport. Set the value **0.1** in the **Chamfer-Amount** spinner and then choose the **OK** button to close this caddy control; two edge loops are added, as shown in Figure 3-33.

9. Select the **Polygon** sub-object level from the **Selection** rollout. Next, select the polygons in the Right, Front, and Back viewports using the CTRL key, as shown in Figure 3-34. Now, delete these polygons.

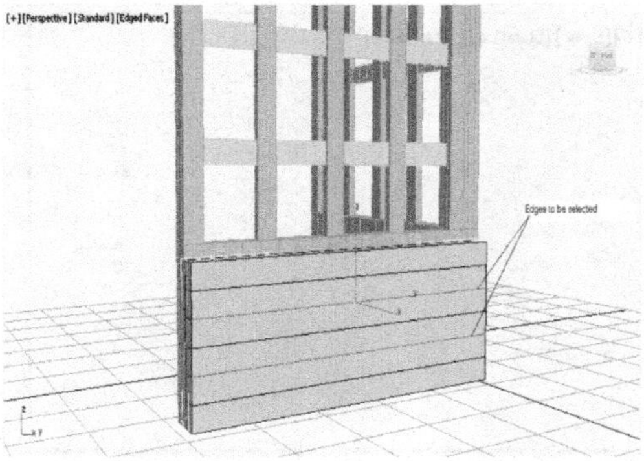

Figure 3-32 Two edges selected

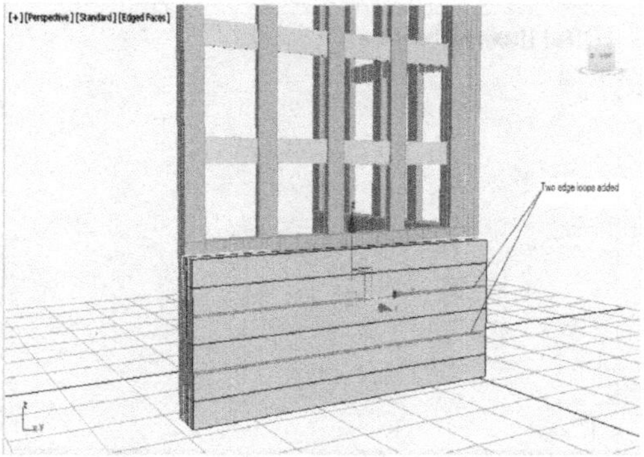

Figure 3-33 Two edge loops added

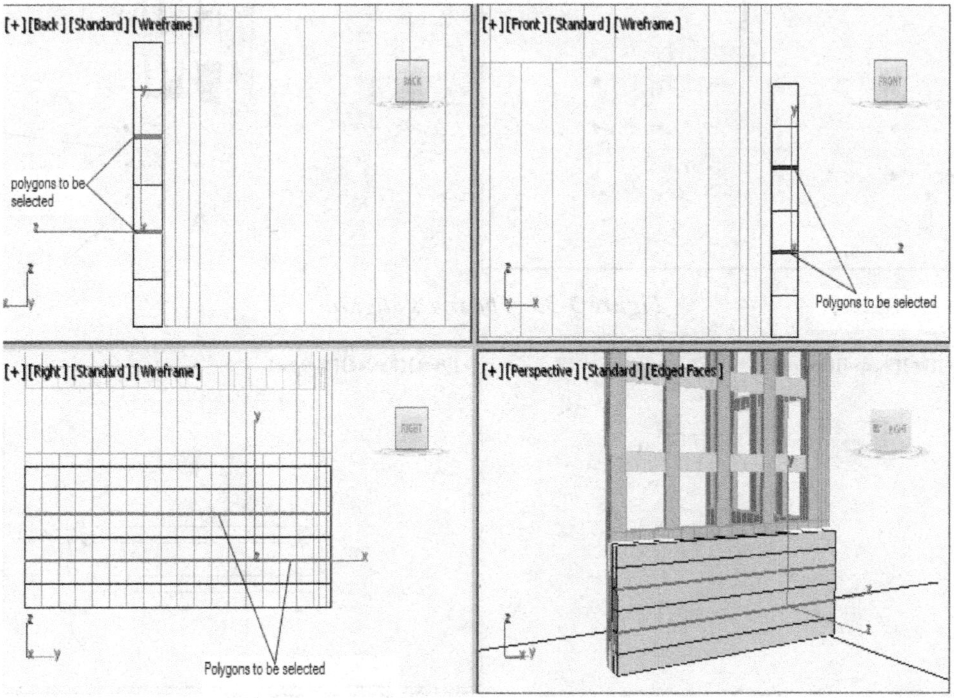

Figure 3-34 The selected polygons

Next, you need to create rivets to join the *support1* with *side01*.

10. Create rivets, as discussed in Tutorial 2 of Chapter 2. Next, align them, as shown in Figure 3-35.

11. Select *support1* as well as all rivets in the scene, and then choose **Group > Group** from the menu bar; the **Group** dialog box is displayed. In this dialog box, enter **support01** in the **Group name** text box and choose the **OK** button to close the dialog box; *support01* group is created.

12. Create a copy of *support 01* with the name *support02*. Now, align and rotate it, as shown in Figure 3-36.

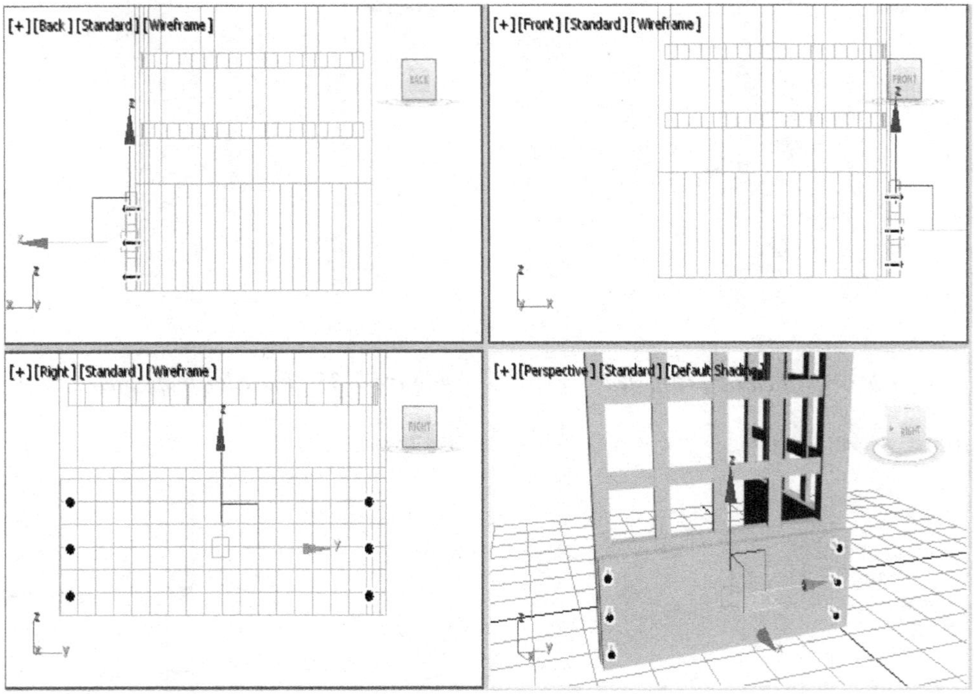

Figure 3-35 *The rivets aligned*

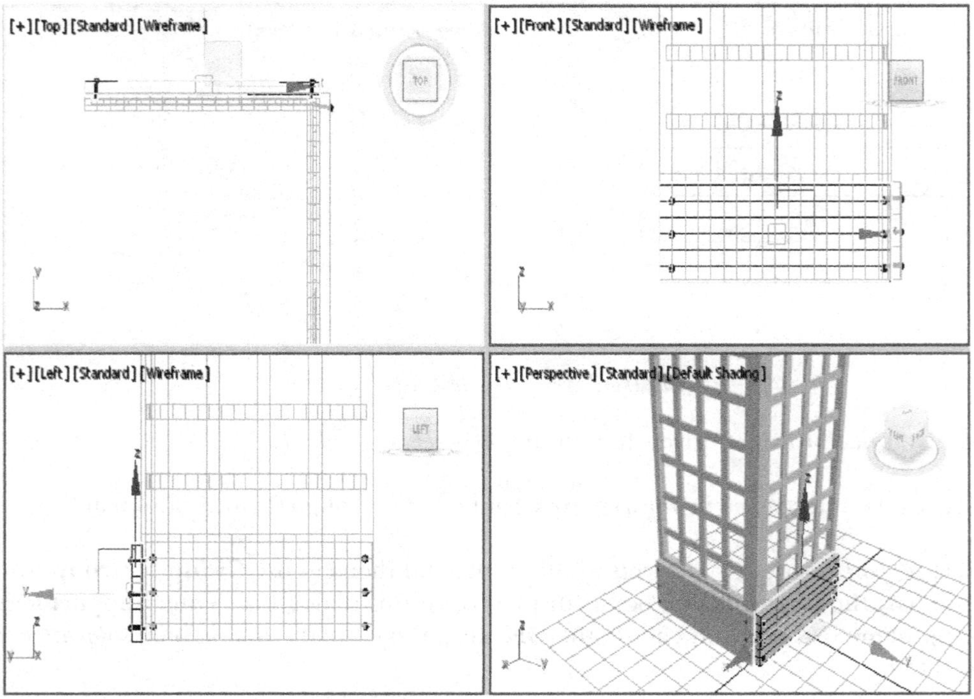

Figure 3-36 *The support02 aligned*

Next, you need to create one more support similar to *support01*.

13. Activate the Top viewport and choose **Create > Geometry** in the **Command Panel**; the **Standard Primitives** option is displayed in the drop-down list. Next, invoke the **Box** tool from the **Object Type** rollout; various rollouts are displayed in the **Command Panel**.

14. Expand the **Keyboard Entry** rollout and set the values as given next:

    Length: **69** Width: **2.3** Height: **17.5**

15. In the **Parameters** rollout, set the value in the **Height Segs** spinner to **6** and **Width Segs** spinner to **1**.

16. Choose the **Create** button from the **Keyboard Entry** rollout; a box is created. Enter **support03** in the **Name and Color** rollout; *support03* is created.

17. Repeat steps 6 to 9 to chamfer the edges and delete the polygons for *support03*. Next, align *support03* using the **Select and Move** and **Select and Rotate** tools from the **Main Toolbar**, as shown in Figure 3-37.

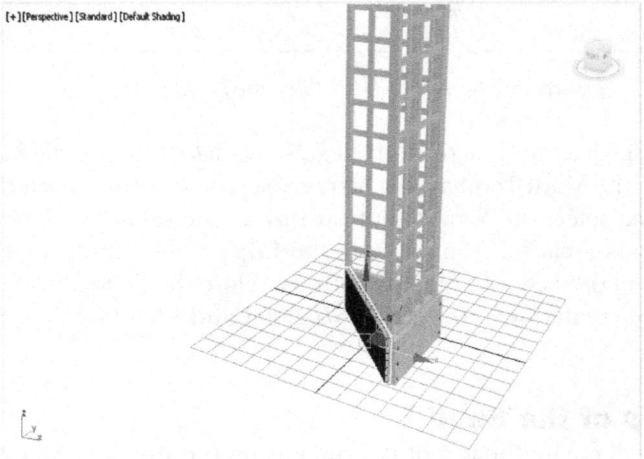

***Figure 3-37** The support03 aligned*

18. Select all objects in the scene and choose **Group > Group** from the menu bar; the **Group** dialog box is displayed. In this dialog box, enter **SIDE01** in the **Group name** text box and choose the **OK** button to close the dialog box; *SIDE01* group is created.

Now, you will create other sides of the shed using the **Mirror** tool.

19. Make sure the Top viewport is activated. Select *SIDE01* and invoke the **Mirror** tool from the **Main Toolbar**; the **Mirror: Screen Coordinates** dialog box is displayed. In this dialog box, select the **Y** radio button and set the value in the **Offset** spinner to **-165** in the **Mirror Axis** area. Next, make sure the **Copy** radio button is selected in the **Clone Selection** area and choose the **OK** button to close the dialog box; the copy of *SIDE01* is created with the name *SIDE002* and positioned, as shown in Figure 3-38.

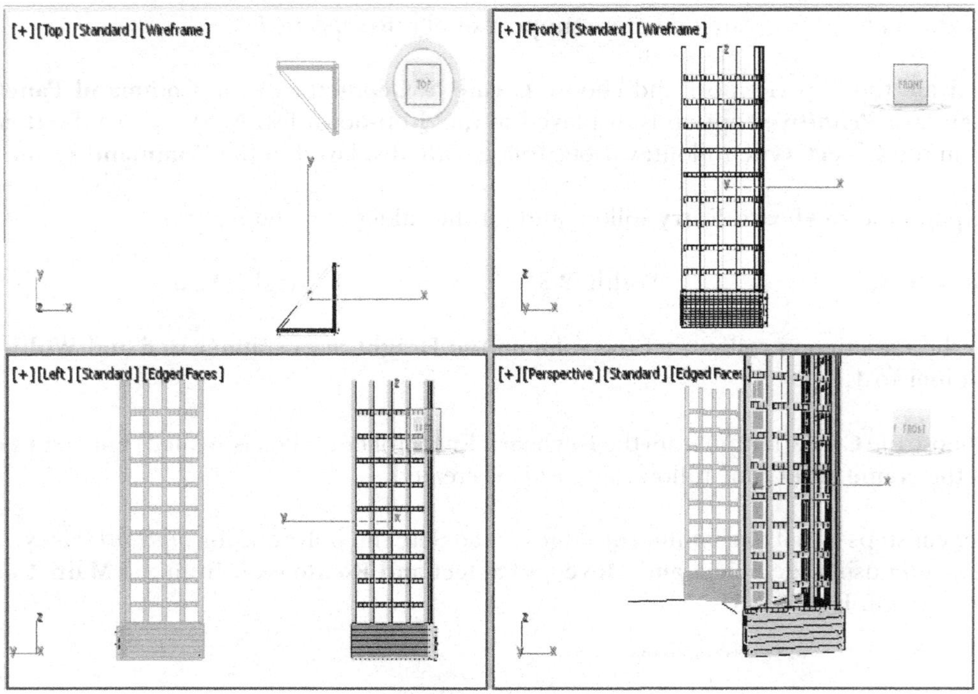

Figure 3-38 The SIDE002 created and positioned

20. Make sure the Top viewport is activated. Next, select *side01* and *side002* and again invoke the **Mirror** tool from the **Main Toolbar**; the **Mirror: Screen Coordinates** dialog box is displayed. In this dialog box, select the **X** radio button and set the value in the **Offset** spinner to **-165** in the **Mirror Axis** area. Next, make sure the **Copy** radio button is selected in the **Clone Selection** area and then choose the **OK** button to close the dialog box; the copies of *SIDE01* and *SIDE002* are created with the name *SIDE003* and *SIDE004* and positioned, as shown in Figure 3-39.

Creating the Top of the Shed

In this section, you will create the top of the shed using the **Box** and **Hedra** tools.

1. Make sure the Top viewport is activated. Choose **Create > Geometry** in the **Command Panel**; **Standard Primitives** is displayed in the drop-down list. Next, invoke the **Box** tool from the **Object Type** rollout; various rollouts are displayed in the **Command Panel**.

2. Expand the **Keyboard Entry** rollout and set the values as given next:

   Length: **205** Width: **195** Height: **4.5**

3. In the **Parameters** rollout, set the values as given next:

   Length Segs: **10** Width Segs: **10** Height Segs: **1**

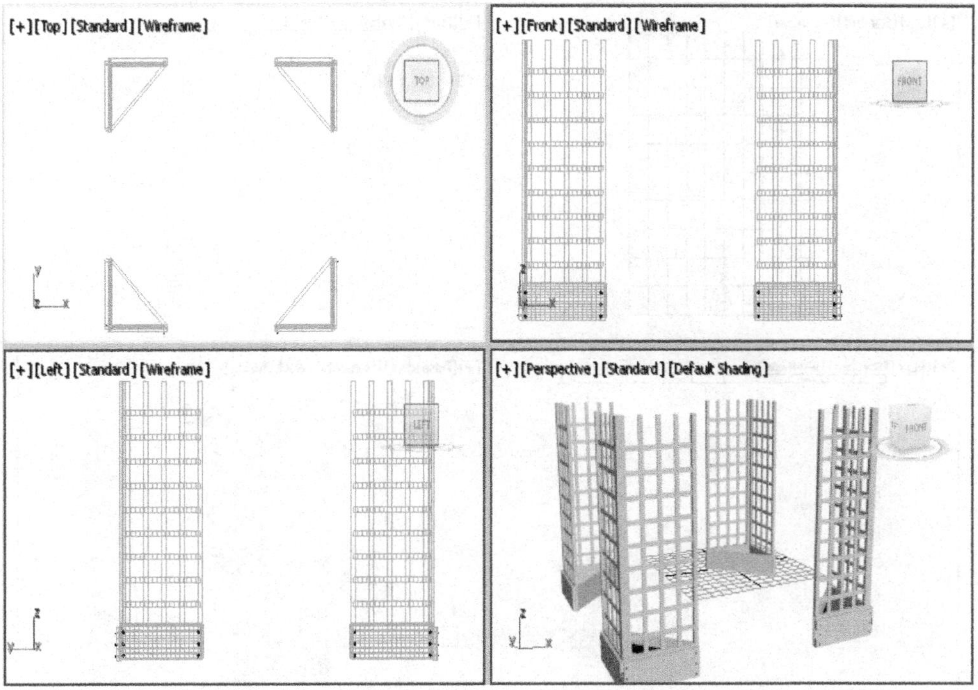

Figure 3-39 *The SIDE003 and SIDE004 positioned*

Choose the **Create** button from the **Keyboard Entry** rollout; a box is created. In the **Name and Color** rollout, enter **Top1** and press ENTER; the box is named as *Top1*.

4. Invoke the **Select and Move** tool from the **Main Toolbar** and align *Top1* in all viewports, as shown in Figure 3-40.

5. Make sure *Top1* is selected and the Top viewport is activated. Next, choose the **Modify** tab in the **Command Panel**. Select **Edit Poly** from the **Modifier List** drop-down list; the **Edit Poly** modifier is applied to *Top1* in the modifier stack.

6. Invoke the **Select Object** tool from the **Main Toolbar** and then select the **Polygon** sub-object level from the **Selection** rollout. Next, select middle polygons of *Top1* by dragging across window around the polygons, as shown in Figure 3-41. Next, delete them.

7. Choose **Create > Geometry** in the **Command Panel**. Select **Extended Primitives** from the drop-down list located below it. Next, invoke the **Hedra** tool from the **Object Type** rollout.

8. Click at a point in the Top viewport, drag the cursor, and then release the left mouse button; the hedra is created, as shown in Figure 3-42 and it is automatically named as *Hedra001*.

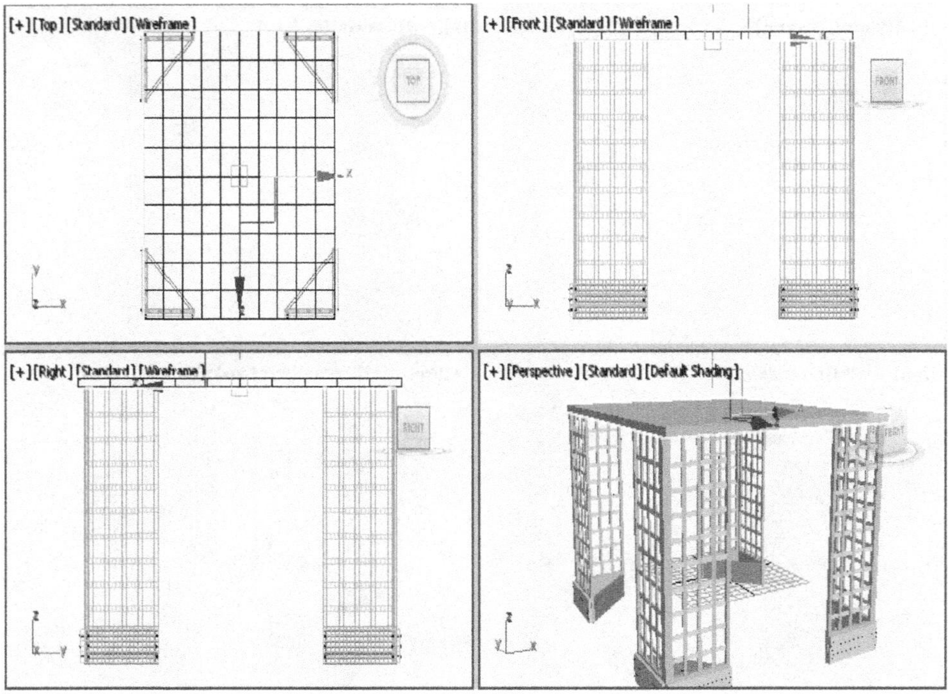

Figure 3-40 *Top 1 aligned in all viewports*

9. Choose the **Modify** tab in the **Command Panel**; various rollouts are displayed in the **Command Panel**. Make sure that the **Tetra** radio button is selected in the **Family** area of the **Parameters** rollout. Next, scroll down in this rollout and set the value of **150** in the **Radius** spinner. Also, rename *Hedra001* as *Top2* in the **Name and Color** rollout.

10. Select the **Edit Poly** modifier from the **OBJECT-SPACE MODIFIER** section in **Modifier List** drop-down list; the **Edit Poly** modifier is displayed in the modifier stack. In the modifier stack, click on the arrow on the left of the **Editable Poly** to view all sub-object levels.

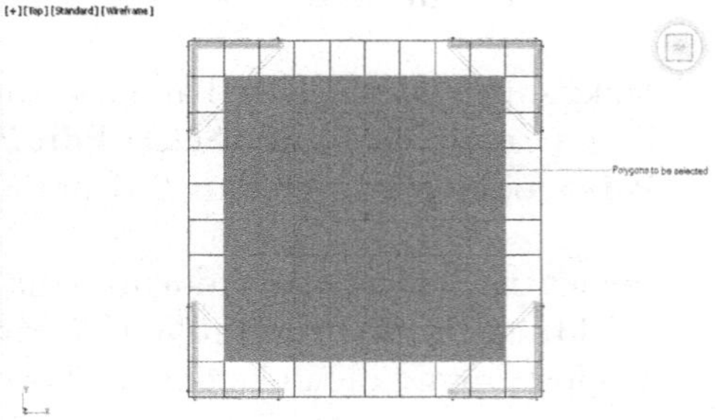

Figure 3-41 *The polygons selected*

11. Activate the Front viewport. Select the **Vertex** sub-object level from the modifier stack. Next, select the vertex of *Top2*, as shown in Figure 3-43. Invoke the **Select and Move** tool from the **Main Toolbar** and move the vertex, as shown in Figure 3-44.

12. Select the **Edge** sub-object level from the **Selection** rollout and then select all edges of *Top2*. Next, choose the **Settings** button on the right of the **Chamfer** button; the **Chamfer** caddy control is displayed, as shown in Figure 3-45. Set the value **2.5** in the **Chamfer-Amount** spinner and make sure that the value in the **Segments** spinner is set to 1. Choose the **OK** button to close the caddy control; all edges of *Top2* are chamfered, refer to Figure 3-46. Select the **Edge** sub-object level in the modifier stack again to deactivate it.

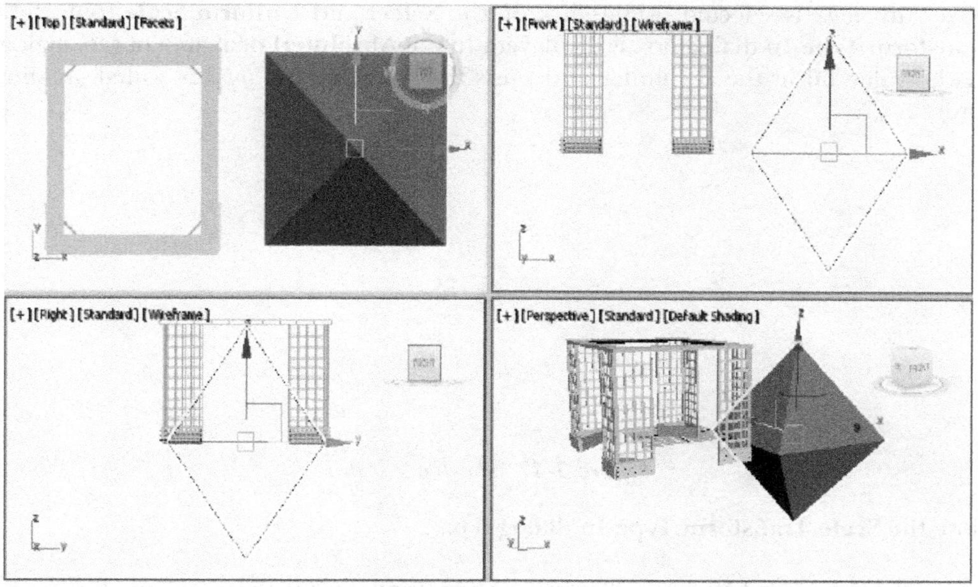

Figure 3-42 *Hedra001 created*

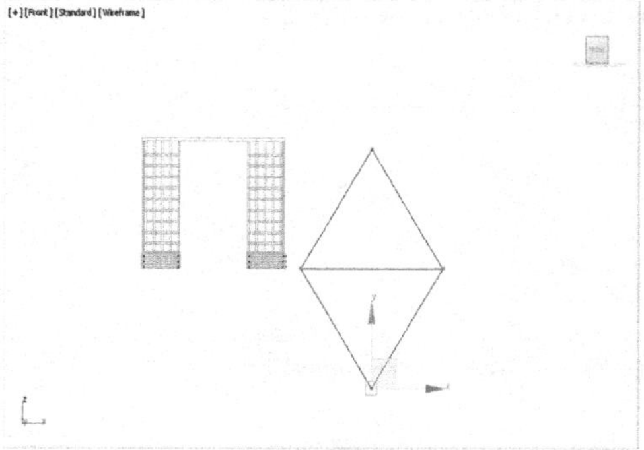

Figure 3-43 *The selected vertex*

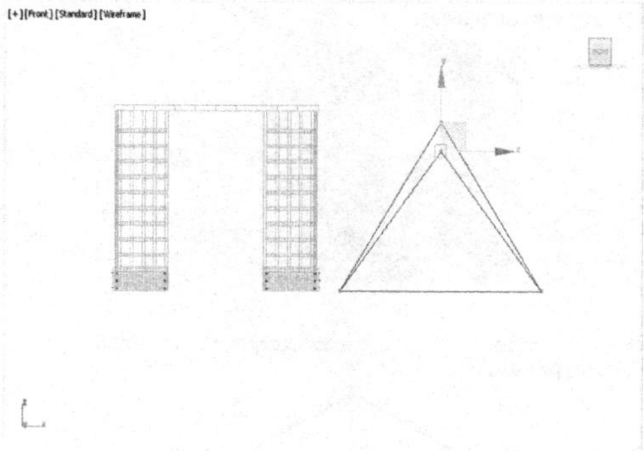

Figure 3-44 *The vertex moved*

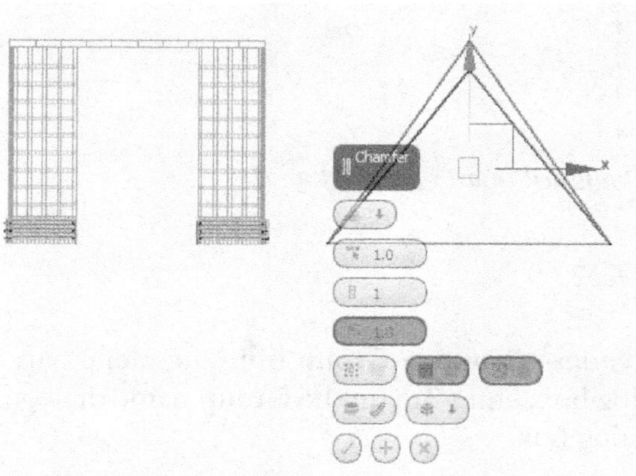

Figure 3-45 *The* ***Chamfer*** *caddy control*

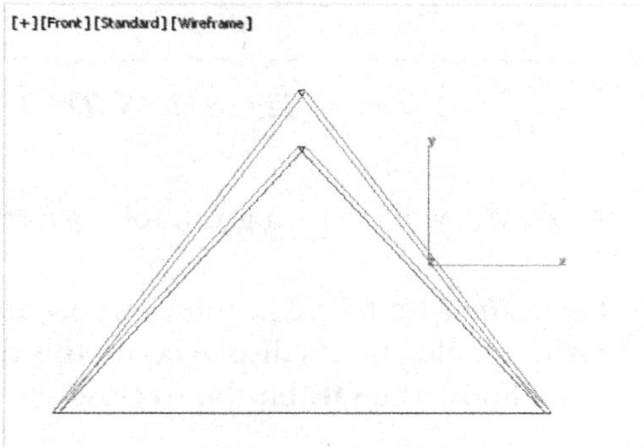

Figure 3-46 *All edges chamfered*

13. Make sure *Top2* is selected. Right-click on the **Select and Uniform Scale** tool; the **Scale Transform Type-In** dialog box is displayed. In the **Absolute:Local** area of this dialog box, set the value **50** in the **Z** spinner and press the ENTER key; *Top2* is scaled, as shown in Figure 3-47.

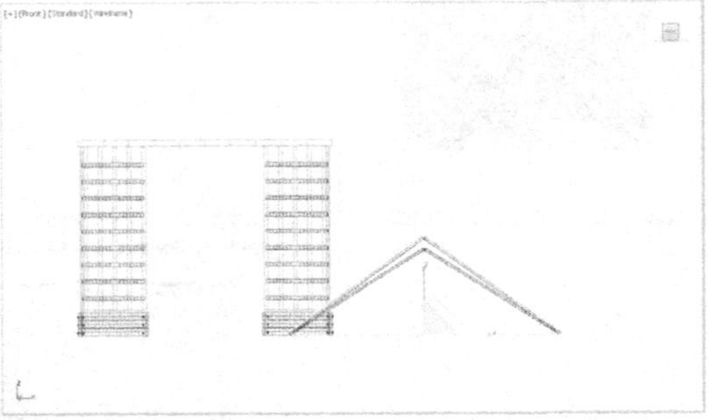

Figure 3-47 *The Top 2 scaled*

14. Close the **Scale Transform Type-In** dialog box.

15. Invoke the **Select and Move** tool from the **Main Toolbar** and align *Top2* in all viewports, as shown in Figure 3-48.

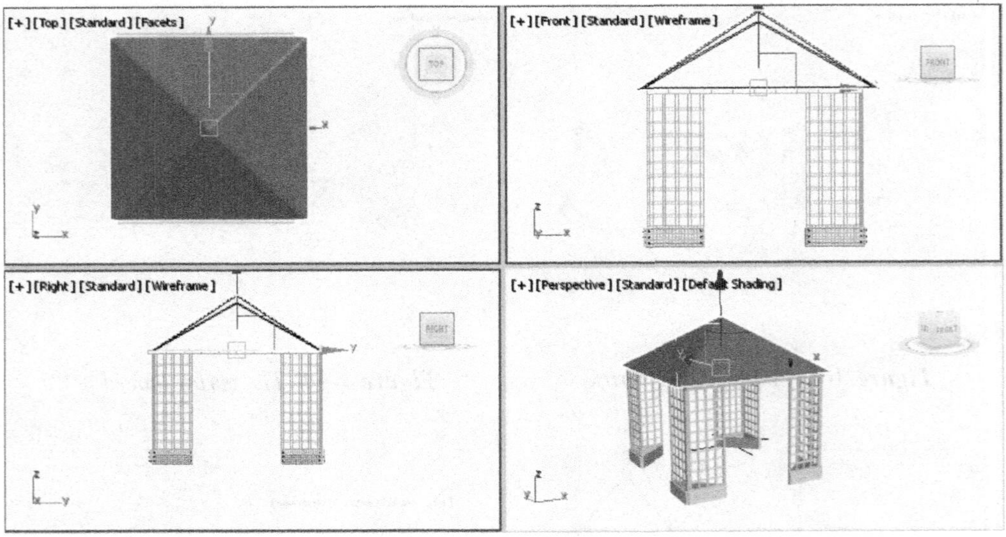

Figure 3-48 *The Top 2 aligned in all viewports*

Next, you will create a group of *Top1* and *Top2*.

16. Select *Top1* and *Top2* in the viewport and choose **Group > Group** from the menu bar; the **Group** dialog box is displayed. In this dialog box, enter **Top** in the **Group name** dialog box and choose the **OK** button to close the dialog box.

Saving and Rendering the Scene

In this section, you will save the scene that you have created and then render it. You can also view the final rendered image of this model by downloading the *c03_3dsmax_2022_rndr.zip* file from *www.cadcim.com*. The path of the file is as follows: *Textbooks > Animation and Visual Effects > Autodesk 3ds Max > Autodesk 3ds Max 2022 for Beginners: A Tutorial Approach.*

1. Change the background color of the scene to white by following the steps given in Tutorial 1 of Chapter 2.

2. Choose **Save** from the **File** menu.

3. Activate the Perspective viewport. Next, invoke the **Render Production** tool from the **Main Toolbar**; the rendered image of a shed is displayed, refer to Figure 3-49.

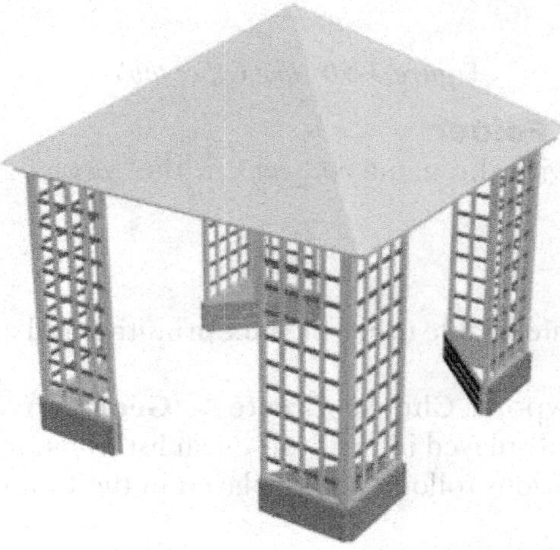

Figure 3-49 *The rendered image of a shed*

Tutorial 3

In this tutorial, you will create the model of an office table using the smart extrude technique, as shown in Figure 3-50. **(Expected time: 25 min)**

The following steps are required to complete this tutorial:

a. Create the project folder.
b. Create a floor.
c. Create a table.
d. Create the drawers.
e. Save and render the scene.

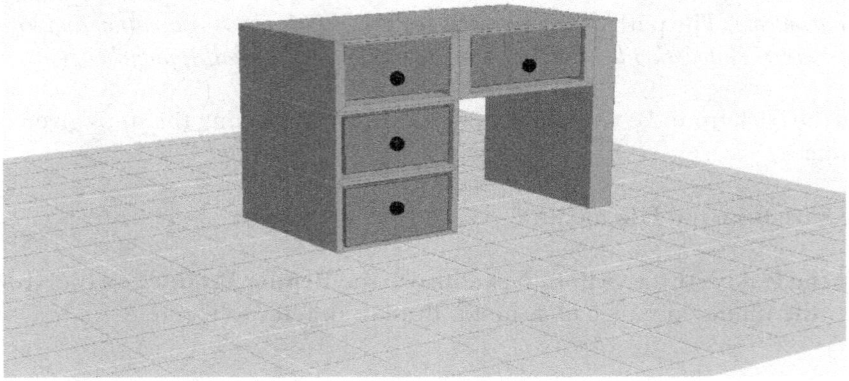

Figure 3-50 The Office table

Creating the Project Folder

Create the project folder with the name *c03_tut3* in the *3dsmax 2022* folder as discussed in Tutorial 1 of Chapter 2.

Creating a Table

In this section, you will create a table using the **box** primitive and apply the smart extrude.

1. Activate the Top viewport. Choose **Create > Geometry** in the **Command Panel**; **Standard Primitives** is displayed in the drop-down list. Next, invoke the **Box** tool from the **Object Type** rollout; various rollouts are displayed in the **Command Panel**.

2. Expand the **Keyboard Entry** rollout and set the values as given next:

   Length: **20** Width: **25** Height: **30**

3. Choose the **Create** button from the **Keyboard Entry** rollout; a box is created. In the **Name and Color** rollout, enter **Officetable** and press ENTER; the box is named as *Officetable*.

4. Choose the **Modify** tab in the **Command Panel**. Next, in the **Parameters** rollout, set the value **3** in **Height Segs**. Next, press F4; edges are displayed on *Officetable*.

   On setting these values, the modified *Officetable* is displayed in all viewports, as shown in Figure 3-51.

   Next, you need to convert *Officetable* to editable poly to modify it at the sub-object level.

5. Make sure *Officetable* is selected and then right-click; a flyout is displayed. Next, choose the **Convert To** option from the flyout; a cascading menu is displayed. Choose the **Convert to Editable Poly** option from the cascading menu; the **Editable Poly** is displayed in the modifier stack. Click on the arrow on the left of the **Editable Poly** in the modifier stack to view all sub-object levels.

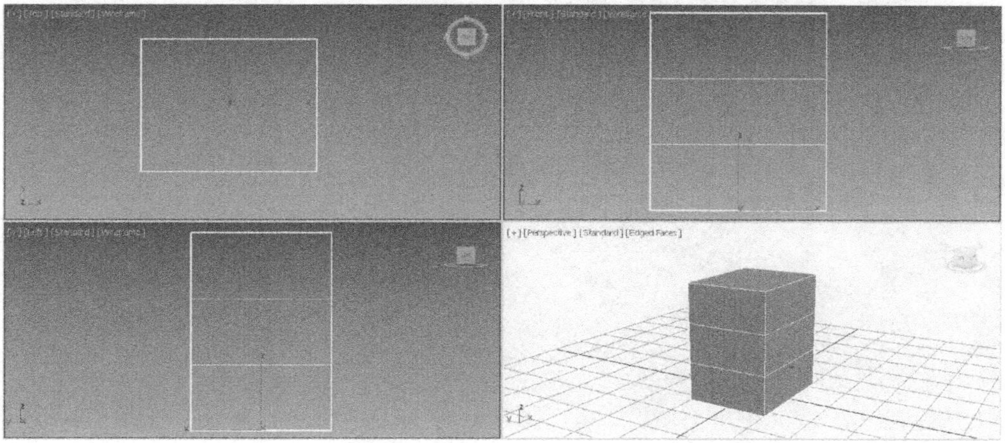

Figure 3-51 The Officetable displayed in all viewports

6. Activate the **Front** viewport. Select the **Polygon** sub-object level. Next, select the polygon of *Officetable*, as shown in Figure 3-52. Make sure the **Select and Move** tool is activated. Now, hold the SHIFT key and drag in the **Perspective** viewport, as shown in Figure 3-53. Make sure the polygon is selected. Again hold the SHIFT key and drag the polygon little bit in the **Perspective** viewport, as shown in Figure 3-54.

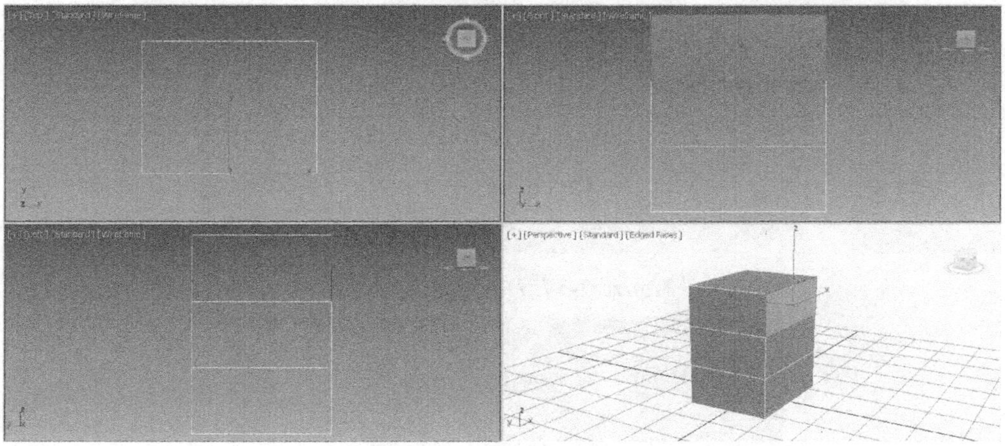

Figure 3-52 Selected polygon

7. In the **Top** viewport choose the **[Top]** tab; a flyout is displayed, as shown in Figure 3-55. Choose **Bottom** from the flyout; the **Bottom** viewport is displayed. Next, choose the bottom polygon, as shown in Figure 3-56. Now, hold the SHIFT key and drag downward in the **Left** viewport, as shown in Figure 3-57. Switch the bottom viewport with the Top viewport.

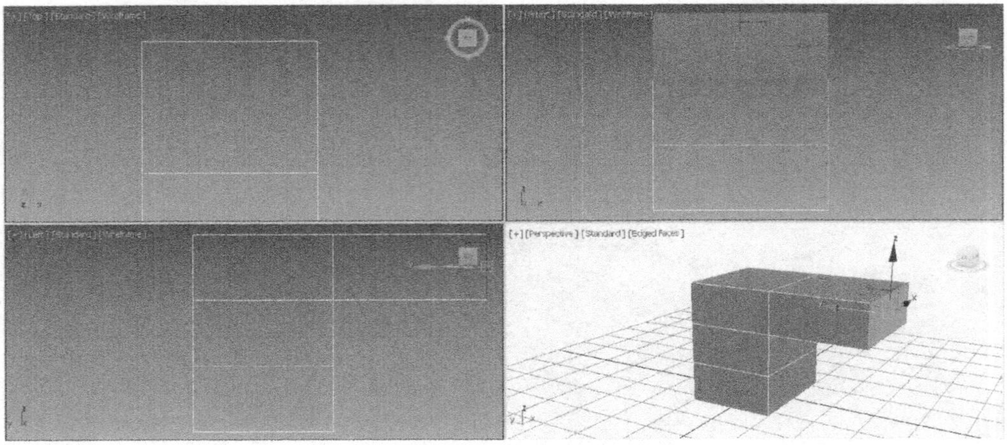

Figure 3-53 Selected polygon dragged

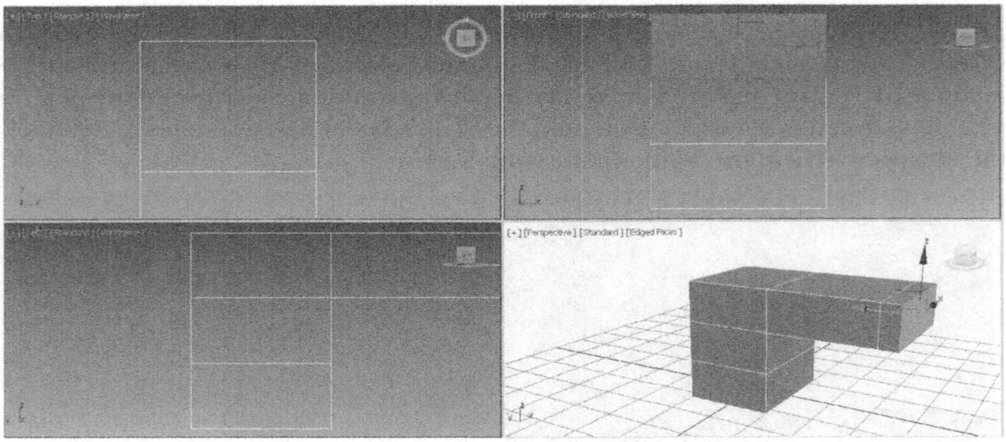

Figure 3-54 Dragged polygon

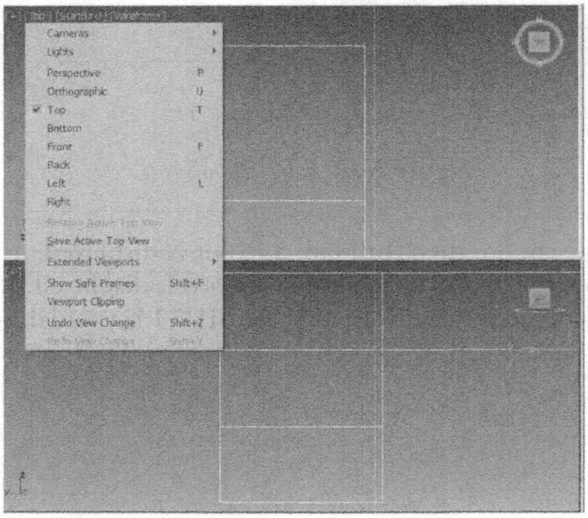

Figure 3-55 The flyout displayed

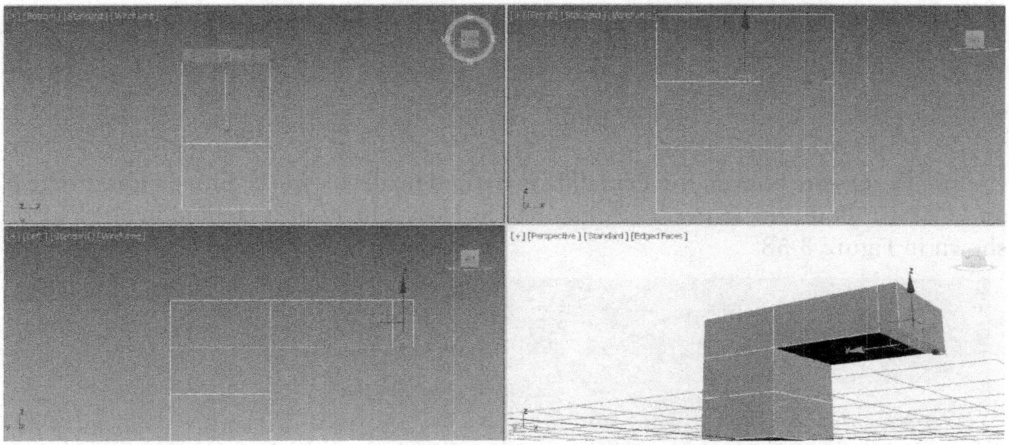

Figure 3-56 Bottom polygon selected

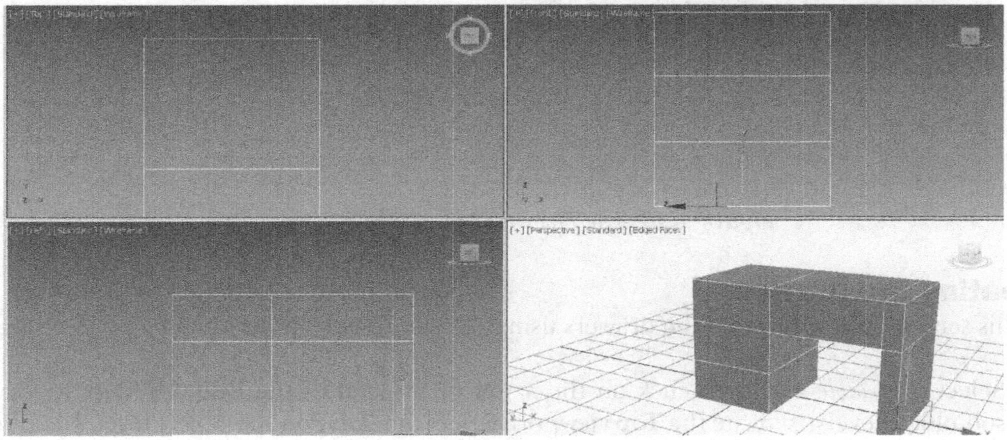

Figure 3-57 Bottom polygon dragged

Creating the Floor

In this section, you will create the floor using the plane primitive.

1. Make sure the **Top** viewport is activated. Choose **Create > Geometry** in the **Command Panel**, if it is not already chosen; different tools and rollouts pertaining to the **Create** command are displayed in the **Command Panel**.

2. Invoke the **Plane** tool from the **Object Type** rollout; the rollouts corresponding to the **Plane** tool such as **Name and Color**, **Creation Method**, **Keyboard Entry**, and **Parameters** are displayed.

   Note that the arrow sign on the left side of each rollout is used to expand or collapse it.

3. Expand the **Keyboard Entry** rollout by clicking on its title bar and set the values as given next.

   Length: **150** Width: **150**

   Choose the **Create** button; the Plane001 is created in all viewports. Enter **Floor** in the **Name and Color** rollout; the *Plane001* is renamed as *Floor*. The *Floor* is created in all viewports, as shown in Figure 3-58.

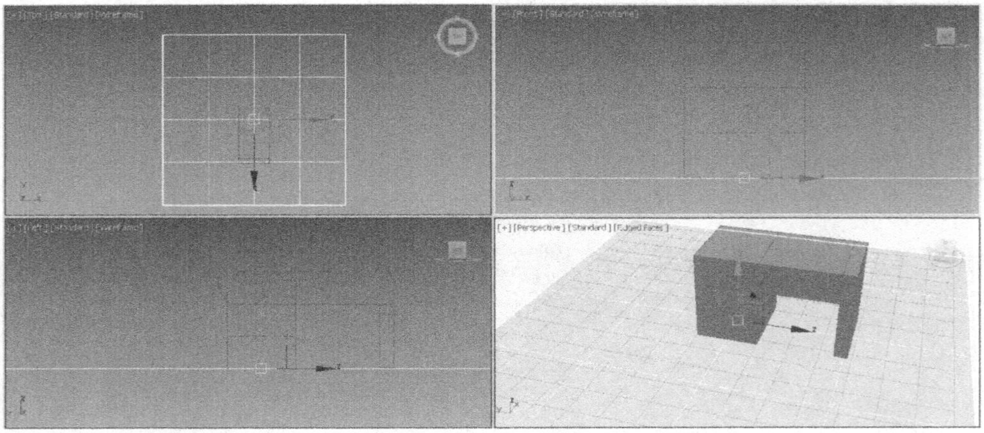

Figure 3-58 *The Floor created in all viewports*

Creating the Drawers

In this section, you will create the drawers using the smart extrude technique.

1. Select *Officetable* and then click on the arrow on the left of the **Editable Poly** to view all sub-object levels. Activate the **Top** viewport. Select the **Polygon** sub-object level. Next, select the polygon of *Officetable*. Activate the **Select and Scale** tool. Now, hold the SHIFT key and drag inward, as shown in Figure 3-59. Make sure the polygon is selected. Now, activate the **Select and Move** tool, hold the SHIFT key and then move the selected polygon inside, as shown in Figure 3-60.

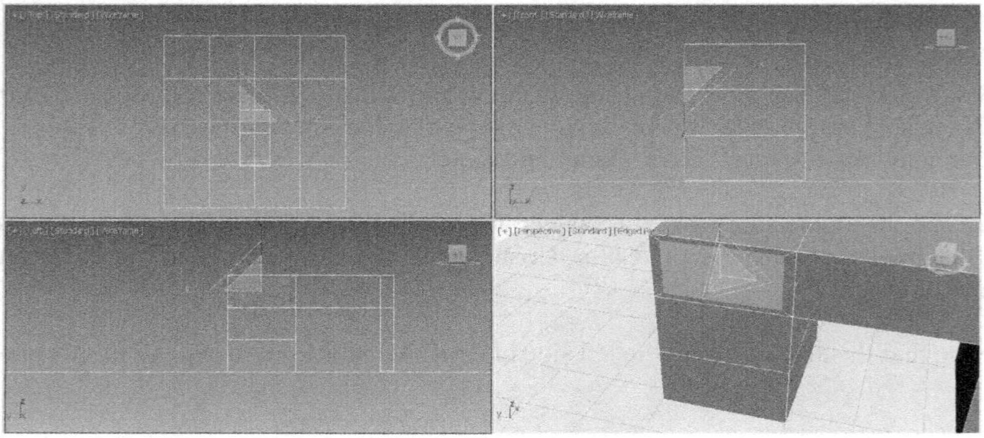

Figure 3-59 *Selected polygon scaled*

2. You need to follow the process mentioned in step 1 to create the remaining three drawers, as shown in Figure 3-61.

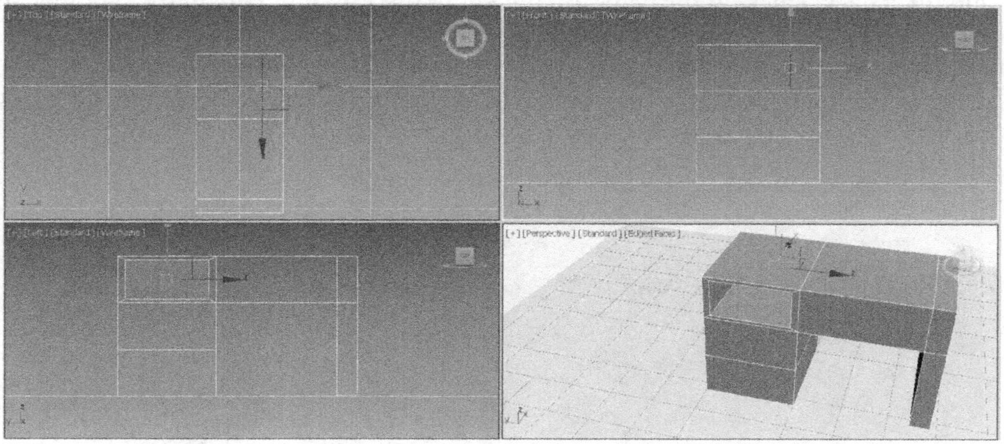

Figure 3-60 *Moving the selected polygon inside*

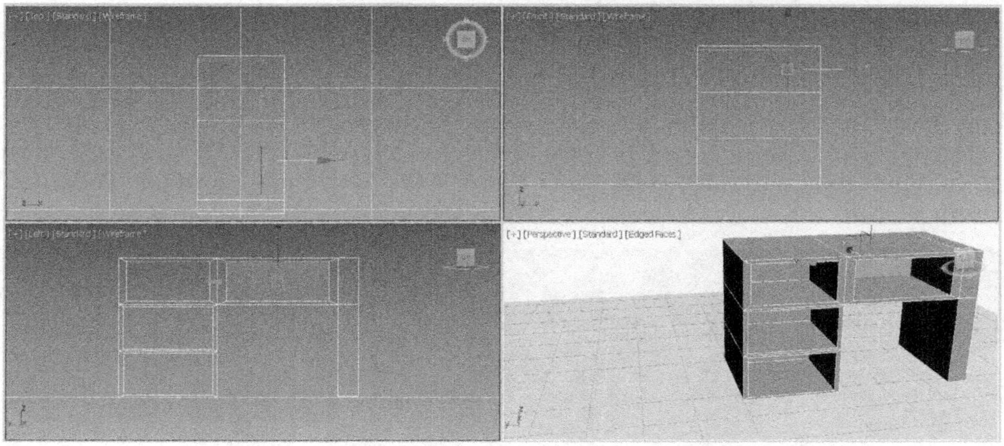

Figure 3-61 *Moving other polygons inside*

3. Activate the Top viewport. Choose **Create > Geometry** in the **Command Panel**; **Standard Primitives** is displayed in the drop-down list. Next, invoke the **Box** tool from the **Object Type** rollout; various rollouts are displayed in the **Command Panel**.

4. Expand the **Keyboard Entry** rollout and set the values as given next:

   Length: **17** Width: **16** Height: **8**

5. Choose the **Create** button from the **Keyboard Entry** rollout; a box is created in all viewports. In the **Name and Color** rollout, enter **Drawer** and then press ENTER; the box is named as *Drawer*.

6. Make sure the *Drawer* is selected. Activate the **Select and Move** tool and move away from *Officetable*, refer to Figure 3-62 Next, right-click; the flyout is displayed. Next, choose the

Convert To option from the flyout; a cascading menu is displayed. Choose the **Convert to Editable Poly** option from the cascading menu; the **Editable Poly** is displayed in the modifier stack. Click on the arrow on the left of the **Editable Poly** in the modifier stack to view all sub-object levels. Select the **Polygon** sub-object level.

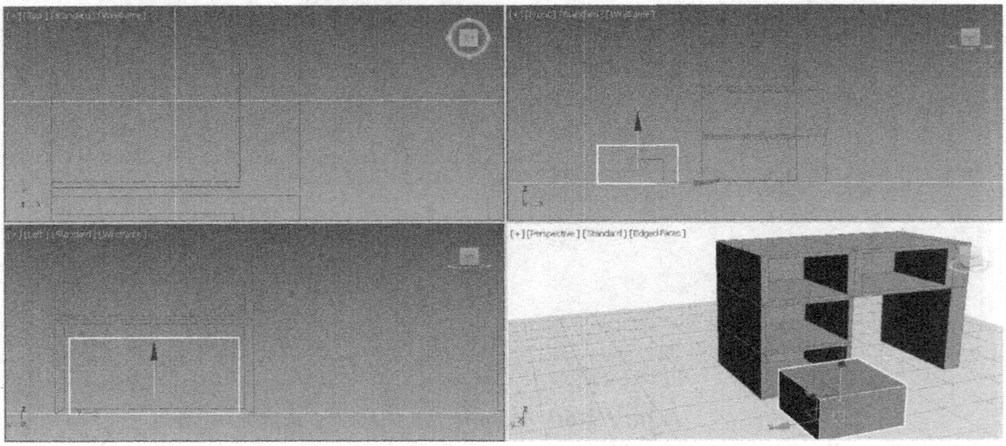

Figure 3-62 *Box created in all viewports*

7. Select the polygon of *Drawer*, as shown in Figure 3-63. Activate the **Select and Scale** tool. Now, hold the SHIFT key and drag inward, as shown in Figure 3-64. Make sure the polygon is selected. Now, activate the **Select and Move** tool and hold the SHIFT key and move the selected polygon downward, as shown in Figure 3-65. Now, choose **Editable Poly** from the modifier stack.

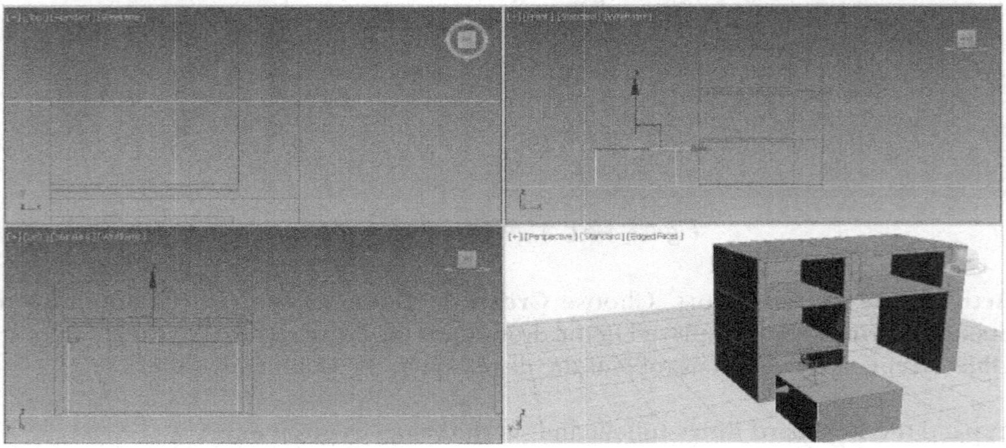

Figure 3-63 *Selected polygon of Drawer*

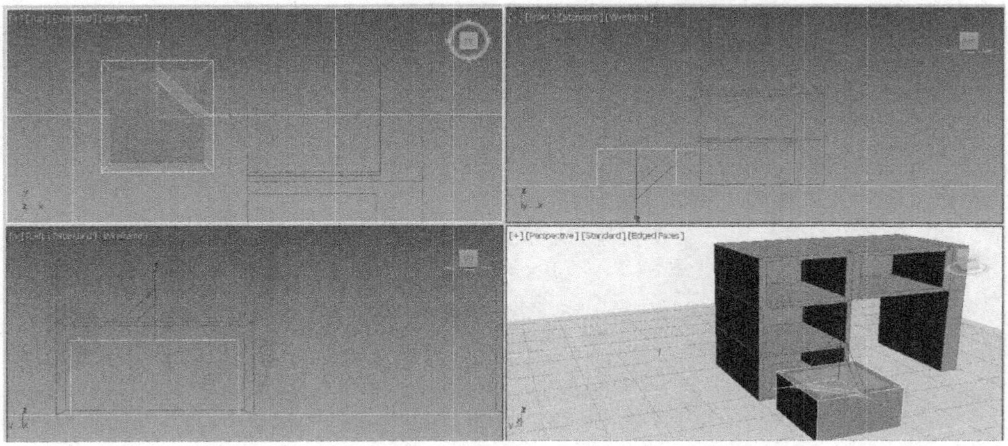

Figure 3-64 *Selected polygon scaled inward*

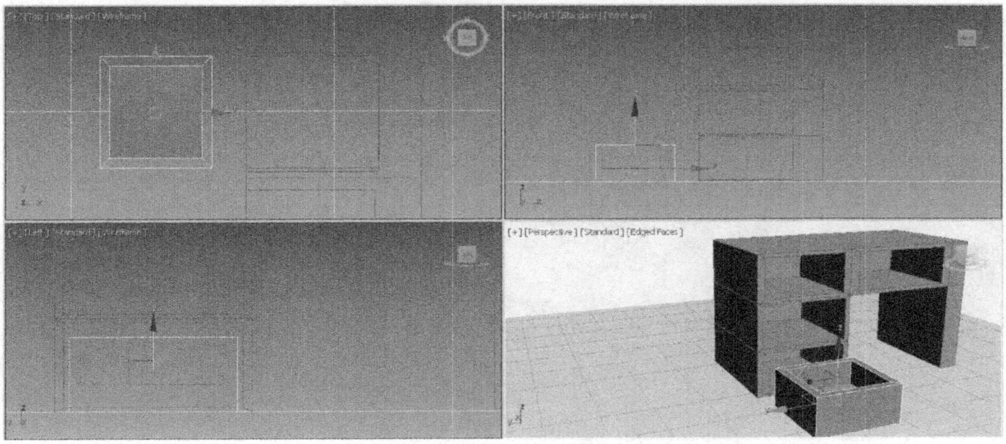

Figure 3-65 *Selected polygon moved downward*

8. Activate the **Select and Move** tool and align the *Drawer* to *Officetable*, as shown in Figure 3-66. Make sure the *Drawer* is selected. Now, hold SHIFT and move upward to the second drawer space; the **Clone Options** dialog box is displayed; enter **2** in the **Number of Copies** slider, as shown in Figure 3-67 and choose the **OK** button; two copies of drawers are created. Now, align them to *Officetable*, as shown in Figure 3-68.

9. Create a copy of drawer to be used as the right drawer. Next, align and scale it, refer to Figure 3-69.

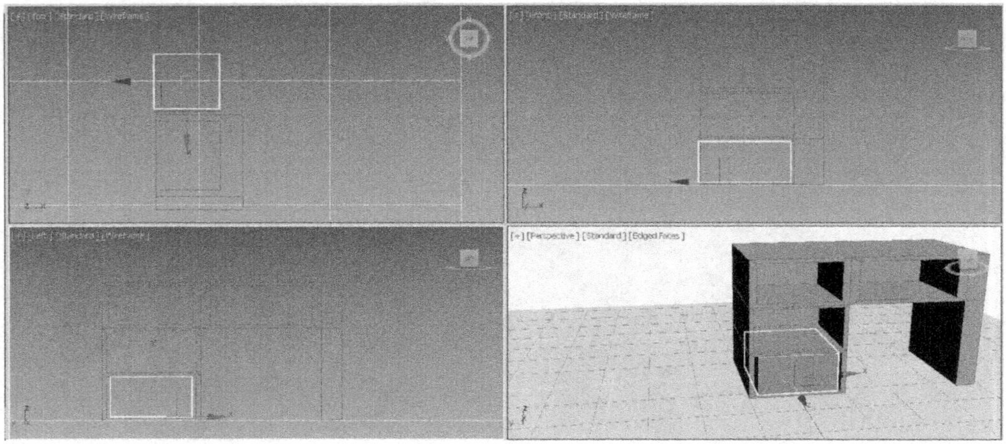

Figure 3-66 *Drawer aligned to Officetable*

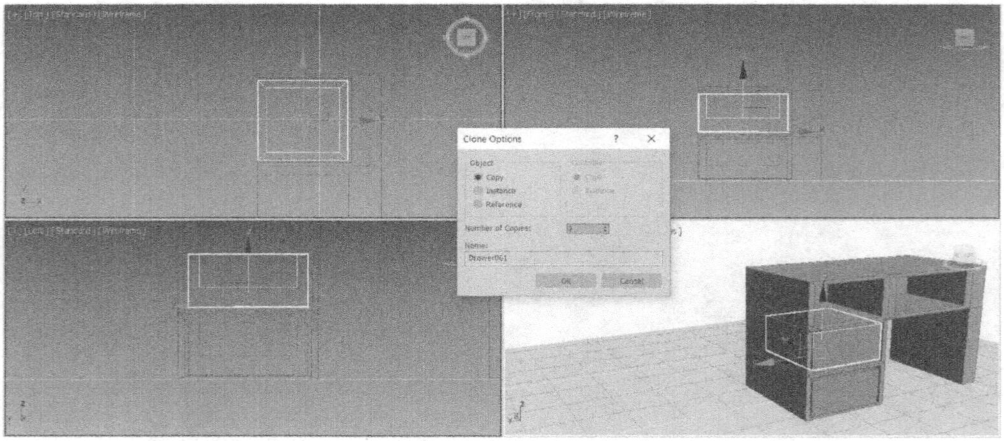

Figure 3-67 *The* ***Clone Options*** *dialog box displayed*

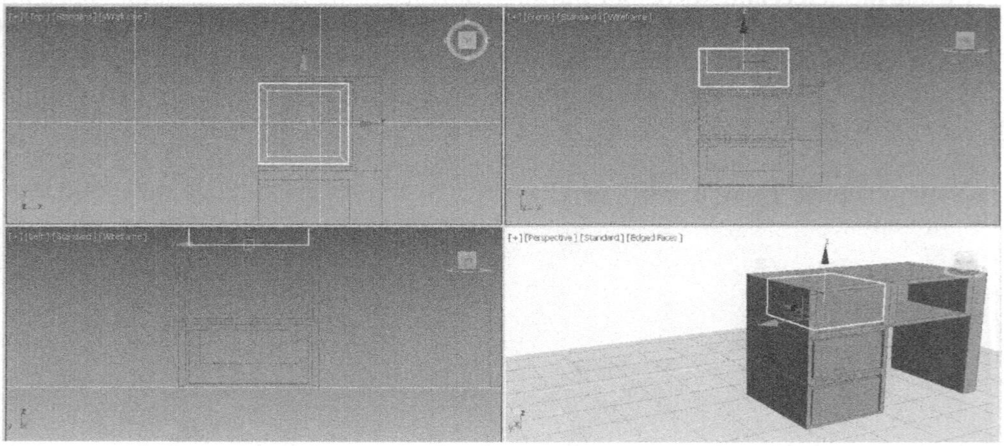

Figure 3-68 *Aligning the drawers*

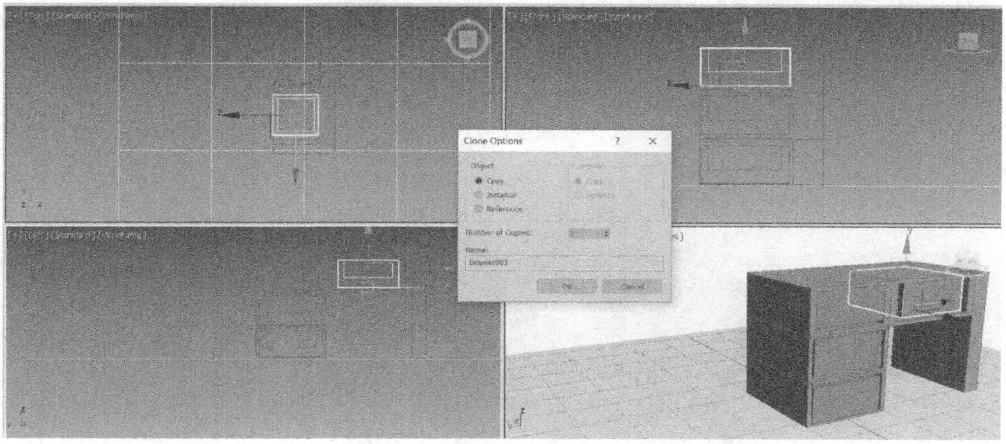

Figure 3-69 *Copy of drawer created*

Now, you will create knobs for drawers.

9. Activate the Top viewport and choose **Create > Geometry** in the **Command Panel**. Now, invoke the **Sphere** tool from the **Object Type** rollout.

10. In the **Keyboard Entry** rollout, set **1** in the **Radius** spinner. In the **Parameters** rollout, set **16** in the **Segments** spinner. Next, choose the **Create** button in the **Keyboard Entry** rollout; a sphere is created. Enter **knob** in the **Name and Color** rollout. Right-click in the viewport to exit the tool. Make sure the **Select and Move** tool is activated. Now, select *knob* and align it, as shown in Figure 3-70.

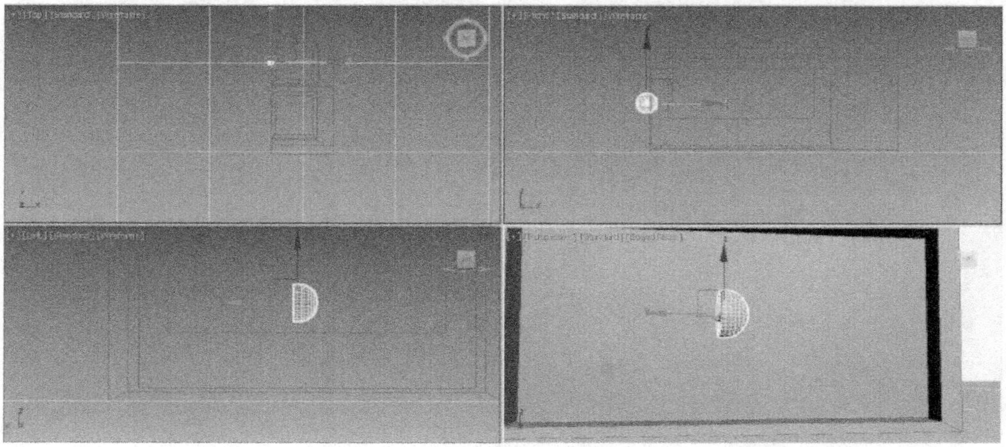

Figure 3-70 *Knob aligned to drawer*

11. Select knob in the Modifier Stack and enter 0.445 value in the Hemisphere slider. Now, choose the Select and Rotate Tool and then rotate the knob, as shown in Figure 3-71.

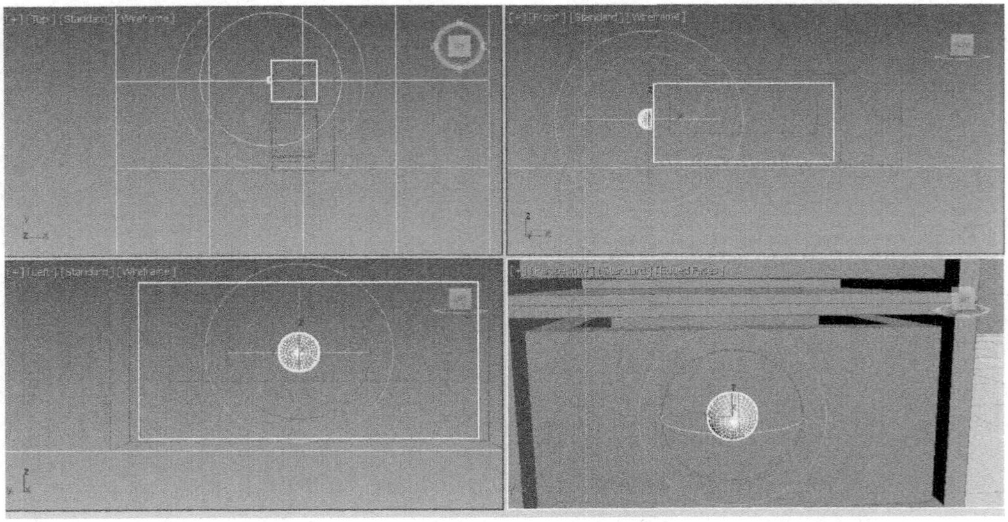

Figure 3-71 Rotating knob

12. Make sure the *knob* is selected in the viewport. Activate the **Select and Move** tool, hold the SHIFT key and then move the selected *knob* upward; a copy of knob is created and the **Clone Options** dialog box is displayed. Enter the value **2** in the **Number of Copies** edit box, as shown in Figure 3-72; two copies of knob are created in the viewport, as shown in Figure 3-73. Create one more knob for the side drawer, as shown in Figure 3-74.

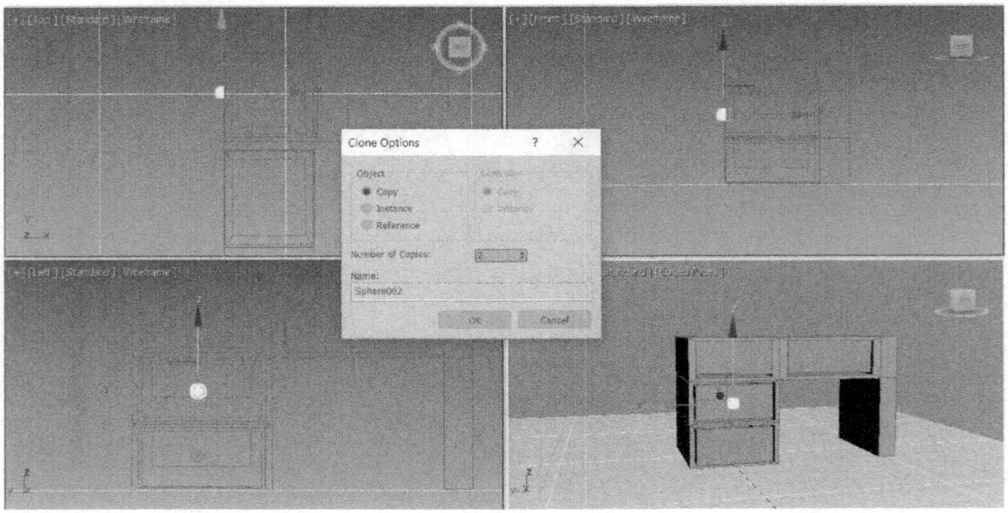

Figure 3-72 *Entering value* ***2*** *in the* ***Number of Copies*** *edit box*

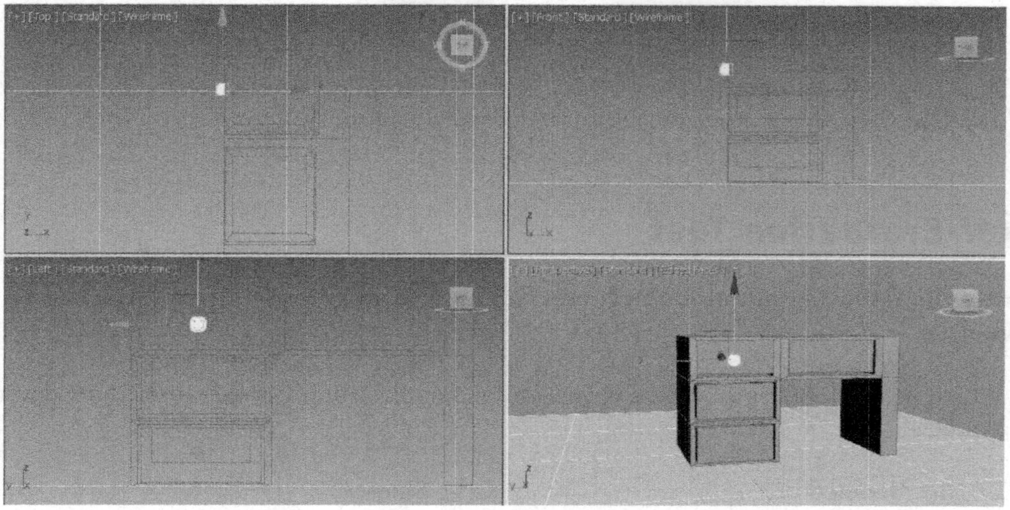

Figure 3-73 Copies of knob created

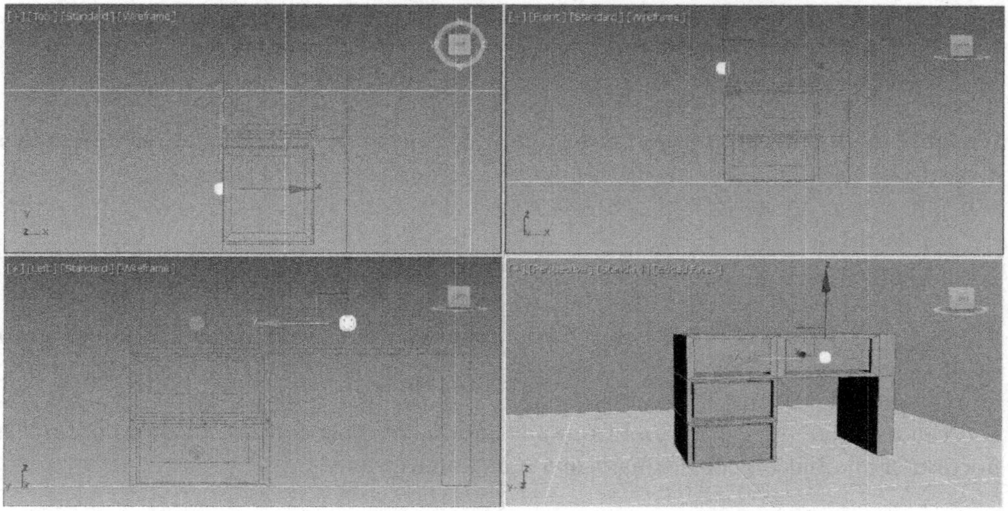

Figure 3-74 Copy of knob created for the side drawer

Saving and Rendering the Scene

In this section, you will save the scene that you have created and then render it. You can also view the final rendered image of this model by downloading the *c03_3dsmax_2022_rndr.zip* file from *www.cadcim.com*. The path of the file is as follows: *Textbooks > Animation and Visual Effects > Autodesk 3ds Max > Autodesk 3ds Max 2022 for Beginners: A Tutorial Approach.*

1. Change the background color of the scene to white by following the steps given in Tutorial 1 of Chapter 2.

2. Choose **Save** from the **File** menu.

3. Activate the Perspective viewport. Next, invoke the **Render Production** tool from the **Main Toolbar**; the rendered image is displayed, refer to Figure 3-50.

Self-Evaluation Test

Answer the following questions and then compare them to those given at the end of this chapter:

1. Which of the following buttons is chosen by default in the **Create** tab of the **Command Panel**?

   (a) **Geometry** (b) **Shapes**
   (c) **Lights** (d) **Helpers**

2. Which of the following tools in the **Main Toolbar** is used to rotate the object in increments?

   (a) **Spinner Snap Toggle** (b) **Percent Snap Toggle**
   (c) **Angle Snap Toggle** (d) None of these

3. Which of the following options is selected by default in the **Reference Coordinate System** drop-down list?

   (a) **World** (b) **View**
   (c) **Screen** (d) **Grid**

4. The __________ tool in the viewport navigation controls is used to view all objects properly in all viewports.

5. You can use the __________ tool from navigation viewport controls to zoom in on a small portion of an object by enclosing it in a selection window.

6. Choose **Array** from the __________ menu to display the **Array** dialog box.

7. Choose **Group** from the __________ menu to display the **Group** dialog box.

8. You can use the **Render Production** tool from the **Main Toolbar** to render the active viewport. (T/F)

Review Questions

Answer the following questions:

1. Which of the following dialog boxes is invoked by right-clicking on the **Select and Move** tool?

    (a) **Scale Transform Type-In** (b) **Rotate Transform Type-In**
    (c) **Move Transform Type-In** (d) None of these

2. Which of the following dialog boxes is displayed by right-clicking on the **Snaps Toggle** tool?

    (a) **Select from Scene** (b) **Grid and Snap Settings**
    (c) **Preference Settings** (d) **Environment and Effects**

3. The __________ tool is used to zoom an object equally in all viewports.

4. The __________ tool is used to create the duplicate copies of the other sides of objects.

5. The __________ is used to select the objects in the scene.

6. The **Axis Constraints** tools in the **Axis Constraints** toolbar do not normally work with snaps on. (T/F)

7. The **Snaps Toggle** tool is used to rotate the object in the increments you set. (T/F)

EXERCISE

The rendered output of the model used in the following exercise can be accessed by downloading the *c03_3dsmax_2022_exr.zip* from *www.cadcim.com*. The path of the file is as follows: *Textbooks > Animation and Visual Effects > Autodesk 3ds Max > Autodesk 3ds Max 2022 for Beginners: A Tutorial Approach*

Exercise 1

Create the scene of an umbrella and table on the ground, as shown in Figure 3-75.

Figure 3-75 *The scene of an umbrella and a table*

Answers to Self-Evaluation Test

1. a, **2.** c, **3.** b, **4. Zoom Extents All**, **5. Zoom Region**, **6. Tools**, **7. Group**, **8.** T

Chapter 4

Working with Splines-I

Learning Objectives

After completing this chapter, you will be able to:

- *Create 2D shapes*
- *Modify splines*
- *Change bezier splines*
- *Work with transform centers*

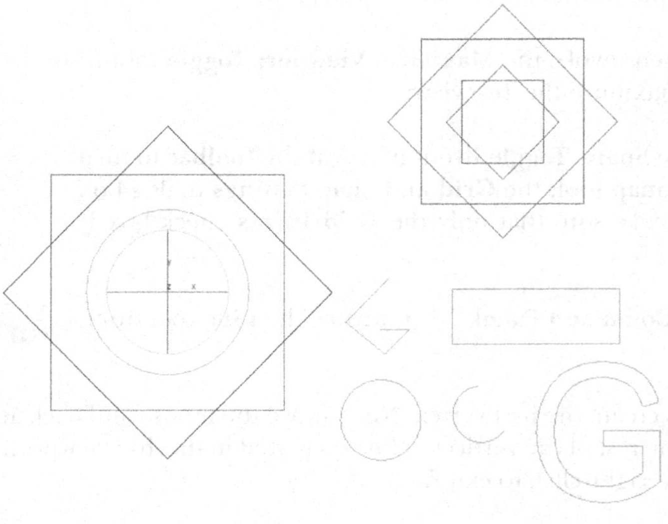

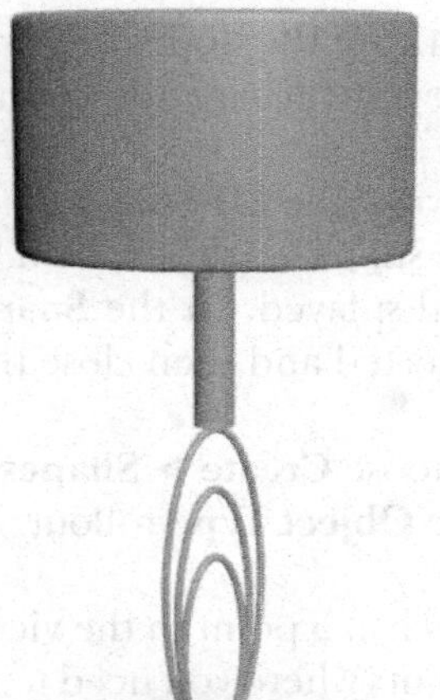

INTRODUCTION

In 3ds Max, splines are used to create various shapes. The advantage of using splines in place of primitive objects is that they require less number of segments to create an object. In this chapter, you will learn to create and modify different types of splines using the **Edit Spline** modifier.

TUTORIALS

Tutorial 1

In this tutorial, you will create different 2D splines such as a line, circle, rectangle, arc, and text. You will modify the vertices and segments, and also adjust the splines using bezier handles. In addition, you will transform the shapes using different transform centers.

(Expected time: 25 min)

The following steps are required to complete this tutorial:

a. Create the project folder.
b. Create 2D Shapes.
c. Modify vertices, segments, and splines.
d. Change bezier splines.
e. Lock handles.
f. Create and modify splines.
g. Copy an object.

Creating the Project Folder

Create a project folder with the name *c04_tut1* in the *3dsmax 2022* folder as discussed in Tutorial 1 of Chapter 2.

Creating 2D Shapes

In this section, you will create 2D shapes in the form of spline polygons.

1. Activate the Top viewport and then invoke the **Maximize Viewport Toggle** tool from the viewport navigation controls to maximize the Top viewport.

2. Invoke the **2D Snap** tool from the **Snaps Toggle** flyout in the **Main Toolbar** to turn on snaps. Right-click on the **2D Snap** tool; the **Grid and Snap Settings** dialog box is displayed. In the **Snaps** tab, make sure that only the **Grid Points** check box is selected and then close the dialog box.

3. Choose **Create > Shapes** in the **Command Panel**. Next, invoke the **Line** tool from the **Object Type** rollout.

4. Click at a point in the viewport to create the first vertex. Next, move the mouse and click at points where you need to create the rest of the vertices; a line is created in the Top viewport, refer to Figure 4-1 (top-left). Now, right-click to exit the tool.

5. Invoke the **Zoom** tool from the viewport navigation controls. Click near the center of the viewport and drag the cursor vertically downward. Release the mouse button after increasing the drawing display area.

6. Invoke the **Rectangle** tool from the **Object Type** rollout.

7. Press and hold the mouse button at a point and then drag the cursor to select the diagonally opposite corner; a rectangle is created, refer to Figure 4-1. You can press and hold the CTRL key to make a square. Right-click to exit the tool.

8. Invoke the **Circle** tool from the **Object Type** rollout. Next, press and hold the mouse button at a point as the center of the circle and then drag the cursor to set the radius, refer to Figure 4-1.

9. Invoke the **Arc** tool from the **Object Type** rollout. Press and hold the mouse button at a point to create the first endpoint of the arc. Next, drag to create the second point and then release the left mouse button. Now, drag and click again in the viewport to specify the third and fourth points to set the radius of the arc, refer to Figure 4-1.

***Figure 4-1** Different spline shapes created in the Top viewport*

10. Invoke the **Text** tool from the **Object Type** rollout. In the **Parameters** rollout, select **Arial** from the drop-down list, if it is not already selected. Next, select **MAX Text** from the **Text** edit box and then enter **G**.

11. Click at a point in the viewport to position the text, refer to Figure 4-1. Invoke the **Select and Move** tool from the **Main Toolbar**. Next, press and hold the left mouse button and drag the cursor to move the text and place it on the screen at a proper position. You can also scale the text as required by using the **Select and Uniform Scale** tool.

12. Click on the empty area of the viewport to deselect any selected shape.

13. Invoke the **Zoom Extents** tool from the viewport navigation controls to view the text clearly in the viewport. Next, choose **Edit > Hold** from the menu bar to copy the current scene to a buffer. This can be retrieved later using the **Fetch** option.

Modifying Vertices, Segments, and Splines

In this section, you will convert the alignment of vertices from linear to curved and vice versa.

1. Select the text object and then select the circle using the CTRL key.

2. Choose the **Modify** tab in the **Command Panel**. Select **Edit Spline** from the **Modifier List** drop-down list. Next, choose the **Vertex** button in the **Selection** rollout; the **Edit Spline** modifier is applied to the shapes and the **Vertex** sub-object mode is activated. The vertices of both the shapes are displayed in the viewport. 

3. Right-click on the vertex on the right side of the circle; a quad menu is displayed. Choose **Corner** from the upper-left quadrant of the quad menu, as shown in Figure 4-2.

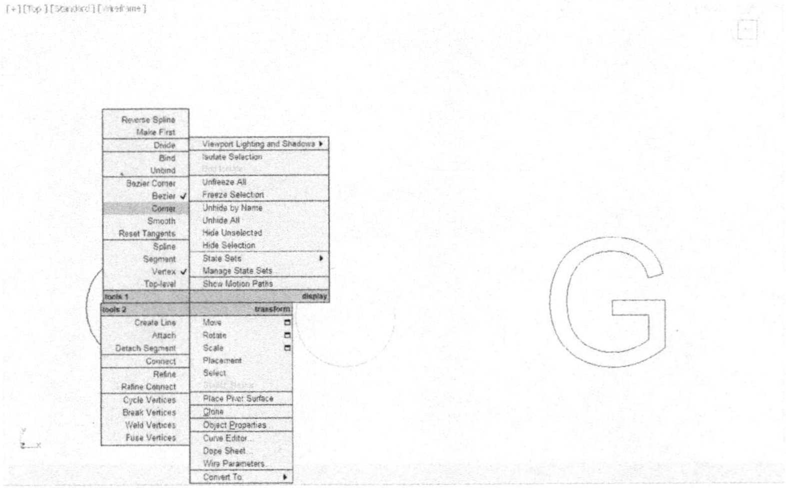

Figure 4-2 Choosing ***Corner*** *from the upper-left quadrant of the quad menu*

4. Make sure the text is selected and right-click on the vertex on the left side of the text. Choose **Corner** from the upper-left quadrant of the quad menu displayed; the curved vertices are changed to corner, as shown in Figure 4-3.

5. Right-click on the vertex on the right side of the circle; the quad menu is displayed. Choose **Smooth** from the upper-left quadrant of the quad menu to make the circle smooth. Choose the **Vertex** button to exit the sub-object mode. Now, select the rectangle.

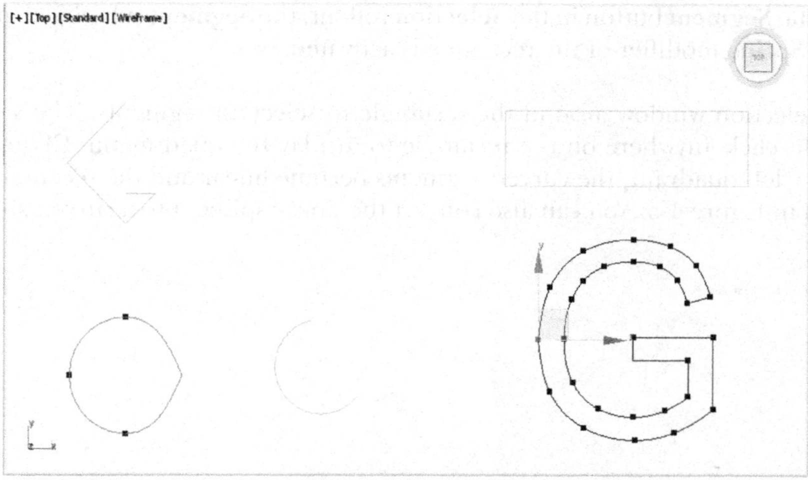

Figure 4-3 The curved vertices changed to corner

Note

You cannot select a new shape when you are in the sub-object mode.

6. Select **Edit Spline** from the **OBJECT-SPACE MODIFIERS** section in the **Modifier List** drop-down list. Next, choose the **Vertex** button in the **Selection** rollout; the **Edit Spline** modifier is applied to the rectangle and the **Vertex** sub-object mode is activated. Also, the vertices are displayed in the viewport.

7. Right-click on the lower-left vertex of the rectangle; quad menu is displayed. Choose **Smooth** in the upper-left quadrant of the quad menu. Now, right-click on the upper-right vertex of the rectangle and then choose **Smooth** from the quad menu displayed; the two linear vertices become curved resulting in a curved shape, as shown in Figure 4-4. Similarly, you can convert the linear segments into curved segments and vice versa.

Figure 4-4 Two corners of the rectangle changed to smooth curves

8. Choose the **Segment** button in the **Selection** rollout; the **Segment** sub-object mode of the **Edit Spline** modifier of the rectangle is activated.

9. Drag a selection window around the rectangle to select all segments. The segments turn red. Right-click anywhere on the rectangle to display the quad menu. Choose **Line** from the upper-left quadrant; the curved segments become linear and the rectangle is restored, as shown in Figure 4-5. You can also convert the linear spline into a curved spline and vice versa.

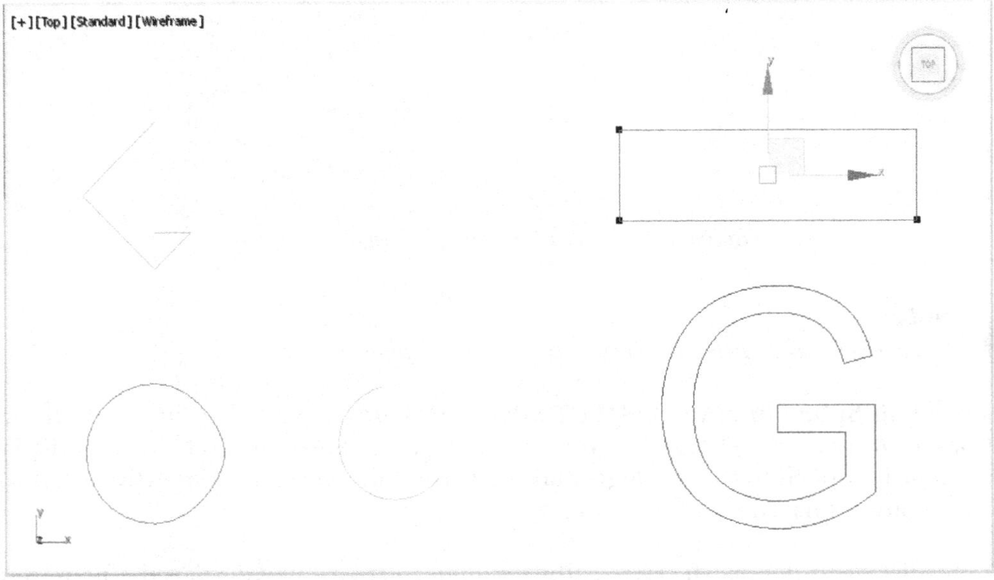

Figure 4-5 *The rectangle restored to its original shape*

10. Choose the **Segment** button in the **Selection** rollout to exit the sub-object mode.

11. Select the object created using the **Line** tool. Next, make sure the **Modify** tab is chosen in the **Command Panel**. In the **Selection** rollout, choose the **Spline** button to enter the **Spline** sub-object mode.

Note

*There is no need to apply the **Edit Spline** modifier to the shapes that were created using the **Line** tool.*

12. Right-click on the spline; the quad menu is displayed. Choose **Curve** from the upper-left quadrant of the quad menu; the whole spline becomes curved, as shown in Figure 4-6.

Changing Bezier Splines

There are two types of bezier vertices: the **Bezier** vertex and the **Bezier Corner** vertex. In this section, you will adjust spline curvature with the help of bezier handles.

1. Choose **Edit > Fetch** from the menu bar; the **About to Fetch. OK?** message box is displayed, as shown in Figure 4-7. Choose the **Yes** button from the message box to restore the scene from buffer.

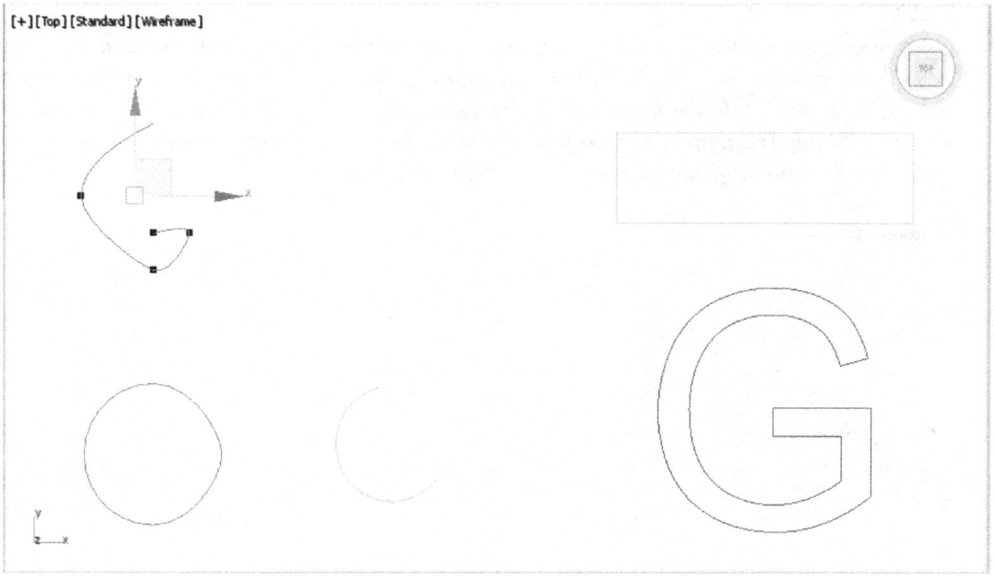

Figure 4-6 *The curved line spline*

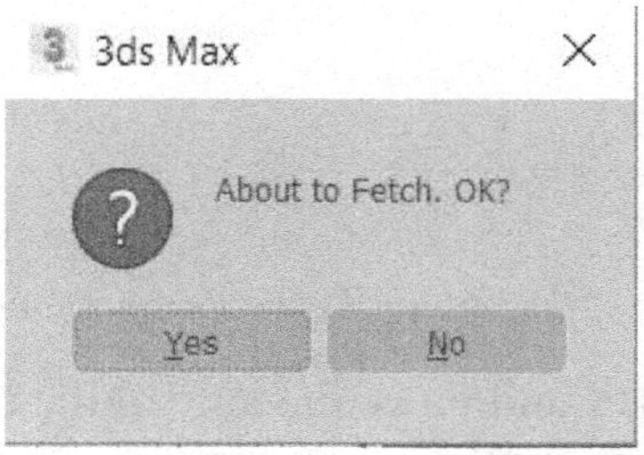

Figure 4-7 *The* ***About to Fetch. OK?*** *message box*

2. Select the rectangle in the viewport and then make sure the **Modify** tab is chosen in the **Command Panel**. Select **Edit Spline** from the **OBJECT-SPACE MODIFIERS** section in the **Modifier-List** drop-down list; the **Edit Spline** modifier is applied. Next, choose the **Vertex** button in the **Selection** rollout to enter into the **Vertex** sub-object mode.

3. Right-click on the top-left vertex of the rectangle; a quad menu is displayed. Choose **Bezier** from the upper-left quadrant of the quad menu; two vector handles are displayed in a straight line forcing the left segment into a curve.

4. Invoke the **Select and Move** tool from the **Main Toolbar**. Next, invoke the **2D Snap** tool again from the **Snaps Toggle** flyout in the **Main Toolbar** to turn it off. Drag one of the handles to form the shape shown in Figure 4-8; both handles move such that the vertex remains tangent to the segments.

Note

On choosing the transform gizmo button, an axis constraint is set. Therefore, it is often difficult to move the handles as desired while the transform gizmo is active. You may need to turn off the transform gizmo while working with bezier handles. To turn off the transform gizmo, choose ***Views > Show Transform Gizmo*** *from the menu bar. On choosing* ***Show Transform Gizmo*** *again, the transform gizmo will be restored in the viewport.*

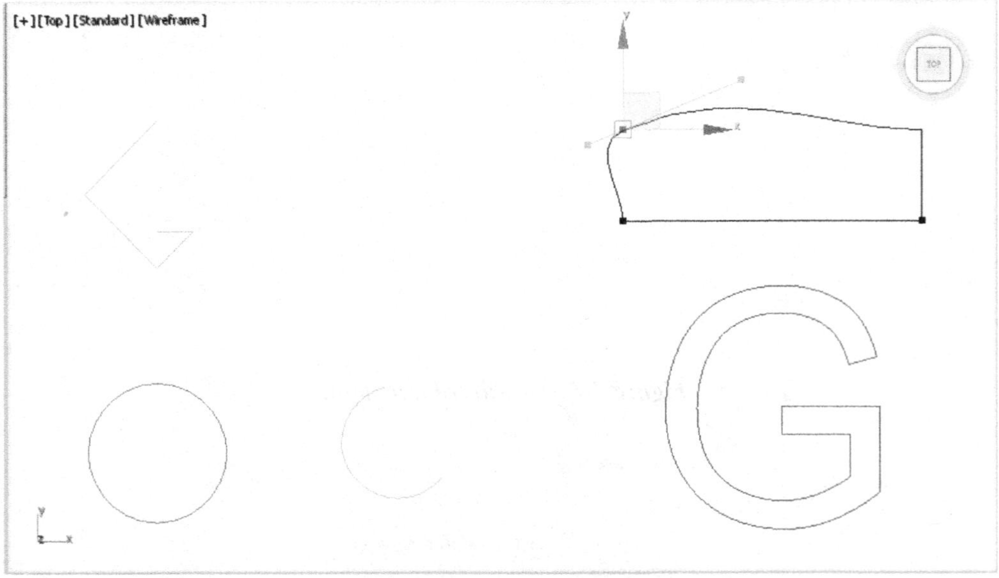

Figure 4-8 *Shape of the rectangle after dragging one of the handles*

5. Press and hold the SHIFT key and move the left handle up toward the left direction; the SHIFT key breaks the tangent by allowing you to move the selected handle converting the bezier vertex into a bezier corner vertex; only the segment associated with this handle is affected whereas the other handle remains unchanged, as shown in Figure 4-9.

6. Right-click on the same vertex again to display the quad menu and choose **Bezier** from it; the changed segment automatically assumes its previous shape forcing the moved handle to be in a straight line with the unchanged handle.

Locking Handles

For any vertex, there are two bezier handles that are used to set the tangency. By selecting the **Lock Handles** check box in the **Selection** rollout and selecting the **Alike** radio button, you can simultaneously transform all similar bezier handles (all lead-in handles or all lead-out handles) for the selected vertices. When the **All** radio button is active, you can simultaneously move both handles for all the selected vertices. In this section, you will explore the use of the **Lock Handles** check box and the radio buttons discussed above.

1. Choose **Edit > Fetch** from the menu bar; the **About to Fetch. OK?** message box is displayed.

2. Choose the **Yes** button in this message box; the original shape of the rectangle is restored.

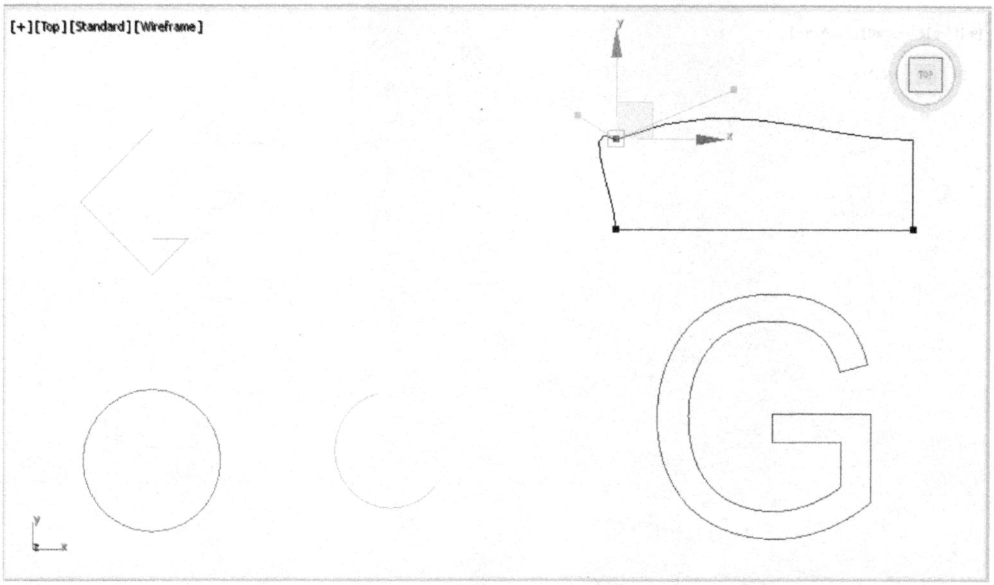

Figure 4-9 Curved line segments of the spline

3. Select the rectangle and apply the **Edit Spline** modifier, as discussed earlier.

4. Choose the **Vertex** button in the **Selection** rollout and select all the vertices in the shape by dragging a selection window around the rectangle; the bezier handles are displayed on all the vertices.

5. In the **Selection** rollout, clear the **Lock Handles** check box if it is selected.

6. Invoke the **Select and Move** tool. Next, select the lower handle of the top-left vertex. Now, drag it to left; only the selected handle is affected and only the segment connected to it changes, as shown in Figure 4-10.

7. Choose the **Undo** button to reverse the changes. In the **Selection** rollout, select the **Lock Handles** check box. Also, make sure the **Alike** radio button is selected. Select all vertices in the rectangle in the viewport.

8. Select the lower handle of the top-left corner vertex and drag it to left; the selected handle and all other lead-in handles for the selected vertices are affected, and all segments connected to them are changed, as shown in Figure 4-11.

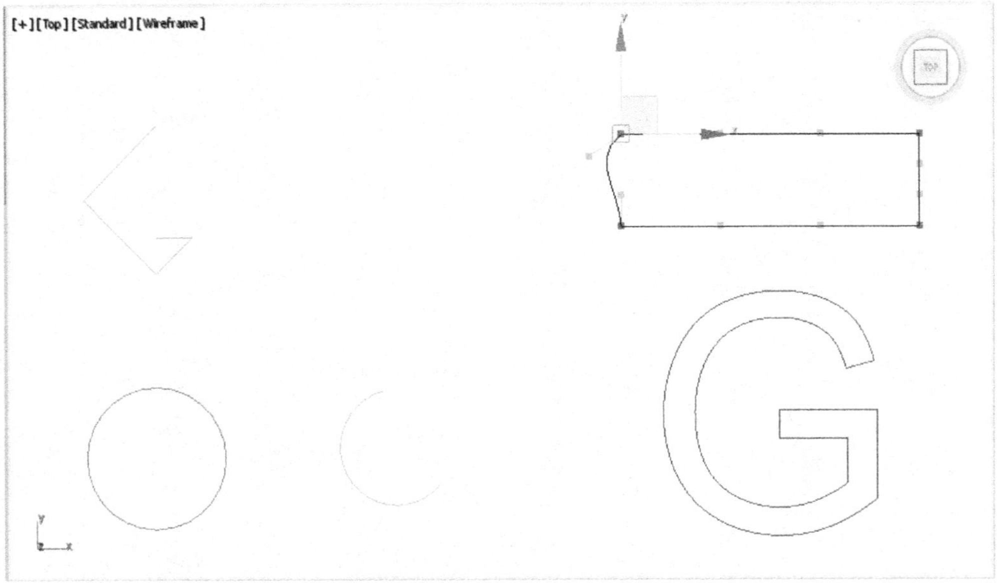

Figure 4-10 *Adjusting the bezier handles*

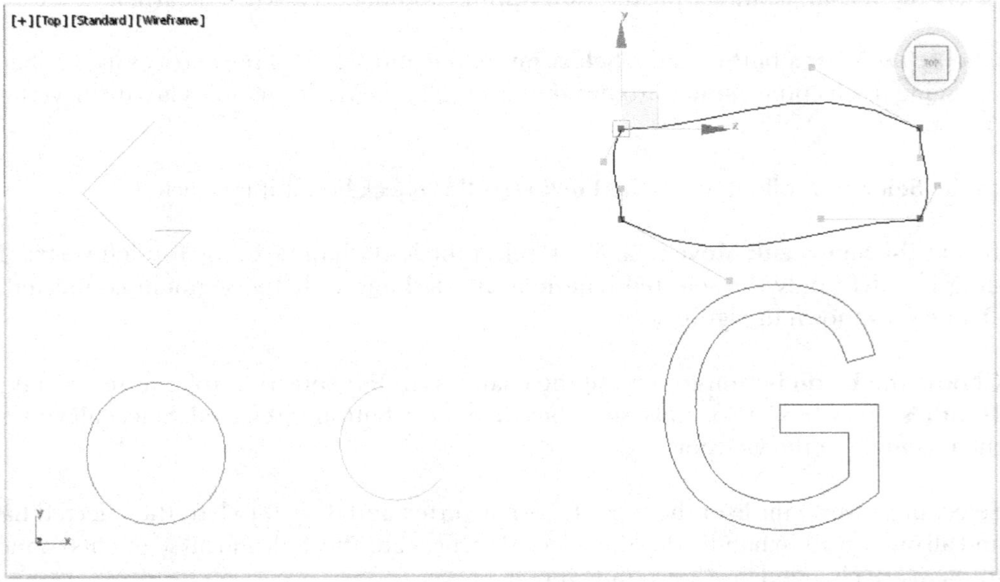

Figure 4-11 *All similar bezier handles affected by the* ***Lock Handles*** *check box and the* ***Alike*** *radio button*

9. Press CTRL + Z to reverse the changes. In the **Selection** rollout, select the **All** radio button.

10. Select the lower handle of the top-left corner vertex and drag it to left. Release the left mouse button; all the handles are affected and all the segments are changed, refer to Figure 4-12.

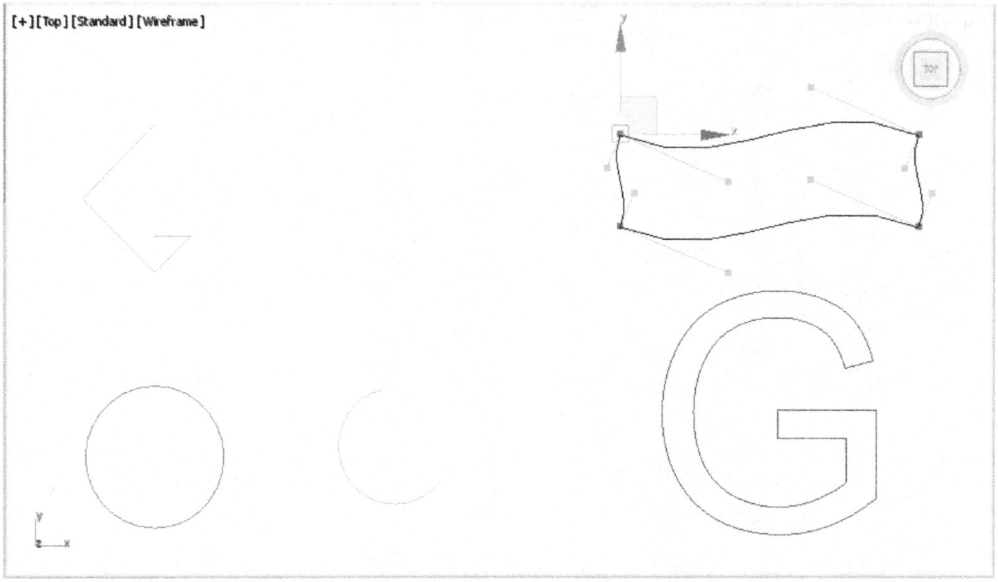

*Figure 4-12 All bezier handles affected on selecting the **All** radio button*

Creating and Modifying Splines

In this section, you will create some more splines and modify them using options in the rollouts.

1. Choose **File > New > New All** from the **File** menu; the **Autodesk 3ds Max 2022** message box is displayed, as shown in Figure 4-13. Choose the **Don't Save** button; all shapes are removed and a blank screen is displayed. But the viewport layout, which contains maximized Top viewport, is retained.

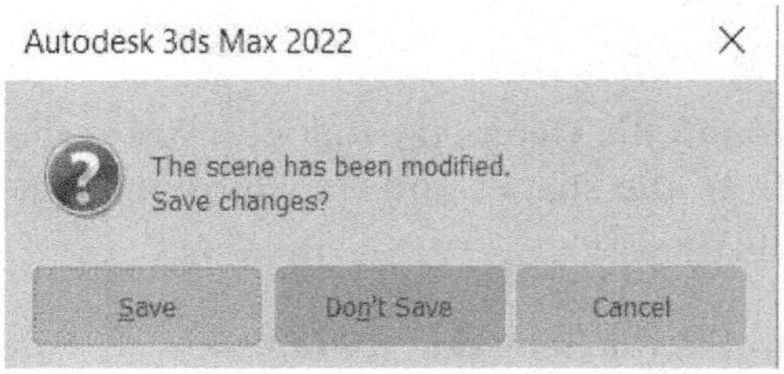

*Figure 4-13 The **Autodesk 3ds Max 2022** message box*

2. Choose **Create > Shapes** in the **Command Panel**. Next, invoke the **Ellipse** tool from the **Object Type** rollout. Create an ellipse in the Top viewport.

3. Invoke the **Rectangle** tool from the **Object Type** rollout. Press and hold the CTRL key and draw a square, as shown in Figure 4-14.

4. Invoke the **Select Object** tool, press and hold the CTRL key, and select the ellipse and square.

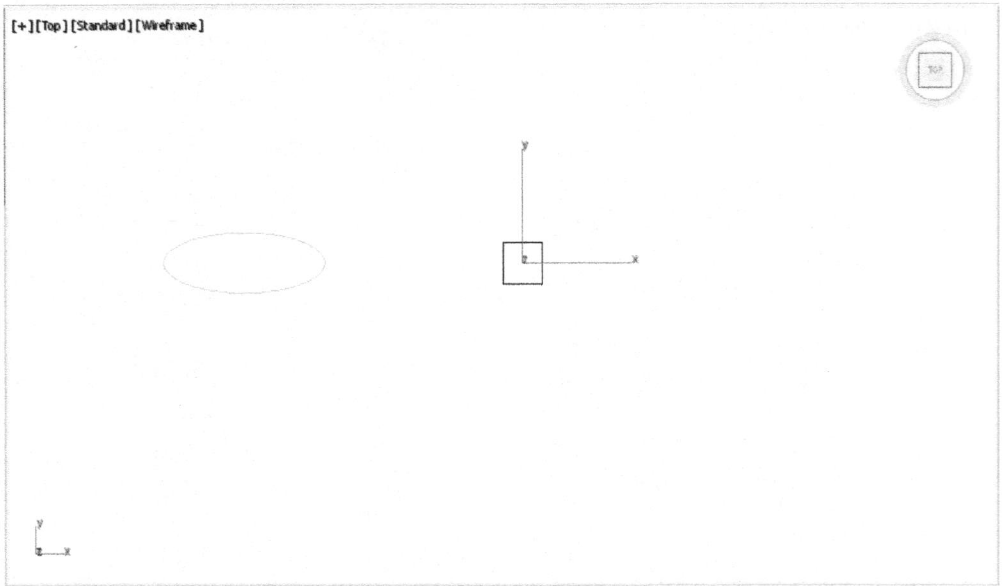

Figure 4-14 The ellipse and square created in the Top viewport

5. Choose the **Modify** tab in the **Command Panel**. Select **Edit Spline** from the **OBJECT-SPACE MODIFIERS** section in the **Modifier List** drop-down list; the **Edit Spline** modifier is applied to both shapes.

6. Choose the **Segment** button in the **Selection** rollout.

7. Select the lower-right segment of the ellipse. Press and hold the CTRL key and select the left and lower side of the square as well; all three segments turn red indicating that they are selected.

8. Choose the **Delete** button in the **Geometry** rollout or press the DELETE key; the selected segments are deleted from the shapes and the ellipse and the square become two open splines, as shown in Figure 4-15.

9. Choose the **Spline** button in the **Selection** rollout.

10. Click anywhere on the remaining portion of the square; the spline turns red indicating selection. Notice that the ellipse is not selected.

11. Choose the **Close** button in the **Geometry** rollout; the end vertices of the two remaining segments are joined by a curved segment and the spline is closed, as shown in Figure 4-16.

12. Choose the **Undo** button to reverse the changes. Next, choose the **Vertex** button in the **Selection** rollout.

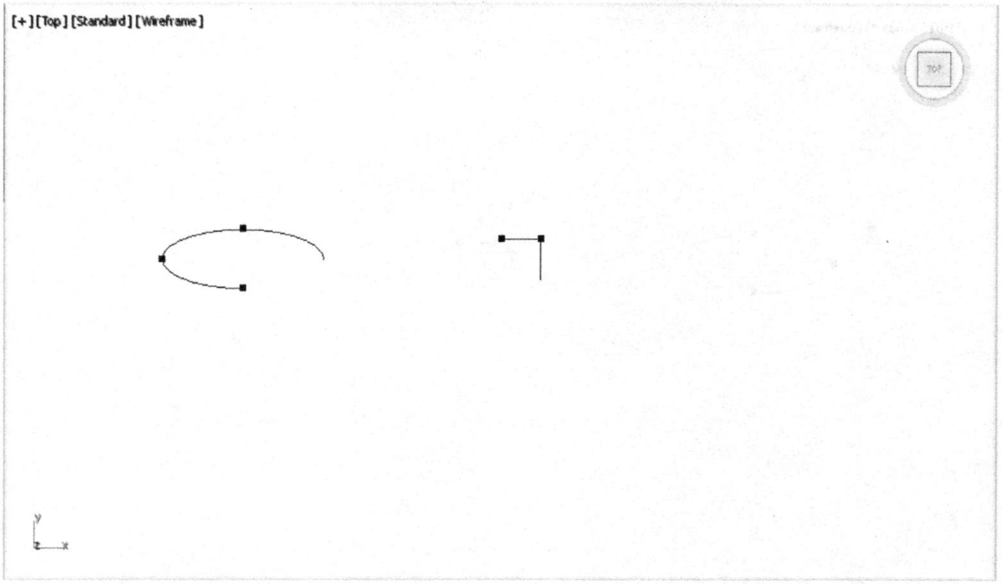

Figure 4-15 Selected segments deleted from the shapes

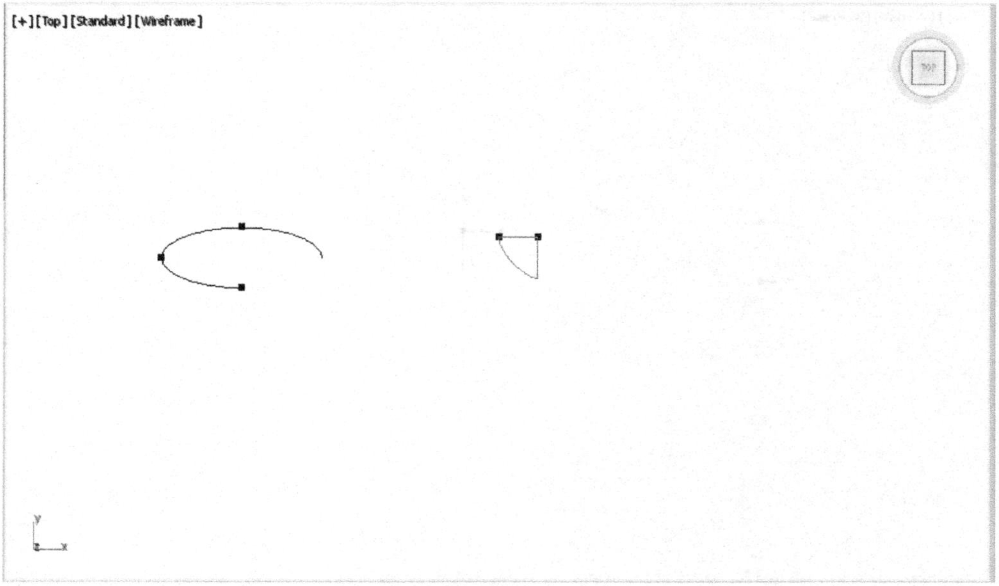

Figure 4-16 The closed spline

13. Choose the **Connect** button in the **Geometry** rollout. Next, select the end vertex of the square segment and then drag the cursor toward the other end vertex. When you move the cursor over the other end vertex, the cursor icon gets changed. Also, a dashed line is attached to the cursor while dragging, as shown in Figure 4-17.

14. Release the mouse button; the two vertices are joined by a straight segment, as shown in Figure 4-18.

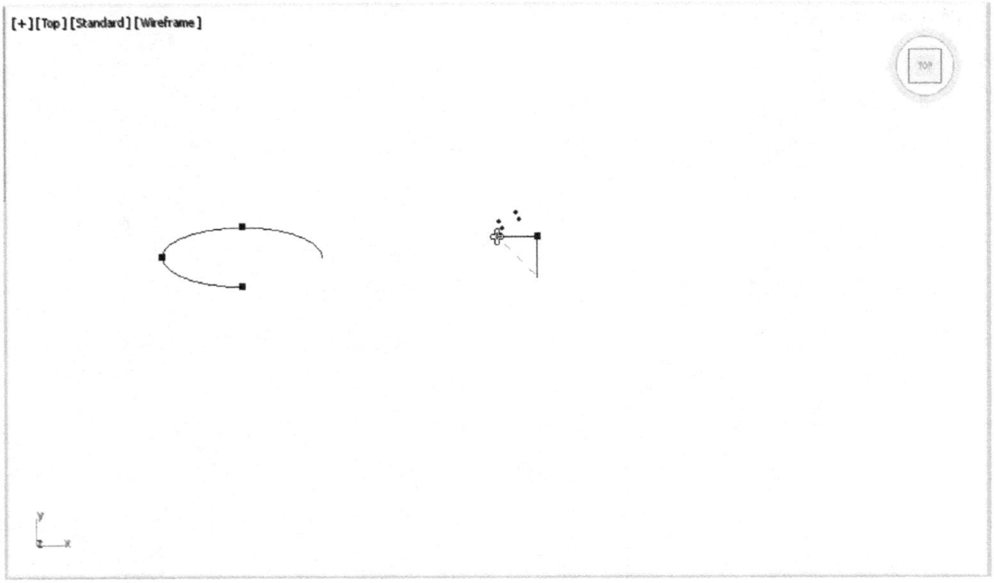

Figure 4-17 *A dashed line attached to the end vertices*

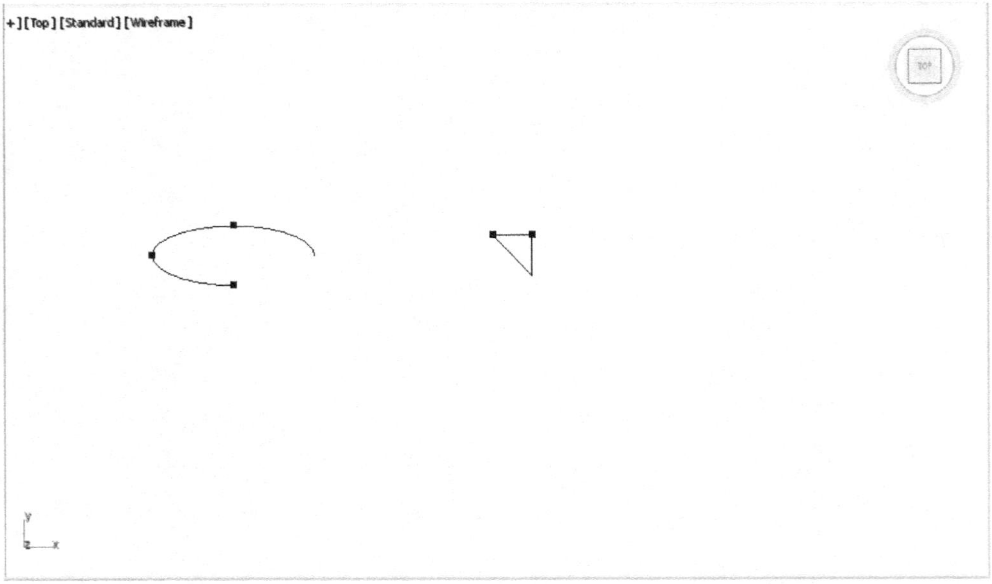

Figure 4-18 *The two vertices joined by a straight segment*

15. Choose the **Undo** button to reverse the changes. Next, choose the **Insert** button in the **Geometry** rollout.

16. Move the cursor and place it near the middle of the top segment of the remaining square; the cursor icon changes.

17. Click anywhere in the top segment to insert the vertex and then move the cursor down. Click again to position the vertex.

18. Move the cursor to the left and click to place another vertex, refer to Figure 4-19. Right-click to complete the command.

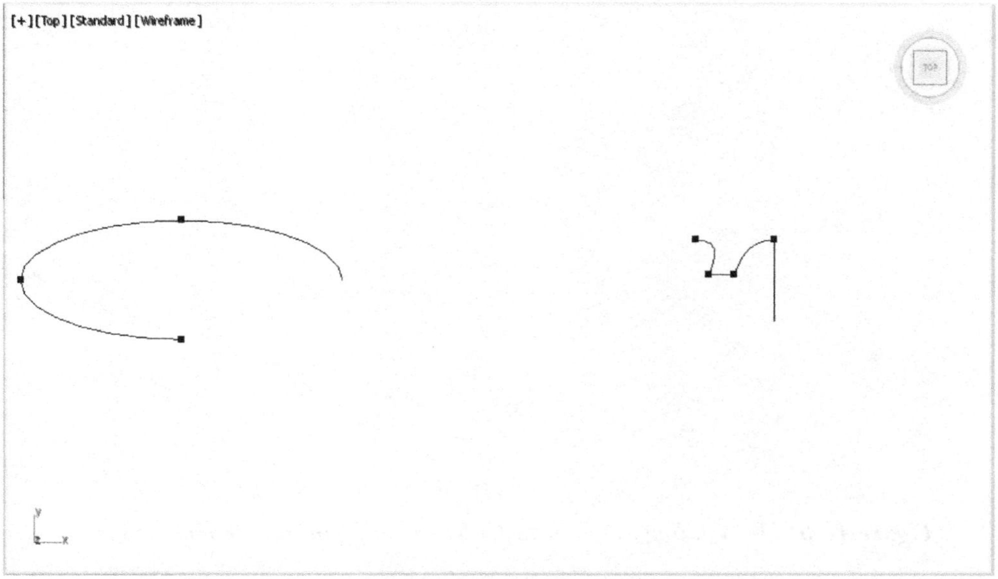

Figure 4-19 *New vertices added to the segment*

19. Choose the **Undo** button to reverse the changes. Next, invoke the **Select Object** tool  and select the white colored end vertex. The direction of the spline runs from the first vertex to the last vertex. The first vertex is yellow in color.

20. Choose the **Make First** button in the **Geometry** rollout; the selected vertex becomes yellow in color indicating that now it is the first vertex. Since the vertex is selected, it is red in color.

21. Choose the **Insert** button in the **Geometry** rollout. Next, move the cursor and place it near the middle of the top segment of the remaining portion of the square; the cursor icon changes.

22. Click to place the vertex and then move it down. Click again to position the vertex.

23. Move the cursor to the right and click again to place another vertex. Right-click to complete the command.

    As the direction of the spline has changed, the resulting shape is different from the one created previously by adding vertices, refer to Figure 4-20.

    Next, you will work with transform centers. There are three types of transform centers: pivot point, selection, and transform coordinate. The effect on objects due to rotation and scale transforms differs depending on the type of transform center selected.

24. Choose **File > New > New All** from the **File** menu; the **Autodesk 3ds Max 2022** message box is displayed. Choose the **Don't Save** button; all shapes are removed.

***Figure 4-20** The resulting shape created after changing the direction of the spline*

25. Invoke the **Zoom Extents** tool. Next, choose **Create > Shapes** in the **Command Panel**. Next, invoke the **Rectangle** tool. Draw a small rectangle near the upper-left corner of the Top viewport.

26. Invoke the **Ellipse** tool and draw a small ellipse close to the rectangle, as shown in Figure 4-21.

***Figure 4-21** A rectangle and an ellipse created*

27. Choose **Edit > Hold** from the menu bar to copy the current state to a buffer. This can be retrieved later using the **Fetch** option.

28. Make sure the **Snaps Toggle** tool is turned off. Next, invoke the **Select and Rotate** tool and make sure the **Transform Gizmo Z Constraint** button is chosen.

29. Make sure the ellipse is selected and then invoke the **Angle Snap Toggle** tool to turn the angle snap on.

    The **Use Pivot Point Center** tool in the **Main Toolbar** is invoked by default.

30. Rotate the ellipse until 0.00, 0.00, 90.0 is displayed in the coordinate display; the ellipse is rotated about its pivot point which is at the center of the ellipse, as shown in Figure 4-22.

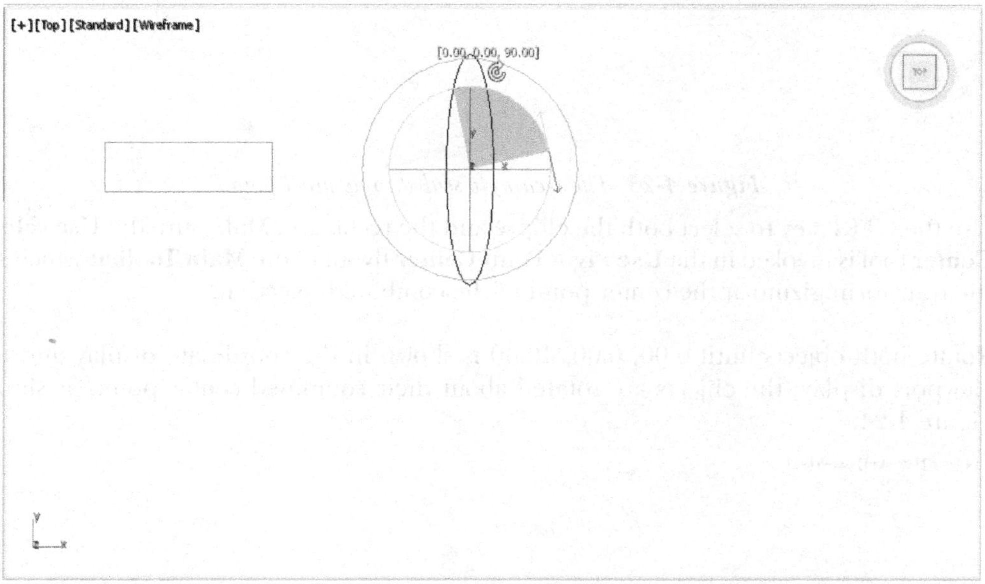

Figure 4-22 *The ellipse rotated to 90 degrees*

31. Invoke the **Select and Uniform Scale** tool from the **Main Toolbar** and make sure the **Transform Gizmo XY Plane Constraint** button is chosen in the **Axis Constraints** toolbar. Select the rectangle and scale it until 50, 50, 50 is shown in the coordinate display; the rectangle is scaled about its pivot point which is at the center of the rectangle, as shown in Figure 4-23.

32. Choose **Edit > Fetch** from the menu bar to retrieve the original drawing; the **About to Fetch. OK** message box is displayed. Choose the **Yes** button in this dialog box; the original shape of the rectangle and the ellipse is restored.

33. Invoke the **Select and Rotate** tool and make sure the **Transform Gizmo Z Constraint** button is chosen. Make sure the **Snaps Toggle** tool is deactivated and the **Angle Snap Toggle** tool is invoked.

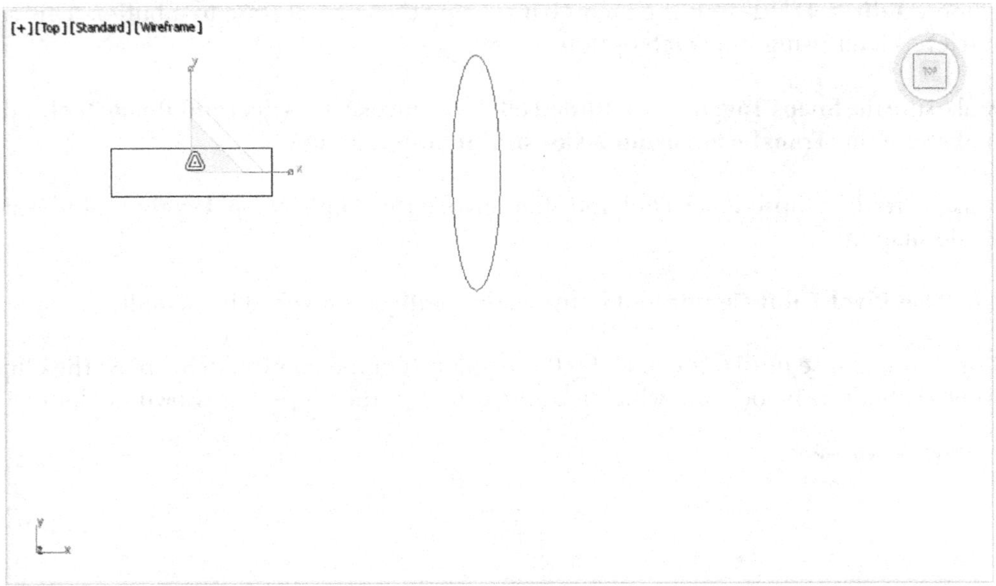

Figure 4-23 *The rectangle scaled to its pivot point*

34. Use the CTRL key to select both the ellipse and the rectangle. Make sure the **Use selection Center** tool is invoked in the **Use Pivot Point Center** flyout of the **Main Toolbar** which shows the transform gizmo at the center point of the combined selection.

35. Rotate both objects until 0.00, 0.00, 40.00 is shown in the coordinate display and in the viewport display; the objects are rotated about their combined center point, as shown in Figure 4-24.

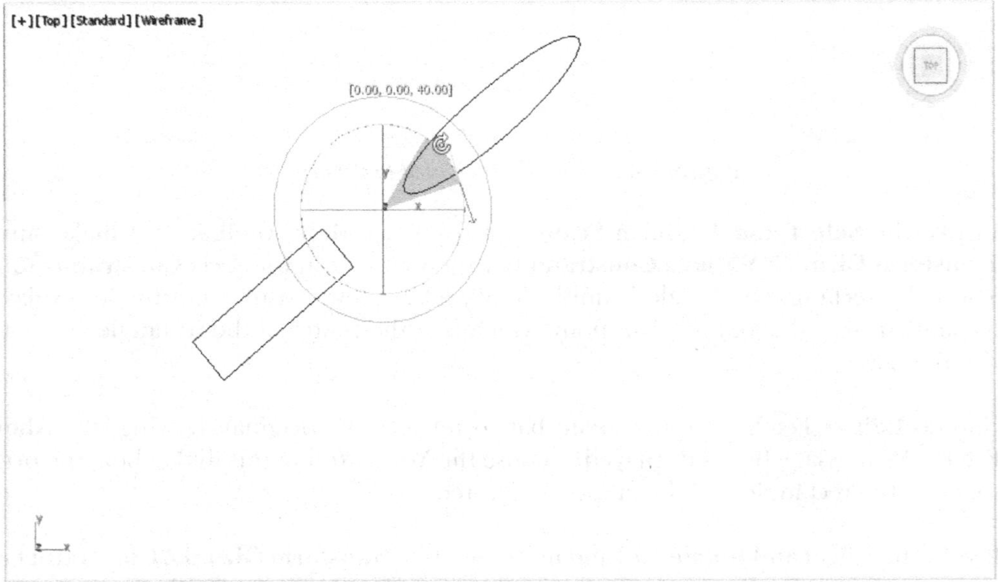

Figure 4-24 *Both objects rotated around the center of their combined center point*

36. Restore the original drawing by choosing the **Undo** button. Make sure the **Select and Rotate** tool is invoked in the **Main Toolbar**.

37. Select only the ellipse. Next, invoke the **Use Transform Coordinate Center** tool from the **Use Selection Center** flyout. Also, select **World** from the **Reference Coordinate System** drop-down list located next to the **Select and Place** tool. Note that the axis tripod moves to the world origin, which is probably near the center of the viewport. 

38. Make sure that the **Angle Snap Toggle** tool is invoked in the **Main Toolbar** and then rotate the ellipse until 0.00, 0.00, -90.00 is shown in the coordinate display; the ellipse is rotated with respect to the origin of the world space, as shown in Figure 4-25.

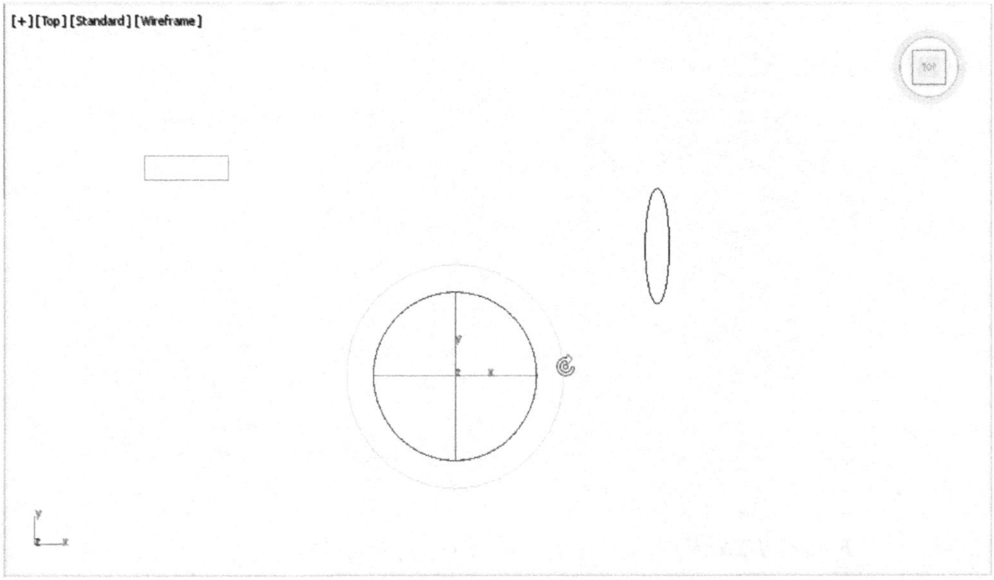

Figure 4-25 *The rotated ellipse after selecting* ***World*** *from the* ***Reference Coordinate System*** *drop-down list*

Copying an Object

In this section, you will use the SHIFT key with the transform commands to make a copy of the original object.

1. Choose **File > New > New All** from the **File** menu; the **Autodesk 3ds Max 2022** message box is displayed. Choose the **Don't Save** button; all the shapes are removed and a blank Top viewport is displayed.

2. Choose **Create > Shapes** in the **Command Panel**. Next, invoke the **Rectangle** tool from the **Object Type** rollout. Expand the **Keyboard Entry** rollout and set the values as given next:

   Length: **100** Width: **100**

   Choose the **Create** button; a square is created in the center of the Top viewport.

3. Invoke the **Select and Rotate** tool and make sure the **Transform Gizmo Z Constraint** button is chosen. Select **View** from the **Reference Coordinate System** drop-down list if it is not already selected. Make sure the **Use Pivot Point Center** tool is also invoked. Also, ensure that the snap is turned off and the **Angle Snap Toggle** tool is invoked.

4. Press and hold the SHIFT key and make sure the square is selected and then rotate it, as shown in Figure 4-26. Next, release the mouse button and the SHIFT key when the square is rotated about 45 degrees; the **Clone Options** dialog box is displayed. Notice that the original square remains at its position, while its copy is rotated about its pivot point.

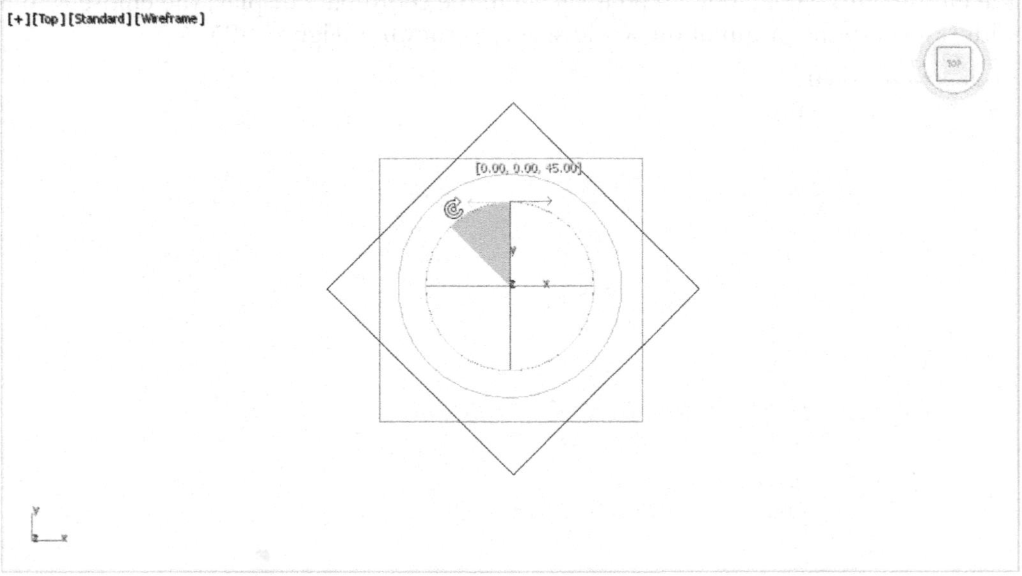

Figure 4-26 Rotating the square at an angle of 45 degrees

5. Make sure that the **Copy** radio button is selected in the **Object** area of the **Clone Options** dialog box. Choose the **OK** button in the dialog box; a copy of the square rotated at an angle of 45 degrees is created, as shown in Figure 4-27.

6. Invoke the **Select and Uniform Scale** tool from the **Main Toolbar**. Next, press and hold the left mouse button and the SHIFT key. Now, select the original square and scale it up to 50 percent of its original size and release the left mouse button and the SHIFT key; the **Clone Options** dialog box is displayed.

7. Choose the **OK** button in the **Clone Options** dialog box; a copy of the square scaled down to half of its original size is created inside the two overlapping squares.

8. Select the first square that you cloned, press and hold the SHIFT key, and scale the square to 50 percent of its original size. Choose **OK** in the **Clone Options** dialog box; a copy of the rotated square, scaled down to half its size, is created, as shown in Figure 4-28.

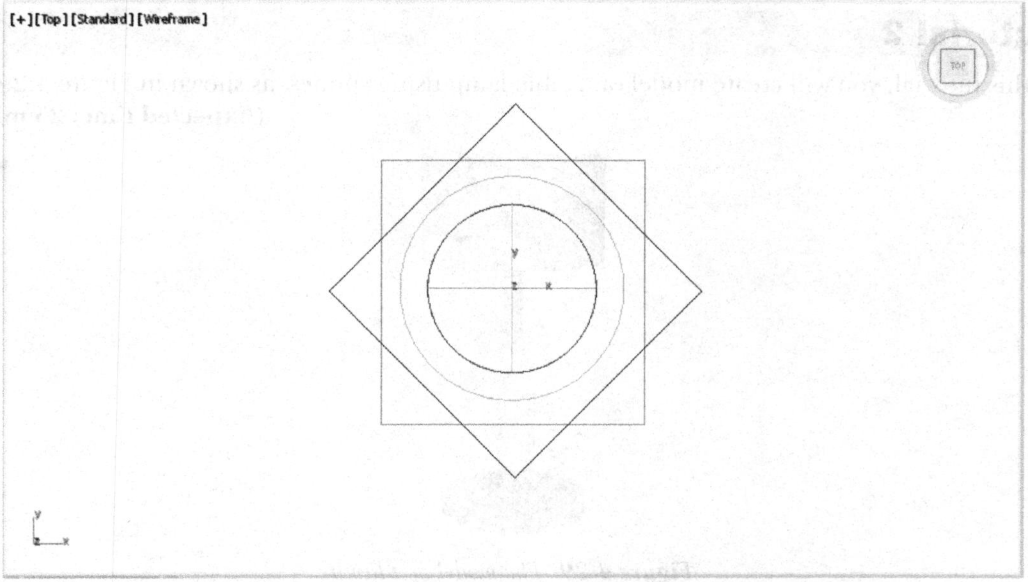

Figure 4-27 A copy of the square created by rotating it at 45 degrees

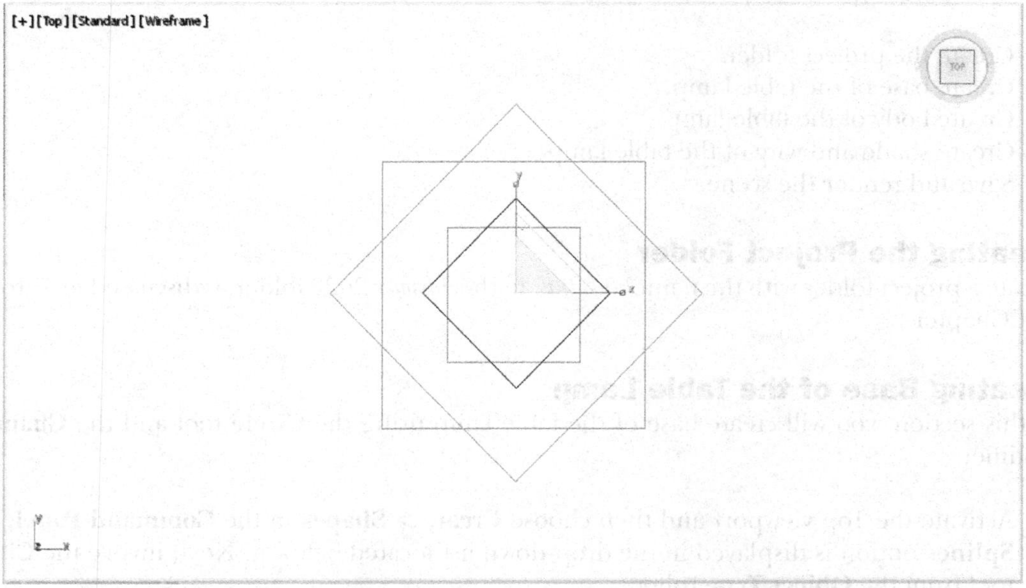

Figure 4-28 Both squares copied and scaled down to 50 percent of their original size

Tutorial 2

In this tutorial, you will create model of a table lamp using splines, as shown in Figure 4-29.

(Expected time: 25 min)

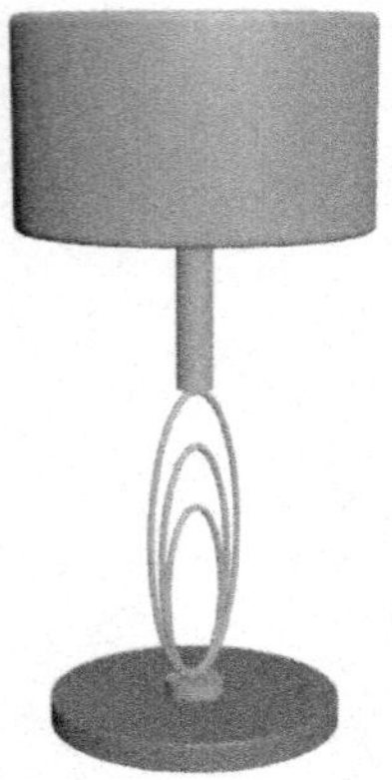

Figure 4-29 *The model of a lamp*

The following steps are required to complete this tutorial:

a. Create the project folder.
b. Create base of the table lamp.
c. Create body of the table lamp.
d. Create shade and wire of the table lamp.
e. Save and render the scene.

Creating the Project Folder

Create a project folder with the name *c04_tut2* in the *3dsmax 2022* folder as discussed in Tutorial 1 of Chapter 2.

Creating Base of the Table Lamp

In this section, you will create base of the table lamp using the **Circle** tool and the **Chamfer** modifier.

1. Activate the Top viewport and then choose **Create > Shapes** in the **Command Panel**; the **Splines** option is displayed in the drop-down list located below it. Next, invoke the **Circle** tool from the **Object Type** rollout.

2. Expand the **Keyboard Entry** rollout and set the value **10** in the **Radius** spinner. Next, expand the **Interpolation** rollout and set the value **16** in the **Steps** spinner. Now, choose the **Create** button in the **Keyboard Entry** rollout; a circle is created in all viewports.

3. Enter **base** in the **Name and Color** rollout. Invoke the **Zoom Extents All** tool; *base* is zoomed in all viewports, as shown in Figure 4-30.

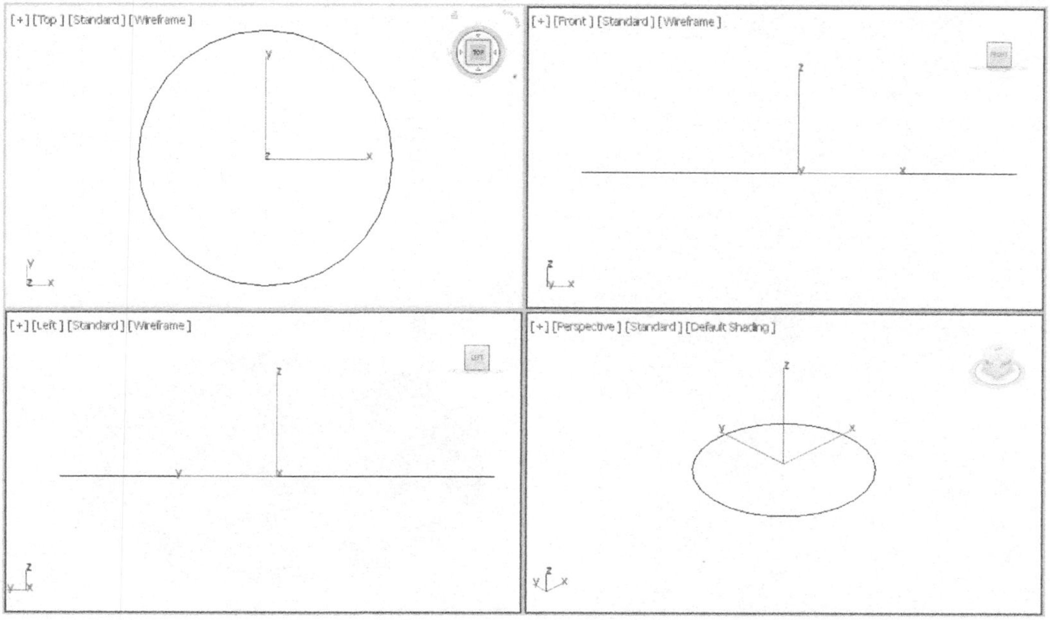

Figure 4-30 *The base zoomed in all viewports*

4. Choose the **Modify** tab in the **Command Panel**. Next, select the **Extrude** option from the **OBJECT-SPACE MODIFIERS** section in the **Modifier List** drop-down list; the **Extrude** modifier is applied to *base* and displayed in the modifier stack. Also, various rollouts of this modifier are displayed in the **Command Panel**.

5. In the **Parameters** rollout, set the value **2** in the **Amount** spinner and **4** in the **Segments** spinner. Next, press the ENTER key; *base* is extruded, as shown in Figure 4-31.

6. Activate the Perspective viewport and press F4. Notice that edges of *base* are now visible in the Perspective viewport.

7. Make sure the **Modify** tab is chosen and then select the **Chamfer** modifier from the **OBJECT-SPACE MODIFIERS** section in the **Modifier List** drop-down list; the **Chamfer** modifier is displayed in the modifier stack and the **Parameters** rollout is displayed in the **Command Panel**.

8. In the **Chamfer Options** area of the **Parameters** rollout, set **0.2** in the **Amounts** spinner and **2** in the **Segments** spinner; *base* is smoothened, as shown in Figure 4-32.

Creating Body of the Table Lamp

In this section, you will create body of the table lamp using the **Ellipse** and **Circle** tools.

1. Activate the Top viewport and then choose **Create > Shapes** in the **Command Panel**; the **Splines** option is displayed in the drop-down list located below it. Next, invoke the **Circle** tool from the **Object Type** rollout.

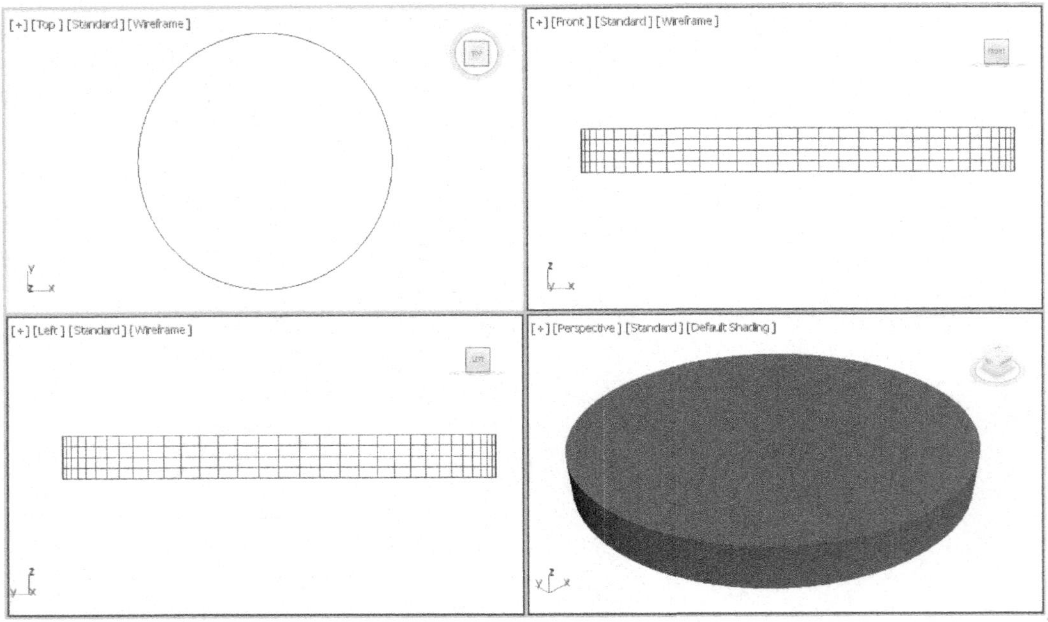

***Figure 4-31** The base zoomed in all viewports*

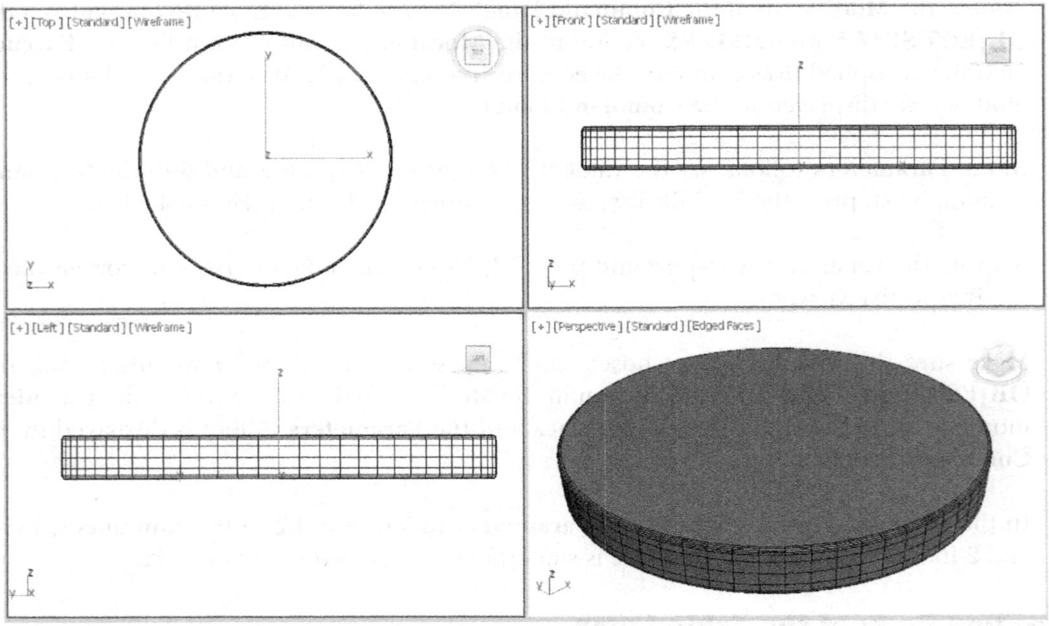

***Figure 4-32** The base smoothened*

2. Expand the **Keyboard Entry** rollout and set the value **2** in the **Radius** spinner. Next, expand the **Interpolation** rollout and make sure **16** is displayed in the **Steps** spinner. Now, choose the **Create** button in the **Keyboard Entry** rollout; a circle is created in all viewports with the name *Circle001*.

3. Choose the **Modify** tab in the **Command Panel**. Next, select the **Extrude** option from the **OBJECT-SPACE-MODIFIERS** section in the **Modifier List** drop-down list; the **Extrude** modifier is applied to *Circle001* and displayed in the modifier stack. Also, various rollouts of this modifier are displayed in the **Command Panel**.

4. In the **Parameters** rollout, set the value **1** in the **Amount** spinner and make sure **4** is displayed in the **Segments** spinner. Next, press the ENTER key; *Circle001* is extruded.

5. Activate the Front viewport. Next, choose the **Select and Move** tool from the **Main Toolbar** and move *Circle001*, as shown in Figure 4-33.

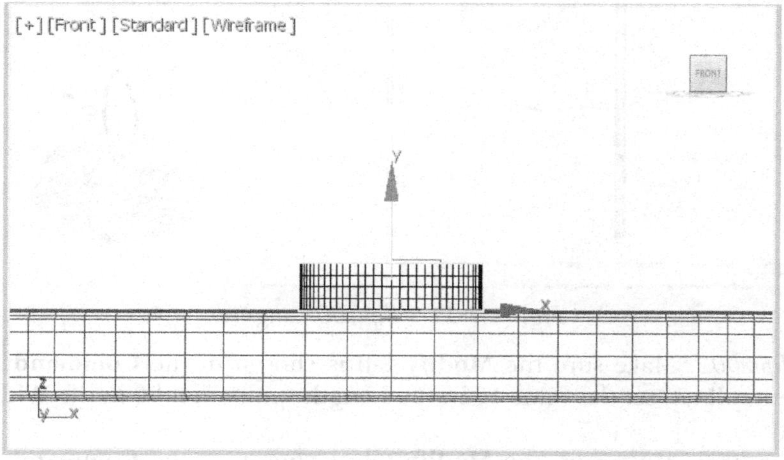

Figure 4-33 *The Circle001 object moved*

6. Make sure the Front viewport is activated. Next, choose **Create > Shapes** in the **Command Panel**; the **Splines** option is displayed in the drop-down list located below it. Next, invoke the **Ellipse** tool from the **Object Type** rollout.

7. Expand the **Keyboard Entry** rollout and set the value **12** in the **Length** spinner and **5** in the **Width** spinner. Next, expand the **Interpolation** rollout and make sure **16** is displayed in the **Steps** spinner. Now, choose the **Create** button in the **Keyboard Entry** rollout; an ellipse is created in all viewports with the name *Ellipse001*.

8. Expand the **Rendering** rollout and select the **Enable in Renderer** and **Enable in Viewport** check boxes. Also, make sure that the **Radial** radio button is selected and set the value **0.5** in the **Thickness** spinner.

9. Align *Ellipse001* on the top of *Circle001,* as shown in Figure 4-34.

10. Make sure *Ellipse001* is selected. Next, press and hold the SHIFT key and drag the cursor in the Y direction. Now, release the left mouse button; the **Clone Options** dialog box is displayed. In this dialog box, make sure the **Copy** radio button is selected. Also, set the value in the **Number of Copies** spinner to **2** and then choose the **OK** button; the copies of *Ellipse001* are created with the names *Ellipse002* and *Ellipse003*.

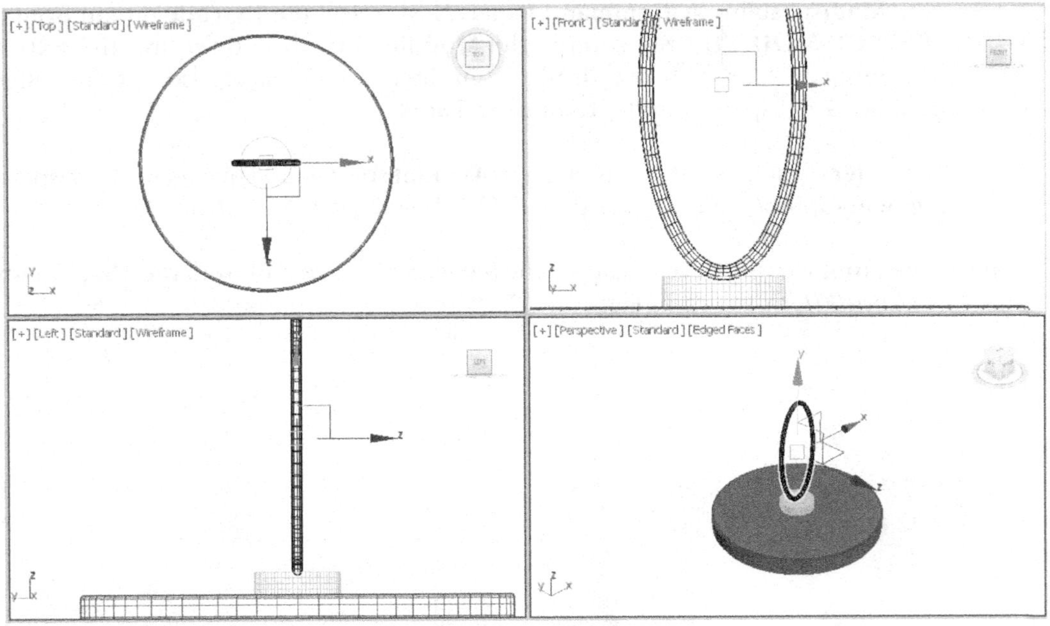

Figure 4-34 Ellipse001 aligned

11. Select *Ellipse002*. Make sure the **Modify** tab is chosen in the **Command Panel**. In the **Parameters** rollout, set the value **16** in the **Length** spinner and **6** in the **Width** spinner.

12. Select *Ellipse003*. Make sure the **Modify** tab is chosen in the **Command Panel**. In the **Parameters** rollout, set the value **20** in the **Length** spinner and **8** in the **Width** spinner.

13. Align *Ellipse002* and *Ellipse003* in all viewports, as shown in Figure 4-35.

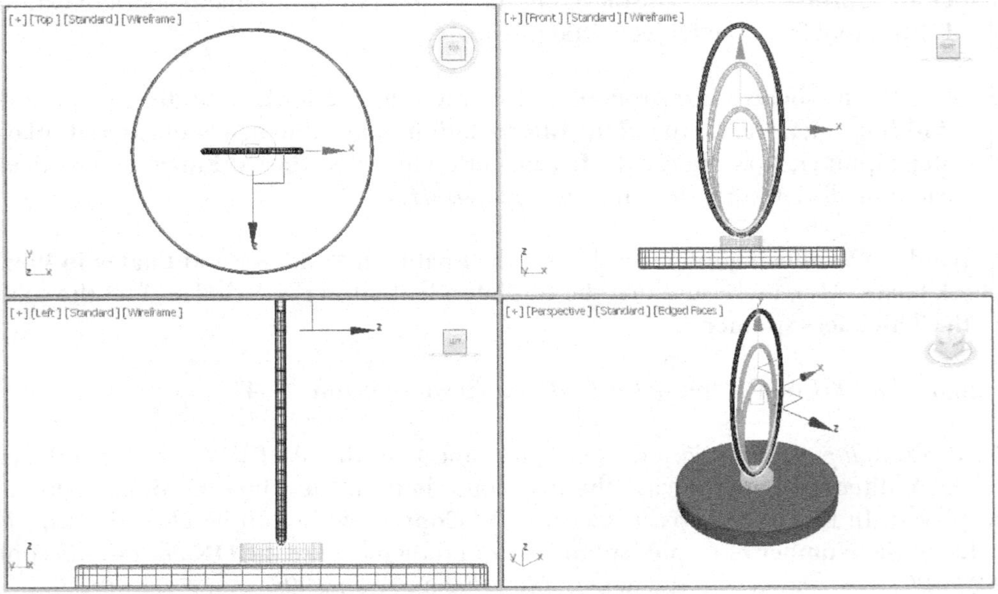

Figure 4-35 Ellipse002 and Ellipse003 aligned

14. Activate the Top viewport and then choose **Create > Shapes** in the **Command Panel**; the **Splines** option is displayed in the drop-down list located below it. Next, invoke the **Circle** tool from the **Object Type** rollout.

15. Expand the **Keyboard Entry** rollout and set the value **1.25** in the **Radius** spinner. Next, expand the **Interpolation** rollout and make sure **16** is displayed in the **Steps** spinner. Now, choose the **Create** button in the **Keyboard Entry** rollout; a circle is created in all viewports. with the name *Circle002*.

16. Choose the **Modify** tab in the **Command Panel**. Next, select the **Extrude** option from the **OBJECT-SPACE MODIFIERS** section in the **Modifier List** drop-down list; the **Extrude** modifier is applied to *Circle002* and displayed in the modifier stack. Also, various rollouts of this modifier are displayed in the **Command Panel**.

17. In the **Parameters** rollout, set the value **12** in the **Amount** spinner and make sure **4** is displayed in the **Segments** spinner. Next, press the ENTER key; *Circle002* is extruded.

18. Activate the Front viewport. Next, choose the **Select and Move** tool from the **Main Toolbar** and align *Circle002*, as shown in Figure 4-36.

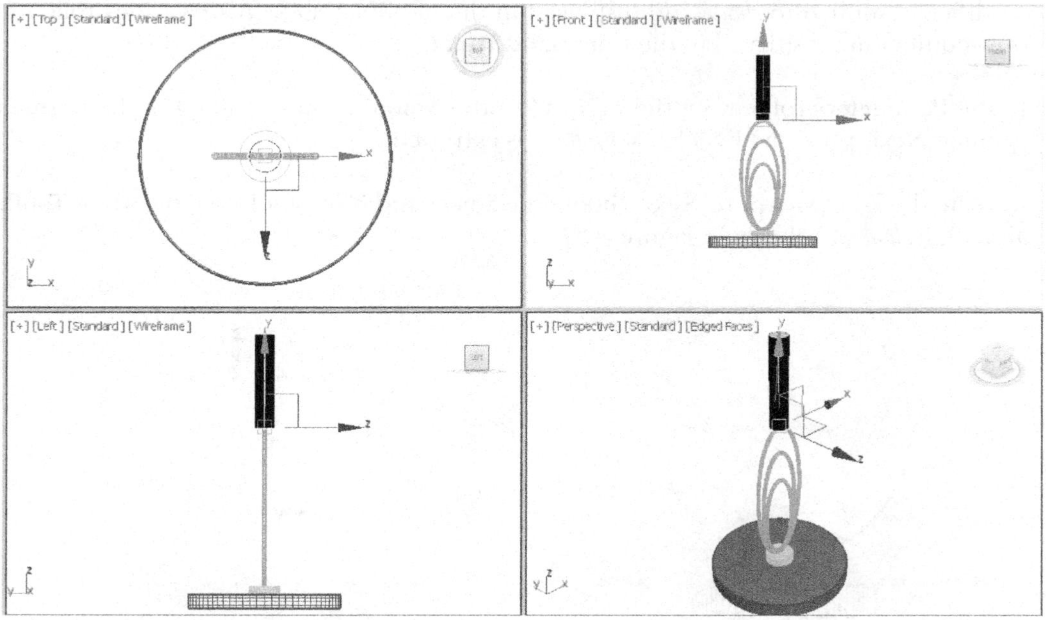

Figure 4-36 *Circle002 aligned*

Next, you need to select all the objects excluding *base* and create a group.

19. Select *Circle001, Circle002, Ellipse001, Ellipse002, and Ellipse003* using the CTRL key. Next, choose **Group > Group** from the menu bar; the **Group** dialog box is displayed. Enter **body** in the **Group name** edit box and choose the **OK** button to close the dialog box.

20. Make sure that the group *body* is selected. Change the color of *body* to grey or black by using the color swatch in the **Name and Color** rollout of the **Command Panel**.

Creating Shade of the Table Lamp

In this section, you will create shade using the **Circle** tool.

1. Activate the Top viewport and then choose **Create > Shapes** in the **Command Panel**; the **Splines** option is displayed in the drop-down list located below it. Next, invoke the **Circle** tool from the **Object Type** rollout.

2. Expand the **Keyboard Entry** rollout and set the value **12** in the **Radius** spinner. Next, expand the **Interpolation** rollout and make sure **16** is displayed in the **Steps** spinner. Now, choose the **Create** button in the **Keyboard Entry** rollout; a circle is created in all viewports. with the name *Circle003*.

3. Enter **shade** in the **Name and Color** rollout to rename it.

4. Choose the **Modify** tab in the **Command Panel**. Next, select the **Extrude** option from the **OBJECT-SPACE MODIFIERS** section in the **Modifier List** drop-down list; the **Extrude** modifier is applied to *shade* and displayed in the modifier stack. Also, various rollouts of this modifier are displayed in the **Command Panel**.

5. In the **Parameters** rollout, set the value **14** in the **Amount** spinner and **4** in the **Segments** spinner. Next, press the ENTER key; *shade* is extruded.

6. Activate the Front viewport. Next, choose the **Select and Move** tool from the **Main Toolbar** and align *shade*, as shown in Figure 4-37.

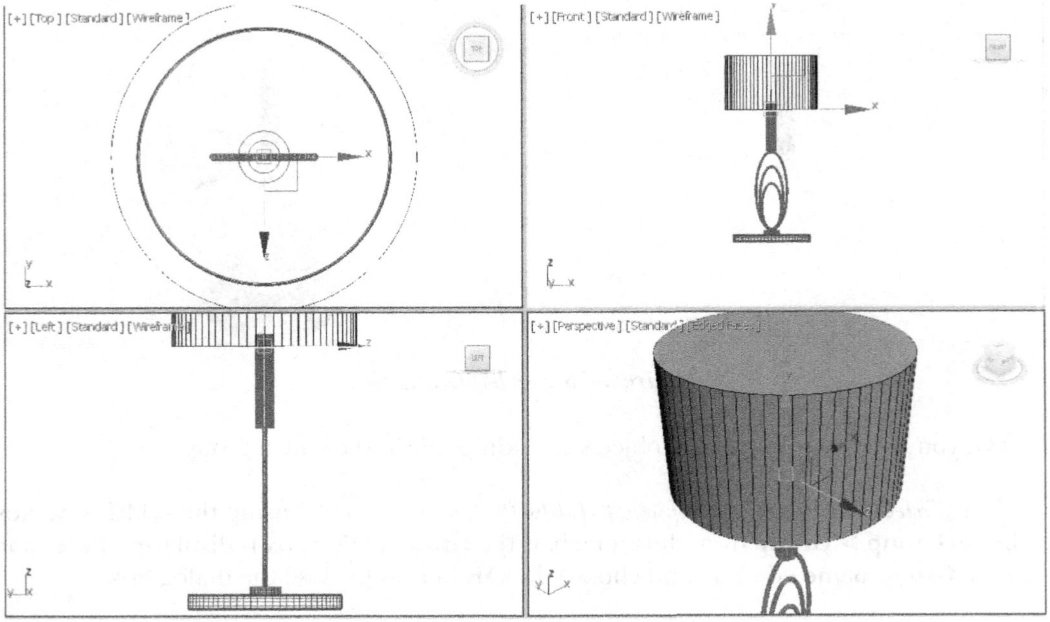

Figure 4-37 *The shade object aligned*

7. Make sure the **Modify** tab is chosen and then select the **Chamfer** modifier from the **OBJECT-SPACE MODIFIERS** section in the **Modifier List** drop-down list; the **Chamfer** modifier is displayed in the modifier stack and the **Parameters** rollout is displayed in the **Command Panel**.

8. In the **Chamfer Options** area of the **Parameters** rollout, set **0.65** in the **Amounts** spinner and **2** in the **Segments** spinner; *shade* is smoothened, as shown in Figure 4-38.

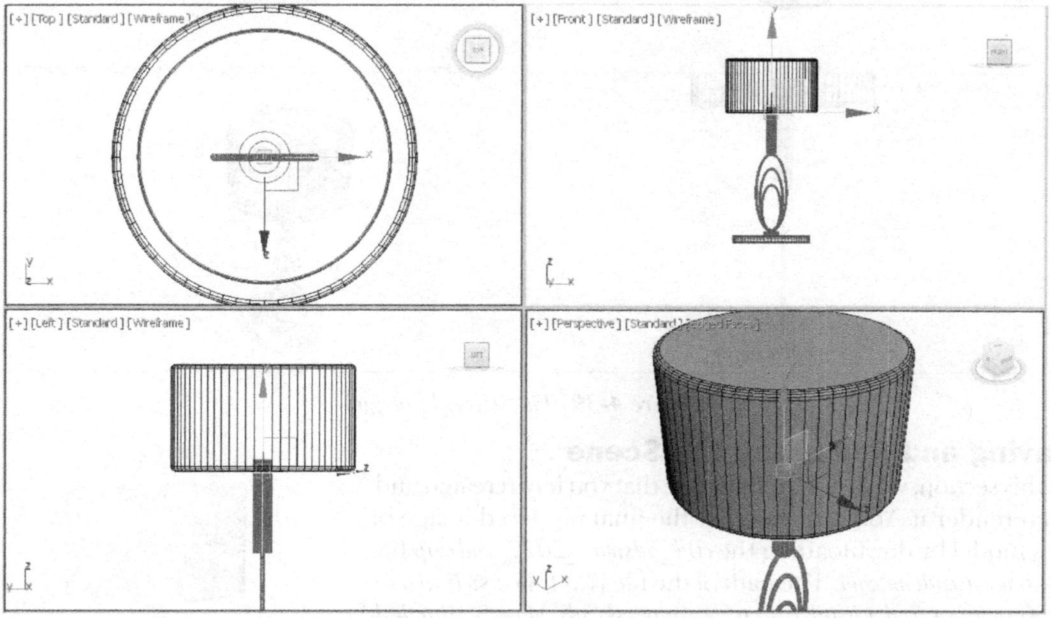

Figure 4-38 *The shade object smoothened*

9. Make sure the **Modify** tab is chosen and then select the **Edit Poly** modifier from the **OBJECT-SPACE MODIFIERS** section in the **Modifier List** drop-down list; the **Edit Poly** modifier is displayed in the modifier stack and various rollouts are displayed in the **Command Panel**.

10. In the modifier stack, click on the arrow sign on the left of the **Edit Poly** modifier to view all the sub-object levels and then select the **Polygon** sub-object level.

    Next, you will activate the Bottom viewport.

11. Click on the POV viewport label in the Top viewport; a flyout is displayed. Choose the **Bottom** option; the Top viewport is switched to the Bottom viewport. Now, select the polygon in the Bottom viewport and delete it, refer to Figure 4-39.

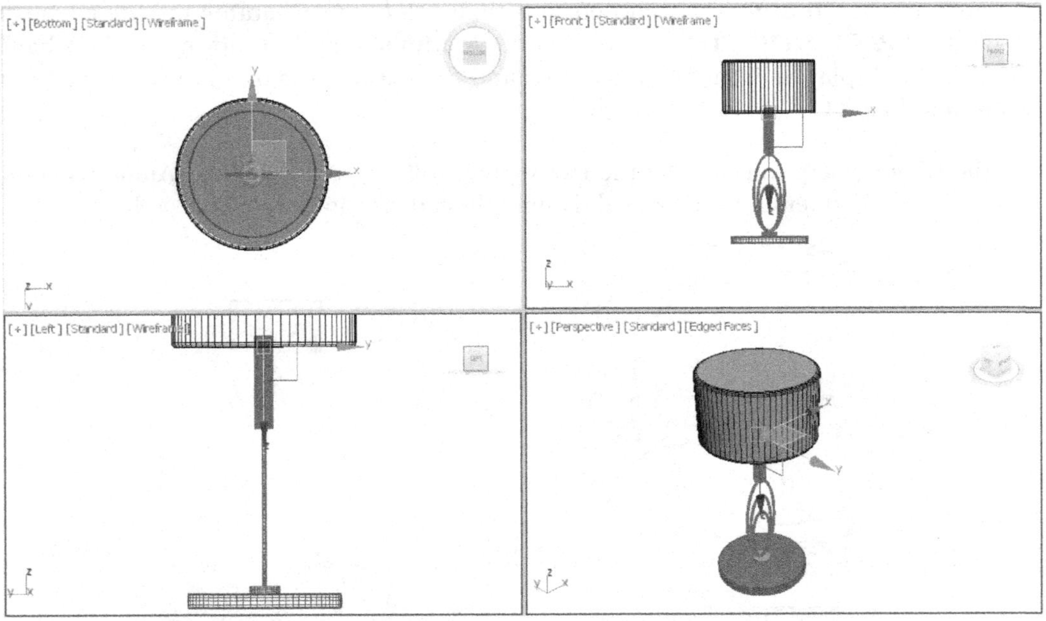

Figure 4-39 The selected polygon

Saving and Rendering the Scene

In this section, you will save the scene that you have created and then render it. You can also view the final rendered image of this model by downloading the *c04_3dsmax_2022_rndr.zip* file from *www.cadcim.com*. The path of the file is as follows: *Textbooks > Animation and Visual Effects > Autodesk 3ds Max > Autodesk 3ds Max 2022 for Beginners: A Tutorial Approach*

1. Change the background color of the scene to white by following the steps as given in Tutorial 1 of Chapter 2.
2. Choose **Save** from the **File** menu.
3. Activate the Perspective viewport. Next, invoke the **Render Production** tool from the **Main Toolbar**; the rendered image is displayed, refer to Figure 4-40.

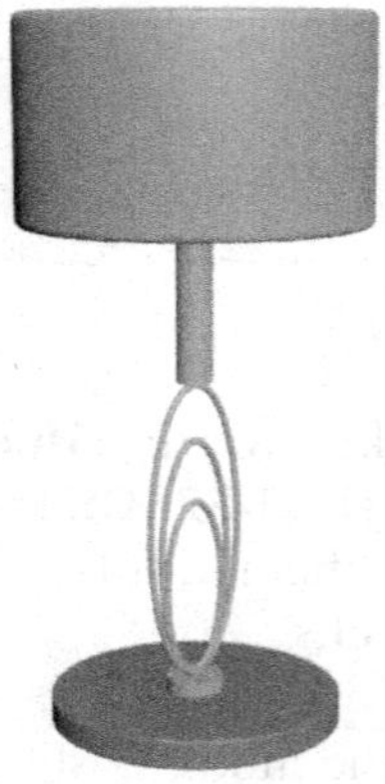

Figure 4-40 The rendered image

Tutorial 3

In this tutorial, you will create the model of a vase by converting a 2D spline into a 3D mesh object and then place a flower plant in it, as shown in Figure 4-41. **(Expected time: 25 min)**

The following steps are required to complete this tutorial:

a. Create the project folder.
b. Create a line spline for the vase.
c. Apply the **Lathe** modifier.

d. Assign the material to the vase.
e. Create a stand for the vase.
f. Save and render the scene.

Figure 4-41 *The model of a vase with flower plant*

Creating the Project Folder

Create a project folder with the name *c04_tut3* in the *3dsmax 2022* folder as discussed in Tutorial 1 of Chapter 2.

Creating a Line Spline for the Vase

In this section, you will create a profile similar to one-half of the vase using the **Line** tool.

1. Activate the Front viewport and choose **Create > Shapes** in the **Command Panel**; the **Splines** option is selected by default in the drop-down list below the **Shapes** button. Now, choose the **Line** tool from the **Object Type** rollout.

2. In the **Creation Method** rollout, select the **Corner** radio button in both the **Initial Type** and **Drag Type** areas.

3. Create a line spline as discussed in Tutorial 1. This spline should be similar to one-half of the vase, refer to Figure 4-42.

4. Choose the **Modify** tab and select the **Vertex** sub-object level in the modifier stack; all vertices of the line spline are displayed in the viewport. Next, select the vertices one by one in the viewport and modify the shape by moving them to get a perfect shape.

Applying the Lathe Modifier

In this section, you will apply the **Lathe** modifier to convert the 2D spline into the 3D mesh object.

1. Select the first vertex of the spline in the Front viewport, as shown in Figure 4-43.

2. Choose **Modifiers > Patch/Spline Editing > Lathe** from the menu bar; the 2D line spline is converted into a 3D mesh object.

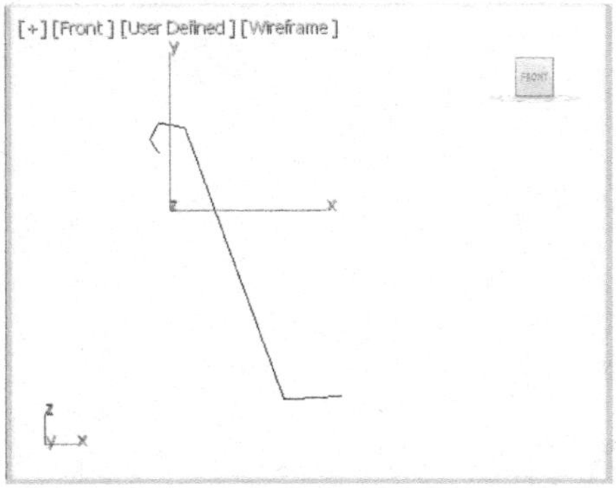

Figure 4-42 *A line spline created for the vase*

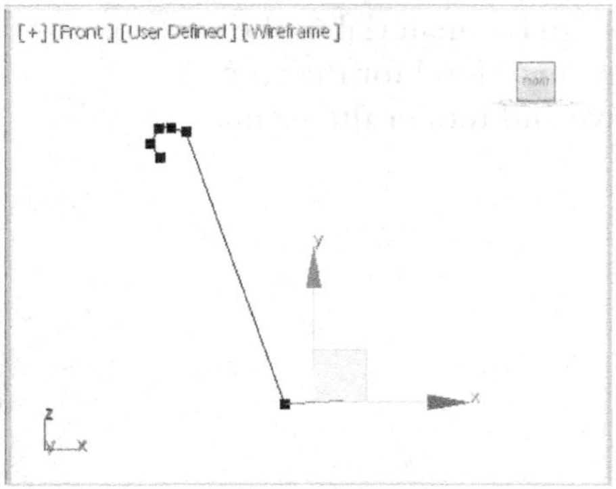

Figure 4-43 *The first vertex of the line spline selected in the Front viewport*

3. Name the new 3D mesh object as *vase*. Also, click anywhere in the viewport to deselect the *vase*.

4. Choose the **Zoom Extents All** tool; the entire *vase* is displayed in the viewports, as shown in Figure 4-44.

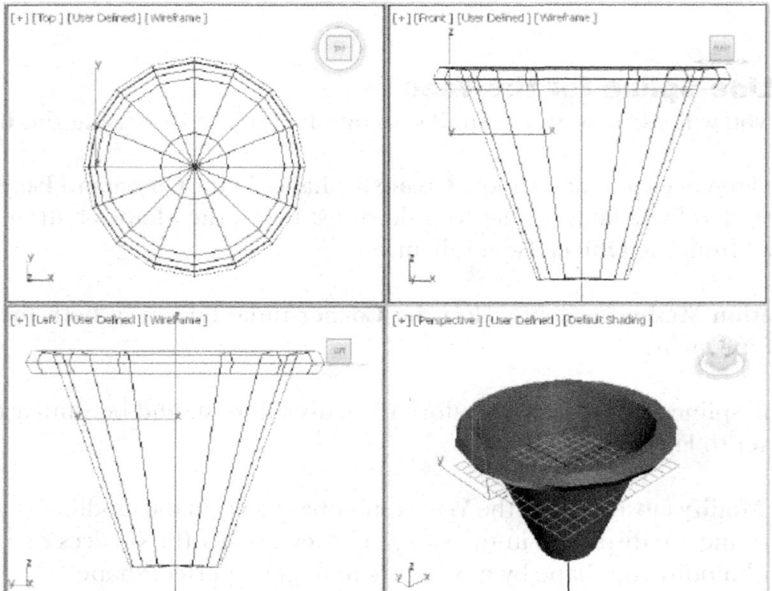

Figure 4-44 *The vase geometry after applying the* ***Lathe*** *modifier*

Assigning Material to Vase

In this section, you will assign the material to *vase* using the **Material Editor** tool as discussed in Tutorial 1 of this chapter.

1. Choose **Rendering > Material Editor > Compact Material Editor** from the menu bar; the **Material Editor** dialog box is displayed.

2. Select the **2-Sided** check box in the **Shader Basic Parameters** rollout of the **Material Editor** dialog box. This enables you to view both sides of *vase* in the final output. Also, in the **Blinn Basic Parameters** rollout, choose the **Diffuse** color swatch; the **Color Selector: Diffuse Color** dialog box is displayed. In this dialog box, enter the values as follows:

   Red: **177** Green: **88** Blue: **27**

   Choose the **OK** button in the **Color Selector: Diffuse Color** dialog box; the selected color is displayed in the **Diffuse** color swatch.

3. Make sure that *vase* is selected in the viewports. Next, choose the **Assign Material to Selection** button in the **Material Editor** dialog box; the selected material is assigned to *vase* in the viewport. Next, choose the **Show Shaded Material in Viewport** button in the same dialog box to view the material in viewport. Next, close the **Material Editor** dialog box.

   Next, you need to use the **Foliage** tool to place a plant in *vase*.

4. Activate the Top viewport. Choose **Create > Geometry** in the **Command Panel**; a drop-down list is displayed below the **Geometry** button. Select **AEC Extended** from the drop-down list and then choose the **Foliage** tool from the **Object Type** rollout.

5. In the **Favorite Plants** rollout, double-click on the **Society Garlic** foliage; the foliage is displayed in the viewports.

6. In the **Name and Color** rollout, change the name of foliage to *plant* and press ENTER.

7. Make sure that *plant* is still selected. Choose the **Modify** tab in the **Command Panel**; the **Parameters** rollout is displayed.

8. In the **Parameters** rollout, set a value in the **Height** spinner to adjust the height of *plant* according to *vase*.

9. Choose the **Select and Move** tool and align *plant* with *vase* in the viewports, refer to Figure 4-45.

10. Select *plant* and *vase* in the viewports and group them as *flower vase*.

   Next, you need to create a stand for *flower vase*.

Creating a Stand for the Flower Vase

In this section, you will create a stand for the *flower vase*.

1. Select *flower vase* and right-click on it; a quad menu is displayed. Choose the **Hide Selection** option from it; *flower vase* disappears temporarily from the viewports. Now, you can work easily in the viewports.

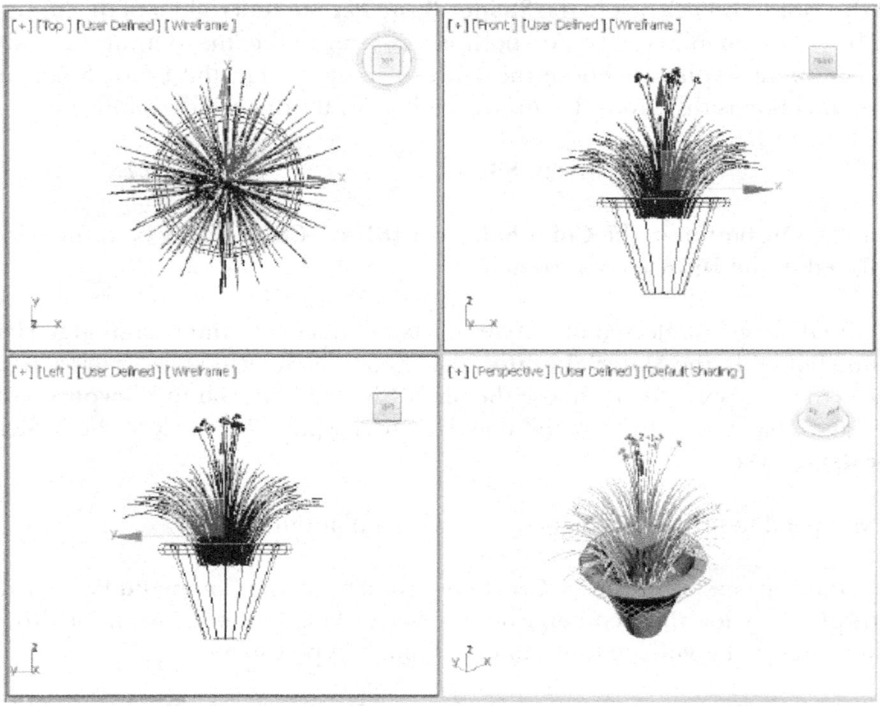

Figure 4-45 Alignment of plant with vase in viewports

2. Activate the Top viewport. Choose **Create > Shapes** in the **Command Panel**; the **Splines** option is selected by default in the drop-down list below the **Shapes** button. Now, choose the **Circle** tool from the **Object Type** rollout and create a circular spline.

3. In the **Parameters** rollout, set the radius of the circle to **11.662** in the **Radius** spinner.

   You may need to modify the radius according to the scene in the later section.

4. Expand the **Rendering** rollout and select the **Enable In Renderer** and **Enable In Viewport** check boxes. Select the **Radial** radio button, if it is not already selected. Also, make sure that the **Auto Smooth** check box is selected. Now, enter the values as follows:

   Thickness: **2.0** Sides: **15** Angle: **0.99**

5. The circular spline is automatically named as *Circle001*. Now, change its color by entering the values as follows:

   Red: **193** Green: **193** Blue: **193**

6. Create another circle in the Top viewport. It is automatically named as *Circle002*.

7. In the **Parameters** rollout, set the radius value of *Circle002*. This radius should be double the size of radius of *Circle001*.

8. Change the color of *Circle002* to black.

9. Choose the **Zoom Extents All** tool and align both the circles using the **Select and Move** tool, as shown in Figure 4-46.

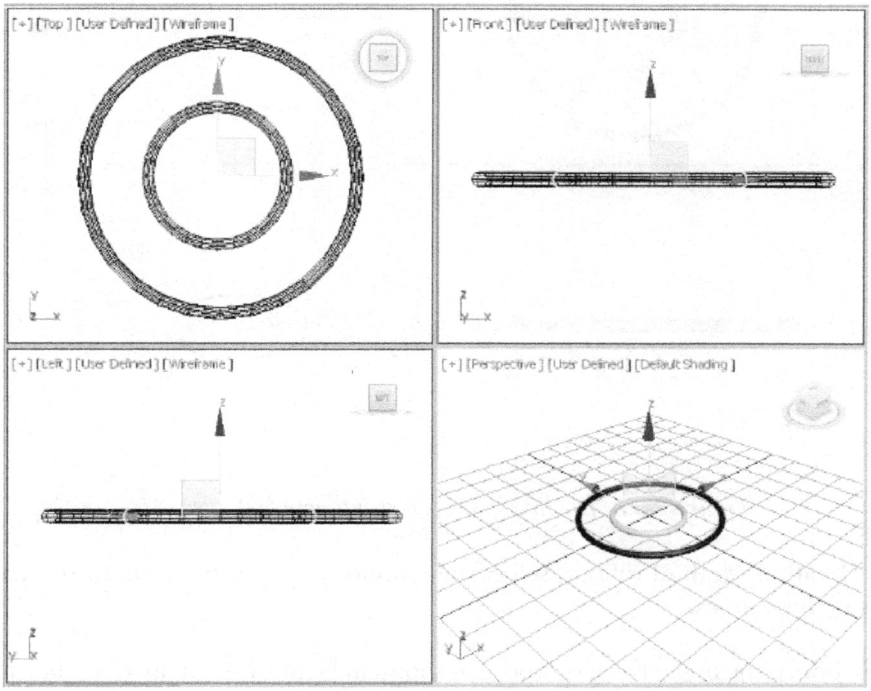

Figure 4-46 *Alignment of the circles in viewports*

10. Choose the **Line** tool from **Create > Shapes > Splines > Object Type** rollout.

11. In the **Creation Method** rollout, select the **Corner** radio button in the **Initial Type** and **Drag Type** areas.

12. Activate the Top viewport. Next, create four line splines to join both the circles, as shown in Figure 4-47.

13. Assign black color to all line splines.

14. Select the line splines and both the circles in any viewport by dragging a selection box around them, and then group them as *upper portion*.

15. Select *upper portion* and hide it temporarily as described earlier.

    Next, you need to create the legs for the stand using the **Line** tool.

16. Activate the Front viewport. Next, choose the **Line** tool from **Create > Shapes > Splines > Object Type** rollout.

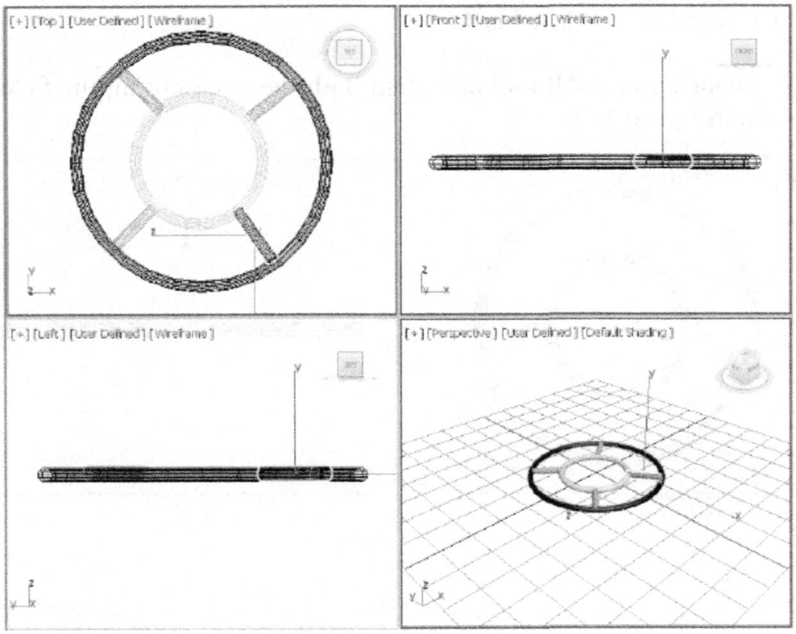

***Figure 4-47** The line splines created to join both the circles*

17. In the **Creation Method** rollout, select the **Smooth** radio button in both the **Initial Type** and **Drag Type** areas.

18. Create a line spline in the Front viewport, as shown in Figure 4-48. It uses the same parameters from the **Rendering** rollout that you had set earlier.

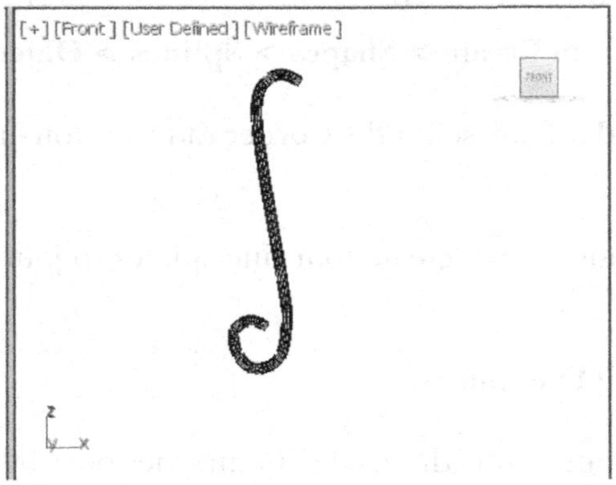

***Figure 4-48** The line spline for the leg of stand*

19. In the **Rendering** rollout, make sure that the **Enable In Renderer** and **Enable In Viewport** check boxes are selected. Select the **Rectangular** radio button and enter the values as follows:

Length: **2.0** Width: **1.5** Angle: **0.0**

20. Name the line spline as *leg01* and change its color to black.

21. Right-click in any of the viewports; a quad menu is displayed. Choose the **Unhide by Name** option; the **Unhide Objects** dialog box is displayed. Select *upper portion* from the list given in this dialog box and choose the **Unhide** button; *upper portion* is now displayed in the viewport. Choose the **Zoom Extents All** tool.

Note

You can resize leg01 according to the size of the upper portion using the ***Select and Uniform Scale*** *tool.*

22. Align *upper portion* and *leg01* in viewports, as shown in Figure 4-49.

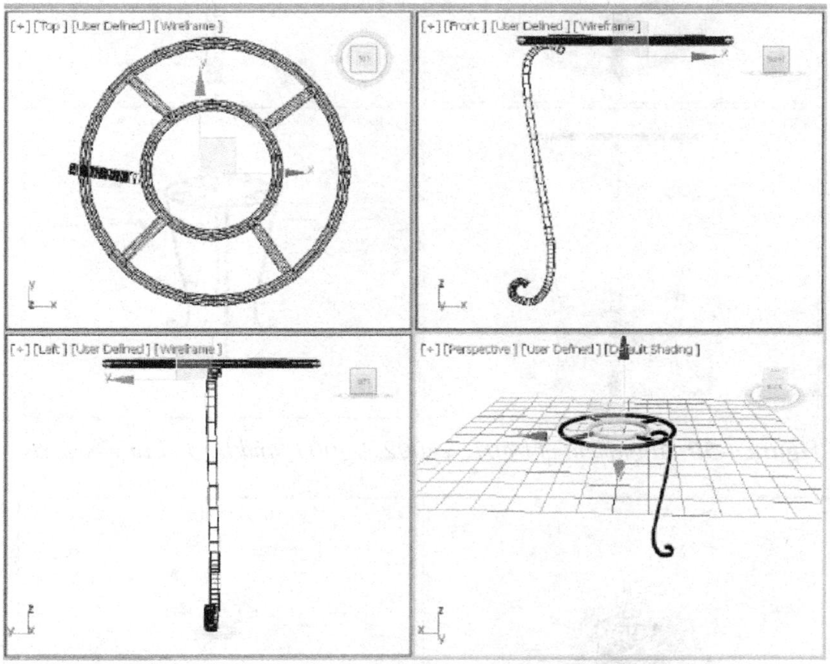

Figure 4-49 *Alignment of upper portion and leg01 in viewports*

Next, you need to copy the *leg01* to create other legs.

23. Create three copies of *leg01*; they are automatically named as *leg002*, *leg003*, and *leg004*.

24. Align *leg01*, *leg002*, *leg003*, and *leg004* using the **Select and Move** and **Select and Rotate** tools, as shown in Figure 4-50.

25. Select all the objects in the viewport and group them as *vase stand*.

26. Right-click on the viewport; a quad menu is displayed. Choose the **Unhide All** option; all the hidden objects are displayed in the viewports.

27. Align *flower vase* and *vase stand* and choose the **Zoom Extents All** tool, refer to Figure 4-51.

Note

*You can adjust the size of the flower vase according to the size of the vase stand using the **Select and Uniform Scale** tool.*

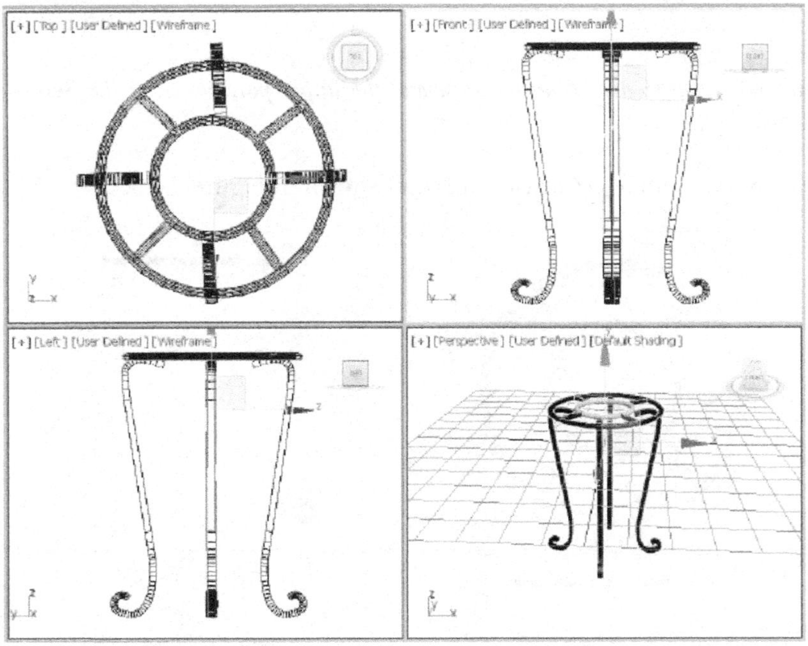

Figure 4-50 *Alignment of leg01, leg002, leg003, and leg004 in viewports*

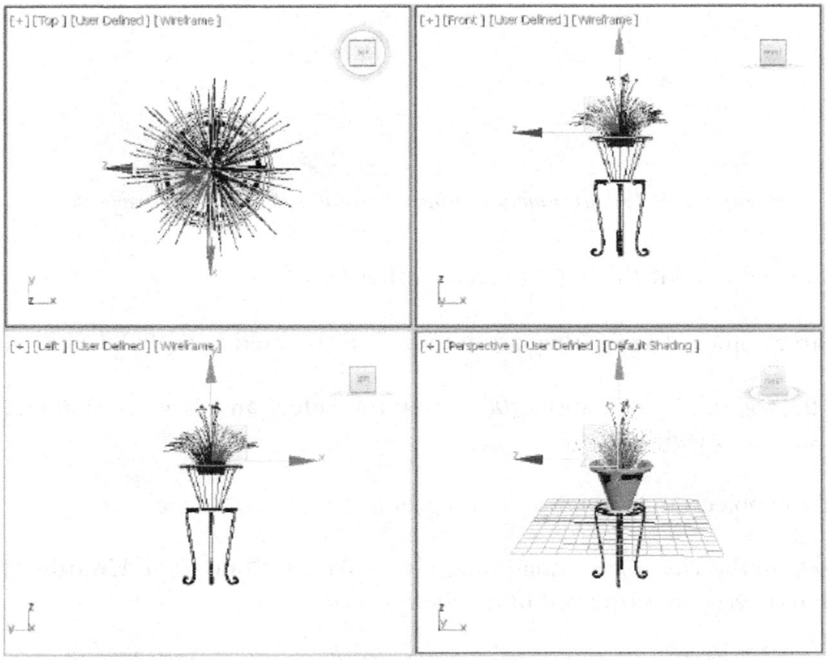

Figure 4-51 *Alignment of flower vase and vase stand in viewports*

Saving and Rendering the Scene

In this section, you will save the scene that you have created and then render it. You can also view the final rendered image of this model by downloading the *c04_3dsmax_2022_rndr.zip* file from *www.cadcim.com*. The path of the file is as follows: *Textbooks > Animation and Visual Effects > Autodesk 3ds Max > Autodesk 3ds Max 2022 for Beginners: A Tutorial Approach*

1. Change the background color of the scene to white by following the steps as given in Tutorial 1 of Chapter 2.

2. Choose **Save** from the **File** menu.

3. Activate the Perspective viewport. Next, invoke the **Render Production** tool from the **Main Toolbar**; the rendered image is displayed, refer to Figure 4-41.

Tutorial 4

In this tutorial, you will create the model of a chain, as shown in Figure 4-52, by converting 2D splines into 3D mesh objects. **(Expected time: 25 min)**

New

Figure 4-52 *The model of a chain*

The following steps are required to complete this tutorial:

a. Create a project folder.
b. Create a circle for chain.
c. Create a chain.
d. Create duplicate copy of the chain.
e. Save and render the scene.

Creating the Project Folder

Create a project folder with the name *c04_tut4* in the *3dsmax 2022* folder as discussed in Tutorial 1 of Chapter 2.

Creating a Circle for the Chain

In this section, you will create a circle.

1. Activate the Top viewport and choose **Create > Shapes** in the **Command Panel**; the **Splines**

option is selected by default in the drop-down list below the **Shapes** button. Now, choose the **Circle** tool from the **Object Type** rollout.

2. Create a circle in the Top viewport. Now, choose the **Modify** tab and in the **Parameters** rollout, set the radius of the circle to **12** in the **Radius** spinner. Now, rename **Circle001** to **Chain**.

3. In the **Modify** tab, make sure *Chain* is selected. Next, right-click on *Chain*; a flyout is displayed. Now, choose **Editable Spline** from the flyout, as shown in Figure 4-53.

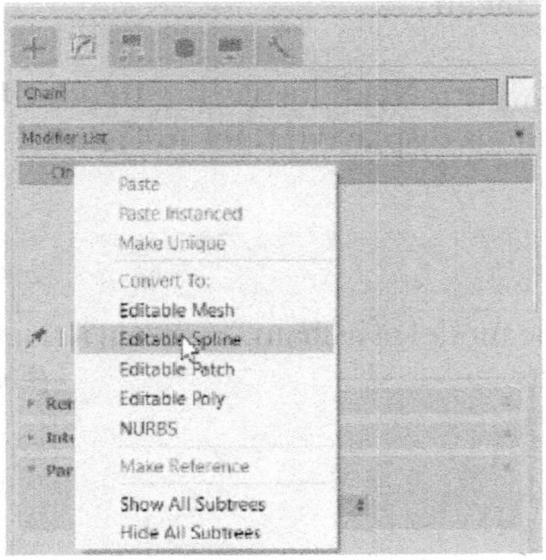

***Figure 4-53** Choosing **Editable Spline** from the flyout*

Creating the Chain

In this section, you will create the shape of a chain.

1. In the Modifier stack, expand **Editable Spline**; a list of sub-objects is displayed. Select **Vertex** from the list of sub-objects level. Now, select the two vertices of the *Chain* from the Top viewport, as shown in Figure 4-54.

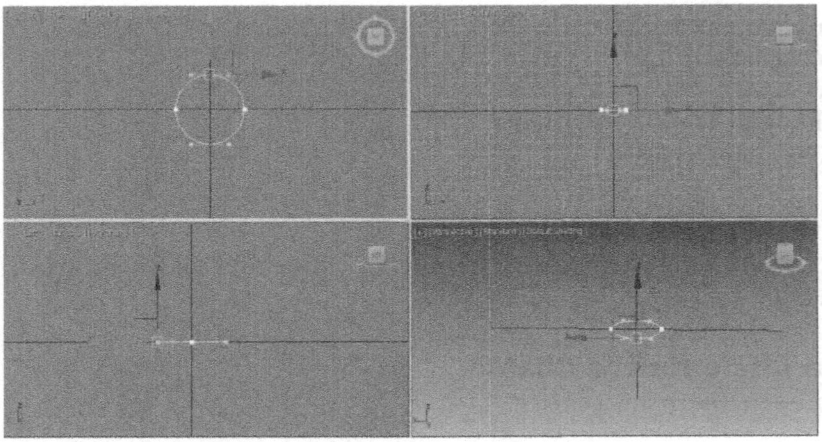

***Figure 4-54** Two vertices selected in the viewport*

2. In the **Geometry** rollout, set the value **7** in the **Chamfer** spinner. Next, select the vertex in the Top viewport, as shown in Figure 4-55.

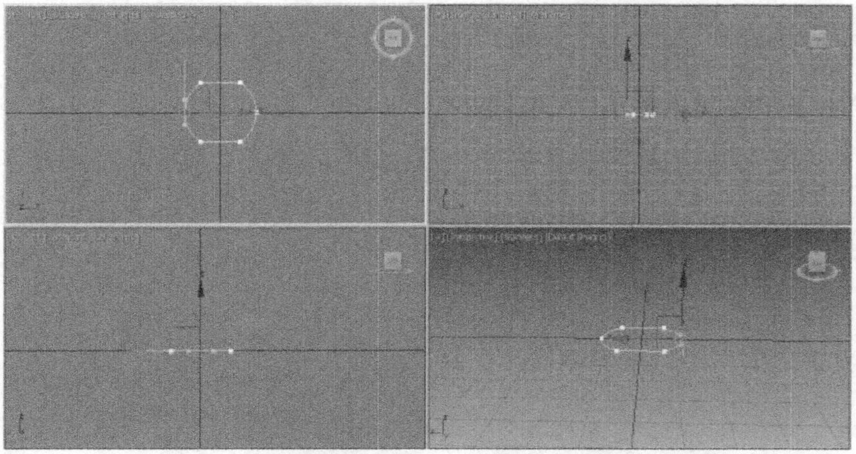

Figure 4-55 *Vertex selected in the viewport*

3. Make sure the vertex is selected in the viewport. Now, in the **Geometry** rollout, choose the **Break** button, as shown in Figure 4-56; the selected vertex is broken into two parts. Next, move the broken vertices, as shown in Figure 4-57.

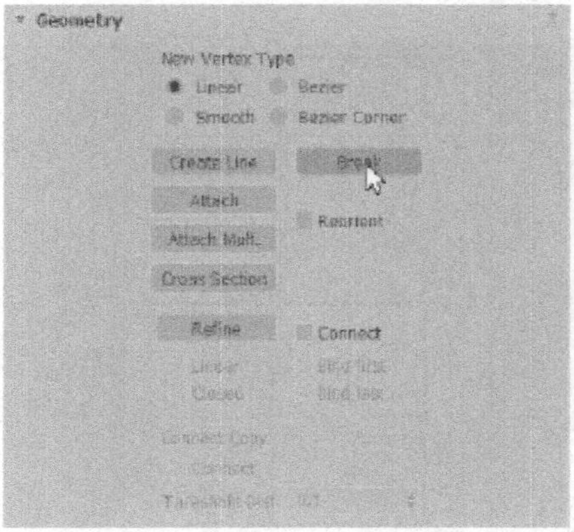

Figure 4-56 *The* ***Break*** *button chosen*

4. In the **Rendering** rollout, select the **Enable In Renderer** and **Enable In Viewport** check boxes. Make sure the **Radial** radio button is selected and then enter the values as follows:

Thickness: **5.0** Sides: **30**

The chain is modified, as shown in Figure 4-58.

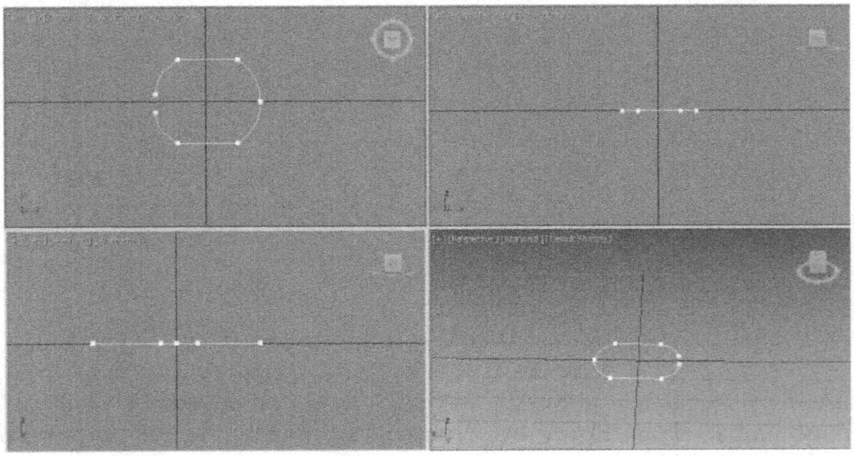

Figure 4-57 *Broken vertices moved*

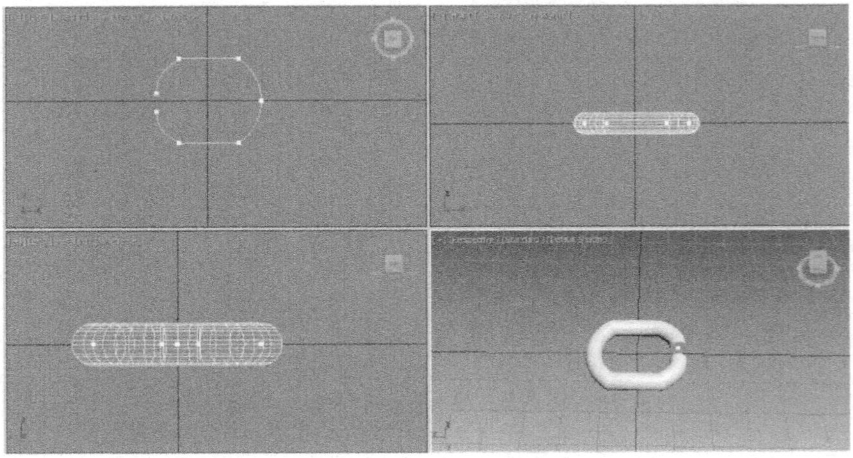

Figure 4-58 *The chain modified*

5. In the Modifier stack, click on **Editable Spline** to exit the sub-object level. Next, press W to activate the **Select and Move** tool. Now, select *Chain* from the viewport and right click; the quad menu is displayed. Choose **Convert > Convert to Editable Poly** from the quad menu.

Creating the Duplicate Copy of the Chain

In this section, you will create a duplicate copy of the chain created in the previous section.

1. Make sure the **Select and Move** tool is activated and *Chain* is selected. Next, press and hold the SHIFT key and move *Chain*; a copy with the name **Chain001** is created and the **Clone Options** dialog box is displayed, as shown in Figure 4-59. Now, close the dialog box.

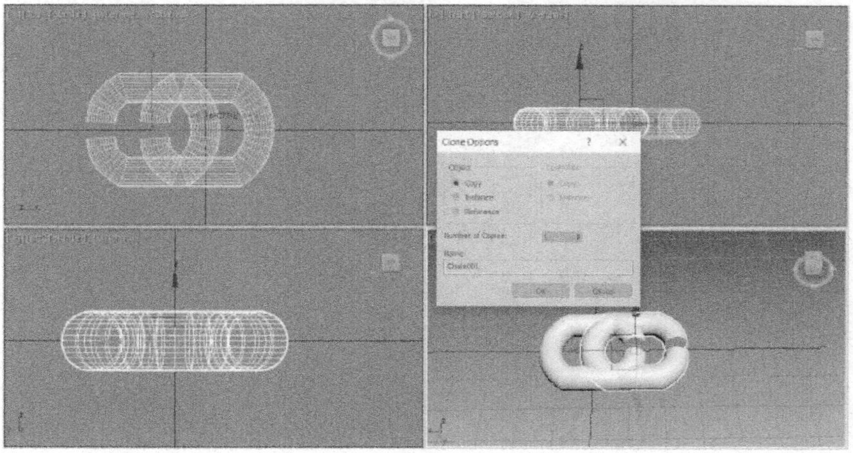

Figure 4-59 *The* ***Clone Options*** *dialog box displayed*

2. Make sure *Chain001* is selected. Next, press E to activate the **Select and Rotate** tool. Next, rotate the selected *Chain001* at 90 degrees, as shown in Figure 4-60.

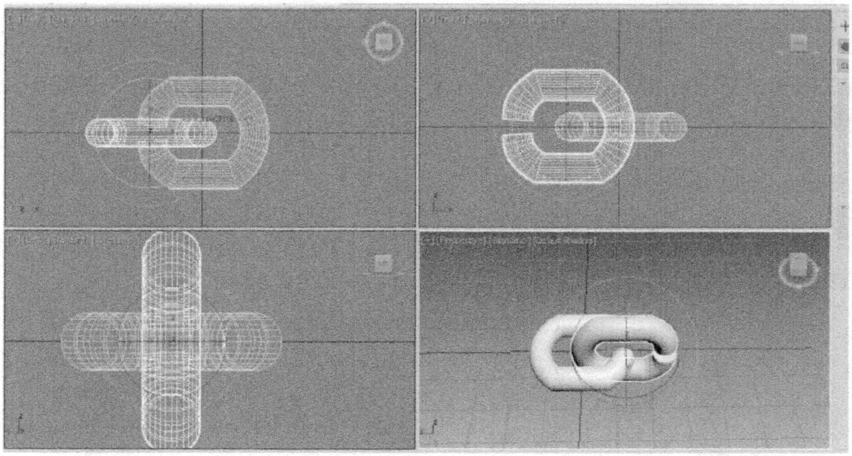

Figure 4-60 *The Chain001 rotated at 90 degrees*

3. By using the CTRL key, select *Chain* and *Chain001* simultaneously. Choose **Group > Group** from the menubar; the **Group** dialog box is displayed. Now, enter **Chaingrp** in the **Group Name** text box and then close the dialog box. Change the color of *Chaingrp* to light grey as discussed in earlier chapters.

4. Make sure the **Select and Move** tool is activated. Choose the *Chaingrp* from the viewport. Press and hold the SHIFT key and move *Chaingrp* slightly to create a chain like structure; the **Clone Options** dialog box is displayed. In this dialog box, enter **15** in the **Number of**

Copies spinner; the copies of *Chaingrp* are created, as shown in Figure 4-61. Next, close the **Clone Options** dialog box.

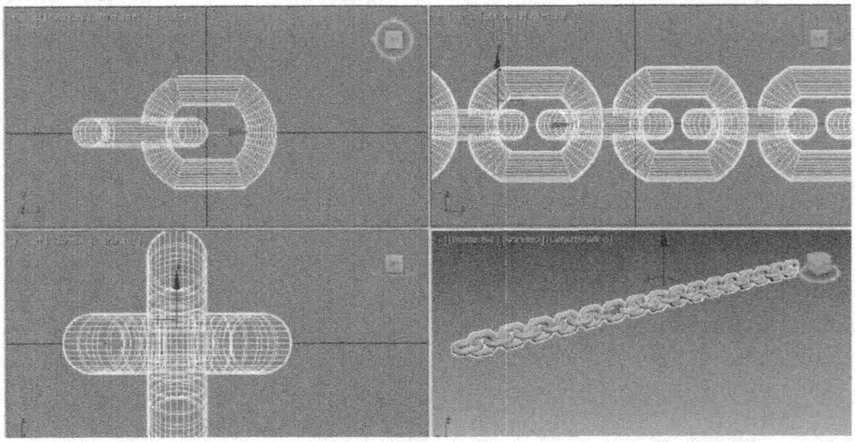

Figure 4-61 The copies of Chaingrp created

Saving and Rendering the Scene

In this section, you will save the scene that you have created and then render it. You can also view the final rendered image of this model by downloading the *c04_3dsmax_2022_rndr.zip* file from *www.cadcim.com*. The path of the file is as follows: *Textbooks > Animation and Visual Effects > Autodesk 3ds Max > Autodesk 3ds Max 2022 for Beginners: A Tutorial Approach*

1. Change the background color of the scene to white by following the steps as given in Tutorial 1 of Chapter 2.

2. Choose **Save** from the **File** menu.

3. Activate the Perspective viewport. Next, invoke the **Render Production** tool from the **Main Toolbar**; the rendered image is displayed, refer to Figure 4-62.

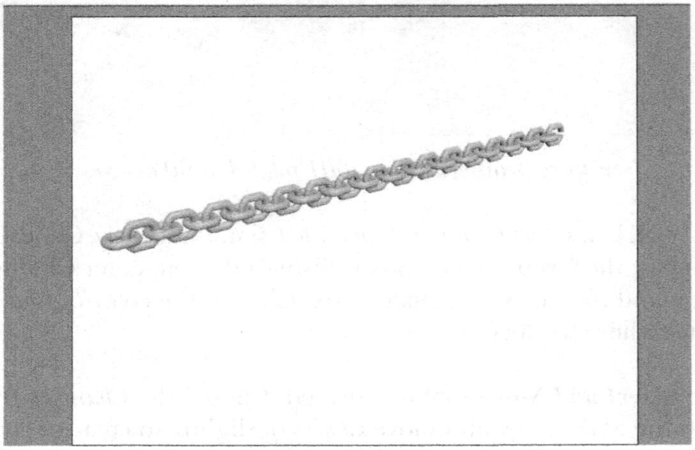

Figure 4-62 The final output after rendering

Self-Evaluation Test

Answer the following questions and then compare them to those given at the end of this chapter:

1. Which of the following options is used to convert a curved spline into a linear spline?

   (a) **Bezier** (b) **Bezier Corner**
   (c) **Corner** (d) **Smooth**

2. Which of the following buttons is chosen to enter the spline sub-object mode?

   (a) **Spline** (b) **Line**
   (c) **Vertex** (d) None

3. You need to choose __________ from the **Edit** menu to copy the existing state on the screen to a buffer.

4. When you choose the **Vertex** button in the **Selection** rollout to turn it on, the __________ of the selected shape are displayed.

5. You can change the curvature through a vertex by manipulating the __________ of the vertex.

6. There are two types of bezier vertices, **Bezier** and __________.

7. The two bezier handles at any vertex are used to move the lead in and out of the vertex. (T/F)

8. When the **Connect** button in the **Geometry** rollout (in the **Vertex** sub-object mode) is used, the two vertices of a spline are joined by a curved segment. (T/F)

9. The first vertex in a spline has white color. (T/F)

10. The **Edit Spline** modifier need not to be applied to the shapes created using the **Line** tool. (T/F)

Review Questions

Answer the following questions:

1. Which of the following tools is used to bring the transform gizmo to the center point of the selected object?

   (a) **Use Transform Coordinate Center** (b) **Use Pivot Point Center**
   (c) **Use Selection Center** (d) **Select and Uniform Scale**

2. Which modifier is used to modify the shape of the splines?

   (a) **Edit Poly** (b) **Edit Spline**
   (c) **Edit Mesh** (d) None of these

3. The __________ option is used to convert a linear spline into a curved spline.

4. The __________ option is used to restore the previously saved state from the buffer.

5. To turn the transform gizmo on and off, choose __________ from the **Views** menu.

6. The __________ button is used to bring the transform gizmo to the center point of the selected object.

7. The **Edit Spline** modifier helps you to modify the shape of splines. (T/F)

8. When the **Use Selection Center** tool becomes active on using the **Rotate** command, the selected objects are rotated along their individual pivot points. (T/F)

9. The first vertex of the spline is always yellow colored. (T/F)

10. You cannot select a new shape when you are in the sub-object mode. (T/F)

EXERCISES

The rendered output of the model used in the following exercises can be accessed by downloading the *c04_3dsmax_2022_exr.zip* from *www.cadcim.com*. The path of the file is as follows: *Textbooks > Animation and Visual Effects > Autodesk 3ds Max > Autodesk 3ds Max 2022 for Beginners: A Tutorial Approach*

Exercise 1

Draw the following shapes by using the dimensions of your choice, refer to Figure 4-63. Create their copies and transform them as required. **(Expected time: 20 min)**

Figure 4-63 *Drawing different shapes*

Exercise 2

Create the model of a street lamp, as shown in Figure 4-64, by applying dimensions of your choice. **(Expected time: 30 min)**

Figure 4-64 *The model of a street lamp*

Answers to Self-Evaluation Test

1. c, **2.** a, **3. Hold**, **4.** vertices, **5.** bezier handles, **6. Bezier Corner**, **7.** T, **8.** F, **9.** F, **10.** T

Chapter 5

Working with Splines-II

Learning Objectives

After completing this chapter, you will be able to:
- *Create objects by using different shapes*
- *Create objects by revolving a shape*
- *Create objects by lofting a shape along a path*

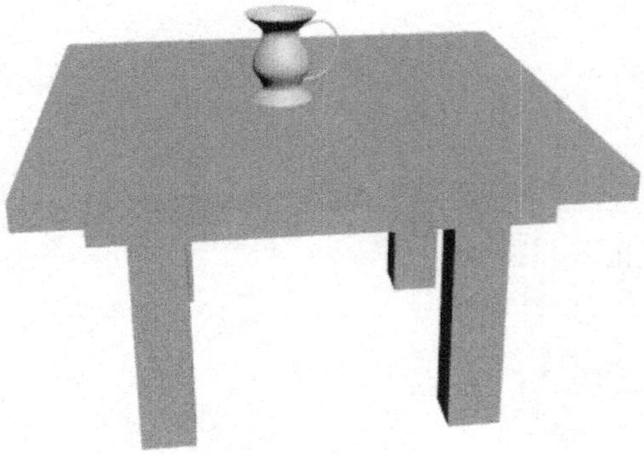

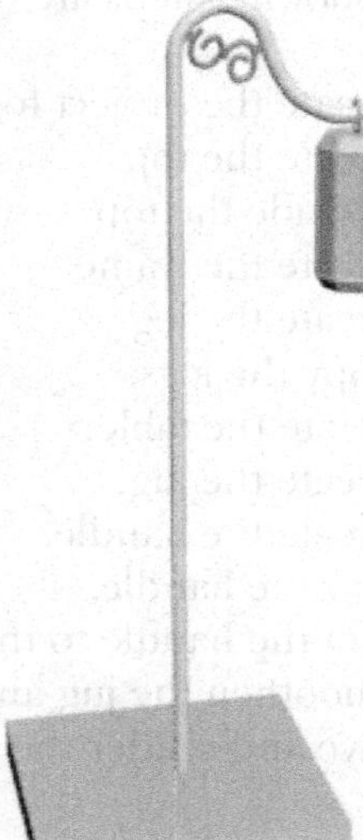

INTRODUCTION

In the previous chapter, you learned to create and modify different shapes using splines. In this chapter, you will learn to create 3D objects using splines. You will also learn the use of the **Lathe** modifier.

TUTORIALS

Tutorial 1

In this tutorial, you will create a model of a table consisting of a top, frame, and legs by extruding simple shapes. You will also create a jug with a handle and place it on the table, refer to Figure 5-1. **(Expected time: 30 min)**

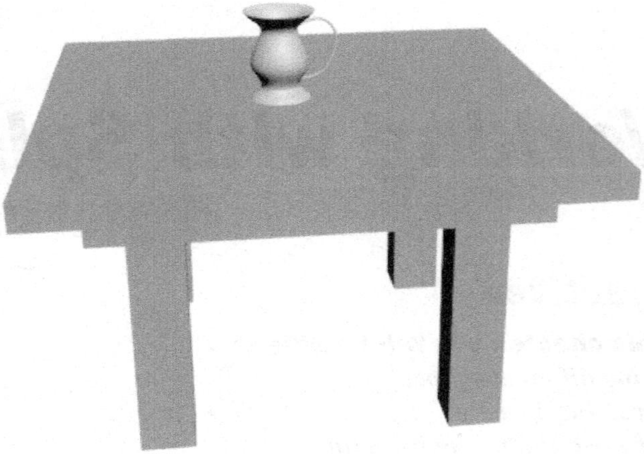

Figure 5-1 *The model of a table and jug*

The following steps are required to complete this tutorial:

a. Create the project folder.
b. Create the top.
c. Extrude the top.
d. Create the frame.
e. Create the leg.
f. Copy the legs.
g. Rotate the table.
h. Create the jug.
i. Create the handle.
j. Loft the handle.
k. Join the handle to the jug.
l. Smoothen the jug and add thickness to it.
m. Save and render the scene.

Creating the Project Folder

Create the project folder with the name *c05_tut1* in the *3dsmax 2022* folder, as discussed in Tutorial 1 of Chapter 2.

Creating the Top

In this section, you will create a top of the table using the **Rectangle** tool.

1. Choose **Create > Shapes** in the **Command Panel**; the **Splines** option is displayed in the drop-down list located below the **Shapes** button. 

2. Invoke the **Rectangle** tool from the **Object Type** rollout.

3. Activate the Top viewport.

4. Expand the **Keyboard Entry** rollout and set the value **200** in both the **Length** and **Width** spinners. Next, choose the **Create** button; a rectangle is created.

5. Invoke the **Zoom Extents All** tool from viewport navigation controls to zoom the rectangle to its extents in all viewports, as shown in Figure 5-2.

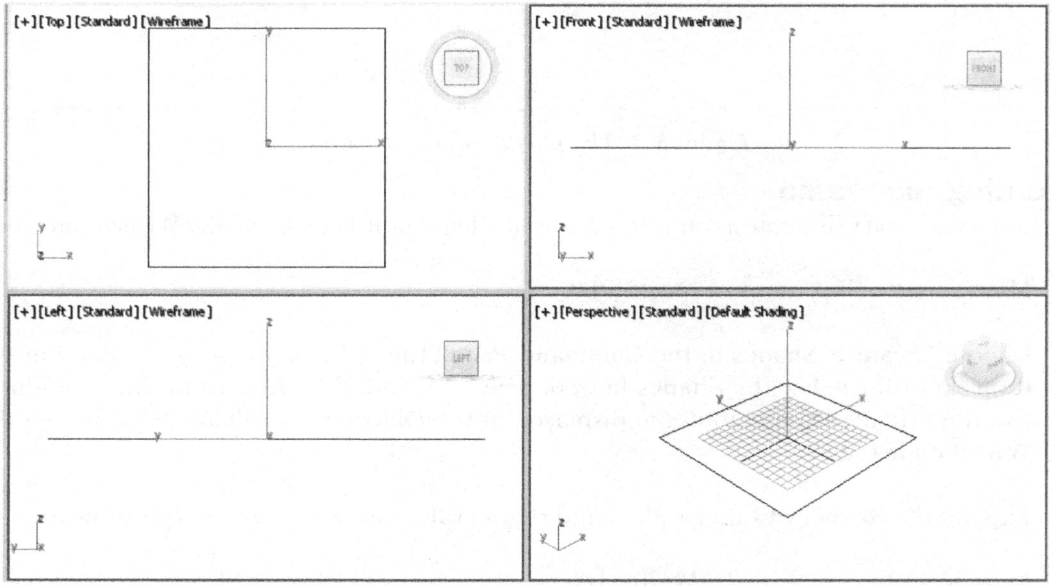

Figure 5-2 *The rectangle created for the table top*

6. In the **Name and Color** rollout, enter **Top** as the name of the object.

Extruding the Top

In this section, you will extrude *Top* to give it a thickness.

1. Make sure *Top* is selected. Next, choose the **Modify** tab in the **Command Panel.** 

2. Select **Extrude** from the **OBJECT-SPACE-MODIFIERS** section in the **Modifier List** drop-down list; the **Extrude** modifier is applied to the rectangle.

3. In the **Parameters** rollout, set the value **10** in the **Amount** spinner and press the ENTER key; *Top* is extruded by 10 units, as shown in Figure 5-3.

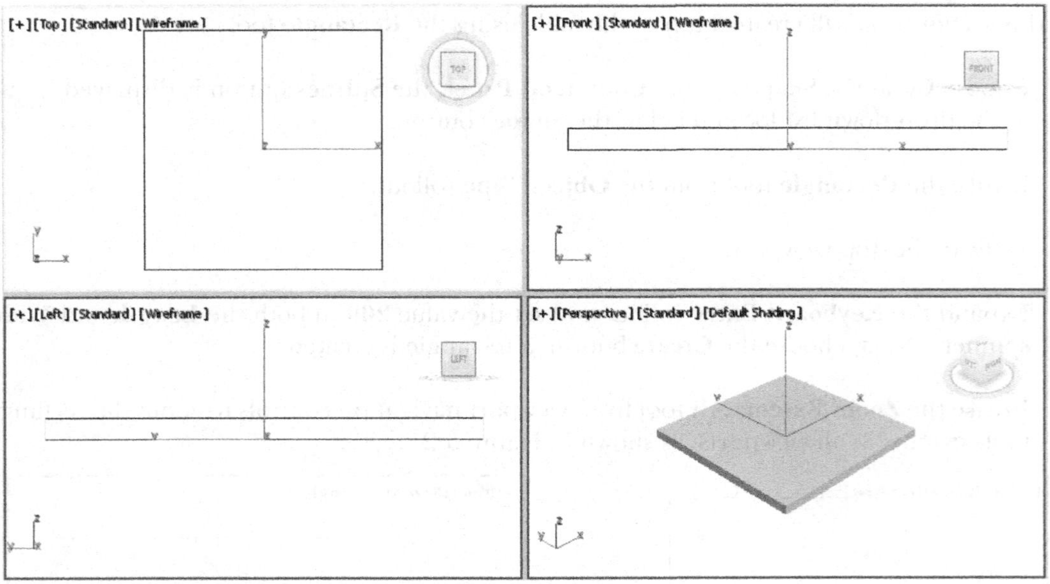

Figure 5-3 *The Top extruded by 10 units*

Creating the Frame

In this section, you will create a frame to support the legs of the table using the **WRectangle** tool.

1. Make sure the Top viewport is activated.

2. Choose **Create > Shapes** in the **Command Panel**; the **Splines** option is displayed in the drop-down list below the **Shapes** button. Select **Extended Splines** from this drop-down list; the extended spline tools are displayed in the **Object Type** rollout. Next, invoke the **WRectangle** tool.

3. Expand the **Keyboard Entry** rollout and set the values in the spinners as given next:

   Length: **160**   Width: **160**   Thickness: **12**

   Choose the **Create** button; *WRectangle001* spline is created.

4. In the **Name and Color** rollout, enter **Frame** as the name of the object.

   Next, you need to extrude *Frame*.

5. Make sure *Frame* is selected. Next, choose the **Modify** tab in the **Command Panel**. 

6. Select **Extrude** from the **OBJECT-SPACE-MODIFIERS** section in the **Modifier List** drop-down list; the **Extrude** modifier is applied to *Frame*.

7. In the **Parameters** rollout, set the value **15** in the **Amount** spinner; *Frame* is extruded by 15 units, as shown in Figure 5-4.

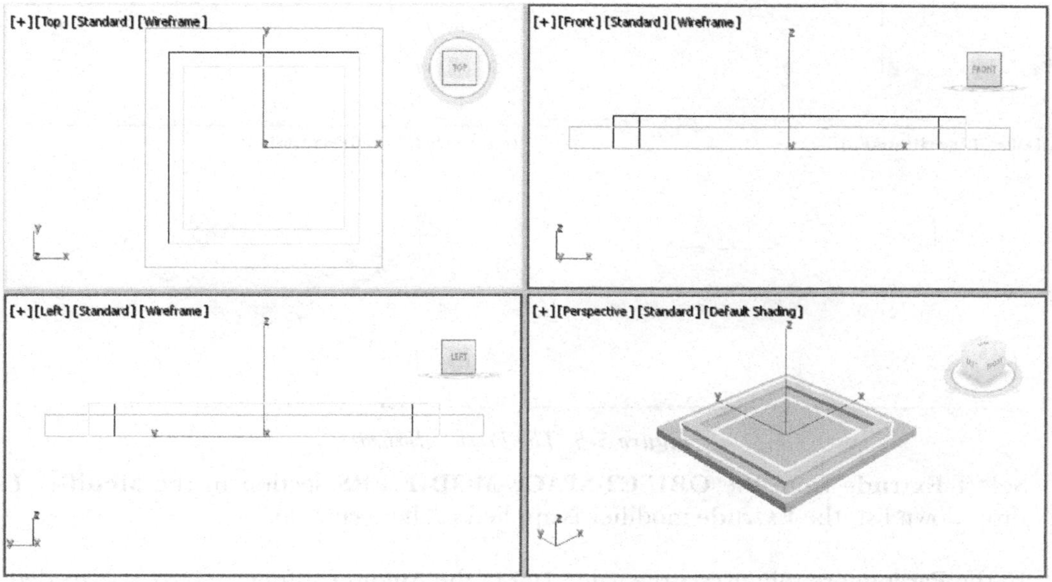

***Figure 5-4** The Frame extruded by 15 units*

8. Invoke the **Select and Move** tool from the **Main Toolbar** and then choose the **Transform Gizmo Y Constraint** button from the **Axis Constraints** toolbar.

9. In the Front viewport, move *Frame* so that its bottom touches the upper part of *Top*, as shown in Figure 5-5.

Creating the Leg

In this section, you will create a leg at one of the corners of *Frame* using the **Rectangle** tool.

1. Activate the Top viewport and then choose **Create > Shapes** in the **Command Panel**.

2. Select **Splines** from the drop-down list located below the **Shapes** button and then invoke the **Rectangle** tool from the **Object Type** rollout.

3. Expand the **Keyboard Entry** rollout and set the value **20** in the **Length** and **Width** spinners each. Next, choose the **Create** button; the rectangle is created at the center of the table.

4. In the **Name and Color** rollout, enter **Leg01** as the name of the object.

5. Make sure *Leg01* is selected. Next, choose the **Modify** tab in the **Command Panel**. 

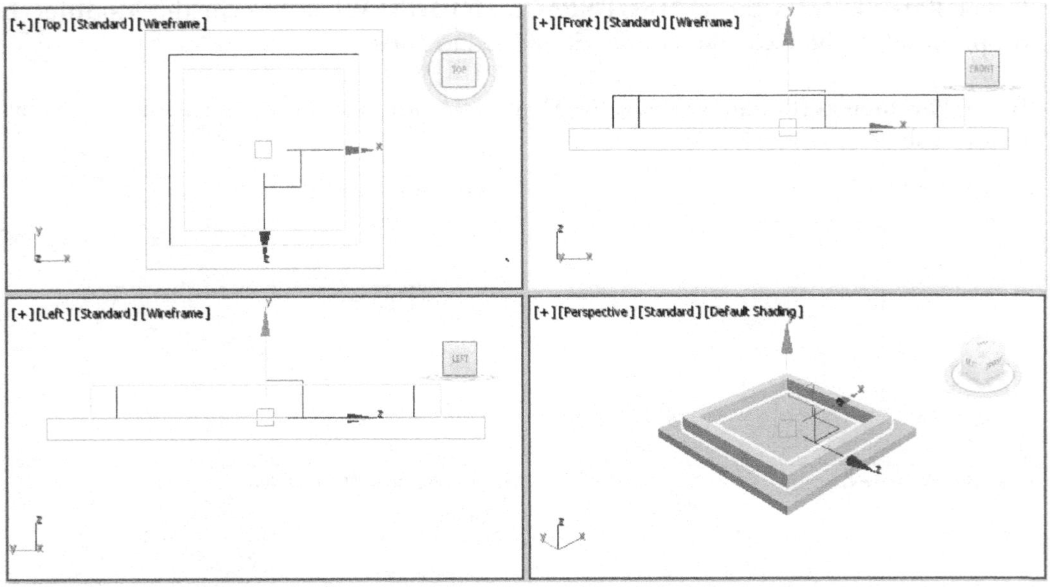

Figure 5-5 The Frame aligned

6. Select **Extrude** from the **OBJECT-SPACE-MODIFIERS** section in the **Modifier List** drop-down list; the **Extrude** modifier is applied to the rectangle.

7. In the **Parameters** rollout, set the value **100** in the **Amount** spinner; *Leg01* is extruded by 100 units.

8. Invoke the **Select and Move** tool from the **Main Toolbar** and then choose the **Transform Gizmo XY Plane Constraint** button from the **Axis Constraints** toolbar.

9. Move *Leg01* to the lower-left inner corner of *Frame*.

10. Activate the Front viewport. Choose the **Transform Gizmo Y Constraint** button from the **Axis Constraints** toolbar and move *Leg01* up so that its bottom touches the upper part of *Top* (not *Frame*).

11. Invoke the **Zoom Extents All** tool from the viewport navigation control to view all objects properly.

Copying the Legs

In this section, you will copy *Leg01* to the other three corners of *Frame*.

1. Activate the Top viewport. Make sure the **Select and Move** tool is invoked and then choose the **Transform Gizmo Y Constraint** button from the **Axis Constraints** toolbar, if it is not already invoked.

2. Select *Leg01* and then choose **Edit > Clone** from the menu bar; the **Clone Options** dialog box is displayed.

3. In this dialog box, enter **Leg02** in the **Name** text box and make sure the **Copy** radio button is selected, as shown in Figure 5-6. Next, choose the **OK** button; *Leg02* is created over *Leg01*.

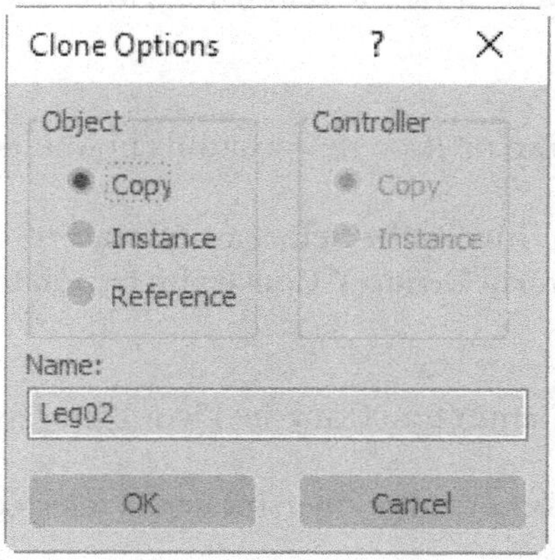

Figure 5-6 *The* ***Clone Options*** *dialog box*

4. Move *Leg02* to the upper-left corner of the frame.

5. Select *Leg02* and then choose **Edit > Clone** from the menu bar; the **Clone Options** dialog box is displayed. In this dialog box, enter **Leg03** in the **Name** text box and then choose the **OK** button. Next, choose the **Transform Gizmo X Constraint** button. Next, move *Leg03* to the upper-right corner of the frame.

6. Similarly, create a copy of *Leg03*. Next, name it as **Leg04** and place it at the lower-right corner of *Frame*; the four legs of table are created and aligned, as shown in Figure 5-7.

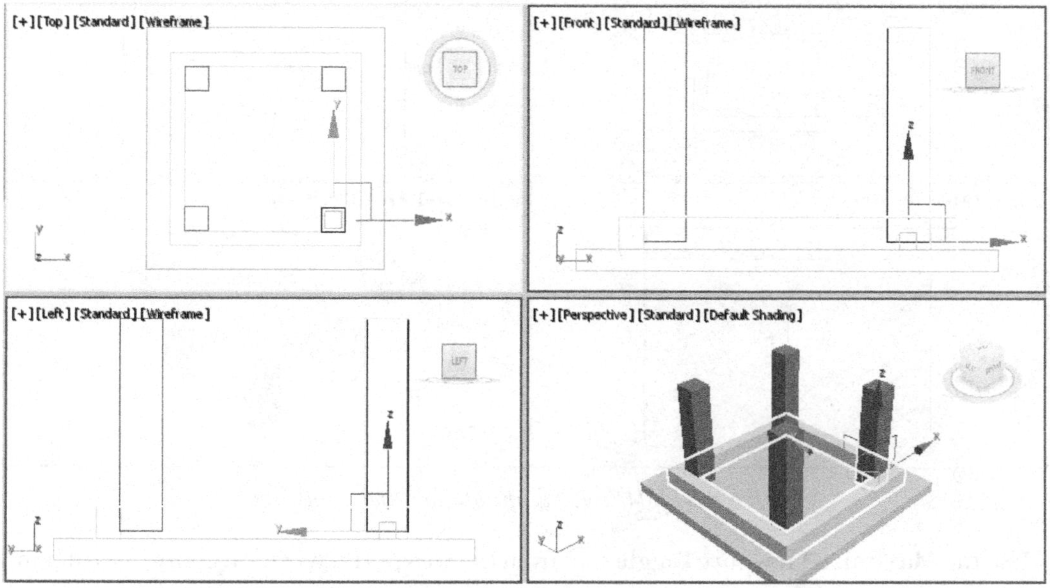

Figure 5-7 *The four legs of the table created and aligned to their correct places*

7. Select all objects in the viewport. Choose **Group > Group** from the menu bar; the **Group** dialog box is displayed. In this dialog box, enter **Table** in the **Group name** text box and then choose the **OK** button to close the dialog box; *Table* group is created.

Rotating the Table

In this section, you will rotate the table so that it stands up on its legs.

1. Activate the Front viewport. Invoke the **Select and Rotate** tool from the **Main Toolbar** and then choose the **Transform Gizmo Y Constraint** button from the **Axis Constraints** toolbar. 

2. Invoke the **Use Selection Center** tool from the Pivot Point flyout in the **Main Toolbar**. 

   The **Use Selection Center** tool enables you to rotate the selected objects around their collective geometric center.

3. Right-click on the **Select and Rotate** tool in the **Main Toolbar**; the **Rotate Transform Type-In** dialog box is displayed. Set the value **180** in the **Z** spinner in the **Offset:Screen** area of this dialog box and then press the ENTER key; *Table* is rotated in upright position, as shown in Figure 5-8. Next, close this dialog box.

4. Invoke the **Zoom Extents All** tool from viewport navigation controls; *Table* is zoomed to its extents in all four viewports, refer to Figure 5-8.

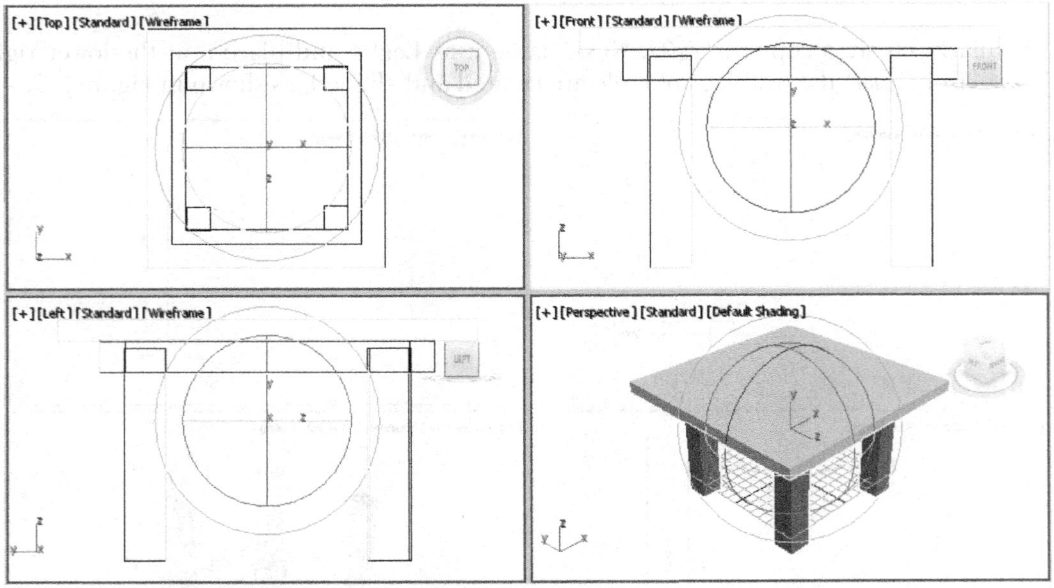

***Figure 5-8** Table rotated upright to stand on its legs*

5. Use the **Maximize Viewport Toggle** tool from the viewport navigation controls to maximize the Front viewport.

Creating the Jug

In this section, you will create a jug on the table by using the **Line** tool and the **Lathe** modifier.

1. Maximize the Front viewport and invoke the **Use Pivot Point Center** tool from the Pivot Point flyout in the **Main Toolbar**.

2. Invoke the **Zoom Region** tool from viewport navigation controls and zoom in on a small area surrounding the center of *Top*, as shown in Figure 5-9. 

Figure 5-9 *The center of Table zoomed in the Front viewport*

3. Choose **Create > Shapes** in the **Command Panel**. Make sure **Splines** is selected in the drop-down list below the **Shapes** button.

4. Invoke the **Line** tool from the **Object Type** rollout.

5. In the Front viewport, place the cursor over the center of the top surface of *Top*. Now, click at this point to specify it as the first vertex of the line.

6. Similarly, click at different points to create the profile of the jug, as shown in Figure 5-10. Note that the profile consists of nearly parallel segments. This will give the jug the required thickness when the **Lathe** modifier is applied to the shape.

7. Right-click to exit the **Line** tool; a double-line profile of the jug is created. Use the **Zoom** and **Pan** tools, if needed, to display the entire profile of the jug, refer to Figure 5-11. 

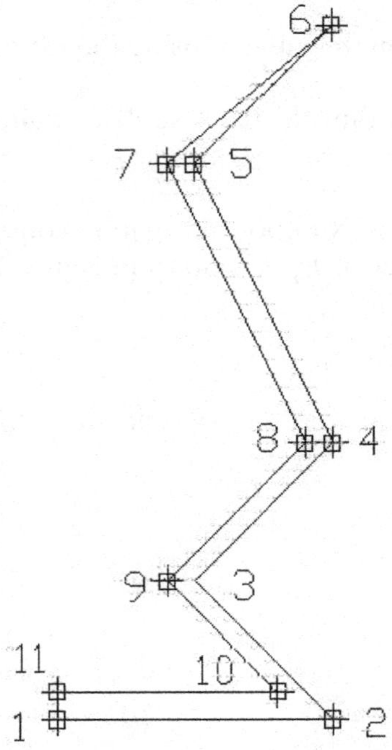

Figure 5-10 The profile curve of the Jug

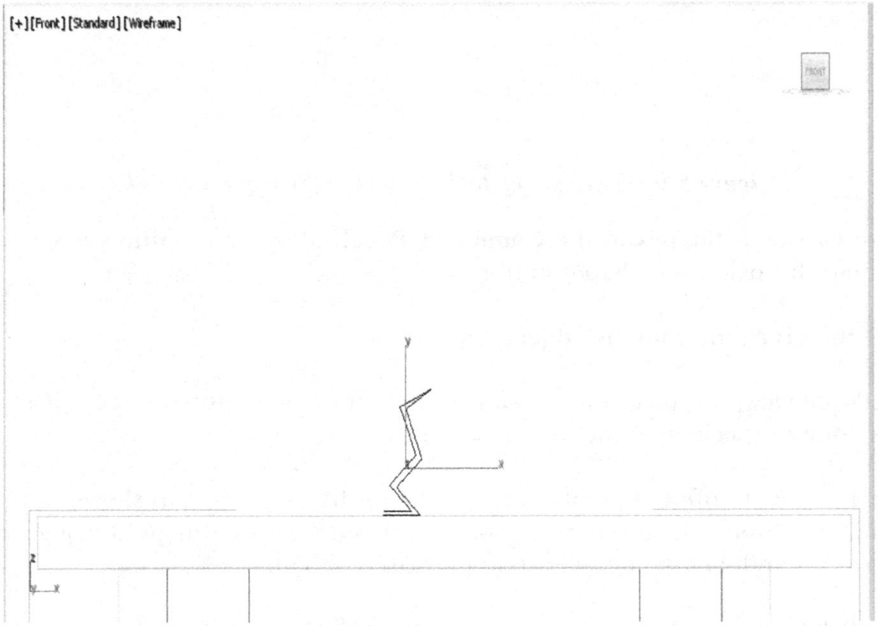

Figure 5-11 The profile of the Jug

8. Enter **Jug** in the **Name and Color** rollout.

Next, you will smoothen *Jug* by modifying its vertices.

9. With *Jug* selected, choose the **Modify** tab in the **Command Panel**. In the **Selection** rollout, choose the **Vertex** button; the **Vertex** sub-object mode is selected and the vertices appear on the profile.

10. Right-click on the inner vertex, refer to Figure 5-12; a quad menu is displayed. Choose **Bezier** from the upper-left quadrant of the quad menu; the spline is curved and the bezier handles are displayed, as shown in Figure 5-12.

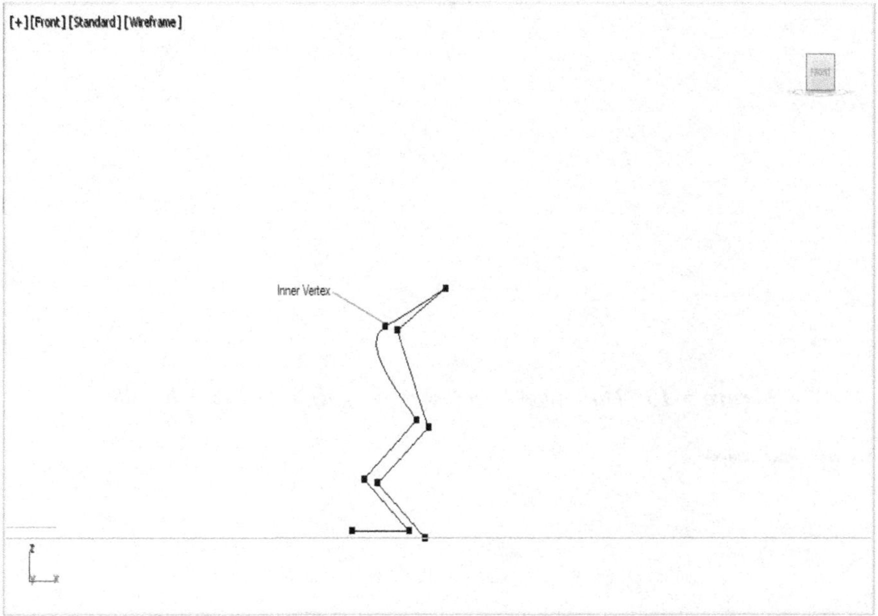

Figure 5-12 *The bezier handles displayed at the vertex*

11. Invoke the **Select and Move** tool from the **Main Toolbar** and choose the **Transform Gizmo XY Plane Constraint** button from the **Axis Constraints** toolbar.

12. Next, hold the upper green handle and drag it to left so that the curve looks similar to the one shown in Figure 5-13. If you make a change by mistake, choose the **Undo Scene Operation** button to undo the previous change.

13. Similarly, change the curvature of the other vertices and move some of them to get the shape of *Jug* similar to the one shown in Figure 5-14. You can move a vertex by directly clicking on it.

Note

You will find snaps useful for selecting vertices, but you may need to disable snaps while moving or changing the curvature of vertices.

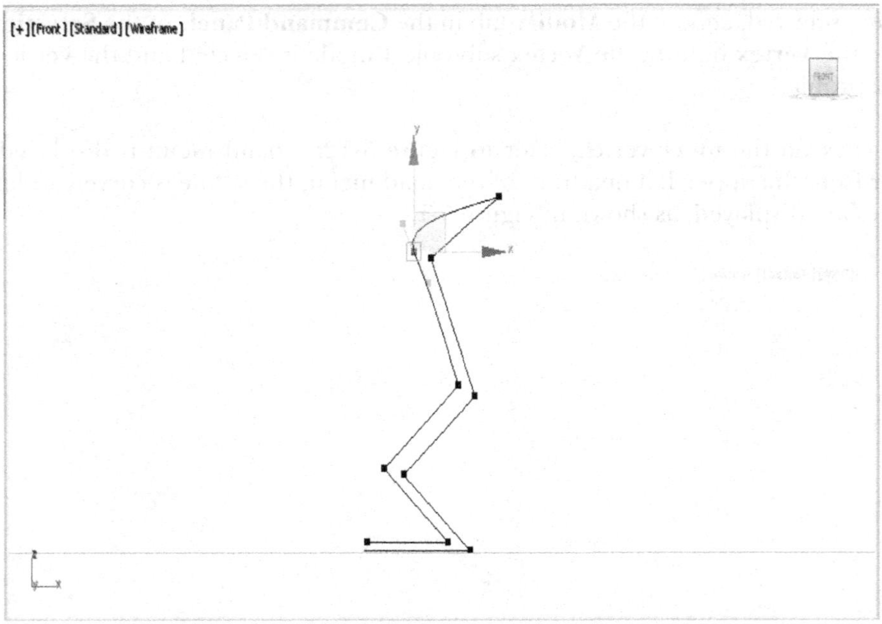

Figure 5-13 The curve reshaped by dragging the bezier handle

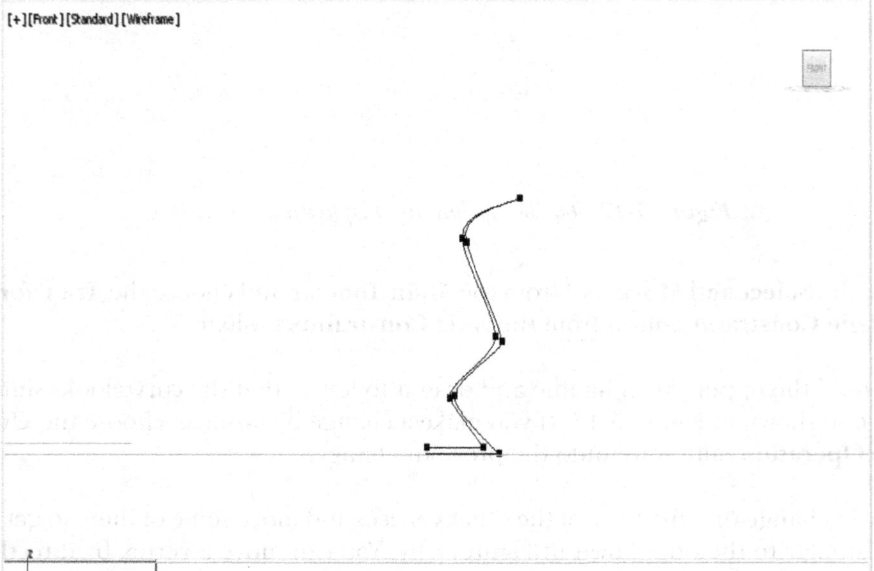

Figure 5-14 The curve edited by moving the bezier handles of the vertices

14. In the **Selection** rollout, choose the **Vertex** button again to exit the sub-object mode.

Next, you will spin the profile of *Jug* around its axis to form a complete jug.

15. Invoke the **Maximize Viewport Toggle** tool from the viewport navigation controls to restore all four viewports. Make sure the profile is selected and invoke the **Zoom Extents All Selected** tool from the viewport navigation controls; the profile is zoomed to its extent in all viewports.

16. Choose the **Modify** tab in the **Command Panel**. Select **Lathe** from the **OBJECT-SPACE-MODIFIERS** section in the **Modifier List** drop-down list; the profile spins around its axis in the middle.

17. Select the **Min** radio button in the **Align** area of the **Parameters** rollout; the profile forms the shape of a jug, refer to Figure 5-15. Invoke the **Zoom Extents All Selected** tool from the viewport navigation controls to zoom the object to its extent.

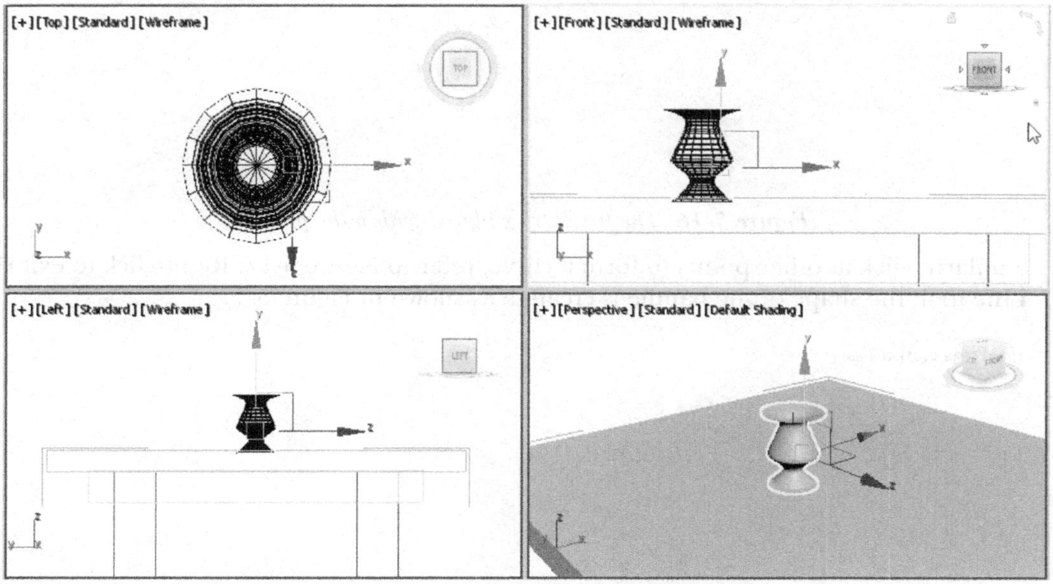

***Figure 5-15** The profile rotated to form the Jug*

18. Invoke the **Zoom Extents All** tool from the viewport navigation controls; all objects are zoomed to their extents in all viewports.

Creating the Handle

In this section, you will create a handle for *Jug* by using the **Circle** and **Loft** tools.

1. Maximize the Front viewport. Invoke the **Zoom Region** tool from the viewport navigation controls and zoom in the right-hand side of *Jug* such that you can create more space to accommodate the handle.

2. Choose **Create > Shapes** in the **Command Panel**. Next, make sure the **Line** tool is invoked from the **Object Type** rollout.

3. Click at a point slightly below the neck and make sure the point is placed inside the outer line of *Jug*, as shown in Figure 5-16.

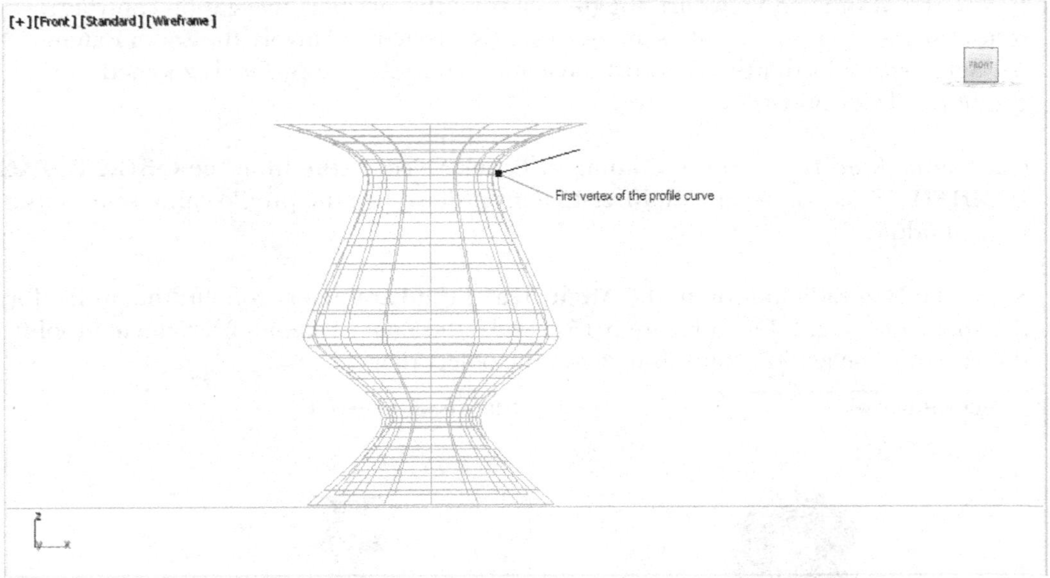

Figure 5-16 *The first vertex placed within the profile*

4. Similarly, click at other points to form a curve, refer to Figure 5-17. Right-click to exit the **Line** tool; the shape of the handle is created, as shown in Figure 5-17.

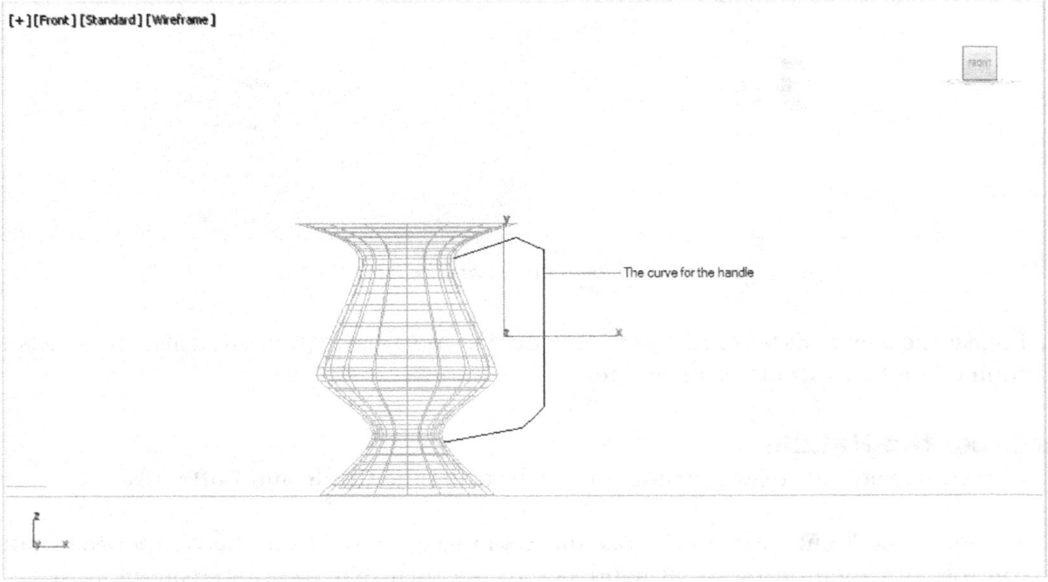

Figure 5-17 *The curve for the handle of the Jug*

You may need to adjust the coordinates given depending on the profile of your *Jug*.

5. In the **Name and Color** rollout, enter **Handle** as the name of the object.

6. With *Handle* selected, choose the **Modify** tab in the **Command Panel**. In the **Selection** rollout, choose the **Vertex** button; the **Vertex** sub-object mode is selected and the vertices appear on the profile.

7. Invoke the **Select and Move** tool from the **Main Toolbar** and right-click on the second vertex. Choose **Bezier** from the quad menu and then move *Handle* to get the desired curve. Similarly, modify the rest of the vertices of *Handle* as needed to form a curved shape, as shown in Figure 5-18.

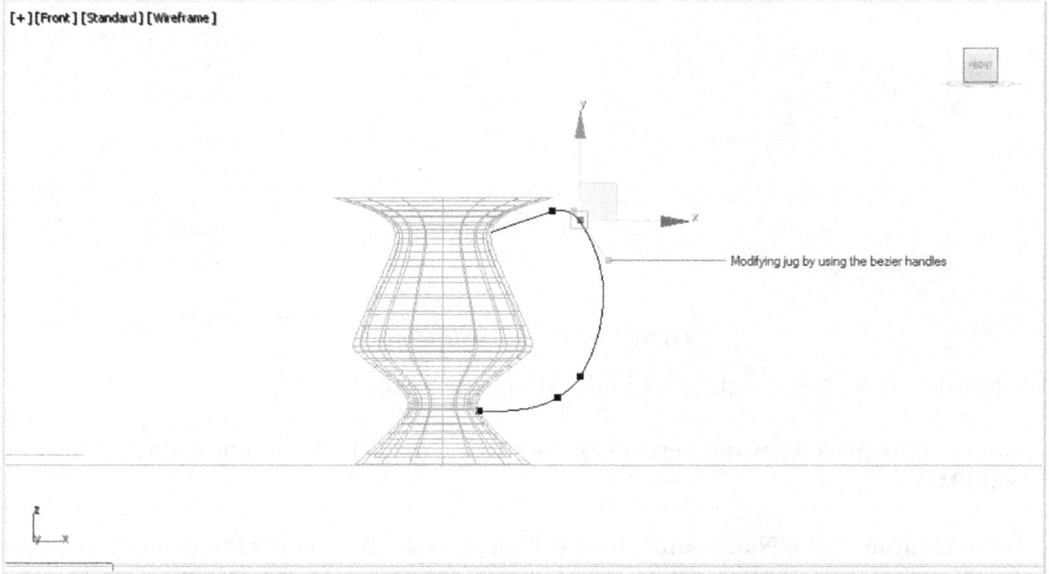

Figure 5-18 *The Handle modified by moving the bezier handles*

8. Choose the **Vertex** button again to exit the vertex selection mode.

Lofting the Handle

In this section, you will add thickness to the handle. This can be done by creating a circle and then lofting it along a path.

1. Choose **Create > Shapes** in the **Command Panel**. Next, invoke the **Circle** tool. Create a circle anywhere in the viewport, as shown in Figure 5-19. Set the value **0.5** in the **Radius** spinner of the **Parameters** rollout.

2. Select *Handle* from the **Scene Explorer** located at the left in the interface.

3. Choose **Create > Geometry** in the **Command Panel**. Select **Compound Objects** from the drop-down list located below it. Next, invoke the **Loft** tool from the **Object Type** rollout; different rollouts for the loft object are displayed.

4. In the **Creation Method** rollout, make sure the **Instance** radio button is selected and then choose the **Get Shape** button; the shape of the cursor changes.

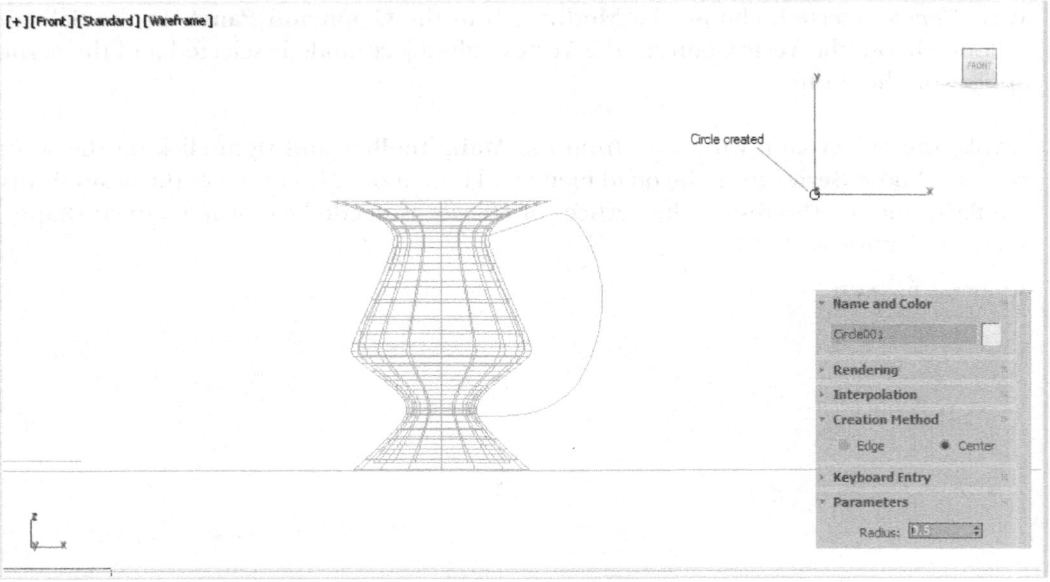

Figure 5-19 The circle created

5. Select the circle. Next, right-click in the viewport to exit the tool.

   The circle is placed on the vertices of the path (*Handle*) and is automatically named as **Loft001**.

6. Enter **Handle** in the **Name and Color** rollout. Invoke the **Select Object** tool to complete the operation.

7. Delete the circle. Next, invoke the **Maximize Viewport Toggle** tool to restore the four-viewport configuration. Now, invoke the **Zoom Extents All** tool to view all objects properly.

Smoothening and Adding Thickness to the Jug

In this section, you will add smoothness and thickness to *Jug* by increasing the segments in the **Lathe** modifier and using the **Shell** modifier.

1. Select *Jug*. Next, select **Lathe** from the modifier stack. In the **Parameters** rollout, set **26** in the **Segments** spinner; *Jug* is smoothened.

2. Make sure the **Modify** tab is chosen in the **Command Panel**. Select **Shell** from the **OBJECT-SPACE-MODIFIERS** section in the **Modifier List** drop-down list; the **Shell** modifier is added to the modifier stack.

3. In the **Parameters** rollout, set **0.65** in the **Outer Amount** spinner. You will notice that thickness is added to *Jug*.

Joining the Handle and the Jug

In this section, you will join *Handle* with *Jug* using the **Boolean** tool.

1. Invoke the **Zoom Region** tool from the viewport navigation controls to zoom in the Top viewport and make sure that *Handle* is centered on *Jug*. If not, move *Handle* as required.

2. Select *Handle*. Next, select **Compound Objects** from the drop-down list below the **Geometry** button. Next, invoke the **Boolean** tool from the **Object Type** rollout; different rollouts for a boolean object are displayed.

3. Choose the **Add Operands** button in the **Boolean Parameters** rollout. Next, choose the **Union** button in the **Operand Parameters** rollout and select *Jug* in any viewport. Now, invoke the **Select Object** tool from the **Main Toolbar** to complete the operation; *Jug* and *Handle* are joined to make a single object, as shown in Figure 5-20.

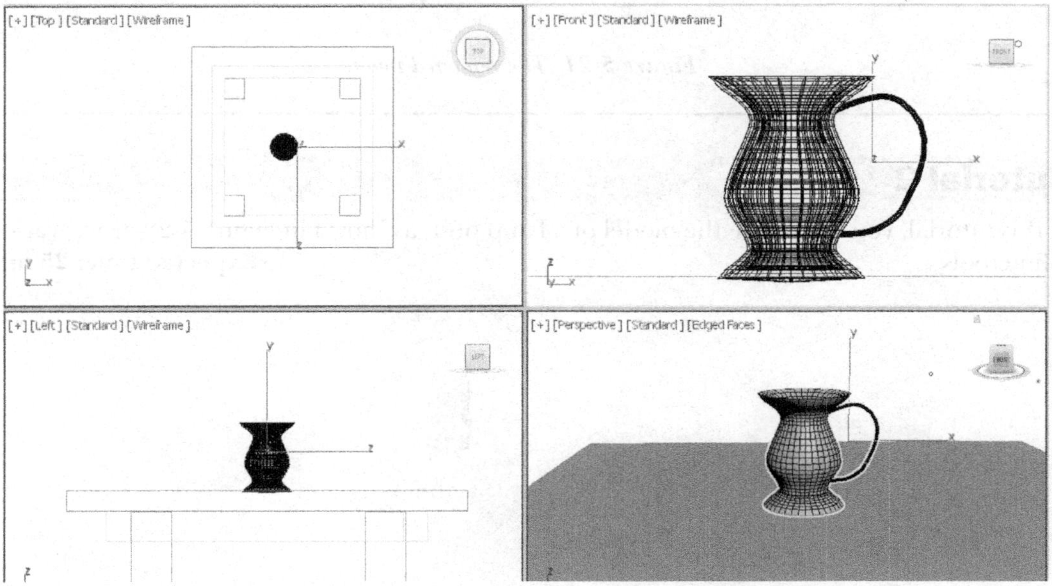

Figure 5-20 *Using the boolean operation to join the Handle and Jug*

4. In the **Name and Color** rollout, enter the name of the object as **jug**. Invoke the **Zoom Extents All** tool from viewport navigation controls to view the objects to their extents.

5. Select *Table*. In the **Name and Color** rollout, change the color of *Table* using the color swatch.

Saving and Rendering the Scene

In this section, you will save the scene that you have created and then render it. You can also view the final rendered image of this scene by downloading the *c05_3dsmax_2022_rndr.zip* file from *www.cadcim.com*. The path of the file is as follows: *Textbooks > Animation and Visual Effects > Autodesk 3ds Max > Autodesk 3ds Max 2022 for Beginners: A Tutorial Approach*

1. Choose **Save** from the **File** menu.

2. Change the background color to white as discussed in the earlier chapters.

3. Activate the Perspective viewport. Next, invoke the **Render Production** tool from the **Main Toolbar**; the rendered image is displayed, refer to Figure 5-21.

Figure 5-21 *The rendered image*

Tutorial 2

In this tutorial, you will create the model of a lamp post, as shown in Figure 5-22, using various spline tools. **(Expected time: 25 min)**

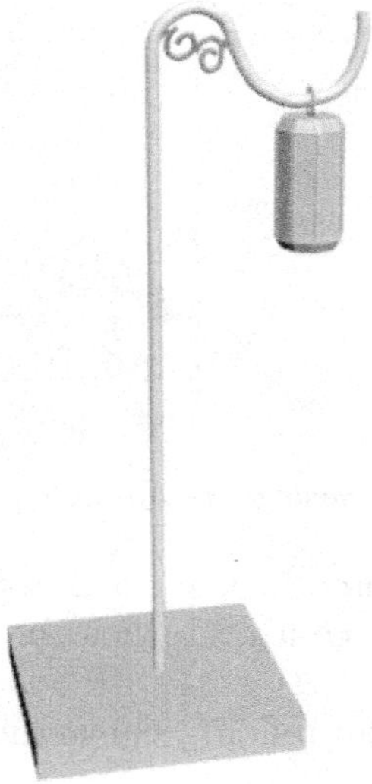

Figure 5-22 *The model of a lamp post*

The following steps are required to complete this tutorial:

a. Create the project folder.
b. Create the base.
c. Create the pillar.

d. Create the joint.
e. Create the lamp holder.
f. Add details.
g. Save and render the scene.

Creating the Project Folder

Create a project folder with the name *c05_tut2* in the *3dsmax 2022* folder as discussed in Tutorial 1 of Chapter 2.

Creating the Base

In this section, you will create the base of the lamp post by using the **Box** tool.

1. Activate the Top viewport and then choose **Create > Shapes** in the **Command Panel**; the **Splines** option is displayed in the drop-down list located below it. Next, invoke the **Rectangle** tool from the **Object Type** rollout.

2. Expand the **Keyboard Entry** rollout and set the values in the spinners as given next:

   Length: **3** Width: **3**

   Choose the **Create** button; a square is created in all viewports.

3. Enter **base** in the **Name and Color** rollout. Invoke the **Zoom Extents All** tool; *base* is zoomed in all viewports, as shown in Figure 5-23.

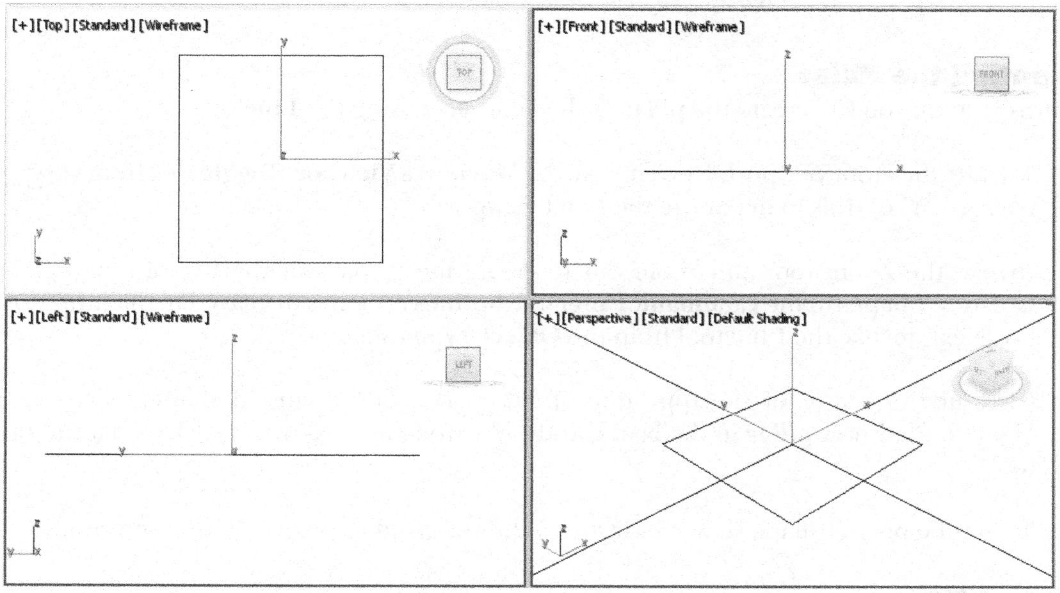

***Figure 5-23** The base zoomed in all viewports*

4. Choose the **Modify** tab in the **Command Panel**. Next, select the **Extrude** option from the **OBJECT-SPACE-MODIFIERS** section in the **Modifier List** drop-down list; the **Extrude**

modifier is applied to *base* and displayed in the modifier stack. Also, various rollouts of this modifier are displayed in the **Command Panel**.

5. In the **Parameters** rollout, set the value **0.4** in the **Amount** spinner and press the ENTER key; *base* is extruded, as shown in Figure 5-24.

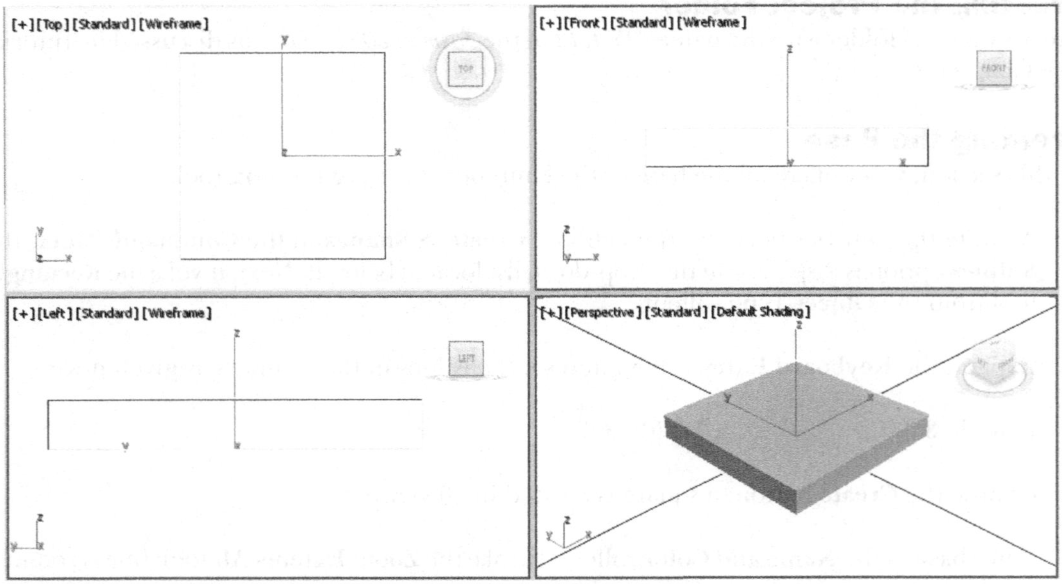

***Figure 5-24** The base extruded in all viewports*

Creating the Pillar

In this section, you will create the pillar of the lamp post using the **Line** tool.

1. Activate the Front viewport. Next, invoke the **Maximize Viewport Toggle** tool from viewport navigation controls to maximize the Front viewport.

2. Invoke the **Zoom** tool and zoom out to create more space at the top of *base*. Choose **Create > Shapes** in the **Command Panel**; the **Splines** option is displayed in the drop-down list. Next, invoke the **Line** tool from the **Object Type** rollout.

3. Click at the center of the top edge of *base* and create a curved shape, as shown in Figure 5-25. Enter **pillar** in the **Name and Color** rollout. Next, right-click to exit the **Line** tool.

   To get the perfect shape shown in Figure 5-25, you need to follow the steps given next.

Note

To achieve the smoothness in the curve, the number of vertices should be more.

4. Choose the **Modify** tab and then in the modifier stack, click on the arrow on the left of the **Line** to view all sub-object levels. Next, select the **Vertex** sub-object level. Now, select

the vertices one by one and move them by using the **Select and Move** tool from the **Main Toolbar** to get the desired shape.

5. Select the upper vertices of *pillar*, as shown in Figure 5-26 and then right-click; a quad menu is displayed. Choose **Smooth** from the quad menu; the upper portion of *pillar* is smoothened. Select the **Vertex** sub-object level again to exit the sub-object mode.

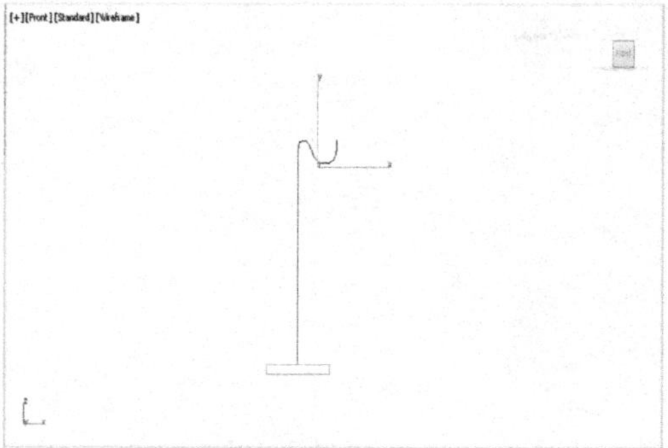

Figure 5-25 *The line created*

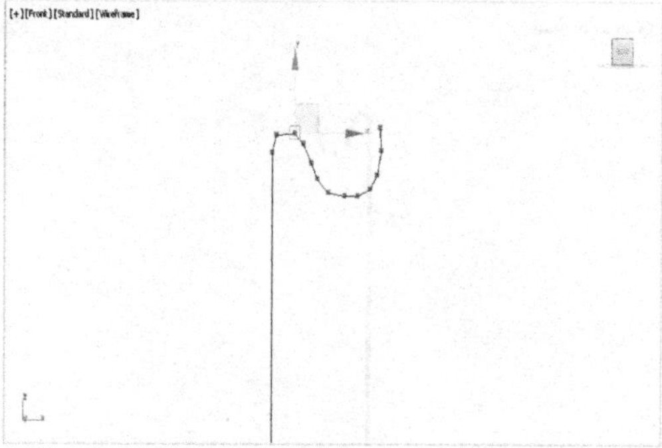

Figure 5-26 *The selected vertices*

Next, you will add thickness to *pillar*.

6. Expand the **Rendering** rollout and select the **Enable In Renderer** and **Enable In Viewport** check boxes.

7. Make sure the **Radial** radio button is selected and set the value **0.15** in the **Thickness** spinner; the thickness is added to *pillar*, as shown in Figure 5-27.

Creating the Joint

In this section, you will create a joint between the pillar and the lamp holder using the **Ellipse** tool.

1. Activate the Left viewport. Next, invoke the **Maximize Viewport Toggle** tool from viewport navigation controls to maximize the Left viewport.

2. Choose **Create > Shapes** in the **Command Panel**; the **Splines** option is displayed in the drop-down list below the **Shapes** button. Next, invoke the **Ellipse** tool from the **Object Type** rollout.

3. Expand the **Keyboard Entry** rollout and set the values of the parameters as given next:

   Length: **0.3** Width: **0.2**

   Choose the **Create** button; an ellipse is created in the viewport. Enter **joint** in the **Name and Color** rollout.

4. In the **Rendering** rollout, make sure that the **Enable In Renderer** and **Enable In Viewport** check boxes are selected. Also, make sure that the **Radial** radio button is selected. Next, set

the value **0.05** in the **Thickness** spinner and press the ENTER key; *joint* is displayed at the bottom of the viewport, as shown in Figure 5-28.

5. Invoke the **Maximize Viewport Toggle** tool to display all viewports. Make sure the **Select and Move** tool is invoked in the **Main Toolbar** and align *joint* in all viewports, as shown in Figure 5-29.

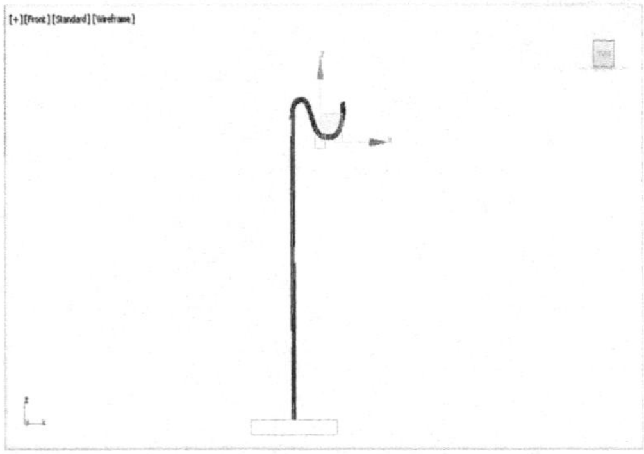

Figure 5-27 *The thickness added to pillar*

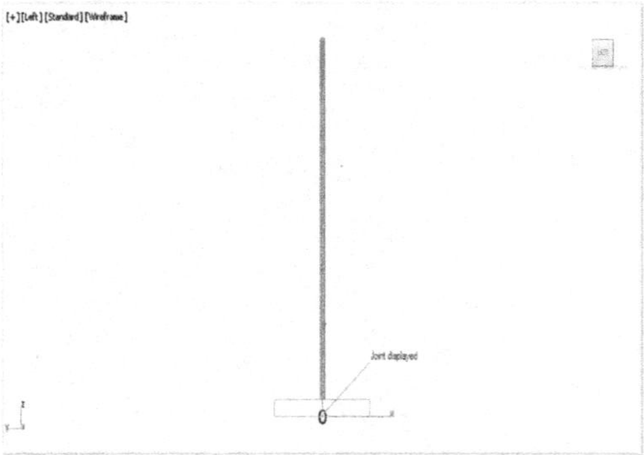

Figure 5-28 *The joint displayed*

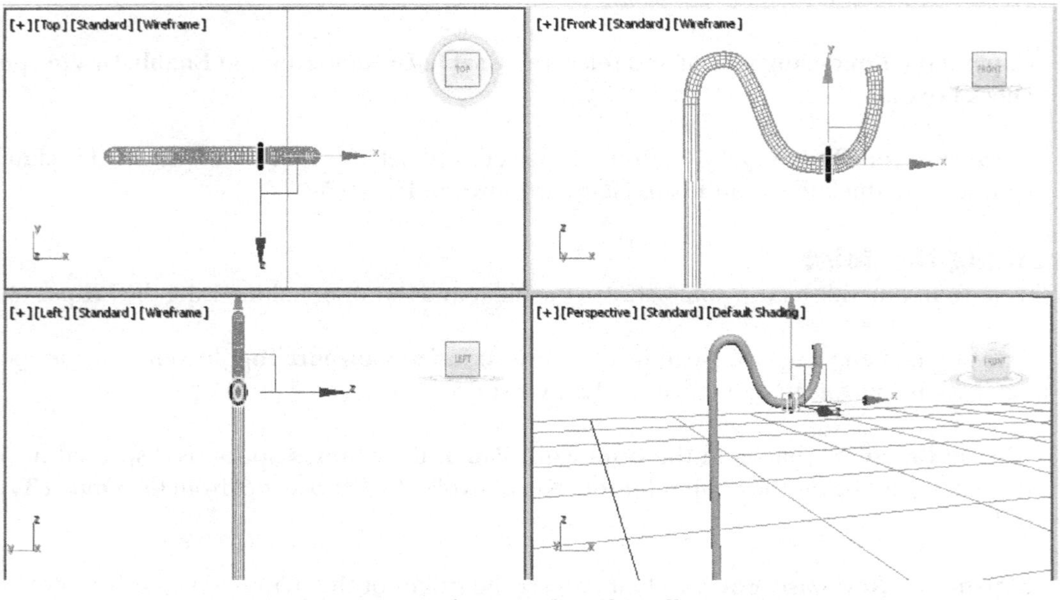

Figure 5-29 *The joint aligned in all viewports*

Creating the Lamp Holder

In this section, you will create the lamp holder using the **NGon** tool and the **Edit Poly** modifier.

1. Activate the Top viewport. Choose **Create > Shapes** in the **Command Panel**; the **Splines** option is displayed in the drop-down list below the **Shapes** button. Next, invoke the **NGon** tool from the **Object Type** rollout.

2. Expand the **Keyboard Entry** rollout and set the value **0.4** in the **Radius** spinner. Choose the **Create** button; *NGon001* is created in the viewport. Enter **lamp holder** in the **Name and Color** rollout. In the **Parameters** rollout, set the value **8** in the **Sides** spinner.

3. Choose the **Modify** tab in the **Command Panel**. Next, select **Extrude** from the **Modifier List** drop-down list; the **Extrude** modifier is applied to *lamp holder*.

4. Set the value **1.2** in the **Amount** spinner of the **Parameters** rollout and press the ENTER key; *lamp holder* is extruded. Next, align it in all viewports, as shown in Figure 5-30.

   The *lamp holder* is aligned slightly below *joint* because you will modify its shape in the following steps by applying the **Edit Poly** modifier.

5. Select **Edit Poly** from the **OBJECT-SPACE-MODIFIERS** section in the **Modifier List** drop-down list; the **Edit Poly** modifier is applied to *lamp holder*.

6. Click on the arrow in the modifier stack; all sub-object levels are displayed. Select the **Edge** sub-object level. Now, activate the Front viewport and select the edge, as shown in Figure 5-31.

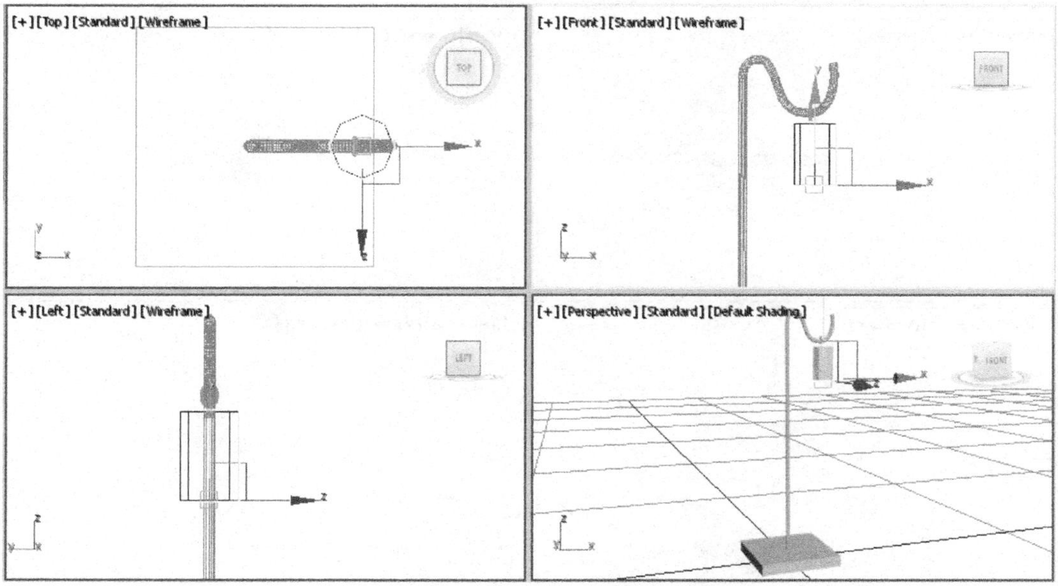

Figure 5-30 *The lamp holder aligned in all viewports*

7. In the **Selection** rollout, choose the **Ring** button; the ring of edges are selected. Next, in the **Edit Edges** rollout, choose the **Settings** button on the right of the **Chamfer** button; the **Chamfer** caddy control is displayed in the viewport. Set the value **0.03** in the **Chamfer-Amount** spinner and **1** in the **Segments** spinner and then choose the **OK** button to close the caddy control; the selected edges are chamfered, as shown in Figure 5-32.

8. Select the **Polygon** sub-object level from the **Selection** rollout and then select all polygons between the chamfered edges, as shown in Figure 5-33.

9. In the **Edit Polygons** rollout, choose the **Settings** button on the right of the **Bevel** button; the **Bevel** caddy control is displayed in the viewport. Set the value **-0.01** in the **Bevel-Height** spinner and **-0.02** in the **Bevel-Outline** spinner and then choose the **OK** button to close the caddy control; the selected polygons are beveled, as shown in Figure 5-34.

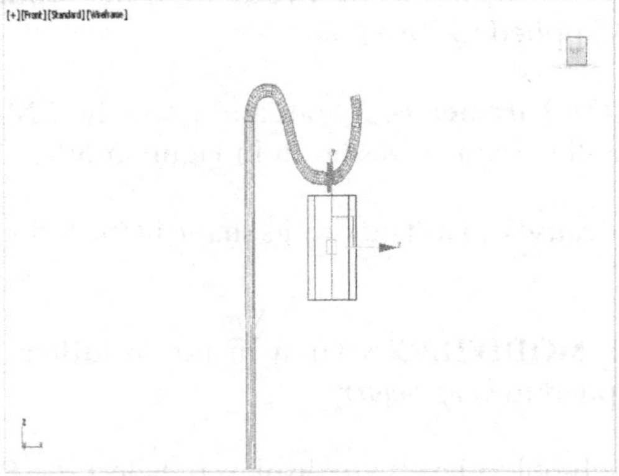

Figure 5-31 *The edge selected*

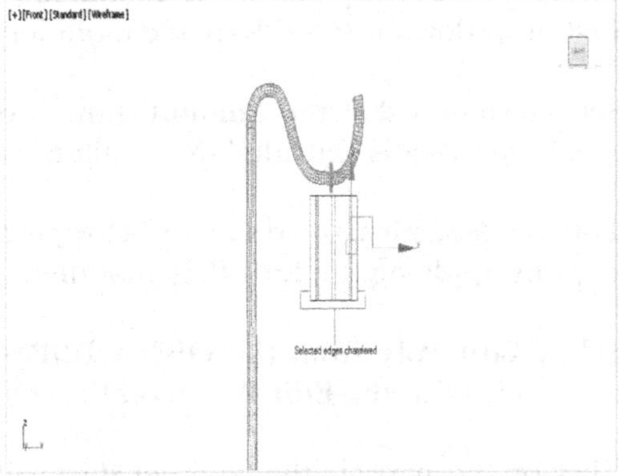

Figure 5-32 *Selected edges chamfered*

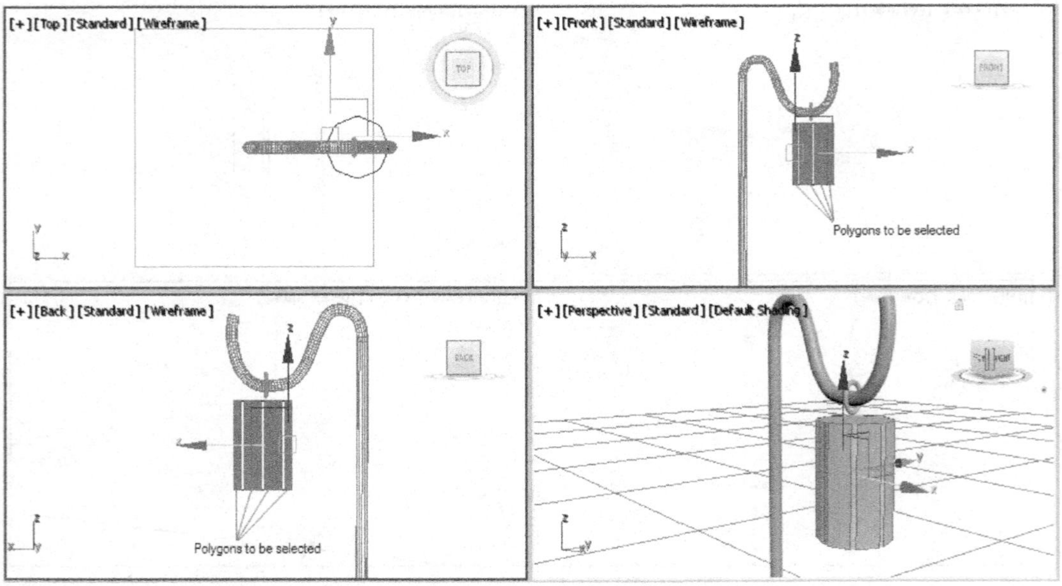

Figure 5-33 *The selected polygons between chamfered edges*

10. Select the uppermost and lowermost polygons, as shown in Figure 5-35.

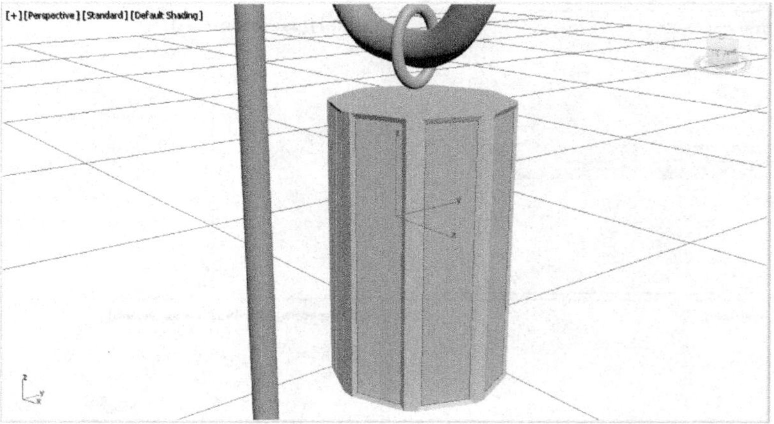

Figure 5-34 The selected polygons beveled

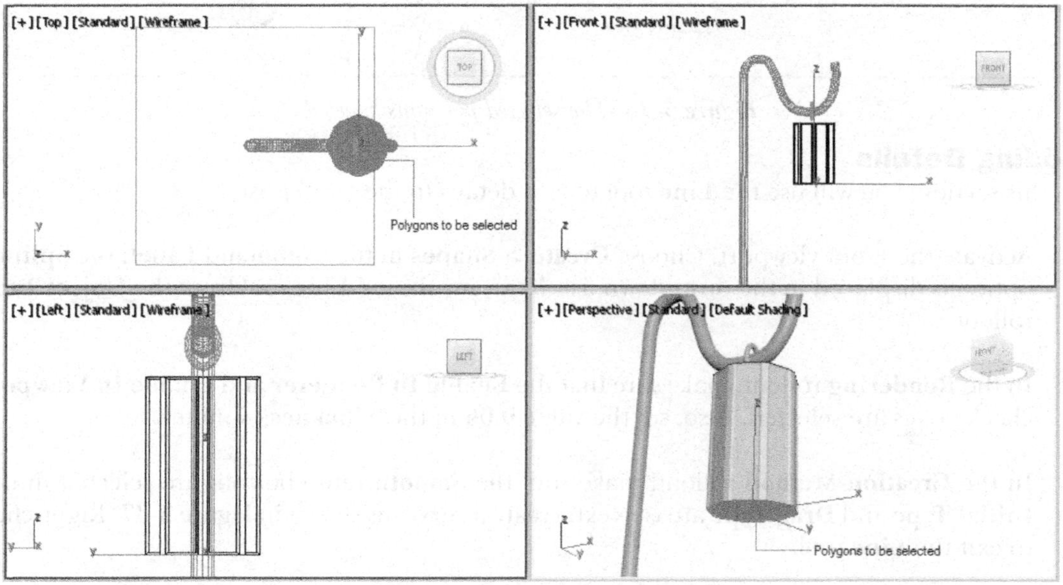

Figure 5-35 The selected polygons

11. In the **Edit Polygons** rollout, choose the **Settings** button on the right of the **Bevel** button; the **Bevel** caddy control is displayed in the viewport. Set the value **0.13** in the **Bevel-Height** spinner and **-0.08** in the **Bevel-Outline** spinner and then choose the **OK** button to close the caddy control; the selected polygons are beveled, as shown in Figure 5-36.

12. Select the **Polygon** sub-object level again in the **Selection** rollout to deactivate it.

13. Invoke the **Select and Move** tool from the **Main Toolbar** and align *lamp holder*, refer to Figure 5-36.

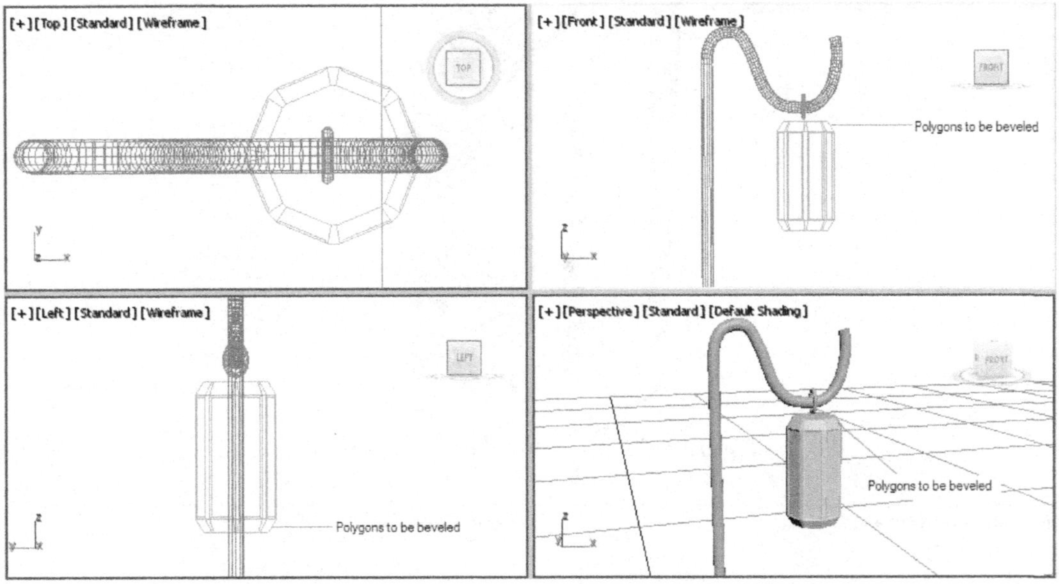

Figure 5-36 The selected polygons beveled

Adding Details

In this section, you will use the **Line** tool to add details to the lamp post.

1. Activate the Front viewport. Choose **Create > Shapes** in the **Command Panel**; the **Splines** option is displayed in the drop-down list. Next, invoke the **Line** tool from the **Object Type** rollout.

2. In the **Rendering** rollout, make sure that the **Enable In Renderer** and **Enable In Viewport** check boxes are selected. Also, set the value **0.06** in the **Thickness** spinner.

3. In the **Creation Method** rollout, make sure the **Smooth** radio buttons are selected in the **Initial Type** and **Drag Type** areas. Next, create a curve, as shown in Figure 5-37. Right-click to exit the **Line** tool.

4. Repeat steps 1, 2, and 3 and create a second curve, as shown in Figure 5-38.

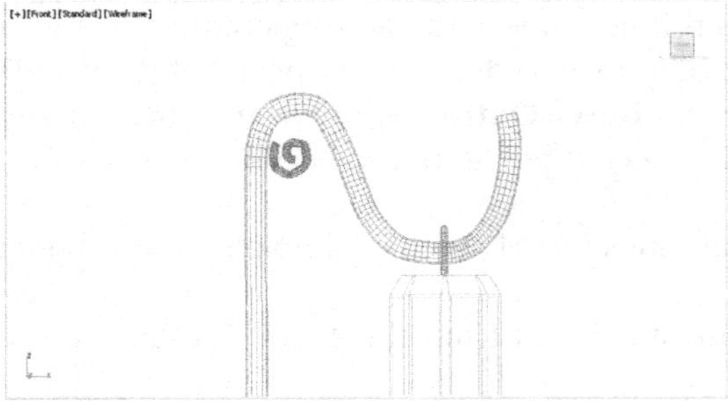

Figure 5-37 The curve created

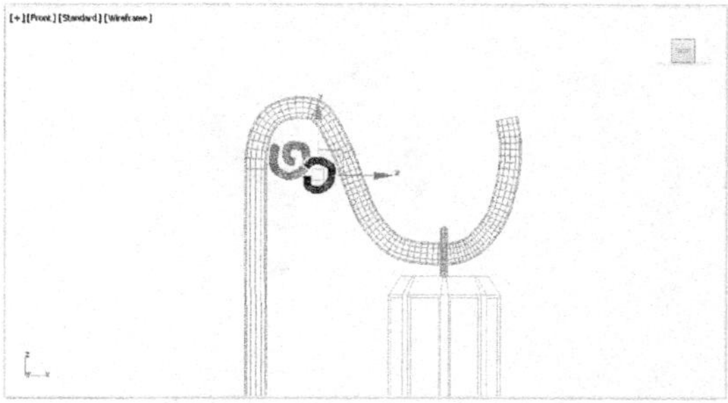

Figure 5-38 The second curve created

Next, you will create two joints between the curves and *pillar*.

5. Make sure the **Line** tool is activated. In the **Rendering** rollout, make sure the **Enable in Renderer** and **Enable In Viewport** check boxes are selected. Also, set the value **0.03** in the **Thickness** spinner. Next, create two joints, as shown in Figure 5-39.

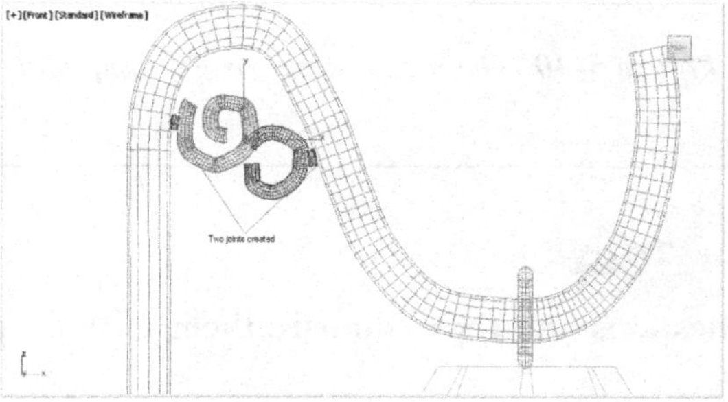

Figure 5-39 Two joints created

Saving and Rendering the Scene

In this section, you will save the scene that you have created and then render it. You can also view the final rendered image of this model by downloading the *c05_3dsmax_2022_rndr.zip* file from *www.cadcim.com*. The path of the file is as follows: *Textbooks > Animation and Visual Effects > Autodesk 3ds Max > Autodesk 3ds Max 2022 for Beginners: A Tutorial Approach*

1. Change the background color of the scene to white, as discussed in Tutorial 1 of Chapter 2.

2. Choose **Save** from the **File** menu.

3. Activate the Perspective viewport. Next, invoke the **Render Production** tool from the **Main Toolbar**; the rendered image of a lamp post is displayed, refer to Figure 5-40.

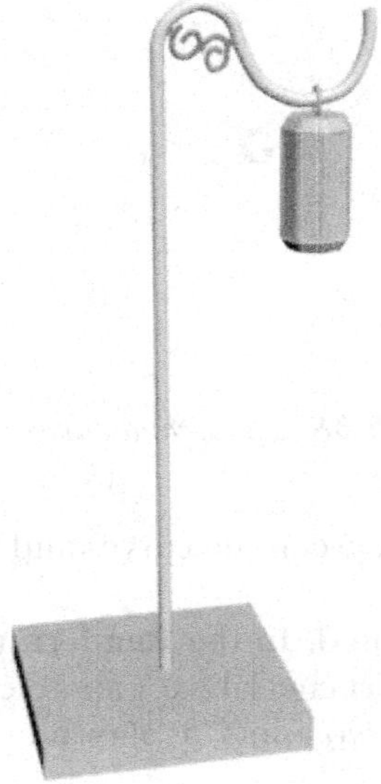

Figure 5-40 The rendered image of a lamp post

Self-Evaluation Test

Answer the following questions and then compare them to those given at the end of this chapter:

1. Which of the following tools is used to render a scene?

   (a) **Rendered Frame Window** (b) **Render Production**
   (c) **Render Setup** (d) None of these

2. You need to choose __________ from the **Edit** menu to invoke the **Clone Options** dialog box.

3. You can use the __________ tool to switch between the single and four-viewports configuration.

4. The __________ button is used to reverse the latest changes made.

5. When you choose **Bezier** from the quad menu, the spline becomes a straight line at the selected vertex along with the display of a single Bezier handle. (T/F)

6. You need to increase the number of vertices in a curve to achieve smoothness. (T/F)

Review Questions

Answer the following questions:

1. Which of the following tools is used to arrange the objects, lights, and cameras in a viewport?

   (a) **Select and Uniform Scale** (b) **Select and Move**
   (c) **Select and Rotate** (d) All of these

2. The __________ button in the **Align** area of the **Parameters** rollout is used to properly align a shape created using the **Lathe** modifier.

3. The __________ tool enables you to rotate the selected objects around their collective geometric center.

4. The __________ modifier is used to spin a spline around its axis to form an object.

5. You can draw a square by invoking the **Rectangle** tool and then dragging the cursor while holding the CTRL key. (T/F)

EXERCISES

The rendered output of the model used in the following exercises can be accessed by downloading *c05_3dsmax_2022_exr.zip* from *www.cadcim.com.* The path of the file is as follows: *Textbooks > Animation and Visual Effects > Autodesk 3ds Max > Autodesk 3ds Max 2022 for Beginners: A Tutorial Approach*

Exercise 1

Create a cup and a saucer, as shown in Figure 5-41, by using the profile curves shown in Figure 5-42. **(Expected time: 25 min)**

Figure 5-41 *The cup and saucer*

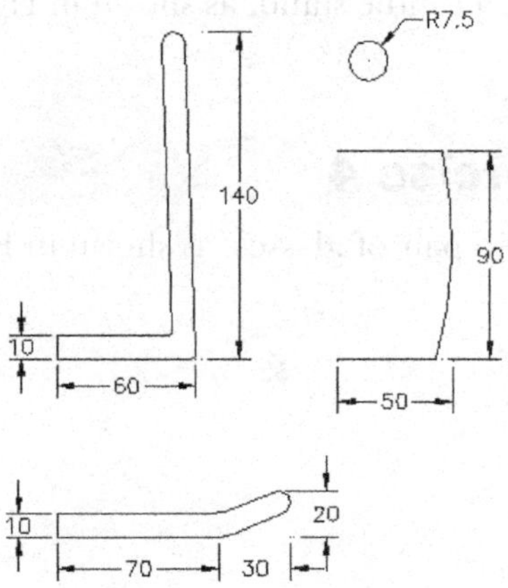

Figure 5-42 *Profile curves of the cup and saucer*

Exercise 2

Create a pair of cocktail glasses, as shown in Figure 5-43, by applying dimensions of your choice.
(Expected time: 20 min)

Figure 5-43 *A pair of cocktail glasses*

Exercise 3

Create a candle stand, as shown in Figure 5-44, by applying dimensions of your choice.
(Expected time: 20 min)

Exercise 4

Create a pair of glasses, as shown in Figure 5-45, by applying dimensions of your choice.
(Expected time: 20 min)

Figure 5-44 *A candle stand*

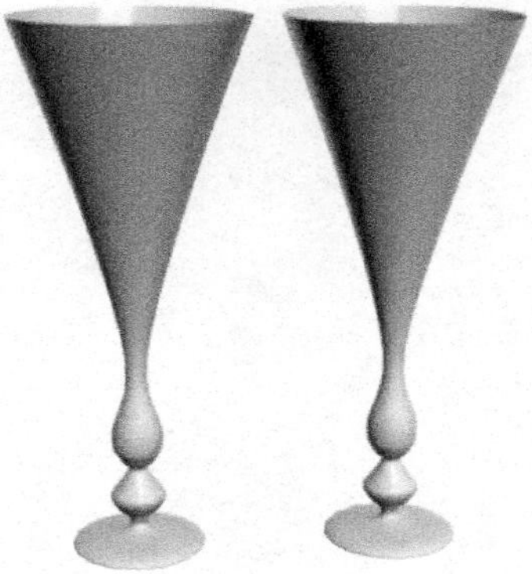

Figure 5-45 *A pair of glasses*

Exercise 5

Create a pair of wine glasses, as shown in Figure 5-46, by applying dimensions of your choice.

(Expected time: 20 min)

Figure 5-46 *A pair of wine glasses*

Answers to Self-Evaluation Test

1. b, **2. Clone**, **3. Maximize Viewport Toggle**, **4. Undo**, **5.** F, **6.** T

Chapter 6

Lofting, Twisting, and Deforming Objects

Learning Objectives

After completing this chapter, you will be able to:

- *Loft spline objects along a path*
- *Change the skin display*
- *Change the path levels*
- *Add different shapes to the loft object*
- *Remove twists from twisted loft objects*
- *Use deformation tools on a loft object*

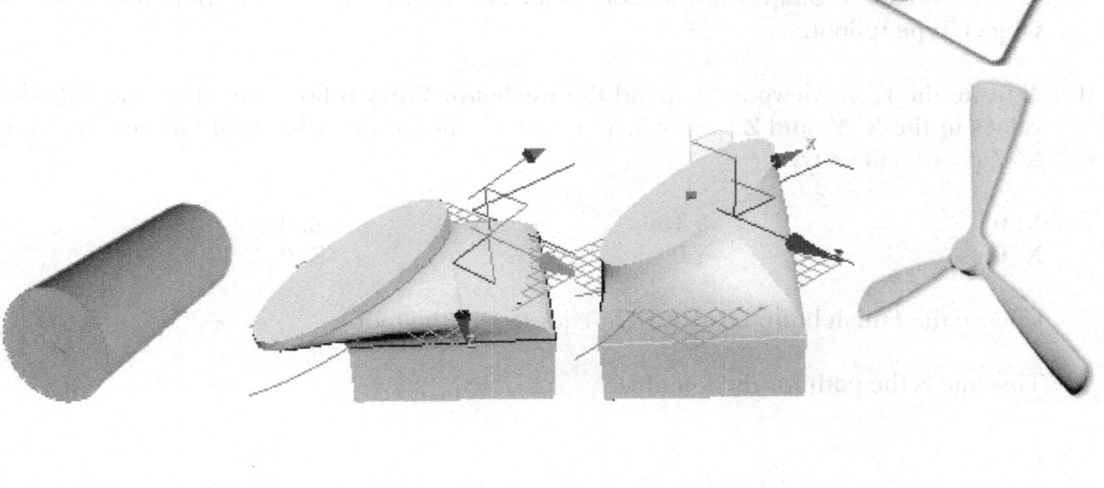

INTRODUCTION

Lofting is a technique which is used to create a 3D object by repeating a shape along a path. In this chapter, you will learn to loft splines, twist the lofted objects, and use various deformation tools to modify them.

TUTORIALS

Tutorial 1

In this tutorial, you will create an ellipse and then loft it along a path to create a 3D object. You will also change various skin parameters to get desired shapes. **(Expected time: 20 min)**

The following steps are required to complete this tutorial:

a. Create the project folder.
b. Loft spline objects.
c. Change the shape of the lofted object.
d. Cap loft objects.

Creating the Project Folder

Create the project folder with the name *c06_tut1* in the *3dsmax 2022* folder, as discussed in Tutorial 1 of Chapter 2.

Lofting Spline Objects

In this section, you will create a line and an ellipse and then loft the line using the **Loft** tool.

1. Invoke the **Zoom** tool from the viewport navigation controls. Zoom in on the Top, Front, and Left viewports to display the subgrid, if it is not already displayed.

2. Invoke the **2D Snap** tool from the **Snaps Toggle** flyout in the **Main Toolbar** to turn on snapping. Right-click on the **2D Snap** tool; the **Grid and Snap Settings** dialog box is displayed. In the **Snaps** tab, ensure that only the **Grid Points** check box is selected. Next, close the dialog box.

3. Choose **Create > Shapes** in the **Command Panel**. Next, invoke the **Line** tool from the **Object Type** rollout.

4. Activate the Front viewport. Expand the **Keyboard Entry** rollout and enter the following values in the **X**, **Y**, and **Z** spinners. You need to choose the **Add Point** button after every XYZ coordinate entry:

   X: **0** Y:**100** Z: **0**
   X: **0** Y:**0** Z: **0**

   Choose the **Finish** button. Next, right-click to exit the tool.

   This line is the path for the loft object.

5. Invoke the **Ellipse** tool from the **Object Type** rollout.

6. Activate the Top viewport. In the **Keyboard Entry** rollout, set **50** in the **Length** spinner and **100** in the **Width** spinner. Next, choose the **Create** button; an ellipse is created in all viewports. Right-click to exit the tool. Now, align the shapes, as shown in Figure 6-1 and invoke the **Zoom Extents All** tool from the viewport navigation controls to view the created shapes properly, refer to Figure 6-1.

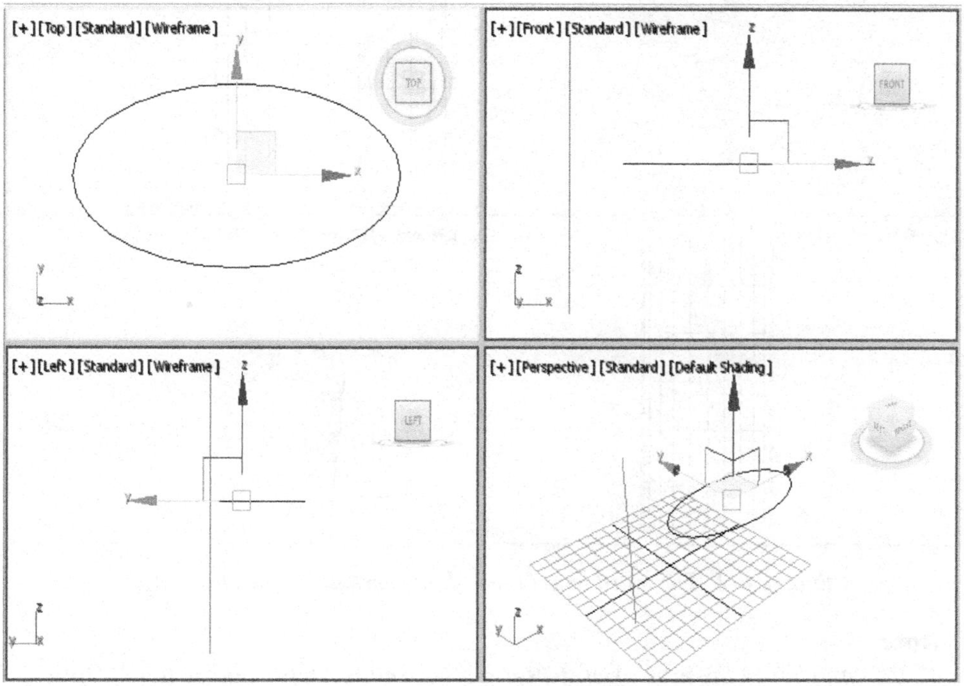

***Figure 6-1** A straight line and an ellipse*

The ellipse is the shape for the loft object.

7. Choose the **2D Snap** tool from the **Main Toolbar** to decativate it. Next, select the line in any viewport.

8. Choose **Create > Geometry** in the **Command Panel**. Next, select **Compound Objects** from the drop-down list below the **Geometry** button. Next, invoke the **Loft** tool from the **Object Type** rollout.

9. In the **Creation Method** rollout, make sure the **Instance** radio button is selected. Then, choose the **Get Shape** button.

10. Select the ellipse in any viewport. Next, right-click to exit the creation of the loft.

The ellipse (shape) is placed along the line (path) at its first vertex. The first vertex of the path is indicated by a small yellow cross mark.

11. With the lofted object selected, invoke the **Zoom Extents All Selected** tool from the viewport navigation controls; the lofted object is zoomed to its extents in all viewports, as shown in Figure 6-2.

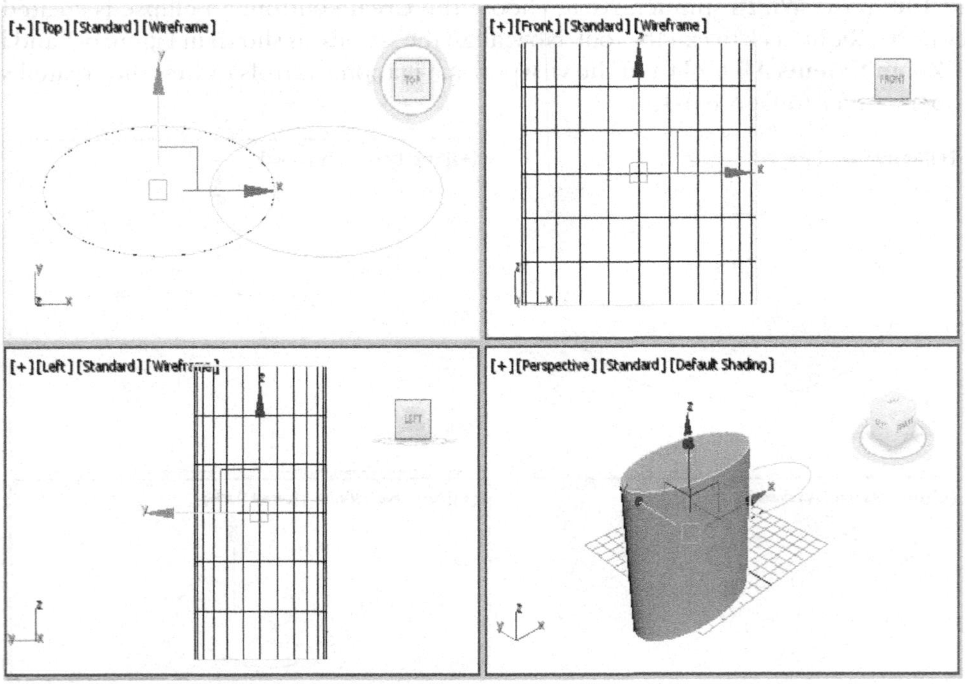

***Figure 6-2** The lofted object zoomed to its extents in all viewports*

Note
*1. You can change the number of steps using the **Shape Steps** spinner in the **Options** area of the **Skin Parameters** rollout to increase or decrease the smoothness of the loft object around (not along) the path.*

2. You can add different shapes at different levels of the path to create a complex loft object. Use of multiple shapes on a path for creating a loft object is explained in detail in Tutorial 3.

Changing the Shape of the Lofted Object

In this section, you will change the shape of the lofted object by using different shapes along its path.

1. Invoke the **2D Snap** tool from the **Snaps Toggle** flyout in the **Main Toolbar** to turn on the snapping mode.

2. Choose **Create > Shapes** in the **Command Panel**. Next, invoke the **Rectangle** tool from the **Object Type** rollout.

3. In the Top viewport, draw a rectangle that fits inside the original ellipse and then right-click to exit the tool.

4. Select the lofted object in any viewport and choose the **Modify** tab in the **Command Panel**.

5. In the **Path Parameters** rollout, select the **On** check box to activate the **Snap** spinner. Make sure the **Percentage** radio button is selected.

6. Set the value **50** in the **Path** spinner; the small yellow cross mark indicating the current level moves to the middle, or 50% of the entire path.

Note

*If the yellow cross mark is not at the center of the path when the **Path** spinner is set to **50**, the path may have been created with one or more bezier vertices. To rectify this, click on the plus sign next to **Loft** in the modifier stack and select the **Path** level from the sub-object hierarchy. Expand the **Line** level and select the **Vertex** sub-object mode. Right-click on the vertices of each of the paths and make sure that the **Corner** option is chosen in the quad menu displayed. When you are done, click on the **Loft** twice in the modifier stack to exit the sub-object mode and then repeat step* 7.

7. Choose the **Get Shape** button in the **Creation Method** rollout. Then, select the rectangle in any viewport. Choose the **Get Shape** button again to deactivate it; the rectangle is added as a shape at the 50% mark and the lofted object is updated, as shown in Figure 6-3.

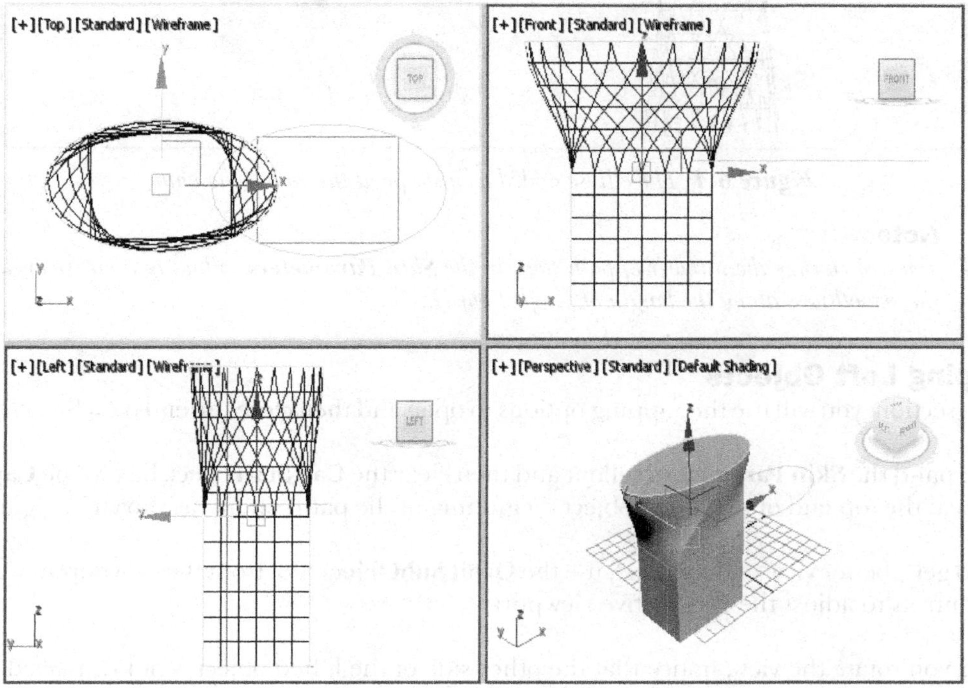

Figure 6-3 *A rectangular section created in the middle of the loft*

8. Now, set the value **100** in the **Path** spinner. In the **Creation Method** rollout, choose the **Get Shape** button. Then, select the original ellipse in any viewport; the ellipse is added as a shape at the end of the path, as shown in Figure 6-4.

The lofted object now starts as an ellipse, tapers to a rectangle, and then tapers back to an ellipse.

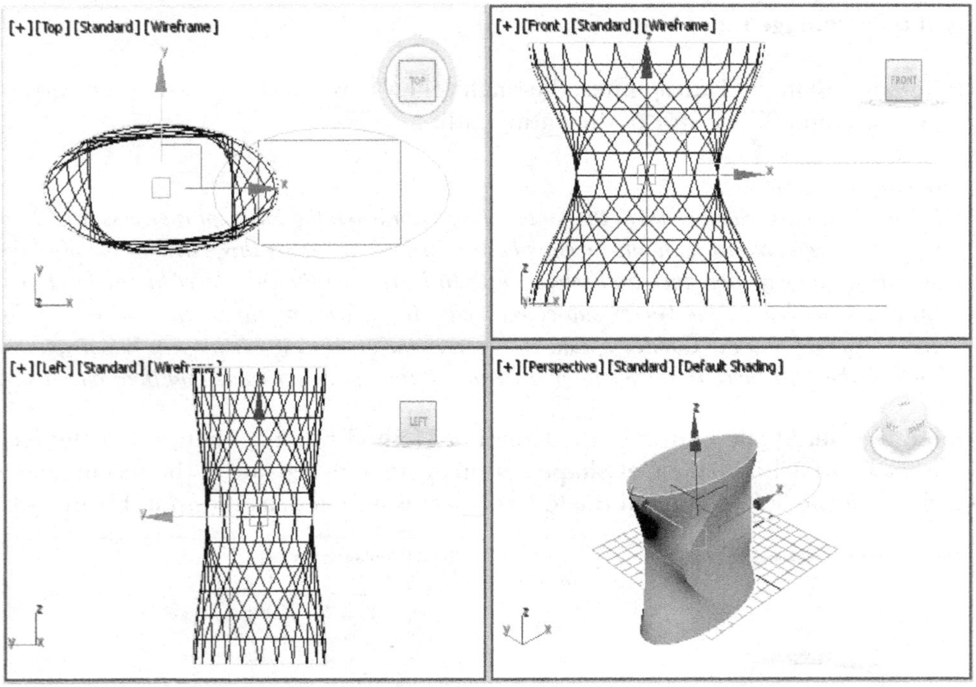

Figure 6-4 *The ellipse added as a shape at the end of the path*

Note

*You can change the number of path steps in the **Skin Parameters** rollout to increase or decrease the smoothness along the length of a lofted object.*

Capping Loft Objects

In this section, you will use the capping options to open and then close the ends of a lofted object.

1. Expand the **Skin Parameters** rollout and then clear the **Cap Start** check box in the **Capping** area; the top end of the lofted object (beginning of the path) opens, as shown in Figure 6-5.

2. To get a better view of the model, use the **Orbit SubObject** tool from the viewport navigation controls to adjust the Perspective viewport.

   As you rotate the view, notice that the other side of the lofted object is not displayed in the shaded viewports. This is because the skin is one-sided. A two-sided material can be applied to see the other side of the lofted object.

3. Clear the **Cap End** check box from the **Capping** area in the **Skin Parameters** rollout; the bottom end of the loft object opens and a hollow cylinder is created because the skin is generated on an open path.

4. Choose **Save** from the **File** menu to save the scene.

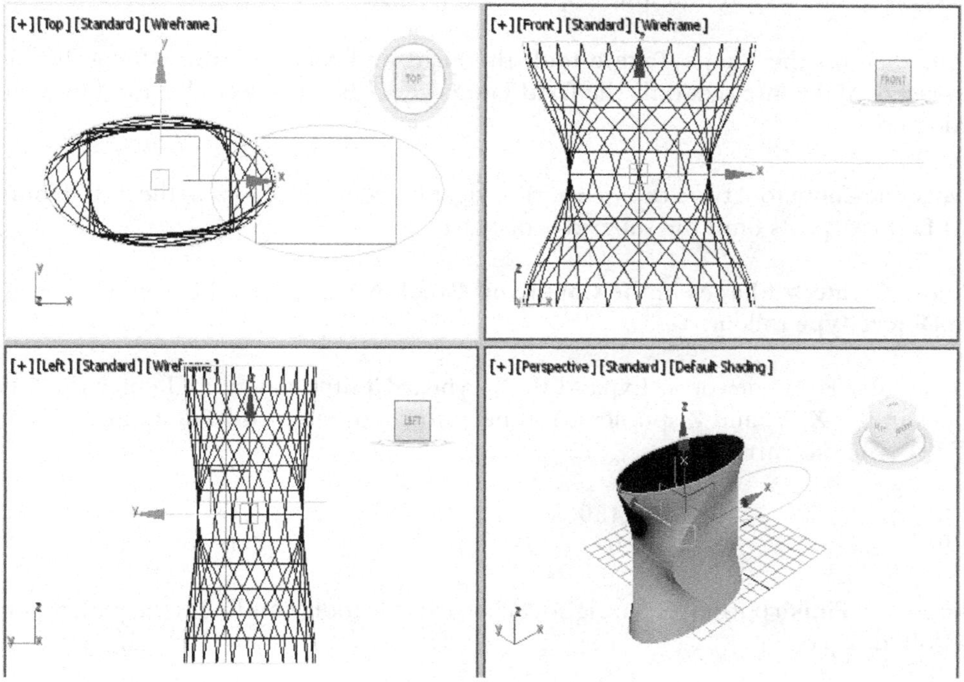

Figure 6-5 The top end of the loft object opened

Tutorial 2

In this tutorial, you will create a lofted object using the **Loft** tool and position the shape at different levels in a path. You will also align the shape vertices to remove the twist from the loft. In addition, you will move the shapes to different levels on the path and then make the path curved. **(Expected time: 25 min)**

The following steps are required to complete this tutorial:

a. Create the project folder.
b. Loft multiple objects.
c. Remove the twist.
d. Move and modify shapes.
e. Change the shape of the path.
f. Move shapes with respect to the path.

Creating the Project Folder

Create the project folder with the name *c06_tut2* in the *3dsmax 2022* folder, as discussed in Tutorial 1 of Chapter 2.

Lofting Multiple Objects

In this section, you will loft multiple shapes along a path.

1. Right-click on the **Snaps Toggle** tool; the **Grid and Snap Settings** dialog box is displayed. Make sure that only the **Grid Points** check box is selected. Next, close the dialog box.

2. Invoke the **Zoom** tool from the viewport navigation controls. Zoom in the Top, Front, and Left viewports until the subgrid is displayed.

3. Choose **Create > Shapes** in the **Command Panel**. Next, invoke the **Line** tool from the **Object Type** rollout.

4. Activate the Front viewport. Expand the **Keyboard Entry** rollout and enter the following values in the **X**, **Y**, and **Z** spinners. You need to choose the **Add Point** button after every XYZ coordinate entry:

X: **0**	Y: **130**	Z: **0**
X: **0**	Y: **0**	Z: **0**

Choose the **Finish** button. Next, right-click to exit the tool. This line is the path for creating the lofted object.

5. Activate the Top viewport. Invoke the **Ellipse** tool from the **Object Type** rollout. In the **Keyboard Entry** rollout, set **60** in the **Length** spinner and **100** in the **Width** spinner. Next, choose the **Create** button; an ellipse is created in all viewports. Right-click to exit the tool.

6. Invoke the **Rectangle** tool from the **Object Type** rollout. In the **Keyboard Entry** rollout, set **130** in the **Length** and **Width** spinners. Next, choose the **Create** button; a rectangle is created in all viewports. Right-click to exit the tool. Now, align the shapes, as shown in Figure 6-6 and invoke the **Zoom Extents All** tool. The rectangle is the second shape to be lofted. The line, ellipse, and rectangle are created, refer to Figure 6-6.

7. Select the line in any viewport.

8. Choose **Create > Geometry** in the **Command Panel**. Select **Compound Objects** from the drop-down list below the **Geometry** button. Next, invoke the **Loft** tool from the **Object Type** rollout.

9. In the **Creation Method** rollout, make sure that the **Instance** radio button is selected. Next, choose the **Get Shape** button.

10. Select the ellipse in any viewport; an instance of the ellipse (shape) is placed at the first vertex of the line (path). The current level, indicated by a small yellow cross, is positioned at the first vertex of the path. Since the line was drawn from the top to the bottom, the first vertex is at the top of the viewport. Next, right-click to exit the tool. The lofted object is shown in Figure 6-7.

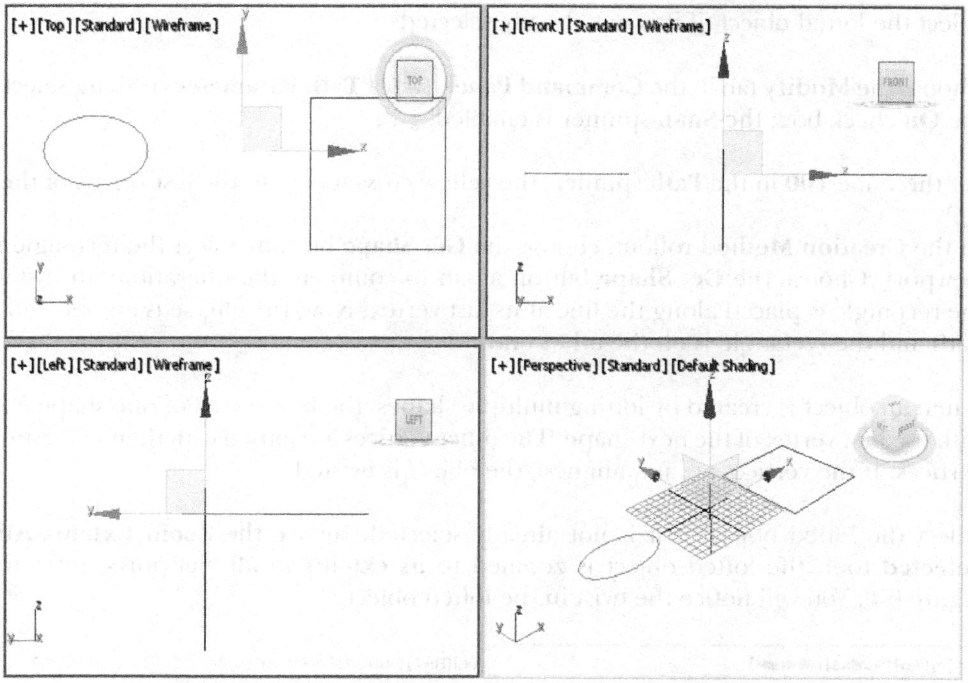

Figure 6-6 *The ellipse, rectangle, and line created*

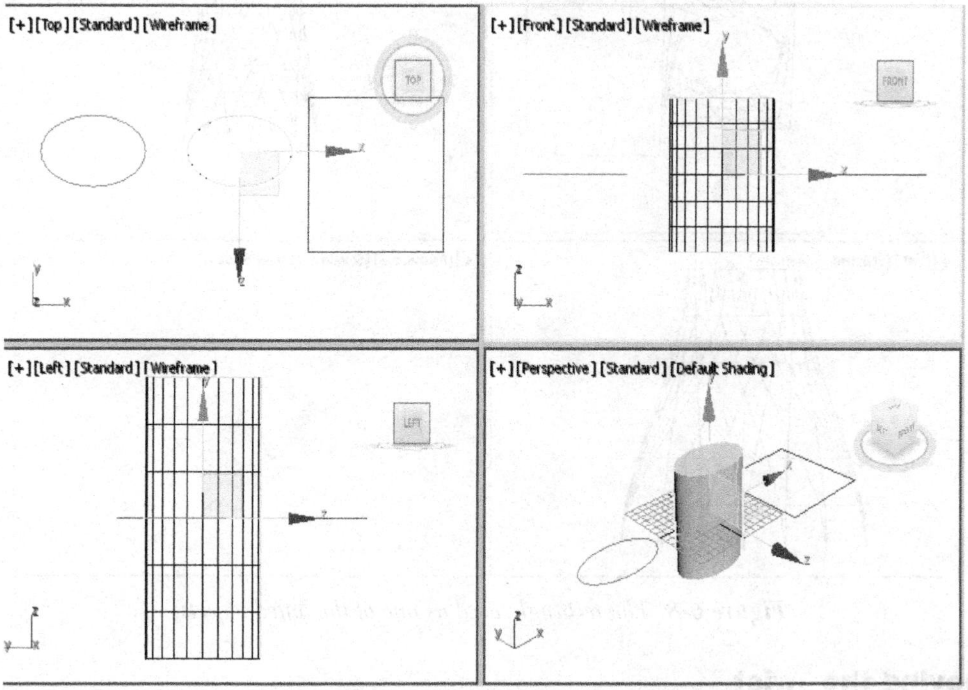

Figure 6-7 *The lofted object and the shapes zoomed to their extents in all viewports*

11. Select the lofted object, if it is not already selected.

12. Choose the **Modify** tab in the **Command Panel**. In the **Path Parameters** rollout, select the **On** check box; the **Snap** spinner is enabled. 

13. Set the value **100** in the **Path** spinner; the yellow cross moves to the last vertex of the path.

14. In the **Creation Method** rollout, choose the **Get Shape** button. Select the rectangle in any viewport. Choose the **Get Shape** button again to complete the operation; an instance of the rectangle is placed along the line at its last vertex. Now, the ellipse is on one end of the path and the rectangle is on the other end.

    When an object is created by lofting multiple shapes, the first vertex of one shape is joined with the first vertex of the next shape. The other vertices are joined with their corresponding vertices. If the vertices are not aligned, the object is twisted.

15. Select the lofted object, if it is not already selected. Invoke the **Zoom Extents All Selected** tool; the lofted object is zoomed to its extents in all viewports, refer to Figure 6-8. You will notice the twist in the lofted object. 

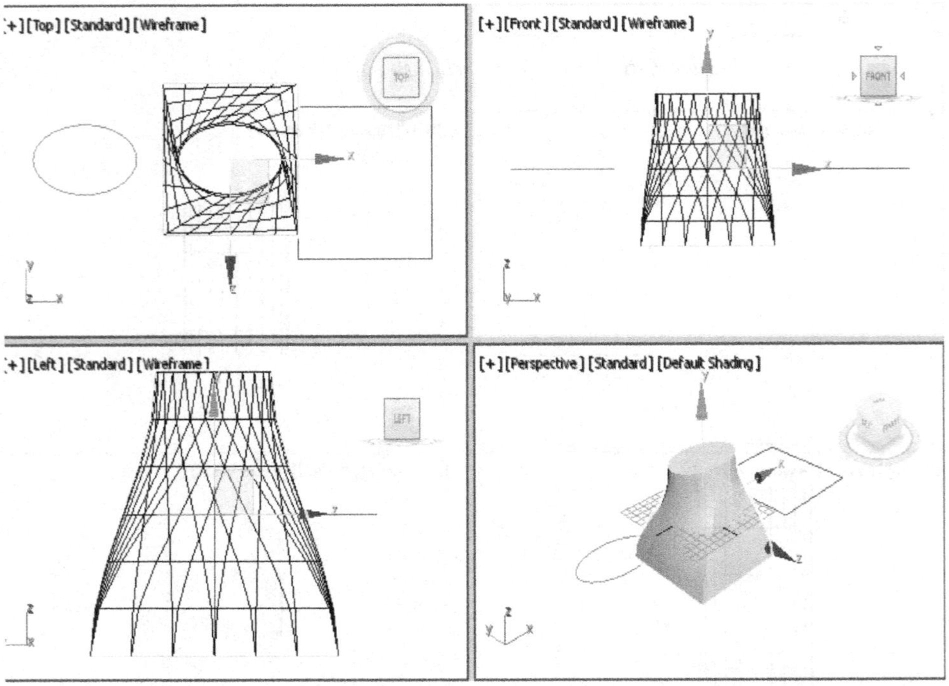

***Figure 6-8** The rectangle used as one of the lofted objects*

Removing the Twist

In this section, you will remove the twist in the lofted object. To avoid twisting of lofted object, the vertices of the shapes should be aligned. This can be done by using the **Compare** button in the **Shape Commands** rollout. This rollout is available when the **Shape** sub-object mode is active for the lofted object.

1. Make sure the lofted object is selected. Next, choose **Loft** in the modifier stack from the **Modify** tab in the **Command Panel**. The modifier stack appears in the window below the **Modifier List** drop-down list.

2. Click on the arrow next to **Loft** to display the sub-object hierarchy. Make sure the **Shape** sub-object level is selected, as shown in Figure 6-9; the **Shape** level is displayed in blue indicating that it is the current sub-object mode. Also, the **Shape Commands** rollout is displayed.

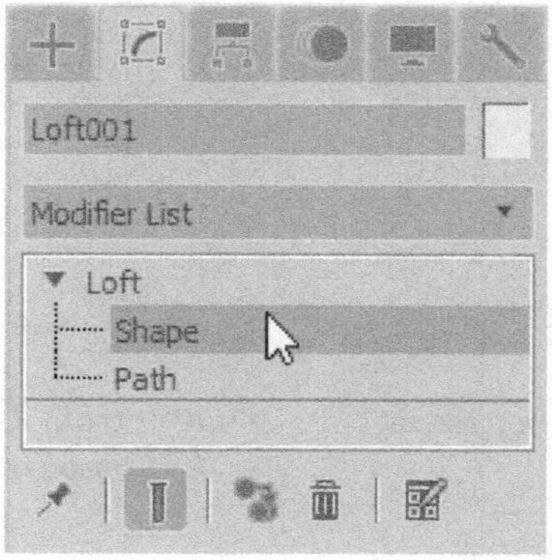

***Figure 6-9** The **Shape** sub-object level selected in the modifier stack*

3. Choose the **Compare** button in the **Shape Commands** rollout; the **Compare** window is displayed, as shown in Figure 6-10.

4. Choose the **Pick Shape** button in the **Compare** window. Next, move the cursor over the ellipse in the lofted object; the shape of the cursor changes. Now, click to select the ellipse in any viewport. 

   The ellipse is displayed in the **Compare** window. The first vertex of the ellipse is displayed as a small square.

5. With the **Pick Shape** button chosen, move the cursor over the rectangle in the lofted object and select it; the rectangle with its first vertex is displayed in the **Compare** window. You can use the **Zoom Extents** tool located at the bottom of the **Compare** window to display both shapes in the window, if needed, as shown in Figure 6-11. 

6. Choose the **Pick Shape** button to deactivate it. Activate the Front viewport and select the rectangle in the lofted object. Next, right-click on the **Select and Rotate** tool from the **Main Toolbar**; the **Rotate Transform Type-In** dialog box is displayed. In this dialog box, set the value **-45** in the **Z** spinner of the **Offset: Local** area and then close the dialog box; the vertices of both the shapes are aligned in the **Compare** window, as shown in Figure 6-12.

7. Close the **Compare** window; the lofted object with aligned sides is displayed in all viewports and the twist is removed from the lofted object, as shown in Figure 6-13.

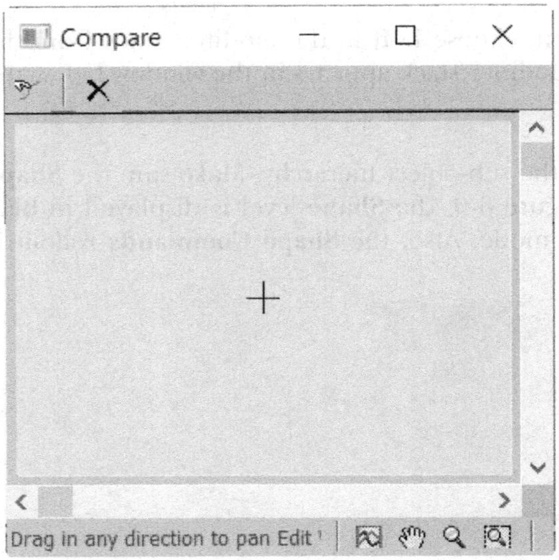

***Figure 6-10** The **Compare** window*

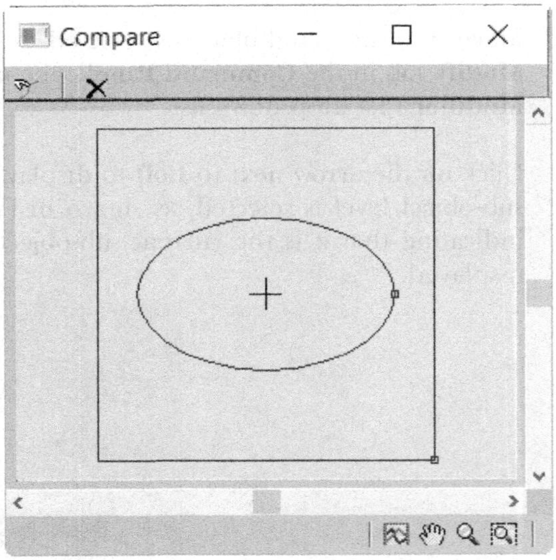

***Figure 6-11** The ellipse and rectangle displayed in the **Compare** window*

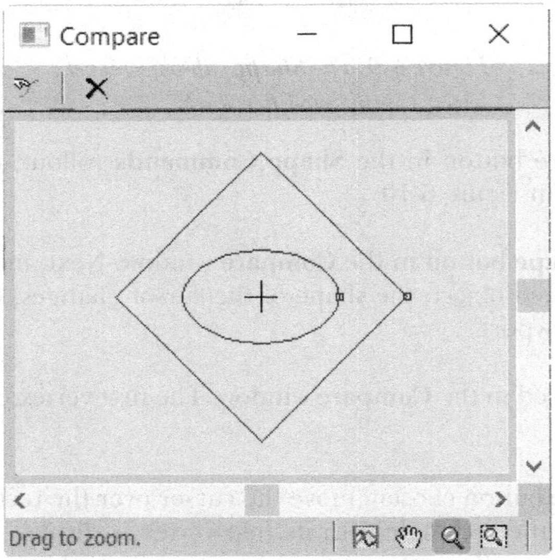

***Figure 6-12** The rectangle rotated to align first vertices of both shapes*

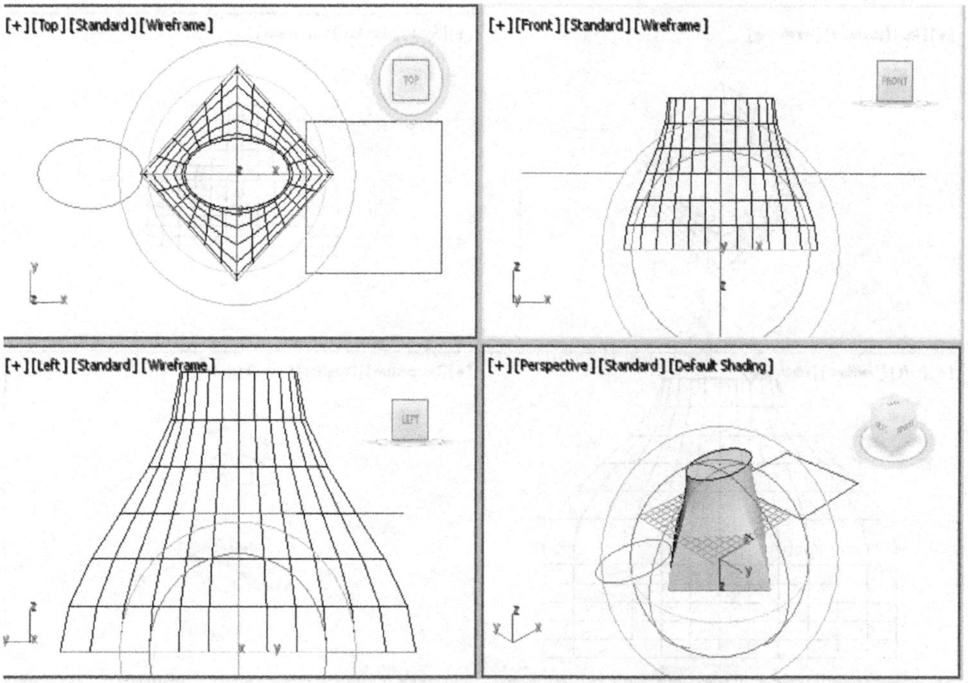

***Figure 6-13** The twist removed from the lofted object by rotating the rectangle shape*

Moving and Modifying Shapes

In this section, you will modify the shape of the lofted object by moving and modifying the shapes along the path.

1. Invoke the **Select Object** tool and select the rectangle in the lofted object, if it is not already selected; the **Rectangle** level appears in the modifier stack below the **Loft** level and its sub-object levels. 

2. In the **Shape Commands** rollout, the **Path Level** spinner displays 100, indicating the level on which the rectangle is placed. Set **60** in this spinner and press the ENTER key; the rectangle moves to the 60% location on the path. The lofted object starts as an ellipse, gradually changes to a rectangle at 60%, and finishes as a rectangle, as shown in Figure 6-14.

3. Select the ellipse on the loft (at 0%). In the modifier stack, select the **Ellipse** level; the rollouts for the ellipse are displayed.

4. In the **Parameters** rollout, set the value **150** in the **Width** spinner and press the ENTER key; the width of the ellipse is increased, as shown in Figure 6-15.

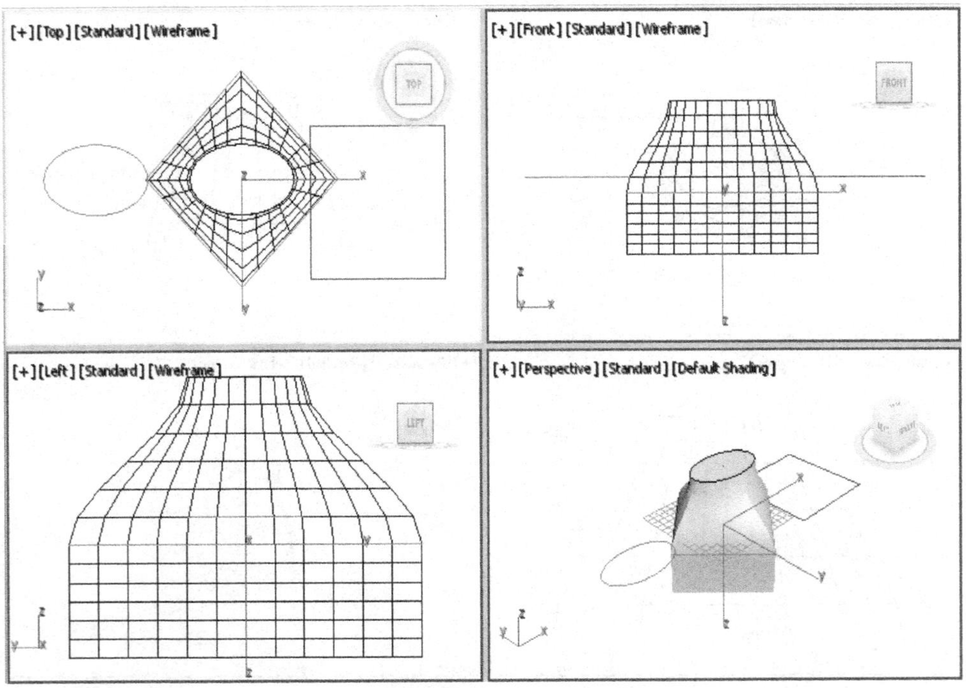

Figure 6-14 *The rectangle shape moved to the 60% position on the path*

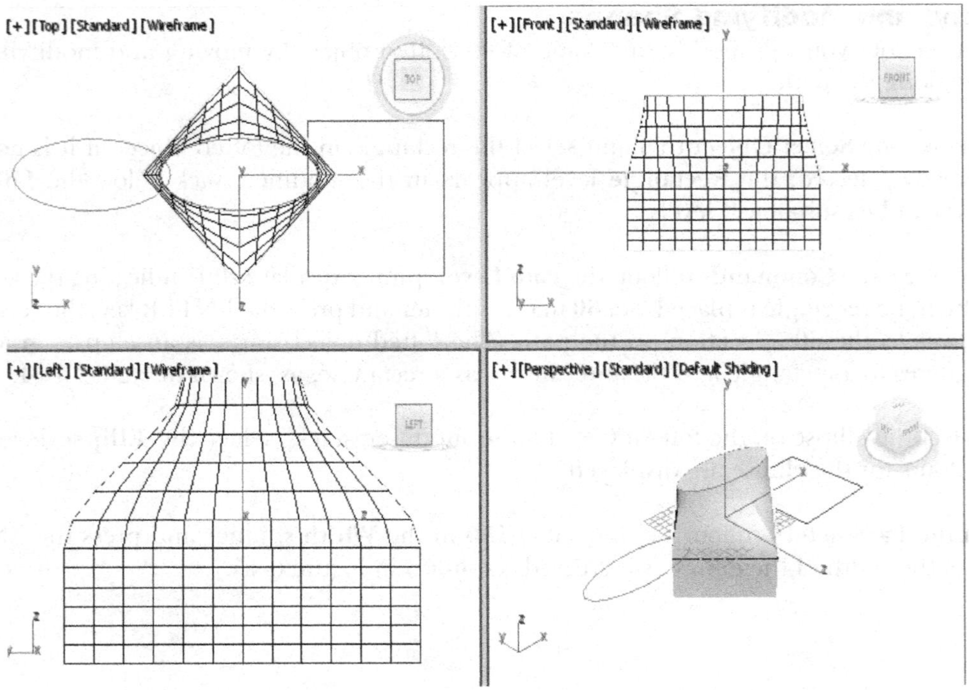

Figure 6-15 *Width of the ellipse increased*

Changing the Shape of the Path

In this section, you will modify the lofted object by changing the shape of the path along which it is lofted.

1. In the modifier stack, select the **Path** level to exit the **Shape** sub-object mode and enter into the **Path** sub-object mode; the **Line** level appears in the modifier stack below the **Path** level of the **Loft** sub-object hierarchy.

2. Select the **Line** level in the modifier stack; the rollouts for a line are displayed. In the **Selection** rollout, choose the **Vertex** button; the **Vertex** sub-object mode is selected for the line.

3. Invoke the **Select and Move** tool from the **Main Toolbar** and make sure the **Transform Gizmo XY Plane Constraint** button is chosen from the **Axis Constraints** toolbar.

4. In the Front viewport, select the first vertex (upper vertex) and move it about 30 units to the left and 20 units down, as shown in Figure 6-16; the lofted object tilts at an angle.

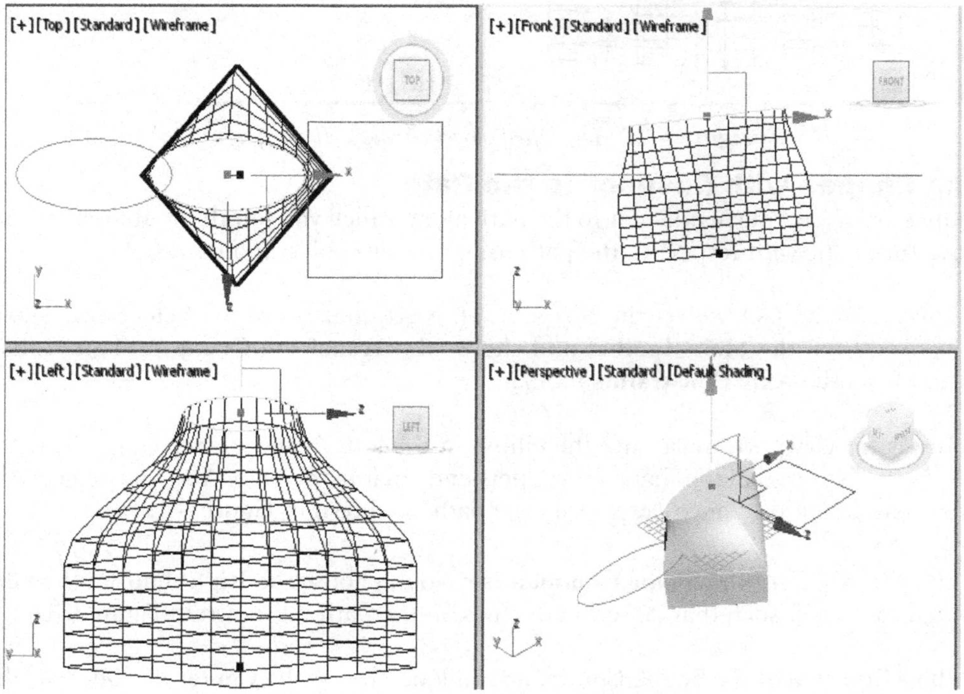

***Figure 6-16** The top vertex of the path moved*

5. Invoke the **Zoom Extents All Selected** tool. Now, in the Front viewport, right-click on the same vertex; the quad menu is displayed. Choose **Bezier** from the upper-left quadrant; the bezier handle is displayed. Since this is an end vertex, there is only one handle.

6. Using the **Select and Move** tool from the **Main Toolbar**, move the bezier handle to the right so that the path is slightly curved; the lofted object curves with the path, as shown in Figure 6-17.

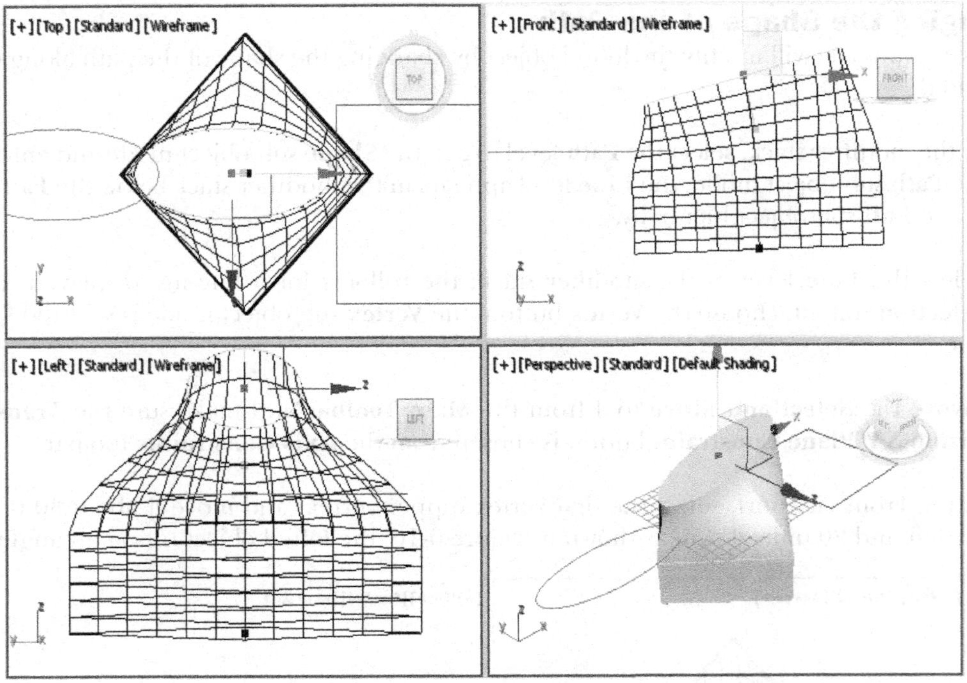

Figure 6-17 The lofted object curves with the path

Moving Shapes with Respect to the Path

The shapes are always perpendicular to the path along which you loft them. In this section, you will move the shape with respect to the path using the sub-object mode.

1. In the modifier stack, select the **Shape** level. Next, make sure the **Select and Move** tool is invoked from the **Main Toolbar** and choose the **Transform Gizmo Z Plane Constraint** button from the **Axis Constraints** toolbar.

2. In the Front viewport, make sure the ellipse in the lofted object is selected. Next, move it downward. Notice that the shape moves perpendicular to the path. Also, the object is slightly skewed as the ellipse moves away from the path, as shown in Figure 6-18.

3. In the **Align** area of the **Shape Commands** rollout, choose the **Right** button; the ellipse is shifted to the left such that its right edge lies on the path, as shown in Figure 6-19.

4. In the **Align** area of the **Shape Commands** rollout, choose the **Center** button; the ellipse is centered on the path.

5. Choose **Save** from the **File** menu to save the scene.

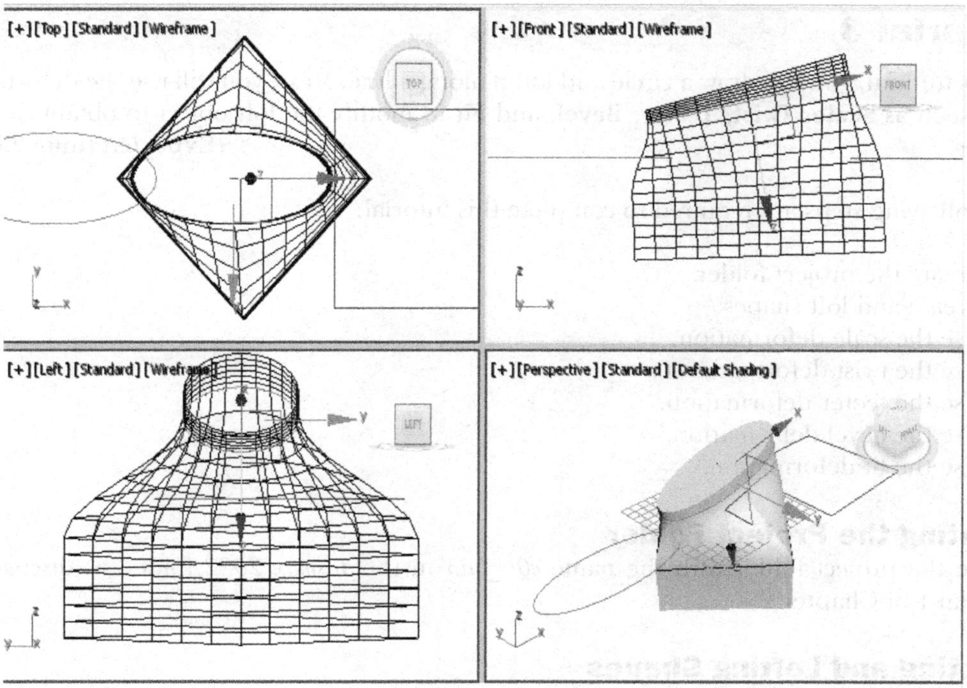

***Figure 6-18** The ellipse moved away from the path (to the right)*

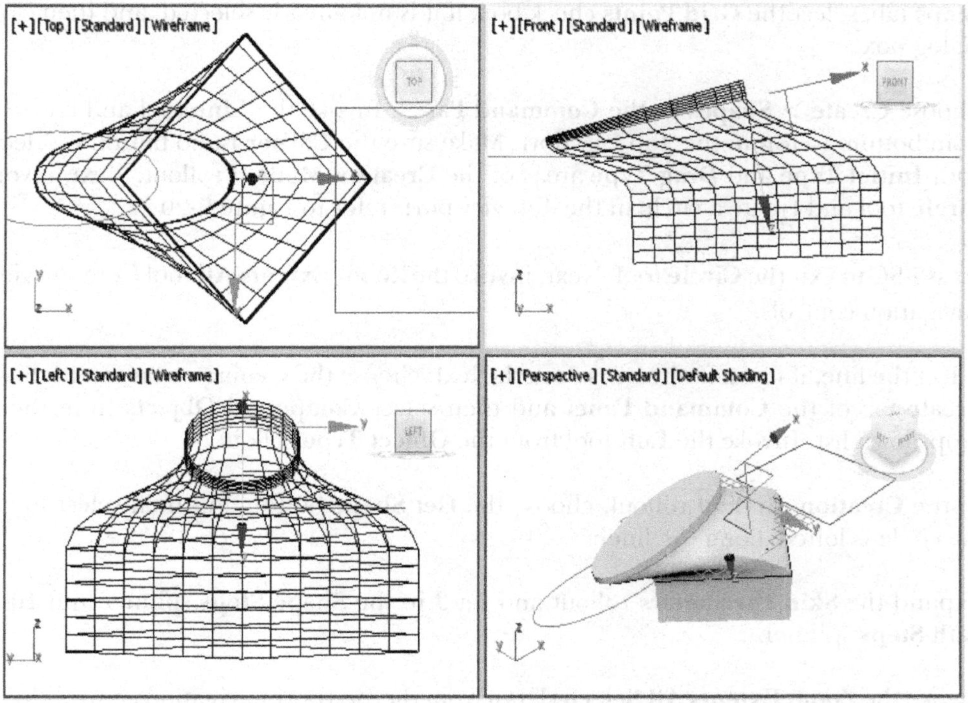

***Figure 6-19** The ellipse aligned on the right by using the **Right** button*

Tutorial 3

In this tutorial, you will draw a circle and loft it along a line. Next, you will use the deformation tools such as **Scale**, **Twist**, **Teeter**, **Bevel**, and **Fit** to modify the loft object to obtain different shapes. **(Expected time: 25 min)**

The following steps are required to complete this tutorial:

a. Create the project folder.
b. Create and loft shapes.
c. Use the scale deformation.
d. Use the twist deformation.
e. Use the teeter deformation.
f. Use the bevel deformation.
g. Use the fit deformation.

Creating the Project Folder

Create the project folder with the name *c06_tut3* in the *3dsmax 2022* folder, as discussed in Tutorial 1 of Chapter 2.

Creating and Lofting Shapes

1. Activate the **Snap Toggle** tool by right-clicking on the **3D Snap** tool from the **Snaps Toggle** flyout in the **Main Toolbar**; the **Grid and Snap Settings** dialog box is displayed. In the **Snaps** tab, select the **Grid Points** check box, if it is not already selected, and then close the dialog box.

2. Choose **Create > Shapes** in the **Command Panel**. Invoke the **Line** tool and create a line from bottom to top in the Top viewport. Make sure the **Corner** radio button is selected in both **Initial Type** and **Drag Type** areas of the **Creation Method** rollout. Next, invoke the **Circle** tool and create a circle in the Top viewport, refer to Figure 6-20.

3. Press ESC to exit the **Circle** tool. Next, invoke the **Zoom Extents All** tool from the viewport navigation controls.

4. Select the line, if it is not already selected. Next, choose the **Geometry** button in the **Create** tab of the **Command Panel** and then select **Compound Objects** from the drop-down list. Invoke the **Loft** tool from the **Object Type** rollout.

5. In the **Creation Method** rollout, choose the **Get Shape** button and then select the circle; the circle is lofted along the line.

6. Expand the **Skin Parameters** rollout and set **3** in the **Shape Steps** spinner and **10** in the **Path Steps** spinner.

7. Invoke the **Zoom Extents All Selected** tool from the viewport navigation controls to zoom the lofted object to its extents, as shown in Figure 6-21.

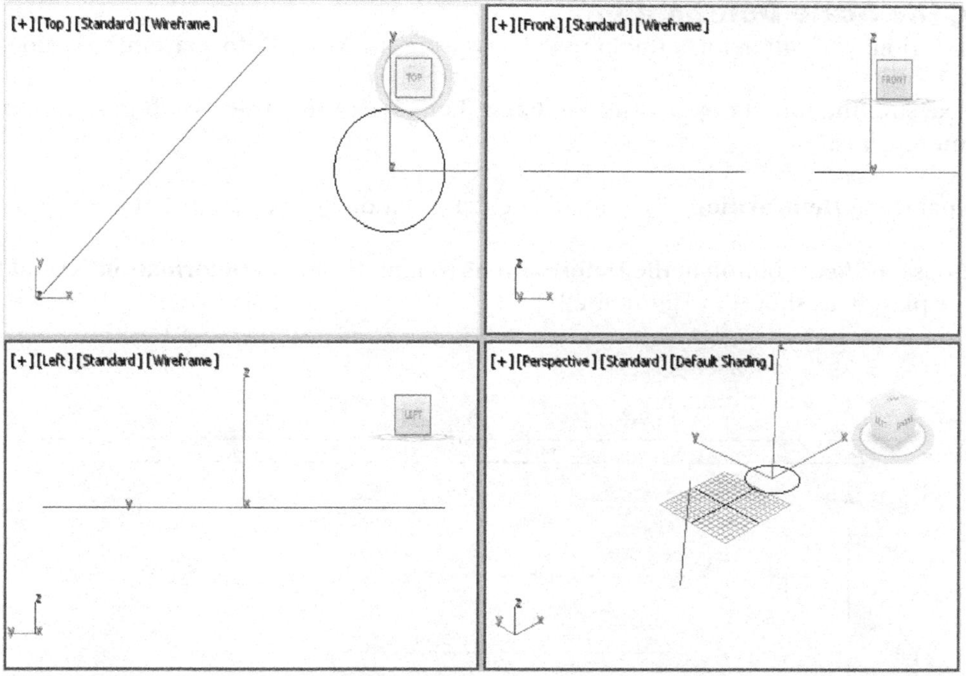

Figure 6-20 *A line and a circle created*

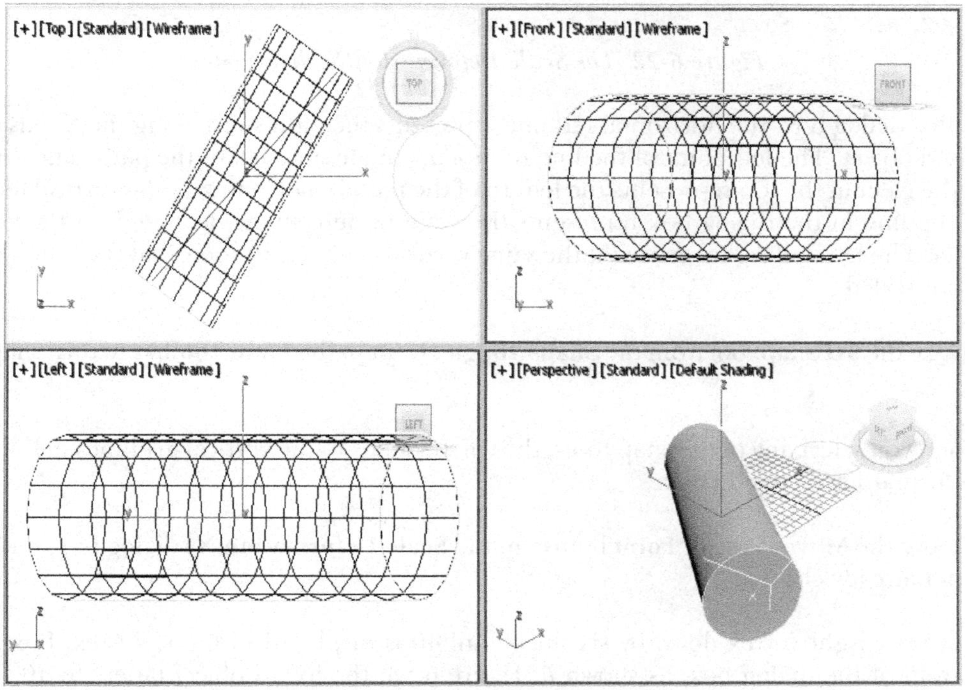

Figure 6-21 *The lofted object zoomed*

Using the Scale Deformation

In this section, you will deform the lofted object using the **Scale Deformation(X)** dialog box.

1. Make sure the lofted object is selected and then choose the **Modify** tab in the **Command Panel**.

2. Expand the **Deformations** rollout; all five deformation buttons are listed in this rollout.

3. Choose the **Scale** button in the **Deformations** rollout; the **Scale Deformation(X)** dialog box is displayed, as shown in Figure 6-22.

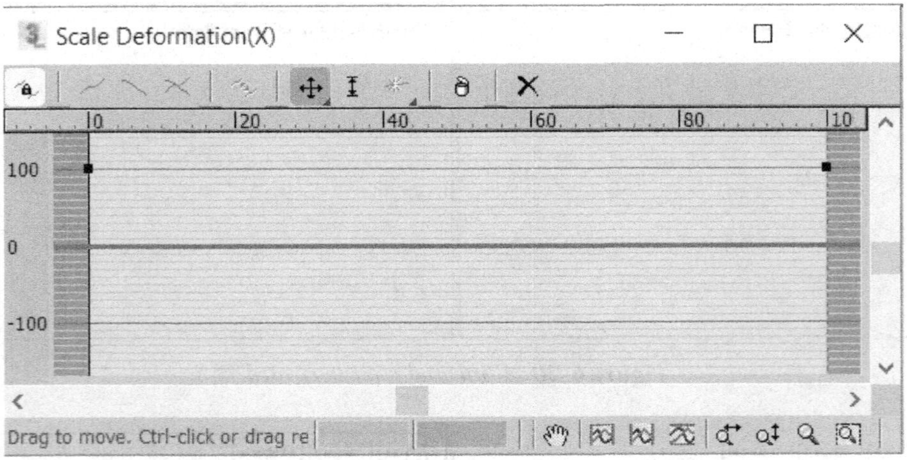

Figure 6-22 *The* ***Scale Deformation(X)*** *dialog box*

In this dialog box, the horizontal red line represents the path steps along the X axis of the lofted object. The left point of the line represents the first vertex of the path. The number in the ruler at the top represents the length of the path in percentage. The vertical location of the line in the dialog box represents the scale in percentage, as noted on the vertical ruler. The left and right vertices at the same vertical point (100) show that the lofted object is not scaled.

4. Invoke the **3D Snap** tool from the **Snaps Toggle** flyout in the **Main Toolbar** to turn snapping on.

When you select any of the snap tools, the cursor snaps to the vertical gridlines in the **Scale Deformation(X)** dialog box.

5. Choose the **Move Control Point** button in the **Scale Deformation(X)** dialog box, if it is not already chosen. 

6. Move the right vertex down by six lines until 40 is displayed in the right text box at the bottom of the dialog box, as shown in Figure 6-23; the lofted object tapers to 40% of its original size at the last vertex, refer to Figure 6-24.

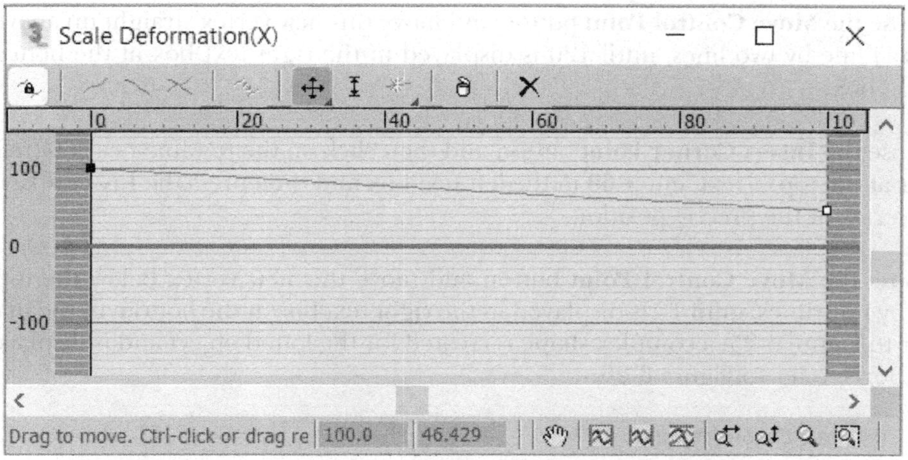

Figure 6-23 The vertex moved in the ***Scale Deformation(X)*** *dialog box*

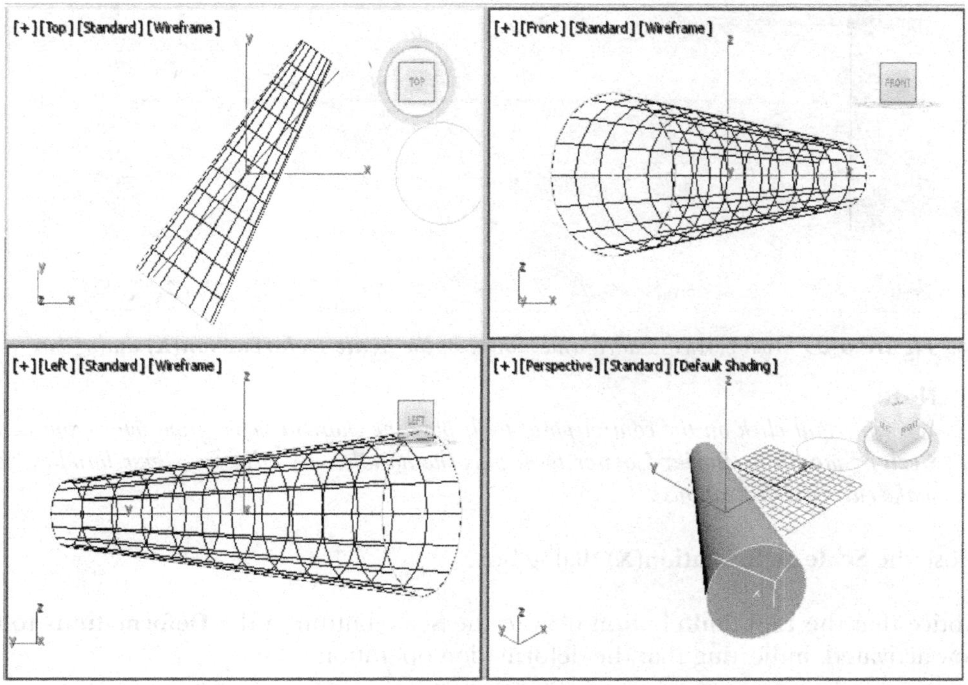

Figure 6-24 *The lofted object tapered after moving the vertex in the* ***Scale Deformation(X)*** *dialog box*

7. Choose the **Insert Corner Point** button in the **Scale Deformation(X)** dialog box and then click on the red line near 40 (refer to the ruler at the top); a vertex is added at this point. To move the vertex to the precise position, enter **40** in the left-hand text box at the bottom of the dialog box and then press the ENTER key. In this way, the corner vertex is placed at the precise position.

8. Choose the **Move Control Point** button and move this new vertex straight up above the top (100%) line by two lines, until 120 is displayed in the right text box at the bottom of the dialog box.

9. Choose the **Insert Corner Point** button and then click on the red line near 60 (refer to the ruler at the top). Next, enter **60** in the left text box and then press the ENTER key to place the vertex at the precise position.

10. Choose the **Move Control Point** button and move this new vertex below the top (100%) line by nine lines, until 10 is displayed in the right text box at the bottom of the dialog box, refer to Figure 6-25; a complex shape is created for the lofted object and is displayed in the viewports, refer to Figure 6-26.

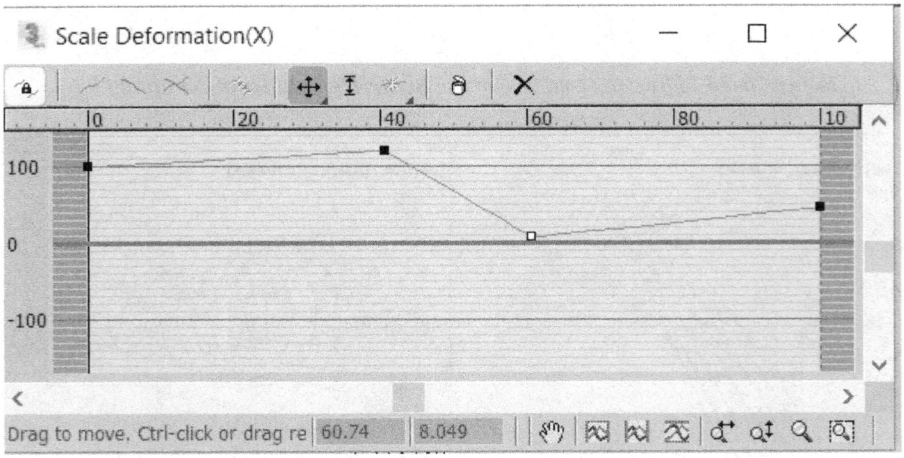

***Figure 6-25** Two vertices added and moved in the **Scale Deformation(X)** dialog box*

Note

*You can right-click on the control point to display the shortcut menu from which you can select **Bezier-Smooth** or **Bezier-Corner** to display the handles. By adjusting these handles, you can make curved deformations.*

11. Close the **Scale Deformation(X)** dialog box.

    Notice that the **Lightbulb** button next to the **Scale** button in the **Deformations** rollout is now activated, indicating that the deformation operation is on.

12. Choose the **Lightbulb** button next to the **Scale** button to turn the deformation off.

Using the Twist Deformation

In this section, you will deform the lofted object using the **Twist Deformation** dialog box.

1. In the **Deformations** rollout, choose the **Twist** button; the **Twist Deformation** dialog box is displayed, which is similar to the **Scale Deformation** dialog box, but with the red horizontal line at the 0 position. The horizontal line represents the degree of twist along the path axis.

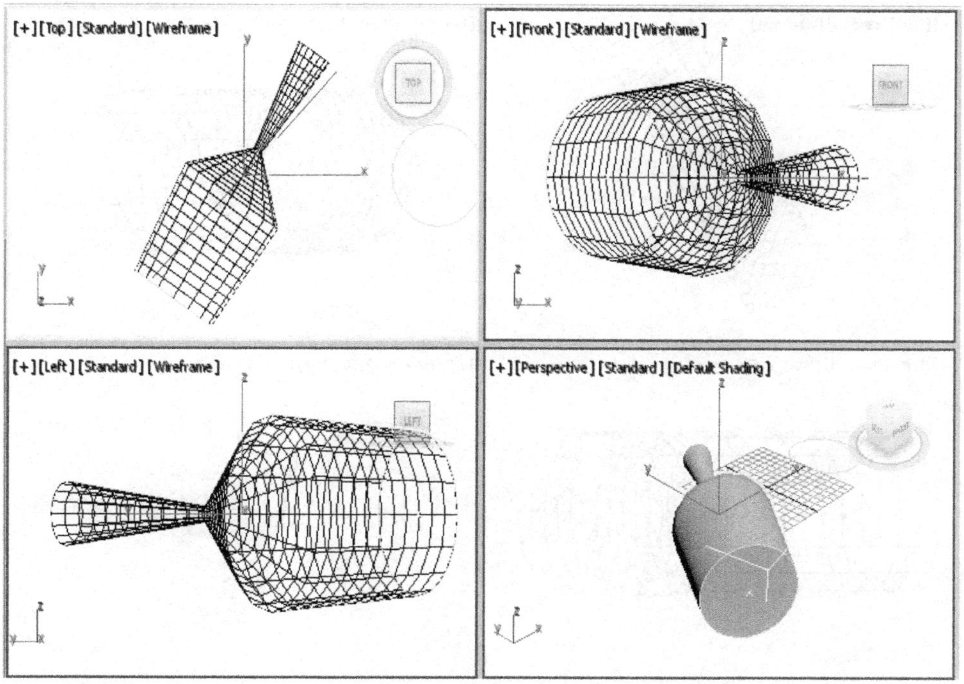

Figure 6-26 *The modified lofted object after adding and moving the vertices*

2. Choose the **Insert Corner Point** button in this dialog box and then click on the red line near 40 (refer to the ruler at the top). Next, enter **40** in the left text box and then press the ENTER key to insert the point at the precise position.

3. Choose the **Move Control Point** button and move the left vertex up to 200 degrees (200 is displayed in the right text box). If 200 is not visible in the graph, use the scroll bar to view the point, refer to Figure 6-27; the lofted object twists 200 degrees between the starting point of the path and a point at 40% of the path's overall length, refer to Figure 6-28.

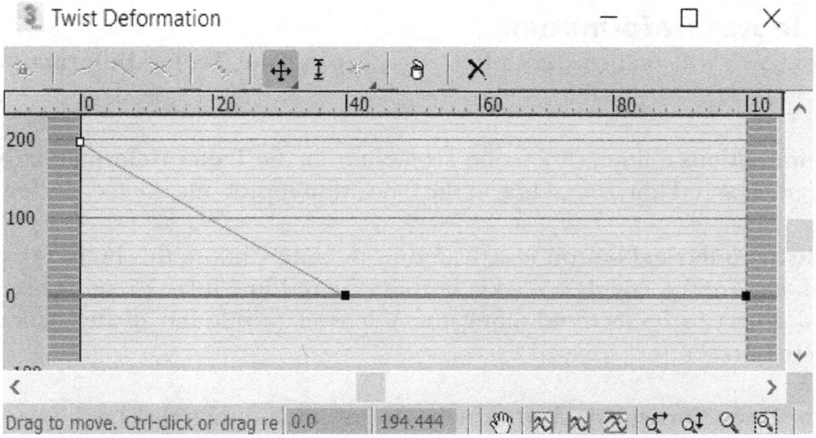

Figure 6-27 *Moving the vertex up to 200 degrees*

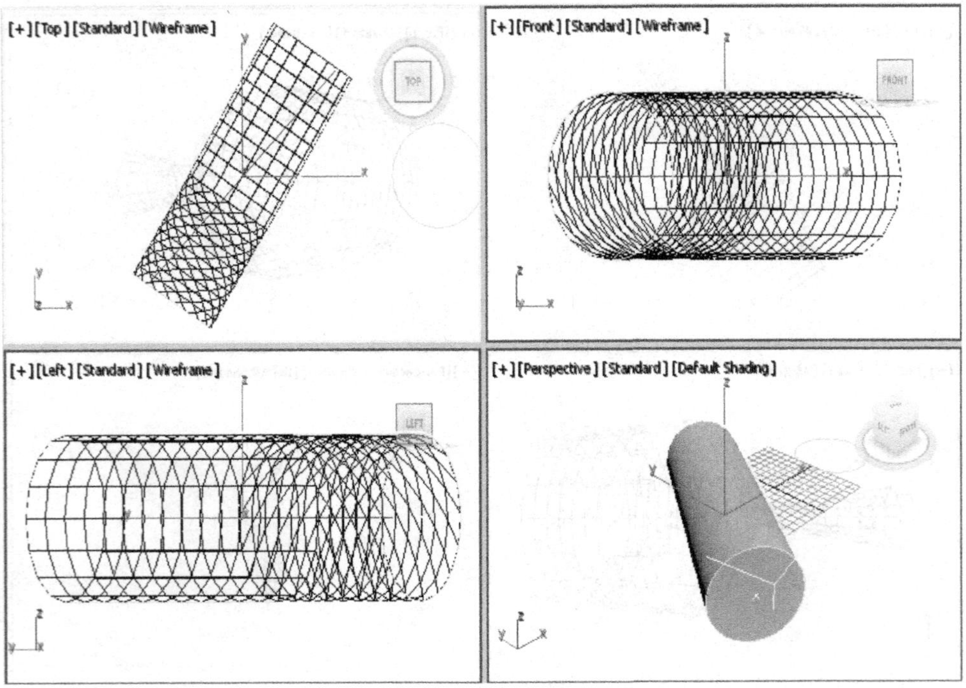

***Figure 6-28** The modified loft object after adding a vertex and moving the left vertex*

4. Close the **Twist Deformation** dialog box.

Note
*If the twist is not readily visible in the shaded viewport, click on the **Realistic** label to display the shortcut menu and then choose **Edged Faces** from the shortcut menu to get a better display.*

5. In the **Deformations** rollout, choose the **Lightbulb** button next to the **Twist** button to turn the deformation off.

Using the Teeter Deformation

In this section, you will deform the lofted object using the **Teeter Deformation(Y)** dialog box.

1. In the **Deformations** rollout, choose the **Teeter** button; the **Teeter Deformation(X)** dialog box is displayed. The red horizontal line at the 0 position represents the teeter along the X axis.

2. The **Make Symmetrical** button is activated by default. Choose this button to turn it off. Then, choose the **Display Y Axis** button; the red line turns green. This allows the lofted object to be teetered along the Y axis. The title bar of the dialog box indicates the presently displayed axis.

3. Choose the **Insert Corner Point** button in the dialog box and then click on the green line (Y axis) near 60. Enter **60** in the left-hand text box and press the ENTER key to place the vertex at the precise position.

4. Choose the **Move Control Point** button in the dialog box and move the right vertex up by five lines so that 50 is displayed in the right text box in the dialog box, as shown in Figure 6-29; the end of the lofted object is tilted 50 degrees along the Y axis, as shown in Figure 6-30.

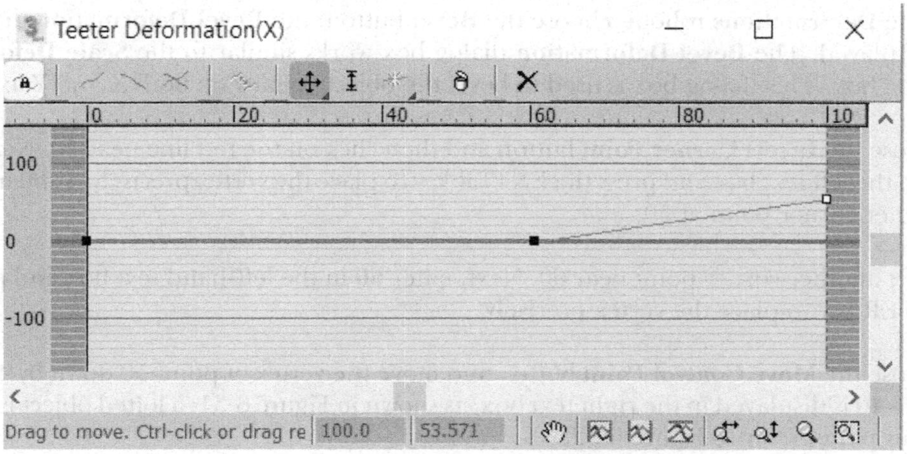

Figure 6-29 *Moving the right-hand vertex up by five lines*

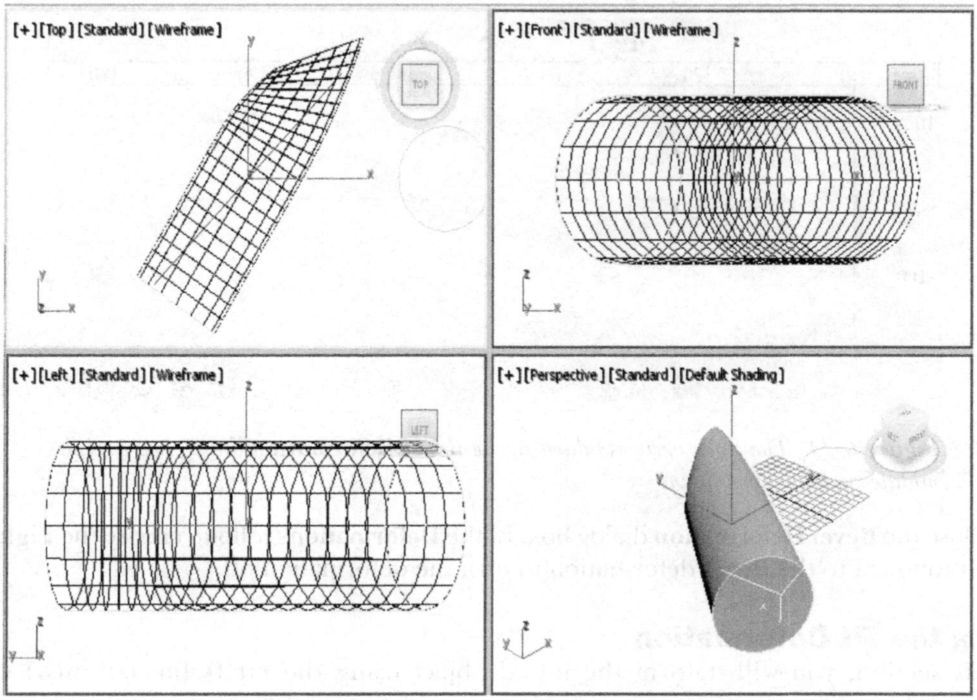

Figure 6-30 *The lofted object tilted along the Y axis*

5. Close the **Teeter Deformation(Y)** dialog box. In the **Deformations** rollout, choose the **Lightbulb** button next to the **Teeter** button to turn the deformation off.

Using the Bevel Deformation

In this section, you will bevel the lofted object using the **Bevel Deformation** dialog box.

1. In the **Deformations** rollout, choose the **Bevel** button; the **Bevel Deformation** dialog box is displayed. The **Bevel Deformation** dialog box works similar to the **Scale Deformation** dialog box. This dialog box is used to bevel the object equally on both X and Y axes.

2. Choose the **Insert Corner Point** button and then click on the red line near 40. Next, enter **40** in the left text box and press the ENTER key to place the vertex precisely. Similarly, place another corner point at 50.

3. Insert another corner point near 60. Next, enter **60** in the left-hand text box and press the ENTER key to place the vertex precisely.

4. Choose the **Move Control Point** button and move the vertex at point 50 down by 4 lines so that – 40 is displayed in the right text box, as shown in Figure 6-31; a lofted object is beveled at this point, refer to Figure 6-32.

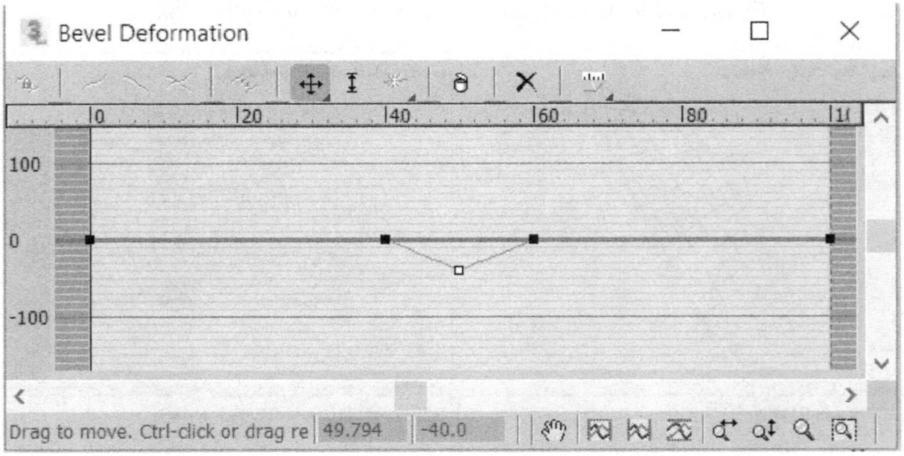

***Figure 6-31** The three vertices added in the **Bevel Deformation** dialog box and the middle vertex moved down*

5. Close the **Bevel Deformation** dialog box. In the **Deformations** rollout, choose the **Lightbulb** button next to the **Bevel** deformation to turn the deformation off.

Using the Fit Deformation

In this section, you will deform the lofted object using the **Fit Deformation(X)** dialog box.

1. Choose **Create > Shapes** in the **Command Panel**. Next, invoke the **Line** tool from the **Object Type** rollout.

2. In the Top viewport, use the **Line** tool to draw a closed shape that represents the way you want the side view of the loft object to appear. Make the shape approximately of the same

length as the loft object. Next, enter **Fit Shape (Side)** in the **Name and Color** rollout to rename the objects, refer to Figure 6-33.

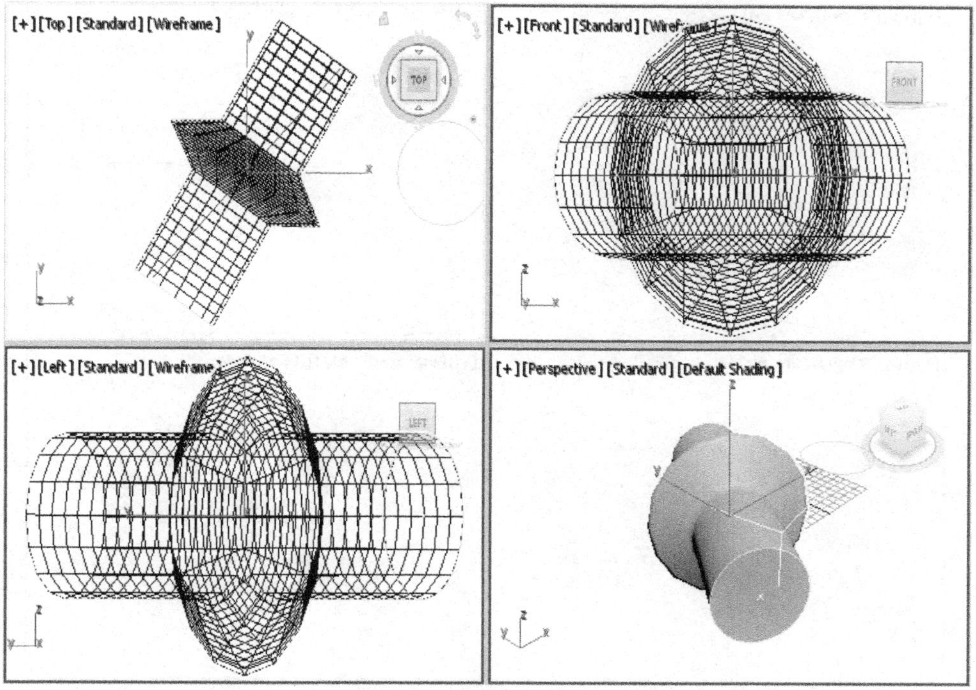

Figure 6-32 *The lofted object beveled after inserting three vertices and moving the middle one*

To draw the fit shape, maximize, zoom, and pan the viewport as needed. Edit the lines at the **Vertex** sub-object level to add smoothness and to get the desired shapes. In order to easily visualize the results of the fit deformation, both the shapes drawn should be of full size, and should have the same length.

3. Draw another closed shape that represents the way you want the loft object to appear when viewed from the top. Next, enter **Fit Shape (Top)** in the **Name and Color** rollout. Invoke the **Zoom Extents All** tool to zoom the objects to their extents, as shown in Figure 6-33.

4. Make sure the object level is activated. Select the lofted object and make sure the **Modify** tab is chosen in the **Command Panel**. In the **Deformations** rollout, choose the **Fit** button; the **Fit Deformation(X)** dialog box is displayed. Unlike the other deformation tools, the **Fit Deformation(X)** dialog box enables you to deform the lofted object by loading the shapes, rather than by editing a baseline.

   The top and side profiles of the lofted object are adjusted so that they can be fitted inside the shapes you select as the fit shapes.

5. In the **Fit Deformation(X)** dialog box, make sure the **Make Symmetrical** button is deactivated to turn off the forced symmetry. Also, make sure the **Display X Axis** button is chosen. Next, choose the **Get Shape** button at the top of the dialog box and select

Fit Shape (Top) in any of the viewports; *Fit Shape (Top)* is displayed in the **Fit Deformation(X)** dialog box. Note that you must select the shape using the cursor.

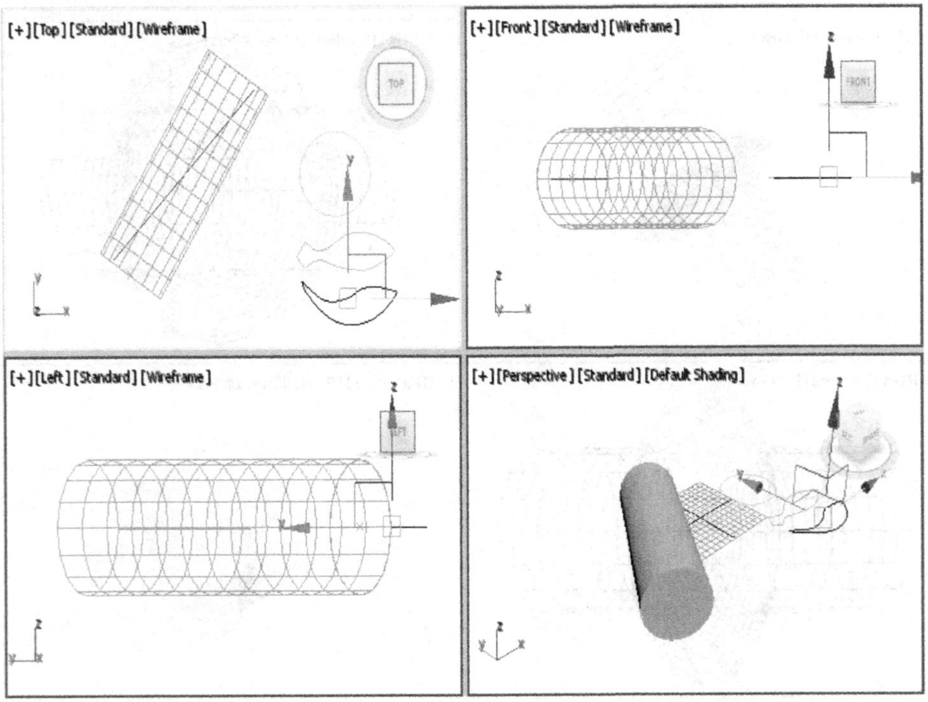

Figure 6-33 *Two closed shapes created in the Top viewport*

6. Invoke the **Zoom** tool and zoom in the graph to view the entire *Fit Shape (Top),* refer to Figure 6-34.

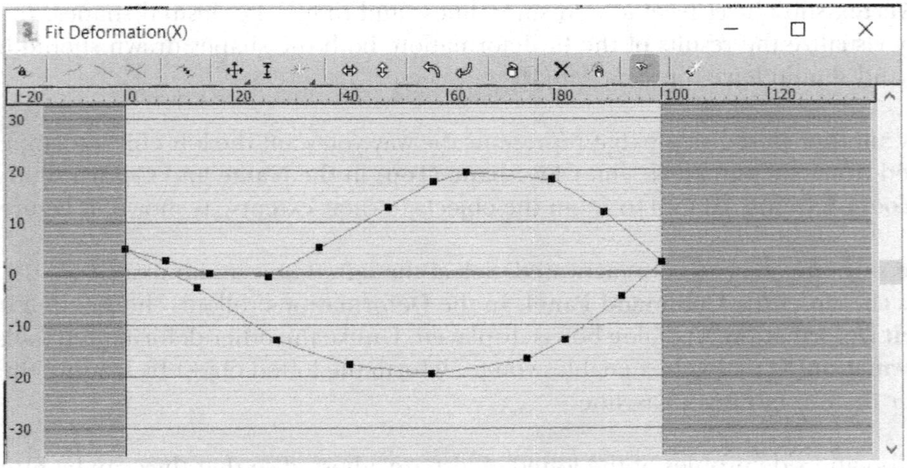

Figure 6-34 *Fit Shape (Top) displayed in the* ***Fit Deformation(X)*** *dialog box*

7. Choose the **Display Y Axis** button. Choose the **Get Shape** button at the top of the dialog box and select *Fit Shape (Side)* in any of the viewports; *Fit Shape (Side)* is displayed in the **Fit Deformation(Y)** dialog box, as shown in Figure 6-35.

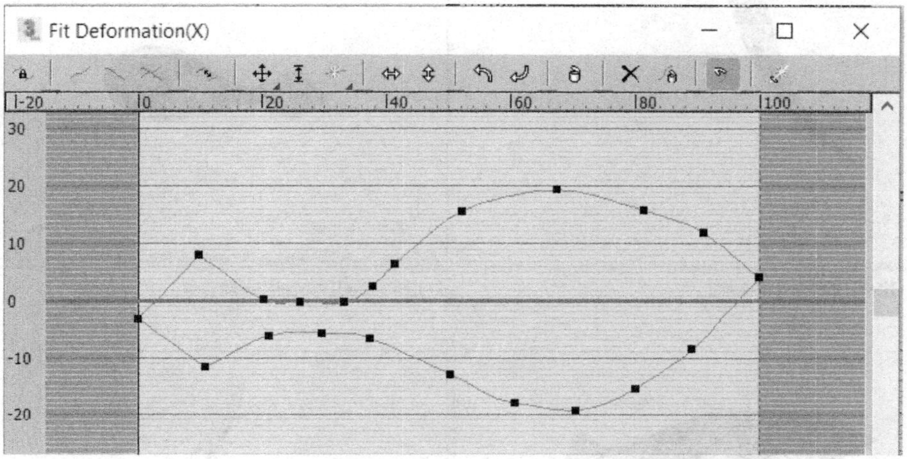

Figure 6-35 *Fit Shape (Side) displayed in the* ***Fit Deformation(Y)*** *dialog box*

Note
When you adjust the fit shapes at the ***Vertex*** *sub-object level in the* ***Fit Deformation*** *dialog box, the lofted object will be automatically updated in the viewports to reflect the change.*

8. Close the **Fit Deformation(Y)** dialog box. Make sure the **Zoom Extents All Selected** tool is invoked from the viewport navigation controls; the lofted object is displayed in the viewports. The shape of the lofted object is clearly determined by the fit shapes, as shown in Figure 6-36.

In the **Deformations** rollout, you may turn on different deformations individually and in combinations by choosing the **Lightbulb** button next to the deformation names.

Note
While applying the ***Scale***, ***Teeter***, *or* ***Fit*** *deformation to a lofted object, you can make changes to the X and Y axes separately by using the* ***Make Symmetrical*** *button.*

9. Choose **Save** from the **File** menu to save the scene.

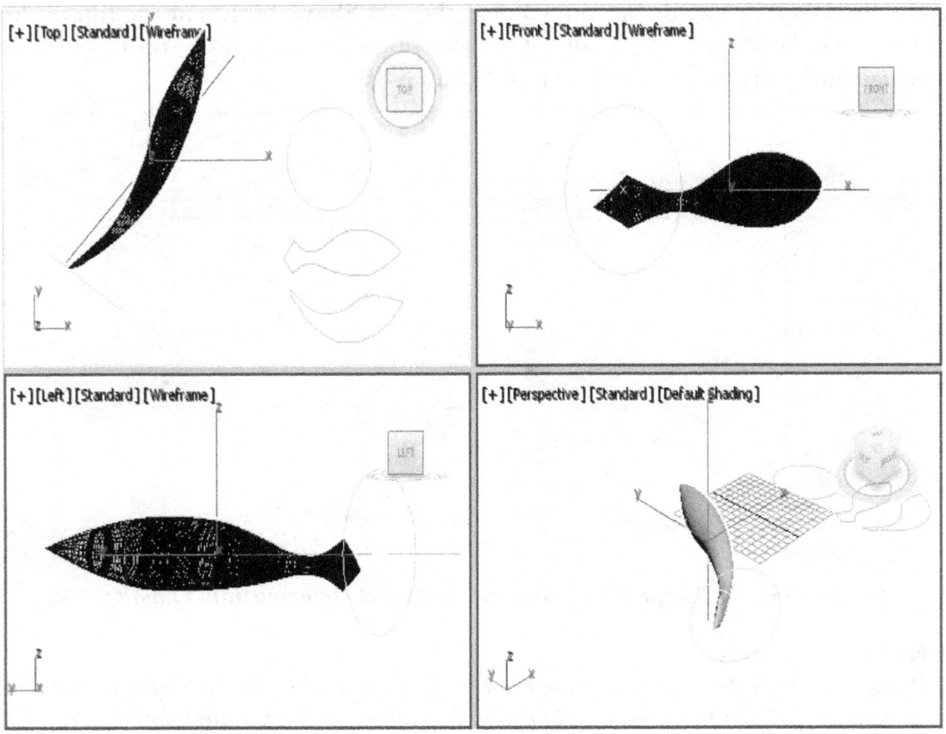

Figure 6-36 *The lofted object showing the effect of fit shapes*

Tutorial 4

In this tutorial, you will create model of a screw driver using the Loft tool, as shown in Figure 6-37. **(Expected time: 20 min)**

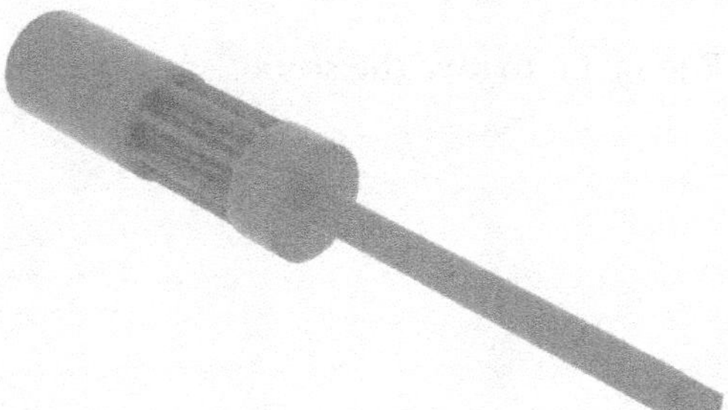

Figure 6-37 *The model of a screw driver*

The following steps are required to complete this tutorial:

a. Create the project folder.
b. Create splines.
c. Create body of the screw driver.
d. Modify the screw driver.
e. Save and render the scene.

Creating the Project Folder

Create a project folder with the name *c06_tut4* in the *3dsmax 2022* folder as discussed in Tutorial 1 of Chapter 2.

Creating Splines

In this section, you will create splines using the **Shape** tool.

1. Activate the Left viewport and then choose **Create > Shapes** in the **Command Panel**; the **Splines** option is displayed in the drop-down list located below it. Next, invoke the **Circle** tool from the **Object Type** rollout.

2. Expand the **Keyboard Entry** rollout and set the value **11** in the **Radius** spinner. Now, choose the **Create** button in the **Keyboard Entry** rollout; a circle is created in all viewports.

3. Enter **base** in the **Name and Color** rollout. Invoke the **Zoom Extents All** tool; *base* is zoomed in all viewports, as shown in Figure 6-38.

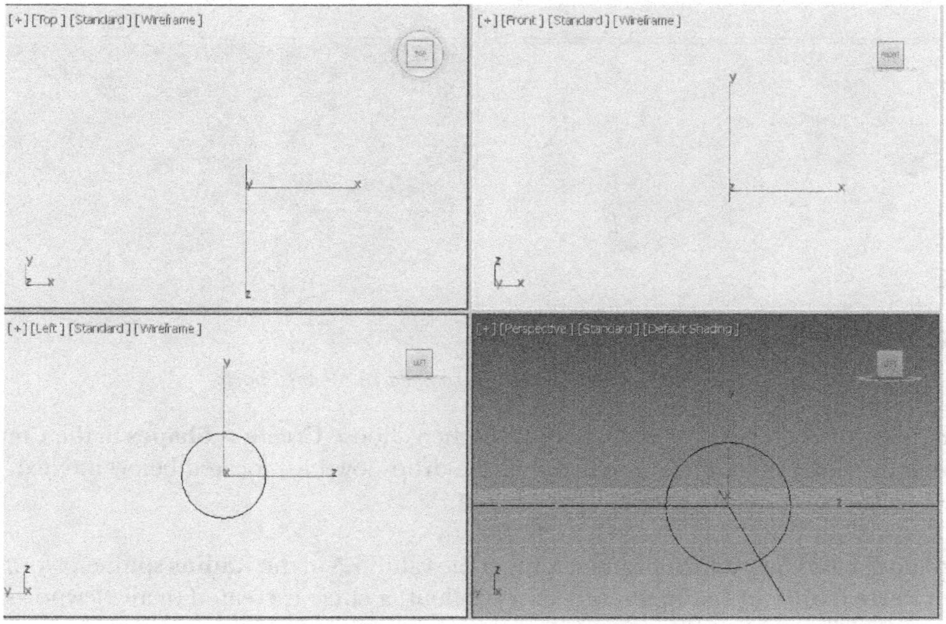

Figure 6-38 *The base zoomed in all viewports*

4. Make sure the Left viewport is activated. Next, invoke the **Star** tool from the **Object Type** rollout.

5. Create a star in the Left viewport and set the properties in the **Parameters** rollout, as shown in Figure 6-39.

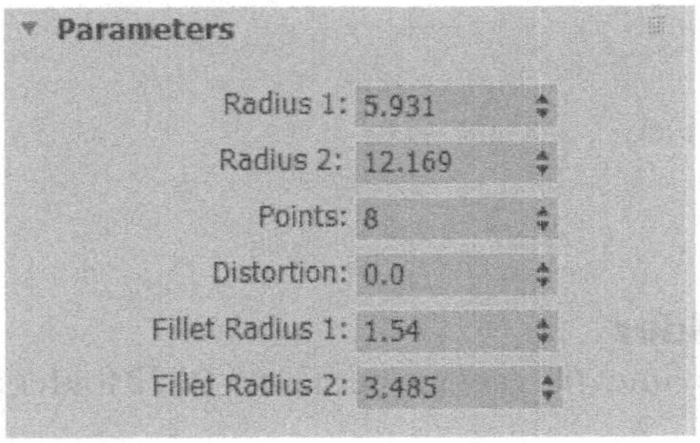

Figure 6-39 *Parameters set in the **Parameters** rollout*

6. Enter **design** in the **Name and Color** rollout. Invoke the **Zoom Extents All** tool; *design* is zoomed in all viewports, as shown in Figure 6-40.

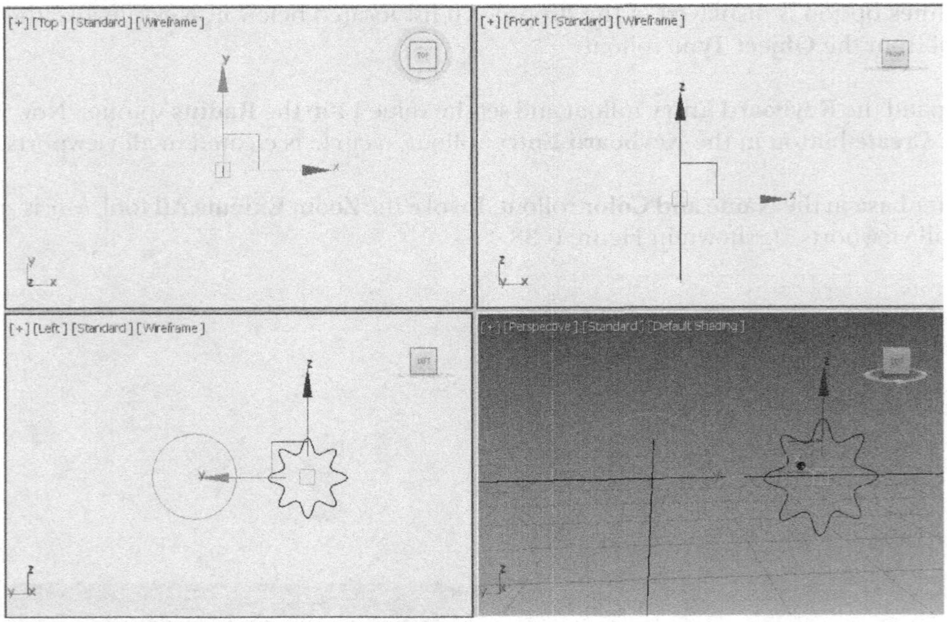

Figure 6-40 *The design zoomed in all viewports*

7. Make sure the Left viewport is activated and then choose **Create > Shapes** in the **Command Panel**; the **Splines** option is displayed in the drop-down list located below it. Next, invoke the **Circle** tool from the **Object Type** rollout.

8. Expand the **Keyboard Entry** rollout and set the value **3.5** in the **Radius** spinner. Now, choose the **Create** button in the **Keyboard Entry** rollout; a circle is created in all viewports.

9. Enter **top_base** in the **Name and Color** rollout. Invoke the **Zoom Extents All** tool; *top_base* is zoomed in all viewports, as shown in Figure 6-41.

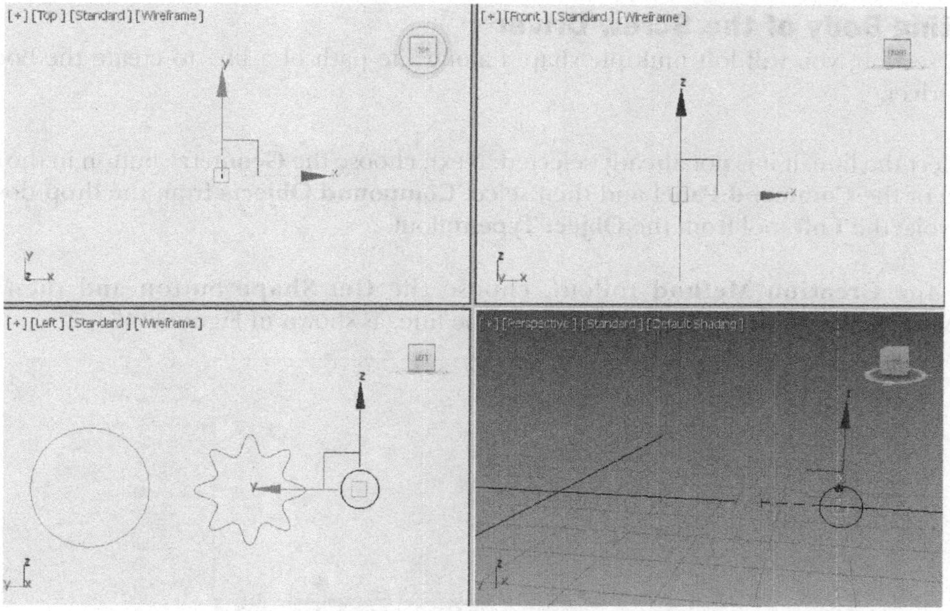

Figure 6-41 *The top_base zoomed in all viewports*

10. Invoke the **Line** tool. Make sure the **Corner** radio button is selected in both the **Initial Type** and **Drag Type** areas of the **Creation Method** rollout. Click on the right side of the viewport. Next, hold the SHIFT key and drag the mouse toward left side and then click again. Now, press ESC to exit the **Line** tool. Line is created in the Left viewport, as shown in Figure 6-42.

Figure 6-42 *The line zoomed in the Left viewport*

Creating Body of the Screw Driver

In this section, you will loft multiple shapes along the path of a line to create the body of a screw driver.

1. Select the line, if it is not already selected. Next, choose the **Geometry** button in the **Create** tab of the **Command Panel** and then select **Compound Objects** from the drop-down list. Invoke the **Loft** tool from the **Object Type** rollout.

2. In the **Creation Method** rollout, choose the **Get Shape** button and then select *base* in the viewport; circle is lofted along the line, as shown in Figure 6-43.

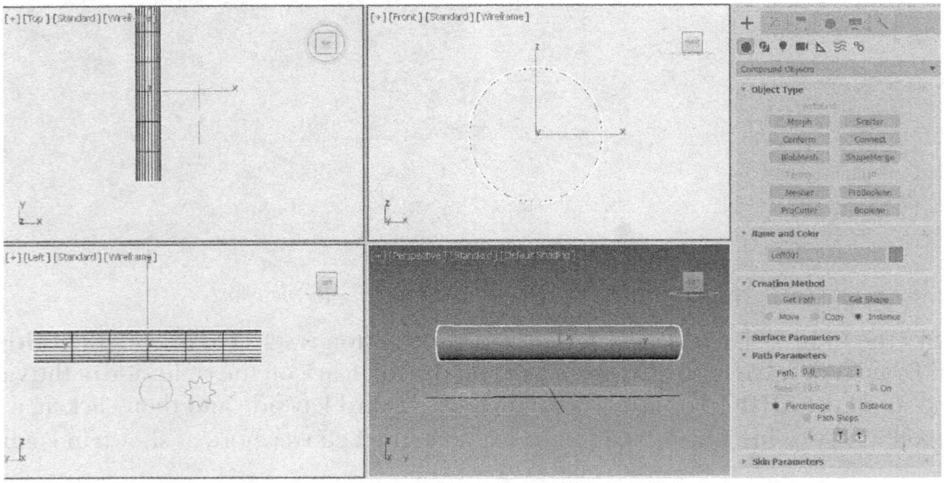

Figure 6-43 *The circle lofted along the line*

3. Expand the **Path Parameters** rollout and set **23** in the **Path** spinner. Next, choose the **Get Shape** button and then select *base*; the path is set, as shown in Figure 6-44.

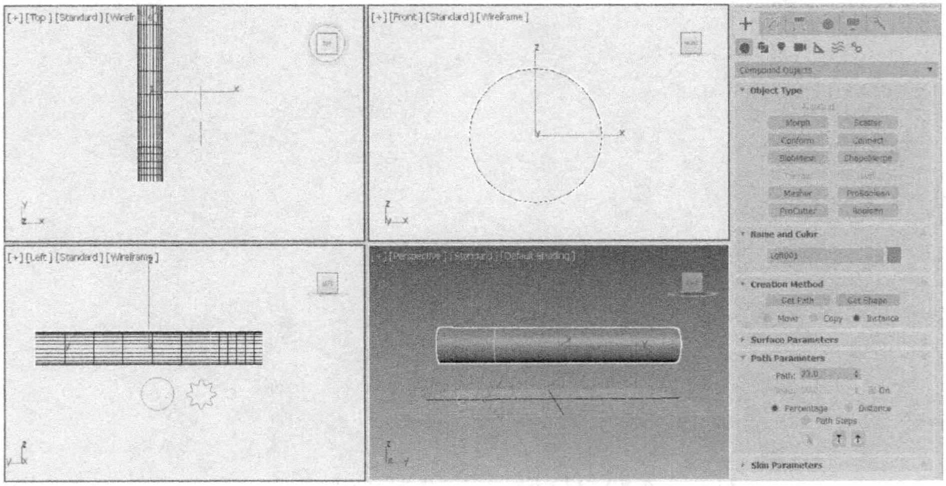

Figure 6-44 *The path set*

4. In the **Path Parameters** rollout, set **24** in the **Path** spinner. Next, choose the **Get Shape** button in the **Creation Method** rollout, and then select design in the viewport; design is lofted along the path, as shown in Figure 6-45.

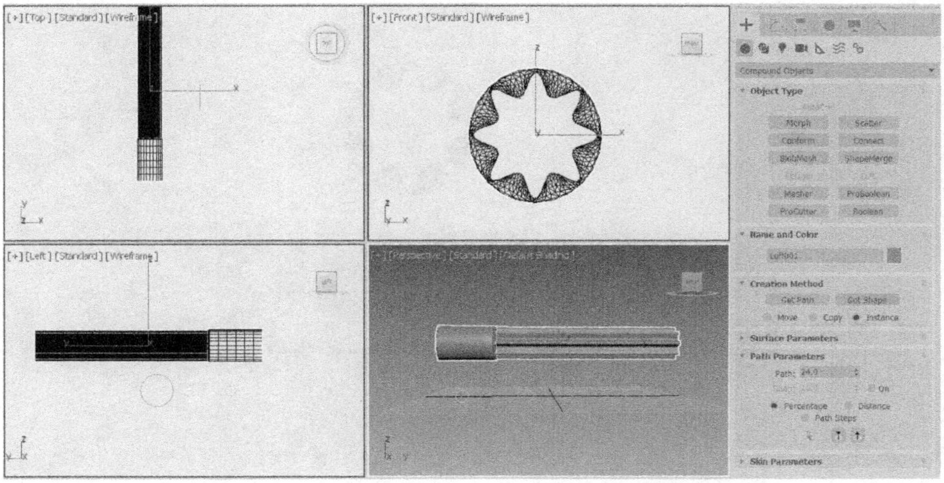

***Figure 6-45** The design lofted along the path*

5. In the **Path Parameters** rollout, set **40** in the **Path** spinner. Next, choose the **Get Shape** button and then select *design*; the path is set, as shown in Figure 6-46.

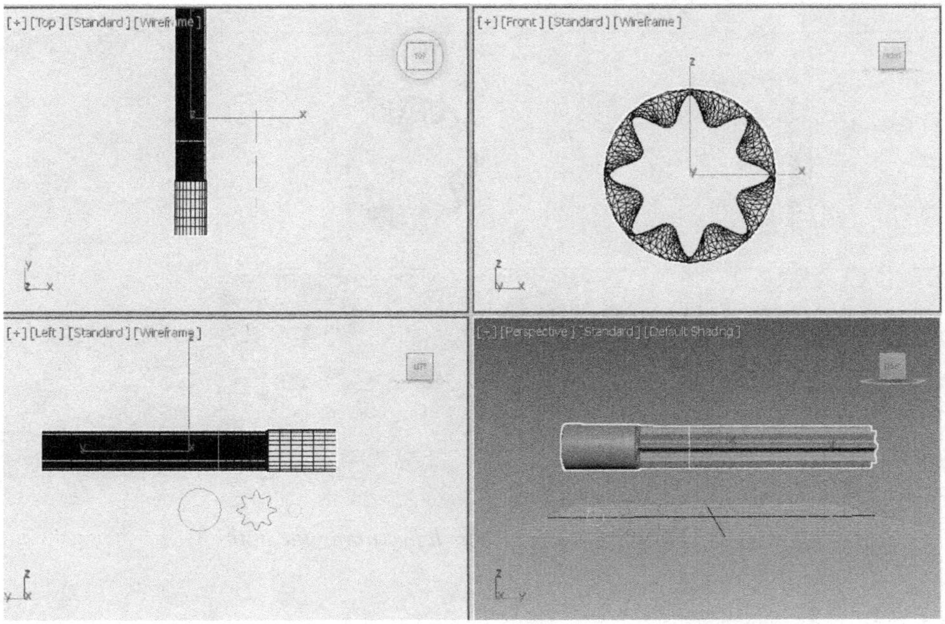

***Figure 6-46** The path set*

6. In the **Path Parameters** rollout, set **41** in the **Path** spinner. Next, choose the **Get Shape** button in the **Creation Method** rollout, and then select *base* in the viewport; *base* is lofted along the path, as shown in Figure 6-47.

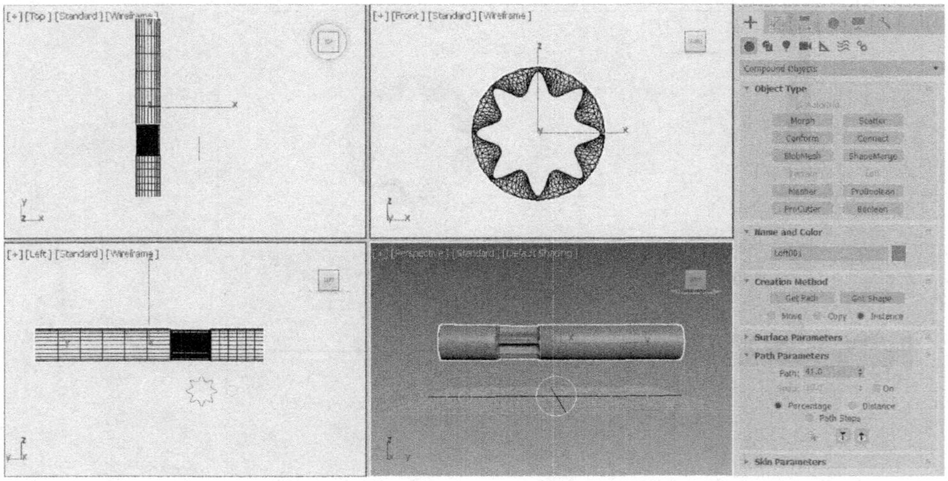

Figure 6-47 *The base lofted along the path*

7. In the **Path Parameters** rollout, set **48** in the **Path** spinner. Next, choose the **Get Shape** button in the **Creation Method** rollout, and then select *base* in the viewport; *base* is lofted along the path, as shown in Figure 6-48.

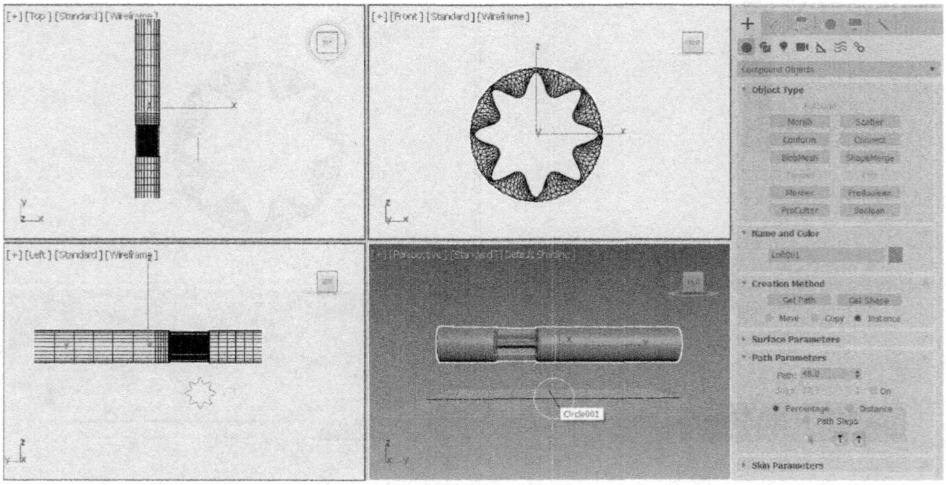

Figure 6-48 *The base lofted along the path*

8. In the **Path Parameters** rollout, set **49** in the **Path** spinner. Next, choose the **Get Shape** button in the **Creation Method** rollout, and then select *top_base* in the viewport; *top_base* is lofted along the path, as shown in Figure 6-49.

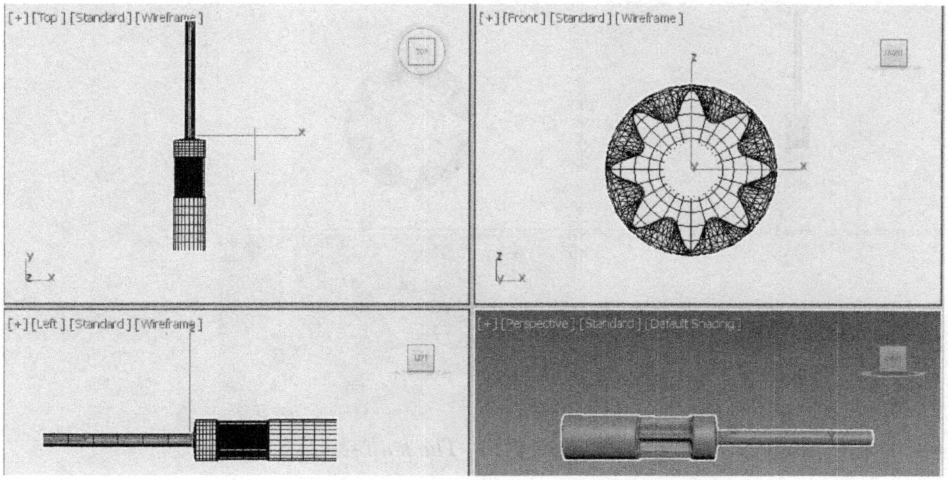

Figure 6-49 *The top_base lofted along the path*

9. Choose **Create > Shapes** in the **Command Panel**. Invoke the **Rectangle** tool and create a small rectangle in the Left viewport, as shown in Figure 6-50. Choose the **Modify** tab in the **Command Panel.**

Figure 6-50 *The rectangle created in the left viewport*

10. Invoke the **Rectangle** tool from the **Object Type** rollout. Expand the **Parameters** rollout and set the values in the spinners as given next:

Length: **1.822** Width: **10.02**

11. In the **Path Parameters** rollout, set **95** in the **Path** spinner. Next, choose the **Get Shape** button in the **Creation Method** rollout, and then select *top_base* in the viewport; the path is set, as shown in Figure 6-51.

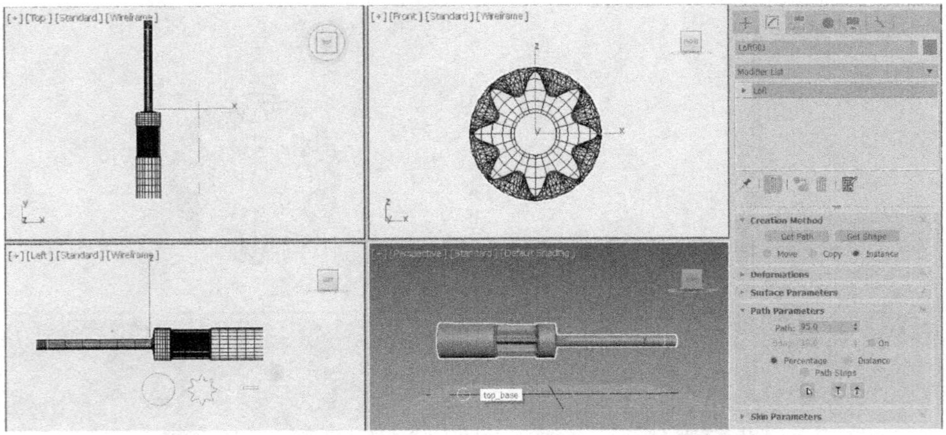

Figure 6-51 *The path set*

12. In the **Path Parameters** rollout, set **96** in the **Path** spinner. Next, choose the **Get Shape** button in the **Creation Method** rollout, and then select rectangle in the viewport; the rectangle is lofted along the path, as shown in Figure 6-52.

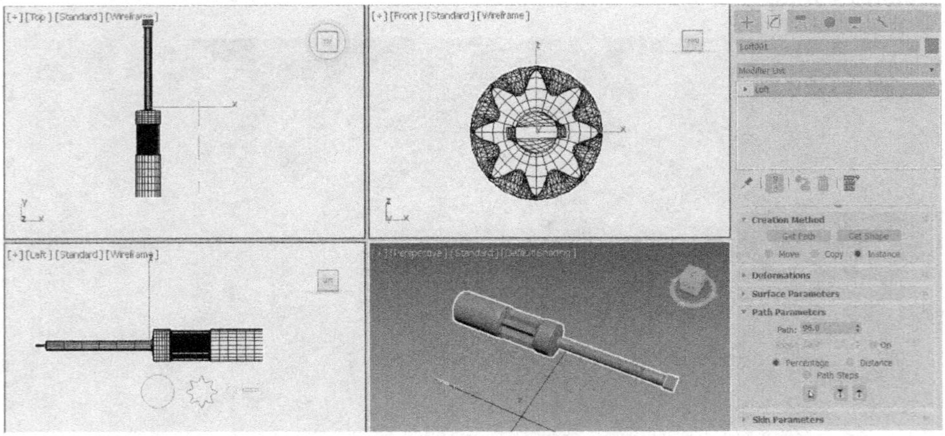

Figure 6-52 *The rectangle lofted along the path*

Modifying Shape of the Screw Driver

In this section, you will modify the lofted object by scaling and rotating the shape.

1. In the modifier stack, expand the **Loft**; the **Loft** sub-object hierarchy is displayed. Next, select the **Shape** sub-object mode of the **Loft** sub-object hierarchy.

2. Maximize the Perspective viewport and zoom it. Select the rectangle shape of the screw driver, as shown in Figure 6-53.

Figure 6-53 *The rectangle shape selected*

3. Invoke the **Rotate** tool from the **Main Toolbar** or press E to slightly rotate the rectangle shape, as shown in Figure 6-54.

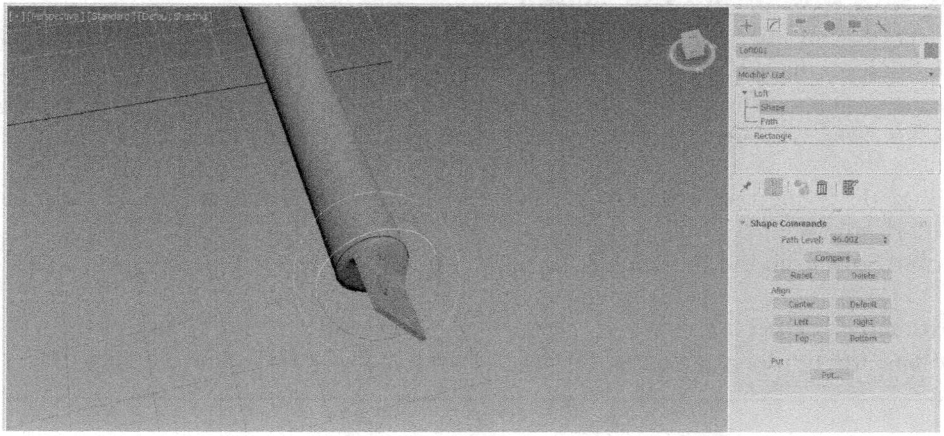

Figure 6-54 *Rotating the rectangle shape*

4. Invoke the **Scale** tool from the **Main Toolbar** or press R to scale the selected shape. You will notice that the tip of the screw driver is properly scaled, as shown in Figure 6-55.

Figure 6-55 The rectangle shape scaled

Note

*To smoothen corners of the rectangle, you can set the value **0.5** in the **Corner Radius** spinner of the **Parameters** rollout.*

Saving and Rendering the Scene

In this section, you will save the scene that you have created and then render it. You can also view the final rendered image of this model by downloading the *c06_3dsmax_2022_rndr.zip* file from *www.cadcim.com*. The path of the file is as follows: *Textbooks > Animation and Visual Effects > Autodesk 3ds Max > Autodesk 3ds Max 2022 for Beginners: A Tutorial Approach*

1. Change the background color of the scene to white, as discussed in Tutorial 1 of Chapter 2.

2. Choose **Save** from the **File** menu.

3. Activate the Perspective viewport. Next, invoke the **Render Production** tool from the **Main Toolbar**; the rendered image of the screw driver is displayed, refer to Figure 6-37.

Self-Evaluation Test

Answer the following questions and then compare them to those given at the end of this chapter:

1. Which of the following shortcut keys is used to invoke the **Angle Snap Toggle** tool?

   (a) A (b) Z
   (c) C (d) None of these

2. Which of the following buttons in the **Creation Method** rollout is used to select the shape of the loft feature?

   (a) **Get Path** (b) **Get Shape**
   (c) both of the above (d) None of the above

3. Which of the following rollouts in the **Command Panel** is available when the **Shape** sub-object mode is activated for the lofted object?

   (a) **Surface Properties** (b) **Shape Commands**
   (c) **Soft Selection** (d) **Geometry**

4. To avoid twisting of multiple shapes while lofting them, _________ of the shapes should be aligned.

5. In the **Teeter Deformation** dialog box, the currently displayed axis is indicated in the _________ of the dialog box.

6. The _________ button in the **Compare** window is used to load shapes into the window.

7. The _________ option in the **Align** area of the **Shape Commands** rollout moves the shape at the center on the path.

8. You can loft a number of different _________ along a single path.

9. Select the _________ check box in the **Skin Parameters** rollout to display the skin of the lofted object.

10. You can add shapes along different _________ of a path.

11. Select the _________ check box in the **Path Parameters** rollout to enable the **Snap** spinner.

12. The **Bevel Deformation** dialog box is used to bevel an object equally on both X and Y axes. (T/F)

13. The lightbulb buttons next to the deformation names are used to turn on or off the effects of deformations. (T/F)

14. When you apply a fit deformation to a lofted object, the shapes used to create the loft are replaced with the fit shapes. (T/F)

15. You can apply a maximum of three deformations to a lofted object. (T/F)

16. When applying the **Scale**, **Teeter**, or **Fit** deformation to a lofted object, you can make the changes to the X and Y axes separately by making the **Make Symmetrical** button inactive. (T/F)

17. The shapes in the lofted object can be moved to different levels of the path. (T/F)

18. Once the lofted object is created, the shape of the path cannot be changed. (T/F)

19. The shapes can be moved with respect to the path using the sub-object mode. (T/F)

20. You can use the capping options to open or close the ends of a lofted object. (T/F)

Review Questions

Answer the following questions:

1. Which of the following buttons in the **Fit Deformation** dialog box is used to move control vertices?

   (a) **Insert corner Point** (b) **Move Control Point**
   (c) **Scale Control Point** (d) **Delete Control Point**

2. The __________ button is used to insert new values in the **Scale Deformation** dialog box.

3. The __________ button is used to create 3D objects by lofting 2D objects along a spline path.

4. If the vertices of the lofted object are not aligned, the object gets __________ .

5. The shapes are always __________ to the path along which you loft them.

6. In the **Creation Method** rollout, you need to make sure that the __________ radio button is selected before lofting the object.

7. In the **Skin Parameters** rollout, clear the __________ check box to open the top end of the lofted object.

8. To get a better view of the model, use the __________ button to adjust the view in the Perspective viewport.

9. In the **Fit Deformation** dialog box, the green horizontal line represents the X axis. (T/F)

10. The **Bevel Deformation** dialog box works similar to the **Scale Deformation** dialog box. (T/F)

11. The shapes in the lofted object can be moved and copied to different levels of the path. (T/F)

12. If you modify the path along which you loft the objects, the lofted object also gets modified. (T/F)

13. You cannot increase or decrease the smoothness of the lofted object around the created path. (T/F)

EXERCISES

The rendered output of the model used in the following exercises can be accessed by downloading the *c06_3dsmax_2022_exr.zip* from *www.cadcim.com*. The path of the file is as follows: *Textbooks > Animation and Visual Effects > Autodesk 3ds Max > Autodesk 3ds Max 2022 for Beginners: A Tutorial Approach*

Exercise 1

In this exercise, you will create a propeller using a circle and an ellipse, refer to Figure 6-56. Both these shapes will be lofted along straight lines, as shown in Figure 6-57. Add material, lights, and camera, and then render the model, refer to Figure 6-58. **(Expected time: 30 min)**

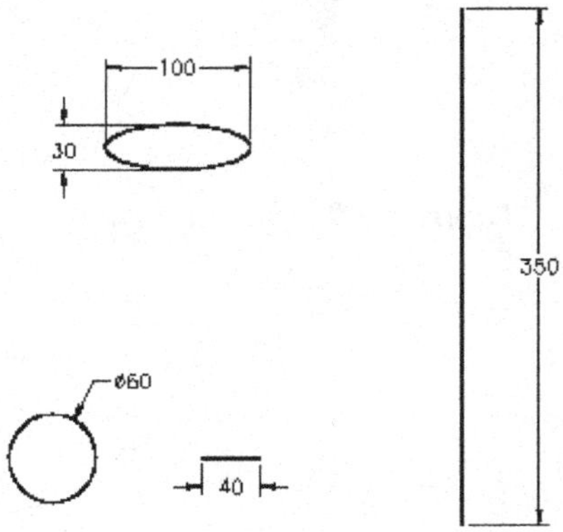

Figure 6-56 *The circular and ellipse shapes*

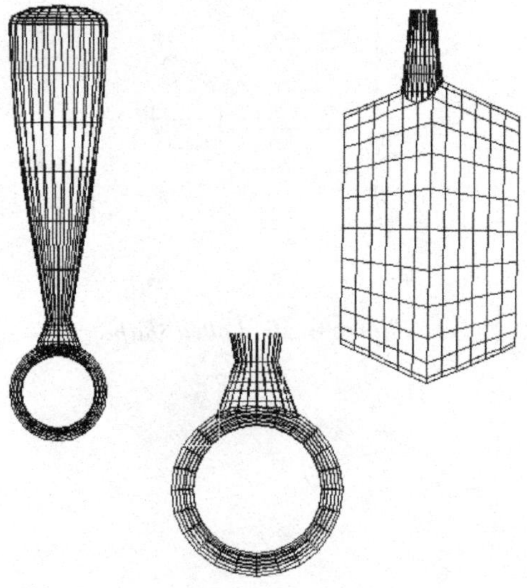

Figure 6-57 *Shapes to be lofted*

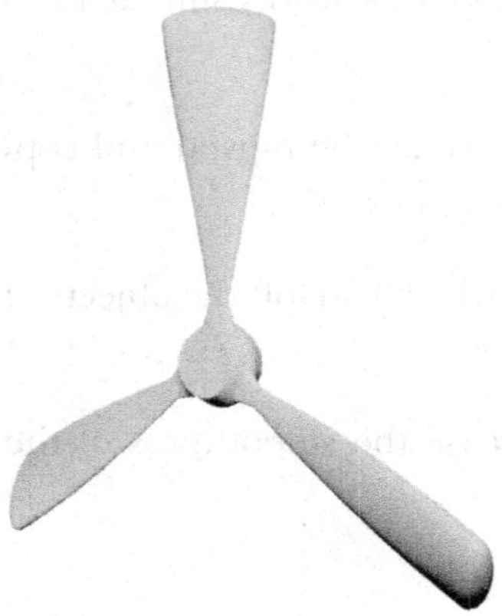

Figure 6-58 The propeller model

Exercise 2

In this exercise, you will draw shapes and loft them to create the travel mug, as shown in Figures 6-59 and 6-60. The lofted objects must be modified to create the final model. Next, add lights and materials to the objects. **(Expected time: 30 min)**

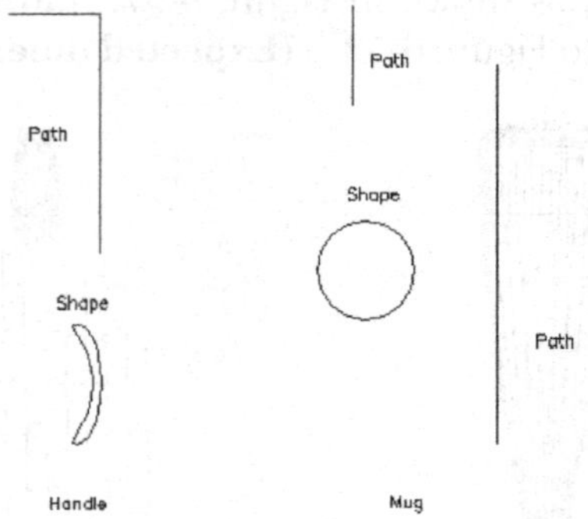

Figure 6-59 *Lofted shapes*

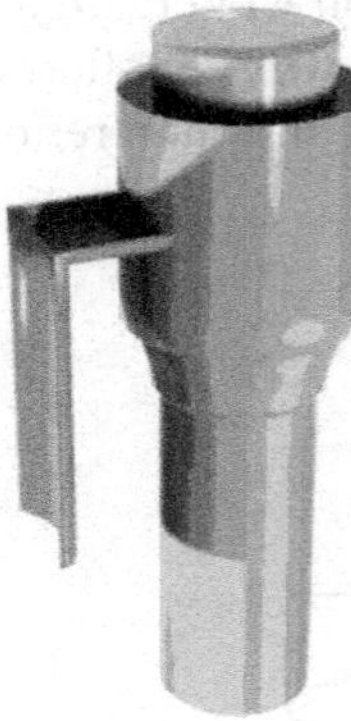

Figure 6-60 *The travel mug*

Answers to Self-Evaluation Test

1. a, **2.** b, **3.** b, **4.** vertices, **5.** title bar, **6. Pick Shape**, **7. Center**, **8.** shapes, **9. Skin**, **10.** levels, **11. On**, **12.** T, **13.** T, **14.** T, **15.** F, **16.** T, **17.** T, **18.** F, **19.** T, **20.** T

Chapter 7

Material Editor: Creating Materials

Learning Objectives

After completing this chapter, you will be able to:

- *Create materials in the material library*
- *Apply ambient and diffuse properties of the materials*
- *Assign materials to the objects*
- *Use ART renderer and Autodesk materials*
- *Render the scene for the final output*

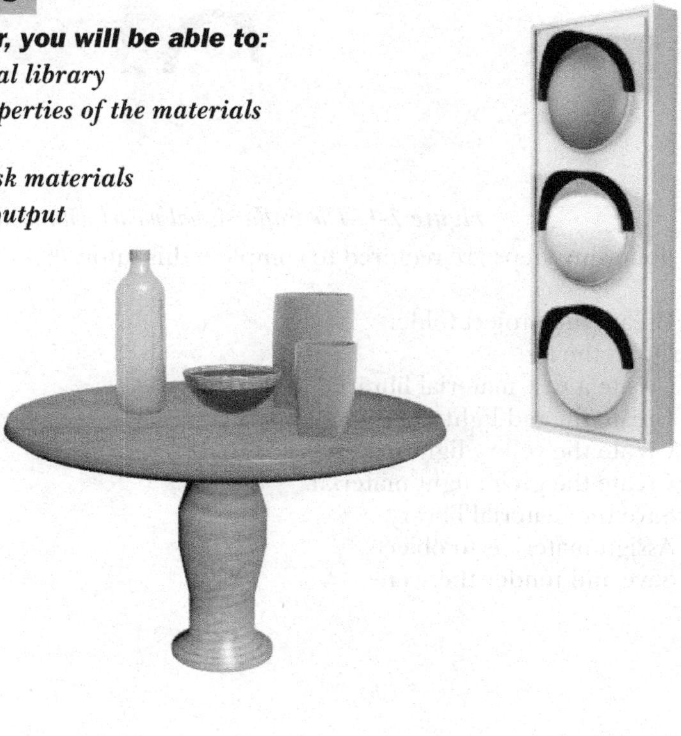

INTRODUCTION

In this chapter, you will create and apply materials to 3D objects. You will use the **Material Editor** dialog box to create materials and then apply them to 3D objects. The materials you create will be stored in a material library. Later, these materials can be assigned to other 3D objects.

TUTORIALS

Before starting the tutorials, you need to download the *c07_3dsmax_2022_tut.zip* file from *www.cadcim.com*. The path of the file is as follows: *Textbooks > Animation and Visual Effects > Autodesk 3ds Max > Autodesk 3ds Max 2022 for Beginners: A Tutorial Approach*

Extract the contents of the zip file and save them in the *Documents* folder.

Tutorial 1

In this tutorial, using the **Material Editor** dialog box, you will create the materials to be applied to a traffic signal model. You will also create shiny red, green, and yellow materials for lights, as shown in Figure 7-1. **(Expected time: 25 min)**

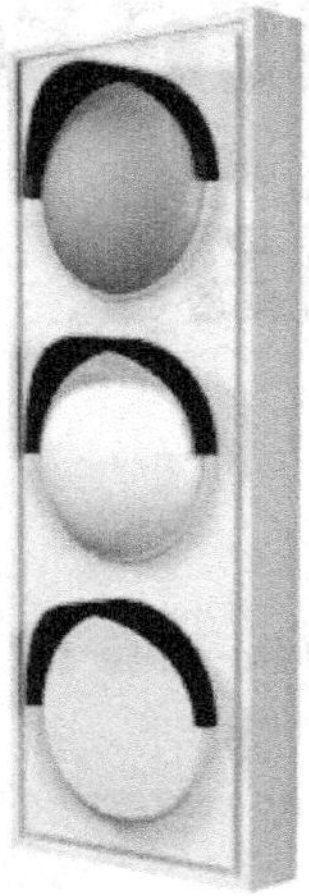

Figure 7-1 *The traffic signal model with materials applied*

The following steps are required to complete this tutorial:

a. Create the project folder.
b. Open the file.
c. Create a new material library.
d. Create the red light material.
e. Create the yellow light material.
f. Create the green light material.
g. Save the material library.
h. Assign materials to objects.
i. Save and render the scene.

Creating the Project Folder

Create the project folder with the name *c07_tut1* in the *3dsmax 2022* folder, as discussed in Tutorial 1 of Chapter 2.

Opening the File

1. Choose **Open** from the **File** menu; the **Open File** dialog box is displayed. In this dialog box, browse to the location *\Documents\c07_3dsmax_2022_tut.zip* and select the **c07_tut1_start.max** file from it. Choose the **Open** button to open this file; a traffic signal model is displayed in the perspective viewport with predefined materials assigned to the case and three hoods, as shown in Figure 7-2.

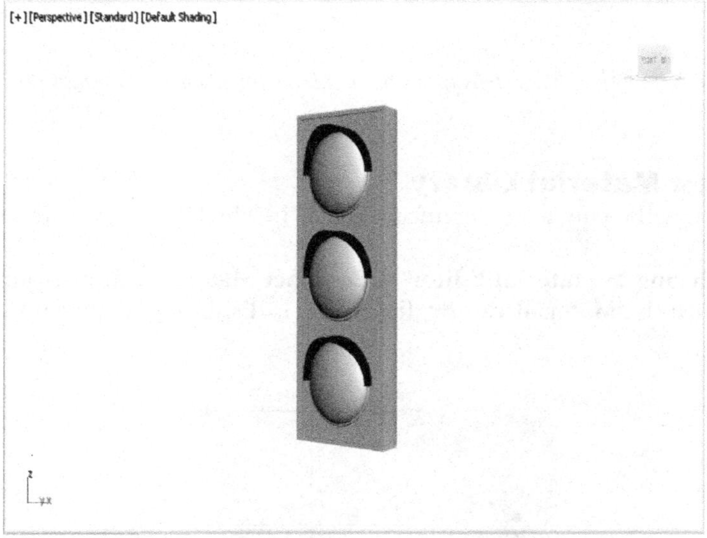

Figure 7-2 *The c07_tut1_start file*

Note

*While opening a Max file, if the **Project Mismatch** dialog box is displayed, select the **Change the Project** radio button and then choose the **Continue** button. Next, if the **Invalid Paths** message box is displayed, choose **OK** in this message box. On doing so, the **Missing External Files** message box is displayed. In this message box, choose the **Continue** button. Next, press SHIFT+T; the **Asset Tracking** window will be displayed. In this window, right-click on the file name whose status is displayed as **File Missing**; a quad menu will be displayed. Choose **Browse** from it; the **Browse Image File** dialog box will be displayed. Navigate to the folder where you have saved the images and then choose the **Open** button; you will notice that the status of the selected file is changed to **Found** in the **Asset Tracking** window. Repeat this process for other such files as well and then close this window.*

2. Choose **Save As** from the **File** menu; the **Save File As** dialog box is displayed. Browse to the location *\Documents\3dsmax 2022\c07_tut1\scenes*. Save the file with the name *c07tut1.max* at this location.

You need to confirm whether the Scanline Renderer is set up.

3. Choose the **Render Setup** tool and do one of the following steps:

a. If the **Render Setup: Arnold** dialog box is displayed, select **Scanline** from the **Renderer** drop-down list; the renderer is changed to Scanline and the name of the dialog box is modified to **Render Setup: Scanline Renderer** dialog box. Close this dialog box and proceed further.

b. If the **Render Setup: Scanline Renderer** dialog box is displayed, close the dialog box and proceed further as discussed next.

Note

You need to follow the above steps to set the renderer for opening the start files in the forthcoming chapters.

Creating a New Material Library

In this section, you will create a new material library in which you can store new materials.

1. Choose **Rendering > Material Editor > Compact Material Editor** from the menu bar or press the M key; the **Material Editor** dialog box is displayed, as shown in Figure 7-3.

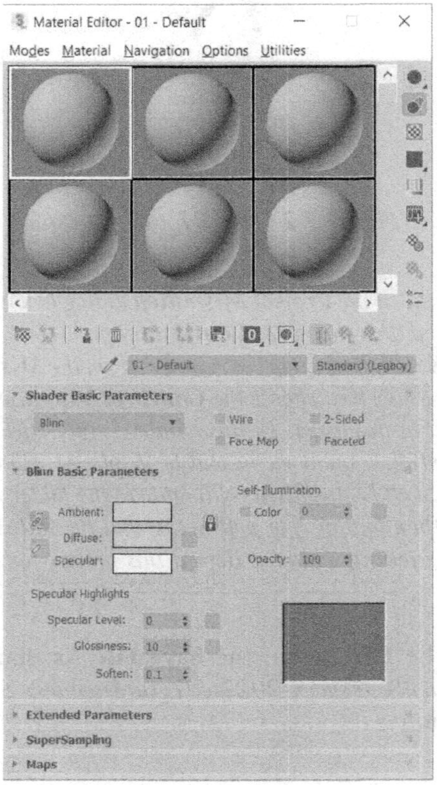

Figure 7-3 *The* ***Material Editor*** *dialog box*

2. Choose the **Get Material** button in the **Material Editor** dialog box; the **Material/Map Browser** dialog box is displayed. 

3. Choose the **Material/Map Browser Options** button in the **Material/Map Browser** dialog box; the **Material/Map Browser Options** flyout is displayed, as shown in Figure 7-4. Select **New Material Library** from the **Material/Map Browser Options** flyout; the **Create New Material Library** dialog box is displayed, as shown in Figure 7-5.

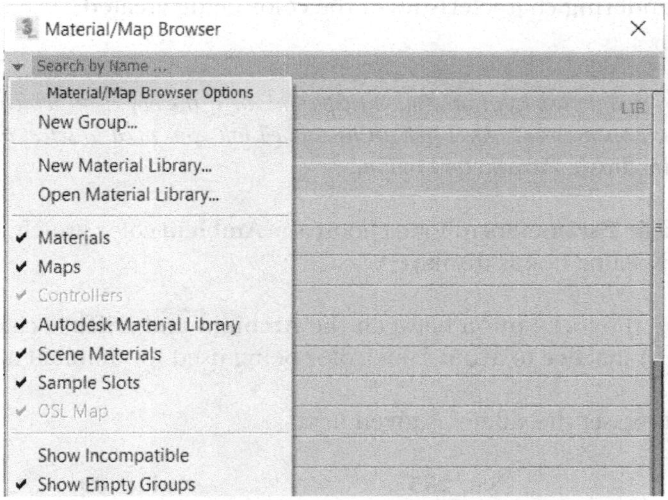

***Figure 7-4** The **Material/Map Browser Options** flyout*

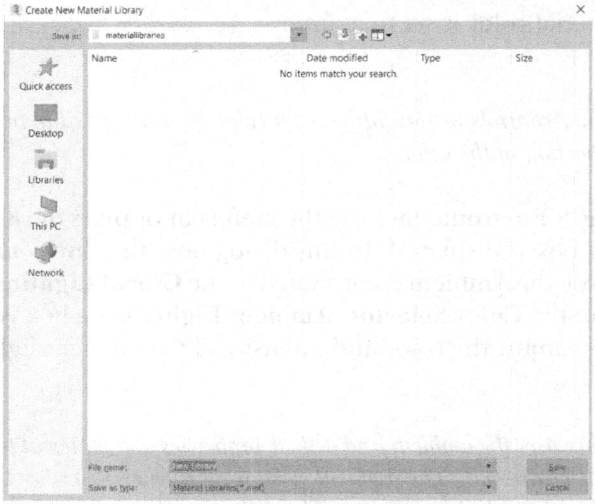

***Figure 7-5** The **Create New Material Library** dialog box*

4. Enter **signal** as the new name for the material library and choose the **Save** button; the new **signal** material library is displayed in the **Material/Map Browser** dialog box.

5. Close the **Material/Map Browser** dialog box to return to the **Material Editor** dialog box.

Creating the Red Light Material

In this section, you will create material for the red light of the signal.

1. Select the first material sample slot in the **Material Editor** dialog box to make it the current sample; the properties for the material are displayed below the material samples.

2. In the **Shader Basic Parameters** rollout, make sure **Blinn** is selected in the drop-down list. This sets the rendering characteristics of the color being created.

Note

*If the material is transparent or semi transparent, or if the object on which the material is to be assigned is an open-ended object like an uncapped loft, you need to select the **2-Sided** check box in the **Shader Basic Parameters** rollout.*

3. In the **Blinn Basic Parameters** rollout, choose the **Ambient** color swatch; the **Color Selector: Ambient Color** dialog box is displayed.

   Click and make the lock button between the **Ambient** and **Diffuse** color swatches (to the left of the labels) inactive to avoid same color being used for ambient and diffuse settings.

4. In this dialog box, set the values as given next:

   Hue: **255** Sat: **255** Value: **89**

   Choose the **OK** button to exit the **Color Selector**: **Ambient Color** dialog box; the ambient portion of the material color is set to maroon.

Note

*The **Value** spinner controls the brightness of a color. Lower the value in this spinner, darker will be the ambient portion of the color.*

5. Choose **Rendering > Environment** from the menu bar or press the 8 key; the **Environment and Effects** dialog box is displayed. In this dialog box, the **Environment** tab is chosen by default. Next, choose the **Ambient** color swatch in the **Global Lighting** area of the **Common Parameters** rollout; the **Color Selector: Ambient Light** dialog box is displayed. Using this dialog box, you can adjust the color and intensity of the ambient light in the scene.

Note

Autodesk 3ds Max uses the ambient and diffuse properties of a material to shade an object to make it appear three-dimensional. The ambient portion of a color describes the part of an object that is not under direct light but in shadow. The diffuse portion of a color describes the part that is in normal lighting conditions. By adjusting the values of these portions, a matte or a non-glossy material is created.

6. In the **Color Selector: Ambient Light** dialog box, set the value **29** in the **Value** spinner. Note the change in the material sample in the **Material Editor** dialog box. Choose the **OK** button to exit the **Color Selector: Ambient Light** dialog box. Also, close the **Environment and Effects** dialog box.

Changing the value of the ambient light setting from 0 to 29 adds a small amount of white ambient light to the scene. If the ambient light value is set to 0 (black), the **Ambient** color settings for the material have no effect. As ambient light is increased in the scene, the ambient color portion of the material has more effect on the material's overall appearance.

7. Choose the **Diffuse** color swatch in the **Blinn Basic Parameters** area of the **Material Editor** dialog box; the **Color Selector: Diffuse Color** dialog box is displayed. In this dialog box, set the values as given next:

   Red: **168** Green: **0** Blue: **0**

   This sets the diffuse portion of the color to medium red.

8. Choose the **OK** button to close the **Color Selector: Diffuse Color** dialog box; the new material is displayed in the **Material Editor** dialog box.

   You need to set the specular value for shiny materials. Shiny materials are represented by highlights when rendered. The specular portion of the color describes hue, saturation, and intensity of the highlights visible when a shiny object is rendered. The strength and shape of the highlight is controlled by the specular highlight settings.

   Now, you will make the material shiny.

9. With the first material sample slot selected, choose the **Specular** color swatch in the **Blinn Basic Parameters** rollout; the **Color Selector: Specular Color** dialog box is displayed.

10. In the **Color Selector: Specular Color** dialog box, set the values as given next:

    Red: **255** Green: **200** Blue: **200**

    This will lighten the specular color.

11. Choose the **OK** button to close the **Color Selector: Specular Color** dialog box.

12. In the **Specular Highlights** area of the **Blinn Basic Parameters** rollout, set the value **60** in the **Specular Level** spinner and **20** in the **Glossiness** spinner.

    The **Specular Level** spinner determines the intensity or brightness of the specular highlight. The **Glossiness** spinner determines the size of the specular highlight.

    Now, you need to name the material and put it in the material library. After the material is saved in the library, it can be accessed and assigned to objects in other scenes.

13. Change the name of the material to **Red light** by entering it in the drop-down list located on the left of the **Standard (Legacy)** button in the **Material Editor** dialog box.

    The name of the material is changed and it appears in the **Material Editor** dialog box as well.

14. Choose the **Put To Library** button in the **Material Editor** dialog box; a flyout is displayed. 

15. Choose **signal.mat** (the newly created material library) from the flyout, as shown in Figure 7-6; the **Put To Library** dialog box is displayed, as shown in Figure 7-7.

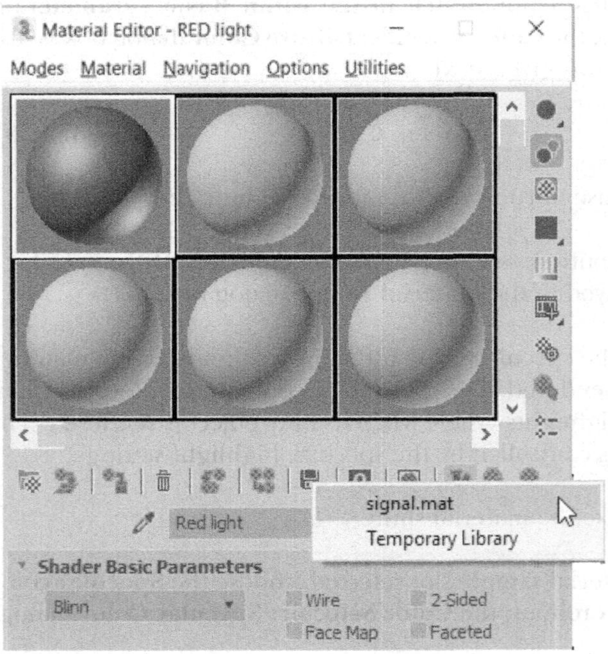

***Figure 7-6** Choosing **signal.mat** from the flyout*

***Figure 7-7** The **Put To Library** dialog box*

The name *Red light* appears in the **Name** text box. You can place this material in the library by either using the default name or renaming it.

16. Choose the **OK** button; the current material is added under the **signal** material library.

17. Choose the **Get Material** button in the **Material Editor** dialog box; the **Material/Map Browser** dialog box is displayed. In this dialog box, click on **signal.mat** to expand it, if not already expanded. You will notice that the *Red light* material is added to the material library, as shown in Figure 7-8.

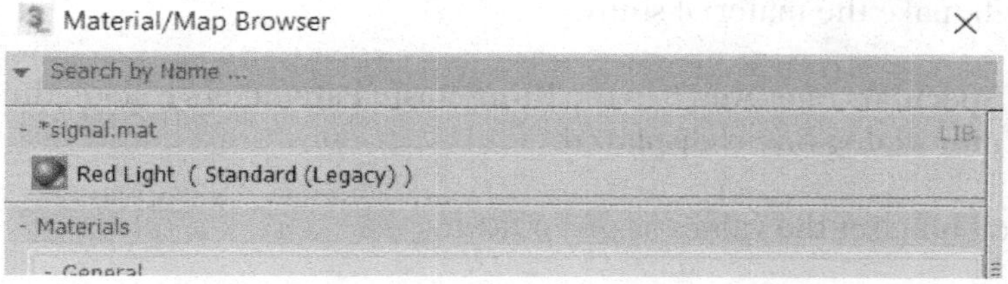

Figure 7-8 The Red light material added to the material library

18. Close the **Material/Map Browser** dialog box to return to the **Material Editor** dialog box.

Note
You can copy material by dragging it from one sample slot to another.

Creating the Yellow Light Material

In this section, you will create material for the yellow light of the signal.

1. Select the second available material sample slot in the **Material Editor** dialog box. Various rollouts are displayed in the **Material Editor** dialog box to modify the **Standard (Legacy)** material.
2. In the **Shader Basic Parameters** rollout, make sure **Blinn** is selected in the drop-down list.
3. In the **Blinn Basic Parameters** rollout, unlock the **Ambient** and **Diffuse** color swatches, if locked. Next, choose the **Ambient** color swatch; the **Color Selector: Ambient Color** dialog box is displayed.
4. In this dialog box, set the values as given next:

   Red: **245** Green: **183** Blue: **15**

5. Choose the **OK** button to close the **Color Selector: Ambient Color** dialog box.

   Since the value of the ambient light setting has already been increased for the scene, you do not need to adjust the value.

6. Choose the **Diffuse** color swatch in the **Material Editor** dialog box; the **Color Selector: Diffuse Color** dialog box is displayed.
7. In the **Color Selector: Diffuse Color** dialog box, set the values as given next:

   Red: **245** Green: **183** Blue: **15**

Note
*The **Red**, **Green**, **Blue** spinners control the amount of color being mixed together. Greater the value, lighter will be the color. Setting all three colors to **0** creates a black material whereas setting all three colors to **255** creates a white material.*

8. Choose the **OK** button to close the **Color Selector: Diffuse Color** dialog box.

Now, you will make the material shiny.

9. Choose the **Specular** color swatch in the **Blinn Basic Parameters** rollout; the **Color Selector: Specular Color** dialog box is displayed.

10. In this dialog box, set the values as given next:

    Red: **255** Green: **255** Blue: **62**

11. Choose the **OK** button to close the **Color Selector: Specular Color** dialog box.

12. In the **Material Editor** dialog box, set the values in the **Specular Highlights** area as given next:

    Specular Level: **60** Glossiness: **20**

    Now, you need to name the material and put it in the material library.

13. Change the name of the material to **Yellow light** by entering it in the drop-down list located on the left of the **Standard (Legacy)** button in the **Material Editor** dialog box.

14. Choose the **Put to Library** button in the **Material Editor** dialog box; a flyout is displayed.

15. Choose **signal.mat** (newly created material library) from the flyout; the **Put To Library** dialog box is displayed.

    The name *Yellow light* is displayed in the **Name** text box. You can place this material in the library by either using the default name or renaming it.

16. Choose the **OK** button; the current material is added under the **signal** material library.

Creating the Green Light Material

In this section, you will create material for the green light of the signal.

1. Select the third material sample slot in the **Material Editor** dialog box. Various rollouts are displayed in the **Material Editor** dialog box to modify the **Standard (Legacy)** material.

2. In the **Shader Basic Parameters** rollout, make sure **Blinn** is selected in the drop-down list.

3. In the **Blinn Basic Parameters** rollout, unlock the **Diffuse** and **Ambient** colors. Choose the **Ambient** color swatch; the **Color Selector: Ambient Color** dialog box is displayed. In this dialog box, set the values as given next:

    Red: **0** Green: **50** Blue: **0**

    The ambience level of the material color is set to dark green.

4. Choose the **OK** button to close the **Color Selector: Ambient Color** dialog box.

5. Choose the **Diffuse** color swatch in the **Material Editor** dialog box; the **Color Selector: Diffuse Color** dialog box is displayed. In this dialog box, set the values as given next:

   Red: **0** Green: **225** Blue: **0**

   The diffuse portion of the material color is set to bright green.

6. Choose the **OK** button to close the **Color Selector: Diffuse Color** dialog box.

   Now, you need to make the material shiny.

7. Choose the **Specular** color swatch in the **Blinn Basic Parameters** rollout; the **Color Selector: Specular Color** dialog box is displayed.

8. In this dialog box, set the values as given next:

   Red: **0** Green: **47** Blue: **0**

9. Choose the **OK** button to close the **Color Selector: Specular Color** dialog box.

10. In the **Specular Highlights** area of the **Blinn Basic Parameters** rollout, set the values as given next:

    Specular Level: **60** Glossiness: **20**

    Now, you need to make the material self-illuminating to make the object appear glowing. It is achieved by replacing a percentage of the ambient color with the diffuse color in the shadowed areas of the object.

11. In the **Self-Illumination** area of the **Blinn Basic Parameters** rollout, set the value in the **Color** spinner to **85**. If you set the value in the **Color** spinner closer to 100, the material appears brighter.

    Now, you need to name the material and put it in the material library.

12. Change the name of the material to **Green light** by entering it in the drop-down list located on the left of the **Standard (Legacy)** button in the **Material Editor** dialog box.

13. Choose the **Put to Library** button in the **Material Editor** dialog box; a flyout is displayed. Choose **signal** from the flyout; the **Put To Library** dialog box is displayed.

14. The name *Green light* is displayed in the **Name** text box. You can place this material in the library by either using the default name or renaming it. To use the current name, choose the **OK** button; the current material is added under the **signal** material library.

Saving the Material Library

The changes made to the material library must be saved. Else, they will be lost when you exit the current file in 3ds Max. In this section, you will save the material library.

1. Choose the **Get Material** button in the **Material Editor** dialog box; the **Material/Map Browser** dialog box is displayed. Next, right-click on the **signal.mat** rollout; a shortcut menu is displayed. Choose **Save** from the shortcut menu, as shown in Figure 7-9.

2. Close the **Material/Map Browser** dialog box.

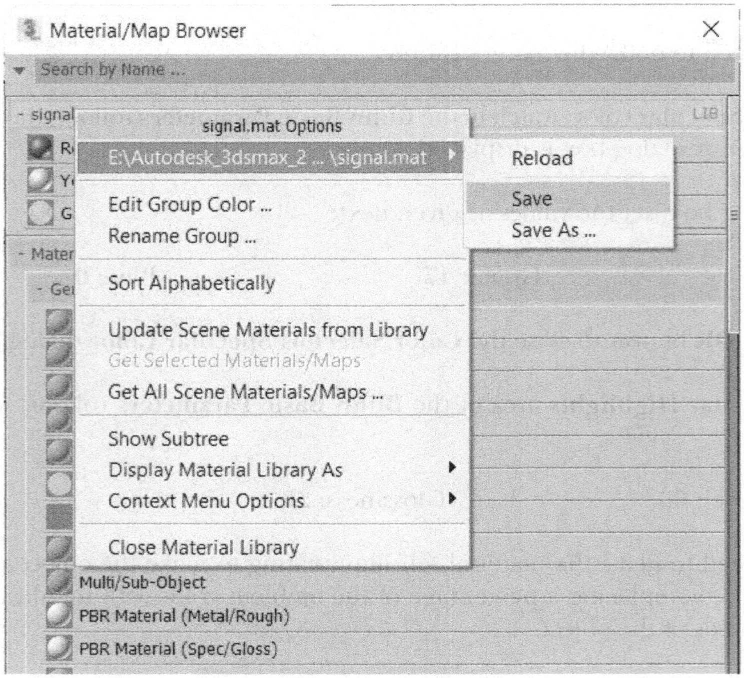

Figure 7-9 *Saving the material library*

Assigning Materials to Objects

The materials have been created and the material library is saved. In this section, you will assign materials to the objects in the scene.

1. Select **Red** from the **Scene Explorer** located on the left of the interface; *Red* is selected in the viewport.

2. In the **Material Editor** dialog box, select the material sample slot which is named as *Red light*; the *Red light* material is set as the current material.

3. Choose the **Assign Material to Selection** button; the currently selected material is assigned to *Red*. Similarly, assign the material *Yellow light* to *Yellow* and *Green light* to *Green*. Next, close the **Material Editor** dialog box.

Note

A material can also be assigned to an object by dragging the material sample slot into the scene and dropping it on the object.

Saving and Rendering the Scene

In this section, you will save the scene that you have created and then render it. You can also view the final rendered image of this model by downloading the *c07_3dsmax_2022_rndr.zip* file from *www.cadcim.com*. The path of the file is as follows: *Textbooks > Animation and Visual Effects > Autodesk 3ds Max > Autodesk 3ds Max 2022 for Beginners: A Tutorial Approach*

1. Choose **Save** from the **File** menu.

2. Activate the Perspective viewport. Next, invoke the **Render Production** tool from the **Main Toolbar**; the rendered image is displayed, refer to Figure 7-10.

Figure 7-10 *The rendered image*

Tutorial 2

In this tutorial, you will create and assign materials to the objects in the scene, refer to Figure 7-11. **(Expected time: 30 min)**

The following steps are required to complete this tutorial:

a. Create the project folder.
b. Open the file.
c. Create material for table.
d. Create material for glasses.
e. Create material for bowl.
f. Create material for bottle.
g. Save and render the scene.

Figure 7-11 The textured scene

Creating the Project Folder

Create the project folder with the name *c07_tut2* in the *3dsmax 2022* folder, as discussed in Tutorial 1 of Chapter 2.

Opening the File

1. Open Windows Explorer and then browse to the *c07_3dsmax_2022_tut* folder and copy all jpeg files from this folder to *\Documents\3dsmax2022\c07_tut2\sceneassets\images.*

2. Choose **Open** from the **File** menu; the **Open File** dialog box is displayed. In this dialog box, browse to the location *\Documents\c07_3dsmax_2022_tut.zip* and select the **c07_tut2_start.max** file from it. Choose the **Open** button to open this file; the scene is displayed in the Perspective viewport with two predefined materials assigned to the *floor* and *wall*, as shown in Figure 7-12.

3. Choose the **Save As** button from the **File** menu; the **Save File As** dialog box is displayed. Browse to the location *\Documents\3dsmax2022\c07_tut2\scenes*. Save the file with the name c07tut2.max at this location.

Creating Material for Table

In this section, you will create material for *table* using the **Standard (Legacy)** material. You will also use the **Poly Select** and **UVW Map** modifiers to properly align the texture.

1. Select **table** from the Scene Explorer. Next, press the M key; the **Material Editor** dialog box is displayed. Select the unused sample slot from the **Material Editor** dialog box. In the **Material Name** drop-down list, enter **table**. Next, various rollouts are displayed in the **Material Editor** dialog box to modify the **Standard (Legacy)** material.

Figure 7-12 The c07_tut2_start file

2. In the **Blinn Basic Parameters** rollout, choose the **Diffuse** map button located next to the **Diffuse** color swatch; the **Material/Map Browser** dialog box is displayed. Choose the **Bitmap** map from the **Maps > General** rollout and choose the **OK** button; the **Select Bitmap Image File** dialog box is displayed. As the project folder is already set, the *images* folder is displayed in the **Look in** drop-down list of this dialog box. Select the file **wood.jpg** and then choose the **Open** button; the image is displayed in the selected sample slot.

3. Choose the **Go to Parent** button. Next, make sure *table* is selected.

4. Set the following values in the **Specular Highlights** area:

   Specular Level: **65** Glossiness: **50**

5. Choose the **Assign Material to Selection** and **Show Shaded Material in Viewport** buttons from the **Material Editor** dialog box; the **wood** material is assigned to *table,* as shown in Figure 7-13.

   You will notice that the **wood** material is not assigned properly on *table*. To properly assign this material, you need to follow the steps given next.

6. Make sure *table* is selected. Choose the **Modify** tab in the **Command Panel**. Next, select the **Poly Select** modifier from the **Modifier List** drop-down list of the **Command Panel**; the **Poly Select** modifier is displayed in the modifier stack and rollouts are displayed below the modifier stack.

7. In the modifier stack, click on the arrow on the left of the **Poly Select** modifier to view all the sub-object levels.

8. Select the **Polygon** sub-object level; it turns green and gets activated.

9. Choose the **Maximize Viewport Toggle** tool. Next, activate the Front viewport and select the polygons, as shown in Figure 7-14.

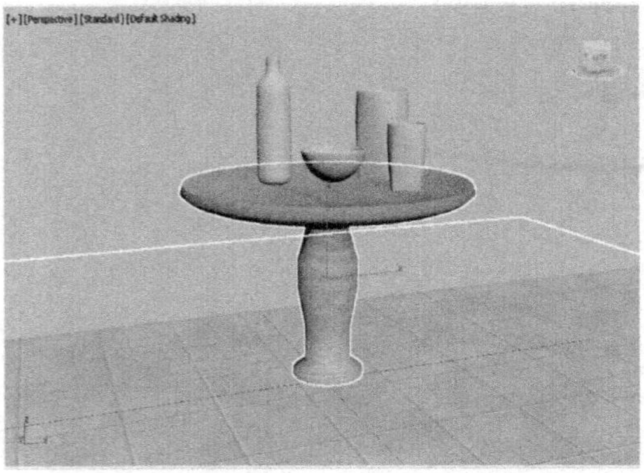

Figure 7-13 *The* ***wood*** *material assigned to table*

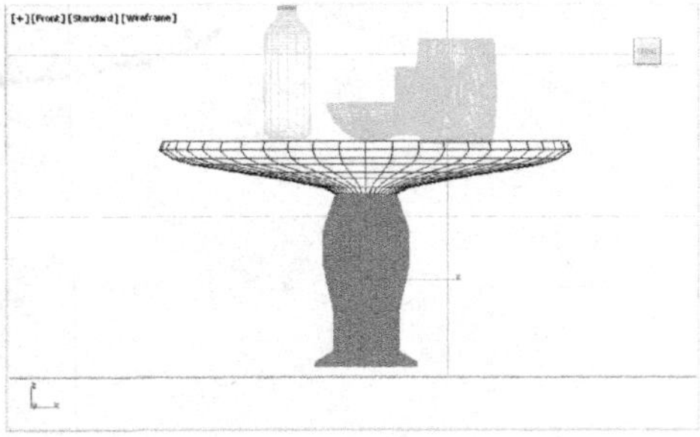

Figure 7-14 *Polygons of table selected*

10. Select the **UVW Map** modifier from the **Modifier List** drop-down list of the **Command Panel**; the **UVW Map** modifier is displayed in the modifier stack.

11. In the **Parameters** rollout, select the **Cylindrical** radio button from the **Mapping** area. Next, make sure the **Z** radio button is selected and then choose the **Fit** button from the **Alignment** area; the texture is properly aligned on the bottom part of *table,* as shown in Figure 7-15.

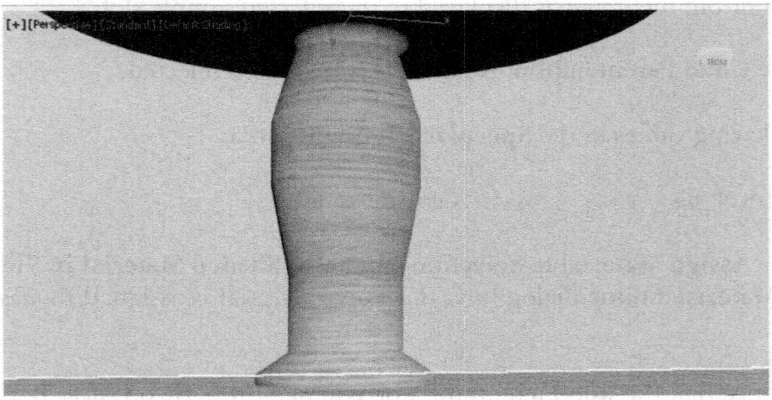

Figure 7-15 *The texture aligned on the bottom part of table*

12. Again, select the **Poly Select** modifier from the **Modifier List** drop-down list of the **Command Panel**; the **Poly Select** modifier is displayed in the modifier stack and rollouts are displayed below the modifier stack.

13. In the modifier stack, click on the arrow on the left of the **Poly Select** modifier to view all the sub-object levels. Next, select the **Polygon** sub-object level.

14. In the Front viewport, select polygons of *table,* as shown in Figure 7-16.

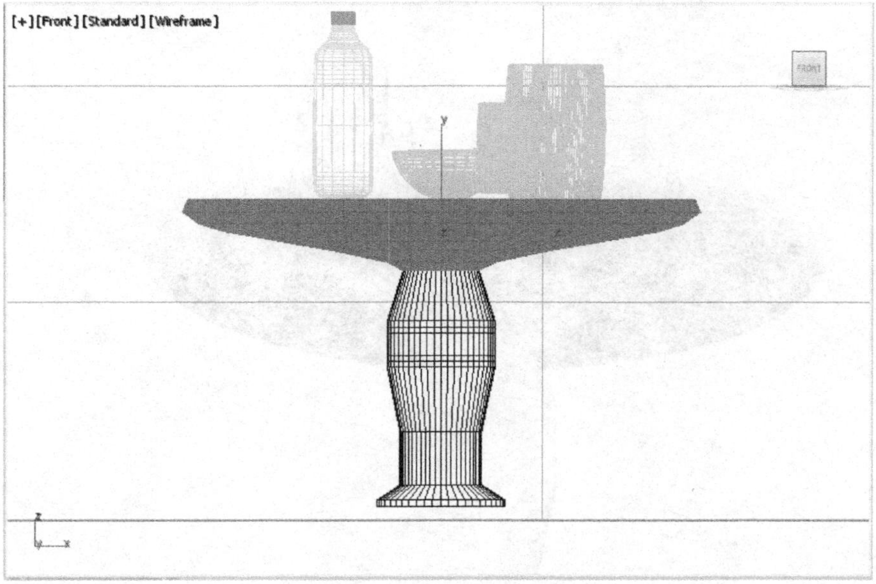

Figure 7-16 Polygons of table selected

15. Select the **UVW Map** modifier from the **Modifier List** drop-down list of the **Command Panel**; the **UVW Map** modifier is displayed in the modifier stack.

16. In the **Parameters** rollout, select the **Cylindrical** radio button from the **Mapping** area. Next, make sure the **Z** radio button is selected and then choose the **Fit** button from the **Alignment** area; the texture is properly aligned on the middle part of *table,* as shown in Figure 7-17.

17. Again, select the **Poly Select** modifier from the **Modifier List** drop-down list of the **Command Panel**; the **Poly Select** modifier is displayed in the modifier stack and rollouts are displayed below the modifier stack.

18. In the modifier stack, click on the arrow on the left of the **Poly Select** modifier to view all the sub-object levels. Next, select the **Polygon** sub-object level.

19. In the Top viewport, select polygon of *table,* as shown in Figure 7-18.

20. Select the **UVW Map** modifier from the **Modifier List** drop-down list of the **Command Panel**; the **UVW Map** modifier is displayed in the modifier stack.

21. In the **Parameters** rollout, make sure the **Planar** radio button is selected in the **Mapping** area. Next, make sure the **Z** radio button is selected and then choose the **Fit** button from the **Alignment** area; the texture is properly aligned on the upper part of *table,* as shown in Figure 7-19.

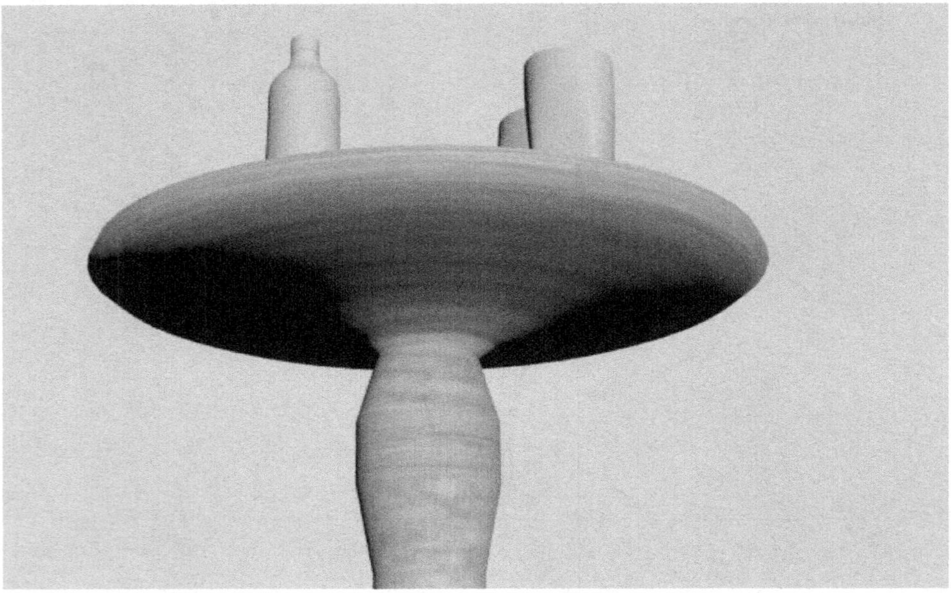

Figure 7-17 *The texture aligned on the middle part of table*

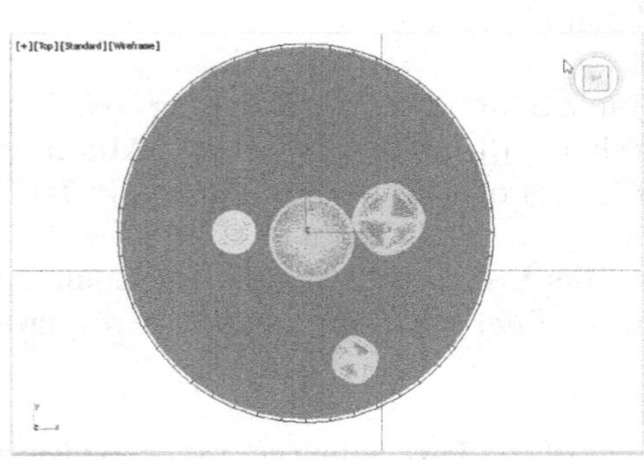

Figure 7-18 *Selected polygon of table*

Figure 7-19 *The texture aligned on the upper part of table*

Creating Material for Glasses

In this section, you will create plastic material for glasses by using material from the Autodesk Material Library. You will also change the default scanline renderer to ART renderer to use the material from the Autodesk Material Library.

1. Select the next unused sample slot from the **Material Editor** dialog box. In the **Material Name** drop-down list, enter **plastic glass**.

2. Choose the **Render Setup** tool from the **Main Toolbar**; the **Render Setup: Scanline Renderer** dialog box is displayed.

3. In this dialog box, select **ART Renderer** from the **Renderer** drop-down list; the **Scanline Renderer** is changed to **ART Renderer**.

4. Choose the **Standard (Legacy)** button available in the **Material Editor** dialog box; the **Material/Map Browser** dialog box is displayed. In this dialog box, select **Smooth - Navy** from the **Autodesk Material Library > Plastic** rollout and choose **OK**; the **Autodesk Plastic/Vinyl** material is displayed in the **Material Editor** dialog box.

5. Choose the color swatch from the **Plastic** rollout of the **Material Editor** dialog box; the **Color Selector: Plastic_Color** dialog box is displayed. In this dialog box, set the values as follows:

   Red: **0.047** Green: **0.349** Blue: **0.659**

   Choose **OK** to close the dialog box.

6. Select **Matte** from the **Finish** drop-down list. Next, select *glass* and *glass1* from the Scene Explorer.

7. Assign the **plastic glass** material to *glass* and *glass1*, as discussed earlier.

8. Render the Perspective viewport; the rendered image is displayed, refer to Figure 7-20.

Figure 7-20 *The rendered image*

Creating Material for Bowl

In this section, you will create material for bowl using the **Multi/Sub-Object** material. You will also use material from the Autodesk material library as a sub-object material in the **Multi/Sub-Object** material.

1. Select next unused sample slot from the **Material Editor** dialog box. In the **Material Name** drop-down list, enter **bowl**.

2. Choose the **Standard (Legacy)** button available in the **Material Editor** dialog box to change the material type; the **Material/Map Browser** dialog box is displayed. In this dialog box, select **Multi/Sub-Object** from the **Materials > General** rollout and then choose the **OK** button; the **Replace Material** message box is displayed. In this message box, select the **Discard old material?** radio button and choose **OK**; the **Standard (Legacy)** material is replaced by the **Multi/Sub-Object** material in the **Material Editor** dialog box. Also, various rollouts are displayed in the **Material Editor** dialog box to modify the **Multi/Sub-Object** material.

3. In the **Multi/Sub-Object Basic Parameters** rollout, choose the **Set Number** button; the **Set Number of Materials** dialog box is displayed. In this dialog box, enter **2** in the **Number of materials** spinner and choose **OK**; sub-materials in the **Multi/Sub-Object Basic Parameters** rollout are reduced to 2.

4. Enter **body** in the **Name** text box of first sub-material and choose the **None** button at the right of this text box; the **Material/Map Browser** dialog box is displayed. In this dialog box, select **Navy Blue** from the **Autodesk Material Library > Ceramic > Porcelain** rollout and choose **OK**; the **Autodesk Ceramic** material is displayed in the **Material Editor** dialog box.

5. Choose the color swatch from the **Ceramic** rollout of the **Material Editor** dialog box; the **Color Selector: Ceramic_Color** dialog box is displayed. In this dialog box, set the values as follows:

   Red: **0.031** Green: **0.031** Blue: **0.2**

   Choose **OK** to close the dialog box.

6. Select the **Enable** check box and enter **1** in the **Amount** spinner of the **Finish Bumps** area.

7. Choose the **Go to Parent** button. Next, click and drag material from the first sub-material to the second sub-material; the **Instance (Copy) Material** dialog box is displayed. In this dialog box, select the **Copy** radio button from the **Method** area and choose **OK**; material in the first sub-material is copied to the second sub-material.

8. Click on the second sub-material. Next, select **Use Map** from the **Color** drop-down list; the color swatch below the **Color** drop-down list is changed to the **No Map** button.

9. Choose the **No Map** button; the **Material/Map Browser** dialog box is displayed. In this dialog box, select **Bitmap** from the **Maps > General** rollout and then choose the **OK** button; the **Select Bitmap Image File** dialog box is displayed.

10. As the project folder is already set, the *images* folder is displayed in the **Look in** drop-down list of this dialog box. Select the file **ceramic.jpg** and then choose the **Open** button; the image is displayed in the selected sample slot.

11. In the **Coordinates** rollout, enter **4** in the **U: Tiling** and **V: Tiling** spinners.

12. Choose the **Go to Parent** button twice. Next, enter **design** in the **Name** text box of second sub-material.

13. Select *bowl* from the Scene Explorer. Next, press ALT +Q; *bowl* is isolated. Now, choose the **Maximize Viewport Toggle** tool; all viewports are displayed.

14. Activate the Front viewport and choose the **Modify** tab from the **Command Panel**. Next, click on the arrow on the left of the **Editable Poly** in the modifier stack and select the **Polygon** sub-object level; it turns green and gets activated.

15. Select the polygons of *bowl,* as shown in Figure 7-21. Next, enter **1** in the **Set ID** spinner of **Polygon: Material IDs** area.

16. Select rest of the polygons of *bowl,* as shown in Figure 7-22. Next, enter **2** in the **Set ID** spinner of the **Polygon: Material IDs** area.

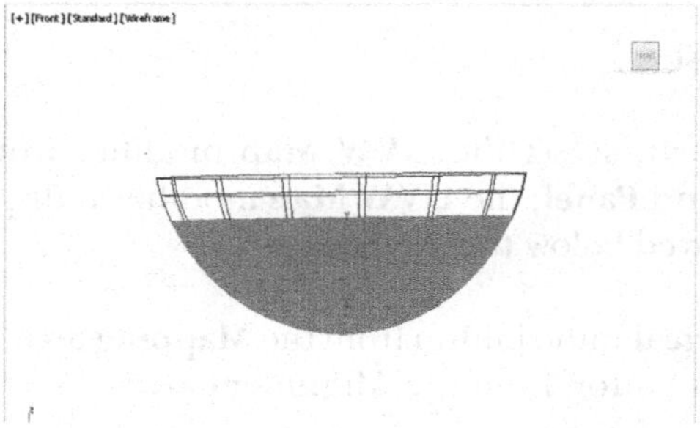

Figure 7-21 *Selected polygons of bowl*

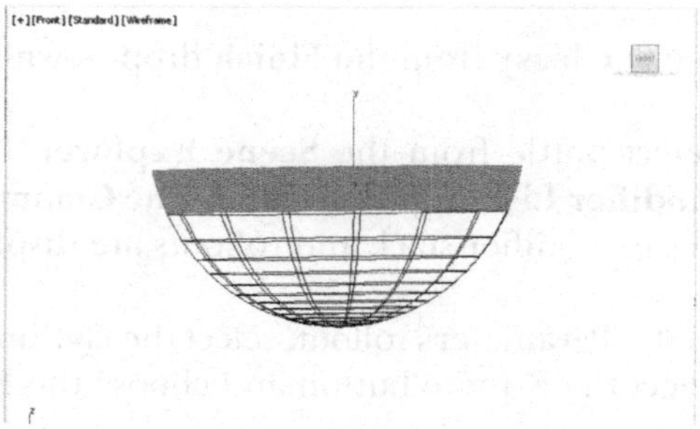

Figure 7-22 *Rest of the polygons selected*

17. In the modifier stack, select the **Polygon** sub-object level again to deactivate it.

18. Select the **UVW Map** modifier from the **Modifier List** drop-down list of the **Command Panel**; the **UVW Map** modifier is displayed in the modifier stack and the rollouts are displayed below the modifier stack.

19. In the **Parameters** rollout, select the **Spherical** radio button from the **Mapping** area. Next, select the **X** radio button and choose the **Fit** button from the **Alignment** area.

20. Make sure *bowl* is selected. Next, choose the **Assign Material to Selection** and **Show Shaded Material in Viewport** buttons from the **Material Editor** dialog box; the **bowl** material is assigned to *bowl,* refer to Figure 7-23. Render the Perpective viewport. You will notice that the assigned material is displayed properly on *bowl.*

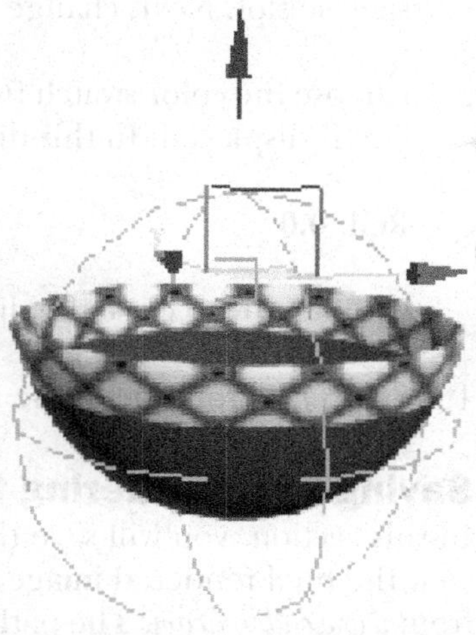

Figure 7-23 *The* ***bowl*** *material applied to bowl*

21. Choose the **Isolate Selection Toggle** button from the status bar; all objects are displayed in the viewports.

Creating Material for Bottle

In this section, you will create material for bottle using material from the Autodesk Material Library.

1. Select the **plastic glass** material from the **Material Editor** dialog box and drag it to the unused sample slot. Next, change the name of the material to **plastic bottle** in the **Material Name** drop-down list.

2. In the **Plastic** rollout, select **Plastic (Transparent)** from the **Type** drop-down list.

3. Choose the color swatch from the **Plastic** rollout; the **Color Selector: Plastic_Color** dialog box is displayed. In this dialog box, set the values as follows:

   Red: **0.039** Green: **1.0** Blue: **0.808**

   Choose **OK** to close the dialog box.

4. Select **Glossy** from the **Finish** drop-down list.

5. Select bottle from the **Scene Explorer**. Next, select the **UVW Map** modifier from the **Modifier List** drop-down list of the **Command Panel**; the **UVW Map** modifier is displayed in the modifier stack and rollouts are displayed below the modifier stack.

6. In the **Parameters** rollout, select the **Cylindrical** radio button from the **Mapping** area. Next, select the **X** radio button and choose the **Fit** button from the **Alignment** area.

7. Assign the **plastic bottle** material to bottle.

8. Select the **plastic glass** material from the **Material Editor** dialog box and drag it to unused sample slot. Next, change the name in the **Material Name** drop-down list to **cap**.

9. Choose the color swatch from the **Plastic** rollout; the **Color Selector: Plastic_Color** dialog box is displayed. In this dialog box, set the values as follows:

   Red: **0.0** Green: **0.439** Blue: **0.298**

   Choose **OK** to close the dialog box.

10. Select *cap* from the Scene Explorer. Next, assign the **cap** material to it.

Saving and Rendering the Scene

In this section, you will save the scene that you have created and then render it. You can also view the final rendered image of this model by downloading the *c07_3dsmax_2022_rndr.zip* file from *www.cadcim.com*. The path of the file is as follows: *Textbooks > Animation and Visual Effects > Autodesk 3ds Max > Autodesk 3ds Max 2022 for Beginners: A Tutorial Approach*

1. Choose **Save** from the **File** menu.

2. Activate the Perspective viewport. Next, invoke the **Render Production** tool from the **Main Toolbar**; the rendered image is displayed, refer to Figure 7-24.

Figure 7-24 The rendered image

Self-Evaluation Test

Answer the following questions and then compare them to those given at the end of this chapter:

1. Which of the following buttons is used to invoke the **Material/Map Browser** dialog box?

   (a) **Put to Library** (b) **Get Material**
   (c) **Go to Parent** (d) None of these

2. In 3ds Max, which of the following material types in the **Material Editor** dialog box is available by default?

   (a) **Architectural** (b) **Physical**
   (c) **Multi/Sub-Object** (d) **Arch & Design**

3. Which of the following shortcut keys is used to invoke the **Asset Tracking** window to re-link the missing materials?

   (a) SHIFT + M (b) SHIFT + L
   (c) SHIFT + I (d) SHIFT + T

4. The __________ dialog box displays all materials available in the currently open material library.

5. The __________ portion of a color describes the part of an object that is not under direct light but in shadow.

6. By adjusting the __________ and __________ portions of a material, you can create a matte, or a non-glossy material.

7. While defining a material, if you set the value of the red, green, and blue colors to __________ in the **Diffuse Color** swatch, black material will be created.

8. The __________ portion of the color describes the highlights of a shiny material.

9. The appearance of the highlights in a material is determined by using the __________ and __________ spinners.

10. Once the material is saved in the library, it can be assigned to objects. (T/F)

Review Questions

Answer the following questions:

1. Which of the following color swatches in the **Blinn Basic Parameters** rollout specifies the part of an object under normal light conditions?

   (a) **Ambient** (b) **Diffuse**
   (c) **Specular** (d) None of these

2. The 3ds Max uses the __________ and __________ properties of a material to shade an object.

3. If an object is transparent, semitransparent, or open ended, you need to select the __________ check box in the **Shader Basic Parameters** rollout.

4. If the lock button between the **Ambient** and **Diffuse** color swatches is turned on then different colors will get applied for ambient and diffuse settings. (T/F)

5. The lower the value of the ambient color, the darker will be the colored portion. (T/F)

6. Which of the following shortcut keys is used to isolate objects in a viewport?

   (a) ALT+I (b) ALT+P
   (c) ALT+Q (d) ALT+S

7. In which of the following locations, the **Isolate Selection Toggle** button is located?

(a) Menu bar (b) Toolbar
(c) **Axis Constraint Toolbar** (d) Status bar

8. Which of the following path sequence is used to invoke the **Select Bitmap Image File** dialog box from the **Material Editor** dialog box?

(a) Choose **Bitmap** map from the **Maps > Standard** rollout
(b) Choose **Bitmap** map from the **Maps > General** rollout
(c) Choose **Bitmap** map from the **Maps > Scanline** rollout
(d) Choose **Bitmap** map from the **Maps > Scene materials** rollout

EXERCISES

The rendered output of the model used in the exercises given next can be accessed by downloading the *c07_3dsmax_2022_exr.zip* from *www.cadcim.com*. The path of the file is as follows: *Textbooks > Animation and Visual Effects > Autodesk 3ds Max > Autodesk 3ds Max 2022 for Beginners: A Tutorial Approach*

Exercise 1

Create the materials needed to texture a scene containing five crayons, as shown in Figure 7-25. You will need five colors of your choice for the wrappers of crayons and five shiny colors for the crayons. **(Expected time: 25 min)**

Figure 7-25 *The crayons*

Exercise 2

Create the model of cups and glasses shown in Figure 7-26 using the **Line** tool and **Lathe** modifier. To create the steel material, you need to use the **Oren-Nayar-Blinn shader** in the **Standard (Legacy)** material and the **Reflection** map in the **Maps** rollout. **(Expected time: 20 min)**

Figure 7-26 *The model of cups and glasses*

Exercise 3

Create the model of a tea cup in the viewport, as shown in Figure 7-27. Also, you need to create copper material using the **Phong** shader from the **Standard (Legacy)** material and assign it to the object. **(Expected time: 20 min)**

Answers to Self-Evaluation Test

1. b, **2.** b, **3.** d, **4. Material/Map Browser**, **5.** ambient, **6.** ambient, diffuse, **7.** 0, **8.** specular, **9. Specular Level, Glossiness**, **10.** T

Chapter 8

Material Editor: Texture Maps-I

Learning Objectives

After completing this chapter, you will be able to:
- *Apply texture maps*
- *Apply bump maps*

INTRODUCTION

In this chapter, you will learn to create materials using texture maps. Multiple texture maps can be used to create complex materials. In addition, you will also learn to use the bump maps.

TUTORIAL

Before starting the tutorial, you need to download the *c08_3dsmax_2022_tut.zip* file from *www.cadcim.com*. The path of the file is as follows: *Textbooks > Animation and Visual Effects > Autodesk 3ds Max > Autodesk 3ds Max 2022 for Beginners: A Tutorial Approach*

Extract the contents of the zip file and save them in the *Documents* folder.

Tutorial 1

In this tutorial, you will create different types of materials and then apply them to a checkerboard, as shown in Figure 8-1. **(Expected time: 20 min)**

Figure 8-1 *The checkerboard after applying materials*

The following steps are required to complete this tutorial:

a. Create a project folder.
b. Open the file.
c. Create a new material library.
d. Create checkerboard border material.
e. Create the board material.
f. Create the red checker material.
g. Create the black checker material.
h. Save the material library.
i. Assign materials.
j. Save and render the scene.

Creating the Project Folder

Create a project folder with the name *c08_tut1* in the *3dsmax 2022* folder, as discussed in earlier chapters.

Opening the File

1. Open Windows Explorer and then browse to *\Documents\c08_3dsmax_2022_tut.* Next, copy the *checkerb.jpg, checkerr.jpg, checker1.jpg,* and *marble.gif* files to *\Documents\3dsmax 2022\c08_tut1\sceneassets\images.*

2. Choose **Open** from the **File** menu; the **Open File** dialog box is displayed. In this dialog box, browse to the location *\Documents\c08_3dsmax_2022_tut.zip* and select the **c08_tut1_start.max** file from it. Choose the **Open** button to open the file, refer to Figure 8-2.

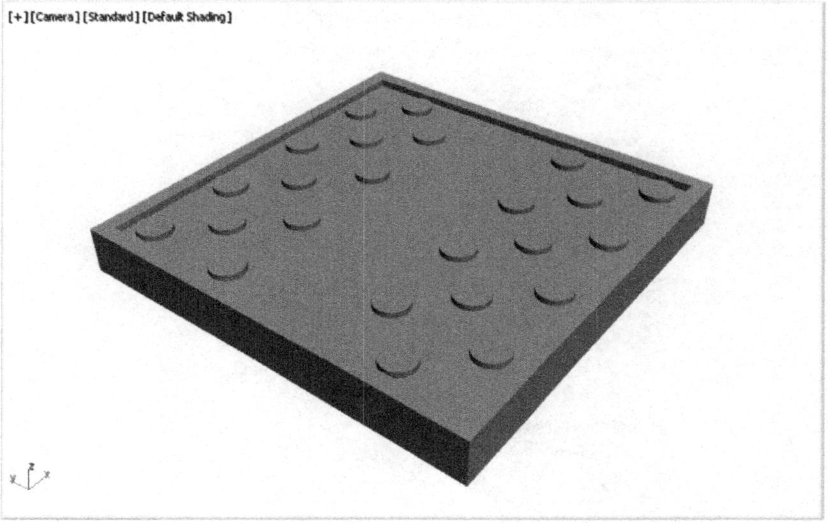

Figure 8-2 *The c08_tut1_start file*

3. Choose **Save As** from the **File** menu; the **Save File As** dialog box is displayed. Browse to the location *\Documents\3dsmax 2022\c08_tut1\scenes.* Save the file with the name *c08tut1.max* at this location.

Creating a New Material Library

In this section, you will create a new material library by using the **Material Editor** dialog box.

1. Choose **Rendering > Material Editor > Compact Material Editor** from the menu bar or press the M key; the **Material Editor** dialog box is displayed, as shown in Figure 8-3.

2. In this dialog box, choose the **Get Material** button; the **Material/Map Browser** dialog box is displayed. 

3. Choose the **Material/Map Browser Options** button from the **Material/Map Browser** dialog box; a flyout is displayed. Choose **New Material Library** from the flyout, as shown in Figure 8-4; the **Create New Material Library** dialog box is displayed, as shown in Figure 8-5.

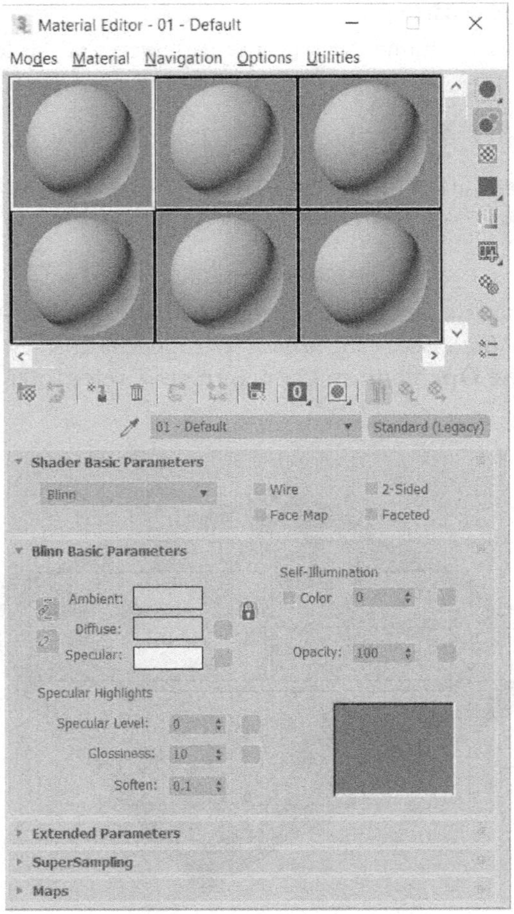

Figure 8-3 *The* ***Material Editor*** *dialog box*

As the project folder is already set, the *materiallibraries* folder of this project is displayed in the **Save in** drop-down list.

4. In this dialog box, enter **Checker Board** in the **File name** text box and then choose the **Save** button; the new material library named **Checker Board** is created in the **Material/Map Browser** dialog box.

5. Close the **Material/Map Browser** dialog box to return to the **Material Editor** dialog box.

Creating Checker Board Border Material

In this section, you will create a checker board material.

1. Select the first material sample slot in the **Material Editor** dialog box to make it the current sample slot.

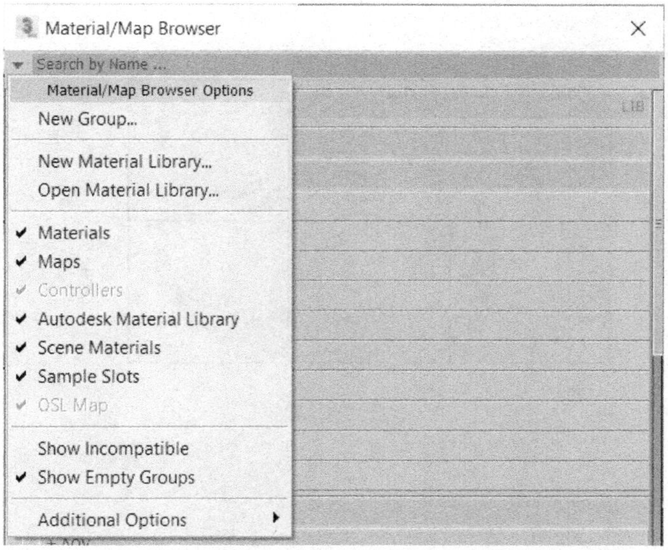

*Figure 8-4 Choosing **New Material Library** from the flyout*

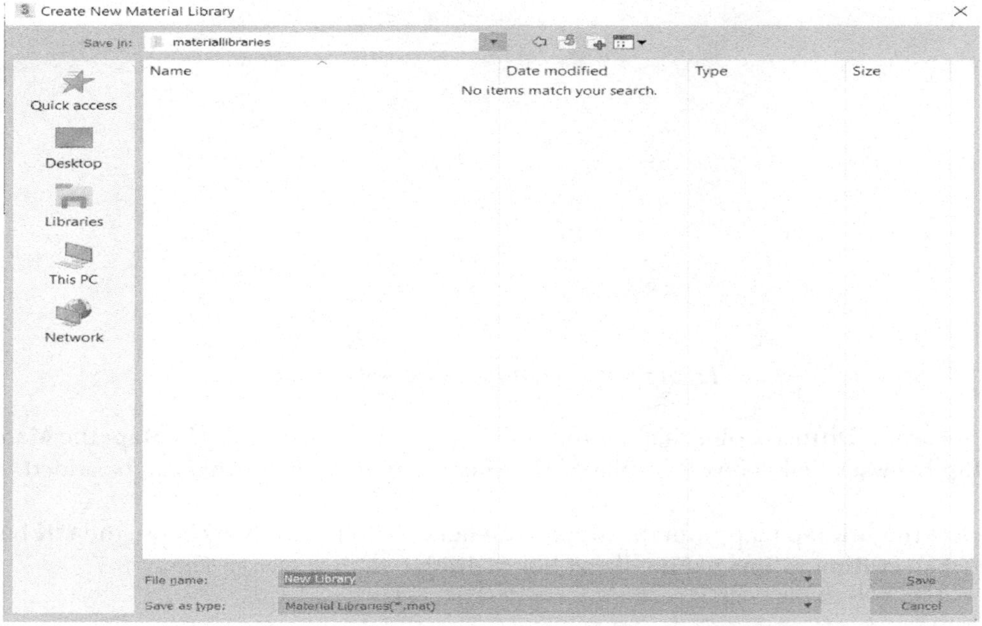

*Figure 8-5 The **Create New Material Library** dialog box*

2. In the **Shader Basic Parameters** rollout, make sure **Blinn** is selected in the drop-down list. Next, expand the **Maps** rollout, refer to Figure 8-6.

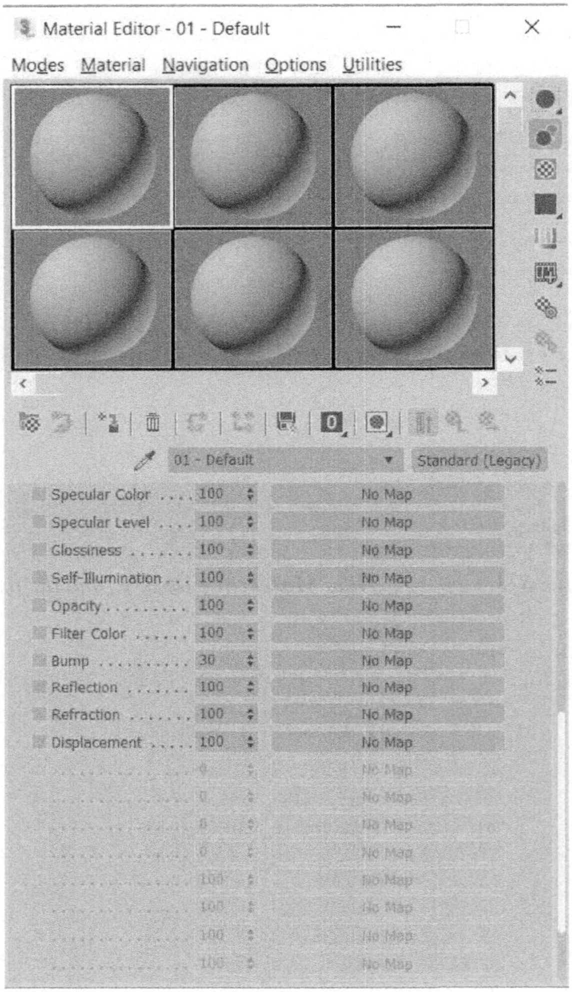

*Figure 8-6 The **Maps** rollout expanded*

3. Choose the **Diffuse Color** map button which is currently labeled as **No Map**; the **Material/ Map Browser** dialog box is displayed showing the types of maps that can be added.

4. Select the **Bitmap** map from the **Maps > General** rollout and then choose the **OK** button; the **Select Bitmap Image File** dialog box is displayed, as shown in Figure 8-7.

5. In this dialog box, make sure **All Formats** is displayed in the **Files of type** drop-down list. As the project folder is already set, the *images* folder of this project is displayed in the **Look in** drop-down list. Select the **marble.gif** file and then choose the **Open** button; the image file is loaded and the parameters are displayed for the map in the **Material Editor** dialog box.

6. Click and hold the **Sample Type** button located on the right in the **Material Editor** dialog box; a flyout is displayed. Choose the **Square** button to switch the sample from sphere to square, refer to Figure 8-8.

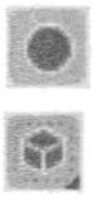

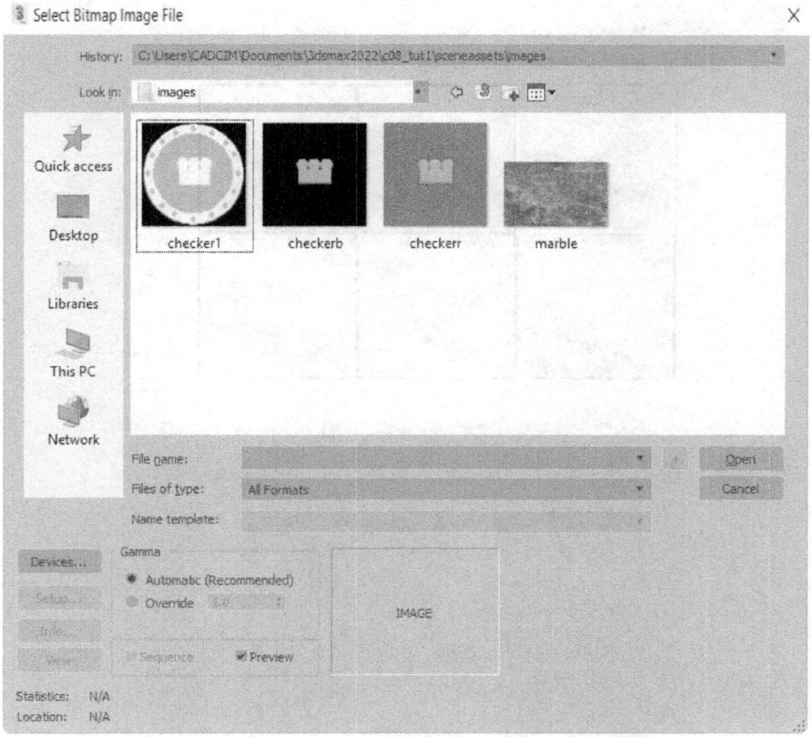

Figure 8-7 *The* ***Select Bitmap Image File*** *dialog box*

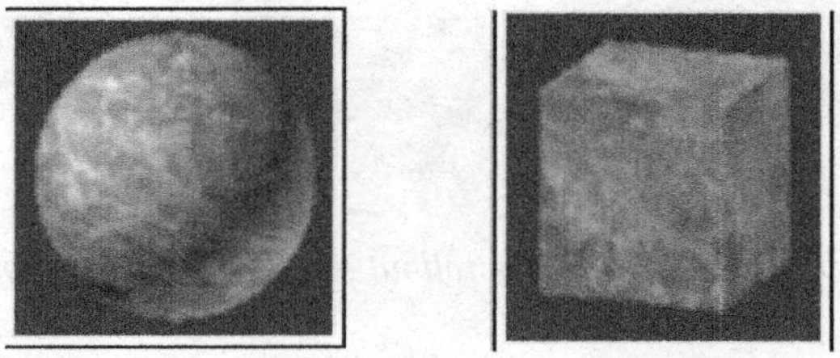

Figure 8-8 *The sample switched from sphere to square in the* ***Material Editor*** *dialog box*

7. Choose the **Show Shaded Material in Viewport** button in the **Material Editor** dialog box. This will make the bitmap image visible in the shaded viewport when you assign material to the object. 

8. In the **Coordinates** rollout, make sure the **Use Real-WorldScale** check box is cleared. Also, make sure **1** is set in the **U:Tiling** and **V:Tiling** spinners, as shown in Figure 8-9.

9. Choose the **Go To Parent** button to return to the root level.

   Next, you will name the material created.

10. Choose the **Put to Library** button in the **Material Editor** dialog box; a flyout is displayed. 

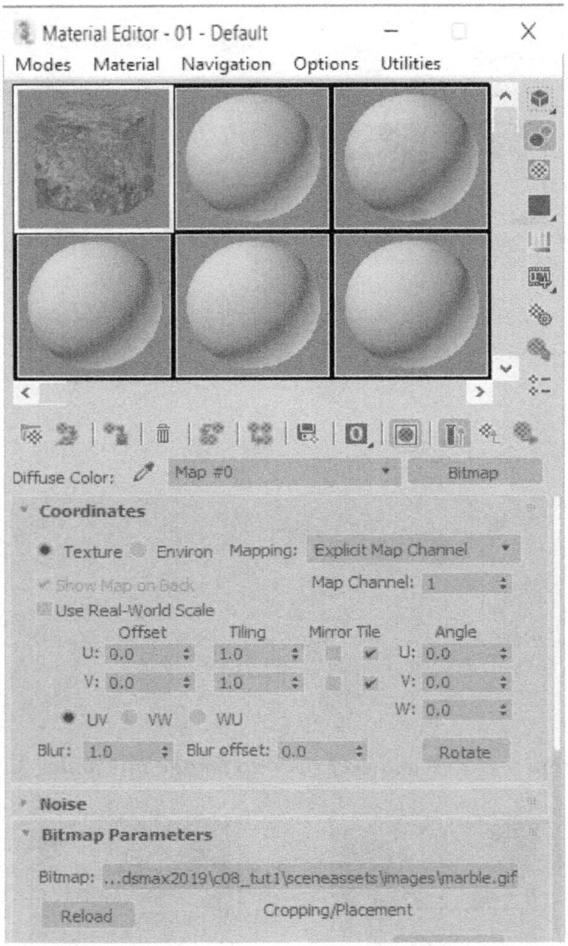

Figure 8-9 *The* ***Coordinates*** *rollout after modifying the parameters*

11. Choose the **Checker Board.mat** option from the flyout; the **Put To Library** dialog box is displayed.

12. Enter **Checker Board border** as the name of the new material in the **Put To Library** dialog box and choose the **OK** button; the material is assigned the name *Checker Board border* and is added to the **Checker Board** material library.

13. Select **border** from the **Scene Explorer**.

14. Select the **UVW Map** modifier from the **Modifier list** drop-down list of the **Command Panel**; the **UVW Map** modifier is applied and displayed in the modifier stack.

15. In the **Parameters** rollout, select the **Box** radio button from the **Mapping** area and enter **108** in the **Length** spinner. Now, the texture will be properly aligned to *border* after assigning material to it.

Creating the Board Material

In this section, you will create the board material.

1. In the **Material Editor** dialog box, select the second material sample slot.

2. Expand the **Maps** rollout and then choose the **Diffuse Color** map button that is currently labeled as **No Map**; the **Material/Map Browser** dialog box is displayed showing the types of maps that can be added.

3. Select the **Checker** map from the **Maps > General** rollout and choose the **OK** button; various rollouts for the **Checker** map are displayed in the **Material Editor** dialog box.

4. Make sure the **Use Real-World Scale** check box is cleared in the **Coordinates** rollout. Next, set the value **1** in the **U:Tiling** and **V:Tiling** spinners.

   The material currently resembles a portion of a black and white checkerboard.

5. Choose the **Show Shaded Material in Viewport** button in the **Material Editor** dialog box. Next, choose the **Go To Parent** button to return to the root level.

   Next, you will name the material created.

6. Choose the **Put to Library** button in the **Material Editor** dialog box; a flyout is displayed.

7. Choose the **Checker Board.mat** option from the flyout; the **Put To Library** dialog box is displayed.

8. Enter **Board** as the name of the new material in the **Put To Library** dialog box. Choose the **OK** button; the material is assigned the name *Board* and is added to the **Checker Board** material library.

Creating the Red Checker Material

In this section, you will create a red checker material with diffuse color and bump map. Also, you will assign the image to the bump map to make the surface of the material appear raised and lowered. Bump maps use a selected image file to make the surface of a material appear raised and lowered. Lighter pixels in the map will make the corresponding areas on the object appear to be raised while darker pixels on the map will make the corresponding areas on the object appear lowered. The value in the **Bump** spinner controls the intensity of the bumps.

1. In the **Material Editor** dialog box, select the third material sample slot.

2. Expand the **Maps** rollout. Next, choose the **Diffuse Color** map button that is currently labeled as **No Map**; the **Material/Map Browser** dialog box is displayed showing the types of maps that can be added.

3. Select the **Bitmap** map from the **Maps > General** rollout and choose the **OK** button; the **Select Bitmap Image File** dialog box is displayed.

4. In this dialog box, make sure **All Formats** is selected in the **Files of type** drop-down list. As the project folder is already set, the *images* folder of this project is displayed in the **Look in** text box of this dialog box. Select the **checkerr.jpg** file and choose the **Open** button to open the image file.

5. In the **Coordinates** rollout, make sure the **Use Real-WorldScale** check box is cleared. Also, make sure **1** is set in the **U:Tiling** and **V:Tiling** spinners.

6. Click and hold the **Sample Type** button on the right side of the **Material Editor** dialog box and choose the **Square** button to switch the sample from a sphere to a square, as done earlier.

   The material currently resembles a red checker with a golden crown at the center. To make the checker appear three-dimensional, you need to apply a bump map.

7. Choose the **Go To Parent** button in the **Material Editor** dialog box; the root level in the material definition is displayed.

8. In the **Maps** rollout, set the value in the **Bump** spinner to **90**.

   Now, a bump map needs to be assigned.

9. Choose the **Bump** map button currently labeled as **No Map**; the **Material/Map Browser** dialog box is displayed again.

10. Select the **Bitmap** map from the **Maps > General** rollout and choose the **OK** button; the **Select Bitmap Image File** dialog box is displayed.

11. In this dialog box, make sure **All Formats** is selected in the **Files of type** drop-down list. As the project folder is already set, the *images* folder of this project is displayed in the **Look in** text box of this dialog box. Select the **checker1.jpg** file and choose the **Open** button; the selected image is displayed in the selected sample slot.

    You will notice that the preview displayed in the **Select Bitmap Image File** dialog box is a grayscale bitmap. The white area will be raised the most while the black area will not be raised at all.

12. In the **Coordinates** rollout, make sure the **Use Real-WorldScale** check box is cleared. Also, make sure **1** is set in the **U:Tiling** and **V:Tiling** spinners.

13. Choose the **Show Shaded Material in Viewport** button in the **Material Editor** dialog box.

    Next, you will make the material shiny.

14. Choose the **Go To Parent** button in the **Material Editor** dialog box to return to the root level in the material definition and then collapse the **Maps** rollout.

15. Choose the **Specular** color swatch in the **Blinn Basic Parameters** rollout; the **Color Selector: Specular Color** dialog box is displayed.

16. In this dialog box, set the value **255** in the **Red**, **Green**, and **Blue** spinners; the specular color is set to white.

17. Choose the **OK** button to close the **Color Selector: Specular Color** dialog box.

18. In the **Specular Highlights** area of the **Blinn Basic Parameters** rollout, set the value **60** in the **Specular Level** spinner.

19. Set the value **20** in the **Glossiness** spinner.

    Next, you will name the material created.

20. Choose the **Put to Library** button in the **Material Editor** dialog box; a flyout is displayed.

21. Choose the **Checker Board.mat** from the flyout; the **Put To Library** dialog box is displayed.

22. Enter **Checker Red** as the name of the new material in the dialog box and choose the **OK** button; the material is added to the **Checker Board** material library.

Creating the Black Checker Material

In this section, you will create the black checker material.

1. Select the material sample slot next to the **Checker Red** material type.

2. Expand the **Maps** rollout and then choose the **Diffuse Color** map button currently labeled **No Map**; the **Material/Map Browser** dialog box is displayed, showing the types of maps that may be added.

3. Select the **Bitmap** map from the **Maps > General** rollout and choose the **OK** button; the **Select Bitmap Image File** dialog box is displayed.

4. In this dialog box, make sure **All Formats** is selected in the **Files of type** drop-down list. As the project folder is already set, the *images* folder of this project is displayed in the **Look in** text box of this dialog box. Select the **checkerb.jpg** file and choose the **Open** button to open the image file; the selected image is displayed in the selected sample slot.

5. In the **Coordinates** rollout, make sure the **Use Real-WorldScale** check box is cleared. Also, make sure **1** is set in the **U:Tiling** and **V:Tiling** spinners.

6. Click and hold the **Sample Type** button in the **Material Editor** dialog box and choose the **Square** button to switch the sample from a sphere to a square; the material currently resembles a black checker with a gold crown in the center.

7. Choose the **Go To Parent** button in the **Material Editor** dialog box; you will return to the root level in the material definition.

8. In the **Maps** rollout, set the value **90** in the **Bump** spinner; the intensity of the bump map is set.

   Now, a bump map needs to be assigned.

9. Choose the **Bump** map button, currently labeled as **No Map**.

10. Select the **Bitmap** map from the **Maps > General** rollout and choose the **OK** button; the **Select Bitmap Image File** dialog box is displayed.

11. In this dialog box, select **All Formats** in the **Files of type** drop-down list. As the project folder is already set, the *images* folder of this project is displayed in the **Look in** text box of this dialog box. Select the **checker1.jpg** file and choose the **Open** button; a bump map is assigned to the material.

12. In the **Coordinates** rollout, make sure the **Use Real-WorldScale** check box is cleared. Also, make sure **1** is set in the **U:Tiling** and **V:Tiling** spinners.

13. Choose the **Show Shaded Material in Viewport** button in the **Material Editor** dialog box.

    Next, you will make the material shiny.

14. Choose the **Go To Parent** button in the **Material Editor** dialog box to return to the root level in the material definition. Next, collapse the **Maps** rollout.

15. Choose the **Specular** color swatch in the **Blinn Basic Parameters** rollout; the **Color Selector: Specular Color** is displayed.

16. In this dialog box, set the value **255** in the **Red**, **Green**, and **Blue** spinners; the specular color is set to white. Now, choose the **OK** button to close the **Color Selector: Specular Color** dialog box.

17. In the **Specular Highlights** area of the **Blinn Basic Parameters** rollout, set the value **60** in the **Specular Level** spinner.

18. Set the value **15** in the **Glossiness** spinner.

    Next, you will name the material created.

19. Choose the **Put to Library** button in the **Material Editor** dialog box; a flyout is displayed.

20. Choose the **Checker Board.mat** option (new material library that you have created); the **Put To Library** dialog box is displayed.

21. Enter **Checker Black** as the name of the new material in the **Put To Library** dialog box and choose the **OK** button; the material is assigned the name **Checker Black** and is added to the current material library.

Note

While creating dark-colored material, you can choose the ***Background*** *button in the* ***Material Editor*** *dialog box. You can also change the sample material window background color from black to a multicolored pattern.*

Saving the Material Library

In this section, you will save the material library.

1. Choose the **Get Material** button in the **Material Editor** dialog box; the **Material/Map Browser** dialog box is displayed.

2. Right-click on **CheckerBoard.mat** in the **Material/Map Browser** dialog box; a flyout is displayed. Choose **Save** from the flyout, as shown in Figure 8-10.

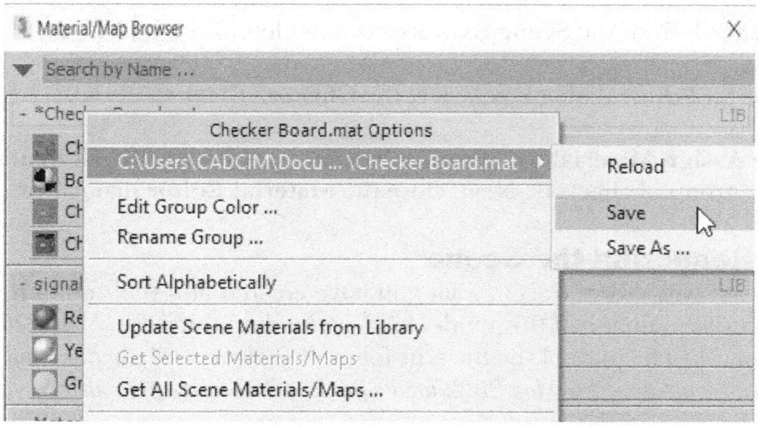

Figure 8-10 *Saving the Material Library*

3. Close the **Material/Map Browser** dialog box.

Assigning Materials

In this section, you will assign materials to all objects in the scene.

1. Make sure the Camera viewport is activated. Next, select **border** from the **Scene Explorer** located on the left in the interface.

2. In the **Material Editor** dialog box, select the **Checker Board border** material. Next, choose the **Assign Material to Selection** button; the currently selected object is assigned the **Checker Board border** material.

   Next, you will assign material to *Board*.

3. Select *Board* from the **Scene Explorer**; *Board* is highlighted and the object name *Board* is displayed in the **Name and Color** rollout.

4. In the **Material Editor** dialog box, select the **Board** material.

5. Choose the **Assign Material to Selection** button; the **Board** material is assigned to the board part of the checker board.

   Next, you will assign the **Checker Red** material to the **Checker-r** group.

6. Select **Checker-r** from the **Scene Explorer** located on the left in the interface.

7. In the **Material Editor** dialog box, select the **Checker Red** material.

8. Choose the **Assign Material to Selection** button; the **Checker Red** material is assigned to one group of the checkers, as discussed earlier.

   Next, you will assign the **black** checker material to the **Checker-b** group.

9. Select **Checker-b** from the **Scene Explorer** located left of the interface.

10. In the **Material Editor** dialog box, select the **Checker Black** material.

11. Choose the **Assign Material to Selection** button; the **Checker Black** material is assigned to the other group of checkers. Next, close the **Material Editor** dialog box.

Saving and Rendering the Scene

In this section, you will save the scene that you have created and then render it. You can also view the final rendered image of this model by downloading the *c08_3dsmax_2022_rndr.zip* file from *www.cadcim.com*. The path of the file is as follows: *Textbooks > Animation and Visual Effects > Autodesk 3ds Max > Autodesk 3ds Max 2022 for Beginners: A Tutorial Approach*

1. Choose **Save** from the **File** menu.

2. Make sure the Camera viewport is activated. Next, invoke the **Render Production** tool from the **Main Toolbar**; the rendered image is displayed, as shown in Figure 8-11.

Figure 8-11 *The rendered scene after assigning materials*

Self-Evaluation Test

Answer the following questions and then compare them to those given at the end of this chapter:

1. Which of the following buttons in the **Material Editor** dialog box is used to display the material in the viewport?

   (a) **Show Shaded Material in Viewport** (b) **Assign Material to Selection**
   (c) **Show End Result** (d) **Put Material to Scene**

2. Which of the following buttons in the **Material Editor** dialog box is used to save the current material in a material library?

   (a) **Get Material** (b) **Assign Material to Selection**
   (c) **Put To Library** (d) **Go To Parent**

3. Which of the following buttons in the **Material Editor** dialog box is used to change the sample material window background color from black to a multicolored pattern?

   (a) **Background** (b) **Sample Type**
   (c) **Get Material** (d) **Assign Material to Selection**

4. Which of the following tools is used to invoke the **Material Editor** dialog box from the **Main Toolbar**?

   (a) **Render Setup** (b) **Render Production**
   (c) **Material Editor** (d) **Manage Layers**

5. The __________ maps can be assigned to create complex materials.

6. A texture map will not be displayed in the viewports unless the __________ button is active.

7. Lighter pixels in an image used as a bump map will appear to be __________, while the darker pixels will appear to be __________.

8. In 3ds Max, you can assign more than one texture map to each material. (T/F)

9. The **Standard(Legacy)** material type is selected by default in the **Material Editor** dialog box. (T/F)

10. The **Scene Explorer** is used to select and highlight the objects in the viewports. (T/F)

Review Questions

Answer the following questions:

1. Which of the following areas of the **Blinn Basic Parameters** rollout is used to control the intensity and shape of highlights on a glossy material?

   (a) **Self-Illumination** (b) **Specular Highlights**
   (c) Both of these (d) None of these

2. The __________ button is used to create a new material library.

3. The value in the __________ spinner controls the intensity of the bumps.

4. The __________ button is used to put a material in the library.

5. The **Get Material** button helps you display a material in the viewport. (T/F)

EXERCISES

The rendered image of the scene used in the following exercises can be accessed by downloading the *c08_3dsmax_2022_exr.zip* from *www.cadcim.com*. The path of the file is as follows: *Textbooks > Animation and Visual Effects > Autodesk 3ds Max > Autodesk 3ds Max 2022 for Beginners: A Tutorial Approach*

Exercise 1

Extract the contents of the *c08_3dsmax_2022_exr.zip* and then open the *c08_exr1_start* file from it. Next, create materials and then apply them to a company logo. Figure 8-12 shows the final rendered image of the company logo. **(Expected time: 20 min)**

***Figure 8-12** Company Logo*

Exercise 2

Open the *c08_exr2_start* file. Next, create and apply materials needed for the model of a vise, refer to Figure 8-13. **(Expected time: 25 min)**

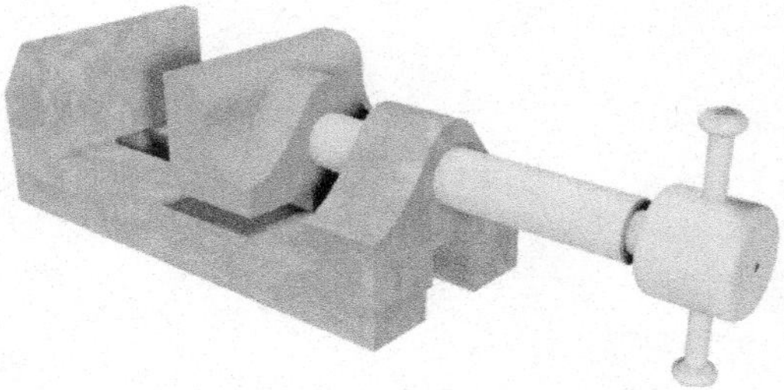

Figure 8-13 *The model of a Vise*

Answers to Self-Evaluation Test

1. a, **2.** c, **3.** a, **4.** c, **5.** Texture, **6.** **Show Shaded Material in Viewport**, **7.** raised, lowered, **8.** T, **9.** F, **10.** T

Chapter 9

Material Editor: Texture Maps-II

Learning Objectives

After completing this chapter, you will be able to:

- *Create transparent materials*
- *Apply transparency maps to objects*
- *Create procedural maps*

INTRODUCTION

In this chapter, you will create materials using the opacity and procedural maps. An opacity map is used to control the transparency of a material. It can be applied to a simple 3D object to give it the appearance of a complex shape. Unlike a bitmap which contains a fixed number of pixels, a procedural map is generated in two or three dimensions by using mathematical algorithm.

TUTORIAL

Before starting the tutorial, you need to download the *c09_3dsmax_2022_tut.zip* file from *www.cadcim.com.* The path of the file is as follows: *Textbooks > Animation and Visual Effects > Autodesk 3ds Max > Autodesk 3ds Max 2022 for Beginners: A Tutorial Approach*

Extract the contents of the zip file and save them in the *Documents* folder.

Tutorial 1

In this tutorial, you will create materials using special texture maps. You will use the opacity and diffuse color maps to make a square block look like a flower, refer to Figure 9-1.

(Expected time: 20 min)

Figure 9-1 *Materials applied to models*

The following steps are required to complete this tutorial:

a. Create the project folder.
b. Open the file.
c. Create a new material library.
d. Create the vase material.
e. Create the flower material.
f. Create the wood block material.
g. Assign the material to the objects.
h. Save and render the scene.

Creating the Project Folder

Create the project folder with the name *c09_tut1* in the *3dsmax 2022* folder, as discussed in earlier chapters.

Opening the File

1. Open Windows Explorer and then browse to *\Documents\c09_3dsmax_2022_tut*. Next, copy all the image files to *\Documents\3dsmax 2022\c09_tut1\sceneassets\images*.

2. Choose **Open** from the **File** menu; the **Open File** dialog box is displayed. In this dialog box, browse to the location *\Documents\c09_3dsmax_2022_tut* and select the **c09_tut1_start.max** file from it. Choose the **Open** button to open the file, refer to Figure 9-2.

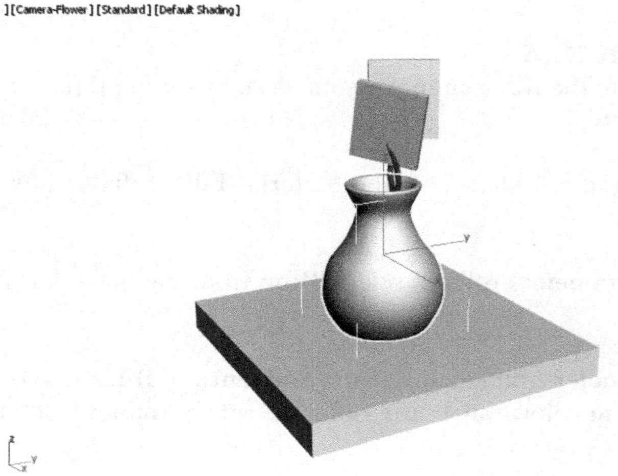

Figure 9-2 *The c09_tut1_start file*

3. Choose **Save As** from the **File** menu; the **Save File As** dialog box is displayed. Browse to the location *\Documents\3dsmax 2022\c09_tut1\scenes*. Save the file with the name *c09tut1.max* at this location.

4. Choose the **Render Setup** tool from the **Main Toolbar** and make sure the **Render Setup: Scanline Renderer** dialog box is displayed. Next, close this dialog box.

Creating a New Material Library

In this section, you will create a new material library which contains materials created in the scene.

1. Choose **Rendering > Material Editor > Compact Material Editor** from the menu bar; the **Material Editor** dialog box is displayed. You will notice that in the first slot there is a material named *Stem* which is applied to the flower stems.

2. Choose the **Get Material** button in the **Material Editor** dialog box; the **Material/Map Browser** dialog box is displayed.

3. Choose the **Material/Map Browser Options** button from the **Material/Map Browser** dialog box; a flyout is displayed. Choose **New Material Library** from this flyout; the **Create New Material Library** dialog box is displayed.

   As the project folder is already set, the *materiallibraries* folder of this project is displayed in the **Save in** drop-down list of this dialog box.

4. Enter **Still-Life** in the **File name** text box as the new name for the material library and choose the **Save** button; the new **Still-Life** material library is created in the **Material/Map Browser** dialog box.

5. Close the **Material/Map Browser** dialog box to return to the **Material Editor** dialog box.

Creating the Vase Material

In this section, you will create the transparent vase material by setting different parameters in the **Material Editor** dialog box.

1. Select the second material sample slot in the **Material Editor** dialog box to make it the current sample.

2. In the **Shader Basic Parameters** rollout, select **Blinn** from the drop-down list, if it is not already selected.

3. In the **Blinn Basic Parameters** rollout, unlock the **Ambient** and **Diffuse** color swatches. Next, choose the **Ambient** color swatch; the **Color Selector: Ambient Color** dialog is displayed. 

4. In the **Color Selector: Ambient Color** dialog box, set the values as given next:

   Red: **133** Green: **166** Blue: **47**

   The ambient color is set to green. Choose the **OK** button to close the **Color Selector: Ambient Color** dialog box.

5. Choose **Rendering > Environment** from the menu bar or press 8; the **Environment and Effects** dialog box is displayed. In the **Global Lighting** area of the **Common Parameters** rollout, choose the **Ambient** color swatch; the **Color Selector: Ambient Light** dialog box is displayed. In this dialog box, set the value **75** in the **Value** spinner and then choose the **OK** button to close the dialog box. Next, close the **Environment and Effects** dialog box.

   This ambient light will bring out the ambient portion of the material color, which was added in the previous steps. If the ambient light value remained at 0, the ambient color settings would have no effect on the appearance of the material.

6. In the **Material Editor** dialog box, choose the **Diffuse** color swatch; the **Color Selector: Diffuse Color** dialog box is displayed.

7. In this dialog box, set the values as given next and then choose the **OK** button to close this dialog box:

   Red: **153** Green: **175** Blue: **255**

8. In the **Material Editor** dialog box, choose the **Specular** color swatch; the **Color Selector: Specular Color** dialog box is displayed.

9. In this dialog box, set the values as given next:

   Red: **255** Green: **255** Blue: **255**

10. Close the **Color Selector: Specular Color** dialog box.

11. In the **Specular Highlights** area of the **Blinn Basic Parameters** rollout, set the value **60** in the **Specular Level** spinner.

12. Set the value **40** in the **Glossiness** spinner and **30** in the **Opacity** spinner.

    Note that smaller the opacity value, more transparent the material will be.

13. Choose the **Background** button located in the right side of the **Material Editor** dialog box; the background is displayed in the selected sample slot. This helps to view the effect of transparency settings in a better way; refer to Figure 9-3.

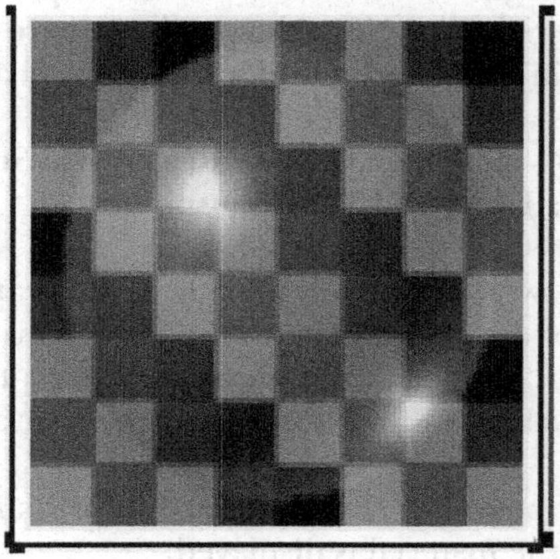

***Figure 9-3** The background displayed in the material sample slot in the **Material Editor** dialog box*

Next, you will name the material and save it in the material library.

14. Choose the **Put to Library** button in the **Material Editor** dialog box; a flyout is displayed. Choose **Still-Life.mat** from the flyout; the **Put To Library** dialog box is displayed. If the **Material Editor** message box is displayed prompting to put the entire material/map tree to the library, choose the **Yes** button. If you are at the root level of the material definition, the **Material Editor** message box does not appear because the entire tree is added by default.

15. Enter **Vase** as the name of the new material in the **Put To Library** dialog box and choose the **OK** button; the material is assigned the name *Vase* and is added to the **Still-Life** material library.

Creating the Flower Material

In this section, you will create the flower material using the opacity maps.

1. In the **Material Editor** dialog box, select the third material sample slot. Various rollouts are displayed in the **Material Editor** dialog box to modify the **Standard (Legacy)** material.

2. Expand the **Maps** rollout. Next, choose the **Diffuse Color** map button, currently labeled **No Map;** the **Material/Map Browser** dialog box is displayed showing the types of maps that may be added.

3. Choose the **Bitmap** map from the **Maps > General** rollout and choose the **OK** button; the **Select Bitmap Image File** dialog box is displayed.

4. In this dialog box, make sure **All Formats** is selected in the **Files of type** drop-down list. As the project folder is already set, the *images* folder of this project is displayed in the **Look in** drop-down list of this dialog box. Select the **daisy.tif** file and choose the **Open** button; the image file is loaded and the parameters are displayed for the map in the **Material Editor** dialog box.

5. In the **Material Editor** dialog box, click and hold the **Sample Type** button located on the right side of the material slots. Now, choose the **Square** button from the flyout displayed; the sample slot in the **Material Editor** dialog box changes into square shape.

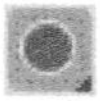

6. In the **Coordinates** rollout, make sure the **Use Real-World Scale** check box is cleared and the value **1** is set in the **U: Tiling** and **V: Tiling** spinners.

   The material uses daisy flower image. When the material is applied to an object, you want the background portion of the image to be transparent. This can be done with an opacity map. The map consists of another image of the daisy with the daisy portion completely white (opaque) and the background portion completely black (transparent).

7. Choose the **Go to Parent** button in the **Material Editor** dialog box to move one level up; the root level of the material is displayed.

8. In the **Maps** rollout, choose the **Opacity** map button that is currently labeled as **No Map**; the **Material/Map Browser** dialog box is displayed showing the types of maps that can be added.

9. Select the **Bitmap** map from the **Maps > General** rollout and choose the **OK** button; the **Select Bitmap Image File** dialog box is displayed.

10. In this dialog box, make sure **All Formats** is selected in the **Files of type** drop-down list. As the project folder is already set, the *images* folder of this project is displayed in the **Look in**

drop-down list of this dialog box. Select the **daisy-o.tga** file and choose the **Open** button; the image file is loaded and the parameters are displayed for the map in the **Material Editor** dialog box.

You will notice that the preview of image displayed in the **Select Bitmap Image File** dialog box has black and white areas. The white area will be opaque while the black area will be transparent when the material is rendered.

11. In the **Coordinates** rollout, make sure the **Use Real-World Scale** check box is cleared and set the value **1.0** in the **U: Tiling** and **V: Tiling** spinners.

12. Choose the **Background** button in the **Material Editor** dialog box; the material sample slot reflects the effect of opacity map, as shown in Figure 9-4.

Figure 9-4 *The material sample slot displayed in the **Material Editor** dialog box*

Note

*The transparent area of the material will still accept specular highlights. In other words, it will look like a clear glass. If you want the transparent areas of the material to be completely invisible, you can set **0** in the **Specular Level** spinner of the **Blinn Basic Parameters** rollout or add a specular level map that matches the opacity map.*

Next, you will name the material and save it in the material library.

13. Choose the **Put to Library** button in the **Material Editor** dialog box; a flyout is displayed. Next, choose **Still-Life.mat** from the flyout; the **Put To Library** dialog box is displayed.

    If the **Material Editor** message box is displayed prompting to put the entire material/map tree to the library, choose the **Yes** button, refer to Figure 9-5. If you are at the root level of the material definition, the **Material Editor** message box does not appear because the entire tree is added by default.

14. Enter **Daisy** as the name of the new material in the **Put To Library** dialog box and then choose the **OK** button; the material is assigned the name *Daisy* and is added to the **Still-Life** material library.

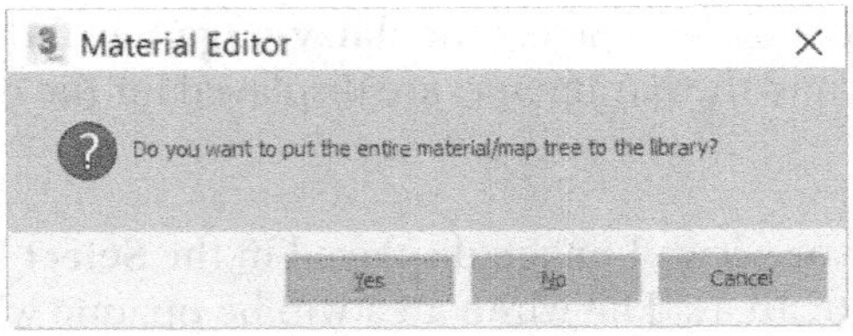

*Figure 9-5 The **Material Editor** message box*

Creating the Wood Block Material

In this section, you will create a wood procedural map.

1. Select the fourth material sample slot in the **Material Editor** dialog box.

2. Expand the **Maps** rollout. Next, choose the **Diffuse Color** map button that is currently labeled as **No Map**; the **Material/Map Browser** dialog box is displayed, showing the types of maps that can be added.

3. Choose **Advanced Wood** from the **Maps > General** rollout and choose the **OK** button; the **Material/Map Browser** dialog box is closed and the parameters for the *Advanced Wood* map are displayed in the **Material Editor** dialog box.

   Various settings in the rollouts of the **Material Editor** dialog box allow you to control the appearance of the wood material.

   You can set the parameters in these rollouts as per your requirement.

4. Make sure *Base* is selected. Choose the **Modify** tab. Next, apply the **UVW Map** modifier to it. In the **Parameters** rollout, select the **Box** radio button. Next, enter **400** in the **Length** spinner.

   As you change parameters, the material sample is dynamically updated.

   Next, you will name the material and save it in the material library.

5. Choose the **Go to Parent** button twice in the **Material Editor** dialog box to move one level up; the root level of the material is displayed. Next, choose the **Put to Library** button in the **Material Editor** dialog box; a flyout is displayed.

6. Choose **Still-Life.mat** from the flyout; the **Put To Library** dialog box is displayed.

7. Enter **Wood** as the name of the new material in the **Put To Library** dialog box and choose the **OK** button; the material is assigned the name *Wood* and is added to the **Still-Life** material library.

   Next, you will save the **Still-Life** material library.

8. Choose the **Get Material** button in the **Material Editor**; the **Material/Map Browser** dialog box is displayed.

9. Right-click on the **Still-Life.mat** in the **Material/Map Browser** dialog box; a flyout is displayed. Choose **Save** from the flyout, as shown in Figure 9-6.

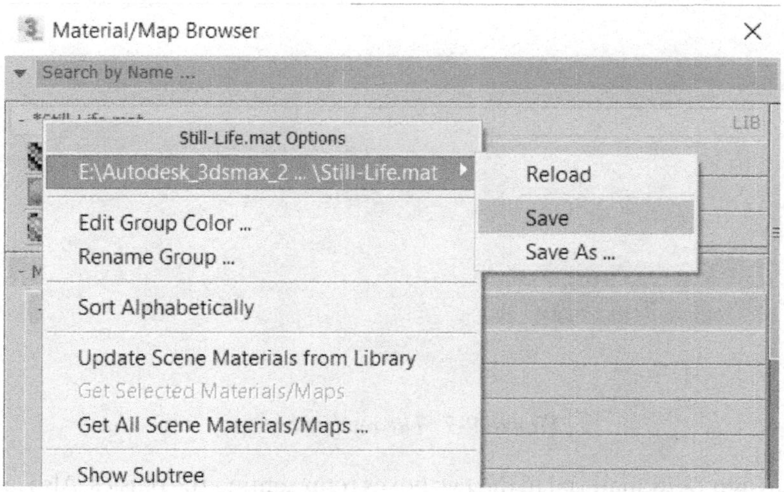

Figure 9-6 *Saving the material library*

10. Close the **Material/Map Browser** dialog box.

Assigning Materials to the Objects

In this section, you will assign materials to the objects in the scene.

1. Activate the Camera-Flower [Standard] [Default-Shading] viewport.

2. Invoke the **Render Production** tool from the **Main Toolbar**; the rendered image is displayed, refer to Figure 9-7.

   You will notice that all the objects except the stems are currently assigned a default material. Notice that the flowers are created as simple 3D boxes. The opacity map on *Daisy* material will give the flowers their shape. The stems already have an appropriate material assigned. Also, the required mapping coordinates have already been applied to the objects.

3. Close the rendered image. Next, invoke the **Select Object** tool from the **Main Toolbar**. Next, select *Vase* in the Camera-Flower [Standard] [Default Shading] viewport; the object is highlighted and *Vase* is displayed in the **Name and Color** rollout.

4. In the **Material Editor** dialog box, select the material sample slot which is named as *Vase* and then choose the **Assign Material to Selection** button; the *Vase* material is assigned to *Vase*. Next, choose the **Show Shaded Material in Viewport** button to display the material on *Vase* in the viewport.

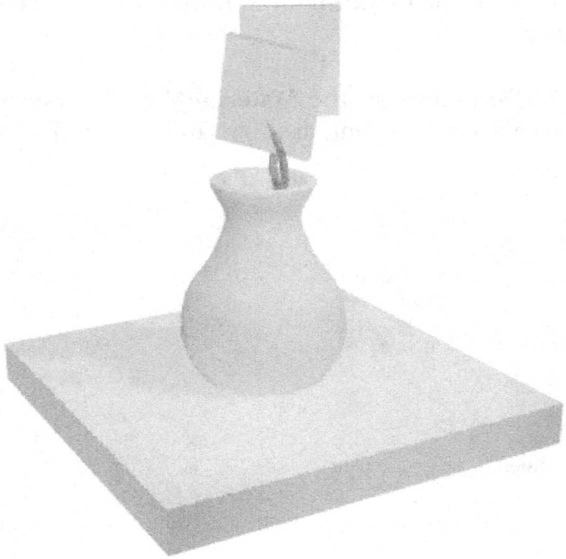

Figure 9-7 *The rendered image*

5. Similarly, assign *Daisy* material to the two boxes representing the daisies. Also, assign *Advanced Wood* material to *Base*. Next, choose the **Show Shaded Material in Viewport** button to display the material on *Base* in the viewport.

6. Select the first material sample slot in the **Material Editor** dialog box.

7. Choose the **Diffuse Map** button that is labelled as M.

8. Select the field located next to **Bitmap** in the **Bitmap Parameters** area; the **Select Bitmap Image File** dialog box is displayed.

9. In this dialog box, navigate to the *C:\Users\CADCIM\Documents\3dsmax 2022\c09_tut1\sceneassets\images* folder and select **EVERGREEN** from it. Next, choose the **Open** button; the map for the stem material is relinked.

10. Close the **Material Editor** dialog box.

Saving and Rendering the Scene

In this section, you will save the scene that you have created and then render it. You can also view the final rendered image of this model by downloading the *c09_3dsmax_2022_rndr.zip* file from *www.cadcim.com*. The path of the file is as follows: *Textbooks > Animation and Visual Effects > Autodesk 3ds Max > Autodesk 3ds Max 2022 for Beginners: A Tutorial Approach*

1. Change the background color of the scene to light grey.

2. Choose **Save** from the **File** menu.

3. Make sure the Camera-Flower [Shaded] viewport is activated. Next, invoke the **Render Production** tool from the **Main Toolbar**; the rendered image is displayed, refer to Figure 9-8.

Figure 9-8 *The rendered image with materials assigned to the objects*

Self-Evaluation Test

Answer the following questions and then compare them to those given at the end of this chapter:

1. Which of the following maps is used to control transparency of the material?

   (a) **Opacity** (b) **Procedural**
   (c) **Mix** (d) None of these

2. Which of the following maps can be created to represent a solid piece of wood?

   (a) **Opacity** (b) **Procedural**
   (c) **Mix** (d) None of these

3. Which of the following buttons is used to move one root level up in the **Material Editor** dialog box?

   (a) **Go to Parent** (b) **Get Material**
   (c) **Material/Map Navigator** (d) **Put Material to Scene**

4. In an opacity map, the __________ pixels will be completely transparent and the __________ pixels will be opaque.

5. The lower the value in the **Opacity** spinner, the higher will be the transparency of the material. (T/F)

6. As you change the parameters of the material types in the **Material Editor** dialog box, the material sample is dynamically updated. (T/F)

Review Questions

Answer the following questions:

1. Which of the following buttons in the **Material Editor** is used to display the opacity effect?

   (a) **Background** (b) **Make Preview**
   (c) **Backlight** (d) None of these

2. The opacity maps help control the __________ of a material.

3. The __________ button in the Main Toolbar is used to render the current frame.

4. If you apply a procedural map to an object with a whole cut in it, the texture will still appear correct. (T/F)

EXERCISES

The rendered output of the models used in the following exercises can be accessed by downloading the *c09_3dsmax_2022_exr.zip* from *www.cadcim.com*. The path of the file is as follows: *Textbooks > Animation and Visual Effects > Autodesk 3ds Max > Autodesk 3ds Max 2022 for Beginners: A Tutorial Approach*.

Exercise 1

Extract the contents of the *c09_3dsmax_2022_exr.zip* and then open *c09_exr01_start.max*. Next, create materials and assign them to three glasses placed on a tray, as shown in Figure 9-9.

Hint: You will require three transparent materials with different opacity settings.

(Expected time: 20 min)

Figure 9-9 *Three glasses with assigned materials*

Exercise 2

Open *c09_exr02_start.max*. Create materials for a scene that displays a broken piece of marble and a wooden board with a notch cut in it, as shown in Figure 9-10. In this exercise, you need to create two procedural map materials. **(Expected time: 15 min)**

Figure 9-10 *Broken piece of marble and a wooden board*

Answers to Self-Evaluation Test

1. a, **2.** b, **3.** a, **4.** black, white, **5.** T, **6.** T

Chapter 10

Material Editor: Controlling Texture Maps

Learning Objectives

After completing this chapter, you will be able to:

- *Modify mapping parameters*
- *Apply decal texture maps*
- *Apply reflection maps*
- *Set up background bitmaps*

INTRODUCTION

In this chapter, you will learn to modify texture maps by setting the mapping parameters in the **Material Editor** dialog box. You will also learn to use reflection maps to provide a photo-realistic appearance to a scene.

TUTORIAL

Before starting the tutorial, you need to download the *c10_3dsmax_2022_tut.zip* file from *www.cadcim.com*. The path of the file is as follows: *Textbooks > Animation and Visual Effects > Autodesk 3ds Max > Autodesk 3ds Max 2022 for Beginners: A Tutorial Approach*

Extract the content of the zip file and save it in the *Documents* folder.

Tutorial 1

In this tutorial, you will create and apply different texture maps to the objects in the scene, as shown in Figure 10-1. **(Expected time: 20 min)**

Figure 10-1 *The textured scene*

The following steps are required to complete this tutorial:

a. Create the project folder.
b. Open the file.
c. Create a new material library.
d. Create material for mug.
e. Create material for mug handle and saucer.
f. Create material for tabletop.
g. Set the environment background.

h. Create material for paper cups.
i. Save the material library.
j. Save and render the scene.

Creating the Project Folder

Create a project folder with the name *c10_tut1* in the *3dsmax 2022* folder, as discussed in earlier chapters.

Opening the File

1. Open Windows Explorer and then browse to *\Documents\c10_3dsmax_2022_tut*. Next, copy the *Logo.tga, Marbtea.gif,* and *Foliage.jpg* files to *\Documents\3dsmax 2022\c10_tut1\sceneassets\images.*

2. Choose **Open** from the **File** menu; the **Open File** dialog box is displayed. In this dialog box, browse to the location *\Documents\c10_3dsmax_2022_tut* and select the *c10_tut1_start.max* file from it. Choose the **Open** button to open the file, refer to Figure 10-2.

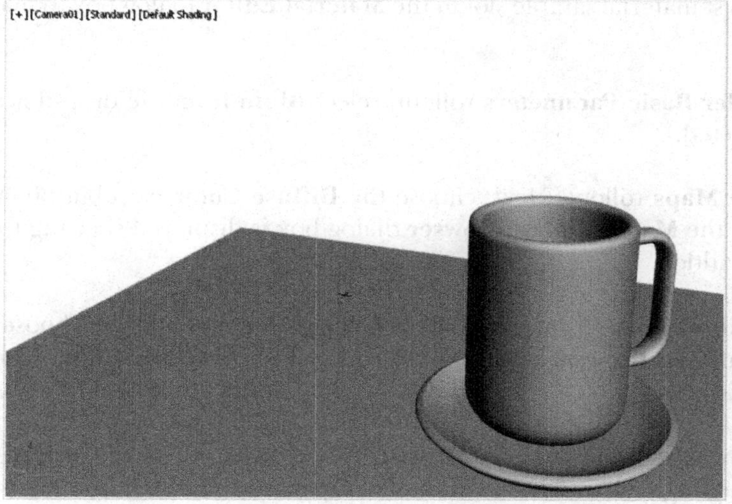

Figure 10-2 *The c10_tut1_start file*

3. Choose **Save As** from the **File** menu; the **Save File As** dialog box is displayed. Browse to the location *\Documents\3dsmax 2022\c10_tut1\scenes*. Save the file with the name *c10tut1.max* at this location.

Creating a New Material Library

In this section, you will create a new material library in which new materials can be stored.

1. Choose **Rendering > Material Editor > Compact Material Editor** from the menu bar; the **Material Editor** dialog box is displayed.

2. Choose the **Get Material** button in the **Material Editor** dialog box; the **Material/Map Browser** dialog box is displayed.

3. Choose the **Material/Map Browser Options** button from the **Material/Map Browser** dialog box; a flyout is displayed. Choose **New Material Library** from this flyout; the **Create New Material Library** dialog box is displayed. 

   As the project folder is already set, the *materiallibraries* folder in this project is displayed in the **Save in** drop-down list.

4. Enter **MugLogo** in the **File name** text box and choose the **Save** button; the new material library **MugLogo.mat** is created at the specified location and displayed in the **Material/Map Browser** dialog box.

5. Close the **Material/Map Browser** dialog box to return to the **Material Editor** dialog box.

Creating Material for Mug

In this section, you will create a new material for *Mug*.

1. Select the first material sample slot in the **Material Editor** dialog box to make it the current sample slot.

2. In the **Shader Basic Parameters** rollout, select **Blinn** from the drop-down list, if it is not already selected.

3. Expand the **Maps** rollout. Next, choose the **Diffuse Color** map button currently labeled as **No Map**; the **Material/Map Browser** dialog box is displayed showing the types of maps that can be added.

4. Select the **Bitmap** map from the **Maps > General** rollout and then choose the **OK** button; the **Material/Map Browser** dialog box is closed and the **Select Bitmap Image File** dialog box is displayed.

5. In this dialog box, make sure **All Formats** is selected in the **Files of type** drop-down list. Select the **logo.tga** file and then choose the **Open** button; the image file is loaded and the parameters for modifying the map are displayed in the **Material Editor** dialog box.

6. In the **Material Editor** dialog box, click and hold the **Sample Type** button located on the right side of the material slots. Now, choose the **Square** button from the flyout displayed; the sample slot in the **Material Editor** dialog box changes into a square shape. 

7. In the **Coordinates** rollout, make sure the **Use Real-World Scale** check box is cleared and the values are set to **1.0** in the **U: Tiling** and **V: Tiling** spinners.

Note

*The **Use Real-World Scale** option is useful if the objects in the scene are drawn to scale, or if the material is to be assigned to multiple objects of different sizes. This feature will be explained later in this chapter.*

8. Choose the **Go to Parent** button in the **Material Editor** dialog box to move one level up. 

   Next, you will assign material to *Mug*.

9. Select **Mug** from the **Scene Explorer**; *Mug* is highlighted.

10. In the **Material Editor** dialog box, make sure the first material sample slot is selected.

11. Choose the **Assign Material to Selection** button; the material is assigned to *Mug*.

12. Choose the **Show Shaded Material in Viewport** button below the material sample slot; the texture map now appears on the object in the shaded viewports.

    Next, you will modify the mapping parameters such as the scale, placement, and rotation angle of the bitmap image. These parameters can be controlled by using the options in the **Coordinates** rollout. You will start by modifying the logo's scale.

13. In the **Maps** rollout, choose the **Diffuse Color** map button which is currently labeled **Map #x (logo.tga)**; you return to the map level of the material definition where the **Coordinates** rollout displays options for the texture map, as shown in Figure 10-3.

14. Set the value **2.0** in the **U: Tiling** spinner; the logo is set horizontally to half of its original size.

    Setting 2.0 in the **U: Tiling** spinner scales down the image so that it can fit twice in its original space.

15. Set the value **2.0** in the **V: Tiling** spinner; the logo is set vertically to half of its original size.

    As the logo is set to half of its original size, it fits better on *Mug*. However, the image is repeating on both the horizontal and vertical axes. To create a decal texture map on *Mug*, the tiling must be turned off.

16. In the **Coordinates** rollout, clear the **U: Tile** check box; the logo no longer repeats on the horizontal axis.

17. Clear the **V: Tile** check box; the logo no longer repeats on the vertical axis.

    To apply the same color on the mug, you need to change the basic color of the mug to dark green.

18. Choose the **Go to Parent** button in the **Material Editor** dialog box to return to the root level of the material definition. 

19. In the **Blinn Basic Parameters** rollout, choose the **Ambient** color swatch; the **Color Selector: Ambient Color** dialog box is displayed.

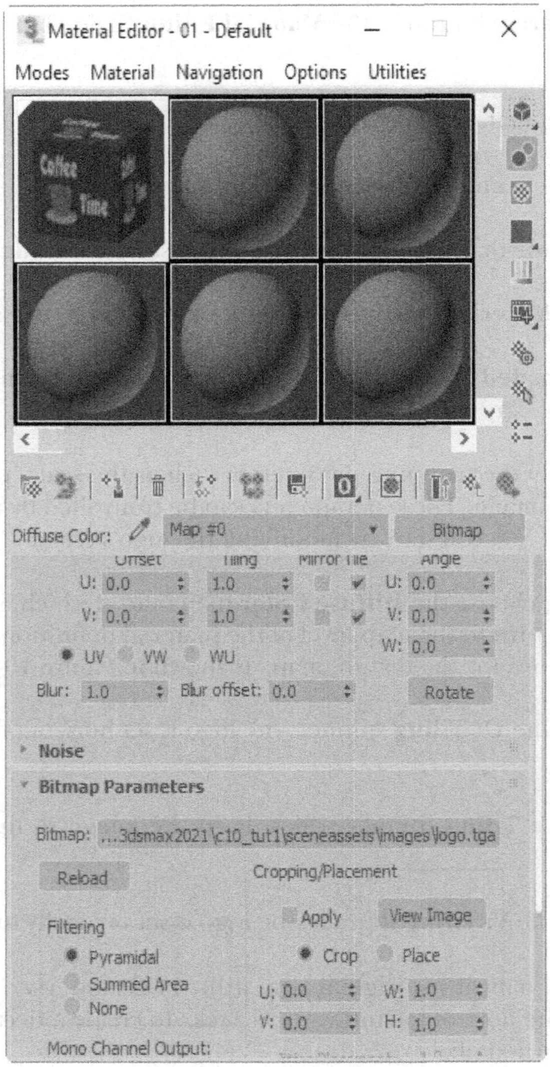

***Figure 10-3** The **Coordinates** rollout in the **Material Editor** dialog box*

20. In the **Color Selector: Ambient color** dialog box, set the values as given next and then choose the **OK** button to close this dialog box:

    Red: **15** Green: **80** Blue: **0**

    The ambient and diffuse colors are currently locked. Therefore, when the ambient color is set to dark green, the diffuse color changes as well.

    You also need to set the specular color to dark green. For this, you need to either choose the **Specular** color swatch and change the color values or you can lock the specular and diffuse (and therefore ambient) colors.

21. Choose the **Lock** button (between the **Diffuse** and **Specular** color swatches). The **Lock Diffuse and Specular?** message box is displayed, as shown in Figure 10-4. Choose the **Yes** button in the message box. Next, invoke the **Render Production** tool from the **Main Toolbar**; the rendered image is displayed and you will notice that a single logo is displayed on the all-green mug.

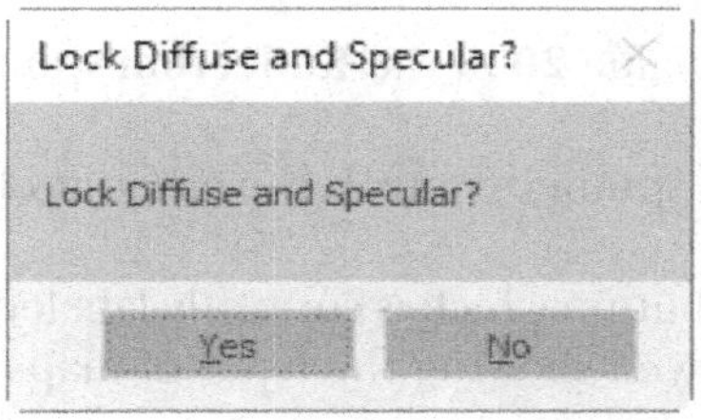

Figure 10-4 *The **Lock Diffuse and Specular?** message box*

Next, you will position the logo. To position the logo on *Mug*, you need to return to the map level of the rollout.

22. Choose the **M** button next to the **Diffuse** color swatch in the **Blinn Basic Parameters** rollout. Alternatively, you can choose the **Diffuse Color** map button in the **Maps** rollout.

23. Set the value **–0.12** in the **U: Offset** spinner; the logo is moved to the left of its original position.

    The distance the logo moves is based on the size of the bitmap image and the tiling setting. When the tiling is set to 1, a value of 0.5 will move any image to the right up to a distance equal to half of the image's width. Since the tiling is set to 2.0, a value of –0.12 in the offset spinner moves the logo to a distance equal to approximately a quarter of its width (–0.12 × 2 = –0.24).

24. Set the value **0.2** in the **V: Offset** spinner; the logo moves up from its original position.

    Next, you will render the scene to view the changes made so far.

25. Make sure the Camera01 viewport is activated, if it is not already activated.

26. Invoke the **Render Production** tool from the **Main Toolbar**; the rendered image is displayed on the screen. Close the render window.

    Next, you will rotate the logo.

27. In the **Coordinates** rollout for the map, set the value **20** in the **W: Angle** spinner; the logo is rotated clockwise.

28. Set the value **–0.1** in the **U: Offset** spinner; the logo moves left.

29. Set the value **0.1** in the **V: Offset** spinner; the logo rotates and reposition itself on *Mug*.

Next, you will add a reflection map. A reflection map makes a 3D scene appear more realistic by reflecting the background image or surrounding elements in the scene.

30. Choose the **Go to Parent** button in the **Material Editor** dialog box; you will return to the root level of the material definition. 

31. In the **Maps** rollout, set the value **20** in the **Reflection** spinner.

    This value in the **Reflection** spinner controls the amount of reflection in the material.

32. Choose the **Reflection** map button which is currently labeled as **No Map**; the **Material/Map Browser** dialog box is displayed showing the types of maps that can be added.

33. Choose **Reflect/Refract** from the **Maps > Scanline** rollout and then choose the **OK** button; various rollouts for the **Reflect/Refract** map are displayed in the **Material Editor** dialog box.

    The **Reflect/Refract** map type automatically generates a reflection from the other elements in the scene and places it on the material during rendering.

34. Choose the **Go to Parent** button to return to the root level of the material definition.

    Next, you will name and save the material.

35. Choose the **Put to Library** button in the **Material Editor** dialog box; a flyout is displayed.

36. Choose **MugLogo.mat** from the flyout; the **Put to Library** dialog box is displayed.

37. Enter **Logo** in the **Name** text box and then choose the **OK** button; the material is assigned the name *Logo* and is added to the **MugLogo** material library.

Creating Material for Mug Handle and Saucer

In this section, you will create a material for mug handle and saucer in the scene. This new material should be identical to the material on the mug but without the logo. For this, you need to copy *Logo* material and modify the copy by removing the bitmap. In this way, all the color and reflection settings for *Logo* material are automatically copied to the new material.

1. Select and hold *Logo* material sample slot. Drag the sample slot to the next unused material sample slot and release the mouse button; a duplicate copy of the material is created.

2. Make sure the copied material is active, not the original one. Also make sure the **Maps** rollout is expanded. Drag and drop the **Specular Color** map button or any other map button labeled **No Map** to the **Diffuse Color** map button; the map is removed from the **Diffuse Color** map button. It means that the logo is removed from the material.

   Next, you will name and save the material.

3. Choose the **Put to Library** button in the **Material Editor** dialog box; a flyout is displayed.

4. Choose **MugLogo.mat** from the flyout; the **Put to Library** dialog box is displayed.

5. Enter **No-Logo** in the **Name** text box and choose the **OK** button; the material is assigned the name *No-Logo* and is added to the **MugLogo** material library.

   Next, you will assign the **No-Logo** material to the handle of the mug.

6. Select *Handle* from the **Scene Explorer** or in the viewport; *Handle* is highlighted.

7. Choose the **Assign Material to Selection** button in the **Material Editor** dialog box; the *No-Logo* material is assigned to *Handle*.

8. Select *Saucer* from the **Scene Explorer**. Next, choose the **Assign Material to Selection** button in the **Material Editor** dialog box; the *No-Logo* material is assigned to *Saucer*.

9. Make sure the [Camera01] [Standard] [Default Shading] viewport is activated. Next, invoke the **Render Production** tool from the **Main Toolbar** to view the new changes after rendering; the rendered image is displayed. Now, close the render window.

Note

Reflection maps are not displayed in the viewport. The scene must be rendered if the user wants to see the effects of a reflection map.

Creating Material for Tabletop

In this section, you will create material for *Tabletop* which should reflect the background and any other object on the tabletop.

1. Activate the next unused material sample slot in the **Material Editor** dialog box.

2. In the **Shader Basic Parameters** rollout, select **Blinn** in the drop-down list, if it is not already selected.

3. Expand the **Maps** rollout. Next, choose the **Diffuse Color** map button which is currently labeled as **No Map**; the **Material/Map Browser** dialog box is displayed showing the types of maps that can be added.

4. Choose **Mix** from the **Maps > General** rollout and then choose the **OK** button; the map type is set as a combination of two colors or images. The **Material/Map Browser** dialog box is closed and the **Mix Parameters** rollout is displayed in the **Material Editor** dialog box.

5. Choose the **Color #1** map button currently labeled as **No Map**; the **Material/Map Browser** dialog box is displayed showing the types of maps that can be added.

6. Choose the **Bitmap** map from the **Maps > General** rollout and choose the **OK** button; the **Material/Map Browser** dialog box is closed and the **Select Bitmap Image File** dialog box is displayed. As the project folder is already set, the *images* folder of this project is displayed in the **Look in** text box of this dialog box.

7. Select the **Marbtea.gif** file and choose the **Open** button; you are moved to the next level of the material definition and various rollouts for the **Bitmap** map are displayed in the **Material Editor** dialog box.

Note

If a file type other than the currently selected file is opened earlier, you need to select ***All Formats*** *in the* ***File of type*** *drop-down list to display the currently selected file.*

8. In the **Coordinates** rollout, select the **Use Real-World Scale** check box. Set the value **250** in the **Width: Size** and **Height: Size** spinners. The material currently resembles a piece of granite. However, there is no reflection.

   Next, to add the reflection, you will set the second color of the mix as a reflection map.

9. Choose the **Go to Parent** button; you move one level up to the mix map level.

   The **Color #1** button is now labeled as **Map #*x* (Marbtea.gif)**.

10. Choose the **Color #2** map button, currently labeled as **No Map**; the **Material/Map Browser** dialog box is displayed showing the types of maps that can be added.

11. Choose the **Reflect/Refract** type from the **Maps > Scanline** rollout and then choose the **OK** button; you are moved to the next level of the material definition and various rollouts for the **Reflect/Refract** map are displayed in the **Material Editor** dialog box.

    You will notice that the first color (the granite bitmap) currently overlays the second color (the reflection map) completely. To avoid this, you need to set the amount of mixing for the two maps in the **Mix Amount** spinner.

12. Choose the **Go to Parent** button; you return to the **Mix** map level.

13. In the **Mix Parameters** rollout, set the value **10** in the **Mix Amount** spinner; 90% of the first color (the granite bitmap) and 10% of the second color (the reflection map) create a combined texture map.

    Next, you will name and save the material.

14. Choose the **Go to Parent** button; you return to the root level of the material definition. Now, choose the **Put to Library** button in the **Material Editor** dialog box; a flyout is displayed.

15. Choose **MugLogo.mat** from the flyout; the **Put to Library** dialog box is displayed.

16. Enter **Table** in the **Put To Library** dialog box and choose the **OK** button; the material is assigned the name *Table* and it is added to the **MugLogo** material library.

    Next, you will assign *Table* material to *Tabletop*.

17. Select *Tabletop* from the **Scene Explorer** or in the viewport; *Tabletop* is highlighted.

18. Choose the **Assign Material to Selection** button in the **Material Editor** dialog box. Next, choose the **Show Shaded Material in Viewport** button; the material *Table* is assigned to *Tabletop*. Render the scene to view the material.

Setting the Environment Background

In this section, you will use a bitmap image as a background for a rendered scene. With reflection maps active, it will appear that the background is actually being reflected on the objects in the scene.

1. Activate the next unused material sample slot in the **Material Editor** dialog box.

2. Select **Blinn** from the drop-down list in the **Shader Basic Parameters** rollout, if it is not already selected.

3. Expand the **Maps** rollout. Next, choose the **Diffuse Color** map button which is currently labeled as **No Map**; the **Material/Map Browser** dialog box is displayed showing the types of maps that can be added.

4. Select the **Bitmap** map from the **Maps > General** rollout and choose the **OK** button; the **Material/Map Browser** dialog box is closed and the **Select Bitmap Image File** dialog box is displayed. As the project folder is already set, the *images* folder of this project is displayed in the **Look in** drop-down list.

5. Select the **Foliage.jpg** file and choose the **Open** button; the image file is loaded and the parameters for the map are displayed in the **Material Editor** dialog box.

6. Click on the drop-down list located below the material sample slots. By default, the name **Map #*x*** is displayed. Rename it to **Foliage** and press the ENTER key.

7. Choose the **Go to Parent** button to return to the root level of the material. Next, choose the **M** button next to the **Diffuse** color swatch in the **Blinn Basic Parameters** rollout.

8. In the **Coordinates** rollout, select the **Environ** radio button. Select **Screen** from the **Mapping** drop-down list, if it is not already selected. Make sure the value **1.0** is set in the **U: Tiling** and **V: Tiling** spinners, as shown in Figure 10-5; the bitmap image will be scaled to fit the screen when it is rendered.

9. Choose the **Go to Parent** button to return to the root level of the material.

   Next, you will name and save the material.

10. Choose the **Put to Library** button in the **Material Editor** dialog box; a flyout is displayed.

11. Choose **MugLogo.mat** from the flyout; the **Put to Library** dialog box is displayed.

12. Enter the name **Background Material** in the **Put To Library** dialog box and choose the **OK** button; the material is assigned the name *Background Material* and is added to the **MugLogo** material library.

13. Choose **Rendering > Environment** from the menu bar; the **Environment and Effects** dialog box is displayed, as shown in Figure 10-6.

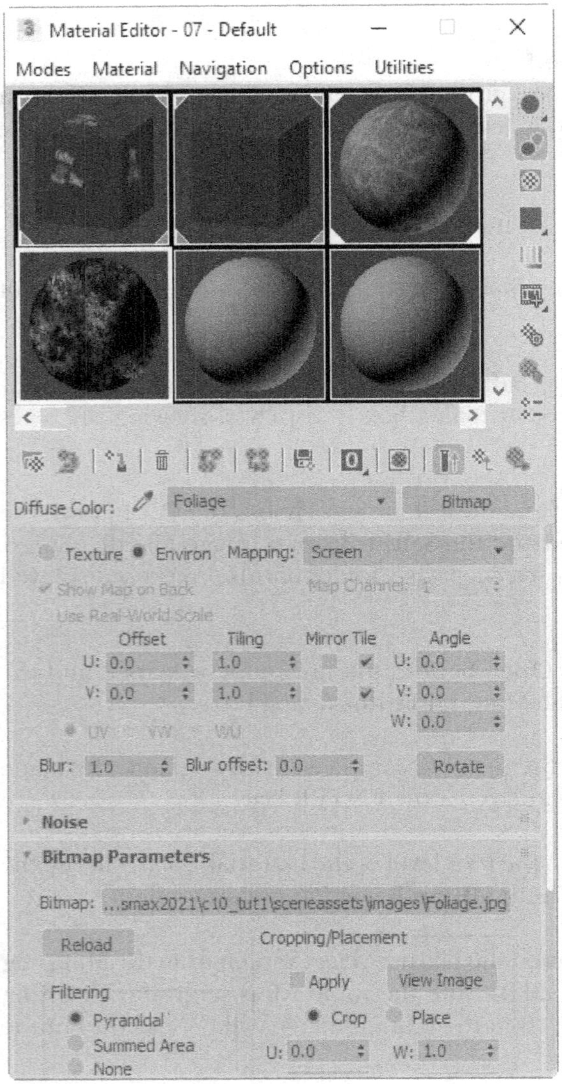

*Figure 10-5 The settings displayed in the **Coordinates** rollout*

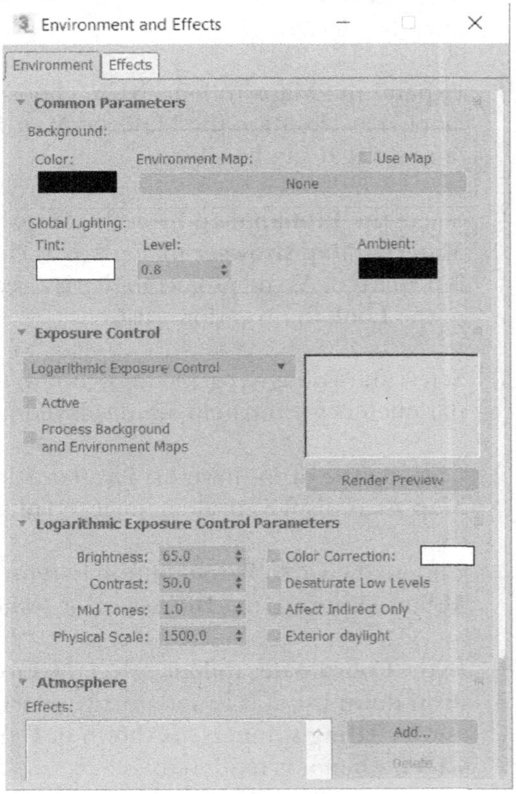

*Figure 10-6 The **Environment and Effects** dialog box*

14. Choose the **Environment Map** button in the **Background** area, currently labeled as **None**; the **Material/Map Browser** dialog box is displayed.

15. Expand the **Sample Slots** rollout, if it is not already expanded and then select **Diffuse Color: Foliage (Foliage.jpg)**, as shown in Figure 10-7. Choose the **OK** button; the **Instance or Copy?** dialog box is displayed, as shown in Figure 10-8.

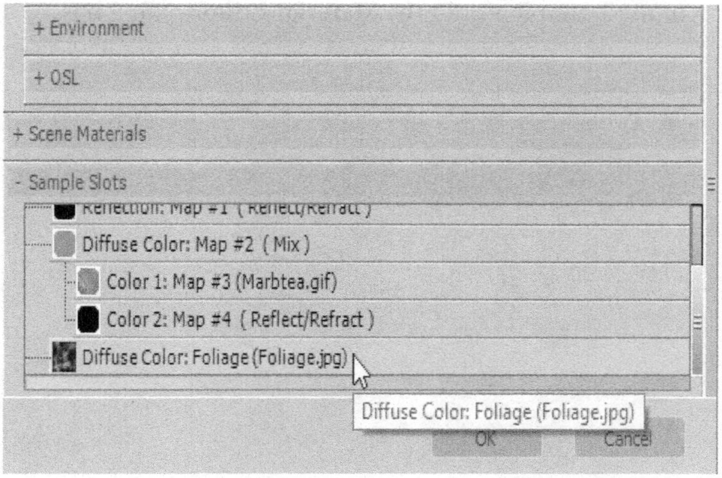

Figure 10-7 *Selecting* ***Diffuse Color: Foliage (Foliage.jpg)*** *from the* ***Sample Slots*** *rollout*

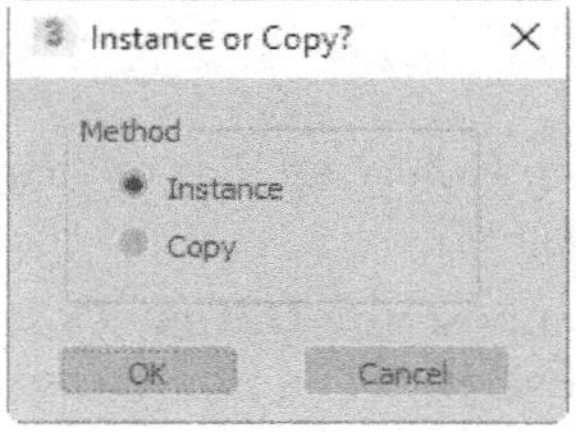

Figure 10-8 *The* ***Instance or Copy?*** *dialog box*

16. Make sure the **Instance** radio button is selected in the **Method** area of the **Instance or Copy?** dialog box and then choose the **OK** button.

    By selecting the **Instance** radio button, any changes made to the map in the **Material Editor** dialog box are automatically transferred to the environment background.

17. Close the **Environment and Effects** dialog box; a bitmap image of bushes will fill the background when the scene is rendered.

Creating Material for Paper Cups

In this section, you will create material for the paper cups using the **Checker** map.

1. Choose the **Display** tab of the **Command Panel**; various rollouts are displayed. In the **Hide** rollout, choose the **Unhide by Name** button; the **Unhide Objects** dialog box is displayed.

2. In this dialog box, select **PaperCup_Lrg** and **PaperCup_Sm**, as shown in Figure 10-9 and then choose the **Unhide** button; *PaperCup_Lrg* and *PaperCup_Sm* are displayed in the active viewport, as shown in Figure 10-10.

3. Select the next unused sample slot in the **Material Editor** dialog box.

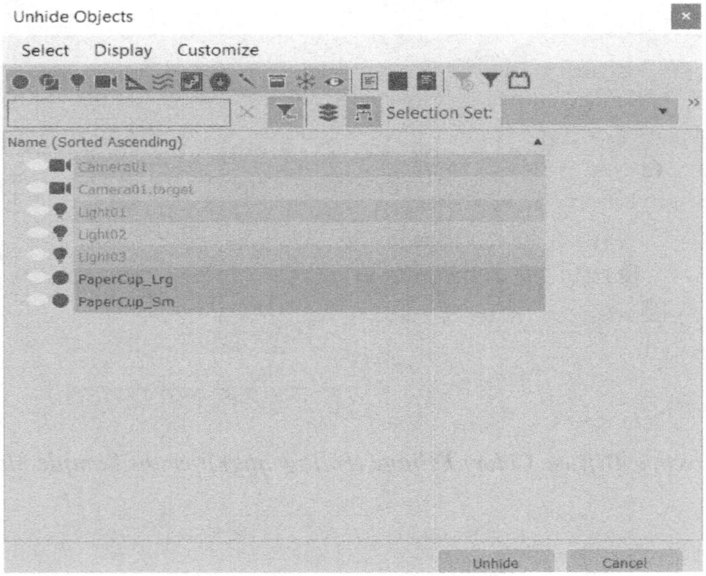

*Figure 10-9 Selecting **PaperCup_Lrg** and **PaperCup_Sm** in the **Unhide Objects** dialog box*

4. Expand the **Maps** rollout and choose the **Diffuse Color** map button, currently labeled as **No Map**; the **Material/Map Browser** dialog box is displayed.

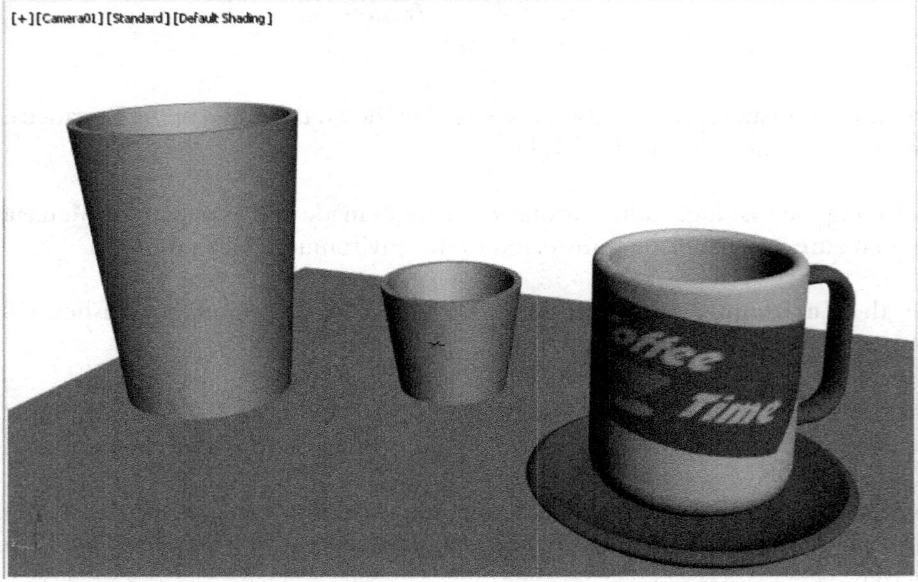

*Figure 10-10 **PaperCup_Lrg** and **PaperCup_Sm** displayed in the viewport*

5. Select the **Checker** map from **Maps > General** rollout in the **Material/Map Browser** dialog box and then choose the **OK** button; the **Material/Map Browser** dialog box is closed and various rollouts for the **Checker** map are displayed in the **Material Editor** dialog box.

6. In the **Material Editor** dialog box, expand the **Checker Parameters** rollout, if it is not already expanded. Next, choose the first **Color** swatch; the **Color Selector: Color 1** dialog box is displayed.

7. In the **Color Selector: Color 1** dialog box, set the values as given next:

   Red: **15** Green: **80** Blue: **0**

   The green color matching the diffuse color of the saucer and the cup is set.

8. In the **Checker Parameters** rollout, choose the second **Color** swatch; the **Color Selector: Color 2** dialog box is displayed, refer to Figure 10-11. In this dialog box, set the values as given next:

   Red: **244** Green: **244** Blue: **0**

***Figure 10-11** The **Color Selector: Color 2** dialog box displayed*

   Close the **Color Selector: Color 2** dialog box; the yellow color matching the yellow color of the logo is set.

9. In the **Coordinates** rollout, make sure the **Use Real-World Scale** check box is cleared. Set the value **8** in the **U: Tiling** and **V: Tiling** spinners and then choose the **Go to Parent** button.

   These settings specify that the map will be repeated 8 times across an object's U dimension, and 8 times across the object's V dimension. The actual size of the map in rendering will depend on the size of the object to which it is assigned.

10. Select *PaperCup_Lrg* and *PaperCup_Sm* in any viewport. Alternatively, you can select them from **Scene Explorer**. Choose the **Assign Material to Selection** button in the **Material Editor** dialog box to assign material to cups. Next, choose the **Show Shaded Material in Viewport** button to display the textured material on the cups in the viewport, as shown in Figure 10-12. Enter the name **Paper Cup** in the drop-down list below the material sample slots; the material created is named as *Paper Cup*.

Figure 10-12 *The textured material displayed on paper cups*

11. Make sure the *Camera01* viewport is activated. Invoke the **Render Production** tool from the **Main Toolbar**.

    Note the difference in the appearance of the material on *PaperCup_Lrg* and *PaperCup_Sm*. *PaperCup_Lrg* is wider and much taller than *PaperCup_Sm*, so the checkerboard pattern in the material is larger. Also, *PaperCup_Lrg* is taller than it is wide, so the material pattern appears stretched out vertically.

12. In the **Material Editor** dialog box, make sure the *Paper Cup* material is selected. Choose the map button (currently labeled as **M**) to the right of the **Diffuse** color swatch; you navigate to the diffuse color map level of the material definition.

13. In the **Coordinates** rollout of the **Material Editor** dialog box, select the **Use Real-World Scale** check box. Set the value **50** in the **Width: Size** and **Height: Size** spinners. Close the **Material Editor** dialog box; you will notice that the map is appearing distorted.

    The **Use Real-World Scale** check box is selected to maintain a consistent appearance when material is applied to objects of different sizes. If you clear this check box, the appearance of the material map during rendering depends on the size of the object on which the map is applied, and the number of times the map is tiled on the object.

14. Select *PaperCup_Sm* in any viewport. Choose the **Modify** tab in the **Command Panel**. Select **Tube** in the modifier stack. In the **Parameters** rollout for the tube, select the **Real-World Map Size** check box.

15. Select *PaperCup_Lrg* in any viewport. Make sure the **Modify** tab is chosen in the **Command Panel**. Select **Tube** in the modifier stack. In the **Parameters** rollout for the tube, select the **Real-World Map Size** check box.

16. Invoke the **Render Production** tool from the **Main Toolbar**.

As the material map size is now based on drawing units instead of object size, the material appears same on both the cups. Next, you will name and save the material.

17. Invoke the **Material Editor** dialog box and select *Paper Cup* material sample slot.

18. Choose the **Go to Parent** button. Next, choose the **Put to Library** button in the **Material Editor** dialog box; a flyout is displayed.

19. Choose **MugLogo.mat** from the flyout; the **Put to Library** dialog box is displayed.

20. Accept the name *Paper Cup* by choosing the **OK** button in the **Put To Library** dialog box; the *Paper Cup* material is added to the **MugLogo** material library.

Saving the Material Library

In this section, you will save the material library. You need to save the changes to the material library before you exit 3ds Max.

1. Choose the **Get Material** button in the **Material Editor** dialog box; the **Material/Map Browser** dialog box is displayed.

2. Right-click on the **MugLogo.mat** rollout; a shortcut menu is displayed. Choose **Save** from the shortcut menu, as shown in Figure 10-13.

3. Close the **Material/Map Browser** dialog box and the **Material Editor** dialog box.

Saving and Rendering the Scene

In this section, you will save the scene that you have created and then render it. You can also view the final rendered image of this model by downloading the *c10_3dsmax_2022_rndr.zip* file from *www.cadcim.com*. The path of the file is as follows: *Textbooks > Animation and Visual Effects > Autodesk 3ds Max > Autodesk 3ds Max 2022 for Beginners: A Tutorial Approach*

1. Choose **Save** from the **File** menu.

2. Activate the *Camera01* viewport. Next, invoke the **Render Production** tool from the **Main Toolbar**; the rendered image is displayed, refer to Figure 10-14.

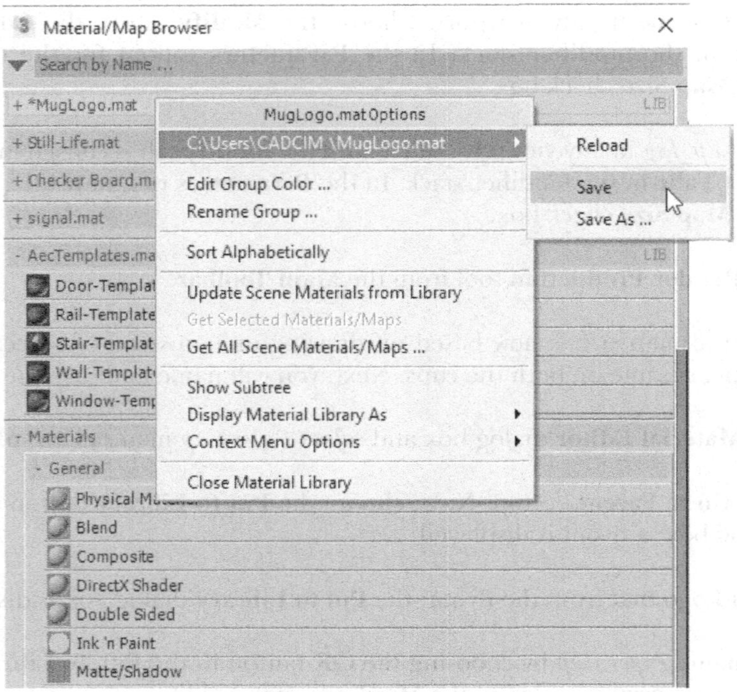

***Figure 10-13** Saving the material library*

***Figure 10-14** The rendered image*

Self-Evaluation Test

Answer the following questions and then compare them to those given at the end of this chapter:

1. Which of the following buttons in the **Material Editor** dialog box is used to assign a material to an object?

   (a) **Show Shaded Material in Viewport** (b) **Assign Material to Selection**
   (c) **Show End Result** (d) **Put Material to Scene**

2. Which of the following rollouts is used to adjust scaling, placement, and rotation angle of a bitmap image?

   (a) **Maps** (b) **Blinn Basic Parameters**
   (c) **Coordinates** (d) None of these

3. You need to turn off the __________ option in the **Coordinates** rollout to stop repetition of a bitmap texture of the object.

4. You need to __________ the scene to see the effects of a reflection map in the scene.

5. The **Environment and Effects** dialog box is used to set the background of a rendered scene. (T/F)

Review Questions

Answer the following questions:

1. Which of the following maps is used to generate reflection from other elements in a scene and then place it on the material during rendering?

   (a) **Reflect/Refract** (b) **Mix**
   (c) **Bitmap** (d) None of these

2. Which of the following buttons in the **Material Editor** dialog box is used to return to the root level of the material?

   (a) **Get Material** (b) **Go to Parent**
   (c) **Put Material to Scene** (d) **Make Material Copy**

3. The ________ map makes a 3D scene appear more realistic by causing objects to reflect a background image or other surrounding elements in a scene.

4. The ________ check box is selected to maintain a consistent appearance when a material is applied to objects of different sizes.

5. The actual size of a map in rendering depends on the size of the _______ to which it is assigned.

6. Bitmap images can be created in a paint program. (T/F)

EXERCISE

The rendered output of the model used in the following exercise can be accessed by downloading the *c10_3dsmax_2022_exr.zip* from *www.cadcim.com*. The path of the file is as follows: *Textbooks > Animation and Visual Effects > Autodesk 3ds Max > Autodesk 3ds Max 2022 for Beginners: A Tutorial Approach*

Exercise 1

Extract the contents of the *c10_3dsmax_2022_exr.zip* and then open *c10_exr01_start.max*. Next, create a decal material to apply on the vise shown in Figure 10-15. **(Expected time: 20 min)**

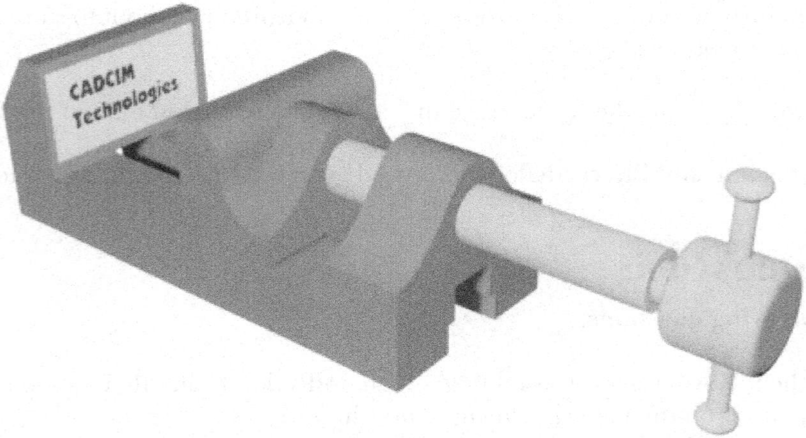

***Figure 10-15** Vise with decal material*

Answers to Self-Evaluation Test

1. b, **2.** c, **3. Tile**, **4.** render, **5.** T

Chapter 11

Material Editor: Miscellaneous Materials

Learning Objectives

After completing this chapter, you will be able to:

- *Use standard Autodesk materials*
- *Apply the UVW Map modifier*
- *Apply materials to different faces of an object*
- *Create background for the scene*
- *Add background to the scene*

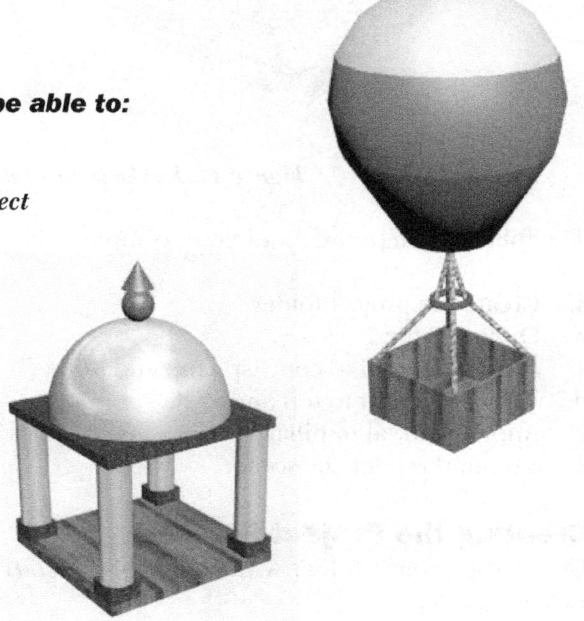

INTRODUCTION

In this chapter, you will assign materials to the objects created in Chapters 2 and 3 using the standard material library *3dsmax.mat*.

TUTORIALS

Before starting the tutorials, you need to download the *c11_3dsmax_2022_tut.zip* file from *www.cadcim.com*. The path of the file is as follows: *Textbooks > Animation and Visual Effects > Autodesk 3ds Max > Autodesk 3ds Max 2022 for Beginners: A Tutorial Approach*

Extract the contents of the zip file and save them in the *Documents* folder.

Tutorial 1

In this tutorial, you will assign different materials to the temple created in Tutorial 1 of Chapter 2, refer to Figure 11-1. **(Expected time: 20 min)**

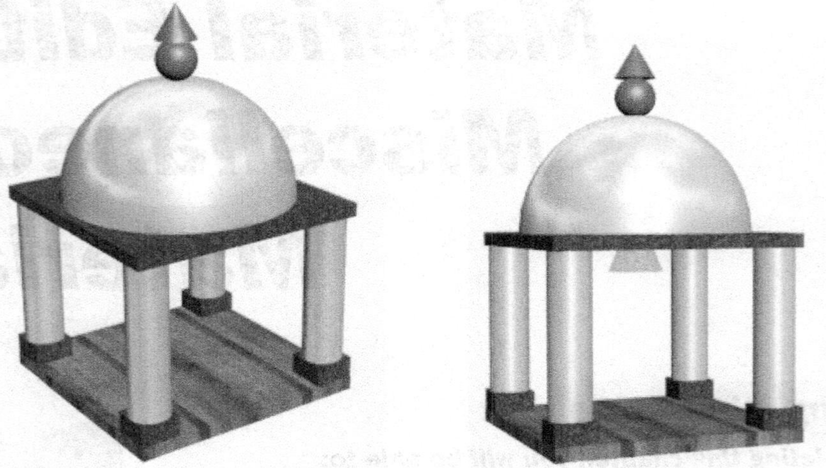

Figure 11-1 *The textured model of the temple*

The following steps are required to complete this tutorial:

a. Create the project folder.
b. Open the file.
c. Assign material to cone, sphere, and dome.
d. Assign material to top and Bell.
e. Apply material to pillars.
f. Save and render the scene.

Creating the Project Folder

Create the project folder with the name *c11_tut1* in the *3dsmax 2022* folder, as discussed in earlier chapters.

Opening the File

1. Open Windows Explorer and then browse to *\Documents\c11_3dsmax_2022_tut.* Next, copy *3dsmax.mat* to *\Documents\3dsmax 2022\c11_tut1\materiallibraries.* Also, copy *cedfence.jpg, lakerem2.jpg, balmoral.jpg,* and *whtgran.jpg* to *\Documents\3dsmax 2022\c11_tut1\sceneassets\images.*

2. Choose **Open** from the **File** menu; the **Open File** dialog box is displayed. In this dialog box, browse to the location *\Documents\c11_3dsmax_2022_tut* and select the **c11_tut1_start.max** file from it. Choose the **Open** button to open the file, refer to Figure 11-2.

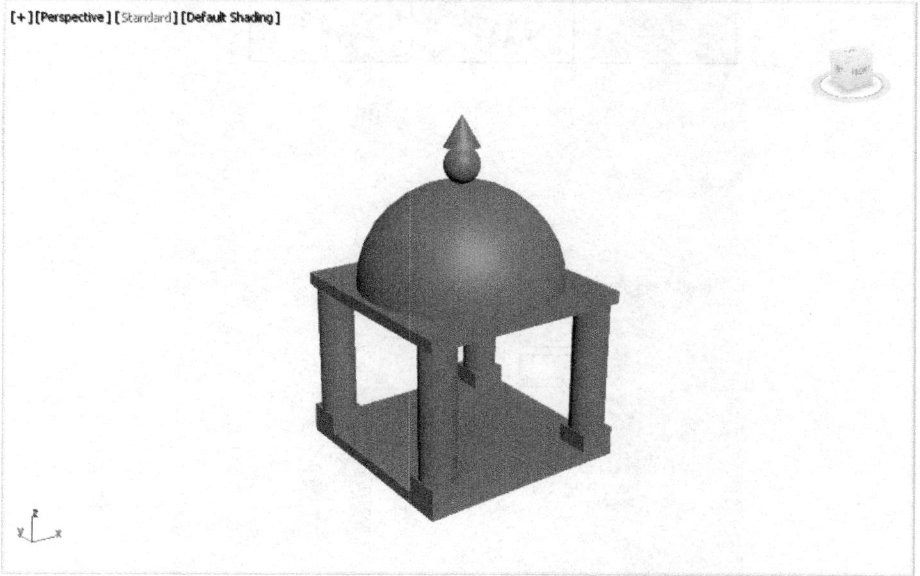

***Figure 11-2** The c11_tut1_start.max file*

Note

Instead of following step 2, you can directly open the file created in Tutorial 1 of Chapter 2 and then follow step 3 given below to save this file.

3. Choose **Save As** from the **File** menu; the **Save File As** dialog box is displayed. Save the file with the name *c11tut1.max.*

   Notice that the saving location of this file is *\Documents\3dsmax 2022\c11_tut1\scenes.*

Applying Material to Cone, Sphere, and Dome

In this section, you will choose materials from the library of materials and then apply the selected materials to cone, sphere, and dome.

1. Select *Cone* and *Sphere* from the **Scene Explorer**. Next, choose **Rendering > Material Editor > Compact Material Editor** from the menu bar; the **Material Editor** dialog box is displayed, as shown in Figure 11-3.

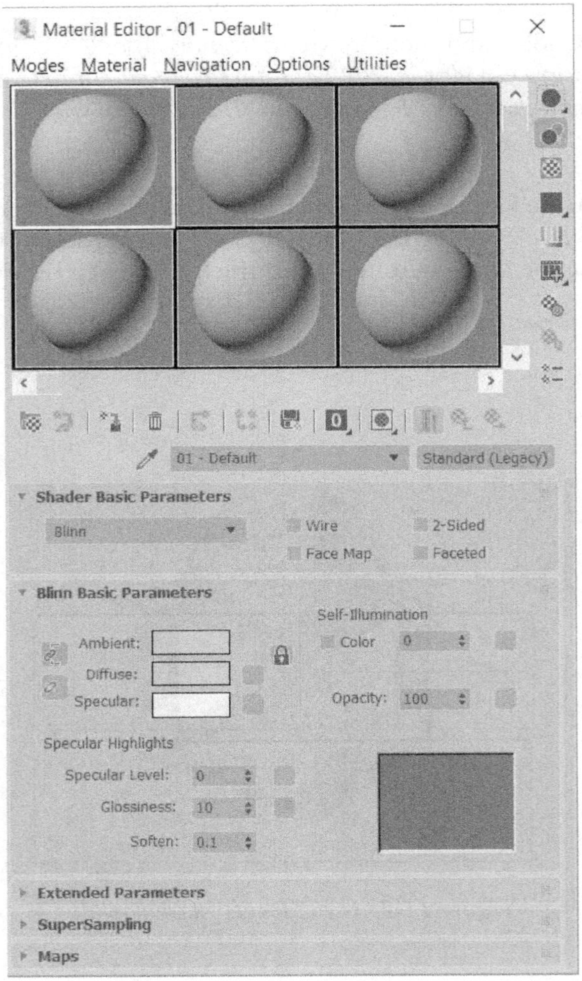

*Figure 11-3 The **Material Editor** dialog box*

2. Make sure the first sample slot is selected, and then choose the **Get Material** button; the **Material/Map Browser** dialog box is displayed, as shown in Figure 11-4. 

3. In this dialog box, choose the **Material/Map Browser Options** button; a flyout is displayed. 

4. Choose **Open Material Library** from this flyout; the **Import Material Library** dialog box is displayed.

   As the project folder is already set, the *materiallibraries* folder of this project is displayed in the **Look in** text box.

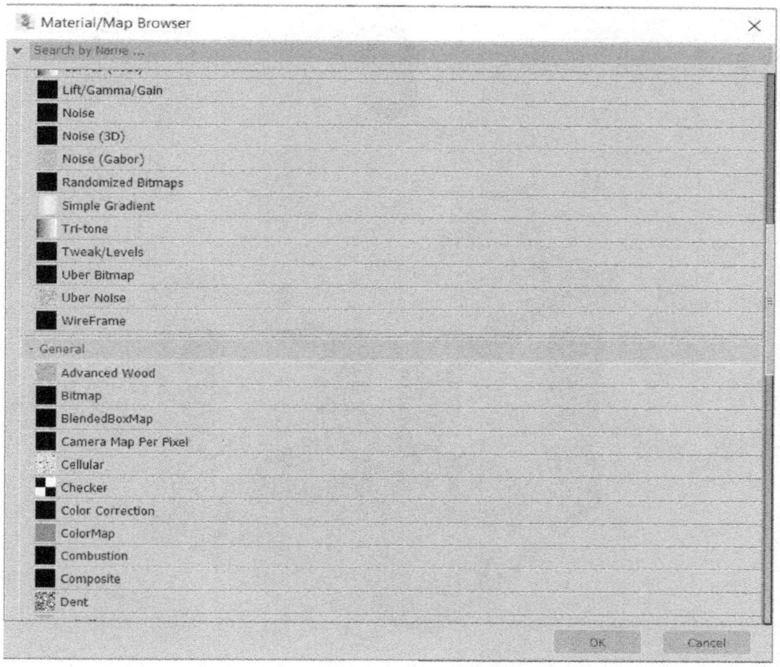

*Figure 11-4 The **Material/Map Browser** dialog box*

5. Now, select **3dsmax** and then choose the **Open** button; different materials present in *3dsmax* library are displayed in the **Material/Map Browser** dialog box.

6. Double-click on **Metal_Black_Plain (Standard)** from the *3dsmax.mat* material library. Make sure *cone* and *sphere* are selected in the viewport and then choose the **Assign Material to Selection** button in the **Material Editor** dialog box; the *Metal_Black_Plain* material is assigned to *cone* and *sphere*. Close the **Material/Map Browser** dialog box.

7. Select *Dome* from the **Scene Explorer** or in the viewport. In the **Material Editor** dialog box, select the second sample slot and choose the **Get Material** button; the **Material/Map Browser** dialog box is displayed.

8. In this dialog box, double-click on the material **Metal_Dark_Gold (Standard)** from the *3dsmax.mat* material library. Next, choose the **Assign Material to Selection** button in the **Material Editor** dialog box; the material *Metal_Dark_Gold* is assigned to *Dome*, as shown in Figure 11-5.

Figure 11-5 The Metal_Dark_Gold material assigned to Dome

9. Close the **Material/Map Browser** dialog box.

Applying Material to Top and Bell

In this section, you will choose materials from the library of materials and then apply the selected materials to *Top* and *Bell*.

1. Select *Top* from the **Scene Explorer** or in the viewport. In the **Material Editor** dialog box, select the third sample slot and choose the **Get Material** button; the **Material/Map Browser** dialog box is displayed.

2. In this dialog box, double-click on the **Stones_Balmoral (Standard)** material from the *3dsmax.mat* material library. Next, choose the **Assign Material to Selection** button in the **Material Editor** dialog box. Minimize the **Material Editor** dialog box and close the **Material/Map Browser** dialog box; the material *Stones_Balmoral* is assigned to *Top*.

3. Make sure *Top* is selected. Next, choose the **Modify** tab in the **Command Panel**. Select **UVW Map** from the **Modifier List** drop-down list; the **UVW Map** modifier is applied to *Top*.

4. In the **Mapping** area of the **Parameters** rollout, make sure the **Real-World Map Size** check box is cleared. Select the **Box** radio button.

5. Set the values in the spinners as given next:

   Height: **360** U Tile: **4** V Tile: **4** W Tile: **4**

The mapping coordinates are applied to *Top*. Note that although *Top* is rectangular, the **UVW Map** modifier gizmo is made cubic. This is done so that the map has the same scale on each side.

If the **UVW Map** modifier gizmo were rectangular, the material would appear squashed on the short sides of the object.

6. Select *Bell* from the **Scene Explorer**. In the **Material Editor** dialog box, select the fourth sample slot and choose the **Get Material** button; the **Material/Map Browser** dialog box is displayed.

7. In this dialog box, double-click on the material **Stones_White_Granite (Standard)** from the *3dsmax.mat* material library. Next, choose the **Assign Material to Selection** button in the **Material Editor** dialog box; the material *Stones_White_Granite* is assigned to *Bell*. Close the **Material/Map Browser** dialog box.

8. In the **Shader Basic Parameters** rollout of the **Material Editor** dialog box, select the **2-Sided** check box and set the value **35** in the **Opacity** spinner; a transparent material is created.

9. Make sure *Bell* is selected and then choose the **Modify** tab in the **Command Panel**. Select **UVW Map** from the **Modifier List** drop-down list; the **UVW Map** modifier is applied to *Bell*.

10. In the **Mapping** area of the **Parameters** rollout, make sure the **Real-World Map Size** check box is cleared and select the **Cylindrical** radio button; the cylindrical mapping coordinates are applied to *Bell*.

Applying Material to Pillars

In this section, you will choose materials from the library of materials and then apply the selected materials to pillars.

1. Select *Pillar01, Pillar02, Pillar03,* and *Pillar04*. In the **Material Editor** dialog box, select the second sample slot.

2. Choose the **Assign Material to Selection** button in the **Material Editor** dialog box; the material *Metal_Dark_Gold* is assigned to *Pillar01, Pillar02, Pillar03,* and *Pillar04*. Click anywhere in the viewport to deselect the pillars.

3. Select *Pillar01* and make sure the **Generate Mapping Coords** check box is selected in the **Parameters** rollout of the Modify panel. Next, clear the **Real-World Map Size** check box.

4. Repeat step 3 for *Pillar02, Pillar03,* and *Pillar04*.

5. Select *Base01, Base02, Base03,* and *Base04*.

6. In the **Material Editor** dialog box, select the sample slot containing the *Stones_Balmoral.* material. Next, choose the **Assign Material to Selection** button; the *Stones_Balmoral* material is assigned to *Base01, Base02, Base03,* and *Base04*.

7. Apply the **UVW Map** modifier to all base objects, as done earlier. In the **Mapping** area in the **Parameters** rollout of the Modify panel, make sure the **Real-World Map Size** check box is cleared. Select the **Box** radio button.

8. Set the values in the respective spinners as given next:

   Length: **360** Width: **360** Height: **360**
   U Tile: **4** V Tile: **4** W Tile: **4**

9. Select *Bottom* from the **Scene Explorer** or in the viewport. In the **Material Editor** dialog box, select the fifth sample slot. Next, choose the **Get Material** button; the **Material/Map Browser** dialog box is displayed.

10. Double-click on the material **Wood_Cedfence (Standard)** from the *3dsmax.mat* material library in the **Material/Map Browser** dialog box. Next, choose the **Assign Material to Selection** button in the **Material Editor** dialog box; the material *Wood_Cedfence* is assigned to *Bottom*.

10. Double-click on the material **Wood_Cedfence (Standard)** from the *3dsmax.mat* material library in the **Material/Map Browser** dialog box. Next, choose the **Assign Material to Selection** button in the **Material Editor** dialog box; the material *Wood_Cedfence* is assigned to *Bottom*.

11. Apply the **UVW Map** modifier to *Bottom*, as done earlier. In the **Mapping** area of the **Parameters** rollout of the **Modify** tab, make sure the **Real-World Map Size** check box is cleared and select the **Box** radio button.

12. In the **Mapping** area of the **Parameters** rollout, set the values in the respective spinners as given next:

    U Tile: **2** V Tile: **2** W Tile: **2**
    Height: **360**

    The box mapping coordinates are applied to *Bottom*.

13. Close the **Material Editor** and **Material/Map Browser** dialog boxes.

Saving and Rendering the Scene

In this section, you will save the scene that you have created and then render it. You can also view the final rendered image of this model by downloading the *c11_3dsmax_2022_rndr.zip* file from *www.cadcim.com*. The path of the file is as follows: *Textbooks > Animation and Visual Effects > Autodesk 3ds Max > Autodesk 3ds Max 2022 for Beginners: A Tutorial Approach*

1. Choose **Save** from the **File** menu.

2. Make sure the Perspective viewport is activated. Next, invoke the **Render Production** tool from the **Main Toolbar**; the rendered image of the temple is displayed, as shown in Figure 11-6.

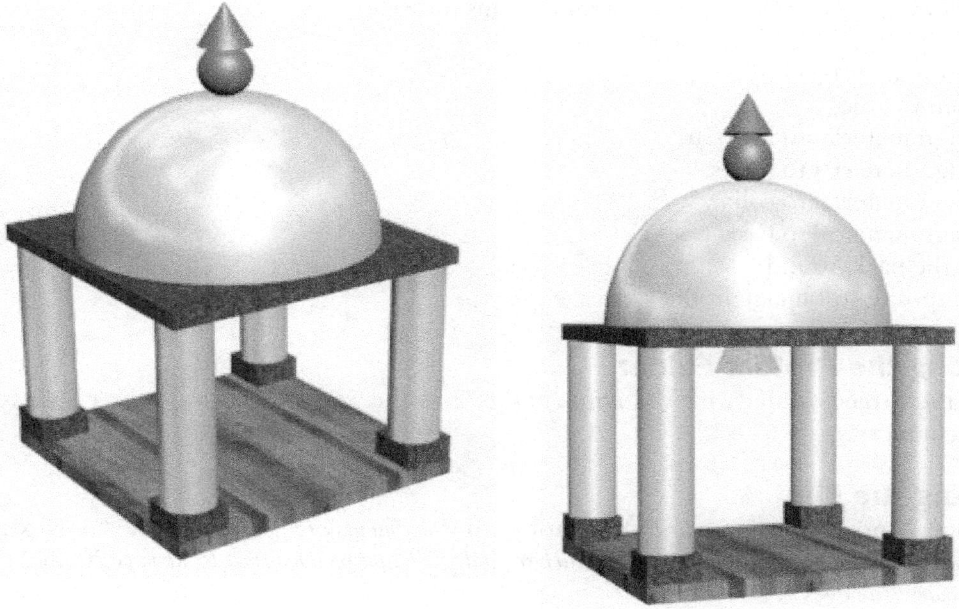

Figure 11-6 *The rendered image of the temple*

Tutorial 2

In this tutorial, you will texture a scene created in Tutorial 1 of Chapter 3, refer to Figure 11-7. You will also composite a still image in the scene using color key OSL shader.

(Expected time: 25 min)

Figure 11-7 *The textured scene with Air balloon*

The following steps are required to complete this tutorial:

a. Create the project folder.
b. Open the file.
c. Assign material to balloon.
d. Assign material to ropes.
e. Assign material to collar.
f. Assign material to basket.
g. Set the background.
h. Composite still image in the scene.

Creating the Project Folder

Create the project folder with the name *c11_tut2* in the *3dsmax 2022* folder, as discussed in earlier chapters.

Opening the File

1. Open Windows Explorer and then browse to *\Documents\c11_3dsmax_2022_tut*. Next, copy *rope.jpg, wood.jpg, car image.jpg and Background_Falls.jpg* to *\Documents\3dsmax 2022\c11_tut2\sceneassets\images*.

2. Choose **Open** from the **File** menu; the **Open File** dialog box is displayed. In this dialog box, browse to the location *\Documents\c11_3dsmax_2022_tut* and select the *c11_tut2_start.max* file from it. Choose the **Open** button to open the file, refer to Figure 11-8.

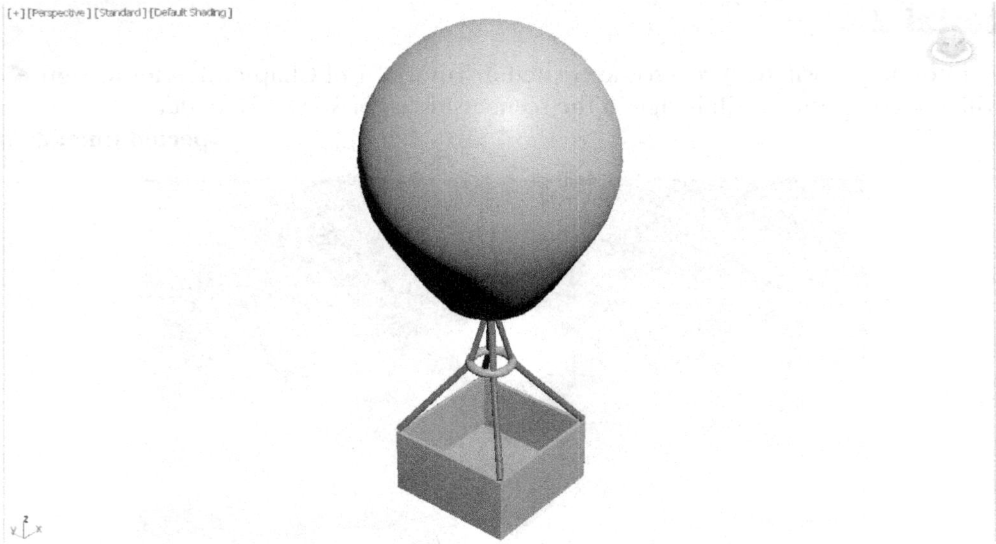

Figure 11-8 *The c11_tut2_start.max file*

3. Choose **Save As** from the **File** menu; the **Save File As** dialog box is displayed. Browse to the location *\Documents\3dsmax 2022\c11_tut2\scenes*. Save the file with the name *c11tut2.max* at this location.

Note

Instead of following step 1, you can directly open the file that you created in Tutorial 1 of Chapter 3 and then follow step 3 to save it.

Assigning Multi/Sub-Object Material to Balloon

In this section, you will assign different materials to different portions of the balloon. This can be accomplished by using the multi/sub-object type of material.

1. Choose **Rendering > Material Editor > Compact Material Editor** from the menu bar; the **Material Editor** dialog box is displayed.

2. Make sure the first material sample slot is selected in this dialog box. Next, choose the **Standard (Legacy)** button; the **Material/Map Browser** dialog box is displayed, as shown in Figure 11-9.

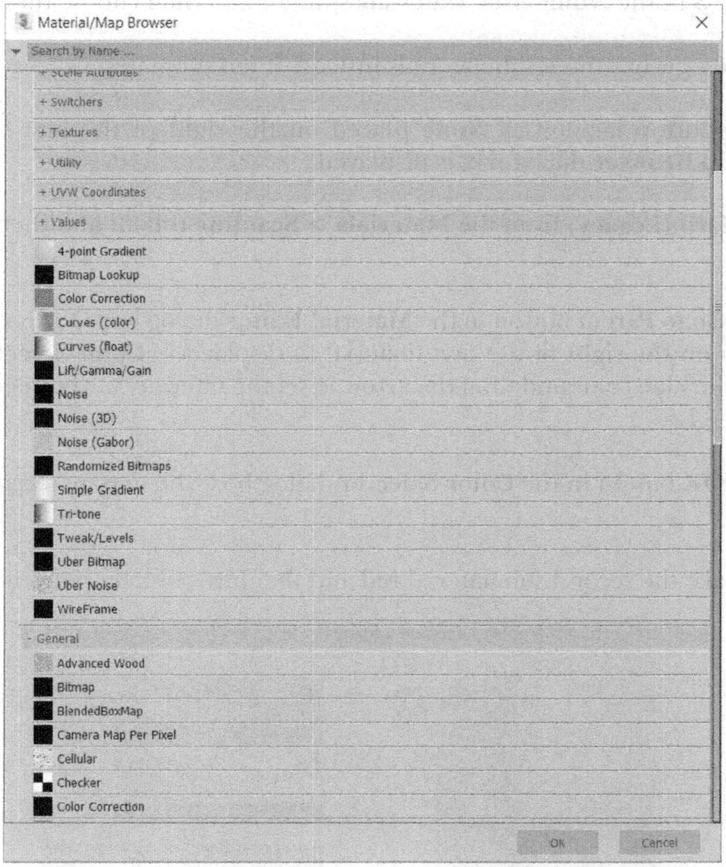

***Figure 11-9** The **Material/Map Browser** dialog box*

3. In this dialog box, select **Multi/Sub-Object** from **Materials > General** rollout and then choose the **OK** button; the **Replace Material** dialog box is displayed. Select the **Discard old material?** radio button.

4. Rename the material as **Balloon Material** in the material name drop-down list located below the material sample slots and then press the ENTER key; the material created is renamed as *Balloon Material*.

5. In the **Multi/Sub-Object Basic Parameters** rollout, choose the **Set Number** button; the **Set Number of Materials** dialog box is displayed, as shown in Figure 11-10.

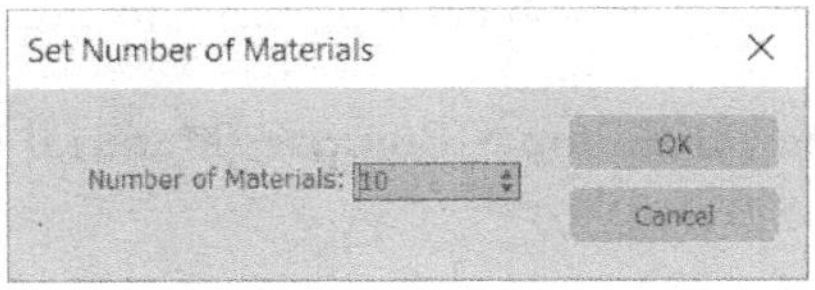

***Figure 11-10** The **Set Number of Materials** dialog box*

6. Set the value **3** in the **Number of Materials** spinner and then choose the **OK** button.

   The material now has three submaterials instead of ten submaterials.

7. Choose the button labeled as **None** placed on the right of the first submaterial; the **Material/Map Browser** dialog box is displayed.

8. Select **Standard (Legacy)** from the **Materials > Scanline** rollout and then choose the **OK** button.

9. Choose the **Go to Parent** button in the **Material Editor** dialog box. Next, choose the color swatch on the right of the first material to display the **Color Selector** dialog box. Select the blue color and drag the arrow to set the intensity of your choice, refer to Figure 11-11.

10. Choose the **OK** button in the **Color Selector** dialog box; the first material changes to blue color.

11. Similarly, make the second submaterial red and the third submaterial yellow.

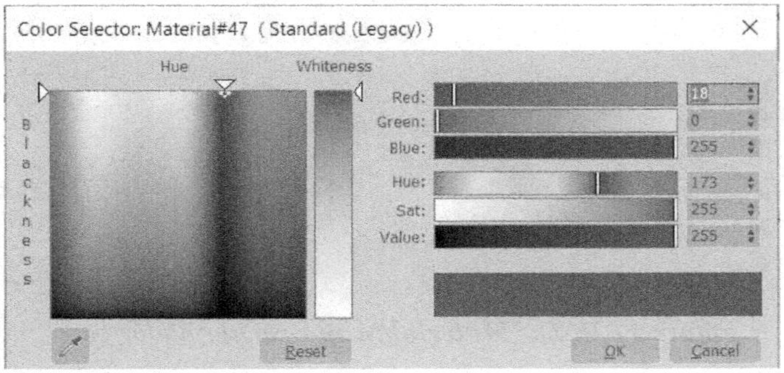

***Figure 11-11** Setting the color of the material to blue using the **Color Selector** dialog box*

12. Select *Balloon* in the Front viewport. Make sure the sample slot containing *Balloon Material* is activated in the **Material Editor** dialog box. Next, choose the **Assign Material to Selection** button in the **Material Editor** dialog box; *Balloon Material* is assigned to *Balloon*.

13. Make sure *Balloon* is selected and then choose the **Modify** tab in the **Command Panel**. In the **Selection** rollout, choose the **Face** button.

14. In the Front viewport, select faces of bottom three rows on *Balloon*, as shown in Figure 11-12.

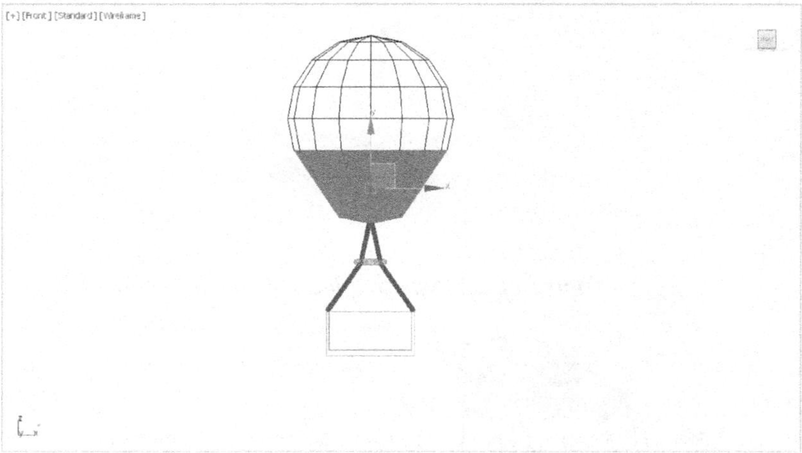

Figure 11-12 *Selecting faces of bottom three rows*

15. Set **1** in the **Set ID** spinner of the **Material** area in the **Surface** properties rollout of the **Modify** panel.

16. Make sure the Front viewport is active. Next, select faces of middle two rows on *Balloon*, as shown in Figure 11-13.

17. Set **2** in the **Set ID** spinner of the **Material** area in the **Surface** properties rollout of the **Modify** panel.

18. Select the remaining faces of top three rows on *Balloon*, as shown in Figure 11-14.

19. Set **3** in the **Set ID** spinner of the **Material** area in the **Surface** properties rollout of the **Modify** panel.

Assigning Material to Balloon

In this section, you will assign materials that have been created for the balloon. This can be accomplished by using the **Material Editor** dialog box.

1. In the **Material Editor** dialog box, make sure the *Balloon Material* is selected. Next, choose the first submaterial button with ID 1; the parameters for the blue material are displayed.

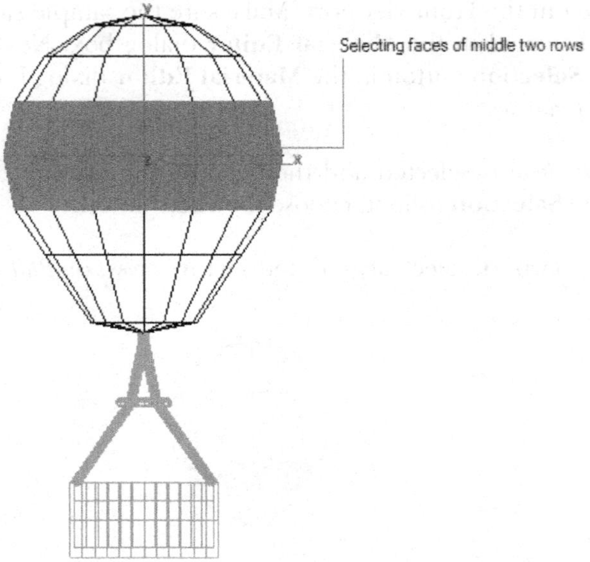

Figure 11-13 Selecting faces of middle two rows

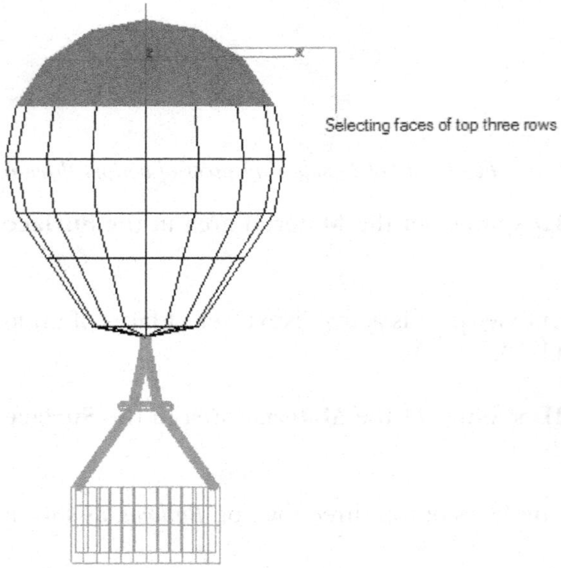

Figure 11-14 Selecting faces of top three rows

2. Enter the name of submaterial as **Blue** in the material name drop-down list located below the material sample slots and then press the ENTER key.

3. In the **Shader Basic Parameters** rollout, select **Phong** from the drop-down list.

4. In the **Specular Highlights** area of the **Phong Basic Parameters** rollout, set the value **50** in the **Specular Level** spinner and **20** in the **Glossiness** spinner.

5. Choose the **Go to Parent** button to return to the root level of the material. Next, choose the second submaterial button with ID 2; the parameters for the red material are displayed. Enter the name of submaterial as **Red** in the **Material name** drop-down list located below the material sample slots and press the ENTER key.

6. Select **Phong** from the drop-down list of the **Shader Basic Parameters** rollout and then using the same process, rename the second submaterial as **Red**. Set the value **50** in the **Specular Level** spinner and **20** in the **Glossiness** spinner.

7. Choose the **Go to Parent** button to return to the root level of the material. Next, choose the third submaterial button with ID 3; the parameters for the yellow material are displayed. Enter the name of submaterial as **Yellow** in the **Material name** drop-down list located below the material sample slots and press the ENTER key.

8. Select **Phong** from the drop-down list of the **Shader Basic Parameters** rollout and then rename the third submaterial as **Yellow**. Set the value **50** in the **Specular Level** spinner and **20** in the **Glossiness** spinner.

9. At *Yellow* submaterial level, select the **Color** check box in the **Self Illumination** area in the **Phong Basic Parameters** rollout; the spinner is replaced by a color swatch.

10. Choose the color swatch to display the **Color Selector: Self-Illum Color** dialog box. Change the color to dark red.

    The **Self-Illumination** color will help give the illusion of a gas burner inside *Balloon*.

11. Return to the parent material level and close the **Material Editor** dialog box.

12. Activate the Perspective viewport and invoke the **Render Production** tool from the **Main Toolbar**; the still image of *Balloon* is rendered in the **Rendered Frame** window.

Assigning Material to Ropes

In this section, you will assign the created materials to ropes to give them a realistic look.

1. Select *Rope01, Rope02, Rope03,* and *Rope04* from the **Scene Explorer**.

2. Invoke the **Material Editor** tool from the **Main Toolbar**; the **Material Editor** dialog box is displayed. Next, select the second material sample slot. In the **Material Name** drop-down list, enter **rope**.

3. Expand the **Maps** rollout. Next, choose the **Diffuse Color** map button currently labeled as **No Map**; the **Material/Map Browser** dialog box is displayed showing the types of maps that can be added.

4. Select the **Bitmap** map from the **Maps > General** rollout and then choose the **OK** button; the **Material/Map Browser** dialog box is closed and the **Select Bitmap Image File** dialog box is displayed.

5. In this dialog box, make sure **All Formats** is selected in the **Files of type** drop-down list. Select the *rope.jpg* file and then choose the **Open** button; the image file is loaded and the parameters for modifying the map are displayed in the **Material Editor** dialog box.

6. In this dialog box, set **0.15** in the **U Tiling** and **V Tiling** spinners and **90** in the **W** spinner.

7. Choose the **Assign Material to Selection** button. Minimize the **Material Editor** dialog box.

8. Make sure the Perspective viewport is activated and then invoke the **Render Production** tool from the **Main Toolbar** to render the scene.

Assigning Material to Collar

In this section, you will assign the material to the collar.

1. Select **Collar** from the **Scene Explorer**.

2. Restore the **Material Editor** dialog box and then drag the second sample slot to third sample slot. In the **Material Name** drop-down list, enter **collar**.

3. In the **Coordinates** rollout, set **0** in the **W** spinner. Next, choose the **Assign Material to Selection** button.

Assigning Material to Basket

In this section, you will assign the material to the basket.

1. Select *Basket01* from the **Scene Explorer**.

2. Restore the **Material Editor** dialog box. Select the fourth sample slot in the **Material Editor** dialog box to make it active. In the **Material Name** drop-down list, enter **basket**.

3. Expand the **Maps** rollout. Next, choose the **Diffuse Color** map button currently labeled as **No Map**; the **Material/Map Browser** dialog box is displayed showing the types of maps that can be added.

4. Select the **Bitmap** map from the **Maps > General** rollout and then choose the **OK** button; the **Material/Map Browser** dialog box is closed and the **Select Bitmap Image File** dialog box is displayed.

5. In this dialog box, make sure **All Formats** is selected in the **Files of type** drop-down list. Select the *basket.jpg* file and then choose the **Open** button; the image file is loaded and the parameters for modifying the map are displayed in the **Material Editor** dialog box.

6. In the **Specular Highlights** area of the **Blinn Basic Parameters** rollout, set **55** in the **Specular Level** spinner and **40** in the **Glossiness** spinner.

7 Make sure the basket is selected and then choose the **Modify** tab in the **Command Panel**. In the **Modifier List** drop-down list, select **UVW Map** in the **OBJECT-SPACE MODIFIERS** area.

8. In the **Mapping** area of the **Parameters** rollout, make sure the **Real-World Map Size** check box is cleared and then select the **Box** radio button.

9. Set the values in the respective spinners as given next:

   Length: **160** Width: **160** Height: **160**

10. Render the Perspective viewport. Close the **Rendered Frame** window after viewing the image.

11. Choose **Create > Cameras** in the **Command Panel**. Next, invoke the **Target** tool from the **Object Type** rollout and place the camera in the Front viewport and position it in all viewports, as shown in Figure 11-15.

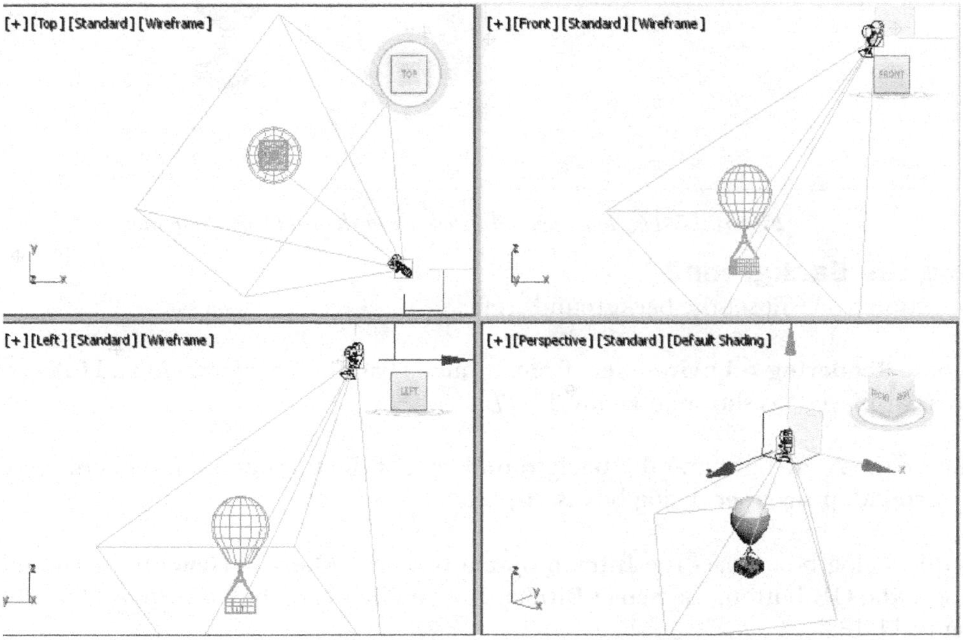

***Figure 11-15** The camera placed and positioned in the Front viewport*

12. Activate the Perspective viewport and then press the C key to make it the Camera viewport.

13. Make sure the **Select and Move** tool is invoked and then move the camera to get the desired view in the Camera viewport. In the Camera viewport, you can use the **Field-of-View** tool and other camera viewport tools to move the camera view closer and get a better view, refer to Figure 11-16. The **Zoom Region** tool is replaced by the **Field-of-View** tool when the Camera viewport is active.

14. Render the Camera viewport.

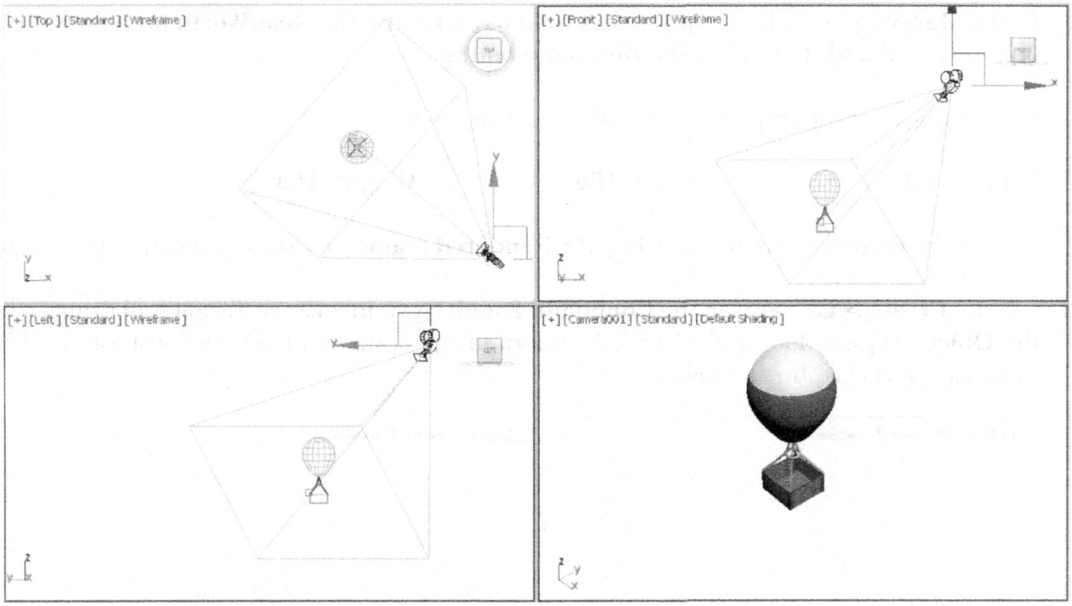

Figure 11-16 The view adjusted using the ***Field-of-View*** *tool*

Setting the Background

In this section, you will set the background scene.

1. Choose **Rendering > Environment** from the menu bar; the **Environment and Effects** dialog box is displayed, as shown in Figure 11-17.

2. Choose the **None** button in the **Background** area of the **Common Parameters** rollout; the **Material/Map Browser** dialog box is displayed.

3. In this dialog box, select the **Bitmap** option from the **Maps > General** rollout and then choose the **OK** button; the **Select Bitmap Image File** dialog box is displayed, as shown in Figure 11-18.

   As the project folder is already set, the *images* folder of this project is displayed in the **Look in** drop-down list.

4. Select **Background_Falls.jpg** and choose the **Open** button; the *Background_Falls.jpg* replaces the **None** label in the **Environment and Effects** dialog box.

5. Make sure the **Use Map** check box is selected in the **Environment and Effects** dialog box. Next, invoke the **Material Editor** dialog box.

6. In the **Environment and Effects** dialog box, press and hold the **Map # X (Background_Falls.jpg)** button and drag it to the fifth sample slot in the **Material Editor** dialog box; the **Instance (Copy) Map** dialog box is displayed. In this dialog box, make sure that the **Instance** radio button is selected. Next, choose the **OK** button. Now, select the fifth sample slot in the **Material Editor** dialog box.

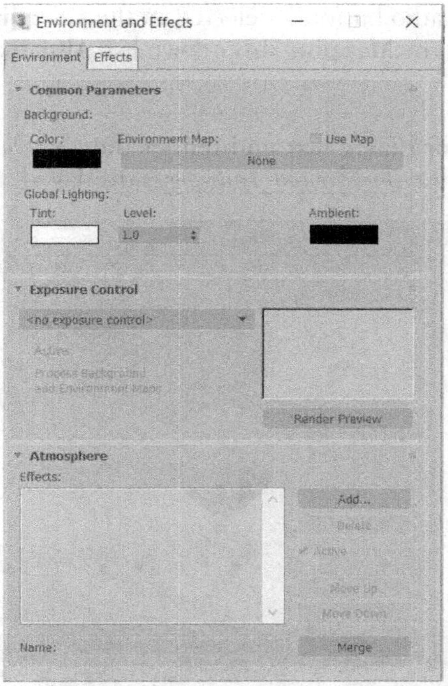

Figure 11-17 The ***Environment and Effects*** *dialog box*

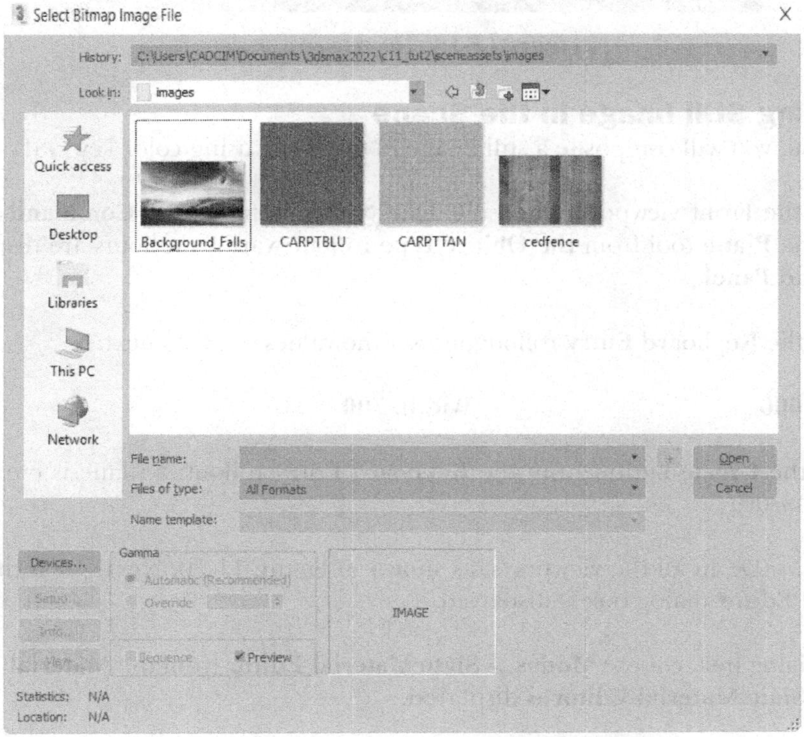

Figure 11-18 The ***Select Bitmap Image File*** *dialog box*

7. Make sure the **Environ** radio button is selected in the **Coordinates** rollout and also make sure **Screen** is selected in the **Mapping** drop-down list. Also, make sure **0.0** is entered in the **Blur offset** spinner.

8. Close the **Material Editor** dialog box and render the *Camera001* viewport; the rendered view of the balloon with the *Background_Falls.jpg* is displayed, as shown in Figure 11-19.

Figure 11-19 *The rendered scene*

Compositing Still Image in the Scene

In this section, you will composite a still image in the scene using color key OSL shader.

1. Activate the Front viewport. Choose **Create > Geometry** in the **Command Panel**. Next, invoke the **Plane** tool from the **Object Type** rollout; various rollouts are displayed in the **Command Panel**.

2. Expand the **Keyboard Entry** rollout and set the values as given next:

   Length: **600** Width: **700**

3. Choose the **Create** button from the **Keyboard Entry** rollout; a plane is created with the name *Plane001*.

4. Align *Plane001* in all the viewports, as shown in Figure 11-20. Next, press the M key; the **Material Editor** dialog box is displayed.

5. In this dialog box, choose **Modes > Slate Material Editor** from the **Material Editor** menu bar; the **Slate Material Editor** is displayed.

6. In the **Slate Material Editor**, select **Materials > General > Physical Material** from the **Material/Map Browser** area and drag it to the **View 1** area, refer to Figure 11-21.

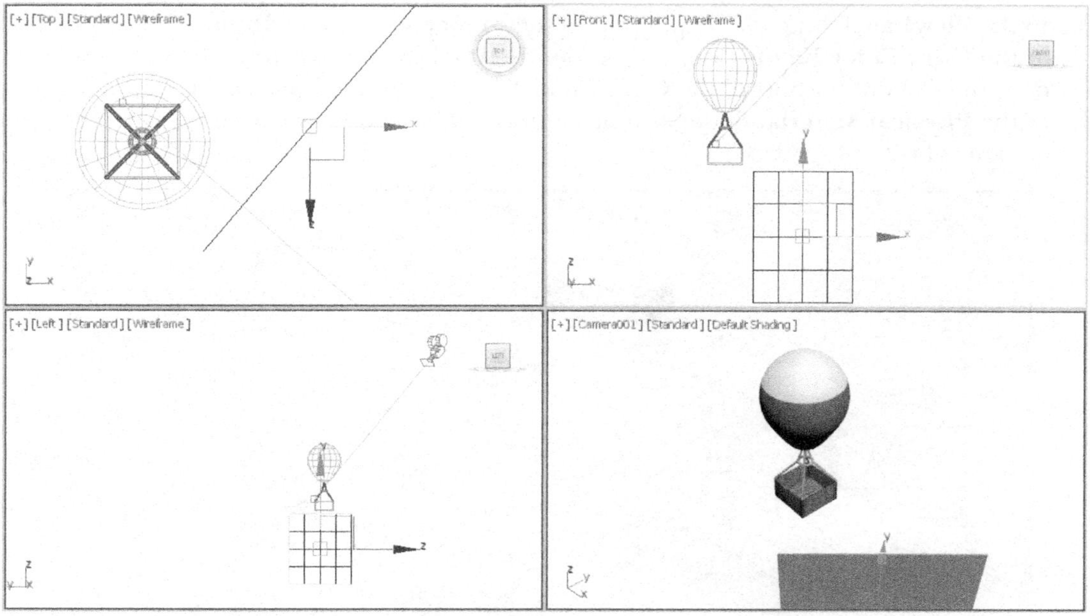

***Figure 11-20** Plane001 aligned in all the viewports*

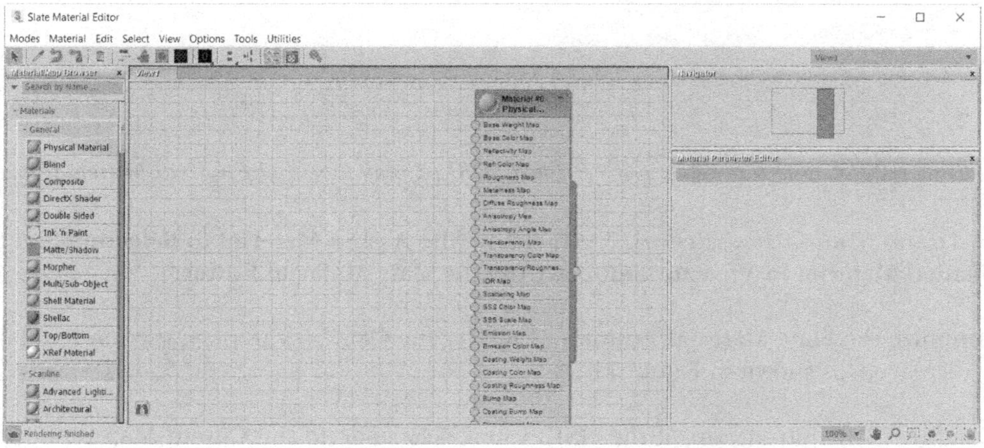

***Figure 11-21** **Physical Material** dragged to the **View1** area*

7. Select **Maps > General > Bitmap** from the **Material/Map Browser** area and drag it to the **View1** area; the **Select Bitmap Image File** dialog box is displayed. As the project folder is already set, the *images* folder of this project is displayed in the **Look in** drop-down list.

8. Select **car image.jpg** and choose the **Open** button; the **Bitmap** node with the car image icon is added to the **View1** area.

9. Select **Maps > OSL > Color Key** from the **Material/Map Browser** area and drag it to the **View1** area; the **OSL: Color Key** node is displayed in the **View1** area.

10. In the **View1** area, drag the output port of the **Bitmap** node to the **Input Image** input port of the **OSL: Color Key** node; a connection is established between these two ports. Next, drag the **Col** output port of the **OSL: Color Key** node to the **Base Color Map** input port of the **Physical Material** node; a connection is established between these two ports, refer to Figure 11-22.

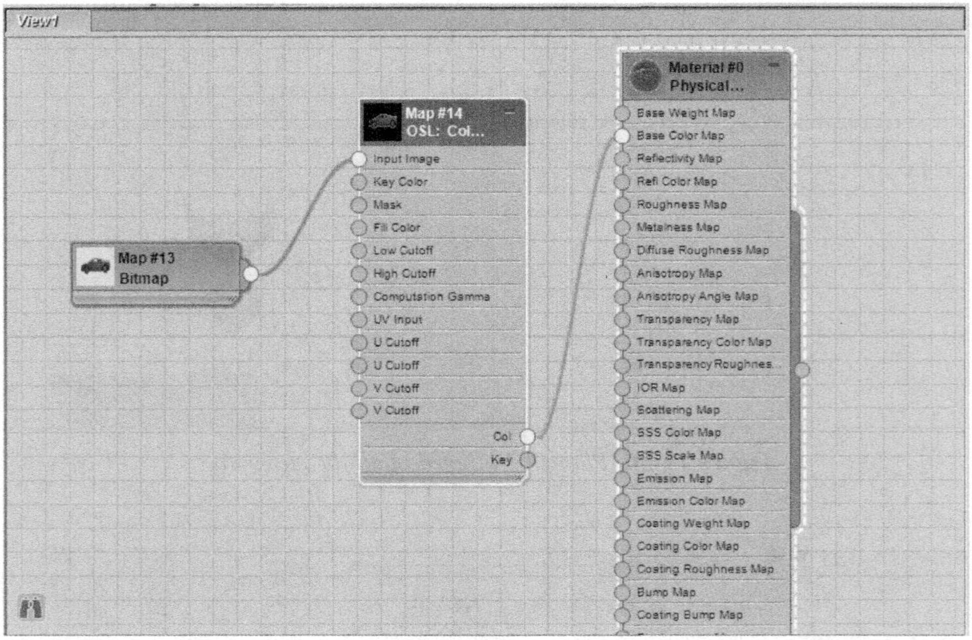

***Figure 11-22** Connection between the **Bitmap**, **OSL: Color Key**, and **Physical Material** nodes*

11. Make sure *Plane001* is selected. Next, choose the **Assign Material to Selection** and **Show Shaded Material in Viewport** buttons from the **Slate Material Editor**.

12. Minimize the **Slate Material Editor** and render the Camera viewport; the rendered image is displayed, as shown in Figure 11-23. `

13. Drag the **Key** output port of the **OSL: Color Key** node to the **Cutout Map** input port of the **Physical Material** node. Next, render the Camera viewport again; the rendered image is displayed, as shown in Figure 11-24. `

Figure 11-23 The rendered image

Figure 11-24 *The rendered image displayed after connecting the* ***OSL: Color Key*** *node to the* ***Physical Material*** *node*

Self-Evaluation Test

Answer the following questions and then compare them to those given at the end of this chapter:

1. Which of the following modifiers is used to apply mapping coordinates to an object?

   (a) **UVW Map** (b) **Extrude**
   (c) **Edit Mesh** (d) **Mesh Select**

2. Which of the following buttons in the **Material Editor** dialog box is used to move one level up in the material definition?

   (a) **Get Material** (b) **Go to Parent**
   (c) **Put to Library** (d) **Go Forward to Sibling**

3. In which of the following sections in the **Modifier** stack, the **UVW Map** is located?

   (a) **OBJECT-SPACE MODIFIERS** (b) **WORLD-SPACE MODIFIERS**
   (c) **Selection Modifiers** (d) None of these

4. You can change the intensity of the omni light by changing the value in the _________ spinner of the **Intensity/Color/Attenuation** rollout.

5. The _________ dialog box is used to set the background scene for an object.

6. The **Assign Material to Selection** button in the **Material Editor** dialog box is used to display the materials on the objects in the viewport. (T/F)

Review Questions

Answer the following questions:

1. The **Get Material** button in the **Material Editor** dialog box is used to display the _________ dialog box.

2. You can apply mapping coordinates to an object by applying the _________ modifier.

3. The _________ button in the **Material Editor** dialog box is used to assign material to an object.

4. You can use the **Multi/Sub Object** material to assign different materials to different portions of an object. (T/F)

EXERCISE

The rendered output of the model used in the following exercise can be accessed by downloading the *c11_3dsmax_2022_exr.zip* from *www.cadcim.com*. The path of the file is as follows: *Textbooks > Animation and Visual Effects > Autodesk 3ds Max > Autodesk 3ds Max 2022 for Beginners: A Tutorial Approach*

Exercise 1

Extract the contents of the *c11_3dsmax_2022_exr.zip* and then open *c11_exr01_start.max*. Next, apply materials to the model of the table, as shown in Figure 11-25. **(Expected time: 20 min)**

Figure 11-25 *The model of the table*

Answers to Self-Evaluation Test

1. a, **2.** b, **3.** a, **4. Multiplier**, **5. Environment and Effects**, **6.** F

Chapter 12

Interior Lighting-I

Learning Objectives

After completing this chapter, you will be able to:

- *Create night interior scene*
- *Create daylight interior scene*

INTRODUCTION

Lights are used to illuminate a scene and make it look more realistic. In Autodesk 3ds Max, there are two main categories of lights: Standard and Photometric. These categories can be further classified into different types. You can select one or more lights based on the requirement of a scene. In this chapter, you will learn to illuminate an interior scene using the standard lights.

TUTORIALS

Before starting the tutorials, you need to download the *c12_3dsmax_2022_tut.zip* file from *www.cadcim.com*. The path of this file is as follows: *Textbooks > Animation and Visual Effects > Autodesk 3ds Max > Autodesk 3ds Max 2022 for Beginners: A Tutorial Approach*

Extract the contents of the zip file and save them in the *Documents* folder.

Tutorial 1

In this tutorial, you will illuminate a night interior scene using standard lights, refer to Figure 12-1. **(Expected time: 45 min)**

Figure 12-1 *The illuminated night interior scene*

The following steps are required to complete this tutorial:

a. Create the project folder.
b. Open the file.
c. Create roof lights.
d. Create wall lights.

e. Create lamp lights.
f. Save and render the scene.

Creating the Project Folder

Create the project folder with the name *c12_tut1* in the *3dsmax 2022* folder, as discussed in earlier chapters.

Opening the File

1. Open Windows Explorer and then browse to *\Documents\c12_3dsmax_2022_tut*. Next, copy all the files to *\Documents\3dsmax 2022\c12_tut1\sceneassets\images*.

2. Choose **Open** from the **File** menu; the **Open File** dialog box is displayed. In this dialog box, browse to the location *\Documents\c12_3dsmax_2022_tut* and select the **c12_tut1_start.max** file from it. Choose the **Open** button to open the file, refer to Figure 12-2.

***Figure 12-2** The c12_tut1_start.max file*

3. Choose **Save As** from the **File** menu; the **Save File As** dialog box is displayed. Browse to the location *\Documents\3dsmax 2022\c12_tut1\scenes*. Save the file with the name *c12tut1.max* at this location.

Creating Roof Lights

In this section, you will create roof lights using the **Omni** tool.

1. Invoke the **Maximize Viewport Toggle** tool and then activate the Top viewport. Choose **Create > Lights** in the **Command Panel**. By default, the **Photometric** option is displayed in the drop-down list. Select the **Standard** option from the drop-down list and then invoke the **Omni** tool from the **Object Type** rollout; various rollouts are displayed in the **Command**

Panel. Next, click at the center of the room; the omni light is displayed in the viewport with the name *Omni001*.

2. Invoke the **Select and Move** tool and align *Omni001* in all viewports, as shown in Figure 12-3.

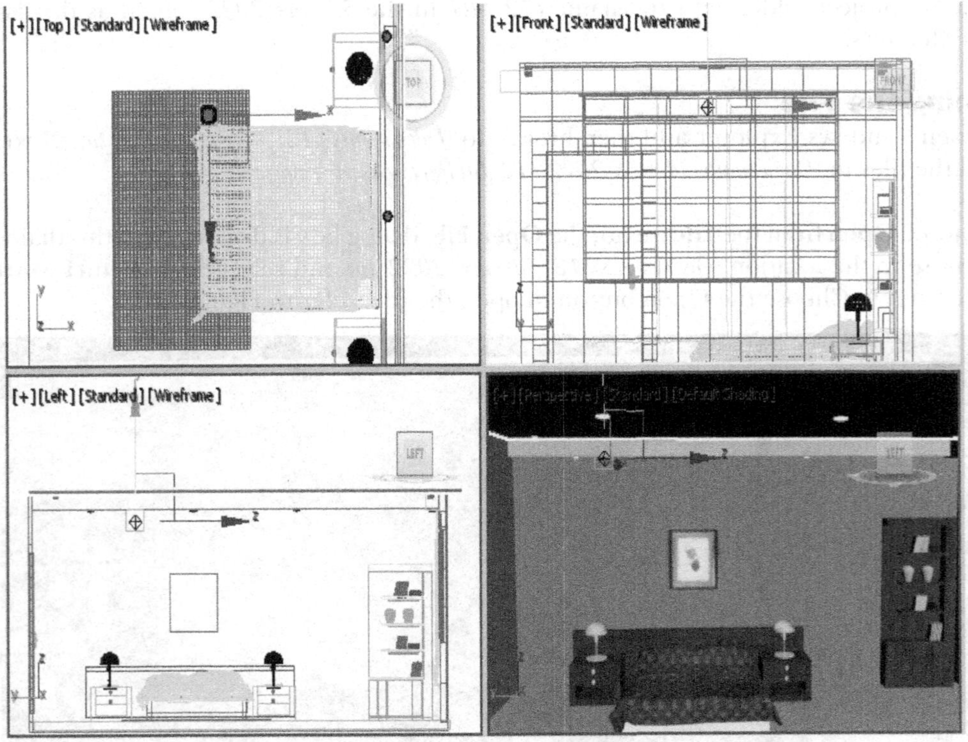

Figure 12-3 *Omni001 aligned in all viewports*

3. In the **Name and Color** rollout, enter the name as **roof Light1**.

4. Choose the **Modify** tab from the **Command Panel**. Now, in the **General Parameters** rollout, select the **On** check box in the **Shadows** area. Next, select **Ray Traced Shadows** from the drop-down list below this check box if not selected.

5. Expand the **Intensity/Color/Attenuation** rollout and set the value **0.5** in the **Multiplier** spinner.

   Next, you will lighten the shadow color.

6. Expand the **Shadow Parameters** rollout and choose the color swatch next to the **Color** parameter in the **Object Shadows** area; the **Color Selector: Shadow Color** dialog box is displayed. In this dialog box, set the values of the **Red**, **Green**, and **Blue** spinners to **52** and then choose the **OK** button.

7. Activate the Left viewport. Make sure *roof light1* is selected and create a copy of it and rename it as **roof light 2**. Now, align it at the position, as shown in Figure 12-4.

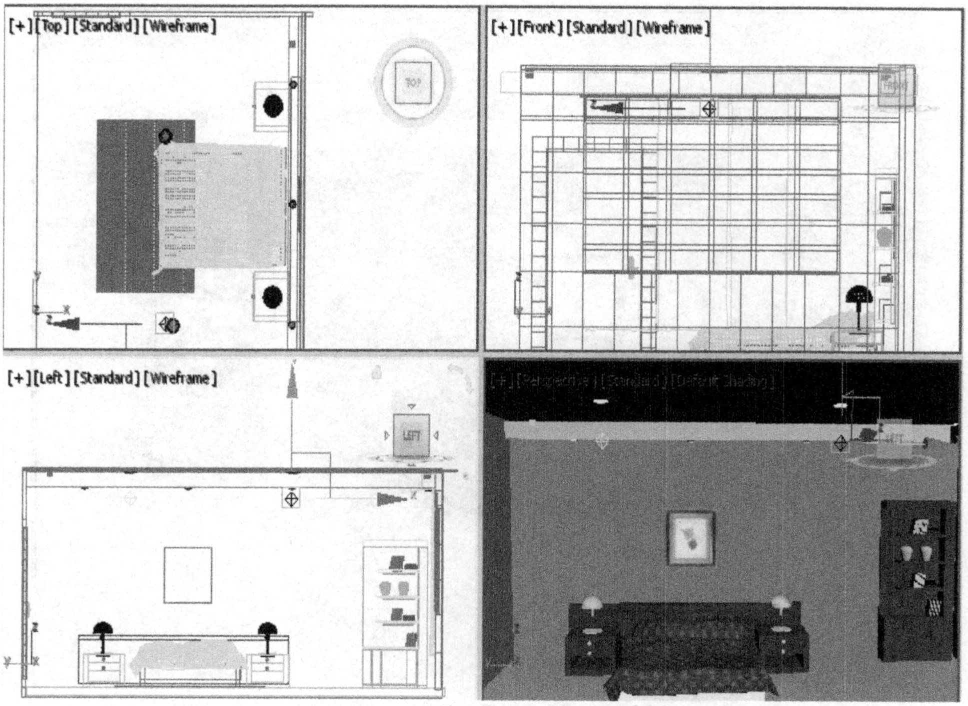

Figure 12-4 The copy of roof light1 aligned in all viewports

8. Set the angle of view in the Perspective viewport and render the Perspective viewport by invoking the **Render Production** tool from the **Main Toolbar**. The rendered image is displayed, refer to Figure 12-5.

Creating Wall Lights

In this section, you will create wall lights using the **Free Spot** and **Omni** tools.

1. Maximize the Top viewport. Choose **Create > Lights** in the **Command Panel**. Make sure **Standard** is selected in the drop-down list. Next, invoke the **Free Spot** tool from the **Object Type** rollout. Next, click and place it near the *fall ceiling1* to create a spot light, as shown in Figure 12-6.

2. Set the value **80** in the **Targeted** spinner in the **General Parameters** rollout. In the **Name and Color** rollout, enter **spot light1**. Next, align the light in all the viewports, as shown in Figure 12-7.

***Figure 12-5** The rendered image*

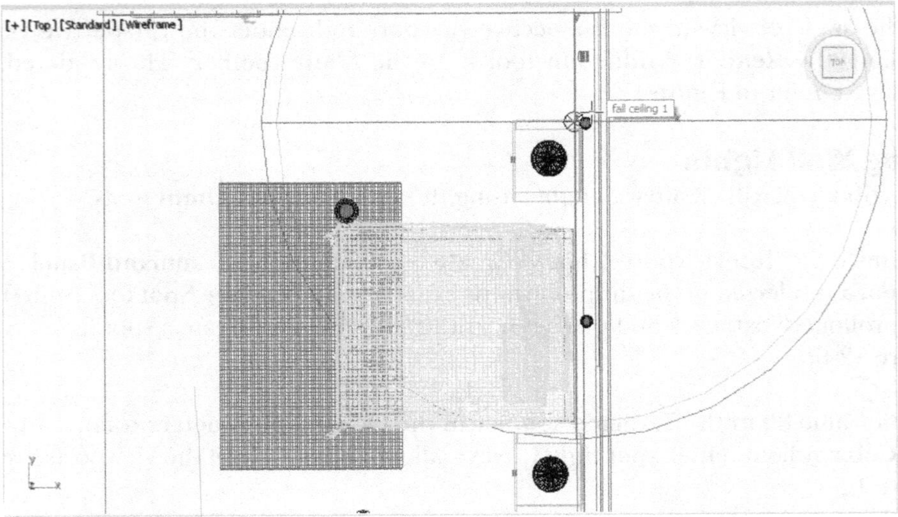

***Figure 12-6** The spot light created*

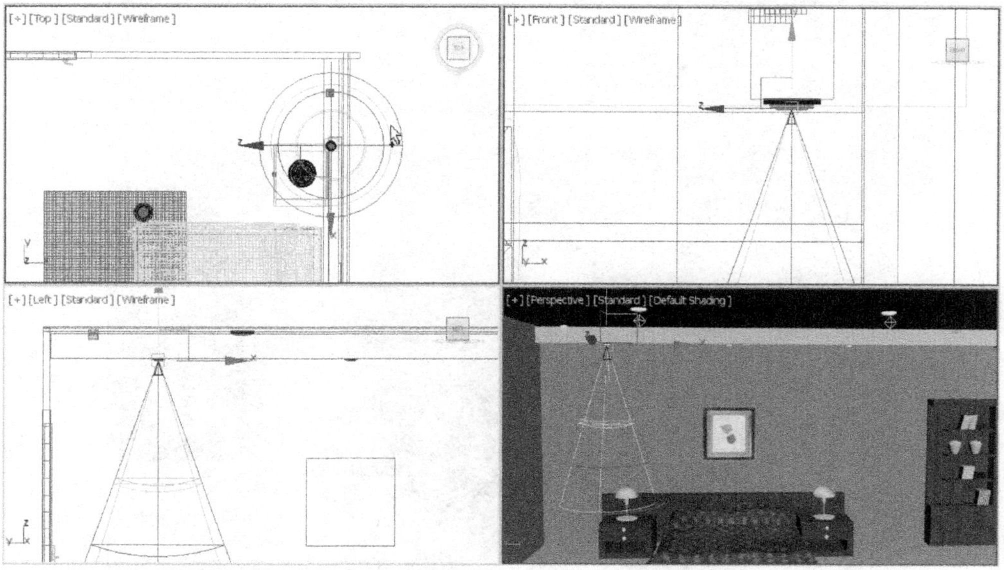

Figure 12-7 *The spot light aligned*

3. Expand the **Intensity/Color/Attenuation** rollout and set the value in the **Multiplier** spinner to **30**. Next, choose the color swatch next to this spinner; the **Color Selector: Light Color** dialog box is displayed. In this dialog box, set the values of the **Red**, **Green**, and **Blue** spinners to **169**, **164**, and **23**, respectively and then choose the **OK** button to close the dialog box.

4. In the **Decay** area, select **Inverse Square** from the **Type** drop-down list. In the **Far Attenuation** area, select the **Use** and **Show** check boxes and then enter **42** and **63** in the **Start** and **End** spinners, respectively.

   Attenuation is the decrease in intensity of light with distance.

5. Expand the **Spotlight Parameters** rollout and set the value of the **Hotspot/Beam** and **Falloff/Field** spinners to **34** and **40**, respectively. Make sure the **Show Cone** check box is selected in the **Light Cone** area of the **Spotlight Parameters** rollout.

6. Set the angle of view in the Perspective viewport and render the Perspective viewport by invoking the **Render Production** tool from the **Main Toolbar**; the rendered image is displayed, refer to Figure 12-8.

Now, you will use the omni light to produce glow at the source of the spot light.

7. Activate the Top viewport. Choose **Create > Lights** in the **Command Panel**. Make sure **Standard** is displayed in the drop-down list located below it. Next, invoke the **Omni** tool from the **Object Type** rollout; various rollouts are displayed in the **Command Panel**. Next, click in the viewport; the omni light is displayed in the viewport. Enter **glow light 1** in the **Name and Color** rollout.

Figure 12-8 *The rendered image*

8. Invoke the **Select and Move** tool and align *glow light 1* in all viewports, refer to Figure 12-9.

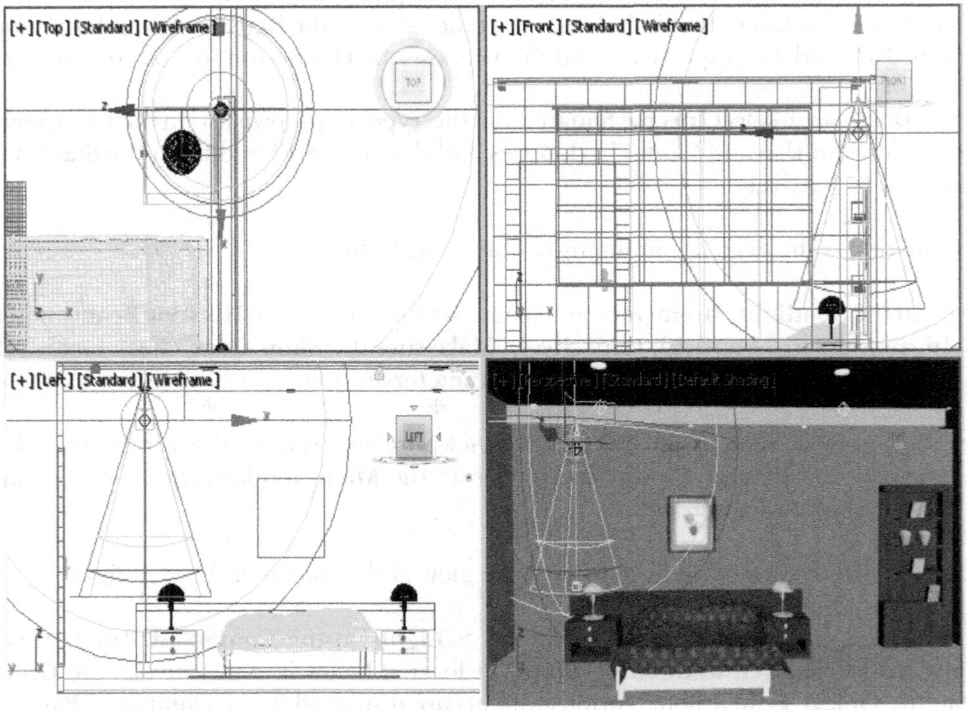

Figure 12-9 *The glow light 1 aligned*

9. Make sure *glow light 1* is selected in the viewport and choose the **Modify** tab in the **Command Panel**. In the **General Parameters** rollout, make sure the **On** check box in the **Shadows** area is cleared.

10. Expand the **Intensity/Color/Attenuation** rollout and set the value in the **Multiplier** spinner to **0.4**. Next, choose the color swatch next to this spinner; the **Color Selector: Light Color** dialog box is displayed. In this dialog box, make sure the values of the **Red**, **Green**, and **Blue** spinners are set to **169**, **164**, and **23**, respectively and then choose the **OK** button to close the dialog box.

11. In the **Decay** area, select **Inverse Square** from the **Type** drop-down list and set the value **10** in the **Start** spinner located below it. Also, select the **Show** check box in this area. Make sure that the **Use** and **Show** check boxes in the **Far Attenuation** and **Near Attenuation** areas are cleared.

12. Set the angle of view in the Perspective viewport and then invoke the **Render Production** tool from the **Main Toolbar**; the rendered image is displayed, refer to Figure 12-10.

Next, you need to create two copies of both *spot light 1* and *glow light 1*.

13. Activate the Left viewport. Select *glow light 1* and *spot light 1* from the **Scene Explorer**. Next, press and hold the SHIFT key and drag the cursor to *light fixture 001*. Release the left mouse button; the **Clone Options** dialog box is displayed. In this dialog box, make sure the **Instance** radio button is selected. Set **2** in the **Number of Copies** spinner and choose the **OK** button; two copies of both *spot light 1* and *glow light 1* are created in all viewports, as shown in Figure 12-11.

Figure 12-10 *The rendered image*

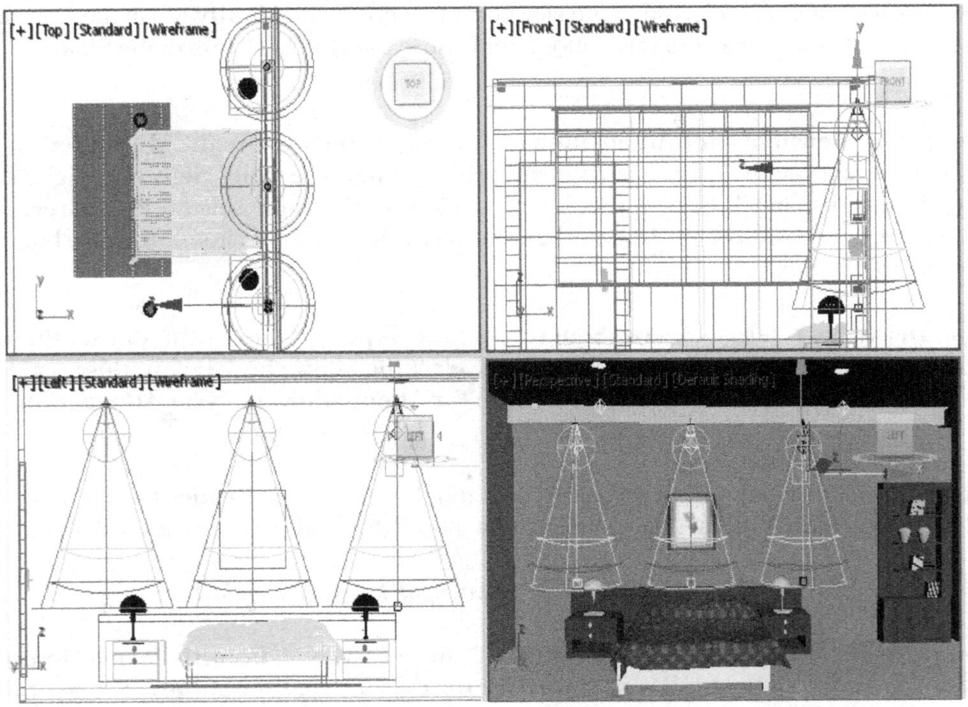

Figure 12-11 The copies of spot light 1 and glow light 1 created

Note

*If you select the **Instance** radio button in the **Clone Options** dialog box, a copy of the object created will be an instance of the original object which means that the changes made in the parameters of one object will be automatically reflected in the other.*

14. Set the angle of view in the Perspective viewport and render it by invoking the **Render Production** tool from the **Main Toolbar**; the rendered image is displayed, refer to Figure 12-12.

Creating Lamp Lights

In this section, you will create lamp lights for the bedside lamps.

1. Maximize the Top viewport. Choose **Create > Lights** in the **Command Panel**. Next, invoke the **Free Spot** tool from the **Object Type** rollout. Next, click and place it near the *fall ceiling1* Now, release the left mouse button; a spot light is created.

2. Set the value **30.0** in the **Targ. Dist** spinner in the **General Parameters** rollout. In the **Name and Color** rollout, enter **lamp light 1** as the name of the light. Next, invoke the **Select and Move** tool and align the *lamp light 1* in all viewports, as shown in Figure 12-13.

3. Select *lamp light 1*. In the **Intensity/Color/Attenuation** rollout, set the value **3** in the **Multiplier** spinner. Next, choose the color swatch next to this spinner; the **Color Selector: Light Color**

dialog box is displayed. In this dialog box, make sure the values of the **Red**, **Green**, and **Blue** spinners are set to **169**, **164**, and **23**, respectively and then choose the **OK** button to close the dialog box.

Figure 12-12 The rendered image

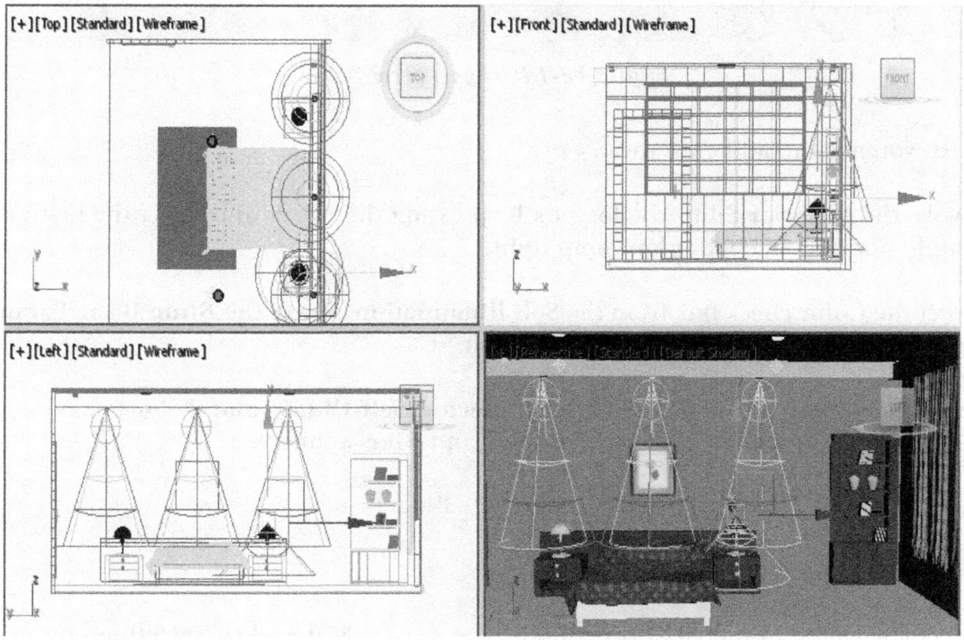

Figure 12-13 The lamp light 1 aligned in all viewports

4. In the **Far Attenuation** area, select the **Use** and **Show** check boxes and set the values of the **Start** and **End** spinners to **10** and **35**, respectively.

5. In the **Spotlight Parameters** rollout, set the value of the **Hotspot/Beam** and **Falloff/Field** spinners to **90** and **95**, respectively.

6. Set the angle of view in the Perspective viewport and render it by invoking the **Render Production** tool from the **Main Toolbar**; the rendered image is displayed, refer to Figure 12-14.

Figure 12-14 *The rendered image*

Next, you need to add glow to *lamp light 1*.

7. Invoke the **Material Editor** dialog box by pressing the M key and select the first material sample slot. Rename it as **glow lamp light**.

8. Select the **Color** check box from the **Self illumination** area of the **Blinn Basic Parameters** rollout; the spinner turns into a color swatch.

9. Select the **Color** color swatch; the **Color Selector: Self-Illum Color** dialog box is displayed. Set the following values in the **Red**, **Green**, and **Blue** spinners:

   Red: **218** Green: **138** Blue: **0**

   Set the value **65** in the **Opacity** spinner

10. Select *lamp top 1* in the viewport. Choose the **Assign Material to Selection** button in the **Material Editor** dialog box to assign material to *lamp top 1*.

11. Set the angle of view in the Perspective viewport and render it by invoking the **Render Production** tool from the **Main Toolbar**; the rendered image is displayed, refer to Figure 12-15.

Figure 12-15 *The rendered image*

Next, you need to create a copy of *lamp light 1* for other bed side lamp.

12. Activate the Left viewport. Select *lamp light 1*. Next, press and hold the SHIFT key and drag the cursor to *lamp top 2*. Release the left mouse button; the **Clone Options** dialog box is displayed. In this dialog box, select the **Copy** radio button. Make sure the value **1** is set in the **Number of Copies** spinner and then choose the **OK** button; a copy of *lamp light1* is created. Next, invoke the **Select and Move** tool and align a copy of *lamp light1* in all the viewports, as shown in Figure 12-16.

13. Assign the *glow lamp light* material to *lamp top 2* as discussed earlier.

Saving and Rendering the Scene

In this section, you will save the scene that you have created and then render it. You can also view the final rendered image of this model by downloading the *c12_3dsmax_2022_rndr.zip* file from *www.cadcim.com*. The path of the file is as follows: *Textbooks > Animation and Visual Effects > Autodesk 3ds Max > Autodesk 3ds Max 2022 for Beginners: A Tutorial Approach*

1. Choose **Save** from the **File** menu.

2. Activate the Perspective viewport. Next, invoke the **Render Production** tool from the **Main Toolbar**; the rendered image is displayed, refer to Figure 12-17.

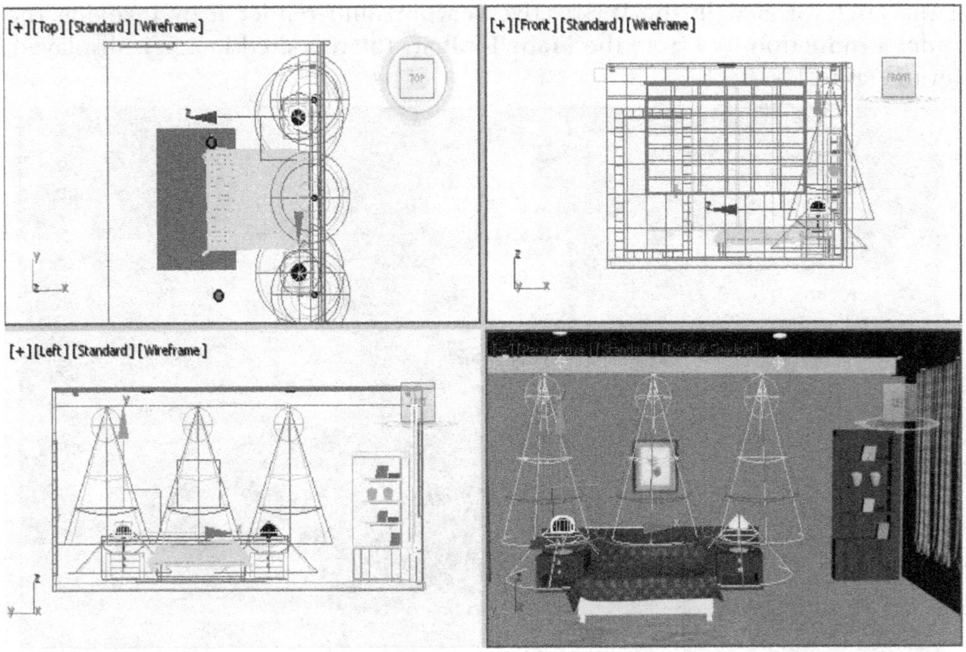

Figure 12-16 *The copy of lamp light 1 aligned*

Figure 12-17 *The rendered image*

Tutorial 2

In this tutorial, you will create a daylight interior scene using the target direct light, as shown in Figure 12-18. **(Expected time: 20 min)**

Figure 12-18 *The daylight interior scene*

The following steps are required to complete this tutorial:

a. Create the project folder.
b. Open the file.
c. Create lights.
d. Save and render the scene.

Creating the Project Folder

Create the project folder with the name *c12_tut2* in the *3dsmax 2022* folder, as discussed in earlier chapters.

Opening the File

1. Open Windows Explorer and then browse to *\Documents\c12_3dsmax_2022_tut*. Next, copy all jpeg files to *\Documents\ 3dsmax 2022\c12_tut2\sceneassets\images*.

2. Choose **Open** from the **File** menu; the **Open File** dialog box is displayed. In this dialog box, browse to the location *\Documents\c12_3dsmax_2022_tut* and open the **c12_tut2_start.max** file, refer to Figure 12-19.

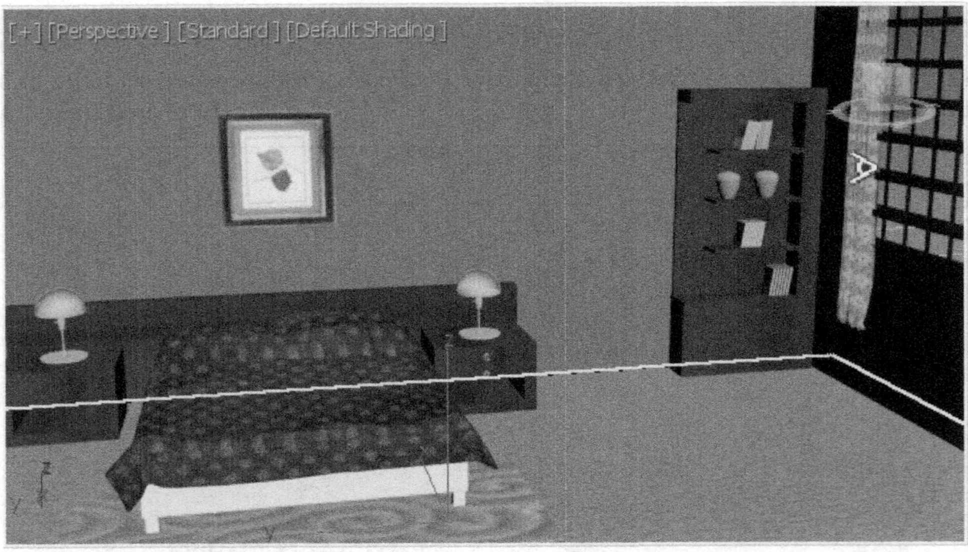

Figure 12-19 The c12_tut2_start file

3. Choose **Save As** from the **File** menu; the **Save File As** dialog box is displayed. Browse to the location *\Documents\3dsmax 2022\c12_tut2\scenes*. Save the file with the name *c12tut2.max* at this location.

Creating Lights

In this section, you will use the **Omni** and **Target Direct** tools to create the daylight effect.

1. Activate the Top viewport. Choose **Create > Lights** in the **Command Panel**. By default, the **Photometric** option is displayed in the **Command Panel**. Select **Standard** from the drop-down list and then invoke the **Omni** tool from the **Object Type** rollout; various rollouts are displayed in the **Command Panel**. Next, click at a point in the room; the omni light is created in the viewport with the name *Omni001*. Next, align it, as shown in Figure 12-20.

2. Make sure *Omni001* is selected. Next, choose the **Modify** tab in the **Command Panel**.

3. In the **General Parameters** rollout, select the **On** check box in the **Shadows** area and also select the **Ray Traced Shadows** option in the drop-down list.

4. Expand the **Intensity/Color/Attenuation** rollout and set the value of the **Multiplier** spinner to **0.5**. Next, choose the color swatch next to this spinner; the **Color Selector: Light Color** dialog box is displayed. In this dialog box, make sure the values of the **Red**, **Green**, and **Blue** spinners are set to **255**, **255**, and **255**, respectively and then choose the **OK** button to close the dialog box.

5. Expand the **Shadow Parameters** rollout and set the value **0.8** in the **Dens** spinner. Next, select the color swatch next to the **Color** parameter; the **Color Selector: Shadow Color** dialog box is displayed. In this dialog box, set the values of the **Red**, **Green**, and **Blue** spinners to **50** and then choose the **OK** button to close the dialog box.

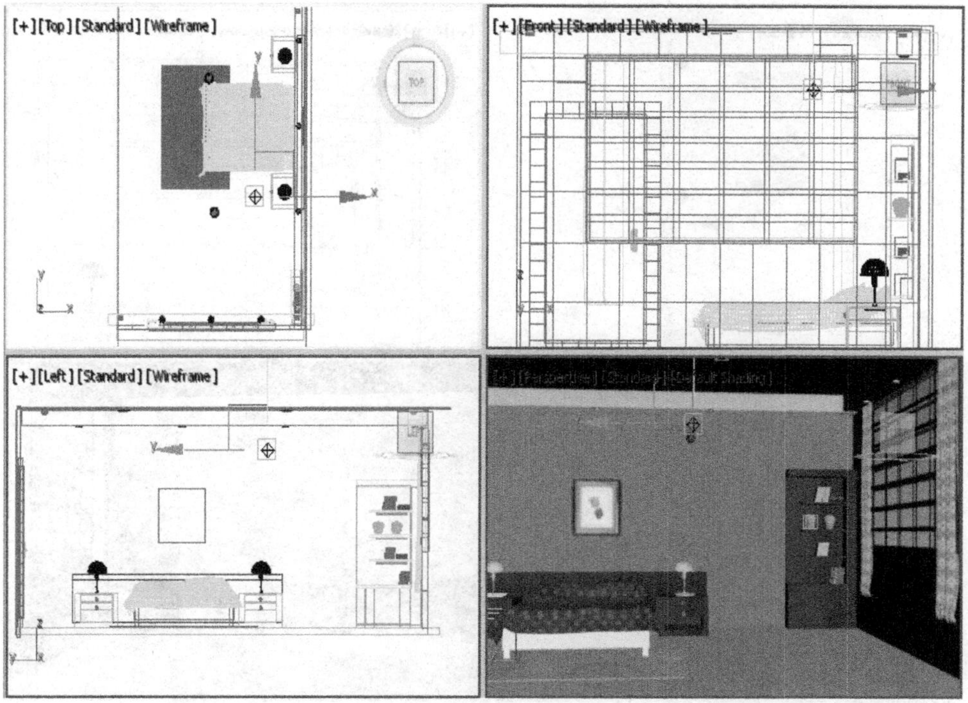

Figure 12-20 *The omni 001 light aligned in all viewports*

6. Similarly, create another omni light as created in step 1 with the name *Omni 002*. Next, align it in all viewports, as shown in Figure 12-21.

7. Make sure *Omni 002* is selected in the viewport. Choose the **Modify** tab in the **Command Panel**. In the **General Parameters** rollout, select the **On** check box in the **Shadows** area and make sure the **Ray Traced Shadows** option is selected in the drop-down list.

8. In the **Intensity/Color/Attenuation** rollout, set the value **0.3** in the **Multiplier** spinner.

9. In the **Shadow Parameters** rollout, make sure the value is set to **0.8** in the **Dens** spinner.

   Next, you will use a direct light to simulate the sunlight coming from the window of the room.

10. Activate the Perspective viewport and invoke the **Render Production** tool from the **Main Toolbar** to view the rendered scene. 

11. Activate and maximize the Left viewport. Invoke the **Target Direct** tool from **Create > Lights > Standard > Object Type** rollout in the **Command Panel**. Move the cursor on the right side of the viewport. Next, press and hold the left mouse button to specify the location of the light and drag the cursor near the front portion of *wall001* to locate the target of the light, refer to Figure 12-22. Release the left mouse button; the light is displayed in all viewports and it is automatically named as *Direct001*.

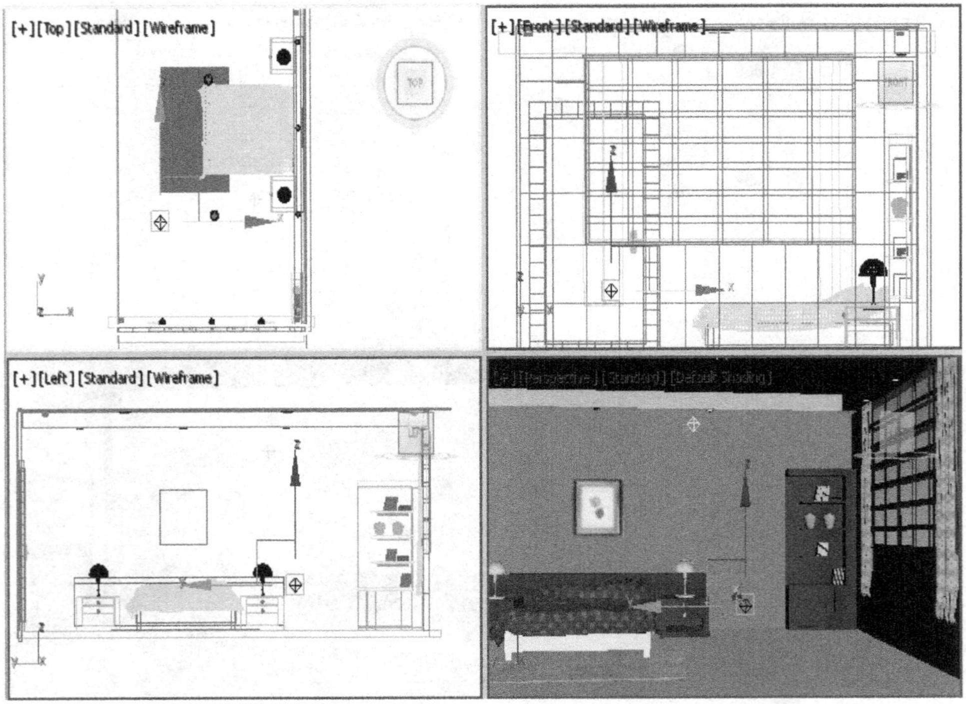

Figure 12-21 *The Omni 002 aligned*

12. Choose the **Modify** tab in the **Command Panel**; various rollouts are displayed to modify *Direct001*.

13. Expand the **General Parameters** rollout and select the **On** check box in the **Shadows** area if not selected.

14. Expand the **Intensity/Color/Attenuation** rollout and set the value of the **Multiplier** spinner to **1.3**. Next, choose the color swatch next to this spinner; the **Color Selector: Light Color** dialog box is displayed. In this dialog box, set the values for the **Red**, **Green**, and **Blue** spinners to **230**, **229**, and **127**, respectively and then choose the **OK** button to close the dialog box.

15. Expand the **Directional Parameters** rollout, select the **Rectangle** radio button and set the values in the respective spinners as given next:

    Hotspot/Beam: **27** Falloff/Field: **29**

16. Invoke the **Select and Move** tool and align *Direct001*, *Direct 001.Target* light in all viewports, as shown in Figure 12-22.

17. Render the scene to view the rendered output.

18. Select *Direct001*. Expand the **Atmospheres & Effects** rollout and choose the **Add** button; the **Add Atmosphere or Effect** dialog box is displayed. Select the **Volume Light** option

and then choose the **OK** button; the **Volume Light** option is displayed in the text area of the **Atmospheres & Effects** rollout.

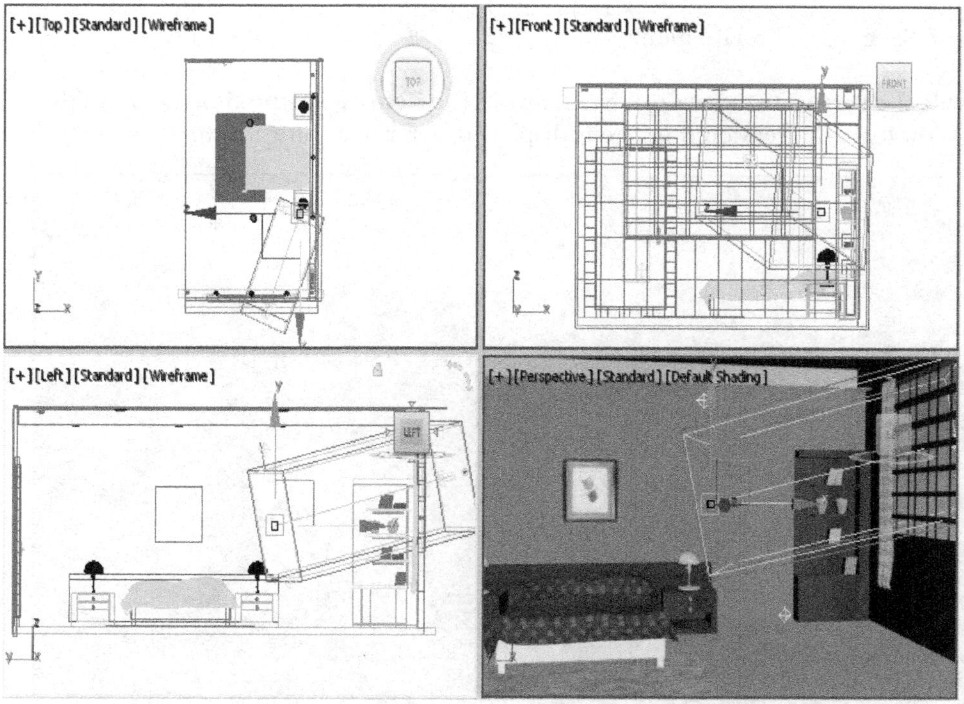

Figure 12-22 *The Direct 001, Direct 001.Target aligned*

Next, you need to modify the parameters of the **Volume Light**.

19. Select the **Volume Light** option in the **Atmospheres & Effects** rollout and choose the **Setup** button; the **Environment and Effects** dialog box is displayed. In this dialog box, the **Environment** tab is chosen by default.

20. In the **Atmosphere** rollout, make sure that the **Volume Light** option is selected in the **Effects** area. Set the value **0.5** in the **Density** spinner of the **Volume** area in the **Volume Light Parameters** rollout. Use the default values for other options and close the **Environment and Effects** dialog box.

Note

You can also use different settings in the ***Volume Light Parameters*** *rollout to apply different effects in the scene.*

21. Render the scene.

Saving and Rendering the Scene

In this section, you will save the scene that you have created and then render it. You can also view the final rendered image of this model by downloading the *c12_3dsmax_2022_rndr.zip* file from *www.cadcim.com*. The path of the file is as follows: *Textbooks > Animation and Visual Effects > Autodesk 3ds Max > Autodesk 3ds Max 2022 for Beginners: A Tutorial Approach*

1. Choose **Save** from the **File** menu.

2. Activate the Perspective viewport. Next, invoke the **Render Production** tool from the **Main Toolbar**; the rendered image is displayed, refer to Figure 12-23.

Figure 12-23 *The rendered image*

Self-Evaluation Test

Answer the following questions and then compare them to those given at the end of this chapter:

1. Which of the following tools is used to create standard lights?

   (a) **Target Direct** (b) **Target Light**
   (c) **Free Light** (d) All of these

2. Which of the following tools is used to create the area spot light?

   (a) **Target Spot** (b) **Free Spot**
   (c) **Skylight** (d) **Omni**

3. The __________ check box in the **General Parameters** rollout is selected to view the shadow of a light.

4. The options in the __________ rollout are used to assign special effects such as fog and glow to the environment on rendering.

5. The **Exclude** button in the **General Parameters** rollout is used to define a number of objects in the scene that will be affected by light on rendering. (T/F)

Review Questions

Answer the following questions:

1. Which of the following spinners is used to specify the intensity of the light in the **Intensity/ Color/Attenuation** rollout?

   (a) **Start** (b) **Color**
   (c) **Multiplier** (d) None of these

2. Which of the following tools is used to create a light that illuminates a scene in all directions?

   (a) **Skylight** (b) **Omni**
   (c) **Free Spot** (d) None of these

3. The parameters in the __________ rollout are used to control the properties of the spotlight.

4. The parameters in the __________ rollout are used to control the properties of the shadow.

5. Attenuation is the effect of light intensity that decreases with distance. (T/F)

EXERCISE

The rendered output of the scene used in the following exercise can be accessed by downloading the *c12_3dsmax_2022_exr.zip* from *www.cadcim.com*. The path of the file is as follows: *Textbooks > Animation and Visual Effects > Autodesk 3ds Max > Autodesk 3ds Max 2022 for Beginners: A Tutorial Approach*

Exercise 1

Extract the contents of the *c12_3dsmax_2022_exr.zip* and then open *c12_exr01_start.max*. Next, illuminate the scene using omni and spot lights, as shown in Figure 12-24.

(Expected time: 20 min)

Figure 12-24 *The illuminated scene*

Answers to Self-Evaluation Test

1. a, **2.** b, **3. On**, **4. Advanced Effects**, **5.** T

Chapter 13

Interior Lighting-II

Learning Objectives

After completing this chapter, you will be able to:
- *Use photometric lights for interior lighting*
- *Use IES files*

INTRODUCTION

In the last chapter, you learned to use standard lights to illuminate interior scene. In this chapter, you will learn to use photometric lights to illuminate an interior scene. Photometric lights are more complex than standard lights but provide more realistic lighting effects.

Additionally, you will create special lighting patterns by using the standard IES (Illuminating Engineering Society) files.

TUTORIALS

Before starting the tutorials, you need to download the *c13_3dsmax_2022_tut.zip* file from *www.cadcim.com*. The path of the file is as follows: *Textbooks > Animation and Visual Effects > Autodesk 3ds Max > Autodesk 3ds Max 2022 for Beginners: A Tutorial Approach*

Extract the contents of the zip file and save them in the *Documents* folder.

Tutorial 1

In this tutorial, you will illuminate a night interior scene using photometric lights, refer to Figure 13-1. **(Expected time: 30 min)**

Figure 13-1 *The illuminated night interior scene*

The following steps are required to complete this tutorial:

a. Create the project folder.
b. Open the file.
c. Create roof light.
d. Create wall lights.
e. Save and render the scene.

Creating the Project Folder

Create a project folder with the name *c13_tut1* in the *3dsmax 2022* folder, as discussed in Tutorial 1 of Chapter 2.

Opening the File

1. Open Windows Explorer and then browse to *\Documents\c13_3dsmax_2022_tut*. Next, copy all jpeg files to *\Documents\3dsmax 2022\c13_tut1\sceneassets\images*. Also, copy *1.IES* file to *\Documents\3dsmax 2022\c13_tut1\sceneassets\photometric*.

2. Choose **Open** from the **File** menu; the **Open File** dialog box is displayed. In this dialog box, browse to the location *\Documents\c13_3dsmax_2022_tut* and select the **c13_tut1_start.max** file from it. Choose the **Open** button to open the file, refer to Figure 13-2.

Figure 13-2 *The c13_tut1_start.max file*

3. Choose **Save As** from the **File** menu; the **Save File As** dialog box is displayed. Browse to the location *\Documents\3dsmax 2022\c13_tut1\scenes*. Save the file with the name *c13tut1.max* at this location.

You need to confirm Scanline Renderer is set up.

4. Choose the **Render Setup** tool from the Main Toolbar and do one of the following two steps:

(a) If the **Render Setup: Arnold** dialog is displayed, select **Scanline** from the **Renderer** drop-down list; the renderer is changed to Scanline and the name of the dialog box is modified to **Render Setup: Scanline Renderer** dialog box. Close this dialog box and proceed further.

(b) If the **Render Setup: Scanline Renderer** dialog box is displayed, close the dialog box and proceed further as discussed next.

Creating Roof Light

In this section, you will create roof light by using the **Free Light** tool.

1. Invoke the **Maximize Viewport Toggle** tool to view the four viewports and activate the Top viewport. Choose **Create > Lights** in the **Command Panel**. By default, the **Photometric** option is displayed in the drop-down list below it.

2. Invoke the **Free Light** tool from the **Object Type** rollout; the **Photometric Light Creation** message box is displayed, as shown in Figure 13-3. Next, choose **Yes** from this message box. Also, various rollouts are displayed in the **Command Panel**.

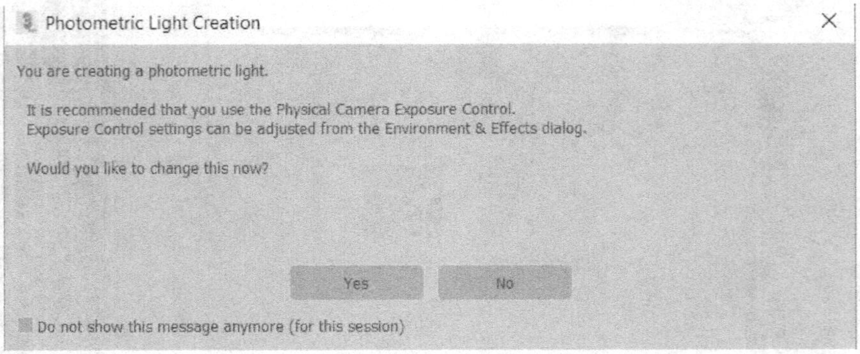

***Figure 13-3** The **Photometric Light Creation** message box*

3. Click at the center of the room; the free light is created with the name *PhotometricLight001*. Enter **rooflight1** in the **Name and Color** rollout.

4. Invoke the **Select and Move** tool and align *rooflight1* in all viewports, as shown in Figure 13-4.

   Next, you will modify the shape of *rooflight1*.

5. Choose the **Modify** tab in the **Command Panel**. Next, in the **Emit Light from (Shape)** area of the **Shape/Area Shadows** rollout, select the **Sphere** option from the drop-down list; the shape of the light changes in the viewport and the **Radius** spinner is displayed

in the **Emit Light from (Shape)** area. Also, the **Shadow Samples** drop-down list is added to the **Rendering** area located below it.

6. Set the value **75.0** in the **Radius** spinner to increase the spread of the light; the shape of *rooflight1* is modified, as shown in Figure 13-5.

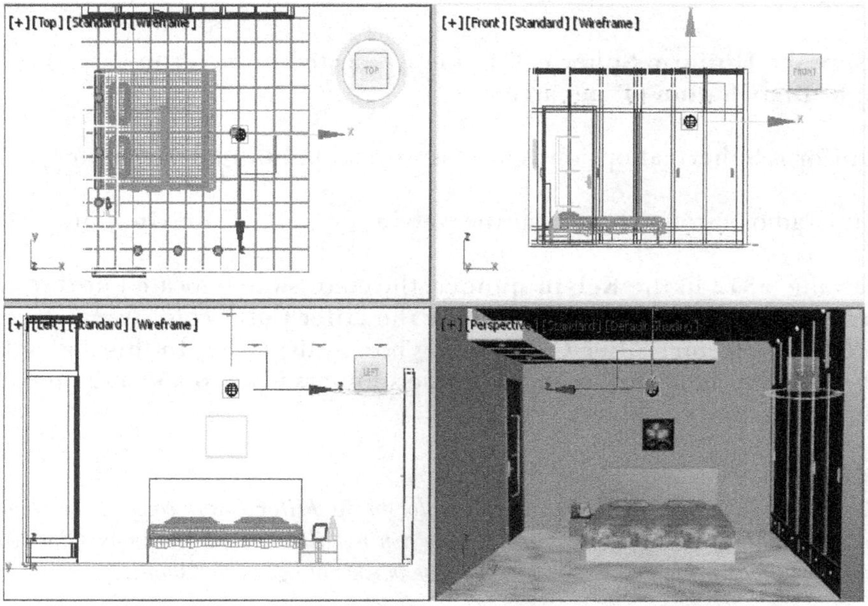

Figure 13-4 *The rooflight1 aligned in all viewports*

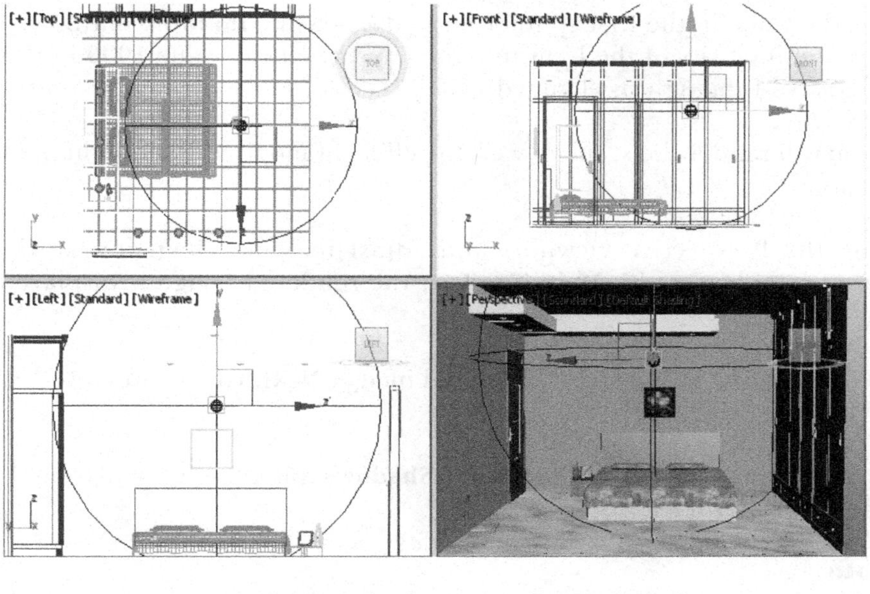

Figure 13-5 *The shape of rooflight1 changed*

Next, you will modify the parameters of *rooflight1* to lit the room with proper intensity, colors, and shadows.

7. In the **General Parameters** rollout, make sure the **On** check box is cleared in the **Shadows** area. Next, select **Ray Traced Shadows** in the drop-down list below this check box.

8. Make sure the **Uniform Spherical** option is selected in the drop-down list located in the **Light Distribution (Type)** area.

   The **Uniform Spherical** option is used to spread light in all directions.

9. Select the radio button located near the **Kelvin** spinner; the **Kelvin** spinner is activated.

10. Set the value **6372** in the **Kelvin** spinner; the color swatch located next to the **Kelvin** spinner is changed to white. Next, choose the **Filter Color** color swatch located below it; the **Color Selector: Filter Color** dialog box is displayed. In this dialog box, make sure the value in the **Red**, **Green**, and **Blue** spinners is set to **255** and then choose the **OK** button to close the dialog box.

Note
*The value in the **Kelvin** spinner and the color in the **Filter Color** color swatch together define the color shade of the light being created. You can use different combinations of the spinner value and color shade in the color swatch to achieve variety of light color shades.*

Next, you will set the intensity of *rooflight1*.

11. In the **Intensity** area of the **Intensity/Color/Attenuation** rollout, select the **lm** radio button; the value in the first spinner located below the **lm** radio button changes and displays the intensity of the light in lumen. Next, set the value **1000** in this spinner; the intensity of the light is changed.

    Now, you will render the scene to view the effect of modifying rooflight parameters in the scene.

12. Activate the Perspective viewport and adjust its view. Next, invoke the **Render Production** tool from the **Main Toolbar**; the rendered image is displayed, refer to Figure 13-6.

    You will notice the grains in the rendered image. Next, you will remove grains in the image by using higher shadow samples.

13. In the **Rendering** area of the **Shape/Area Shadows** rollout, select **256** from the **Shadow Samples** drop-down list.

Note
*If you select higher values for sampling shadows from the **Shadow Samples** drop-down list, the quality of the render improves but rendering time also increases.*

14. Render the Perspective viewport again as discussed earlier. The rendered image is displayed, refer to Figure 13-7.

Figure 13-6 The rendered image with grains

Figure 13-7 The rendered image without grains

Creating Wall Lights

In this section, you will create wall lights by using the **Target Light** tool. You will also use IES file to achieve a specific light pattern.

1. Switch from the Left viewport to the Right viewport. Now, activate the Front viewport. Choose **Create > Lights** in the **Command Panel**. Next, invoke the **Target Light** tool from the **Object Type** rollout; various rollouts are displayed in the **Command Panel**.

2. Click at a point and drag the cursor downward. Next, release the left mouse button; a target light is created. In the **Name and Color** rollout, enter **spot light1**. Next, invoke the **Select and Move** tool and then select *spot light1* and *spot light1.Target*. Now, align them in all viewports, as shown in Figure 13-8. Select *spot light1.Target* and align it with the wall such that the light is visible on upper portion of the wall only.

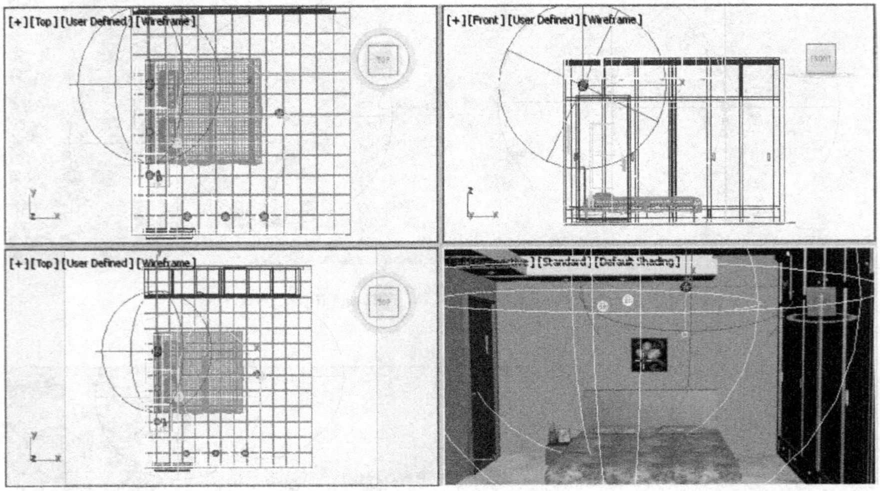

***Figure 13-8** The spot light1 and spot light1.Target aligned in all viewports*

Next, you need to modify the parameters of *spot light1*.

3. Select *spot light1*. Next, choose the **Modify** tab in the **Command Panel**. In the **General Parameters** rollout, select the **On** check box in the **Shadows** area. Next, select the **Photometric Web** option in the drop-down list of the **Light Distribution (Type)** area; the **Distribution (Photometric Web)** rollout is displayed in the **Command Panel**.

   Now, you will use a standard photometric IES file to create a specific lighting pattern. These IES files are standard files provided by light manufacturers and are available on the internet for free downloading.

4. In the **Emit Light from (Shape)** area of the **Shape/Area Shadows** rollout, make sure the **Point** option is selected in the drop-down list.

5. Choose the **< Choose Photometric File >** button from the **Distribution (Photometric Web)** rollout; the **Open a Photometric Web File** dialog box is displayed. As the project folder is already set, the photometric folder of this project is displayed in the **Look**

in drop-down list. Next, select the **1.ies** file from this dialog box and then choose the **Open** button; label of the button is replaced by the name of the selected file. Also, the lighting pattern is displayed in the window located above this button.

Next, you need to set the intensity and color of *spot light1*.

6. Select the radio button located near the **Kelvin** spinner; the **Kelvin** spinner is activated.

7. Choose the **Filter Color** color swatch located below it; the **Color Selector: Filter Color** dialog box is displayed. In this dialog box, set the value in the **Red**, **Green**, and **Blue** spinners as given next:

   Red: **243** Green: **240** Blue: **146**

   Choose the **OK** button to close the dialog box; the color in the **Filter Color** color swatch is changed to yellow.

8. In the **Intensity** area of the **Intensity/Color/Attenuation** rollout, select the **lm** radio button; the value in the first spinner located below this radio button changes and displays the intensity of the light in Lumen. Next, set the value **100** in this spinner; the intensity of the light is increased.

   At this stage, you may render the scene to view the effect of modifying *spot light1* parameters in the scene. Next, you will set the parameters in the **Far Attenuation** area to attenuate *spot light1*.

9. In the **Far Attenuation** area, select the **Use** and **Show** check boxes. Next, set the value **1** and **55** in the **Start** and **End** spinners, respectively.

10. Select *spot light1.Target*. Now, move it toward the wall behind the bed, refer to Figure13-9.

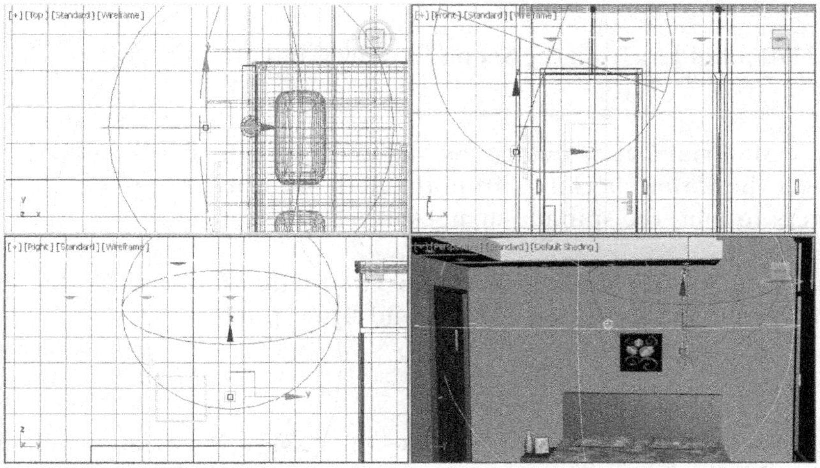

Figure 13-9 *The spot light1.Target moved*

11. Activate the Perspective viewport. Next, invoke the **Render Production** tool from the **Main Toolbar**; the rendered image is displayed, refer to Figure 13-10.

Figure 13-10 *The rendered image*

12. Choose the **Modify** tab in the **Command Panel**. Next, in the **Light Distribution (Type)** area of the **General Parameters** rollout, make sure **Uniform Spherical** is selected in the drop-down list.

13. Select all lamp holders from the **Scene Explorer** and group them with the name **lamp holder**.

14. Invoke the **Material Editor** dialog box and make sure the first sample slot is selected. Now, select the **Color** check box from the **Self Illumination** area of the **Blinn Basic Parameters** rollout; the spinner changes into color swatch.

15. Click on the color swatch; the **Select Color: Self Illum Color** dialog box is displayed. Set the following values in **Red**, **Green**, and **Blue** spinners:

    Red: **225** Green: **220** Blue: **20**

16. Select *lamp holder* in the viewport. Invoke the **Assign Material to Selection** tool in the **Material Editor** dialog box. Invoke the **Show Shaded Material** tool to display the glow of lights in all lamp holders in the *lamp holder* group. Now, close the **Material Editor** dialog box.

17. Activate the Perspective viewport. Next, invoke the **Render Production** tool from the **Main Toolbar**; the rendered image is displayed, refer to Figure 13-11.

Figure 13-11 *The rendered image*

Next, you will make two copies of *spot light1* and place them at the other two spotlights located at its left.

18. Activate the Right viewport. Next, select *spot light1* and *spot light1.Target*. Next, press and hold the SHIFT key and drag the cursor to *new lamp holder001*. Release the left mouse button; the **Clone Options** dialog box is displayed. In this dialog box, select the **Copy** radio button. Set the value **2** in the **Number of Copies** spinner and choose the **OK** button; two copies of *spot light1* and *spot light1.Target* are created in all viewports. Next, invoke the **Select and Move** tool and align these lights in all viewports, as shown in Figure 13-12.

19. Similarly, you can create copies of *spot light1* and *spot light1.Target* and place them at the remaining spot lights of the false ceiling.

Saving and Rendering the Scene

In this section, you will save the scene that you have created and then render it. You can also view the final rendered image of this model by downloading the *c13_3dsmax_2022_rndr.zip* file from *www.cadcim.com*. The path of the file is as follows: *Textbooks > Animation and Visual Effects > Autodesk 3ds Max > Autodesk 3ds Max 2022 for Beginners: A Tutorial Approach*

1. Choose **Save** from the **File** menu.

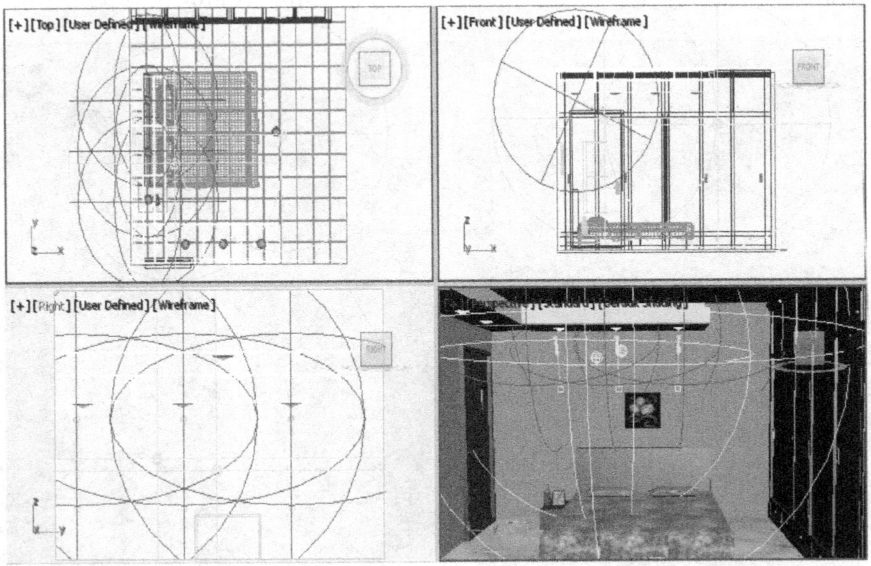

Figure 13-12 *Copies of spot light1* and *spot light1.Target aligned in all viewports*

2. Activate the Perspective viewport. Next, invoke the **Render Production** tool from the **Main Toolbar**; the rendered image is displayed, refer to Figure 13-13.

Figure 13-13 *The rendered image*

Tutorial 2

In this tutorial, you will create false ceiling lights, as shown in Figure 13-14, using the **Free Light** tool. **(Expected time: 20 min)**

Figure 13-14 *The scene with false ceiling lights*

The following steps are required to complete this tutorial:

a. Create the project folder.
b. Open the file.
c. Create the false ceiling lights.
d. Save and render the scene.

Creating the Project Folder

Create the project folder with the name *c13_tut2* in the *3dsmax 2022* folder, as discussed in earlier chapters.

Opening the File

1. Open Windows Explorer and then browse to *\Documents\c13_3dsmax_2022_tut*. Next, copy all jpeg files to *\Documents\3dsmax 2022\c13_tut2\sceneassets\images*.

2. Choose **Open** from the **File** menu; the **Open File** dialog box is displayed. In this dialog box, browse to the location *\Documents\c13_3dsmax_2022_tut* and select the **c13_tut2_start.max** file from it. Next, choose the **Open** button to open the file, refer to Figure 13-15.

You will notice that one light is already there in the scene.

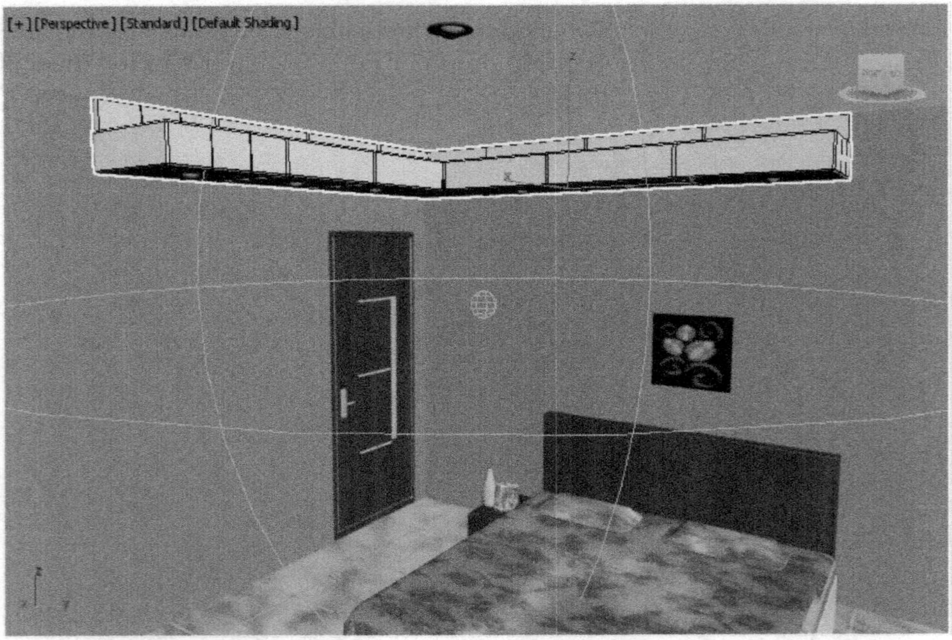

Figure 13-15 *The c13_tut2_start.max file*

3. Choose **Save As** from the **File** menu; the **Save File As** dialog box is displayed. Browse to the location *\Documents\3dsmax 2022\c13_tut2\scenes*. Save the file with the name *c13tut2.max* at this location.

Creating the False Ceiling Lights

In this section, you will create false ceiling lights using the **Free Light** tool.

1. Activate the Top viewport and then choose **Create > Lights** in the **Command Panel**. By default, the **Photometric** option is displayed in the **Command Panel**.

2. Invoke the **Free Light** tool from the **Object Type** rollout. If the **Photometric Light Creation** message box is displayed, choose the **Yes** button from this message box. Now, click at a point; the free light is displayed with the name *PhotometricLight001*. Enter **fceilinglight1** in the **Name and Color** rollout.

3. Choose the **Modify** tab. Now, in the **General Parameters** rollout, select the **On** check box in the **Shadows** area and the **Ray Traced Shadows** option in the drop-down list.

4. In the **Emit light from (Shape)** area of the **Shape/Area Shadows** rollout, select the **Line** option from the drop-down list; the shape of the light changes in the viewport and the **Length** spinner is added to the **Emit Light from (Shape)** area. Make sure the **Uniform Spherical** option is selected in the drop-down list located in the **Light Distribution (Type)** area.

5. Set the value **100** in the **Length** spinner to increase the spread of the light; the shape of *fceilinglight1* is changed to a line shape.

6. Invoke the **Select and Move** tool and align *fceilinglight1* in all viewports, as shown in Figure 13-16.

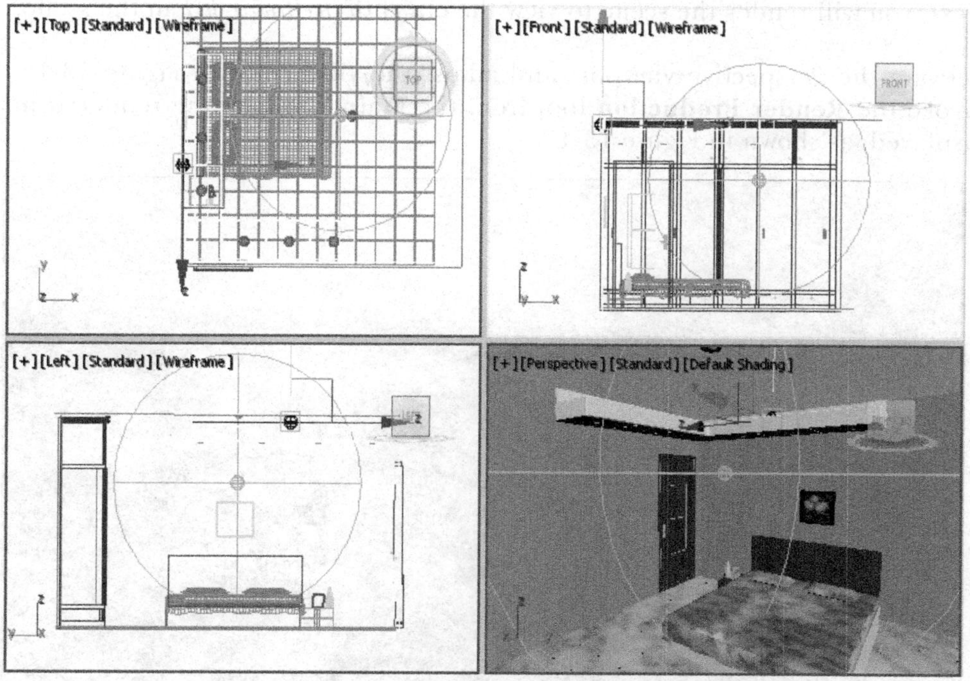

Figure 13-16 *The fceilinglight1 aligned in all viewports*

Make sure that you align *fceilinglight1* close to the wall behind the bed and close to the false ceiling.

Next, you will modify the parameters of *fceilinglight1* to achieve the desired lighting effect.

7. Select the radio button located near the **Kelvin** spinner; the **Kelvin** spinner is activated.

8. Set the value **3386** in the **Kelvin** spinner; the color in the color swatch located next to it is changed.

9. Choose the **Filter Color** color swatch located below it; the **Color Selector: Filter Color** dialog box is displayed. In this dialog box, set the values in the **Red**, **Green**, and **Blue** spinners as given next:

Red: **230** Green: **17** Blue: **32**

Choose the **OK** button to close the dialog box; the color in the **Filter Color** color swatch is changed.

10. In the **Intensity** area of the **Intensity/Color/Attenuation** rollout, select the **lm** radio button; the value in the first spinner located below this radio button changes and displays the intensity of the light in Lumen. Next, set the value **700.0** in this spinner; the intensity of the light is decreased.

    Next, you will render the scene to view the effect of *fceilinglight1* in the scene.

11. Activate the Perspective viewport and adjust the view, refer to Figure 13-16. Next, invoke the **Render Production** tool from the **Main Toolbar**; the rendered image is displayed, as shown in Figure 13-17.

Figure 13-17 *The rendered image*

Next, you need to copy *fceilinglight1* and place the copy at the other side of false ceiling.

12. Activate the Top viewport and select *fceilinglight1*. Now, press and hold the SHIFT key and drag the cursor. Next, release the left mouse button; the **Clone Options** dialog box is displayed. In this dialog box, select the **Copy** radio button. Set the value **1** in the **Number of Copies** spinner and choose the **OK** button; a copy of *fceilinglight1* is created in all viewports with the name *fceilinglight002*. Next, align it in all viewports using the **Select and Move** and **Select and Rotate** tools, as shown in Figure 13-18.

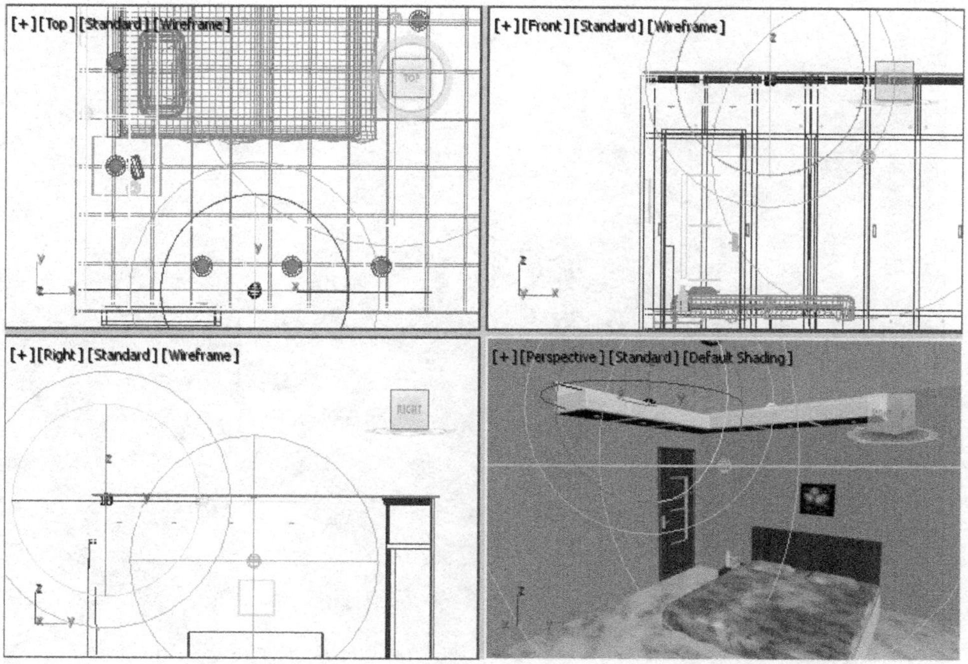

Figure 13-18 *The fceilinglight002 aligned in all viewports*

Make sure that you align *fceilinglight002* close to the wall and also close to the false ceiling.

13. Choose the **Modify** tab in the **Command Panel**. In the **Emit Light from (Shape)** area of the **Shape/Area Shadows** rollout, set the value **96** in the **Length** spinner.

Note
To achieve different lighting effects, you can set different length, color, and intensity values for false ceiling lights.

Saving and Rendering the Scene

In this section, you will save the scene that you have created and then render it. You can also view the final rendered image of this model by downloading the *c13_3dsmax_2022_rndr.zip* file from *www.cadcim.com*. The path of the file is as follows: *Textbooks > Animation and Visual Effects > Autodesk 3ds Max > Autodesk 3ds Max 2022 for Beginners: A Tutorial Approach*

1. Choose **Save** from the **File** menu.

2. Activate the Perspective viewport. Next, invoke the **Render Production** tool from the **Main Toolbar**; the rendered image is displayed, as shown in Figure 13-19.

Figure 13-19 *The rendered image*

Self-Evaluation Test

Answer the following questions and then compare them to those given at the end of this chapter:

1. Which of the following tools is used to create photometric lights?

   (a) **Omni** (b) **Target Spot**
   (c) **Free Light** (d) All of these

2. Which of the following options in the **Light Distribution (Type)** area of the **General Parameters** rollout is used to spread light in all directions?

   (a) **Uniform spherical** (b) **Uniform Distribution**
   (c) **Photometric Web** (d) All of these

3. You can create specific lighting patterns by using the standard __________ files.

4. Higher the value you specify in the **Shadow Samples** drop-down list, more will be the rendering time. (T/F)

5. The **On** check box in the **General Parameters** rollout of the spot light can be selected to view the shadow of a light. (T/F)

Review Questions

Answer the following questions:

1. Which of the following areas of the **Intensity/Color/Attenuation** rollout is used to attenuate the spotlight?

   (a) **Dimming** (b) **Far Attenuation**
   (c) **Intensity** (d) None of these

2. The options in the __________ area are used to change the shape of the light.

3. The value in the __________ spinner and the color in the __________ color swatch together define the color shade of the light being created.

4. Photometric Lights are more complex than standard lights. (T/F)

5. You can improve the quality of rendering by changing the options in the **Shadow Samples** drop-down list. (T/F)

EXERCISE

The rendered output of the scene used in the following exercise can be accessed by downloading the *c13_3dsmax_2022_exr.zip* from *www.cadcim.com*. The path of the file is as follows: *Textbooks > Animation and Visual Effects > Autodesk 3ds Max > Autodesk 3ds Max 2022 for Beginners: A Tutorial Approach*

Exercise 1

Extract the contents of the *c13_3dsmax_2022_exr.zip* and then open the *c13_exer1_start* file from it. Next, illuminate the scene by using the photometric spot lights, as shown in Figure 13-20.

(Expected time: 25 min)

Figure 13-20 The illuminated scene

Answers to Self-Evaluation Test

1. c, **2.** a, **3.** IES, **4.** T, **5.** T

Chapter 14

Animation Basics

Learning Objectives

After completing this chapter, you will be able to:

- *Adjust the number of frames in an animation*
- *Create keyframes*

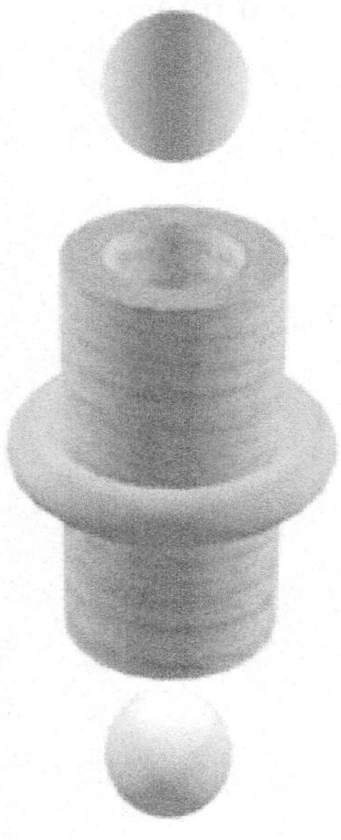

INTRODUCTION

A computer animation is created by defining the transformation of an object at key points during the animation. These defined points are known as keyframes or keys. The interpolation between the keyframes consists of the information of the actions performed between those keyframes. The final animation consists of a series of frames. Each frame of the series has the objects in different positions. When the frames are played quickly, the objects in the viewport appear to be moving. The greater the number of frames, the longer the animation lasts. A computer typically plays frames at the rate of 24 frames per second (1,440 frames per minute). If you want the animation to last for 5 minutes, you need to have 7200 frames (24 frames/second × 60 second/minute × 5 minutes).

TUTORIALS

Before starting the tutorials, you need to download the *c14_3dsmax_2022_tut.zip* file from *www.cadcim.com*. The path of the file is as follows: *Textbooks > Animation and Visual Effects > Autodesk 3ds Max > Autodesk 3ds Max 2022 for Beginners: A Tutorial Approach*

Extract the contents of the zip file and save them in the *Documents* folder.

Tutorial 1

In this tutorial, you will create simple objects and add basic materials to them. You will also impart motion to the objects and save the animation to a file. Figure 14-1 shows the animated scene at frame 12. **(Expected time: 25 min)**

Figure 14-1 *Animated scene at frame 12*

The following steps are required to complete this tutorial:

a. Create the project folder.
b. Create the shaft.

c. Create and position the bearing.
d. Create and position the ball.
e. Assign materials.
f. Add lights.
g. Add cameras.
h. Add animation to the scene.
i. Save and render the animation.

Creating the Project Folder

Create a project folder in the *3dsmax 2022* folder with the name *c14_tut1* as discussed in the previous chapters.

Creating the Shaft

In this section, you will create shaft by using the **Tube** tool.

1. Activate the Top viewport. Choose **Create > Geometry** in the **Command Panel** and then invoke the **Tube** tool from the **Object Type** rollout; various rollouts related to the **Tube** tool are displayed in the **Command Panel**.

2. In the **Parameters** rollout, set the value **20** in the **Sides** spinner and select the **Smooth** check box if not selected.

3. Expand the **Keyboard Entry** rollout. Next, set the values **50** in the **Inner Radius** spinner, **80** in the **Outer Radius** spinner, and **300** in the **Height** spinner. Next, choose the **Create** button in the **Keyboard Entry** rollout; a tube is created and displayed in the viewport.

4. In the **Name and Color** rollout, enter **Shaft** and choose the color swatch; the **Object Color** dialog box is displayed.

5. In this dialog box, select green color and then choose the **OK** button.

6. Activate and maximize the Perspective viewport. Invoke the **Zoom Extents All** tool so that *Shaft* is zoomed to its extents, refer to Figure 14-2.

Creating and Positioning the Bearing

In this section, you will create the bearing by using the **Torus** tool.

1. Restore the four viewport configuration. Activate the Top viewport and then choose **Create > Geometry** in the **Command Panel**. Next, invoke the **Torus** tool from the **Object Type** rollout.

2. Expand the **Keyboard Entry** rollout and set the value **100** in the **Major Radius** spinner and **20** in the **Minor Radius** spinner. Next, choose the **Create** button.

3. In the **Name and Color** rollout, enter **Bearing** and choose the color swatch; the **Object Color** dialog box is displayed. Select blue color from it. Next, choose the **OK** button to close the dialog box.

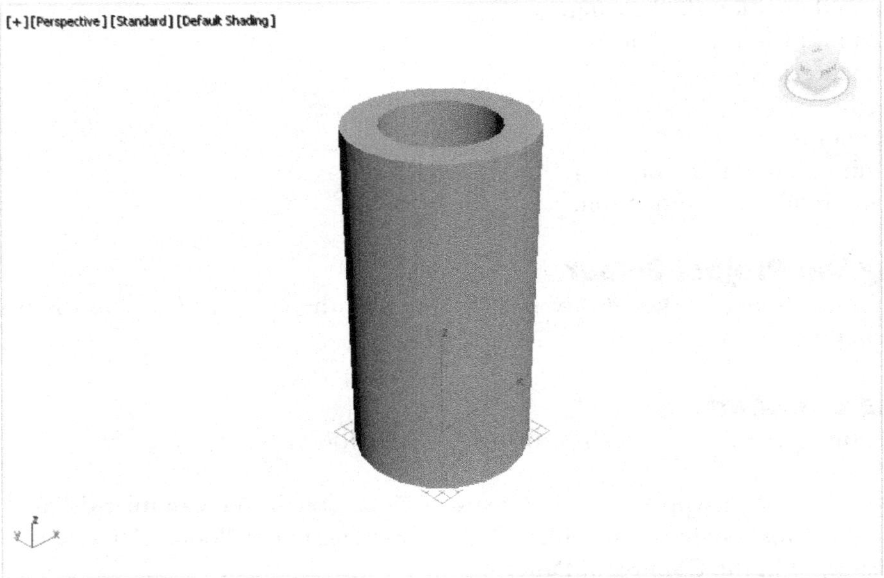

***Figure 14-2** The Shaft zoomed*

4. Invoke the **Zoom Extents All** tool; the objects are zoomed to their extents, as shown in Figure 14-3.

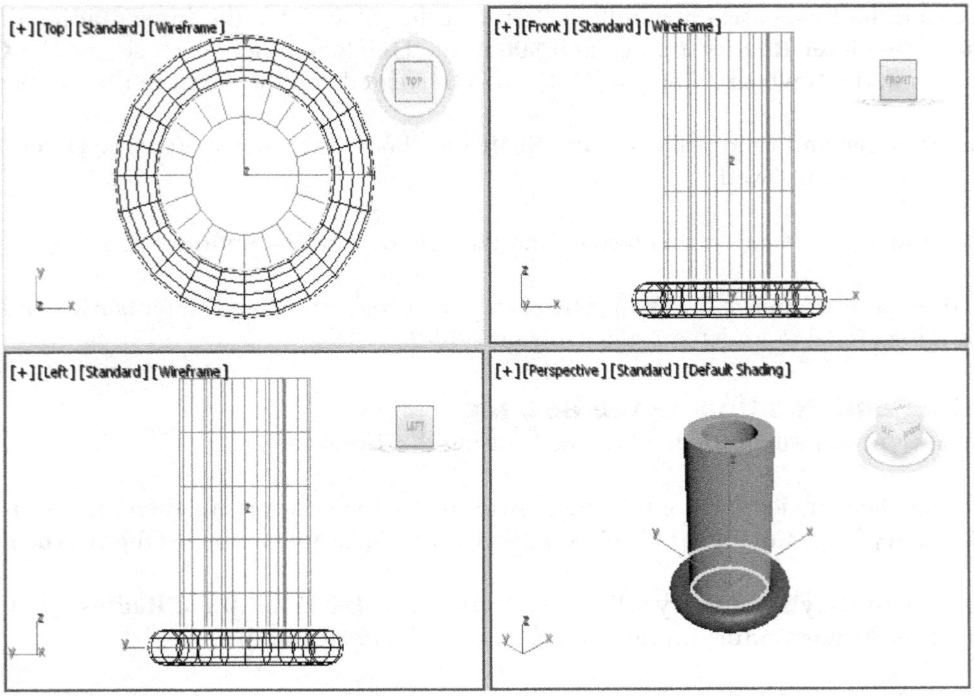

***Figure 14-3** The objects zoomed to their extents*

Bearing is created at the bottom of *Shaft*. You need to move it upward.

5. Invoke the **Select and Move** tool from the **Main Toolbar**.

6. Activate the Front viewport. Drag *Bearing* up until the Y coordinate in the coordinate display reads almost 150 and release the mouse button; *Bearing* is placed in the middle of *Shaft*, as shown in Figure 14-4.

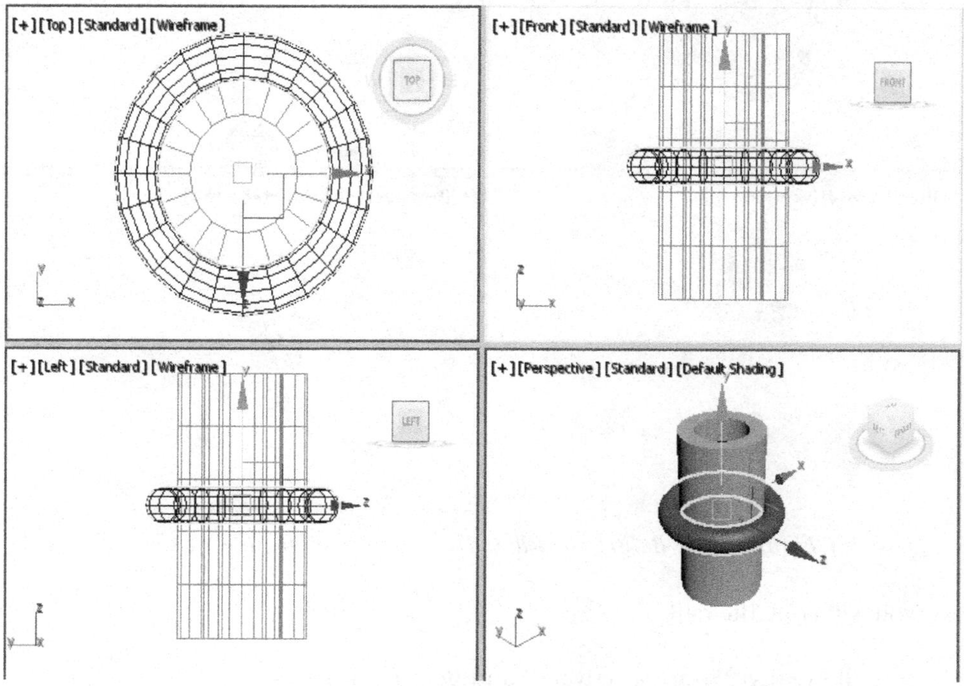

Figure 14-4 Bearing placed near the middle of Shaft

Creating and Positioning the Ball

In this section, you will create the ball using the **Sphere** tool and then move it.

1. Activate the Top viewport and then choose **Create > Geometry** in the **Command Panel**. Next, invoke the **Sphere** tool from the **Object Type** rollout.

2. In the **Parameters** rollout of the **Command Panel**, set the value **30** in the **Segments** spinner and make sure that the **Smooth** check box is selected.

3. Expand the **Keyboard Entry** rollout and set **50** in the **Radius** spinner. Next, choose the **Create** button; the sphere is created near the bottom of *Shaft*.

4. In the **Name and Color** rollout, enter **Ball01**. Next, assign red color to *Ball01* as done earlier.

5. Activate the Left viewport and invoke the **Select and Move** tool.

6. Make sure *Ball01* is selected in the viewport. Now, drag it down until the Y coordinate in the coordinate display reads about –125; *Ball01* moves down along the Y axis, as shown in Figure 14-5.

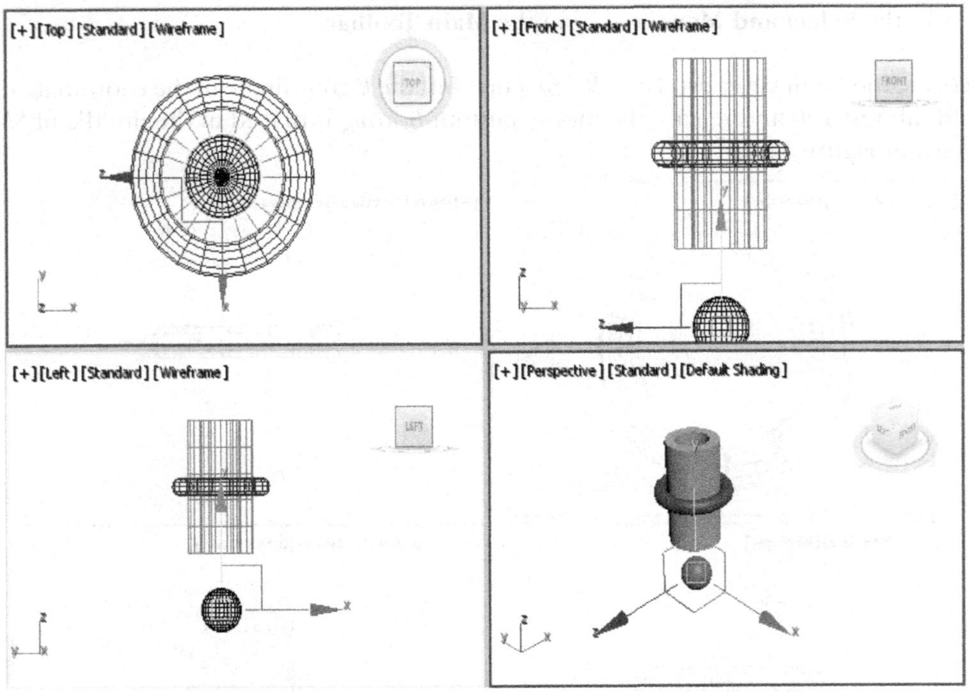

Figure 14-5 *Ball01 created and moved down along the Y axis*

Next, you will copy the ball.

7. Make sure the Left viewport is activated and *Ball01* is selected.

8. Press and hold the SHIFT key and move *Ball01* up until the Y coordinate in the coordinate display reads about 550, and then release the mouse button; the **Clone Options** dialog box is displayed.

9. In this dialog box, enter **Ball02** in the **Name** text box and make sure the **Copy** radio button is selected. Next, choose the **OK** button; a copy of *Ball01* is created and placed at the selected location, as shown in Figure 14-6.

10. Invoke the **Zoom Extents All** tool to zoom the objects to their extents.

Assigning Materials

In this section, you will assign materials to all objects in the scene.

1. Navigate to the *\Documents\c14_3dsmax_2022_tut* folder and copy the *Wood1, Masonry_Stone1.jpg,* and *Metal1* files to *\Documents\3dsmax 2022\c14_tut1\sceneassets\images.*

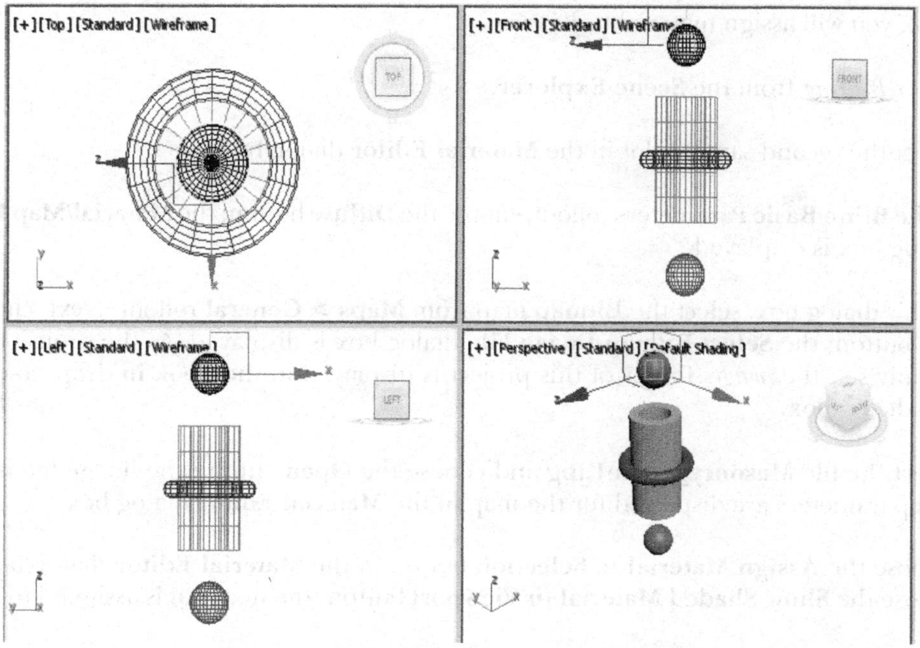

Figure 14-6 Ball02 created above Shaft

2. Choose the **Render Setup** tool and confirm Scanline Renderer is selected.

3. Select *Shaft* from the **Scene Explorer**. Next, choose **Rendering > Material Editor > Compact Material Editor** from the menu bar; the **Material Editor** dialog box is displayed. In this dialog box, select the first sample slot.

4. In the **Blinn Basic Parameters** rollout of the **Material Editor** dialog box, choose the **Diffuse** button; the **Material/Map Browser** dialog box is displayed.

5. In this dialog box, select the **Bitmap** map from **Maps > General** rollout and choose the **OK** button; the **Select Bitmap Image File** dialog box is displayed.

6. In this dialog box, make sure **All Formats** is selected in the **Files of type** drop-down list. As the project folder is already set, the *images* folder of this project is displayed in the **Look in** text box. Select the **wood1.jpg** file and choose the **Open** button; the image file is loaded and the parameters are displayed for the map in the **Material Editor** dialog box.

7. Choose the **Assign Material to Selection** button in the **Material Editor** dialog box. Next, choose the **Show Shaded Material in Viewport** button to display the assigned material in the viewport.
8. Move one level up and enter **Shaft** in the material name drop-down list located below the material sample slots.

9. Choose the **Modify** tab in the **Command Panel**. In the **Parameters** rollout, make sure the **Real-World Map Size** check box is cleared.

Next, you will assign material to *Bearing*.

10. Select *Bearing* from the **Scene Explorer**.

11. Select the second sample slot in the **Material Editor** dialog box.

12. In the **Blinn Basic Parameters** rollout, choose the **Diffuse** button; the **Material/Map Browser** dialog box is displayed.

13. In this dialog box, select the **Bitmap** map from **Maps > General** rollout. Next, choose the **OK** button; the **Select Bitmap Image File** dialog box is displayed. As the project folder is already set, the *images* folder of this project is displayed in the **Look in** drop-down list of this dialog box.

14. Select the file **Masonry_Stone1.jpg** and choose the **Open** button; the image file is loaded and parameters are displayed for the map in the **Material Editor** dialog box.

15. Choose the **Assign Material to Selection** button in the **Material Editor** dialog box. Also, choose the **Show Shaded Material in Viewport** button; the material is assigned to *Bearing*.

16. Move one level up and enter **Bearing** in the material name drop-down list located below the material sample slots.

Next, you will assign material to balls.

17. Select *Ball01* and *Ball02* in the viewport.

18. In the **Material Editor** dialog box, select the third sample slot.

19. In the **Blinn Basic Parameters** rollout, choose the **Diffuse** button; the **Material/Map Browser** dialog box is displayed.

20. Select the **Bitmap** map from the **Maps > General** rollout from the **Material/Map Browser** dialog box. Next, choose the **OK** button; the **Select Bitmap Image File** dialog box is displayed. As the project folder is already set, the *images* folder of this project is displayed in the **Look in** drop-down list.

21. Select the **metal1.jpg** file and choose the **Open** button; the image file is loaded and the parameters are displayed for the map in the **Material Editor** dialog box.

22. Choose the **Assign Material to Selection** button in the **Material Editor** dialog box. Next, choose the **Show Shaded Material in Viewport** button to display the assigned material in the viewport; the material is assigned to *Ball01* and *Ball02*.

23. Move one level up and enter **Ball** in the drop-down list located below the material sample slots.

24. Close the **Material Editor** dialog box.

Adding Lights

In this section, you will create and position two omni lights in the scene. The lights should be positioned away from the object to prevent hot spots. For this reason, you may need to zoom out to create more space around the objects.

1. Invoke the **Zoom** tool. Next, zoom out in the Top viewport until the objects are displayed about one-third of their original size. 

2. Choose the **Lights** button in the **Create** tab of the **Command Panel**. 

3. Select **Standard** from the drop-down list located below it. Invoke the **Omni** tool from the **Object Type** rollout.

4. Click at a point in the lower-left corner of the Top viewport to place a light; an omni light is placed at the selected point.

5. Click at a point near the upper-right corner of the Top viewport to place another omni light; an omni light is placed at the selected point.

6. Invoke the **Zoom Extents All** tool to zoom the objects in the viewports to their extents and invoke the **Select and Move** tool.

7. In the Front viewport, select the light on the left and move it up to a position near the upper left corner of the viewport.

8. Select the light on the right and move it up to a position in the upper right corner of the viewport.

9. Invoke the **Zoom Extents All** tool; the two omni lights are displayed in all viewports, as shown in Figure 14-7.

Adding Camera

In this section, you will add camera to the scene. By adding a camera to the scene, you can look at the object from any direction and from any distance.

1. Activate the Top viewport. Choose **Create > Cameras** in the **Command Panel**. Next, invoke the **Target** tool from the **Object Type** rollout. 

2. Click in the lower-right corner of the viewport to position the camera. Next, press and hold the left mouse button and drag the cursor to the center of the objects, and then release the left mouse button.

   The second point defines the target point for the camera.

3. Activate the Perspective viewport and press the C key to change the viewport to the camera view. Also, press the G key to turn off the grid; the Perspective viewport becomes the Camera viewport.

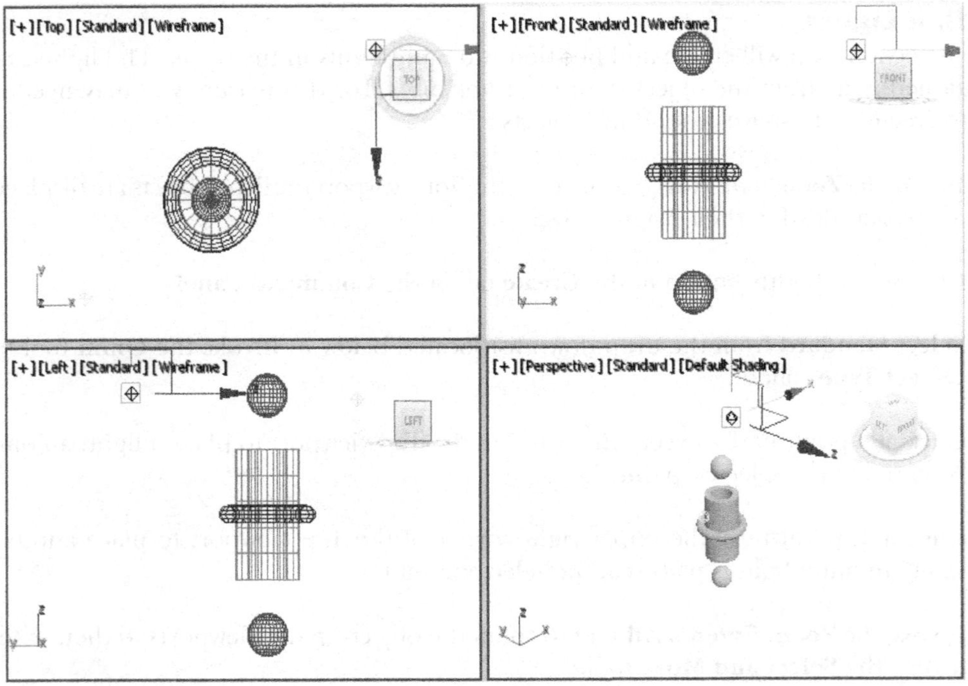

Figure 14-7 The two omni lights displayed in all viewports

4. Activate the Left viewport and invoke the **Select and Move** tool.

5. Select the camera (not its target) and drag it up and select a point that gives you the desired view of the objects in the Camera viewport.

   As you move the camera, the Camera viewport is dynamically updated.

6. Activate the Camera viewport and invoke the **Truck Camera** tool in the viewport navigation controls. 

7. Move the cursor to place the objects at the center of the Camera viewport, as shown in Figure 14-8. Right-click to exit the tool.

8. Invoke the **Render Production** tool from the **Main Toolbar**; the rendered image is displayed in the **Rendered Frame** window, refer to Figure 14-9. Close this window.

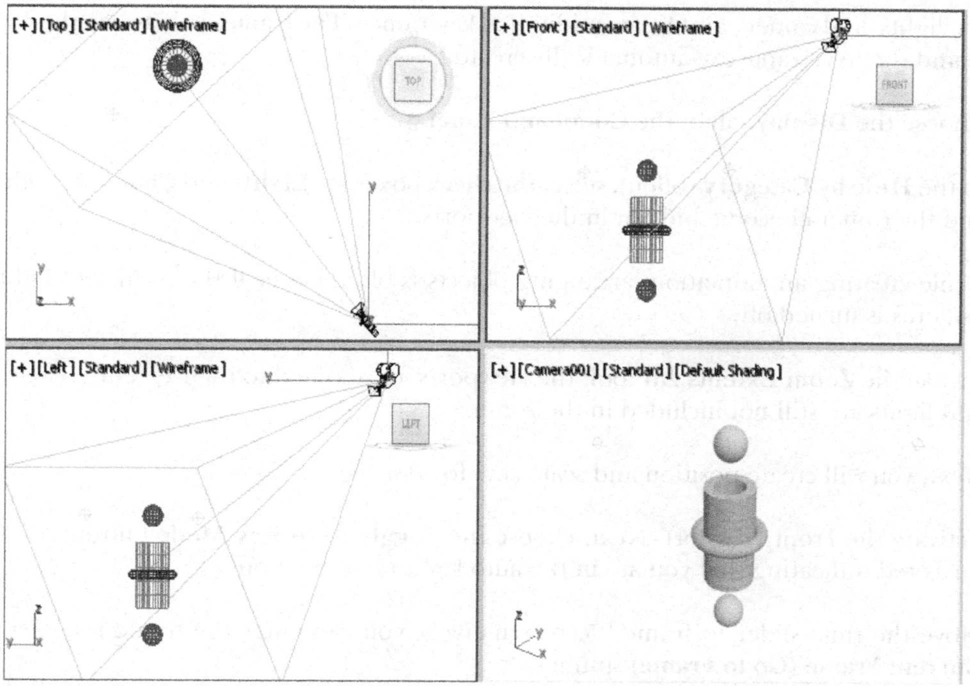

Figure 14-8 Placing the objects at the center of the Camera viewport

Figure 14-9 The rendered image

Creating Animation

In this section, you will create animation for balls and bearing. When you create objects and position lights and camera, you create the first animation scene (frame 0). You can arrange the objects, lights, and camera and create additional keyframes. The frames in between the previous frame and the new frame are automatically created.

1. Choose the **Display** tab in the **Command Panel**.

2. In the **Hide by Category** rollout, select the check boxes for **Lights** and **Cameras**; both lights and the camera become hidden in the viewports.

   While creating an animation, arranging objects is often easier if the display of lights and cameras is turned off.

3. Invoke the **Zoom Extents All** tool; the viewports are zoomed to their extents. The camera and lights are still not included in the zoom.

   Next, you will create position and scale keys for *Bearing*.

4. Activate the Front viewport. Next, choose the **Toggle Auto Key Mode** button; it turns red indicating that you are in the auto key animation mode.

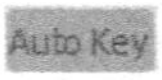

5. Move the time slider to frame 25. Alternatively, you can enter the frame number in the **Current Frame (Go to Frame)** spinner.

6. Invoke the **Select and Move** tool from the **Main Toolbar**.

7. Select *Bearing* from the **Scene Explorer**. Move *Bearing* straight up so that its top edge touches the top edge of *Shaft*.

8. Invoke the **Select and Uniform Scale** tool.

9. Make sure *Bearing* is selected. Now, drag the cursor until the coordinate display reads 125%; *Bearing* is scaled up to 125%.

10. Move to frame 75 and scale *Bearing* down to 80% of its current size.

11. Invoke the **Select and Move** tool and move *Bearing* straight down until its bottom edge touches the bottom edge of *Shaft*.

12. Move to frame 100 and move *Bearing* up until it is in the middle of the tube (its original position).

13. Choose the **Toggle Auto Key Mode** button to exit the auto key animation mode.

14. Activate the Camera viewport.

15. Choose the **Play Animation** button; the animation is previewed in the Camera viewport.

The **Play Animation** button is replaced by the **Stop Animation** button once the animation is started. 

16. Choose the **Stop Animation** button to stop the playback.

    Next, you will create the position keys for *Ball01*.

17. Activate the Front viewport and invoke the **Zoom** tool. Drag the cursor down such that the objects are displayed in half of their original size, and then invoke the **Maximize Viewport Toggle** tool.

18. Choose the **Display** tab in the **Command Panel** and then choose the **Hide by Name** button in the **Hide** rollout; the **Hide Objects** dialog box is displayed, as shown in Figure 14-10.

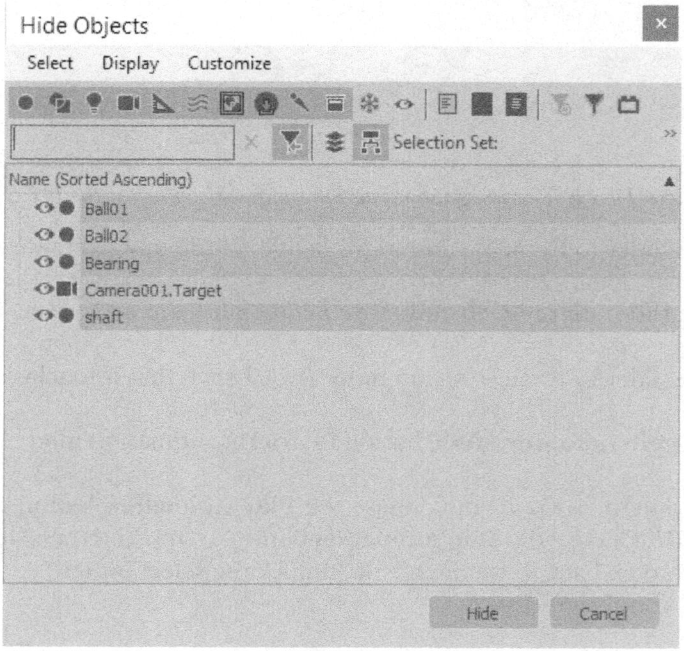

***Figure 14-10** The **Hide Objects** dialog box*

19. Select **Ball02** from the list and then choose the **Hide** button; the dialog box is closed and *Ball02* is hidden.

20. Choose the **Toggle Auto Key Mode** button to activate the auto key animation mode. 

21. Move the time slider to frame 15. Next, invoke the **Select and Move** tool. Select *Ball01* in the viewport and lock the selection set. Move *Ball01* downward toward right. Figure 14-11 shows the suggested paths for the two balls.

22. Move the time slider to frame 30 and move the *Ball01* a little upword toward right.

23. Move the time slider to frame 50. Again, move *Ball01* a little upward toward right so that it is in line with the middle of the tube.

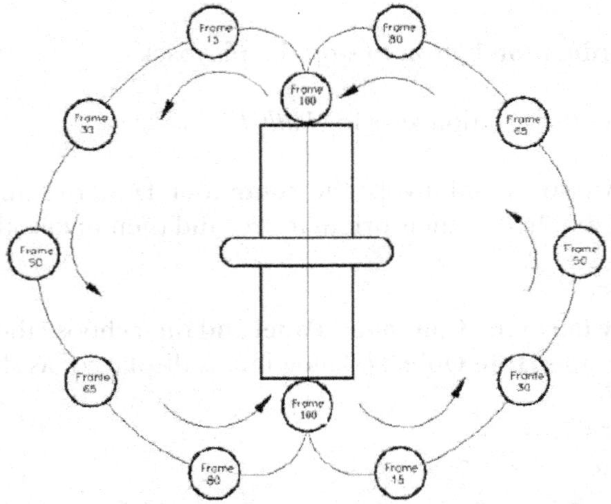

***Figure 14-11** The image showing the suggested paths for the two balls*

24. Move the time slider to frame 65 and move *Ball01* a little upward toward left.

25. Move the time slider to frame 80 and move *Ball01* a little more upward toward left.

26. Move the time slider to frame 100 and move *Ball01* such that it touches the top of *Shaft*.

27. Choose the **Toggle Auto Key Mode** button to exit the animation mode.

28. Activate the Camera viewport and choose the **Play Animation** button to check the motion of *Ball01*. Choose the **Stop Animation** button to stop and then choose the **Go to Start** button to go back to frame 0. Now, unlock the selection set.

Note
The locations of balls are approximate. You can use your own paths or the ones shown in Figure 14-11.

Next, you will create position keys for *Ball02*. You will also change the background color to white.

29. Choose the **Display** tab in the **Command Panel**. Next, choose the **Unhide All** button in the **Hide** rollout.

30. Choose the **Hide by Name** button in the **Hide** rollout; the **Hide Objects** dialog box is displayed.

31. Select **Ball01** in the list and choose the **Hide** button.

32. Choose the **Toggle Auto Key Mode** button to enter the auto key animation mode and move to frame 15.

33. Make sure *Ball02* is selected and then lock the selection set.

34. Activate the Front viewport. Move *Ball02* upward toward left, refer to Figure 14-11.

35. Move to frame 30 and move *Ball02* a little downward toward left.

36. Move to frame 50 and move *Ball02* a little toward left and down such that it is in line with the middle of the tube.

37. Move to frame 65 and move *Ball02* a little downward toward right.

38. Move to frame 80 and move *Ball02* a little more downward toward right.

39. Move to frame 100 and move *Ball02* so that it touches the bottom of the shaft.

40. Choose the **Toggle Auto Key Mode** button to exit the auto key animation mode.

41. Activate the Camera viewport and choose the **Play Animation** button to check the motion of *Ball02*. Next, choose the **Stop Animation** button.

42. In the **Display** tab of the **Command Panel**, choose the **Unhide All** button available under the **Hide** rollout; *Ball01* is displayed.

43. Choose the **Play Animation** button; a full animation is played.

44. Choose the **Stop Animation** button to exit the playback. You can then choose the **Go to Start** button to move to frame 0 or you can simply drag the time slider to frame 0. Now, unlock the selection set.

45. Choose **Rendering > Environment** from the menu bar; the **Environment and Effects** dialog box is displayed.

46. In this dialog box, choose the **Color** swatch in the **Background** area; the **Color Selector: Background Color** dialog box is displayed. Set the color to white and then choose the **OK** button to close this dialog box. Also, close the **Environment and Effects** dialog box.

Note
If all objects are not displayed during animation, invoke the ***Field-of-View*** *tool. Use it in the Camera viewport to increase the field of view so that all objects are easily displayed in the Camera viewport.*

Saving and Rendering the Animation

In this section, you will render the animation to a file. Next, you will play the rendered animation.

1. Choose **Save** from the **File** menu.

2. Make sure the Camera viewport is activated and invoke the **Render Setup** tool; the **Render Setup: Scanline Renderer** dialog box is displayed, as shown in Figure 14-12.

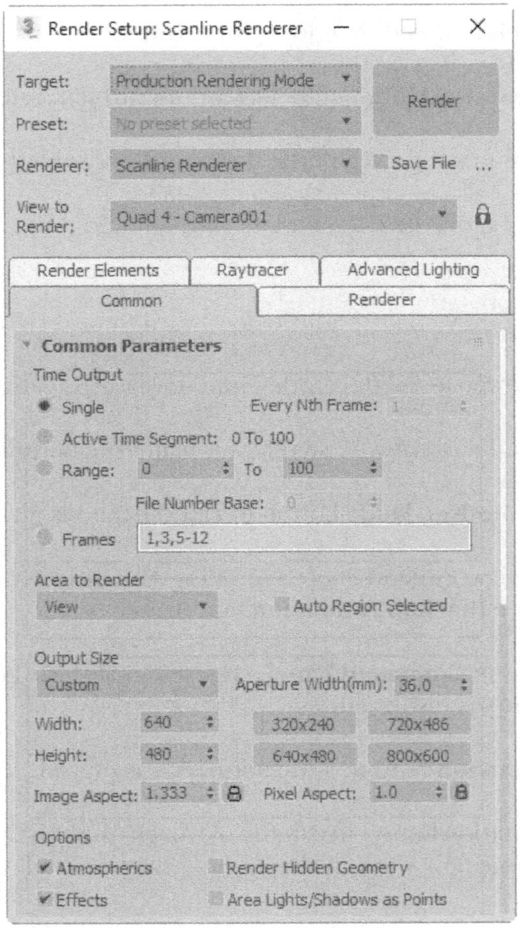

***Figure 14-12** The partial view of **Render Setup: Default Scanline Renderer** dialog box*

3. In the **Time Output** area, select the **Active Time Segment** radio button.

4. In the **Render Output** area, choose the **Files** button; the **Render Output File** dialog box is displayed, as shown in Figure 14-13. As the project folder is already set, the *renderoutput* folder of this project is displayed in the **Save in** text box. Select **AVI File (*.avi)** from the **Save as type** drop-down list.

5. In the **File name** text box, enter **c14tut1render** and choose the **Save** button; the path of the file is displayed in the **Render Output** area of the **Render Setup** dialog box.

   If the **AVI File Compression Setup** dialog box is displayed, choose the **OK** button to close it.

6. Make sure that the **Save File** check box is selected in the **Render Output** area. Now, choose the **Render** button to render the animation and close the **Render Setup** dialog box.

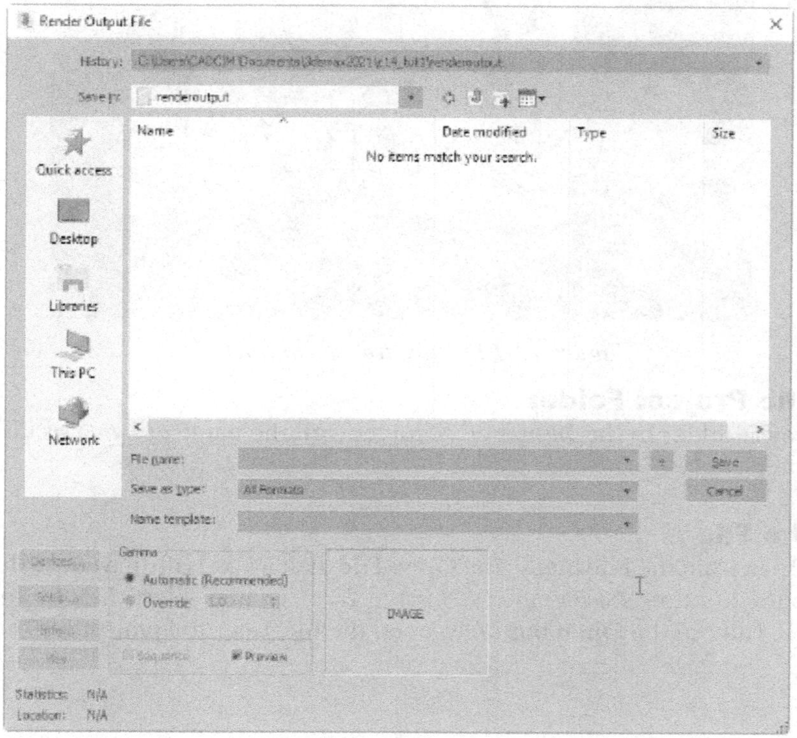

*Figure 14-13 The **Render Output File** dialog box*

It will take several minutes to render the animation frame by frame. After the rendering is complete, close the render window.

7. Browse to the *renderoutput* folder of the *c14_tut1* project and double-click on the **c14tut1render.avi** file; the AVI file is played in Windows Media Player.

Tutorial 2

In this tutorial, you will create an animated logo, refer to Figure 14-14.

(Expected time: 25 min)

The following steps are required to complete this tutorial:

a. Create the project folder.
b. Open the file.
c. Add keyframes to the text.
d. Save and render the animation.

Figure 14-14 Animated logo at frame 39

Creating the Project Folder

Create the project folder in the *3dsmax 2022* folder with the name *c14_tut2* as discussed in the previous chapters.

Opening the File

1. Choose **Open** from the **File** menu; the **Open File** dialog box is displayed. In this dialog box, browse to the location *\Documents\c14_3dsmax_2022_tut* and select the **c14_tut2_start.max** file from it. Choose the **Open** button to open the file, refer to Figure 14-15.

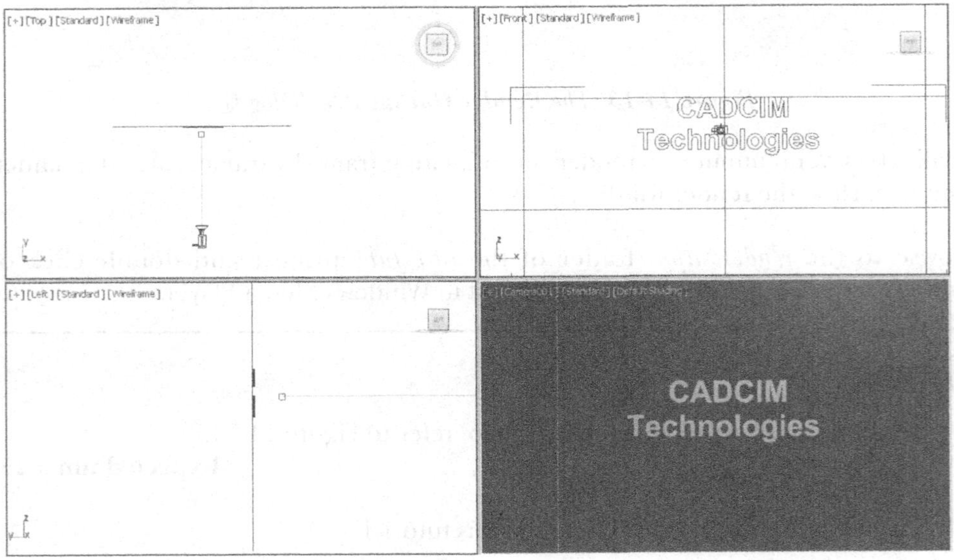

Figure 14-15 The c14_tut2_start.max file

2. Choose **Save As** from the **File** menu; the **Save File As** dialog box is displayed. Browse to the location *\Documents\3dsmax 2022\c14_tut2\scenes*. Save the file with the name *c14tut2.max* at this location.

Adding Keyframes to the Text

1. Choose the **Time Configuration** button in the lower-right corner of the interface; the **Time Configuration** dialog box is displayed, as shown in Figure 14-16. 

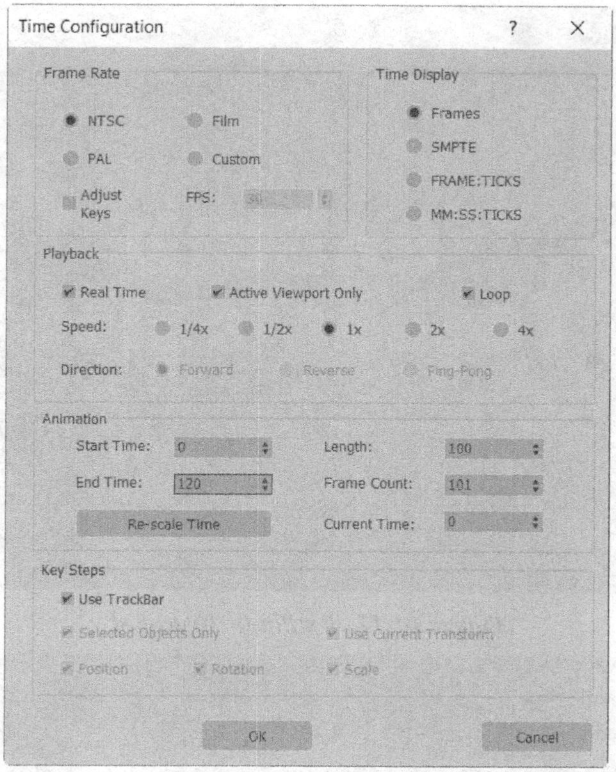

***Figure 14-16** The **Time Configuration** dialog box*

2. In the **Animation** area, set **120** in the **Length** spinner. Next, choose the **OK** button; the total number of frames in the animation are set to 120.

3. Make sure the time slider is set at frame 0. Next, choose the **Toggle Auto Key Mode** button.

4. Move the time slider to frame10 and select *TextPlus01*. Next, choose the **Modify** tab in the **Command Panel**.

5. In the **Parameters** rollout, set **50** in the **Tracking** spinner; red border is displayed around arrows of the **Tracking** spinner and *TextPlus001* is modified, as shown in Figure 14-17. Also, notice that keyframes are created at frame 0 and 10.

6. In the **Geometry** rollout, select the **Apply Bevel** check box from the **Beveling** area; *TextPlus001* is beveled. Next, set **4** and **-0.1** in the **Bevel Depth** and **Bevel Push** spinners, respectively.

7. Move the time slider to frame 20. Next, set **100** in the **Leading** spinner; red border is displayed around arrows of the **Leading** spinner and *TextPlus001* is modified again, as shown in Figure 14-18. Also, notice that keyframe is created at frame 20.

Figure 14-17 *TextPlus001 modified*

Figure 14-18 *Textplus001 modified again*

8. Move the time slider to frame 30. Next, set **450** in the **Size** spinner; red border is displayed around arrows of the **Size** spinner and the size of *TextPlus001* is modified, as shown in Figure 14-19. Also, notice that keyframe is created at frame 30.

9. Move the time slider to frame 40. Next, set **110** in both the **V Scale** and **H Scale** spinners; TextPlus001 is scaled, refer to Figure 14-20.

Figure 14-19 *Size of TextPlus001 modified*

Figure 14-20 *TextPlus001 scaled*

10. Move the time slider to frame 50. Next, set **50** in the **V Scale** and **H Scale** spinners. Next, move the time slider to frame 60. Next, set **0** in the **Leading** spinner.

11. Move the time slider to frame 70. Next, set **0** in the **Tracking** spinner; *TextPlus001* is modified again.

12. Move the time slider to frame 80. In the **Geometry** area, set **11** and **-0.8** in the **Bevel Depth** and **Bevel Push** spinners, respectively; bevel shape of *TextPlus001* is changed. To see the change in bevel shape in *TextPlus001,* switch to the Perspective viewport from any of the viewport and set the angle. Figure 14-21 shows change in bevel shape of *TextPlus001* in the Perspective viewport.

Note

To switch to the Perspective viewport, activate any of the viewport and press P.

Figure 14-21 *Bevel shape of TextPlus001 changed*

13. Make sure the timeslider is at Frame 80. Next, select *Camera001* from the Scene Explorer.

14. Choose the **Toggle Auto Key Mode** button to deactivate it. Next, choose the **Toggle Set Key Mode** and the **Set Keys** buttons; keyframe is created at frame 80 for *Camera001*.

15. Move the time slider to frame 110. Next, activate the Top viewport and move *Camera001* towards *TextPlus001*, refer to Figure 14-22. Now, choose the **Set Keys** button; keyframe is created at frame 110 for *Camera001*.

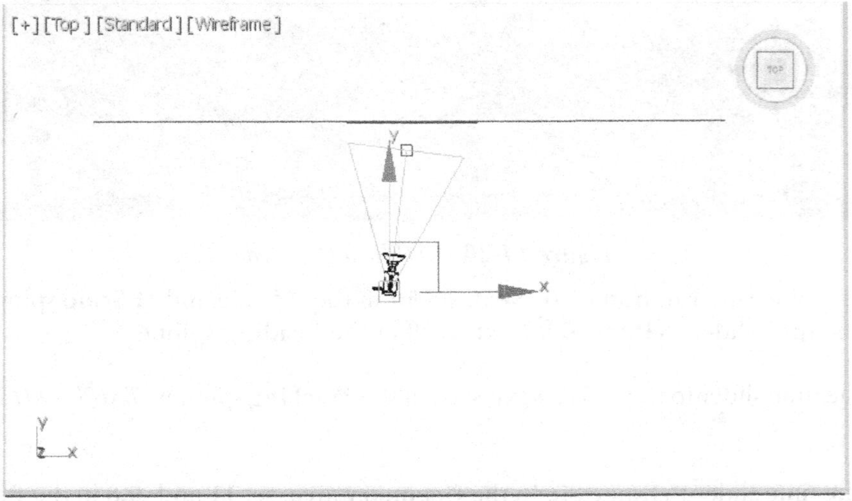

Figure 14-22 *Camera001 moved*

16. Move the time slider to frame 120. Next, choose the **Set Keys** button; keyframe is created at frame 120 for *Camera001*. You will notice that the camera position is same at frames 110 and 120. This is done to increase the timespan of *TextPlus001* when zoomed in.

Saving and Rendering the Animation

In this section, you will render the animation and create an AVI file. Next, you will play the rendered animation.

1. Choose **Save** from the **File** menu.

2. Make sure the Camera viewport is activated and invoke the **Render Setup** tool; the **Render Setup: Default Scanline Renderer** dialog box is displayed.

3. In the **Time Output** area, select the **Active Time Segment** radio button.

4. In the **Render Output** area, choose the **Files** button; the **Render Output File** dialog box is displayed. As the project folder is already set, the *renderoutput* folder of this project is displayed in the **Save in** text box. Select **AVI File (*.avi)** from the **Save as type** drop-down list.

5. In the **File name** text box, enter **c14tut2render** and choose the **Save** button; the path of the file is displayed in the **Render Output** area of the **Render Setup** dialog box.

   If the **AVI File Compression Setup** dialog box is displayed, choose the **OK** button to close it.

6. Make sure that the **Save File** check box is selected in the **Render Output** area. Now, choose the **Render** button to render the animation and close the **Render Setup** dialog box.

   It will take several minutes to render the animation frame by frame. After the rendering is complete, close the render window.

7. Browse to the *renderoutput* folder of the *c14_tut2* project and double-click on the **c14tut2render.avi** file; the AVI file is played in Windows Media Player.

Self-Evaluation Test

Answer the following questions and then compare them to those given at the end of this chapter:

1. Which of the following buttons is used to set the number of frames in animation?

   (a) **Time Configuration** (b) **Toggle Set Key mode**
   (c) **Toggle Auto Key Mode** (d) None of these

2. You can press the __________ key to change the view of the Camera viewport.

3. Computer animation is created by defining __________ and then calculating the movement that occurs between those frames.

4. You can adjust the camera to view an object from any direction and from any distance. (T/F)

5. Autodesk 3ds Max automatically creates frames in between two keyframes. (T/F)

Review Questions

Answer the following questions:

1. Which of the following buttons in the **Command Panel** is used to create keyframes automatically?

   (a) **Time Configuration** (b) **Toggle Auto Key Mode**
   (c) **Set Keys** (d) None of these

2. The __________ rollout in the **Command Panel** is used to set the bevel shape of TextPlus object.

3. The default range available in 3ds Max for animation is __________ frames.

4. The **Play Animation** button is replaced by the **Stop Animation** button once you start playing the animation. (T/F)

5. Animation keys cannot be adjusted, moved, copied, or deleted once they are created. (T/F)

EXERCISES

The rendered sequence of the scenes used in the following exercises can be accessed by downloading the *c14_3dsmax_2022_exr.zip* from *www.cadcim.com*. The path of the file is as follows: *Textbooks > Animation and Visual Effects > Autodesk 3ds Max > Autodesk 3ds Max 2022 for Beginners: A Tutorial Approach*

Exercise 1

Extract the content of the *c14_3dsmax_2022_exr.zip* and then open *c14_exr01_start.max*. Next, animate a chest with a lid and double doors, as shown in Figures 14-23 through 14-26.

(Expected time: 25 min)

Figure 14-23 Animation at frame 15

Figure 14-24 Animation at frame 30

Figure 14-25 Animation at frame 45

Figure 14-26 Animation at frame 60

Exercise 2

Open *c14_exr02_start.max* and animate the company logo, refer to Figures 14-27 through 14-29. **(Expected time: 25 min)**

Figure 14-27 Animation at frame 0

Figure 14-28 Animation at frame 10

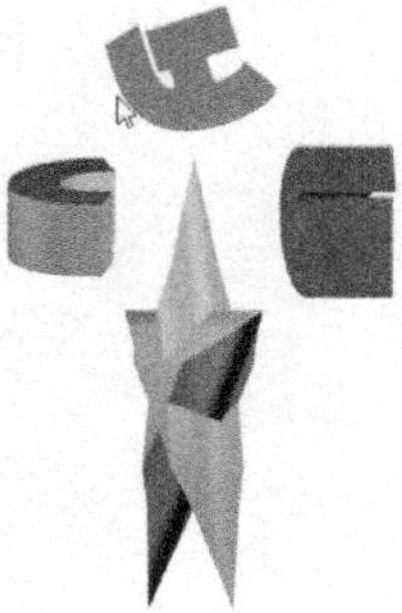

Figure 14-29 Animation at frame 30

Exercise 3

Open *c14_exr03_start.max* and animate the computer disk holder, refer to Figures 14-30 through 14-32. **(Expected time: 20 min)**

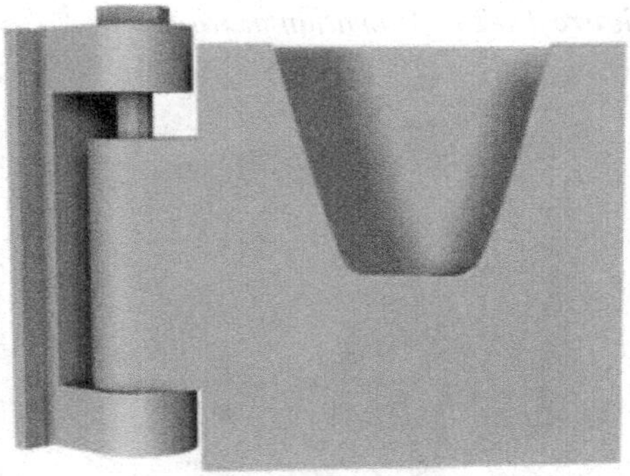

Figure 14-30 Animation at frame 0

Figure 14-31 Animation at frame 30

Figure 14-32 Animation at frame 40

Answers to Self-Evaluation Test

1. a, **2.** C, **3.** keyframes, **4.** T, **5.**T

Chapter 15

Complex Animation

Learning Objectives

After completing this chapter, you will be able to:
- *Create dummy objects*
- *Use graph editor to add, copy, and position keys*
- *Create snapshot objects*
- *Hide objects during animation*
- *Animate and link camera*

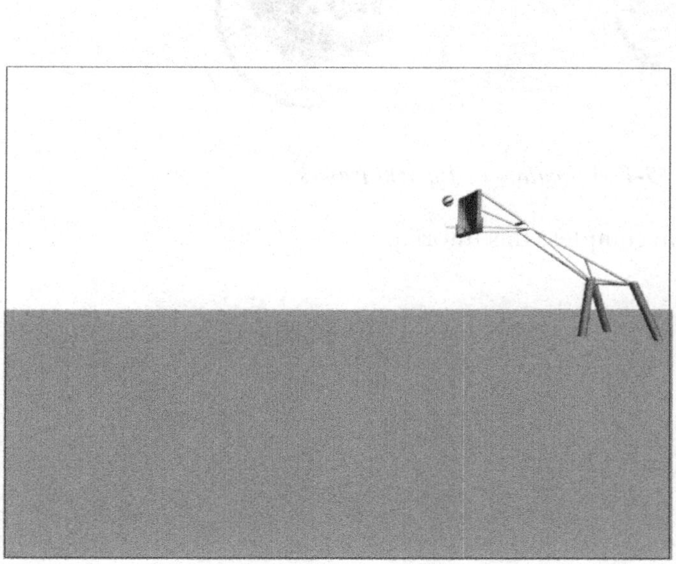

INTRODUCTION

In this chapter, you will learn to animate objects using dummy objects. Dummy objects link the objects to control their movement. If several objects are linked to a dummy object, the dummy can be used to move all objects simultaneously. The objects linked to a dummy can also be moved individually without affecting the dummy object or any other object linked to the dummy object.

TUTORIALS

Before starting the tutorials, you need to download the *c15_3dsmax_2022_tut.zip* file from *www.cadcim.com*. The path of the file is as follows: *Textbooks > Animation and Visual Effects > Autodesk 3ds Max > Autodesk 3ds Max 2022 for Beginners: A Tutorial Approach*

Extract the contents of the zip file and save them in the *Documents* folder.

Tutorial 1

In this tutorial, you will create the animation of a steel manufacturing process. Next, you will create dummy objects to assist in movement during animation, refer to Figure 15-1.

(Expected time: 35 min)

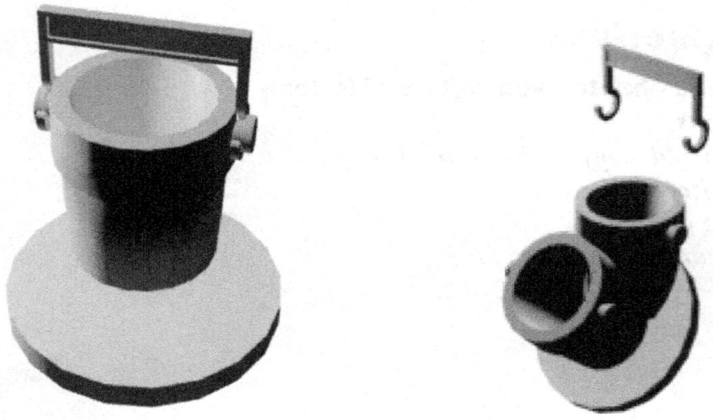

Figure 15-1 *Animation at different frames*

The following steps are required to complete this tutorial:

a. Create the project folder.
b. Open the file.
c. Create dummy objects.
d. Link the objects.
e. Set continuity value.
f. Animate the hook and cauldron.
g. Add a position key.
h. Create a snapshot.

i. Hide the object during animation.
j. Hide the object using object properties.
k. Preview the animation.
l. Set the camera view.
m. Save and preview the animation.

Creating the Project Folder

Create the project folder with the name *c15_tut1* in the *3dsmax 2022* folder, as discussed in earlier chapters.

Opening the File

1. Choose **Open** from the **File** menu; the **Open File** dialog box is displayed. In this dialog box, browse to the location *\Documents\c15_3dsmax_2022_tut* and select the **c15_tut1_start.max** file from it. Choose the **Open** button to open the file, refer to Figure 15-2.

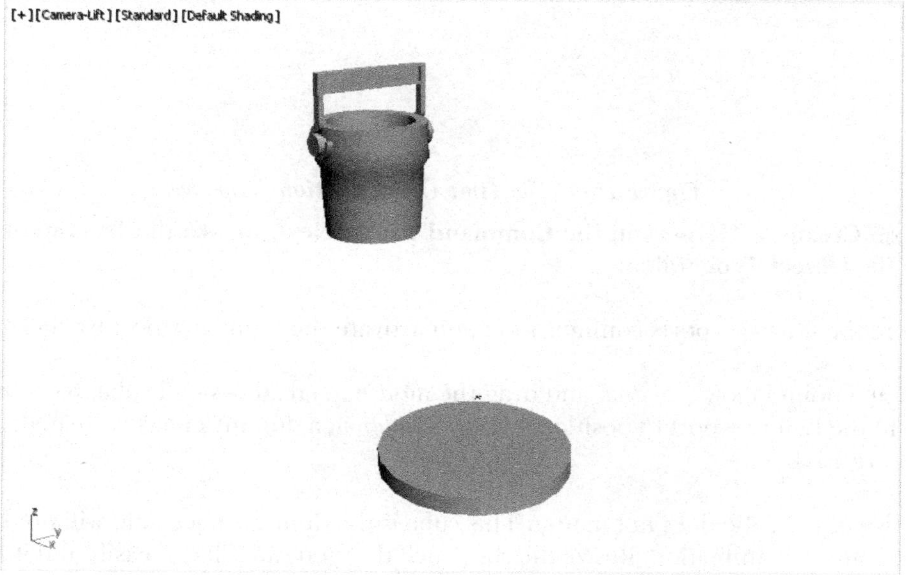

***Figure 15-2** The c15_tut1_start.max file*

2. Choose **Save As** from the **File** menu; the **Save File As** dialog box is displayed. Browse to the location *\Documents\3dsmax 2022\c15_tut1\scenes*. Save the file with the name *c15tut1.max* at this location.

Creating Dummy Objects

In this section, you will set the number of frames in the animation and create dummy objects.

1. Choose the **Time Configuration** button from the animation playback controls; the **Time Configuration** dialog box is displayed, as shown in Figure 15-3.

2. In the **Animation** area, set **150** in the **Length** spinner and choose the **OK** button; the total number of frames for the animation is set to 150.

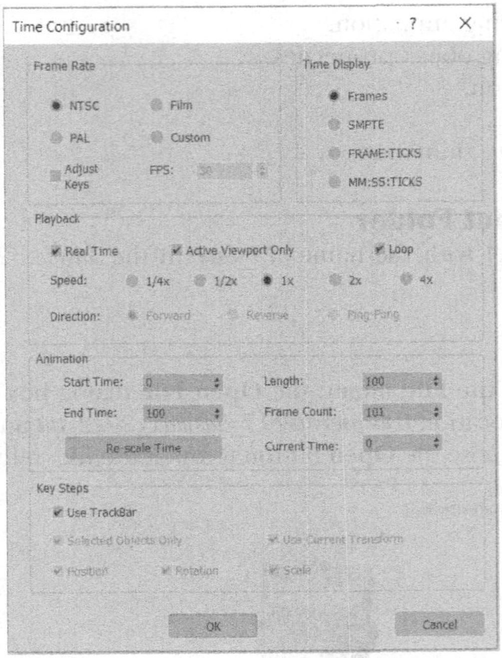

***Figure 15-3** The **Time Configuration** dialog box*

3. Choose **Create > Helpers** in the **Command Panel**. Next, invoke the **Dummy** tool from the **Object Type** rollout. 

4. Restore the four viewports configuration and activate the Front viewport by clicking on it.

5. Click at a point below *Cauldron* and drag the mouse to create a small cube. Next, move the cube in the Left viewport to position it below *Cauldron*; a dummy object is created, as shown in Figure 15-4.

   The size of the cube does not matter. This cube is the dummy object and will not be visible in the rendered animation. Resize the cube such that you can select it easily and it also does not interfere with the selection of other objects in the scene.

6. Enter **Cal-Dummy** in the **Name and Color** rollout.

7. Invoke the **Dummy** tool again from the **Object Type** rollout. In the Front viewport, click at a point below *Base* and drag the cursor to create another small cube. Adjust the position of the cube in the Left viewport to position it below *Base,* as shown in Figure 15-5.

8. Enter **Base-Dummy** in the **Name and Color** rollout.

9. Activate the Top viewport. Next, use the **Align** tool from the **Main Toolbar** to center *Base-Dummy* with the center of the Base object along the X and Y axes. Also, align the center of *Cal-Dummy* with the center of *Cauldron* along the X and Y axes. 

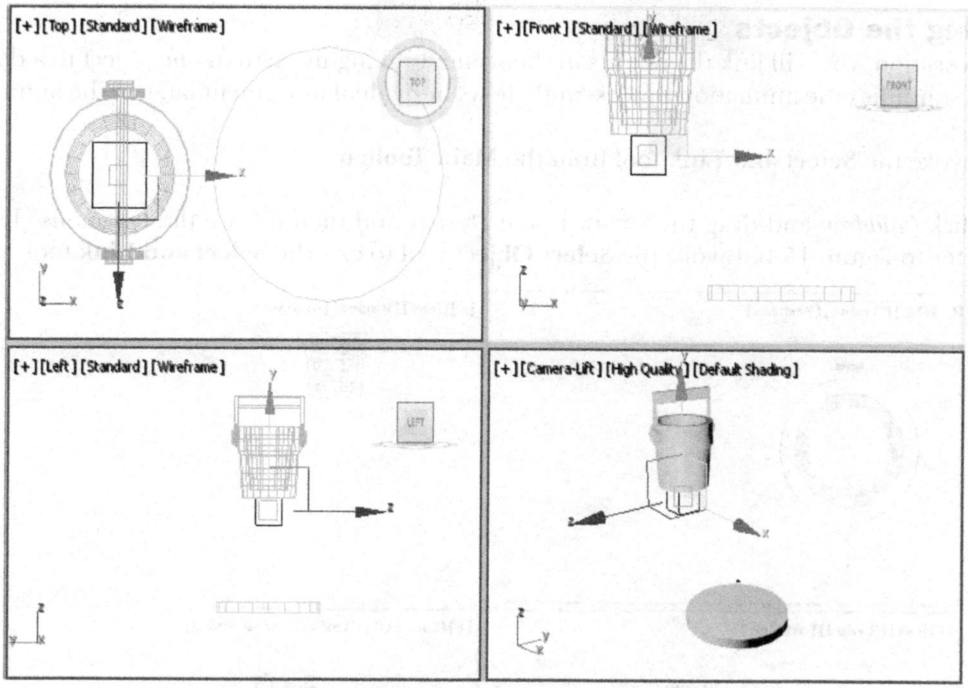

Figure 15-4 A dummy object created

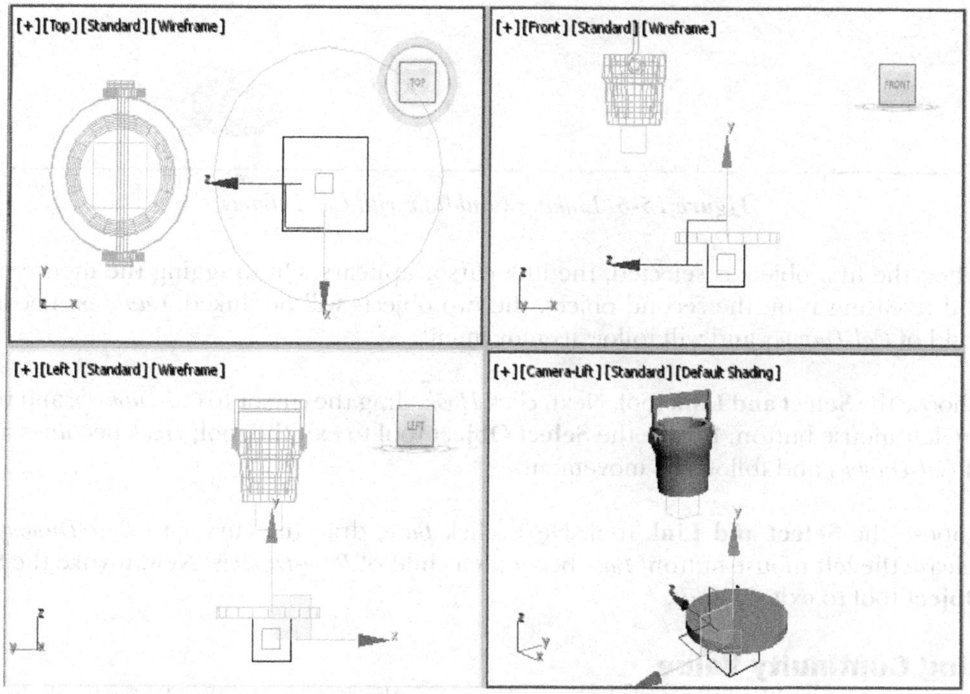

Figure 15-5 Adjusting the position of the cube below Base

Linking the Objects

In this section, you will link the objects in the scene. Linking more than one object to a dummy object simplifies the animation process and allows individual movement later in the animation.

1. Invoke the **Select and Link** tool from the **Main Toolbar**. 

2. Click *Cauldron* and drag the cursor to *Cal-Dummy* and then release the left mouse button, refer to Figure 15-6. Invoke the **Select Object** tool to exit the **Select and Link** tool.

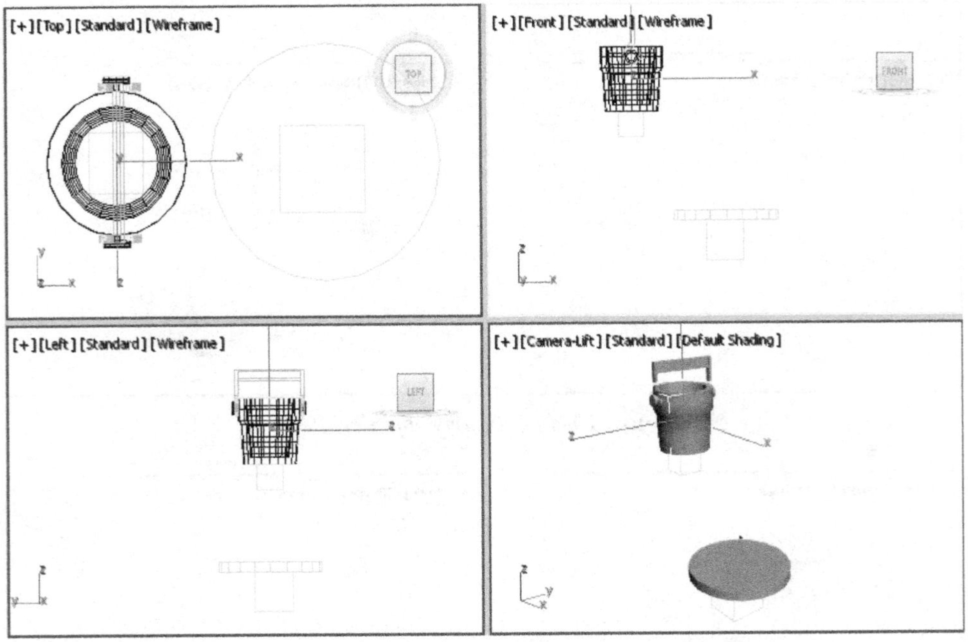

Figure 15-6 Linking Cauldron with Cal-Dummy

When the first object is selected, the link cursor appears. On dragging the mouse button and releasing it on the second object, the two objects will be linked. *Cauldron* becomes a child of *Cal-Dummy* and will follow its movement.

3. Choose the **Select and Link** tool. Next, click *Hook*, drag the cursor to *Cal-Dummy*, and release the left mouse button. Invoke the **Select Object** tool to exit the tool; *Hook* becomes a child of *Cal-Dummy* and follows its movement.

4. Choose the **Select and Link** tool. Next, click *Base*, drag the cursor to *Base-Dummy*, and release the left mouse button; *Base* becomes a child of *Base-Dummy*. Now, invoke the **Select Object** tool to exit the tool.

Setting Continuity Value

In this section, you will set the continuity value using the **Set Controller Defaults** dialog box. Continuity controls the movement of an object through animation keys and creates life-like or mechanical motion.

A continuity default value can be set before creating animation keys so that you need not adjust each key later.

1. Choose **Customize > Preferences** from the menu bar; the **Preference Settings** dialog box is displayed. Next, choose the **Animation** tab in the **Preference Settings** dialog box.

2. In the **Controller Defaults** area, choose the **Set Defaults** button; the **Set Controller Defaults** dialog box is displayed, as shown in Figure 15-7.

3. In this dialog box, select **TCB Float** from the list and then choose the **Set** button; the **TCB Default Key Values** dialog box is displayed, as shown in Figure 15-8.

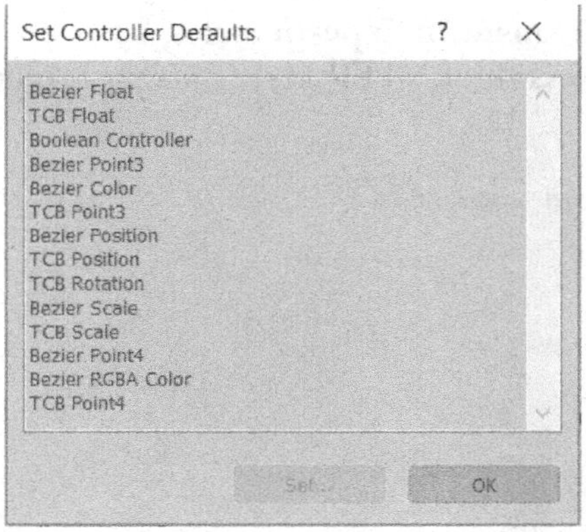

*Figure 15-7 The **Set Controller Defaults** dialog box*

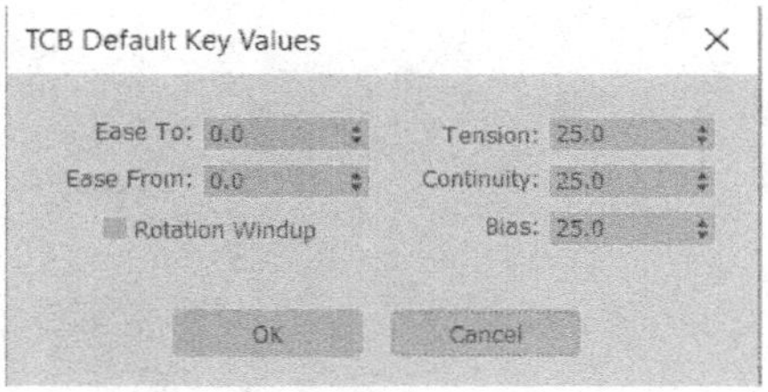

*Figure 15-8 The **TCB Default Key Values** dialog box*

4. In this dialog box, set the value **0** in the **Continuity** spinner and then choose the **OK** button.

5. Choose **OK** in the **Set Controller Defaults** dialog box. Next, choose **OK** in the **Preference Settings** dialog box.

Animating the Hook and Cauldron

In this section, you will animate *Hook* and *Cauldron*. To create an animation, a series of keys are defined.

1. Choose the **Toggle Auto Key Mode** button to turn the animation mode on (red). 

   When the **Toggle Auto Key Mode** button is chosen (red), any transformation creates an animation key on the current frame. If this button is off, the transformation is applied to frame 0 regardless of the current frame.

2. Drag the time slider to frame 25. Alternatively, you can set the value in the spinner at the left side of the **Time Configuration** button as **25**.

3. Activate the Front viewport and click *Cal-Dummy*. Next, right-click on the **Select and Move** tool; the **Move Transform Type-In** dialog box is displayed, refer to Figure 15-9.

4. Set the value **300** in the **X** spinner of the **Offset:Screen** area and press the ENTER key; *Cauldron* is now centered over *Base*.

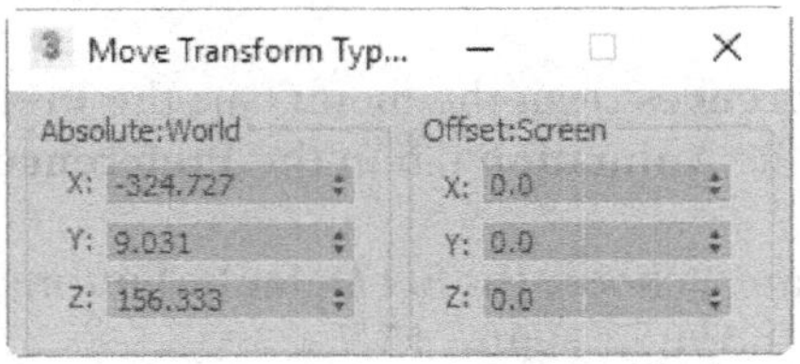

Figure 15-9 *The* ***Move Transform Type-In*** *dialog box*

5. Drag the time slider to frame 50.

6. Make sure *Cal-Dummy* is selected. In the **Move Transform Type-In** dialog box, set **–300** in the **Y** spinner of the **Offset:Screen** area and press the ENTER key; *Cauldron* is placed on *Base*, refer to Figure 15-10.

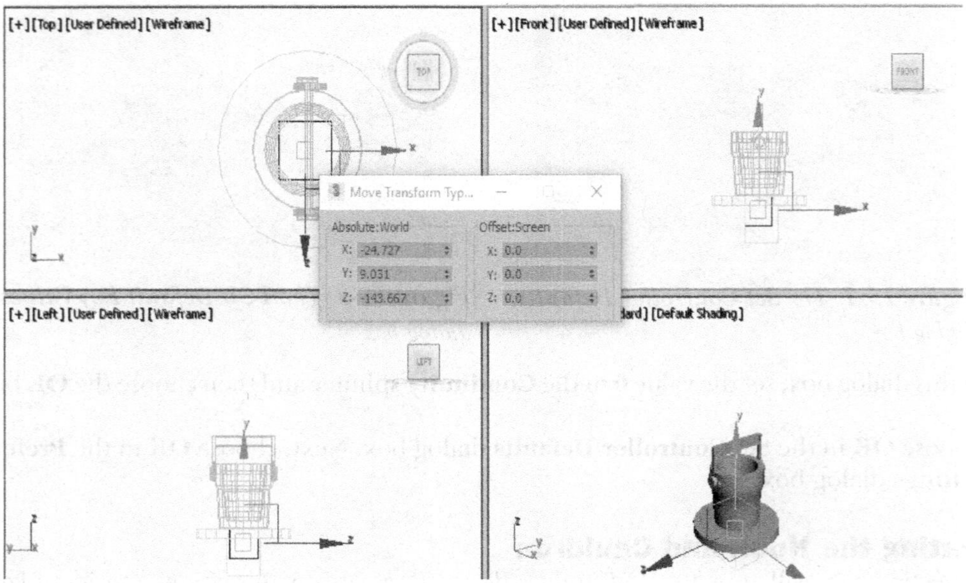

Figure 15-10 *The Cauldron placed on the circular Base*

7. Close the **Move Transform Type-In** dialog box and exit the animation mode by choosing the **Toggle Auto Key Mode** button.

Adding a Position Key

If the movement of *Hook* is animated, the movement would begin at frame 0 and would end at the active frame that was current when *Hook* was moved. As a result, *Hook* would gradually drift away from its parent object. To prevent this, *Hook* needs a position key added at frame 50. The position key locks the *Hook's* position at frame 50. Any movement applied to *Hook* only affects the frames between frame 50 and the current frame. In this section, you will use different frame settings in the **Track View - Dope Sheet** window to create the position key.

1. Select all objects in the viewport. Next, choose **Graph Editors > Track View - Dope Sheet** from the menu bar; the **Track View - Dope Sheet** window is displayed.

2. In this window, click on the plus sign at the left of **Hook** and then click on the plus sign at the left of **Transform**. Notice that currently there are no animation keys for *Hook*.

3. Choose the **Add/Remove Key** button on the upper left of the **Track View - Dope Sheet** window. 

4. Click on the **Position** track for *Hook* at frame 50; a position key for *Hook* is created at frame 50, refer to Figure 15-11. Next, close the **Track View - Dope Sheet** window.

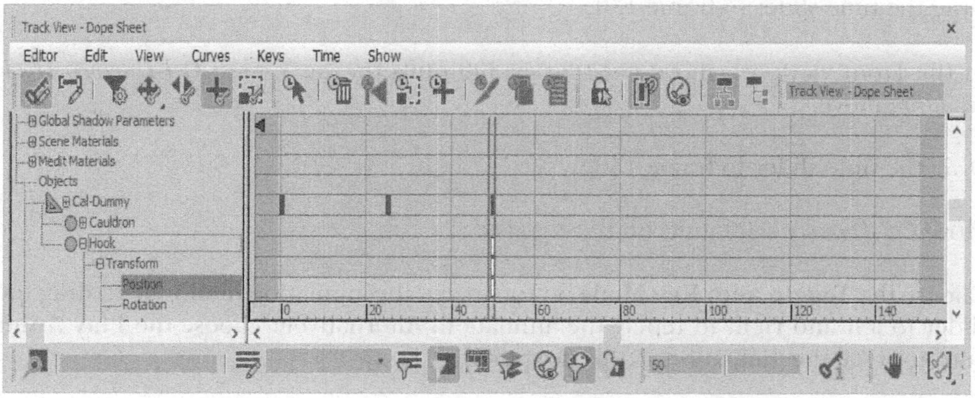

***Figure 15-11** The new key added to the Hook's **Position** track*

5. Drag the time slider to frame 55 and choose the **Toggle Auto Key Mode** button.

6. Select *Hook* in the Front viewport. Move *Hook* –60 units on the X axis and –60 units on the Y axis; *Hook* is detached from *Cauldron*.

7. Drag the time slider to frame 70. Next, move *Hook* 300 units on the Y axis.

8. Choose **Graph Editors > Saved Track Views > Track View - Dope Sheet** from the menu bar; the **Track View-Dope Sheet** window is displayed with the previous settings active.

9. Choose the **Add/Remove Key** button on the upper left of the **Track View-Dope Sheet** window.

10. Zoom in the window and add a key in the **Position** track at frame 90 for *Hook*. Next, close the **Track View-Dope Sheet** window; the current position of *Hook* is recorded in a key at frame 90.

11. Choose the **Absolute Mode Transform Type-In** button; it is replaced by the **Offset Mode Transform Type-In** button.

12. Drag the time slider to frame 105. Next, in the Front viewport, move *Hook* –300 units on the Y axis.

13. Drag the time slider to frame 110. Next, move *Hook* 60 units on the X axis and 60 units on the Y axis; *Hook* is reattached to *Cauldron*.

14. Choose **Graph Editors > Saved Track Views > Track View-Dope Sheet** from the menu bar; the **Track View-Dope Sheet** window is displayed. Next, select *Cal-Dummy* in the viewport and click on the plus sign on the left of **Cal-Dummy** and then click **Transform** in the **Track View-Dope Sheet** window.

15. Choose the **Add/Remove Key** button and add a key to the **Position** track at frame 110 for *Cal-Dummy*. Next, close the **Track View-Dope Sheet** window.

16. Drag the time slider to frame 130.

17. In the Front viewport, move *Cal-Dummy* 300 units on the Y axis; *Cal-Dummy*, *Hook*, and *Cauldron* move.

18. Drag the time slider to frame 150.

19. Move *Cal-Dummy* –300 units on the X axis.

20 Choose the **Toggle Auto Key Mode** button to exit the animation mode. Next, drag the time slider to left and right to replay the animation. Alternatively, choose the **Play Animation** button to see the animation in the viewport.

Creating a Snapshot

The **Snapshot** tool duplicates an object in its current state, but unlike the **Copy** tool, it does not copy any animation keys associated with the object. The snapshot copy can then be used for complex animation sequences. In this section, you will create a snapshot of *Cauldron* and rotate it by 45 degrees.

1. Drag the time slider to frame 50. Next, select *Cauldron*.

2. Choose **Tools > Snapshot** from the menu bar; the **Snapshot** dialog box is displayed, as shown in Figure 15-12.

3. Make sure the **Single** radio button is selected in the **Snapshot** area. Also, select the **Copy** radio button in the **Clone Method** area, and then choose **OK** to close the dialog box.

   A duplicate copy of *Cauldron* is created at the same location as the original. Therefore, the copy is to be moved to frame 50 and is named as *Cauldron001*.

Note

*The **Snapshot** dialog box allows you to create multiple copies over a range of frames using the **Range** radio button or a single copy using the **Single** radio button.*

4. Drag the time slider to frame 0.

There are now two cauldrons in the scene: the original and the copy located at the original's position at frame 50 when the snapshot was created.

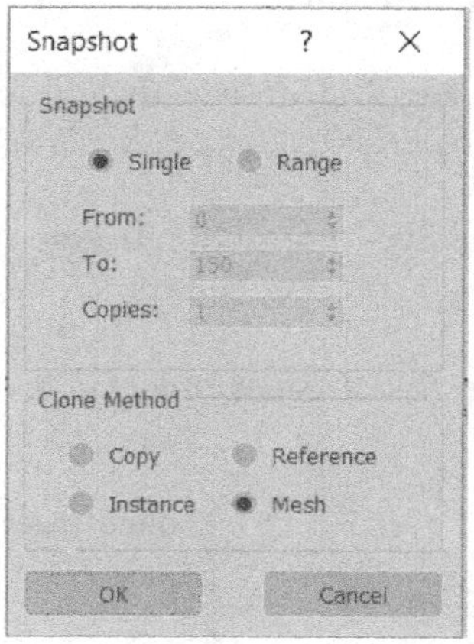

***Figure 15-12** The **Snapshot** dialog box*

5. Invoke the **Select and Link** tool from the **Main Toolbar**.

6. Click and drag *Cauldron001* to *Base Dummy*, as shown in Figure 15-13. Release the left mouse button and invoke the **Select Object** tool to exit the tool; *Cauldron001* is now the child of *Base Dummy* and will follow its movement.

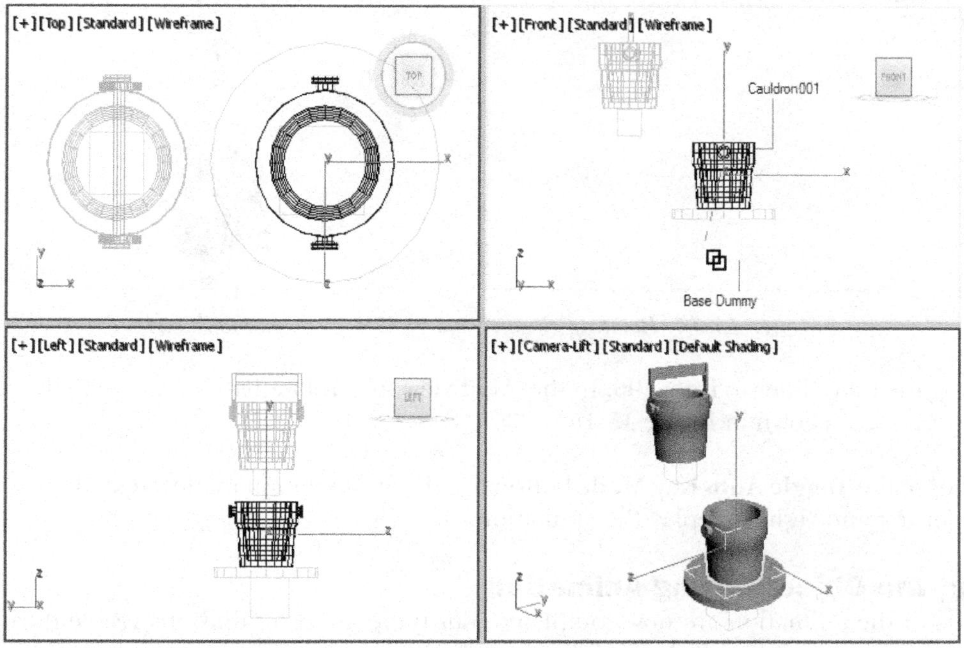

***Figure 15-13** Linking Cauldron with Base-Dummy*

7. Select *Base Dummy* in the viewport.

8. Choose **Graph Editors > Saved Track Views > Track View-Dope Sheet** from the menu bar; the **Track View-Dope Sheet** window is displayed. Next, click on the (+) plus sign at the left of **Base-Dummy** and **Transform**.

9. Make sure the **Add/Remove Key** button is activated and add a key to the **Rotation** track at frame 70 for *Base Dummy*. Next, close the **Track View - Dope Sheet** window.

10. Drag the time slider to frame 80. Choose the **Toggle Auto Key Mode** button to enter the animation mode.

11. In the Front viewport, rotate *Base Dummy* object – 45 degrees on the Z axis, as shown in Figure 15-14.

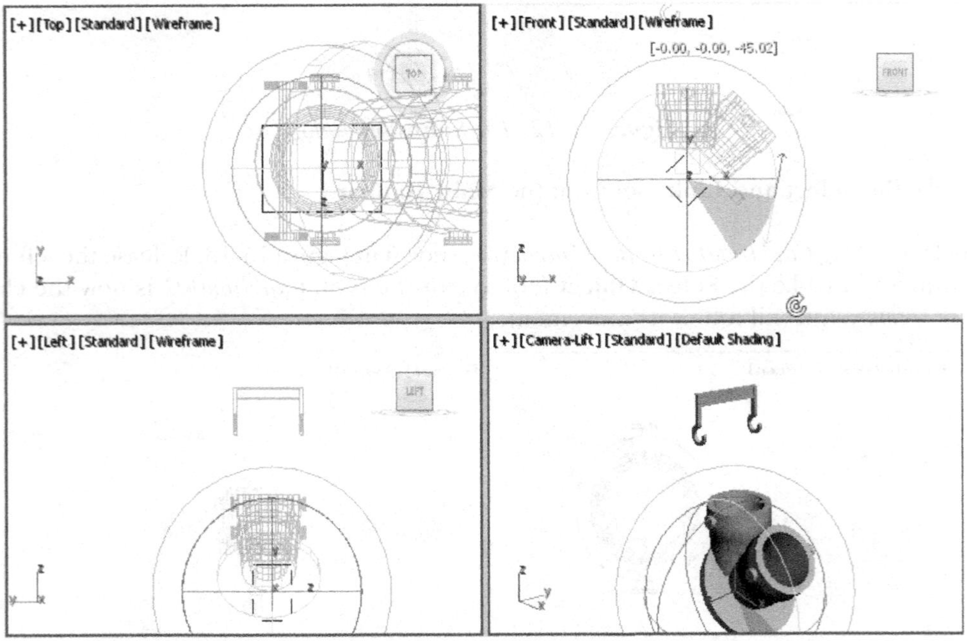

Figure 15-14 *Rotating Base-Dummy -45 degrees on the Z axis*

12. Drag the time slider to frame 90. In the Front viewport, rotate *Base Dummy* +45 degrees on the Z axis, as shown in Figure 15-15.

13. Choose the **Toggle Auto Key Mode** button to exit the animation mode. Next, drag the time slider left and right to replay the animation.

Hiding the Object During Animation

All parts of the animation are now completed, but there are two cauldrons visible during the animation. You must turn off the visibility or hide the cauldrons during different segments of the animation. In this section, you will hide object using the **Track View - Dope Sheet** window. There are two ways to do this. You can use the **Track View - Dope Sheet** window or modify the object's properties.

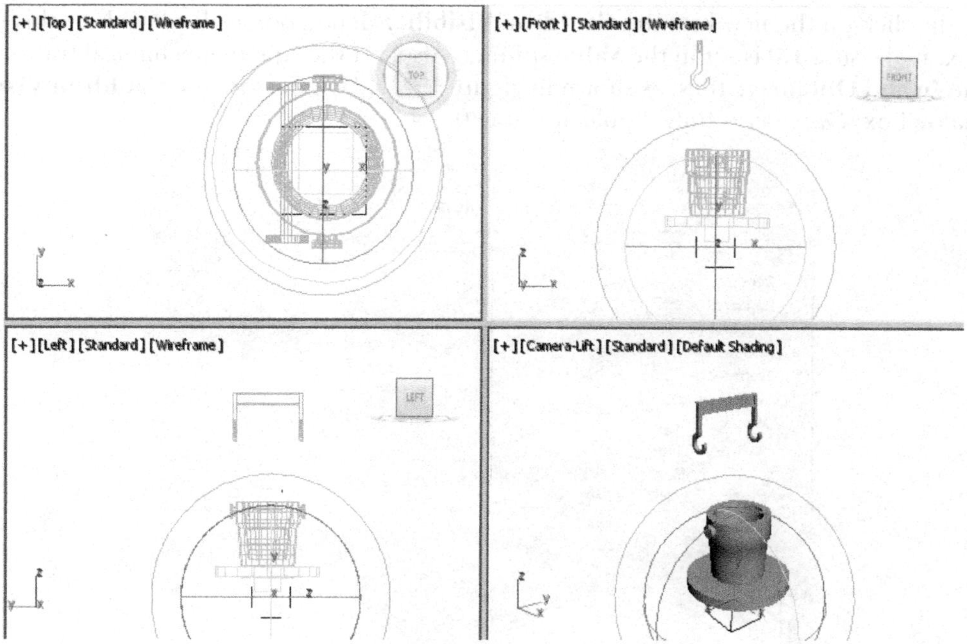

Figure 15-15 *Rotating Base-Dummy +45 degrees on the Z axis*

1. Select *Cauldron* in the viewport. Next, choose **Graph Editors > Saved Track Views > Track View-Dope Sheet** from the menu bar; the **Track View - Dope Sheet** window is displayed.

2. Select **Cauldron** in the **Track View - Dope Sheet** window. Click on the (+) plus sign at the left of *Cauldron*.

3. Choose **Edit > Visibility Tracks > Add** from the menu bar in the **Track View - Dope Sheet** window; a visibility track is added for **Cauldron**.

4. Choose the **Add/Remove Key** button and add a key in the **Visibility** track at frame 0 for *Cauldron*, refer to Figure 15-16.

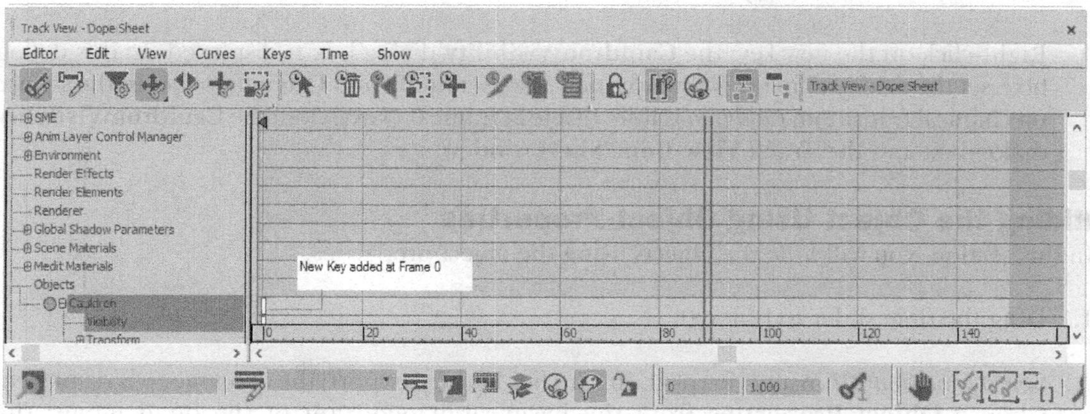

Figure 15-16 *Adding a new key in the* ***Visibility*** *track at frame 0*

5. Right-click on the new key; the **Cauldron\Visibility** dialog box is displayed. In this dialog box, make sure **1.0** is set in the **Value** spinner. Also, set the square mechanical transition in the **In** and **Out** image tiles, as shown in Figure 15-17. Do not close the **Cauldron\Visibility** dialog box; *Cauldron* is fully visible at frame 0.

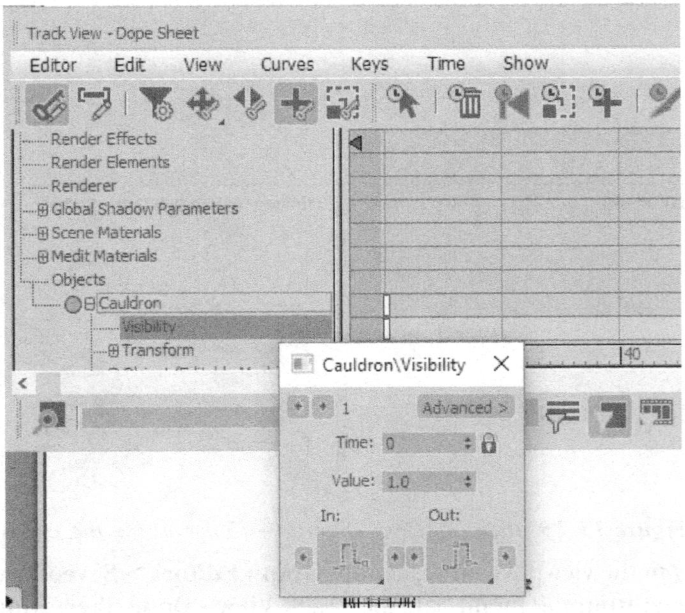

*Figure 15-17 The **Cauldron\Visibility** dialog box*

6. Add a key to the **Visibility** track at frame 50 for *Cauldron*.

7. In the **Cauldron\Visibility** dialog box, set the value **0** in the **Value** spinner. Also, set the **In** and **Out** image tiles to the square mechanical transition; *Cauldron* is invisible at frame 50. Close the **Cauldron\Visibility** dialog box.

8. Add a key in the **Visibility** track at frame 90 for *Cauldron*.

9. Right-click on the new key; the **Cauldron\Visibility** dialog box is displayed. In this dialog box, set the value **1.0** in the **Value** spinner. Also, set the **In** and **Out** image tiles to the square mechanical transition; *Cauldron* is fully visible at frame 0. Next, close the **Cauldron\Visibility** dialog box and the **Track View-Dope Sheet** window.

Hiding the Object Using Object Properties

In this section, you will hide the objects using the object properties.

1. Drag the time slider to frame 0.

2. Select *Cauldron001* in any viewport. Right-click in the viewport; the quad menu is displayed. Choose **Object Properties** from the lower-right quadrant of the quad menu; the **Object Properties** dialog box is displayed, as shown in Figure 15-18.

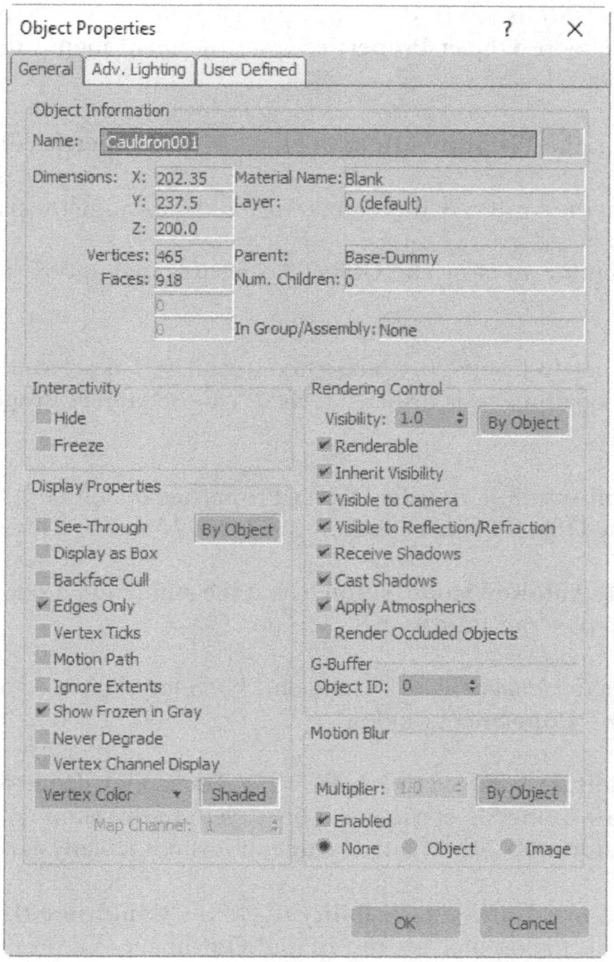

***Figure 15-18** The **Object Properties** dialog box*

3. In this dialog box, make sure the **By Object** button is activated; the options in this area are activated.

Note

*If the **By Layer** button is displayed in the **Object Properties** dialog box, you need to click on this button so that the **By Object** button is displayed.*

4. Set the value **0** in the **Visibility** spinner. Next, choose the **OK** button to close the **Object Properties** dialog box; *Cauldron001* is invisible at this frame (frame 0). This also adds a visibility track in the **Track View - Dope Sheet** window.

5. Drag the time slider to frame 50 and then choose the **Toggle Auto Key Mode** button to enter the animation mode.

6. Make sure *Cauldron001* is selected. Next, right-click in the viewport; the quad menu is displayed. Next, choose **Object Properties** from the quad menu; the **Object Properties** dialog box is displayed.

7. Set the value **1.0** in the **Visibility** spinner of the **Object Properties** dialog box.

   You will notice that the corners of the spinner turn red indicating the parameter is animated.

8. Choose the **OK** button to close the **Object Properties** dialog box; *Cauldron001* becomes visible at frame 50.

9. Drag the time slider to frame 90. Make sure *Cauldron001* is selected. Next, right-click in the viewport; the quad menu is displayed. Next, choose **Object Properties** from the quad menu.

10. Set **0** in the **Visibility** spinner of the **Object Properties** dialog box. Next, choose the **OK** button to close the **Object Properties** dialog box; *Cauldron001* is invisible at frame 90.

11. Choose the **Toggle AutoKey Mode** button to exit the animation mode. Drag the time slider left and right to replay the animation.

    You will notice that *Cauldron001* fades in and out. Therefore, you need to adjust the transition in the **Track View - Dope Sheet** window.

12. Choose **Graph Editors > Saved Track Views > Track View-Dope Sheet** from the menu bar; the **Track View-Dope Sheet** window is displayed. In this window, click on the plus sign on the left of **Cauldron001** to expand the tree, if it is not already expanded.

13. Right-click on the first key in the **Visibility** track; the **Cauldron001/Visibility** dialog box is displayed. In this dialog box, set the **In** and **Out** image tiles to the square mechanical transition.

14. Repeat this for the visibility keys at frame 50 and frame 90. Close the **Cauldron/Visibility** dialog box and the **Track View-Dope Sheet** window.

15. Drag the time slider left and right to preview the animation or choose the **Play Animation** button.

    Notice that *Cauldron001* no longer fades in and out. The visibility changes for both cauldron objects are instantaneous which gives an illusion of only one cauldron in the scene.

Previewing the Animation

In this section, you will preview the animation. The preview animation is a low resolution/low color format used to quickly verify that the animation is working correctly.

1. Make sure the Camera-Lift viewport is activated.

2. Choose **Tools > Preview - Grab Viewport > Create Preview Animation** from the menu bar; the **Make Preview** dialog box is displayed, as shown in Figure 15-19.

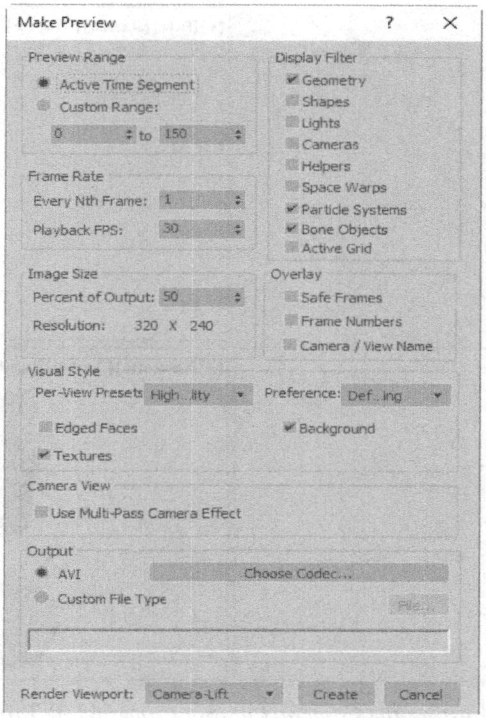

Figure 15-19 The ***Make Preview*** *dialog box*

3. In this dialog box, select the **Frame Numbers** check box in the **Overlay** area. Next, choose the **Create** button. If the **Video Compression** dialog box is displayed, choose **OK** to accept the default settings.

   When the animation is completed, it is played in Windows Media Player. Notice that the invisible cauldrons will not appear clearly. This is just a representation in the preview rendering. In the final animation, the cauldrons are completely invisible.

Setting the Camera View

In this section, you will set the camera view such that all objects are visible throughout the animation.

1. Choose the **Display** tab in the **Command Panel**. 

2. In the **Hide by Category** rollout, clear the **Cameras** check box; the camera named *Camera-Lift* is displayed in the scene.

3. Invoke the **Zoom Extents All** tool; the scene is zoomed out in the viewports so that the camera and target fit on the screen.

4. Set the Camera-Lift and Camera-Lift.Target in all viewports so that all objects in the scene are visible throughout the animation, refer to Figure 15-20.

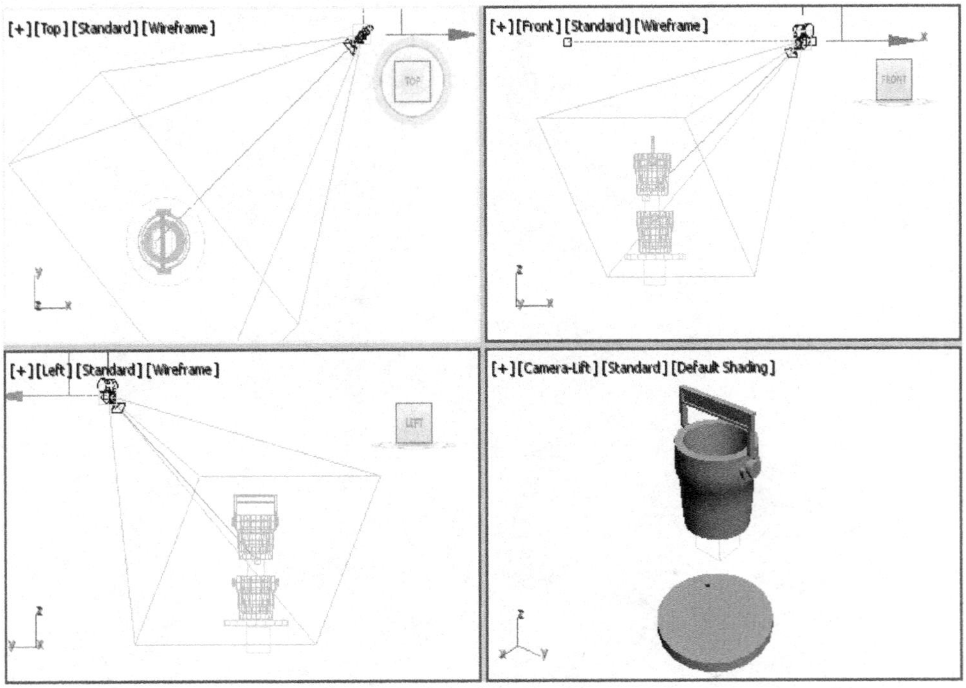

Figure 15-20 *Adjusting the Camera-Lift and Camera-Lift.Target in all viewports*

Next, you will preview the animation to view the effect of moving the camera.

5. Make sure the Camera-Lift viewport is activated. Next, choose **Tools > Preview - Grab Viewport > Create Preview Animation** from the menu bar; the **Make Preview** dialog box is displayed.

6. In this dialog box, choose the **Create** button. If the **Video Compression** dialog box is displayed, choose the **OK** button to accept the default settings.

Saving and Previewing the Animation

In this section, you will create a quick preview of the updated animation to show the movement of the camera target.

1. Change the background color of the scene to white by following the steps given in Tutorial 1 of Chapter 2.

2. Choose **Save** from the **File** menu.

3. Make sure the Camera-Lift viewport is activated.

4. Choose **Tools > Preview - Grab Viewport > Create Preview Animation** from the menu bar; the **Make Preview** dialog box is displayed.

5. Choose the **Create** button. If the **Video Compression** dialog box is displayed, choose the **OK** button to accept the default settings.

   When the animation is completed, it is played in the Windows Media Player.

Tutorial 2

In this tutorial, you will create a basket ball animation, as shown in Figure 15-21.

(Expected time: 35 min)

Figure 15-21 *Basketball animation at frame11*

The following steps are required to complete this tutorial:

a. Create the project folder.
b. Open the file.
c. Animate the basketball.
d. Fine-tune the animation.
e. Save and render the animation.

Creating the Project Folder

Create the project folder with the name *c15_tut2* in the *3dsmax 2022* folder, as discussed in earlier chapters.

Opening the File

1. Choose **Open** from the **File** menu; the **Open File** dialog box is displayed. In this dialog box, browse to the location *\Documents\c15_3dsmax_2022_tut* and select the **c15_tut2_start.max** file from it. Choose the **Open** button to open the file, refer to Figure 15-22.

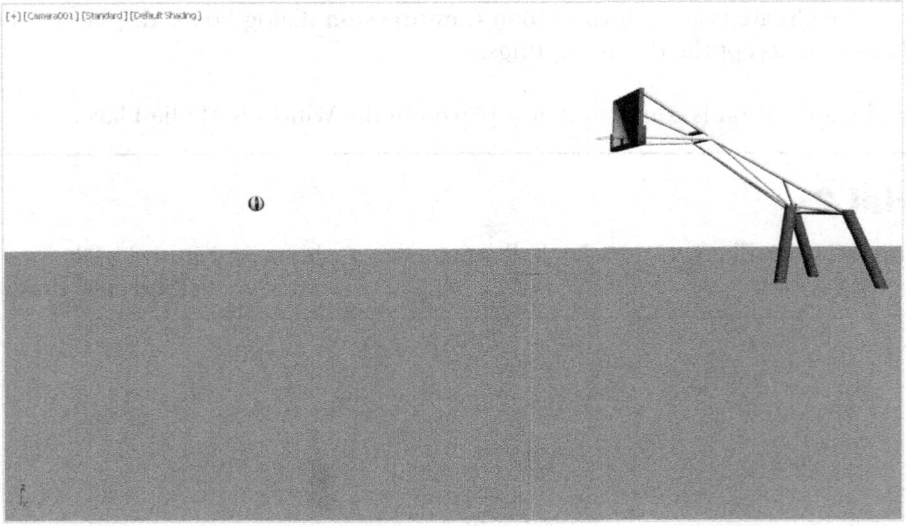

Figure 15-22 The c15_tut2_start.max file

2. Choose **Save As** from the **File** menu; the **Save File As** dialog box is displayed. Browse to the location *\Documents\3dsmax 2022\c15_tut2\scenes*. Save the file with the name *c15tut2.max* at this location.

Animating the Basketball

In this section, you will set the number of frames in the animation and create keyframes.

1. Choose the **Time Configuration** button from the animation playback controls; the **Time Configuration** dialog box is displayed, as shown in Figure 15-23.

2. In the **Animation** area, set **120** in the **Length** spinner and choose the **OK** button; the total number of frames for the animation is set to 120.

3. Make sure the time slider is set at frame 0. Next, choose the **Toggle Auto Key Mode** button.

4. Drag the time slider to frame 5 and select *basketball*. Next, right-click on the **Select and Move** tool in the **Main Toolbar**; the **Move Transform Type-In** dialog box is displayed. In this dialog box, set **555**, **-132**, and **880** in the **X**, **Y**, and **Z** spinners of the **Absolute World** area, refer to Figure 15-24; the *basketball* is positioned, as shown in Figure 15-25 and keyframes are created at frame 0 and 5. Do not close the **Move Transform Type-In** dialog box.

5. In the **Command Panel**, choose the **Display** tab. Next, scroll down and select the **Motion Path** check box from the **Display Properties** area. You will notice that the motion path of *basketball* is visible now, refer to Figure 15-26.

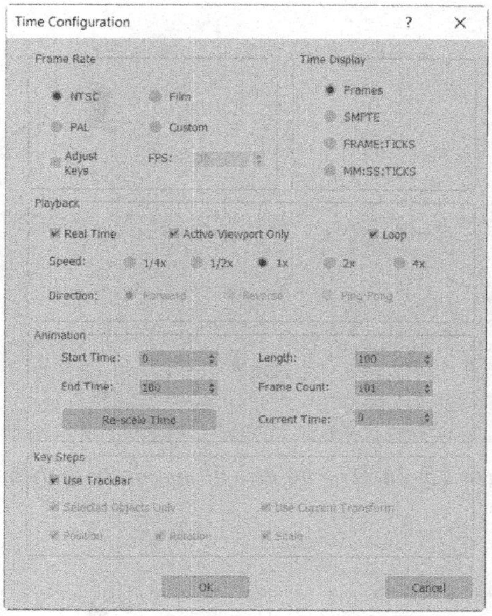

Figure 15-23 *The* ***Time Configuration*** *dialog box*

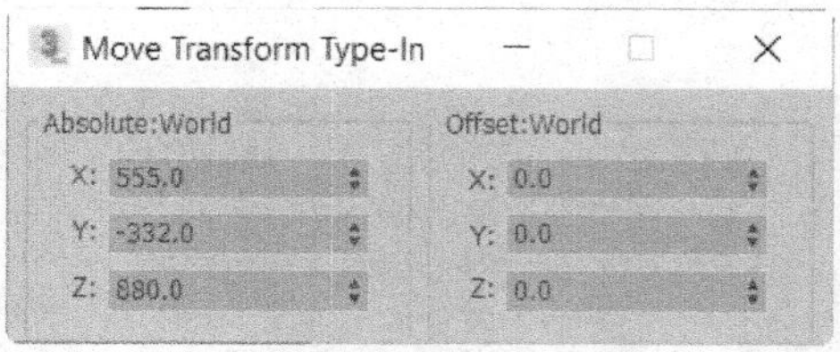

Figure 15-24 *Values in the* ***Move Transform Type-In*** *dialog box*

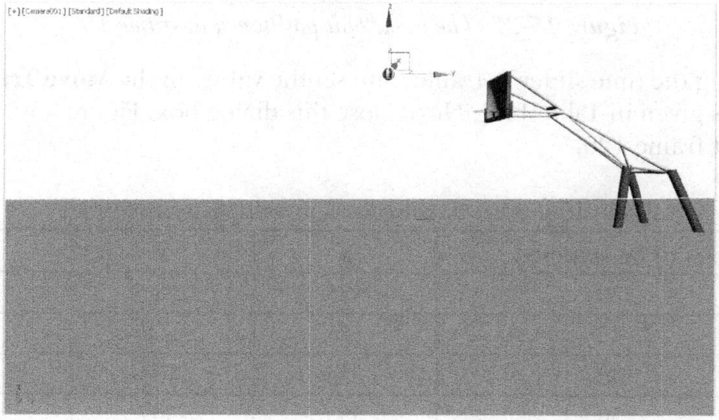

Figure 15-25 *The basketball positioned at frame 5*

6. Drag the time slider to frame 10. In the **Move Transform Type-In** dialog box, set **555**, **147**, and **859** in the **X**, **Y**, and **Z** spinners of the **Absolute World** area; keyframe is created at frame 10 and basketball is positioned, as shown in Figure 15-27.

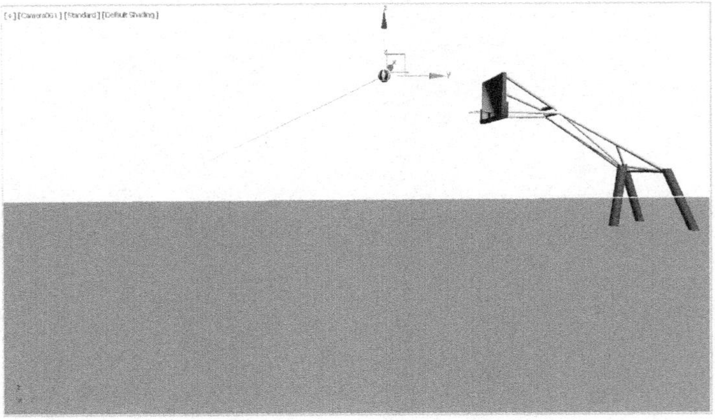

Figure 15-26 *The basketball motion path displayed*

Figure 15-27 *The basketball positioned at frame 10*

7. Similarly, drag the time slider to frames and set the values in the **Move Transform Type-In** dialog box as given in Table 15-1. Next, close this dialog box. Figure 15-28 shows *basketball* positioned at frame 120.

Table 15-1 The transformation values of basketball

Frames to be selected	**X**	**Y**	**Z**
20	555	228.5	503
30	555	111	29
40	555	-134	368
50	555	-347	29
60	555	-508	251
70	555	-647	29
80	555	-780	159
90	555	-1011	29
120	555	-2448	29

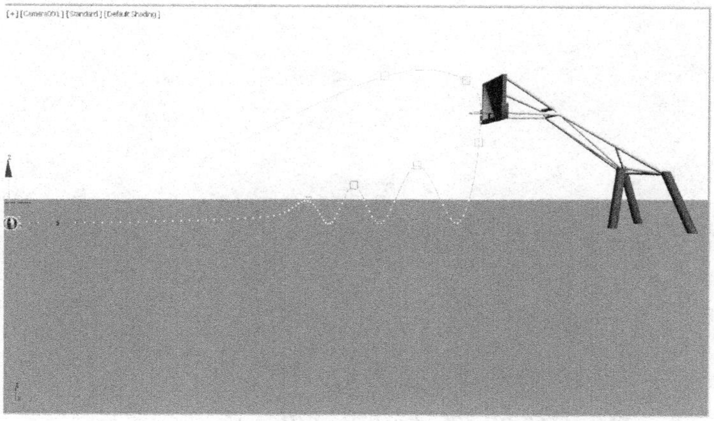

Figure 15-28 *The basketball positioned at frame 120*

8. Choose the **Play Animation** button to play the animation.

   Next, you will rotate *basketball*. To do so, you will use the **Track View Dope Sheet** window.

9. Make sure *basketball* is selected. Next, choose **Graph Editors > Track View - Dope Sheet** from the menu bar; the **Track View Dope Sheet** window is displayed.

10. In this window, click on the plus sign at the left of **basketball** and then click on the plus sign at the left of **Transform**. Notice that currently position keys are assigned to *basketball*.

11. Select the **Rotation** track from the list at the left in the **Track View Dope Sheet** window, refer to Figure 15-29.

12. Choose the **Add/Remove Key** button on the upper left of the **Track View - Dope Sheet** window and then click on the **Rotation** track for *basketball* at frame 0; a keyframe is created at frame 0.

13. Right-click on the keyframe on the **Rotation** track at frame 0; the **Basketball/Rotation** dialog box is displayed. In this dialog box, make sure **0** is displayed in the **Angle** spinner and **1** is displayed in the **X** spinner, as shown in Figure 15-29. Next, close this dialog box.

14. Make sure the **Add/Remove Key** button is chosen on the upper left of the **Track View - Dope Sheet** window and then click on the **Rotation** track for *basketball* at frame 30; a keyframe is created at frame 30.

15. Right-click on the keyframe on the **Rotation** track at frame 30; the **Basketball/Rotation** dialog box is displayed. In this dialog box, set **180** in the **Angle** spinner and make sure **1** is displayed in the **X** spinner. Next, close this dialog box.

16. Choose the Move keys button and then select the keyframe on the **Rotation** track at frame 0. Next, press and hold the SHIFT key and drag the keyframe to frame 51, refer to Figure 15-30; keyframe at frame 0 is copied to frame 51.

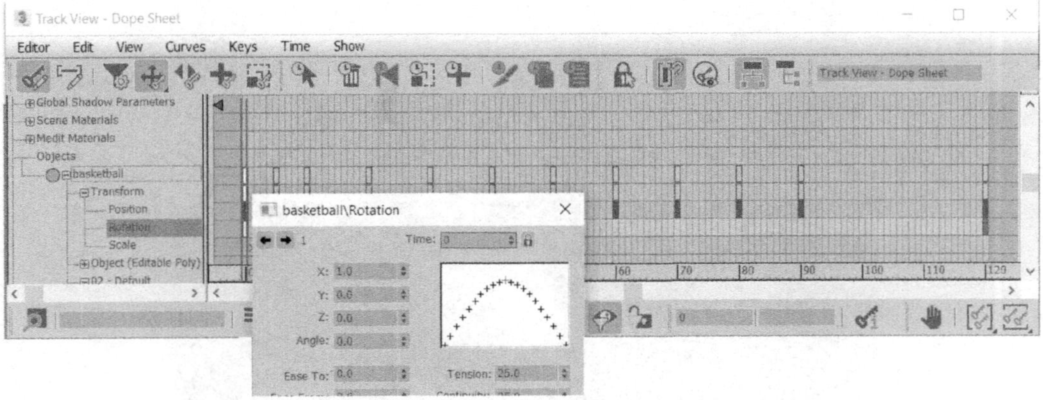

*Figure 15-29 The **basketball/Rotation** dialog box*

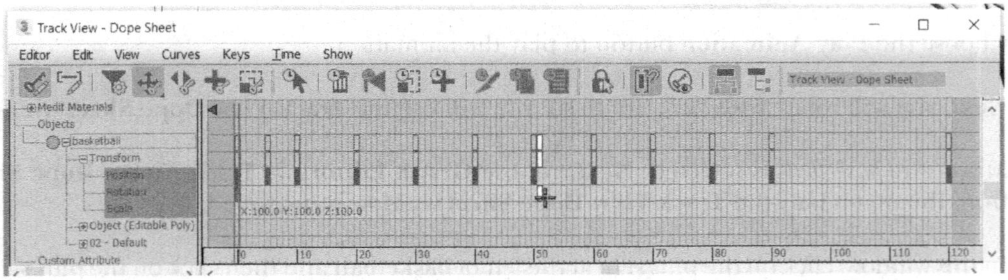

Figure 15-30 Keyframe dragged to frame 51

17. Similarly, copy the keyframe on the **Rotation** track at frame 0 to frames 71 and 90. Next, copy the keyframe on the rotation track at frame 30 to frame 80.

18. Choose the **Add/Remove Key** button on the upper left of the **Track View - Dope Sheet** window and then click on the **Rotation** track for *basketball* at frame 50; a keyframe is created at frame 50.

19. Right-click on the keyframe on the **Rotation** track at frame 50; the **Basketball/Rotation** dialog box is displayed. In this dialog box, set **360** in the **Angle** spinner and make sure **1** is displayed in the **X** spinner. Next, close this dialog box. Now, choose the Move Keys button and copy the keyframe on the **Rotation** track at frame 50 to frames 70 and 120.

    Next, you will create the squash and stretch effect on *basketball* when it touches Floor.

20. Select **Scale** from the list in the **Track View Dope Sheet** window. Next, choose the **Add/ Remove Key** button on the upper left of the **Track View - Dope Sheet** window and then click on the **Scale** track for *basketball* at frame 0; a keyframe is created at frame 0, refer to Figure 15-31.

    Next, you will copy the keyframe at frame 0 to frame 29 so that *basketball* is not scaled till it touches the floor.

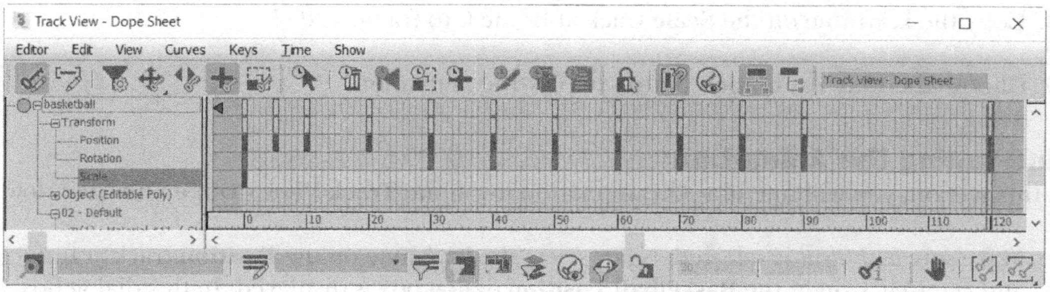

Figure 15-31 *Keyframe created at frame 0*

21. Choose the **Move Keys** button on the upper left of the **Track View - Dope Sheet** window. Next, select the keyframe on the **Scale** track at frame 0 and copy this keyframe to frame 29 as discussed earlier.

22. Make sure the **Add/Remove Key** button is chosen on the upper left of the **Track View - Dope Sheet** window and then click on the **Scale** track for *basketball* at frame 30; a keyframe is created at frame 30.

23. Similarly, create keyframe on the **Scale** track at frame 40.

24. Right-click on the keyframe on the **Scale** track at frame 30; the **Basketball/Scale** dialog box is displayed. Set **85** in the **Z Value** spinner, refer to Figure 15-32, and close the dialog box.

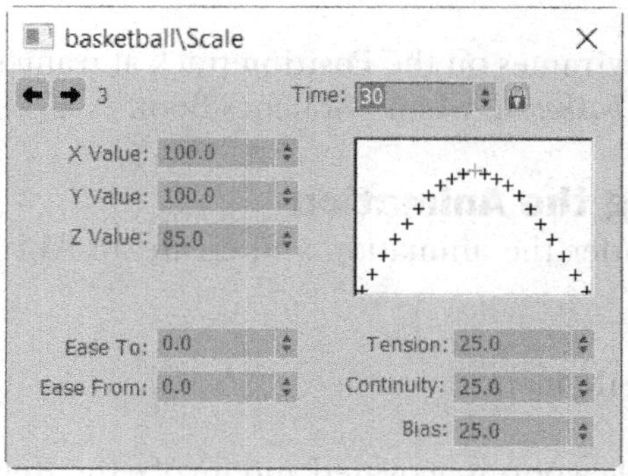

Figure 15-32 *The* ***basketball/Scale*** *dialog box*

25. Right-click on the keyframe on the **Scale** track at frame 40; the **Basketball/Scale** dialog box is displayed. Set **100** in the **Z Value** spinner and **85** in the **Y Value** spinner. Next, close the dialog box.

26. Choose the **Move Keys** button on the upper left of the **Track View - Dope Sheet** window. Next, select the keyframe on the **Scale** track at frame 30. Now, press and hold the SHIFT key and drag the keyframe to frame 50; the keyframe at frame 30 is copied to frame 50.

27. Similarly, copy the keyframe on the **Scale** track at frame 30 to frames 70 and 90. Also, copy the keyframe on the **Scale** track at frame 40 to frames 60 and 80.

28. Copy the keyframe on the **Scale** track at frame 0 to frame 120.

29. Choose the **Play Animation** button to play the animation.

Fine-tuning the Animation

In this section, you will fine tune the animation using the **Track View - Dope Sheet** window.

1. Make sure *basketball* is selected. Next, select the keyframe on the **Position** track at frame 30 and right-click on it; the **Basketball/Position** dialog box is displayed. In this dialog box, set **0** in the **Continuity** spinner, refer to Figure 15-33.

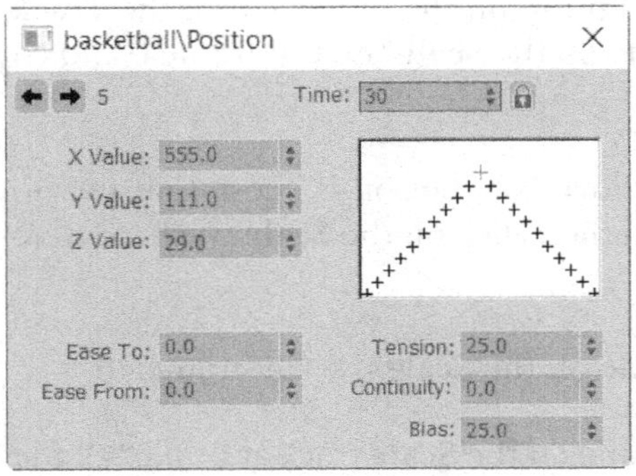

Figure 15-33 *The **basketball/Position** dialog box*

2. Repeat step 1 for the keyframes on the **Position** track at frames 50, 70, and 90. As a result, weight will be added to *basketball* when it touches floor.

Saving and Rendering the Animation

In this section, you will render the animation and create an AVI file. Next, you will play the rendered animation.

1. Choose **Save** from the **File** menu.

2. Make sure the Camera viewport is activated and invoke the **Render Setup** tool; the **Render Setup: Default Scanline Renderer** dialog box is displayed.

3. In the **Time Output** area, select the **Active Time Segment** radio button.

4. In the **Render Output** area, choose the **Files** button; the **Render Output File** dialog box is displayed. As the project folder is already set, the *renderoutput* folder of this project is displayed in the **Save in** text box. Select **AVI File (*.avi)** from the **Save as type** drop-down list.

5. In the **File name** text box, enter **c15tut2render** and choose the **Save** button; the path of the file is displayed in the **Render Output** area of the **Render Setup** dialog box.

If the **AVI File Compression Setup** dialog box is displayed, choose the **OK** button to close it.

6. Make sure that the **Save File** check box is selected in the **Render Output** area. Now, choose the **Render** button to render the animation and close the **Render Setup** dialog box.

   It will take several minutes to render the animation frame by frame. After the rendering is complete, close the render window.

7. Browse to the *renderoutput* folder of the *c15_tut2* project and double-click on the **c15tut2render.avi** file; the AVI file is played in the Windows Media Player.

Self-Evaluation Test

Answer the following questions and then compare them to those given at the end of this chapter:

1. Which of the following tools duplicates an object in its current state, but does not copy any animation key associated with the object?

   (a) **Copy** (b) **Dummy**
   (c) **Snapshot** (d) None of these

2. An object can be linked to a __________ object, and it will then follow the movement of the dummy object.

3. A dummy object will not be visible on __________.

4. The __________ window can be used to create a position key.

5. While adjusting the key information of a key in an object's visibility track, the setting of __________ in the **Value** spinner will make the object completely visible.

6. An object can be made invisible by creating a visibility track in **Track View - Dope Sheet** window, or by adjusting the object's __________.

7. When a camera is animated, both __________ and __________ can be moved or rotated.

8. On selecting __________ in the **Display Properties** rollout of the **Command Panel**, camera's path is displayed in the viewport.

9. Frames that have animation keys for the camera are indicated in the trajectory by white boxes. (T/F)

Review Questions

Answer the following questions:

1. Which of the following dialog boxes is used to set total number of frames in an animation?

   (a) **Set Controller Defaults** (b) **Time Configuration**
   (c) **Object Properties** (d) None of these

2. The __________ objects are used to control the movement of other objects by linking them, but they are not visible in rendering.

3. The __________ tool is used to link one object with another.

4. The __________ button is used to add keys to **Track View - Dope Sheet**.

5. The __________ option verifies the current state of animation.

6. The **Snapshot** tool is similar to the **Copy** command. (T/F)

EXERCISE

The rendered sequence of the scene used in the following exercise can be accessed by downloading the *c15_3dsmax_2022_exr.zip* from *www.cadcim.com*. The path of the file is as follows: *Textbooks > Animation and Visual Effects > Autodesk 3ds Max > Autodesk 3ds Max 2022 for Beginners: A Tutorial Approach*

Exercise 1

Extract the contents of the *c15_3dsmax_2022_exr.zip* and then open *c15_exr01_start.max*. Next, animate the vise, refer to Figures 15-34 through 15-37. **(Expected time: 20 min)**

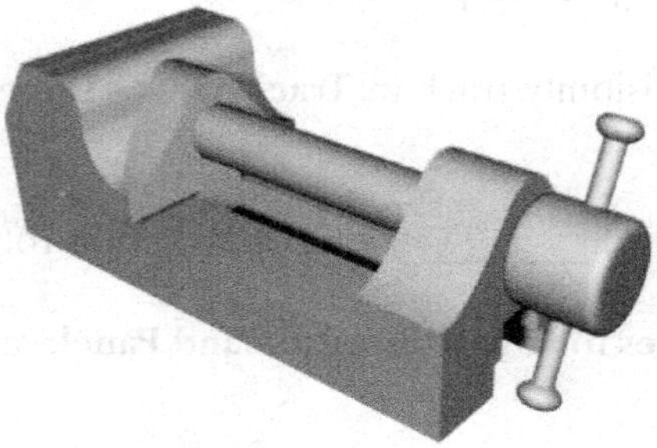

Figure 15-34 *Animation at frame 50*

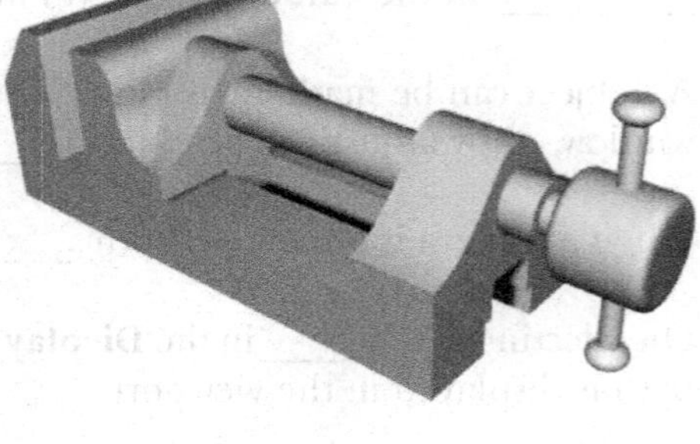

Figure 15-35 *Animation at frame 75*

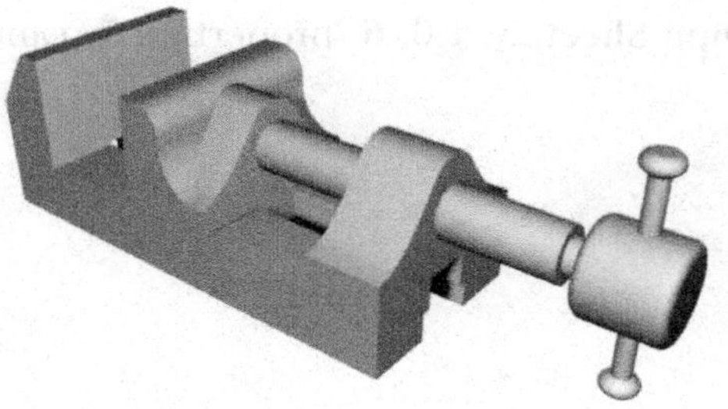

Figure 15-36 *Animation at frame 125*

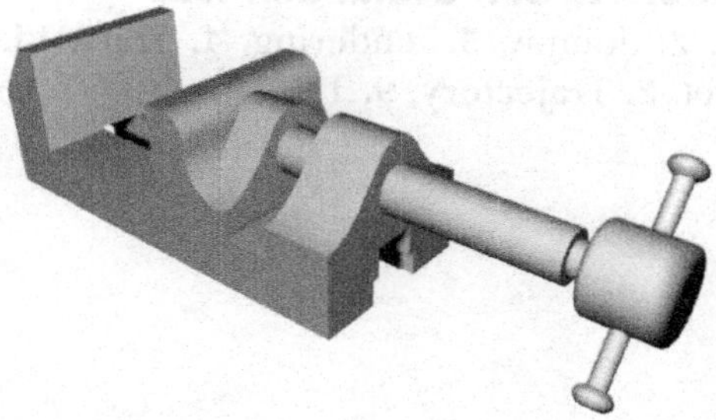

Figure 15-37 *Animation at frame 150*

Answers to Self-Evaluation Test

1. c, **2.** dummy, **3.** rendering, **4. Track View - Dope Sheet**, **5.** 1.0, **6.** properties, **7.** camera, target, **8. Trajectory**, **9.** T

Chapter 16

Arnold Materials, Lights, and Rendering

Learning Objectives

After completing this chapter, you will be able to:
- *Use arnold materials*
- *Set up arnold lights*
- *Modify arnold render settings*

INTRODUCTION

Arnold is a high-end renderer that provides photorealistic renders. It is compatible with physical materials, physical camera, and photometric lights. In this chapter, you will learn to use arnold materials, set up various arnold lights, and modify render settings for a scene.

TUTORIAL

Before starting the tutorial, you need to download the *c16_3dsmax_2022_tut.zip* file from *www.cadcim.com*. The path of the file is as follows: *Textbooks > Animation and Visual Effects > Autodesk 3ds Max > Autodesk 3ds Max 2022 for Beginners: A Tutorial Approach*

Extract the contents of the zip file and save them in the *Documents* folder.

Tutorial 1

In this tutorial, you will assign arnold materials to the objects in the scene, illuminate the scene using arnold quad lights, modify the render settings, and render the scene using arnold renderer, refer to Figure 16-1. You will also simulate sunlight in the scene using arnold distant light, refer to Figure 16-2. **(Expected time: 25 min)**

Figure 16-1 *The illuminated scene with arnold quad lights*

Figure 16-2 *Simulated sunlight with arnold distant light*

The following steps are required to complete this tutorial:

a. Create the project folder.
b. Open the file.
c. Assign arnold materials to the objects.
d. Illuminate the scene using arnold quad light.
e. Simulate sunlight using arnold distant light.
f. Save and render the scene.

Creating the Project Folder

Create the project folder with the name *c16_tut1* in the *3dsmax 2022* folder, as discussed in earlier chapters.

Opening the File

1. Choose **Open** from the **File** menu; the **Open File** dialog box is displayed. In this dialog box, browse to the location *\Documents\c16_3dsmax_2022_tut* and select the **c16_tut1_start.max** file from it. Choose the **Open** button to open the file, refer to Figure 16-3.

Figure 16-3 *The c16_tut1_start.max file*

2. Choose **Save As** from the **File** menu; the **Save File As** dialog box is displayed. Browse to the location *\Documents\3dsmax 2022\c16_tut1\scenes*. Save the file with the name *c16tut1.max* at this location.

Assigning Arnold Materials to the Objects

In this section, you will switch to arnold renderer and assign arnold materials to the objects in the scene.

1. Choose the **Render Setup** tool from the **Main Toolbar**. Make sure the **Render Setup: Arnold** dialog box is displayed, refer to Figure 16-4. Next, close this dialog box.

2. Choose the **Material Editor** tool from the **Main Toolbar** or press M; the **Material Editor** dialog box is displayed.

3. In this dialog box, select first material sample slot. In the **Material Name** drop-down list, enter **wall**.

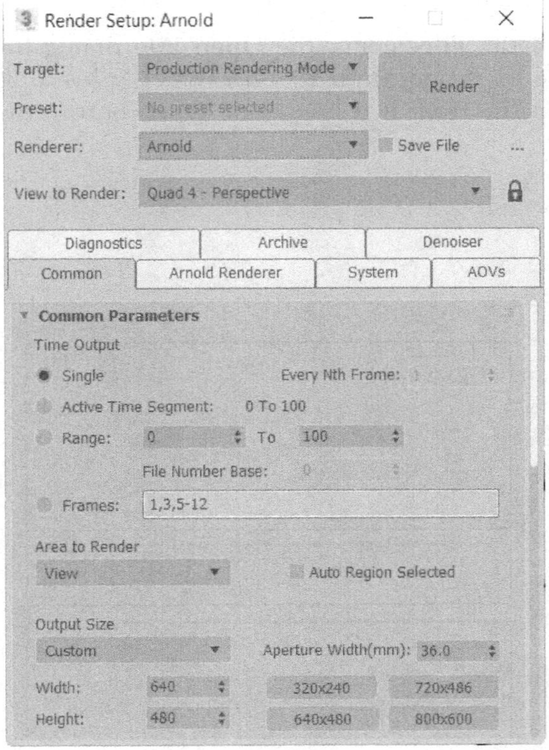

*Figure 16-4 The **Render Setup: Arnold** dialog box*

4. Choose the **Physical Material** button available in the **Material Editor** dialog box; the **Material/Map Browser** dialog box is displayed.

5. In the **Material/Map Browser** dialog box, select **Standard Surface** from the **Materials > Arnold > Surface** rollout and choose **OK**; the **Standard Surface** material is displayed in the **Material Editor** dialog box.

6. In the **Maps** rollout, choose the **No Map** button next to **Base Color**; the **Material/Map Browser** dialog box is displayed. In this dialog box, select **Bitmap** from the **Maps > General** rollout and then choose the **OK** button; the **Select Bitmap Image File** dialog box is displayed.

7. As the project folder is already set, the *images* folder is displayed in the **Look in** drop-down list of this dialog box. Select the file **wall.jpg** and then choose the **Open** button; the image is displayed in the selected sample slot.

8. Choose the **Go to Parent** button. Figure 16-5 shows the **Material Editor** dialog box with the **wall** material.

9. Select *wall* from the Scene Explorer. Next, choose the **Assign Material to Selection** and **Show Shaded Material in Viewport** buttons from the **Material Editor** dialog box; the **wall** material is assigned to *wall*, as shown in Figure 16-6.

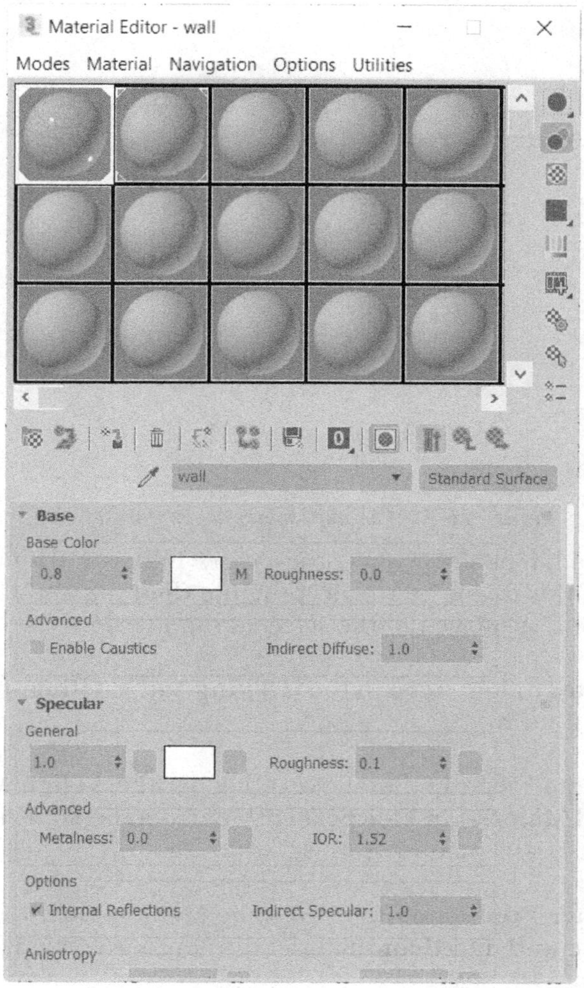

***Figure 16-5** The **Material Editor** dialog box with the **wall** material displayed*

Next, you will assign material to floor.

10. In the **Material Editor** dialog box, select the second material sample slot. In the **Material Name** drop-down list, enter **floor**.

11. Choose the **Physical Material** button available in the **Material Editor** dialog box; the **Material/Map Browser** dialog box is displayed. In this dialog box, select **Standard Surface** from the **Materials > Arnold > Surface** rollout and choose **OK**; the **Standard Surface** material is displayed in the **Material Editor** dialog box.

12. In the **Maps** rollout, choose the **No Map** button next to **Base Color**; the **Material/Map Browser** dialog box is displayed. In this dialog box, select **Bitmap** from the **Maps > General** rollout and then choose the **OK** button; the **Select Bitmap Image File** dialog box is displayed.

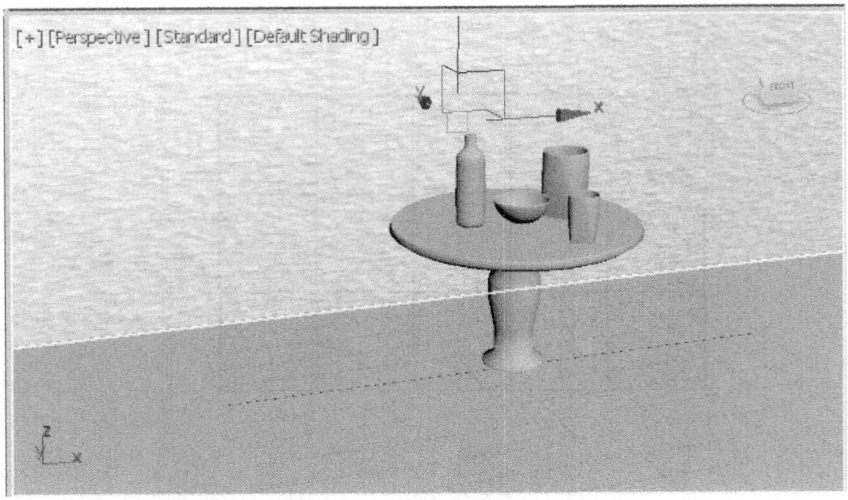

Figure 16-6 *The **wall** material assigned to wall*

13. As the project folder is already set, the *images* folder is displayed in the **Look in** drop-down list of this dialog box. Select the file **floor.jpg** and then choose the **Open** button; the image is displayed in the selected sample slot.

14. In the **Co-ordinates** rollout, set **5** in the **U Tiling** and **V Tiling** spinners. Next, choose the **Go to Parent** button.

15. Select *floor* from the Scene Explorer. Next, choose the **Assign Material to Selection** and **Show Shaded Material in Viewport** buttons from the **Material Editor** dialog box; the **floor** material is assigned to *floor*.

16. Choose the **Render Production** tool from the Main Toolbar; the rendered image is displayed with the **wall** and **floor** material displayed, refer to Figure 16-7.

    Next, you will assign wood material to the table in the scene.

17. In the **Material Editor** dialog box, select the third material sample slot. In the **Material Name** drop-down list, enter **table**.

18. Choose the **Physical Material** button available in the **Material Editor** dialog box; the **Material/Map Browser** dialog box is displayed. In this dialog box, select **Standard Surface** from the **Materials > Arnold > Surface** rollout and choose **OK**; the **Standard Surface** material is displayed in the **Material Editor** dialog box.

19. In the **Maps** rollout, choose the **No Map** button next to **Base Color**; the **Material/Map Browser** dialog box is displayed. In this dialog box, select **Bitmap** from the **Maps > General** rollout and then choose the **OK** button; the **Select Bitmap Image File** dialog box is displayed.

20. As the project folder is already set, the *images* folder is displayed in the **Look in** drop-down list of this dialog box. Select the file **wood.jpg** and then choose the **Open**

button; the image is displayed in the selected sample slot. Choose the **Show Shaded Material in Viewport** button in the **Material Editor** dialog box.

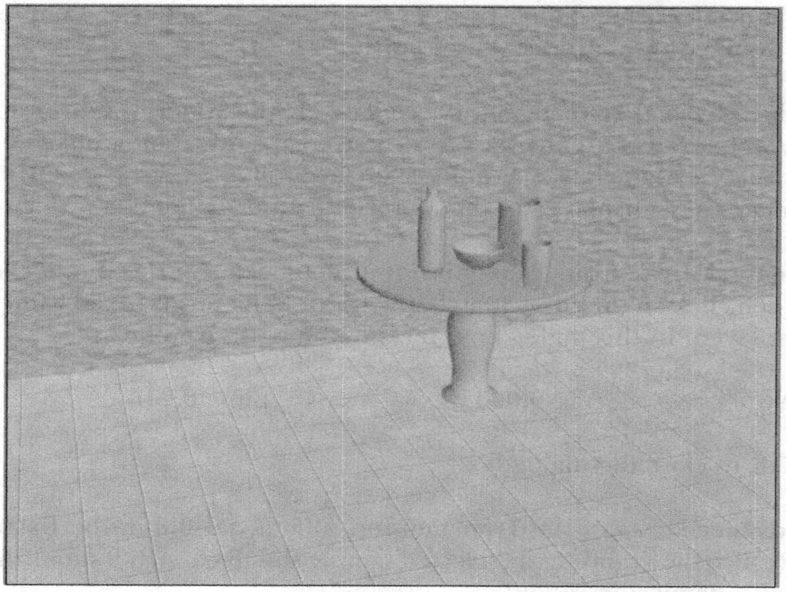

Figure 16-7 *The rendered image with the **wall** and **floor** material displayed*

21. Select *table* from the Scene Explorer. Next, choose the **Assign Material to Selection** and **Show Shaded Material in Viewport** buttons from the **Material Editor** dialog box; the **table** material is assigned to *table*.

22. Choose the **Render Production** tool from the Main Toolbar; the rendered image is displayed with the **table** material displayed, refer to Figure 16-8.

Figure 16-8 *The rendered image with the **table** material displayed*

Next, you will assign colored glass material to the bottle in the scene.

23. In the **Material Editor** dialog box, select the fourth material sample slot. In the **Material Name** drop-down list, enter **bottle**.

24. Choose the **Physical Material** button available in the **Material Editor** dialog box; the **Material/Map Browser** dialog box is displayed. In this dialog box, select **Standard Surface** from the **Materials > Arnold > Surface** rollout and choose **OK**; the **Standard Surface** material is displayed in the **Material Editor** dialog box.

25. In the **General** area of the **Transmission** rollout, set **1** in the spinner located at the left and then click on the color swatch; the **Color Selector: transmission_color** dialog box is displayed. In this dialog box, set the values as given next.

    Red: **0.996** Green: **0.004** Blue: **0.216**

    Choose **OK** to close the dialog box.

26. In the **Advanced** area of the **Transmission** rollout, set **0.6** in the **Extra Roughness** spinner. Notice the grains in the selected sample slot. Figure 16-9 shows the parameters set in the **Transmission** rollout.

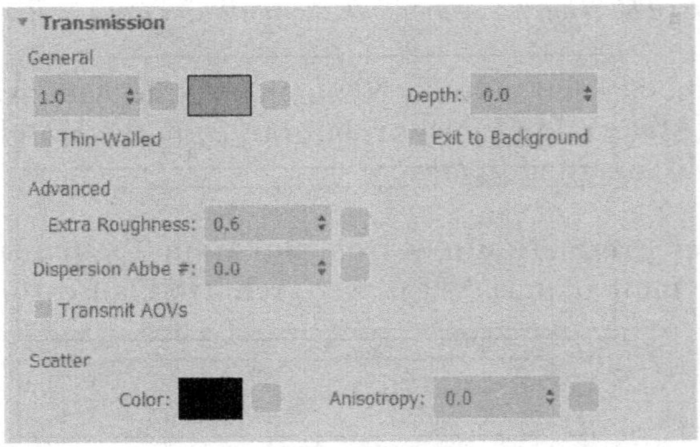

Figure 16-9 *Parameters set in the **Transmission** rollout*

27. Select *bottle* in the viewport. Next, choose the **Assign Material to Selection** and the **Show Shaded Material in Viewport** buttons from the **Material Editor** dialog box; the **bottle** material is assigned to *bottle*.

    Next, you will assign ceramic material to the bowl in the scene.

28. In the **Material Editor** dialog box, select the fifth material sample slot. In the **Material Name** drop-down list, enter **bowl**.

29. Choose the **Physical Material** button available in the **Material Editor** dialog box; the **Material/Map Browser** dialog box is displayed. In this dialog box, select **Standard Surface** from the **Materials > Arnold > Surface** rollout and choose **OK**; the **Standard Surface** material is displayed in the **Material Editor** dialog box.

30. In the **Base Color** area of the **Base** rollout, click on the color swatch; the **Color Selector: base_color** dialog box is displayed. In this dialog box, set the values as given next.

    Red: **0.996** Green: **0.004** Blue: **0.192**

31. Right-click on the color swatch in the **Base** rollout; a flyout is displayed. Choose **Copy** from this flyout. Next, set **0.6** in the spinner located at the left in the **General** area of the **Transmission** rollout. Now, right-click on the color swatch in the **Transmission** rollout; a flyout is displayed. Choose **Paste** from this flyout; the color in the color swatch of the **Base** rollout is copied to the color swatch of the **Transmission** rollout.

32. Select *bowl* in the viewport. Next, choose the **Assign Material to Selection** and **Show Shaded Material in Viewport** buttons from the **Material Editor** dialog box; the **bowl** material is assigned to *bowl*.

33. In the **Material Editor** dialog box, select the sixth material sample slot. In the **Material Name** drop-down list, enter **plastic glass**.

34. Choose the **Physical Material** button available in the **Material Editor** dialog box; the **Material/Map Browser** dialog box is displayed. In this dialog box, select **Standard Surface** from the **Materials > Arnold > Surface** rollout and choose **OK**; the **Standard Surface** material is displayed in the **Material Editor** dialog box.

35. In the **Base Color** area of the **Base** rollout, click on the color swatch; the **Color Selector: base_color** dialog box is displayed. In this dialog box, set the values as given next.

    Red: **0.071** Green: **0.106** Blue: **0.839**

36. Set **0.9** in the spinner located at the left in the **General** area of the **Specular** rollout. Next, set **0.35** in the **Roughness** spinner.

37. Select *glass* and *glass1* in the viewport. Next, choose the **Assign Material to Selection** and **Show Shaded Material in Viewport** buttons from the **Material Editor** dialog box; the **plastic glass** material is assigned to both *glass* and *glass1*.

38. In the **Material Editor** dialog box, click and drag the **plastic glass** material to the seventh material sample slot; the **plastic glass** material is copied to the seventh material sample slot.

39. Make sure the seventh material sample slot is selected. In the **Material Name** drop-down list, enter **bottle cap**.

40. In the **Base Color** area of the **Base** rollout, click on the color swatch; the **Color Selector: base_color** dialog box is displayed. In this dialog box, set **0** in the **Red**, **Green**, and **Blue** spinners and close the dialog box; black color is set in the color swatch.

41. Select *cap* in the viewport. Next, choose the **Assign Material to Selection** and **Show Shaded Material in Viewport** buttons from the **Material Editor** dialog box; the **bottle cap** material is assigned to *cap*.

42. Choose the **Render Production** tool from the Main Toolbar; the rendered image is displayed with all the materials, refer to Figure 16-10.

Figure 16-10 *The rendered image with all the materials displayed*

Illuminating the Scene Using Arnold Quad Light

In this section, you will illuminate the scene using arnold quad lights and modify the render settings.

1. Activate the Top viewport. Choose **Create > Lights** in the **Command Panel**. By default, the **Photometric** option is displayed in the drop-down list below it.

2. Select **Arnold** from the drop-down list and then invoke the **Arnold Light** tool from the **Object Type** rollout; various rollouts are displayed in the **Command Panel**. Next, click at a point in theTop viewport; the arnold light is created in the viewport with the name *ArnoldLight001*. Next, align it in all the viewports, as shown in Figure 16-11.

3. Rename this light as *quadlight01* and activate the Perspective viewport. Next, choose the **Render Production** tool from the **Main Toolbar**; the rendered image is displayed.

   You will notice that the render is almost dark. To rectify it, you need to modify the parameters in the **Modify** tab of the **Command Panel**.

4. Make sure *quadlight01* is selected in the viewport. Next, choose the **Modify** tab in the **Command Panel**.

5. In the **Color/Intensity Parameters** rollout, set **4** in the **Intensity** spinner and **12** in the **Exposure** spinner; the value displayed next to **Resulting Intensity** is changed.

   The **Exposure** spinner acts as a intensity multipler for the selected light.

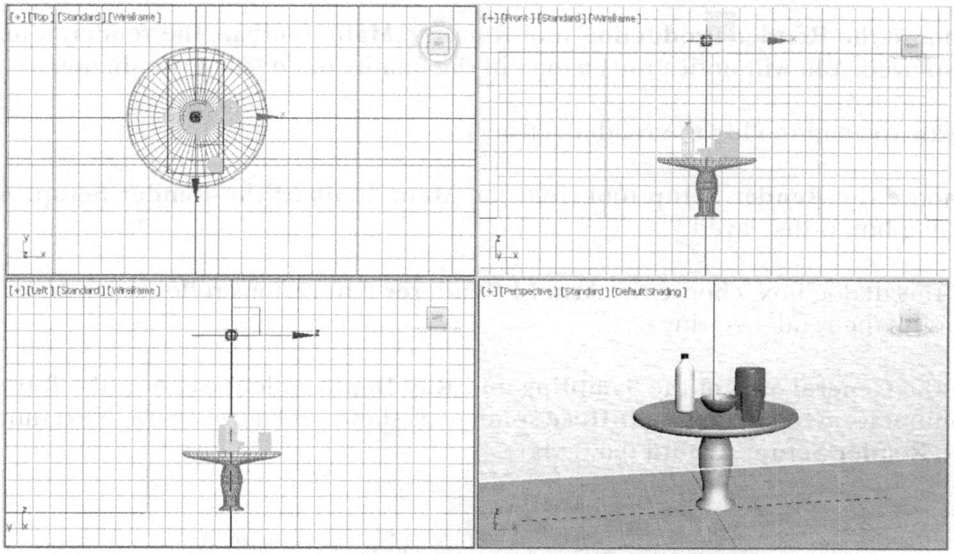

Figure 16-11 *ArnoldLight001 aligned in all the viewports*

6. Choose the **Render Production** tool from the **Main Toolbar**; the rendered image is displayed, as shown in Figure 16-12. Notice that the render is quite improved now.

Figure 16-12 *The rendered image*

Next, you will improve the render further.

7. In the **Rendering** rollout, set **8** in the **Samples** spinner.

   The value in the **Samples** spinner controls the noise quality of the shadows and the specular light.

8. Choose the **Render Production** tool from the **Main Toolbar**; the rendered image is displayed. You will notice the change in the quality of edges of the shadows.

   Next, you will modify the render settings.

9. Choose the **Render Setup** tool from the **Main Toolbar**; the **Render Setup: Arnold** dialog box is displayed.

10. In this dialog box, choose the **Arnold Renderer** tab; various rollouts are displayed to modify the render settings.

11. In the **General** area of the **Sampling and Ray Depth** rollout, set **6** in the **Ray Depth** spinner located next to the **Diffuse** spinner, as shown in Figure 16-13. Next, minimize the **Render Setup: Arnold** dialog box.

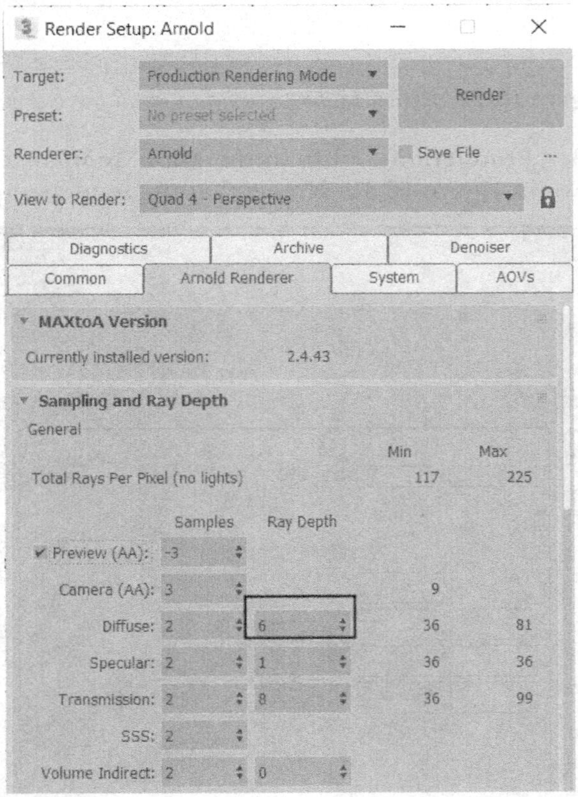

***Figure 16-13** Value set in the **Ray Depth** spinner*

12. Choose the **Render Production** tool from the **Main Toolbar**; the rendered image is displayed. You will notice the change in the overall render quality.

Next, you will add two more quad lights in the scene to remove darkness below the table in the scene.

13. Activate the Front viewport. Make sure *quadlight01* is selected. Next, press and hold the SHIFT key and drag towards left; a copy of *quadlight01* is created. Rename it as *quadlight02*.

14. Align *quadlight02* in all the viewports using the **Select and Move** and **Select and Rotate** tools, as shown in Figure 16-14.

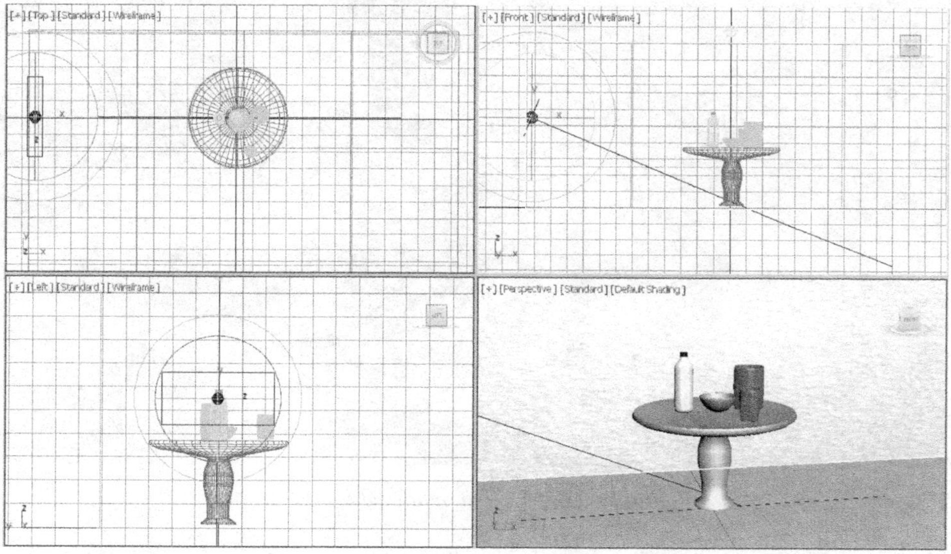

Figure 16-14 *The quadlight02 aligned in all the viewports*

15. Activate the Perspective viewport and then choose the **Render Production** tool from the **Main Toolbar**; the rendered image is displayed, as shown in Figure 16-15.

16. Create one more copy of *quadlight01* and rename it as *quadlight03*. Next, align it in all the viewports, as shown in Figure 16-16.

Next, you will further modify the render settings for global illumination.

17. Maximize the **Render Setup: Arnold** dialog box. In the **General** area of the **Sampling and Ray Depth** rollout, set **6** in the **Diffuse** spinner. Notice the change in the value of the **Min** and **Max** spinners located next to the **Diffuse** spinner.

18. Set **6** in the **Camera(AA)** spinner. You will notice that the values in the **Min** and **Max** spinners located next to **Diffuse**, **Specular**, and **Transmission** spinners are changed.

The value in the **Camera(AA)** spinner acts as a global multipler for improving quality of the rendered image.

19. Save the file with the name *c16tut1_quadlight.max* file.

Figure 16-15 The rendered image

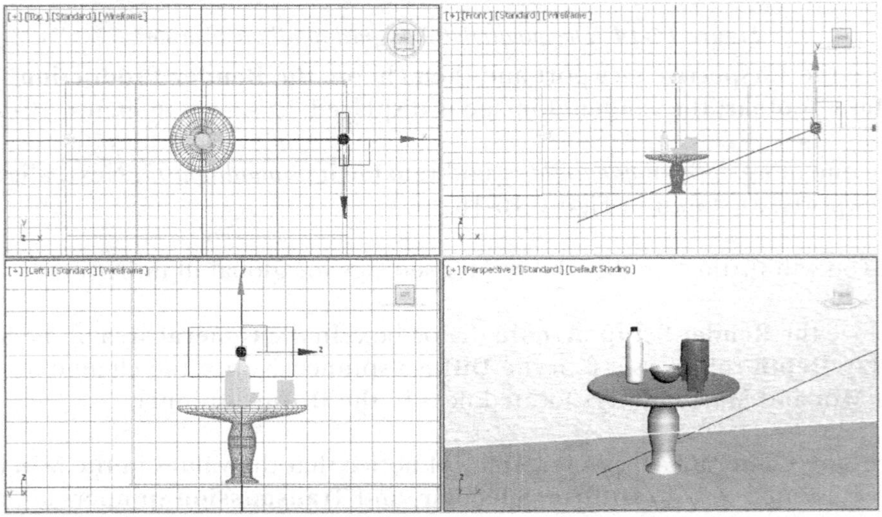

Figure 16-16 The quadlight03 aligned in all the viewports

20. Choose the **Render Production** tool from the **Main Toolbar**; the rendered image is displayed, refer Figure 16-1.

Simulating Sunlight using Arnold Distant Light

1. Delete *quadlight02* and *quadlight03* from the scene. Next, select *quadlight01* in the viewport. Next, make sure the **Modify** tab is chosen.

2. In the **Light Properties** area of the **General** rollout, select the **Targeted** check box; target is added to *quadlight01* with the name *quadlight01.Target*.

3. In the **Emit Light From** area of the **Shape** rollout, select **Distant** from the **Type** drop-down list. Also, rename *quadlight01* as *sunlight*.

4. In the **Intensity** area of the **Color/Intensity** rollout, set **1** in the **Intensity** spinner and **2** in the **Exposure** spinner.

   Next, you will change the color of *sunlight* to light yellow color.

5. In the **Intensity** area of the **Color/Intensity** rollout, click on the color swatch next to **Color**; the **Color Selector: Color** dialog box is displayed. In this dialog box, set the values as given next and then close the dialog box.

   Red: **239** Green: **240** Blue: **193**

6. In the **Rendering** rollout, set **16** in the **Samples** spinner.

7. In the **Shadows** rollout, click on the color swatch next to **Color**; the **Color Selector: Color** dialog box is displayed.

8. In this dialog box, set **87** in the **Red**, **Green**, and **Blue** spinners and then close the dialog box.

Saving and Rendering the Scene

In this section, you will save the scene that you have created and then render it. You can also view the final rendered image of this model by downloading the *c16_3dsmax_2022_rndr.zip* file from *www.cadcim.com*. The path of the file is as follows: *Textbooks > Animation and Visual Effects > Autodesk 3ds Max > Autodesk 3ds Max 2022 for Beginners: A Tutorial Approach*

1. Save the file with the name *c16tut2.max* file.

2. Activate the Perspective viewport. Next, invoke the **Render Production** tool from the **Main Toolbar**; the rendered image is displayed, refer to Figure 16-2.

Self-Evaluation Test

Answer the following questions and then compare them to those given at the end of this chapter:

1. Which of the following rollouts in the **Material Editor** dialog box is significant for creating glass material?

   (a) **Base** (b) **Transmission**
   (c) **Specular** (d) **Subsurface**

2. The __________ tool is used to create various types of arnold lights.

3. The value in the __________ spinner of the **Transmission** rollout is used to display grains in the selected material.

4. The __________ drop-down list in the **Shape** rollout is used to select the type of arnold light.

5. The **Exposure** spinner acts as a intensity multipler for the selected light.(T/F)

Review Questions

Answer the following questions:

1. Which of the following spinners controls the noise quality of the shadows and the specular light?

   (a) **Samples** (b) **Exposure**
   (c) **Density** (d) None of these

2. Which of the following arnold lights is used to simulate sunlight?

   (a) **Quad** (b) **Point**
   (c) **Distant** (d) **Disc**

3. The value displayed next to **Resulting Intensity** changes with the change in the values of the __________ and __________ spinners.

4. The __________ check box in the **General** rollout is used to add target to the arnold light.

5. The value in the **Ray Depth** spinner acts as a global multipler for the rendered image quality. (T/F)

EXERCISE

The rendered sequence of the scene used in the following exercise can be accessed by downloading *c16_3dsmax_2022_exr.zip* from *www.cadcim.com*. The path of the file is as follows: *Textbooks > Animation and Visual Effects > Autodesk 3ds Max > Autodesk 3ds Max 2022 for Beginners: A Tutorial Approach*

Exercise 1

Extract the contents of *c16_3dsmax_2022_exr.zip* and then open *c16_exr01_start.max*. Create a studio light setup using the Arnold lights and render the scene using Arnold renderer, refer to Figures 16-17. **(Expected time: 25 min)**

Figure 16-17 *The final render*

Answers to Self-Evaluation Test

1. b, **2. Arnold Light, 3. Extra Roughness, 4. Type, 5.** T

Chapter 17

Creating a Walkthrough

Learning Objectives

After completing this chapter, you will be able to:

- ***Create a motion path for the camera to follow***
- ***Modify the motion path***
- ***Assign a path constraint controller to the position track of an object***
- ***Adjust camera parameters and animate the camera***
- ***Animate 3ds Max AEC objects***
- ***Render the preview and final walkthrough***

INTRODUCTION

A walkthrough is created by defining a path of the walk by drawing a line from the start point to the end point. A camera is created and assigned to the path which records the animation along the path. A path constraint controller is assigned to the position track of the camera so that the camera follows the path. The path is modified to allow precise control over the animation. In addition, the camera rotation is animated, as it is the movement of 3ds Max AEC objects such as doors and windows. Then, a preview is created before the final animation is rendered.

TUTORIAL

Before starting the tutorial, you need to download the file *c17_3dsmax_2022_tut.zip* from *www.cadcim.com*. The path of the file is as follows: *Textbooks > Animation and Visual Effects > Autodesk 3ds Max > Autodesk 3ds Max 2022 for Beginners: A Tutorial Approach*

Extract the contents of the zip file and save them in the *Documents* folder.

Tutorial 1

In this tutorial, you will create a walkthrough in an architectural setting. You will draw a path and make a camera follow the path, refer to Figure 17-1. **(Expected time: 45 min)**

Figure 17-1 *The final walkthrough at frame 70*

The following steps are required to complete this tutorial:

a. Create the project folder.
b. Open the file.
c. Create the motion path.
d. Specify the length of the animation.
e. Use the walkthrough assistant.

f. Adjust the head tilt angle.
g. Animate objects in the scene.
h. Preview the animation.
i. Render and save the final animation.

Creating the Project Folder

1. Create a new project folder with the name *c17_tut1* at *\Documents\3dsmax2022* and then save the file with the name **c17tut1**, as discussed in Tutorial 1 of Chapter 2.

2. Open the Windows Explorer and then browse to the *c17_3dsmax_2022_tut* folder. Next, copy *floor.jpg*, *wood1.jpg*, and *wall.jpg* to *\Documents\3dsmax2022\c17_tut1\sceneassets\images*.

Opening the File

1. Choose **Open** from the **File** menu; the **Open File** dialog box is displayed. In this dialog box, browse to the location *\Documents\c17_3dsmax_2022_tut* and select the **c17_tut1_start.max** file from it. Choose the **Open** button to open the file, refer to Figure 17-2.

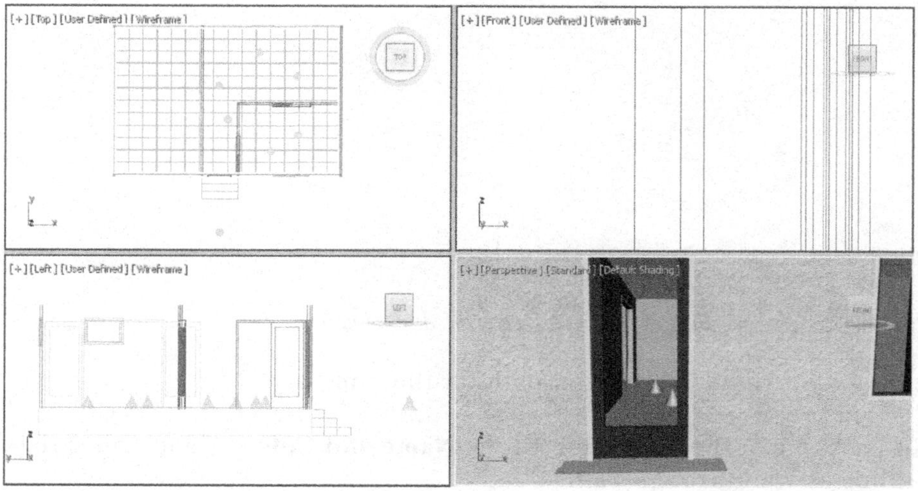

***Figure 17-2** The c17_tut1_start file*

2. Choose **Save As** from the **File** menu; the **Save File As** dialog box is displayed. Browse to the location *\Documents\3dsmax_2022\c17_tut1\scenes*. Save the file with the name *c17tut1.max* at this location.

Creating the Motion Path

In this section, you will create the motion path by using the **Line** tool. This path needs to have round corners to represent the smooth path a person takes as he walks naturally. Also, the path must not lie too close to any of the walls and must pass through the center of door openings.

1. Activate the Top viewport and maximize it by invoking the **Maximize Viewport Toggle** tool. 

2. Invoke the **Zoom Extents All** tool to zoom into the viewports.

3. Choose **Create > Shapes** in the **Command Panel** and make sure **Splines** is selected in the drop-down list below the **Shapes** button, if it is not already selected.

4. Invoke the **Line** tool from the **Object Type** rollout.

   The 2D objects (shapes) available are listed in the **Object Type** rollout. Shapes are used to construct objects such as lofts or are used as motion paths.

5. Select the **Smooth** radio button in the **Initial Type** area from the **Creation Method** rollout.

6. Use the positions of cones as a guide to mark vertices of a line, starting from the bottom cone, as shown in Figure 17-3.

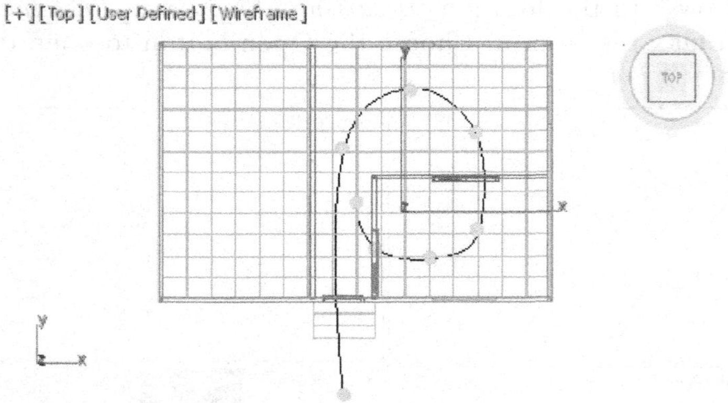

Figure 17-3 *A line created to be used as a path*

   The line will be used as a motion path for the camera.

7. Right-click to exit the command. In the **Name and Color** rollout, enter the name of the spline as **Walkthrough-Path**.

   Next, you will refine Walkthrough-Path.

8. With the line selected, choose the **Modify** tab from the **Command Panel**.

9. In the **Selection** rollout, choose the **Vertex** button to activate the vertex sub-object mode to display each individual vertex of the path.

10. Choose the **Refine** button from the **Geometry** rollout.

11. Add a new vertex where the path enters the building by clicking on the line at that point; a new vertex appears on the path, as shown in Figure 17-4.

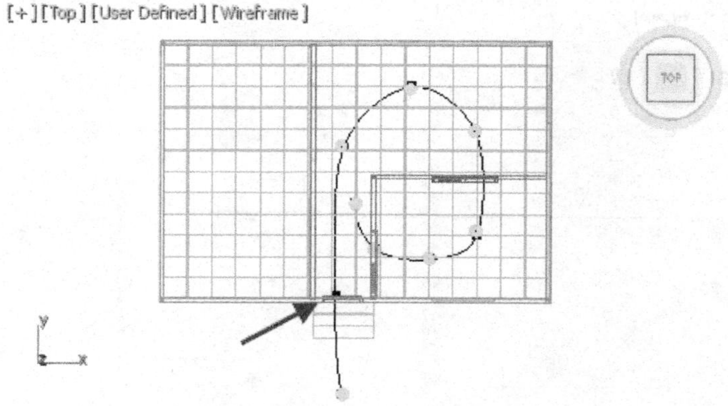

Figure 17-4 A new vertex added to the path

12. Invoke the **Select and Move** tool and select the newly added vertex. Now, move it such that it is centered in the doorway.

13. Invoke the **Maximize Viewport Toggle** tool. Activate the Left viewport and maximize it.

14. Move the first vertex down so that it is at the same level as the bottom of the house, as shown in Figure 17-5.

    The motion path is modified such that it follows the stairs into the building.

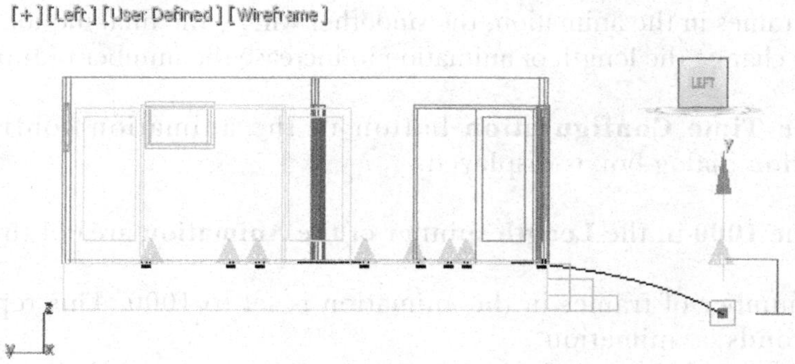

Figure 17-5 The first vertex moved down

15. Choose the **Vertex** button in the **Selection** rollout to exit the **Vertex** sub-object mode.

16. Move *Walkthrough-Path* 66.0 units up on the Y axis.

    The motion path is moved up to the height of an average person's line of sight, as shown in Figure 17-6.

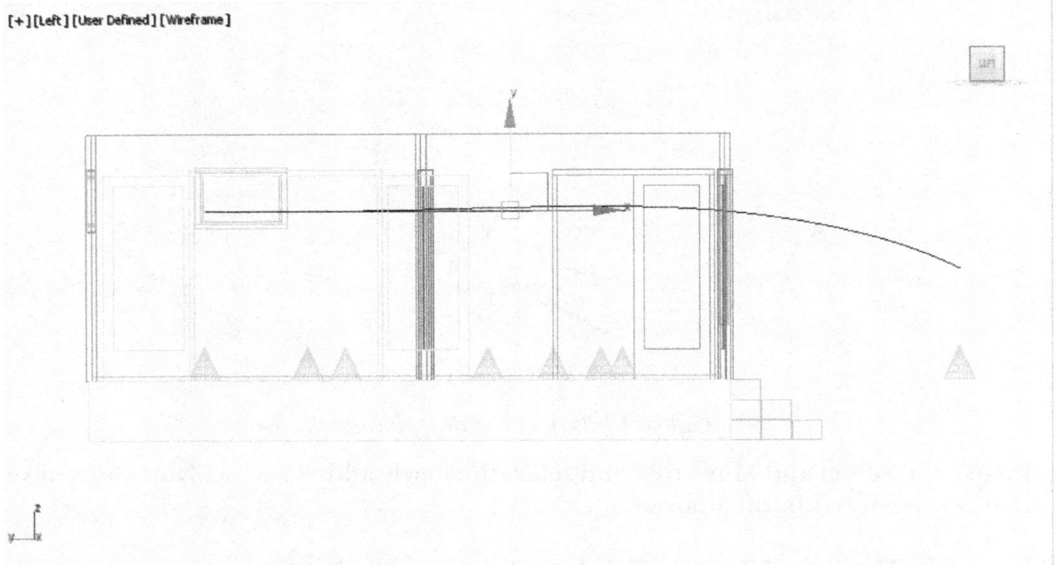

Figure 17-6 The motion path moved up to the height of an average person's line of sight

Specifying the Length of the Animation

Walkthrough animations need a large number of frames. If less number of frames are used for the animation, the camera will appear to be turning at the corners very quickly. In addition, it will appear as if the person is running in the scene, instead of walking. Therefore, the more the number of frames in the animation, the smoother will be the final motion created. In this section, you will change the length of animation to increase the number of frames in it.

1. Choose the **Time Configuration** button in the animation controls; the **Time Configuration** dialog box is displayed.

2. Set the value **1000** in the **Length** spinner of the **Animation** area of this dialog box.

   The total number of frames in the animation is set to 1000. This represents a little over 30 seconds of animation.

3. Choose the **OK** button to close the dialog box.

Using the Walkthrough Assistant

The **Walkthrough Assistant** automates the process of creating a camera and makes it follow the path you created. It provides settings for the camera height and head angles.

1. Choose **Animation > Walkthrough Assistant** from the menu bar; the **Walkthrough Assistant** dialog box is displayed, as shown in Figure 17-7. You can activate any viewport in this case.

2. In the **Camera Creation** area of the **Main Controls** rollout, make sure the **Free** radio button is selected. Next, choose the **Create New Camera** button; a free camera is automatically created with the name *Walkthrough_Cam001*.

Now, the camera must be moved to the start of the path.

3. In the **Path Control** area, select the **Move Path to Eye level** check box. Next, choose the **Pick Path** button and select the line created as the path.

   The camera is positioned and oriented along the path. The camera and motion path are automatically set at average person's eye level.

4. In the **Advanced Controls** rollout, set the value **75** in the **Field of View** spinner and make sure **160** is entered in the **Target Distance** spinner of the **Camera Controls** area.

   This sets the accurate camera view.

5. Activate the Perspective viewport. Change the display to *Walkthrough_Cam001* view by pressing the C key.

6. Drag the time slider to the right and left to replace the animation of the camera following the motion path.

   If the animation does not seem to be correct, you need to check it and remove the flaw. You may need to close the **Walkthrough Assistant** dialog box and delete the camera. You can then start over again by opening the **Walkthrough Assistant** dialog box and creating a new camera.

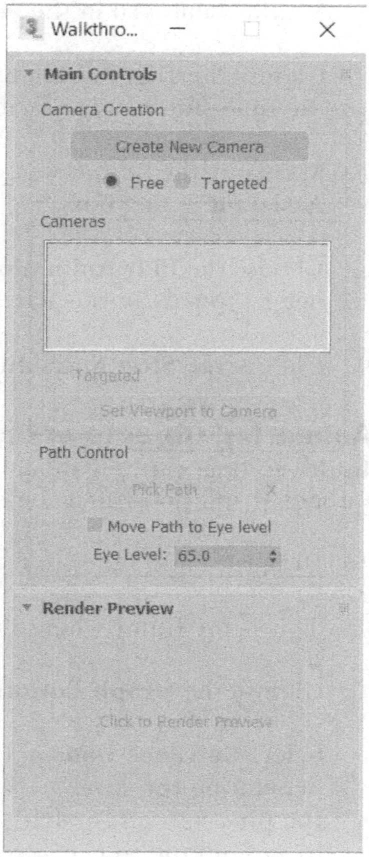

***Figure 17-7** The **Walkthrough Assistant** dialog box*

Adjusting the Head Tilt Angle

As the camera follows the motion path up toward the stairs, it tilts up to follow the path. However, a person walking up to the stairs usually does not tilt his head in this manner. In this section, you will adjust the head tilt angle.

1. Drag the time slider bar to frame 0.

2. Open the **Walkthrough Assistant** dialog box, if it is not already open. Scroll down and set the value **–45** in the **Head Tilt Angle** spinner in the **View Controls** rollout.

   The camera's line of sight is slightly below horizontal in frame 0 of the animation.

3. Choose the **Toggle Auto Key Mode** button to turn on the animation mode.

4. Drag the time slider to the frame in your animation (most probably frame 134) where the camera is halfway through the doorway.

5. Set the value **–10** in the **Head Tilt Angle** spinner.

6. Choose the **Toggle Auto Key Mode** button to turn off the animation mode. Next, drag the time slider bar to frame 0.

7. Make sure the *Walkthrough_Cam001* viewport is activated and then close the **Walkthrough Assistant** dialog box.

8. Choose the **Play Animation** button in the animation playback controls; the animation begins to play in the active viewport.

9. Choose the **Stop Animation** button to stop the animation replay.

Animating Objects in the Scene

In this section, you will animate the entrance door so that it opens when the camera passes through it and closes after some time.

1. Drag the time slider to frame 50.

   This is the frame where the door should begin to open.

2. Choose the **Graph Editors > Track View - Dope Sheet** from the menu bar.

3. Select **Entrance Door** in the viewport or you can select it from the Scene Explorer by expanding the *Exterior Walls* tree. In the **Track View - Dope Sheet** window, click on the plus sign on the left side of **Entrance Door** and then on the **Object (PivotDoor)** to expand the tree.

   The **Track View - Dope Sheet** window displays the **Open (degrees)** track for **Entrance Door**, as shown in Figure 17-8.

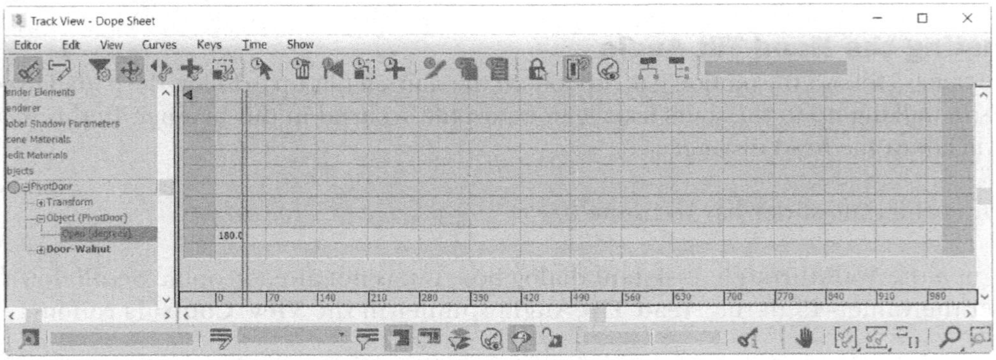

Figure 17-8 *The* ***Track View - Dope Sheet*** *window*

4. Choose the **Add/Remove Key** button on the top of the **Track View-Dope Sheet** window.

5. Click on the **Open (degrees)** track at frame 50 for **Entrance Door**, as shown in Figure 17-9.

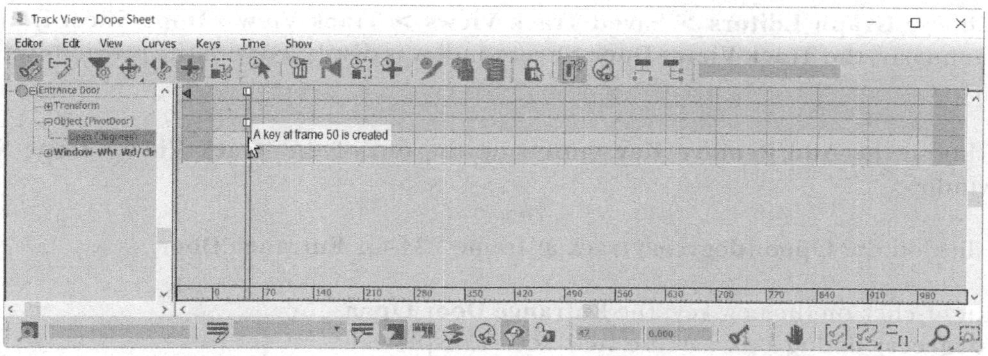

Figure 17-9 The new key created at frame 50 for ***Entrance Door***

A key for **Entrance Door** is created at frame 50. The value of this key is 0 which means the door is closed. This key sets the frame where the door begins to open.

6. Close the **Track View-Dope Sheet** window.

7. Drag the time slider to frame 100 and activate the animation mode, if it is not already active.

8. Make sure **Entrance Door** is selected in any viewport.

9. Choose the **Modify** tab in the **Command Panel**.

10. In the **Parameters** rollout, set the value **90** in the **Open** spinner; **Entrance Door** is opened 90 degrees at frame 100, as shown in Figure 17-10.

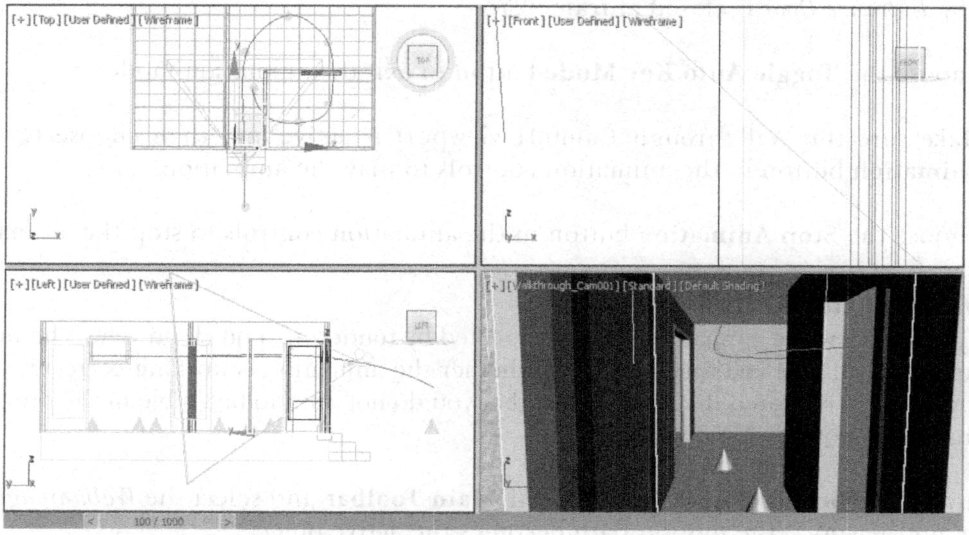

Figure 17-10 The ***Entrance Door*** *is opened 90 degrees at frame 100*

11. Choose **Graph Editors > Saved Track Views > Track View - Dope Sheet** from the menu bar; the **Track View - Dope Sheet** window is displayed with the previous settings active.

12. Choose the **Add/Remove Key** button on the top of the **Track View - Dope Sheet** window.

13. Click on the **Open (degrees)** track at frame 224 for **Entrance Door**.

14. Right-click on the new key; the **Entrance Door\Open (degrees)** dialog box for the new key is displayed, as shown in Figure 17-11.

15. Make sure 90 is displayed as the value in the **Value** spinner.

    The value 90 degrees sets the *Entrance Door* to the fully open position.

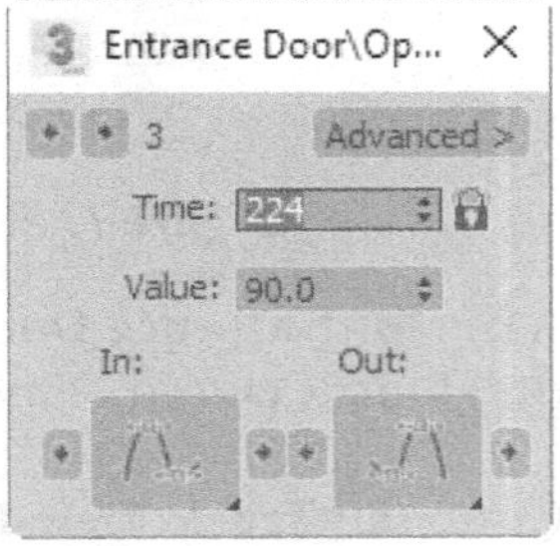

Figure 17-11 *The* ***Entrance Door\ Open (degrees)*** *dialog box*

16. Close both the **Entrance Door\Open (degrees)** dialog box as well as the **Track View - Dope Sheet** window.

17. Drag the time slider to frame 275.

18. Make sure the **Toggle Auto Key Mode** button is chosen. Next, set the value **0** in the **Open** spinner in the **Parameters** rollout.

    The *Entrance Door* is closed at frame 275.

19. Choose the **Toggle Auto Key Mode** button to exit the animation mode.

20. Make sure the Walkthrough_Cam001 viewport is active and then choose the **Play Animation** button in the animation controls to play the animation.

21. Choose the **Stop Animation** button in the animation controls to stop the animation.

Previewing the Animation

Now, you can view the animation you have created by rendering a quick preview. The preview of animation will give you a good idea of whether the animation is working correctly. Before the animation is rendered, hide any object that you do not want to be visible in the preview or the final animation.

1. Invoke the **Select Object** tool from the **Main Toolbar** and select the *Walkthrough-Path* in any viewport; the motion path becomes the active object.

2. Right-click in the viewport to display the quad menu. Choose **Hide Selection** from the upper-right quadrant of this quad menu; the path is hidden and will not be visible in the animation, as shown in Figure 17-12.

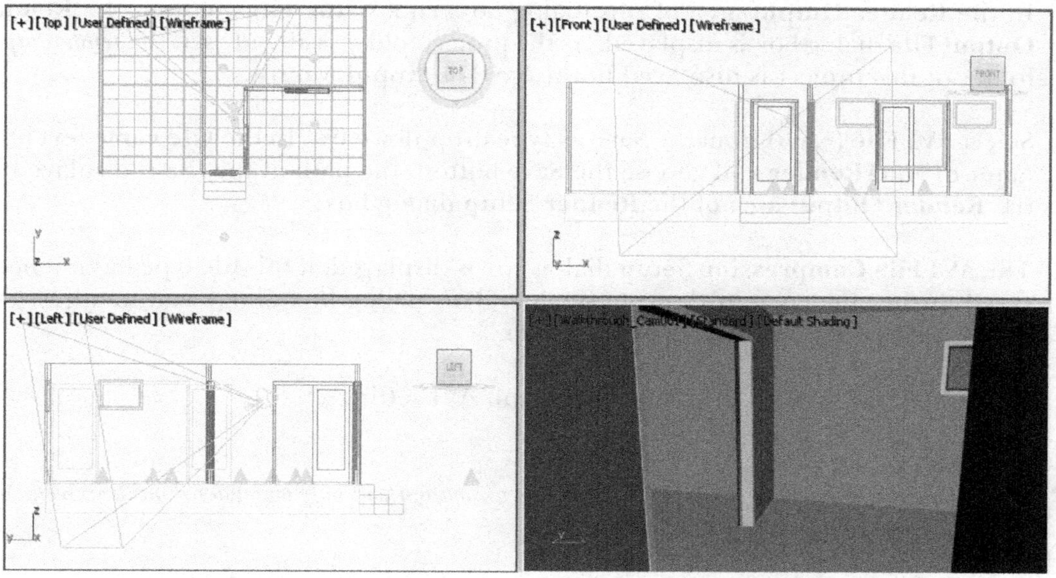

Figure 17-12 The motion path hidden in the viewports

3. Choose **Tools > Preview - Grab Viewport > Create Preview Animation** from the menu bar; the **Make Preview** dialog box is displayed.

4. In this dialog box, choose the **Create** button to accept the default settings and render the preview.

   The preview is generated and then played in Windows Media Player.

Rendering the Final Animation

In this section, we will render the final animation. The final animation contains all movements and is rendered with colors, textures, and shadows. This process can take some time, depending on the speed of the computer's processor and the length, complexity, and resolution of the animation.

1. Activate the Walkthrough_Cam001 viewport and then invoke the **Render Setup** tool from the **Main Toolbar**; the **Render Setup: Scanline Renderer** dialog box is displayed.

2. In the **Time Output** area, select the **Active Time Segment** radio button.

3. In the **Output Size** area of the dialog box, choose the **320×240** button.

   This sets the animation to be rendered at a resolution of 320 × 240 pixels. This animation does not need to be of high resolution. The higher the resolution, the longer it will take to render the animation and also the resulting file size will be large.

4. In the **Render Output** area of this dialog box, choose the **Files** button; the **Render Output File** dialog box is displayed. As the project folder is already set, the *renderoutput* folder of this project is displayed in the **Save in** drop-down list.

5. Select **AVI File (*.avi)** from the **Save as type** drop-down list. In the **File name** text box, enter **c17tut1Render** and choose the **Save** button; the path of the file is displayed in the **Render Output** area of the **Render Setup** dialog box.

   The **AVI File Compression Setup** dialog box is displayed, if this file type has not been saved during the current drawing session. This dialog box allows you to adjust the quality and keyframe rate of the animation.

6. Choose the **OK** button to accept the default AVI settings.

Note
*If the **Save File** check box is not selected, the animation will only be rendered on the screen. You will not be able to replay the animation.*

7. Choose the **Render** button in the **Render Setup: Scanline Renderer** dialog box; the **Rendered Frame** window and the **Rendering** dialog box are displayed. When the rendering is complete, the **Rendering** dialog box will be closed and the last frame of the animation will be displayed. The animation has been saved to the disk and must be viewed to see the results.

8. Close the **Rendered Frame** window and the **Render Setup: Scanline Renderer** dialog box.

9. Navigate *renderoutput* folder of the project folder *c17_tut1* and double-click on **c17tut1Render.avi**; the animation file is opened and played in Windows Media Player. Now, save the scene.

Self-Evaluation Test

Answer the following questions and then compare them to those given at the end of this chapter:

1. Which of the following buttons is used to toggle the animation mode on/off ?

   (a) **Toggle Set Key Mode** (b) **Toggle Auto Key Mode**
   (c) **Isolate Selection Toggle** (d) **Selection Lock Toggle**

2. Which of the following buttons is used to create new keys in the **Track View - Dope Sheet** window?

   (a) **Add Keys** (b) **Edit Keys**
   (c) **Slide Keys** (d) **Show Selected Key Stats**

3. Which of the following cameras is automatically created with the name *Walkthrough_Cam001* while creating a walkthrough?

   (a) **Target** (b) **Free**
   (c) Both (a) and (b) (d) None of these

4. Which of the following is the default renderer in Autodesk 3ds Max?

   (a) **Scanline** (b) **VUE File**
   (c) **NVIDIA mental ray** (d) **Quicksilver Hardware**

5. A __________ can be used as a motion path.

6. The path is followed precisely by using the __________ to represent a person's line of sight.

7. If there are not enough vertices in a line, it can be __________ to add more vertices.

8. A walkthrough animation is created by defining the path of the walk by drawing a line from the start point to end point. (T/F)

9. Walkthrough animations need a large number of frames to make the animation smooth. (T/F)

10. The **Walkthrough Assistant** has a setting that allows the head tilt angle to be modified as the camera moves along the path. (T/F)

Review Questions

Answer the following questions:

1. Which of the following dialog boxes is used to set the length of the animation within the defined frames?

   (a) **Time Configuration** (b) **Walkthrough Assistant**
   (c) **Make Preview** (d) **Render Setup**

2. The ________ button is used to add new vertices to the existing spline.

3. The ________ is a controller assigned to an object to make the object follow a selected path.

4. The vertices in a spline cannot be modified. (T/F)

5. The camera used for a walkthrough assistant is usually a **Target** camera. (T/F)

EXERCISE

The rendered sequence of the scene used in the following exercise can be accessed by downloading the *c17_3dsmax_2022_exr.zip* file from *www.cadcim.com*. The path of the file is as follows: *Textbooks > Animation and Visual Effects > Autodesk 3ds Max > Autodesk 3ds Max 2022 for Beginners: A Tutorial Approach*

Exercise 1

Extract the contents of *c17_3dsmax_2022exr.zip* and then open *c17_exr01_start.max*. Create a walkthrough for the lobby, as shown in Figure 17-13. **(Expected time: 30 min)**

Figure 17-13 *Walkthrough for the lobby*

Answers to Self-Evaluation Test

1. b, **2.** a, **3.** b, **4.** a, **5.** line(Spline), **6.** free camera, **7.** refined, **8.** T, **9.** T, **10.** T

Project 1

Creating a Windmill

PROJECT DESCRIPTION

In this project, you will create the structure of a windmill, as shown in Figure P1-1. You will create the base structure, middle structure, and top structure of the windmill. Next, you will create the deck, deck rails, handrails, and blades of the windmill. You will apply materials to all objects of the windmill, animate the blades of the windmill, and then render the scene for the final output.

Figure P1-1 The model of the windmill

Creating the Project Folder

Create a new project folder with the name *Prj_1* at *\Documents\3dsmax 2022* and then save the file with the name *Prj1*, as discussed in Tutorial 1 of Chapter 2.

Downloading the Files

Before starting the project, you need to download the *prj1_3dsmax_2022.zip* file from *www.cadcim.com*. The path of the file is given next: *Textbooks > Animation and Visual Effects > Autodesk 3ds Max > Autodesk 3ds Max 2022 for Beginners: A Tutorial Approach*

Extract the contents of the zip file to *Documents*. Next, open Windows Explorer and then browse to *\Documents\prj1_3dsmax_2022* and then copy all files to *\Documents\3dsmax 2022\Prj1\sceneassets\images*.

1. Choose the **Render Setup** tool from the **Main Toolbar**; the **Render Setup: Arnold** dialog box is displayed.

2. In this dialog box, select **Scanline Renderer** from the **Renderer** drop-down list; the renderer is changed to Scanline and the name of the dialog box is modified to the **Render Setup: Scanline Renderer** dialog box. Next, close this dialog box.

Creating the Base of the Windmill

In this section, you will create the base structure of the windmill by using the **NGon** tool.

1. Make sure the **Wireframe** option is chosen in the Shaded viewport label in the Top, Front, and Left viewports. Turn the grid off in all viewports by using the G key.

2. Activate the Top viewport and then choose **Create > Shapes** in the **Command Panel**. Next, invoke the **NGon** tool from the **Object Type** rollout; various rollouts are displayed in the **Command Panel**. 

3. In the **Parameters** rollout, make sure the **Inscribed** radio button is selected. Next, set **8** in the **Sides** spinner and then press the ENTER key.

4. Expand the **Keyboard Entry** rollout and set **100** in the **Radius** spinner. Now, choose the **Create** button; the shape is created with the name *NGon001*, as shown in Figure P1-2.

5. Invoke the **3D Snaps Toggle** tool from the **Snaps Toggle** flyout to activate 3D snapping. 

6. Activate the Front viewport and make sure the **Shapes** button is activated in the **Command Panel**. Invoke the **Line** tool from the **Object Type** rollout. Make sure the **Corner** radio button is selected in **Initial Type** and **Drag Type** areas from the **Creation Method** rollout.

7. Expand the **Keyboard Entry** rollout and choose the **Add Point** button to create first point of the line. Next, set **150** as the value in the **Y** spinner and then choose the **Add Point** button; a line of 150 units in length is displayed in the viewports. Right-click in the viewport to exit the tool.

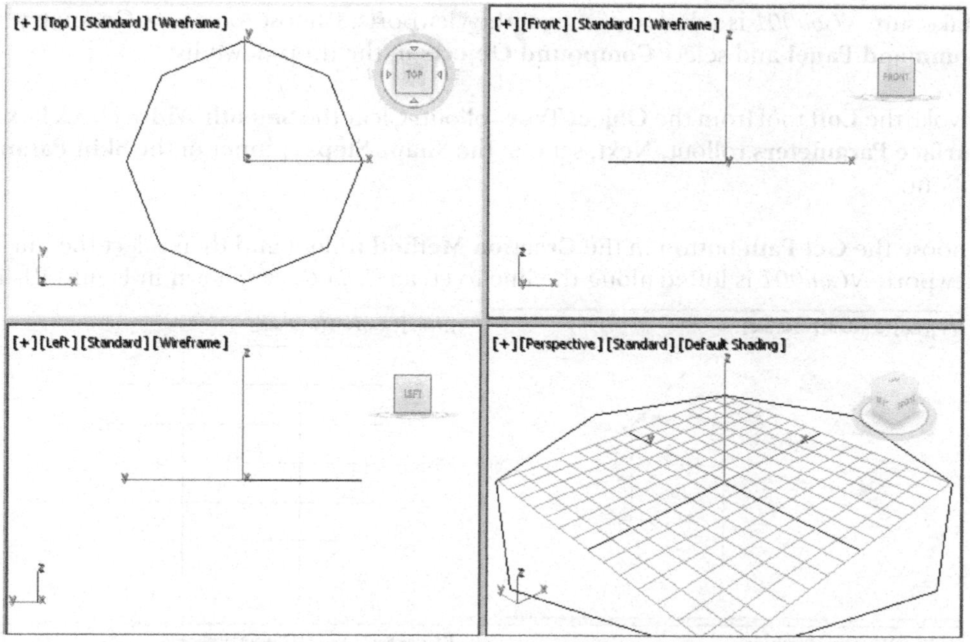

Figure P1-2 *The shape created with the name NGon001*

8. Select *NGon001* and right-click on the **Select and Rotate** tool; the **Rotate Transform Type-In** dialog box is displayed. In this dialog box, set **22.5** as the value in the **Z** spinner in the **Absolute World** area and then press the ENTER key; *NGon001* rotates, as shown in Figure P1-3. Now, close the dialog box and exit the **Select and Rotate** tool.

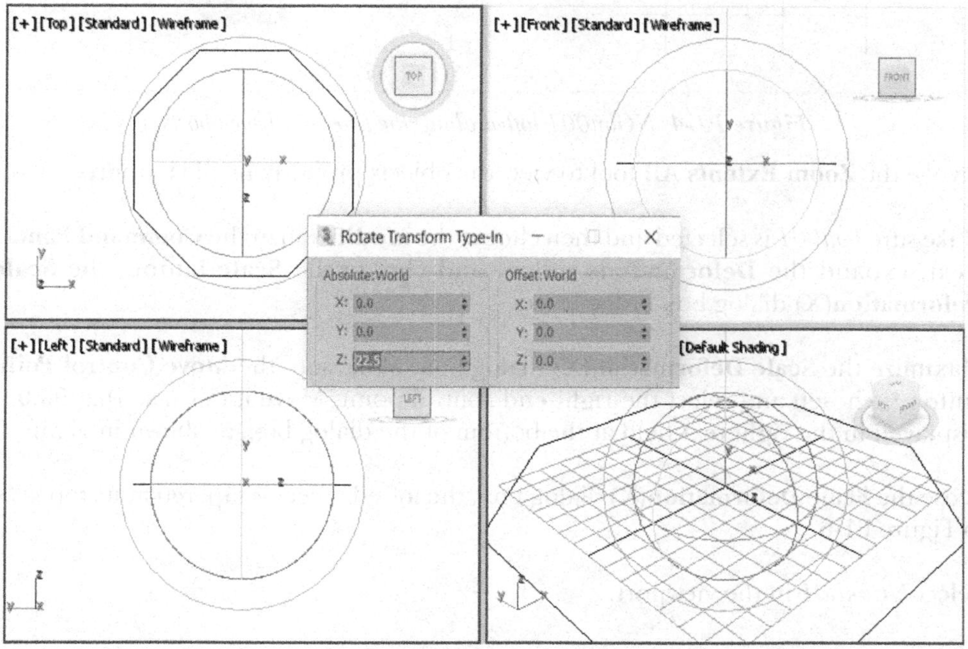

Figure P1-3 *Rotating NGon001*

9. Make sure *NGon001* is selected in any of the viewports. Choose **Create > Geometry** in the **Command Panel** and select **Compound Objects** in the drop-down list.

10. Invoke the **Loft** tool from the **Object Type** rollout. Clear the **Smooth Width** check box in the **Surface Parameters** rollout. Next, set **0** in the **Shape Steps** spinner of the **Skin Parameters** rollout.

11. Choose the **Get Path** button in the **Creation Method** rollout and then select the line in the viewport; *NGon001* is lofted along the line to create *Loft001*, as shown in Figure P1-4.

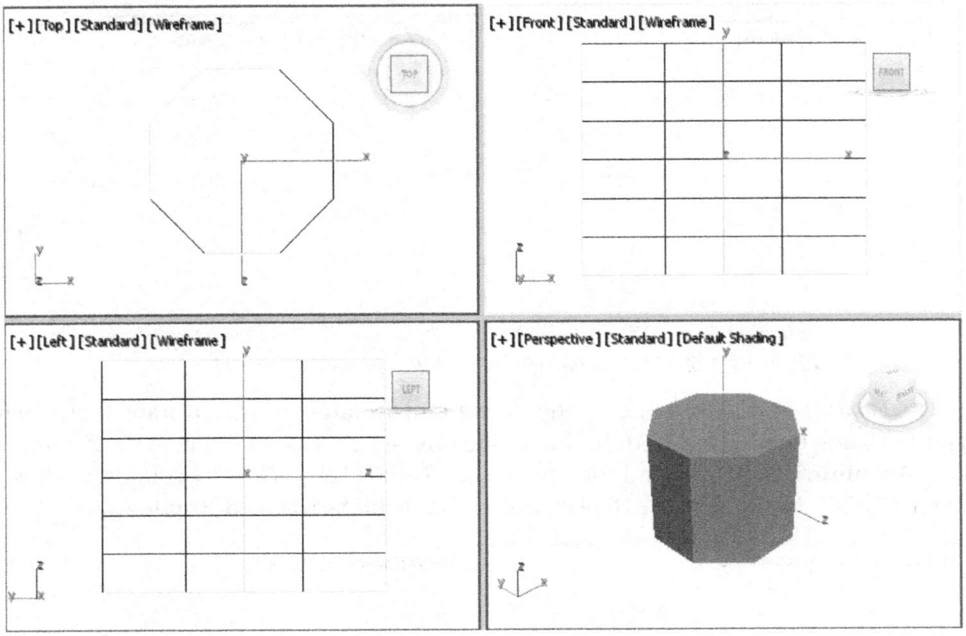

Figure P1-4 *NGon001 lofted along the line in all viewports*

12. Invoke the **Zoom Extents All** tool to view the objects properly in all viewports. 

13. Make sure *Loft001* is selected and then choose the **Modify** tab in the **Command Panel**. Next, expand the **Deformations** rollout and choose the **Scale** button; the **Scale Deformation(X)** dialog box is displayed. 

14. Maximize the **Scale Deformation(X)** dialog box. Make sure the **Move Control Point** button is chosen and select the right end control point. Move it down so that **90.0** is displayed in the right text field at the bottom of the dialog box, as shown in Figure P1-5. 

15. Close the **Scale Deformation(X)** dialog box; the lofted object is tapered at its top, as shown in Figure P1-6.

16. Select *NGon001* in the viewport.

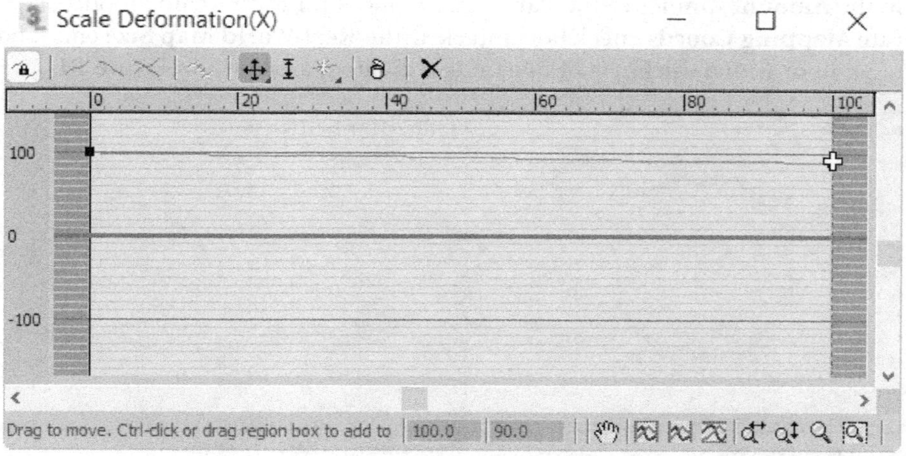

***Figure P1-5** The **Scale Deformation(X)** dialog box*

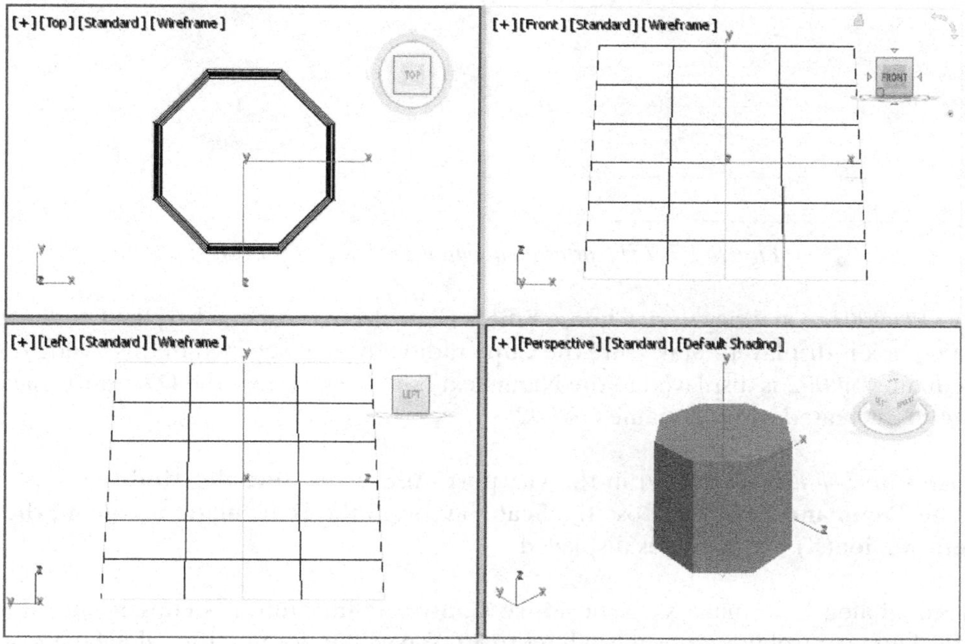

***Figure P1-6** Loft001 tapered at its top*

17. Choose **Edit > Clone** from the menu bar; the **Clone Options** dialog box is displayed. In this dialog box, select the **Copy** radio button in the **Object** area. Next, enter **Floor_lo** in the **Name** text box and choose the **OK** button to close this dialog box; a copy of *NGon001* is created with the name *Floor_lo* at the same location as the original.

18. Make sure *Floor_lo* is selected in the viewport. Expand the **Parameters** rollout and set **99** in the **Radius** spinner. Next, select **Extrude** in the **Modifier List** drop-down list under the **OBJECT-SPACE MODIFIERS** section.

19. Set **5** in the **Amount** spinner of the **Parameters** rollout for the **Extrude** modifier. Select the **Generate Mapping Coords** check box and clear the **Real-World Map Size** check box if it is selected; a floor with a thickness of five units is created, as shown in Figure P1-7.

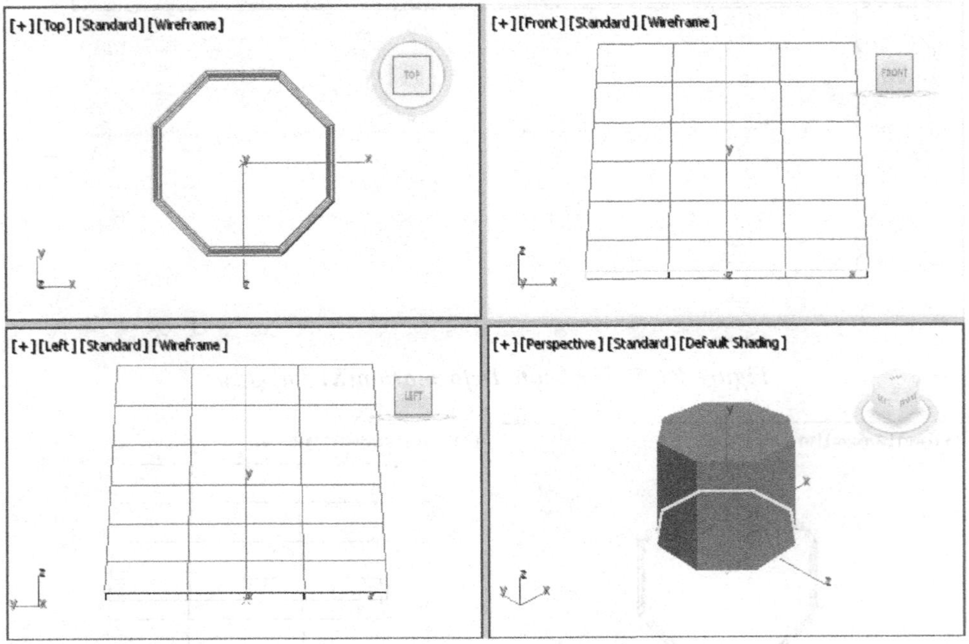

Figure P1-7 *Floor created with a thickness of five units*

20. Select *Loft001* in any viewport. Choose **Edit > Clone** from the menu bar; the **Clone Options** dialog box is displayed. Make sure the **Copy** radio button is selected in this dialog box and the name *Loft002* is displayed in the **Name** text box. Now, choose the **OK** button; a copy of *Loft001* is created with the name *Loft002*.

21. Make sure *Loft002* is selected in the viewport. Also, make sure the **Modify** tab is chosen in the **Command Panel**. Choose the **Scale** button in the **Deformations** rollout; the **Scale Deformation(X)** dialog box is displayed.

22. In this dialog box, make sure the **Move Control Point** button is chosen and move the right-hand control point down two lines to **80**. You will notice that the value 80 is displayed in the right text box at the bottom of this dialog box. Similarly, move the left-hand point down to **90**; the deformation is applied to *Loft002*. Now, close the **Scale Deformation(X)** dialog box.

23. Select *Loft001* from the **Scene Explorer**. Choose **Create > Geometry** in the **Command Panel** and select **Compound Objects** in the drop-down list, if it is not selected. Next, invoke the **Boolean** tool from the **Object Type** rollout.

24. Choose the **Add Operands** button in the **Boolean Parameters** rollout and select *Loft002* in any viewport. Next, choose the **Subtract** button from the **Operand Parameters** rollout; *Loft001* is hollowed out, leaving the walls for the base of the windmill, as shown in Figure P1-8.

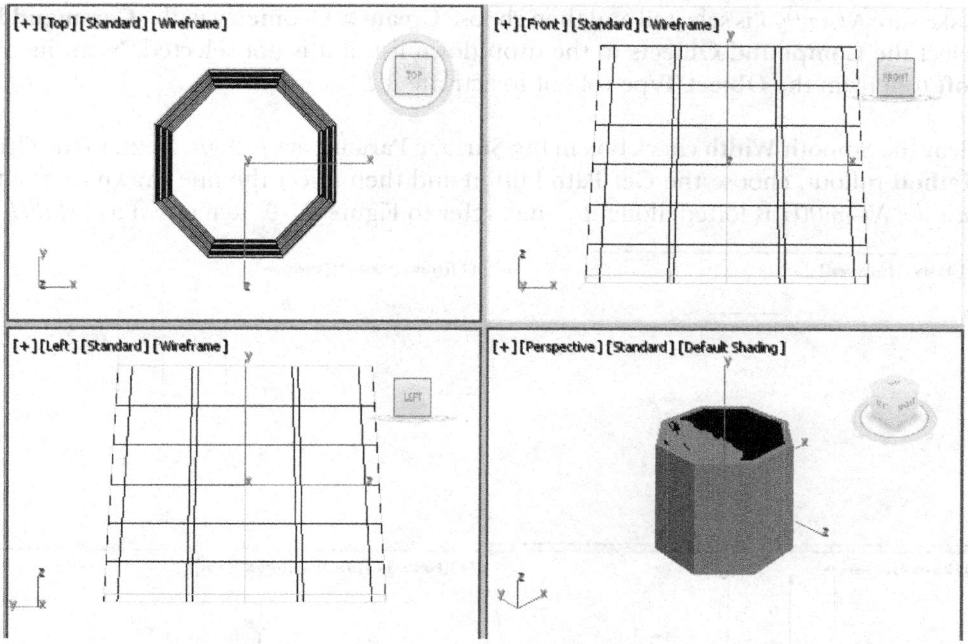

***Figure P1-8** Loft001 displayed after performing the Boolean operation*

25. Invoke the **Select Object** tool from the **Main Toolbar** to exit the Boolean operation.

26. Enter **Base_lo** as the name of the object in the **Name and Color** rollout.

Creating the Middle Structure of the Windmill

In this section, you will create the middle structure of the windmill.

1. Make sure the Front viewport is selected. Invoke the **Pan View** tool to pan the viewport down so that only the top of *Base_lo* is displayed. Make sure the subgrid is displayed.

2. Choose **Create > Shapes** in the **Command Panel.** Invoke the **Line** tool from the **Object Type** rollout to activate it. Make sure the **Corner** radio button is selected in the **Initial Type** and **Drag Type** areas from the **Creation Method** rollout.

3. Create a vertical line **150** units in length starting from the top-center of *Base_lo*. Now, right-click to exit the **Line** tool. Also, invoke the **Select Object** tool.

4. Select *NGon001* in any viewport. Choose **Edit > Clone** from the menu bar; the **Clone Options** dialog box is displayed. Make sure the **Copy** radio button is selected in this dialog box and enter **NGon003** in the **Name** text box. Next, choose the **OK** button; the dialog box is closed and a copy of *NGon001* is created with the name *NGon003*.

5. Invoke the **Align** tool and align *NGon003* with the top of *Base_lo*.

6. Make sure *NGon003* is selected and then choose **Create > Geometry** in the **Command Panel**. Select the **Compound Objects** in the drop-down list, if it is not selected. Next, invoke the **Loft** tool from the **Object Type** rollout to activate it.

7. Clear the **Smooth Width** check box in the **Surface Parameters** rollout. Next, in the **Creation Method** rollout, choose the **Get Path** button and then select the line drawn on the top of *Base_lo*; *NGon003* is lofted along the line, refer to Figure P1-9. Rename it as *Loft001*.

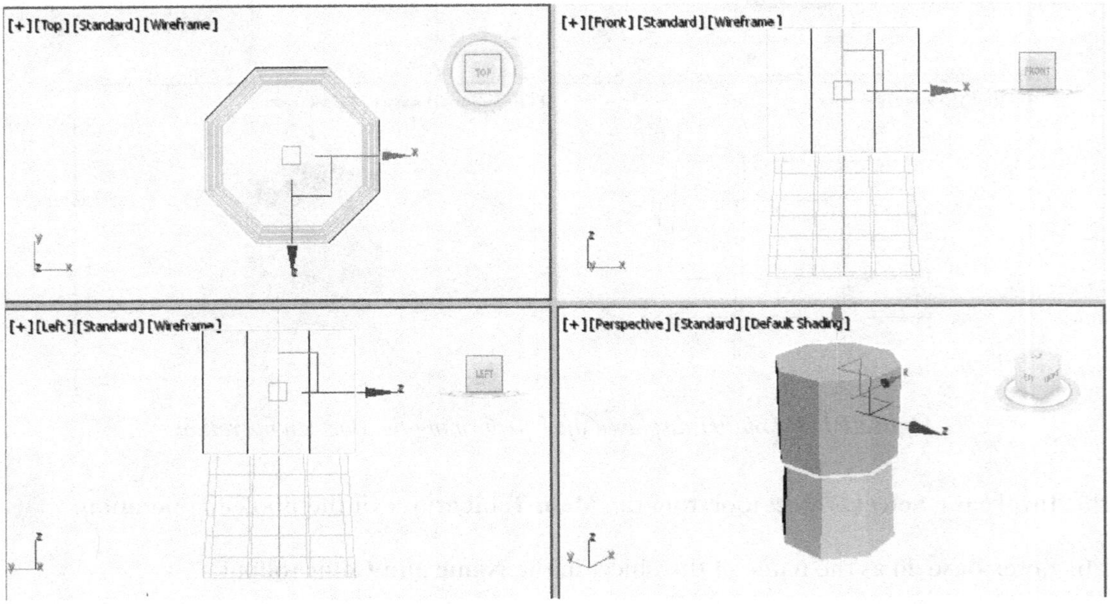

Figure P1-9 *NGon003 lofted along the line*

8. Invoke the **Zoom Extents All** tool to zoom the viewports to their extents.

9. Make sure *Loft001* is selected and then choose the **Modify** tab of the **Command Panel**. Expand the **Skin Parameters** rollout and make sure the value **0** is entered in both the **Shape Steps** and **Path Steps** spinners in the **Options** area.

10. In the **Deformations** rollout, choose the **Scale** button; the **Scale Deformation(X)** dialog box is displayed.

11. In the **Scale Deformation(X)** dialog box, choose the **Insert CornerPoint** button and insert a point at **30** on the horizontal scale. Next, choose the **Move Control Point** button and move the point down to **75**.

Note

You can specify exact values in the text boxes at the bottom of the dialog box. The value entered in the text box on the left is used for positioning the control point in relation to the horizontal scale. The value entered in the text box on the right positions the control point in relation to the vertical scale.

12. Similarly, insert a point at **60** and move it down horizontally to **60**, insert a point at **80** and move it down horizontally to **53**, and move the right corner point (**100**) down horizontally to **50**, as shown in Figure P1-10.

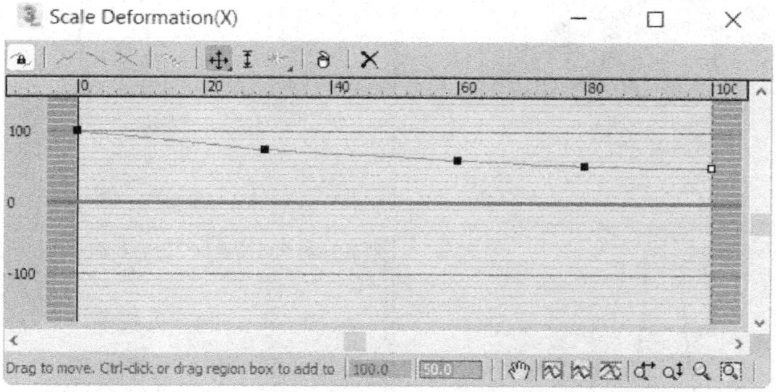

Figure P1-10 *Inserting and moving the control points in the* ***Scale Deformation(X)*** *dialog box*

13. Close the **Scale Deformation(X)** dialog box and invoke the **Zoom Extents All** tool to view the objects in all viewports.

14. Create a copy of *Loft001* with the name **Loft002**.

15. Make sure *Loft002* is selected. Also, make sure the **Modify** tab is chosen in the **Command Panel**. Choose the **Scale** button in the **Deformations** rollout; the **Scale Deformation(X)** dialog box is displayed. Maximize this dialog box and move all control points down to two lines (**–5** units) on the vertical scale. This can be accomplished by selecting the control points one at a time and subtracting **5** from the value displayed in the right-hand text box at the bottom of the dialog box.

16. Close the **Scale Deformation(X)** dialog box and select *Loft001* in any viewport to activate it.

17. Choose **Create > Geometry** in the **Command Panel**. Select the **Compound Objects** in the drop-down list, if it is not selected. Next, invoke the **Boolean** tool from the **Object Type** rollout.

18. Choose the **Add Operands** button from the **Boolean Parameters** rollout and select *Loft002* in any viewport. Next, choose the **Subtract** button from the **Operand Parameters** rollout; the walls for the middle structure of the windmill are created, as shown in Figure P1-11.

19. Invoke the **Select Object** tool to exit the boolean operation.

20. Enter **Base_mid** in the **Name and Color** rollout and save the changes.

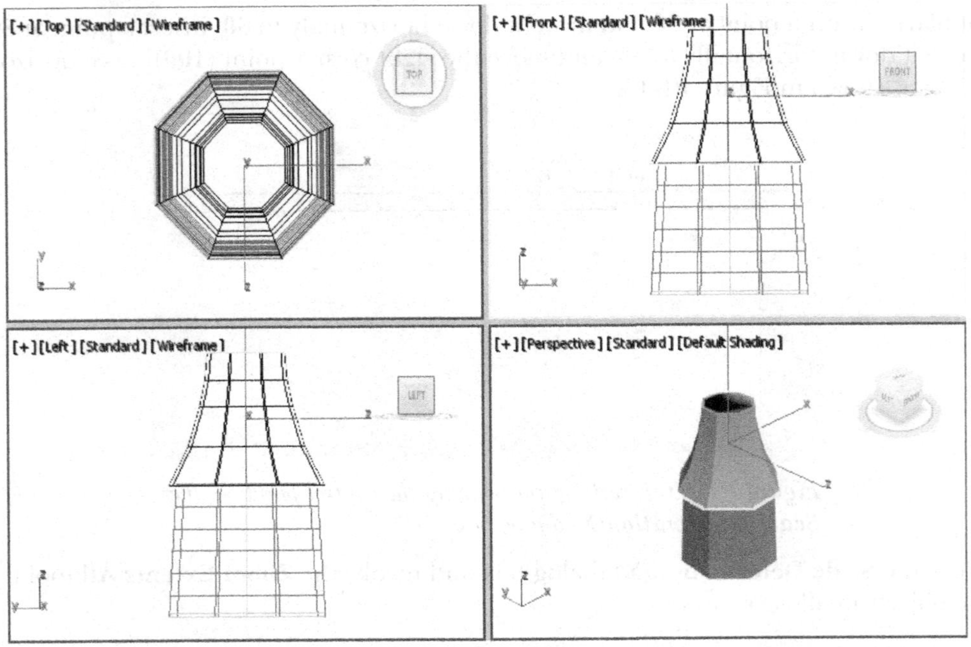

Figure P1-11 Walls created for the middle structure of the windmill

Creating the Top Structure of the Windmill

In this section, you will create the top structure of the windmill. You will then alter the top structure to give it a dome shape.

1. Select *NGon001* and create a copy with the name **NGon002**.

2. Move *NGon002* up in the Front viewport until it is in level with the top of *Base_mid*.

3. Choose the **Modify** tab in the **Command Panel**. In the **Parameters** rollout, set **60** in the **Radius** spinner.

4. Zoom in the Front viewport to display the subgrid. Then, pan the viewport down so that only the upper portion of *Base_mid* is visible.

5. Choose **Create > Shapes** in the **Command Panel**. Invoke the **Line** tool from the **Object Type** rollout. Next, create a vertical line of length **80** units from the top-center of *Base_mid*. Right-click to exit the **Line** tool.

6. Select *NGon002* in any of the viewports. Choose **Create > Geometry** in the **Command Panel** and select **Compound Objects** in the drop-down list, if it is not selected. Next, invoke the **Loft** tool from the **Object Type** rollout.

7. Choose the **Get Path** button in the **Creation Method** rollout, and then select the line object you just created; *NGon002* is lofted along the line, as shown in Figure P1-12.

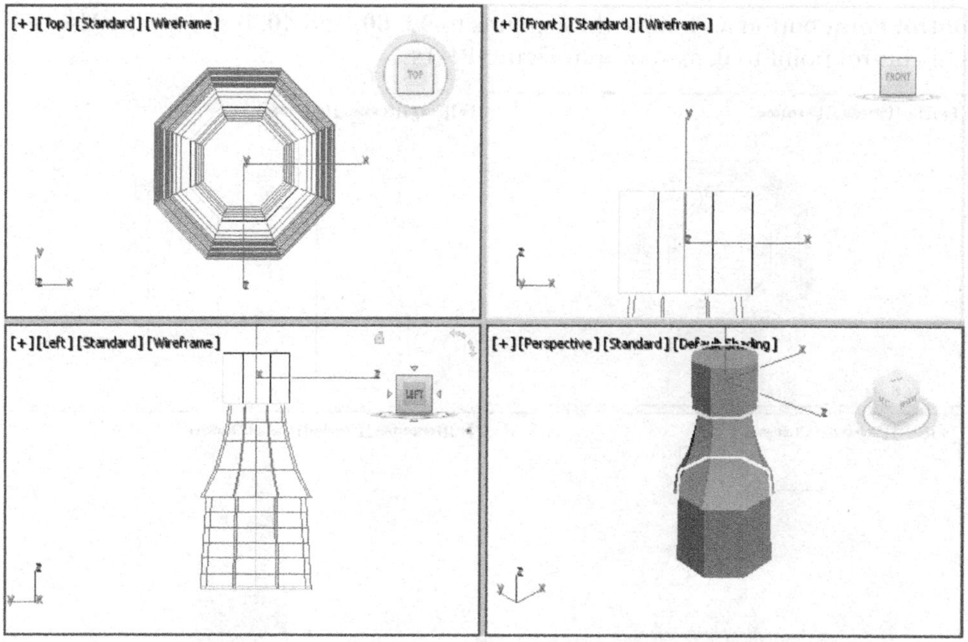

Figure P1-12 *NGon002 lofted along the line*

8. Enter **Base_top** in the **Name and Color** rollout.

9. Create a copy of *Base_top* and name it as **Top_drill**.

10. Make sure *Top_drill* is selected and then choose **Hierarchy > Pivot** in the **Command Panel**. Choose the **Affect Pivot Only** button in the **Move/Rotate/Scale** area of the **Adjust Pivot** rollout. Next, move the pivot so that it is aligned with the bottom of *Top_drill*. Choose the **Affect Pivot Only** button to exit the pivot adjustment mode.

Note
The pivot point may already be at the correct location and may not require any adjustment.

11. Invoke the **Select and Uniform Scale** tool. Also, invoke the **Percent Snap Toggle** tool to turn on the percentage snaps.

12. In the Front viewport, scale *Top_drill* until the coordinate display shows 90.0 in each of the **X**, **Y**, and **Z** spinners; *Top_drill* is scaled uniformly, as shown in Figure P1-13. You can also use the **Scale Transform Type-In** dialog box to scale the *Top_drill*.

13. Make sure *Top_drill* is selected in the viewport and then choose the **Modify** tab in the **Command Panel**. Clear the **Smooth Width** check box in the **Smoothing** area of the **Surface Parameters** rollout. Choose the **Scale** button in the **Deformations** rollout; the **Scale Deformation(X)** dialog box is displayed.

14. Maximize the **Scale Deformation(X)** dialog box, if required. Choose the **Insert Corner Point** button and insert points at **40**, **80**, and **90** on the top scale. Next, choose the **Scale**

Control Point button and move these points to **90**, **60**, and **40**, respectively. Also, move the right control point to **0**, as shown in Figure P1-14.

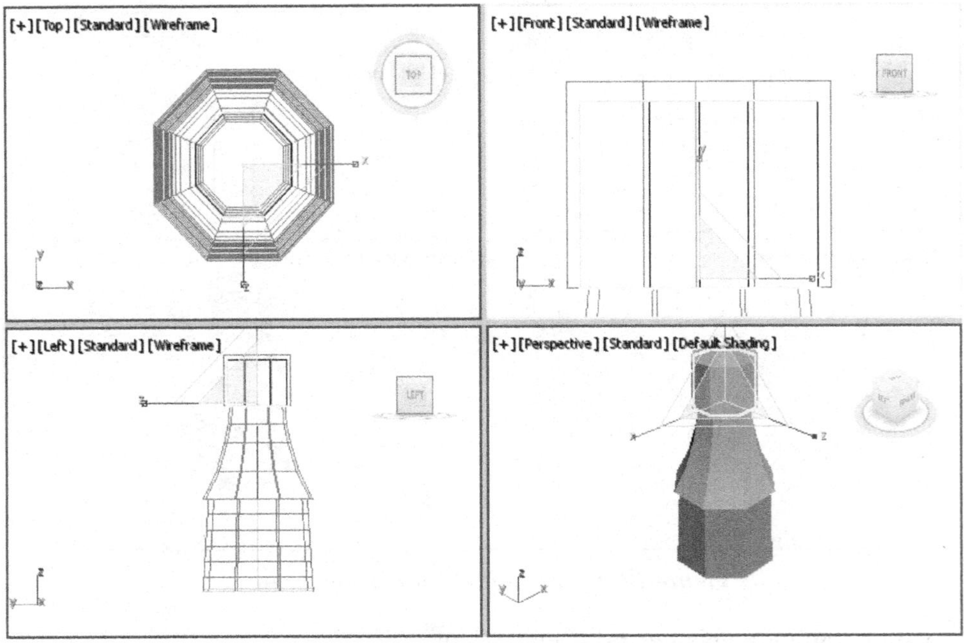

Figure P1-13 *Top_drill scaled uniformly*

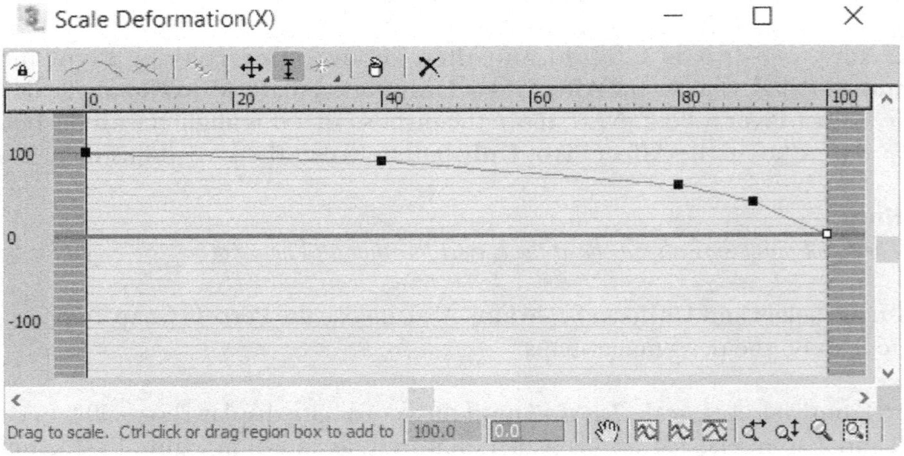

Figure P1-14 *Top_drill modified using the* ***Scale Deformation(X)*** *dialog box*

15. Close the **Scale Deformations(X)** dialog box.

    The shape of drill is altered to a dome shape. However, *Base_top* needs to be changed too.

16. Select *Base_top* and choose the **Modify** tab in the **Command Panel**. Clear the **Smooth Width** check box in the **Surface Parameters** rollout and choose the **Scale** button in the **Deformations** rollout; the **Scale Deformation(X)** dialog box is displayed.

17. In the **Scale Deformation(X)** dialog box, insert and scale the points at same positions as done for *Top_drill*. Refer to step 14. Now, close the **Scale Deformation(X)** dialog box.

18. Select *NGon002* which was used to make the top of the building. Create a copy with the name **Floor_top**.

19. Make sure *Floor_top* is selected in the viewport. Also, make sure the **Modify** tab is chosen in the **Command Panel**. Set **58** in the **Radius** spinner in the **Parameters** rollout. Select **Extrude** in the **Modifier List** drop-down list under the **OBJECT-SPACE MODIFIERS** section.

20. Set **5** in the **Amount** spinner of the **Parameters** rollout; *Floor_top* is extruded, as shown in Figure P1-15.

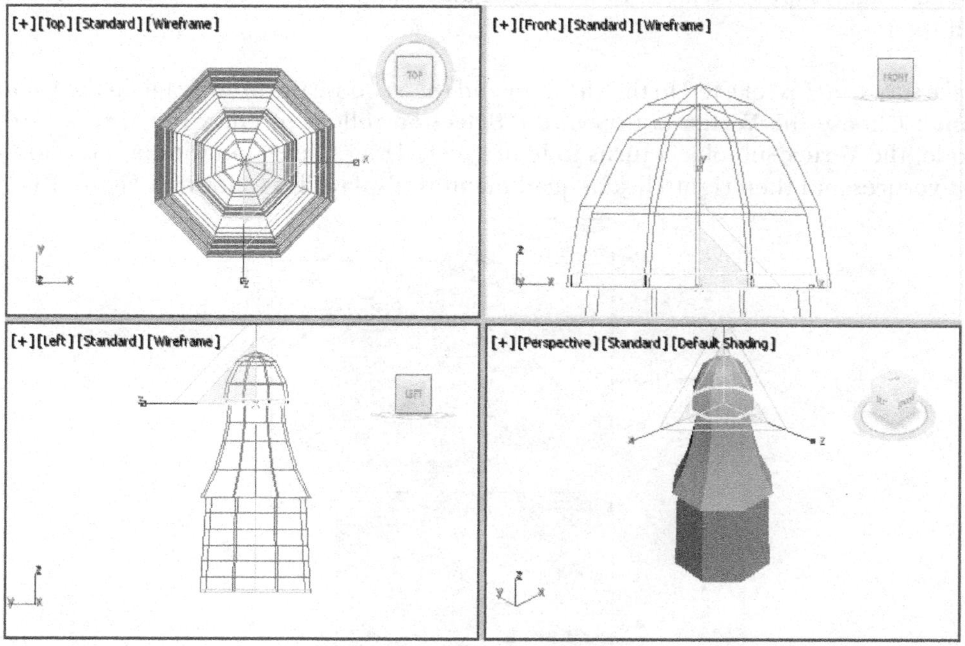

Figure P1-15 *Floor_top extruded*

21. Select *Base_top, Top_drill,* and *Floor_top*.

22. Choose the **Display** tab in the **Command Panel**. Invoke the **Hide Selected** tool in the **Hide** rollout; the top portion of the building is hidden.

Note

An alternative method of hiding the objects would be to select the objects, right-click in the viewport, and choose ***Hide Selection*** *from the quad menu.*

23. Make sure the Front viewport is active and choose **Create > Shapes** in the **Command Panel**. Invoke the **Line** tool from the **Object Type** rollout.

24. Expand the **Keyboard Entry** rollout and enter the following values in the **X**, **Y**, and **Z** spinners. You need to choose the **Add Point** button after every XYZ coordinate entry:

X: **0**	Y: **380**	Z: **0**
X: **–20**	Y: **370**	Z: **0**
X: **–30**	Y: **350**	Z: **0**
X: **–40**	Y: **300**	Z: **0**
X: **40**	Y: **300**	Z: **0**
X: **30**	Y: **350**	Z: **0**
X: **20**	Y: **370**	Z: **0**

25. After the last entry, choose the **Close** button to add the last segment and create a closed line; a dome shape is created. In the **Name and Color** rollout, enter **Ext01**. Right-click to exit the tool.

26. Make sure *Ext01* is selected in the viewport and then choose the **Modify** tab in the **Command Panel**. Choose the **Vertex** button in the **Selection** rollout to enter the **Vertex** sub-object mode; the **Vertex** sub-object turns to light green. Drag a selection window around the top five vertices and then right-click; a quad menu is displayed, as shown in Figure P1-16.

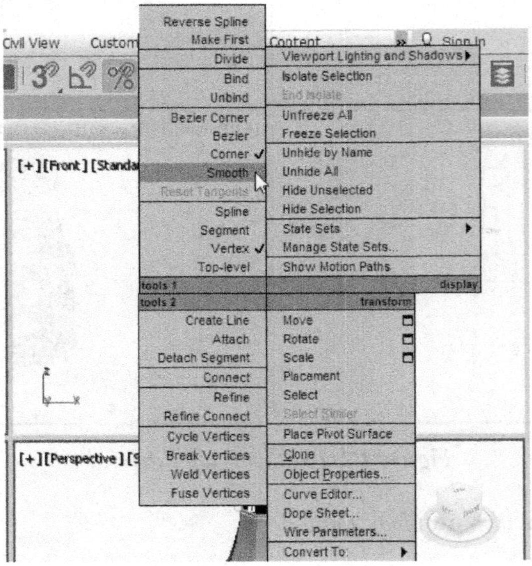

***Figure P1-16** The quad menu displayed*

27. Choose **Smooth** from the upper-left quadrant; the dome shape becomes smooth, refer to Figure P1-17.

28. Choose the **Vertex** button again to exit the **Vertex** sub-object mode.

29. Select **Extrude** in the **Modifier List** drop-down list under the **OBJECT-SPACE MODIFIERS** section. Set **80** in the **Amount** spinner of the **Parameters** rollout for the **Extrude** modifier; *Ext01* is extruded, as shown in Figure P1-18.

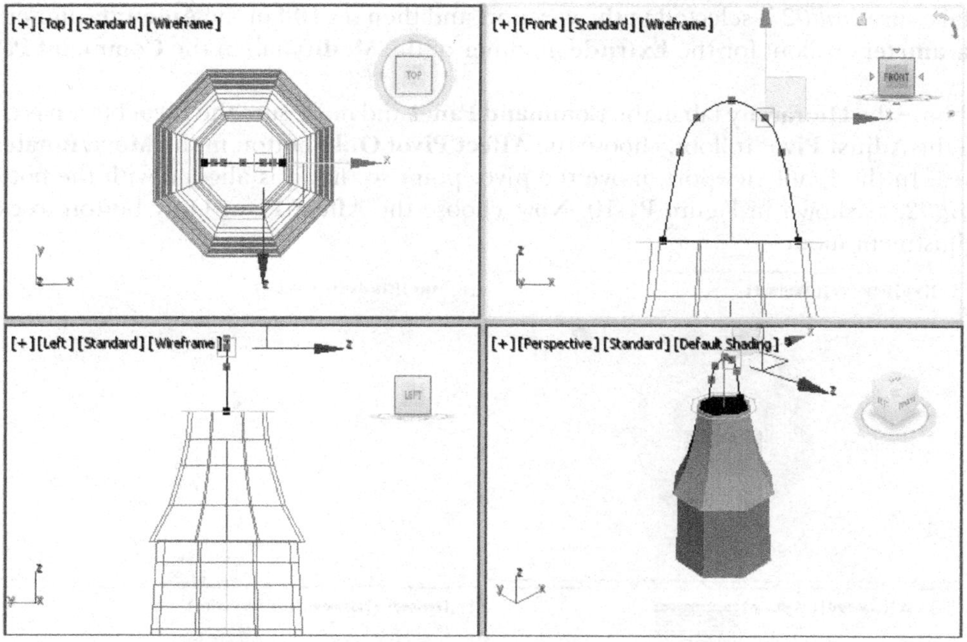

Figure P1-17 *Ext01 line smoothened*

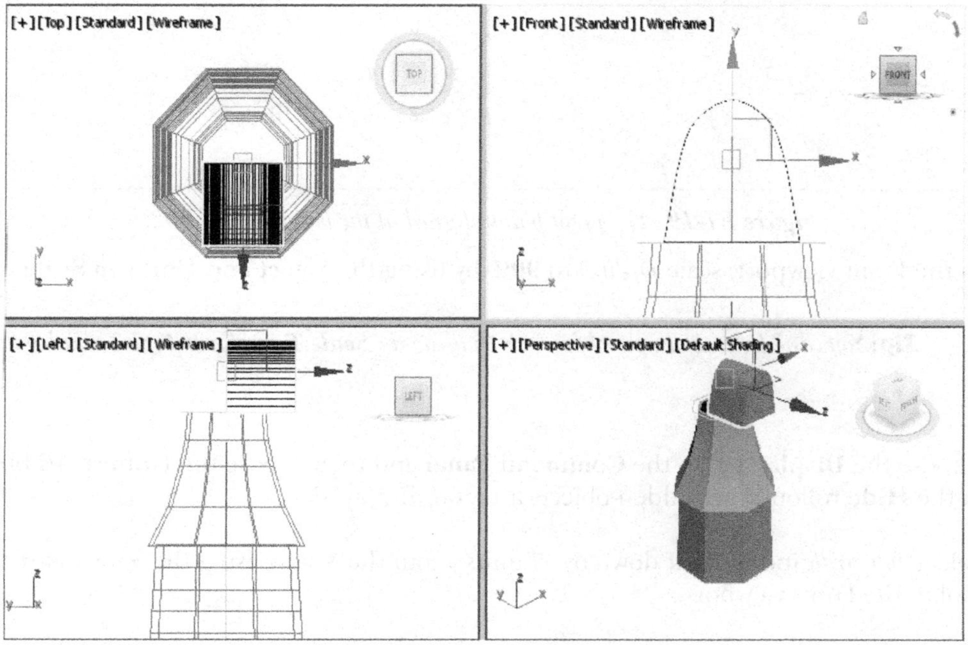

Figure P1-18 *Ext01 line extruded*

30. Create a copy of *Ext01* and name it **Drill02**.

31. Make sure *Drill02* is selected in the viewport and then set **100** in the **Amount** spinner of the **Parameters** rollout for the **Extrude** modifier of the **Modify** tab in the **Command Panel**.

32. Choose the **Hierarchy** tab in the **Command Panel** and make sure the **Pivot** button is chosen. In the **Adjust Pivot** rollout, choose the **Affect Pivot Only** button in the **Move/Rotate/Scale** area. In the Front viewport, move the pivot point so that it is aligned with the bottom of *Drill02*, as shown in Figure P1-19. Now, choose the **Affect Pivot Only** button to exit the adjustment mode.

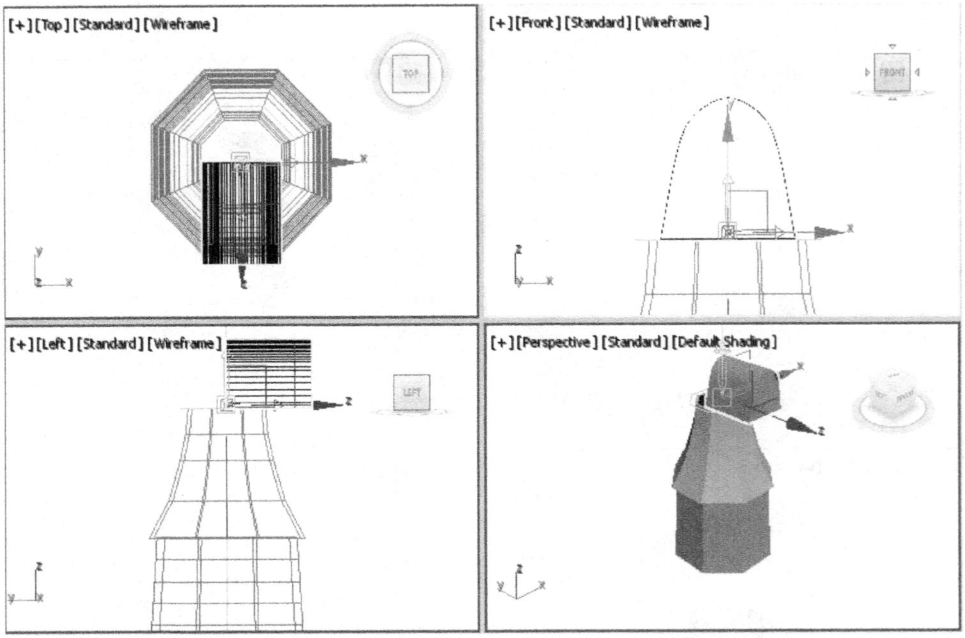

Figure P1-19 *The pivot point aligned at the bottom of Drill02*

33. In the Front viewport, scale *Drill02* to **90%** by using the **Select and Uniform Scale** tool.

Tip: *You can also scale an object precisely by using the* ***Scale Transform Type-In*** *dialog box.*

34. Choose the **Display** tab in the **Command Panel** and then choose the **Unhide All** button in the **Hide** rollout; the hidden objects are now displayed.

35. Select *Top_drill* and move it down by **-5** units about the **Y** axis using the **Select and Move** tool in the Front viewport.

36. Select *Base_top* and choose **Create > Geometry** in the **Command Panel**. Select **Compound Objects** in the drop-down list, if it is not selected.

37. Invoke the **Boolean** tool from the **Object Type** rollout. Choose the **Add Operands** button in the **Boolean Parameters** rollout. Choose the **Subtract** button from the **Operand Parameters**

rollout and select the **Operands** and **Display as Shaded** check boxes in the **Display** area. Now, select *Top_drill* in any viewport.

38. Invoke the **Select Object** tool from the **Main Toolbar** to complete the operation.

39. Select *Ext01* in the viewport. Invoke the **Boolean** tool from the **Object Type** rollout and choose the **Add Operands** button in the **Boolean Parameters** rollout. Select *Drill02* in any of the viewports. Next, choose the **Subtract** button in the **Operand Parameters** rollout.

40. Invoke the **Select Object** tool from the **Main Toolbar** to exit the Boolean operation.

41. Make sure *Base_top* is selected in the viewport and convert it into editable poly. Choose the **Polygon** button from the **Selection** rollout. Now, choose the **Attach** button from the **Edit Geometry** rollout and select *Ext01* in the viewport; *Base_top* and *Ext01* are now attached.

42. Choose the **Edge** button from the **Selection** rollout and then move the edges inside the *Base_top*, as shown in Figure P1-20.

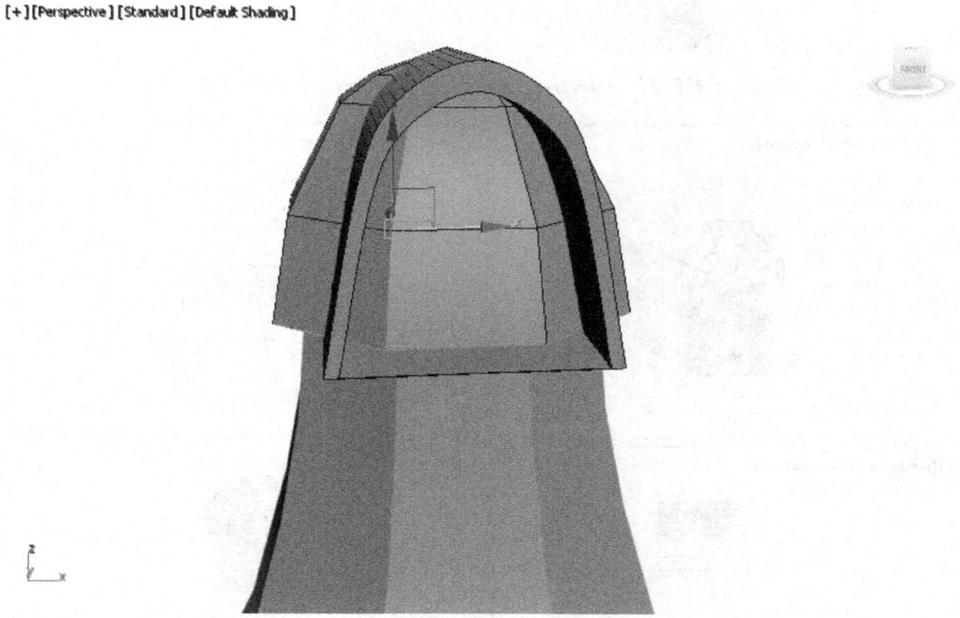

Figure P1-20 *Moving edges of Base_top*

43. Choose the **Polygon** button from the **Selection** rollout and select the inner polygons of the *Base_top*, as shown in Figure P1-21. Next, delete the selected polygons to hollow the *Base_top* from inside, as shown in Figure P1-22.

44. Invoke the **Select Object** tool from the **Main Toolbar** to exit the Boolean operation.

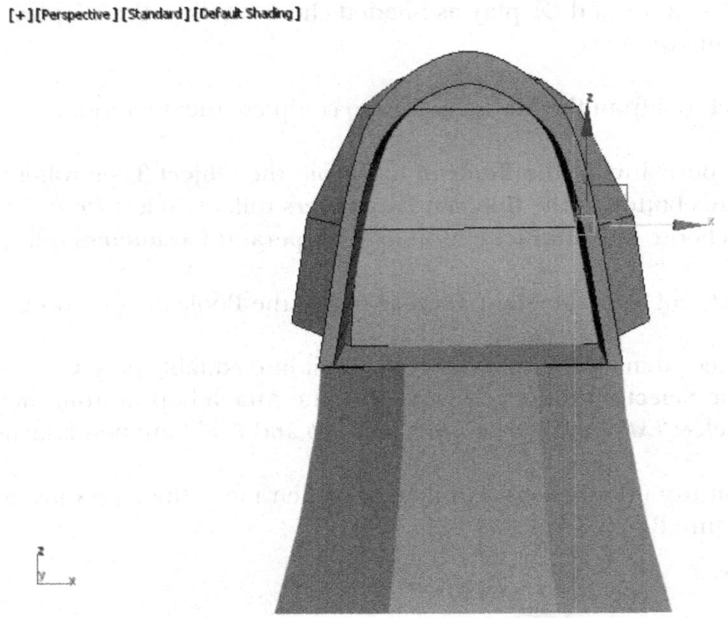

Figure P1-21 *Selecting the inner polygons of Base_top*

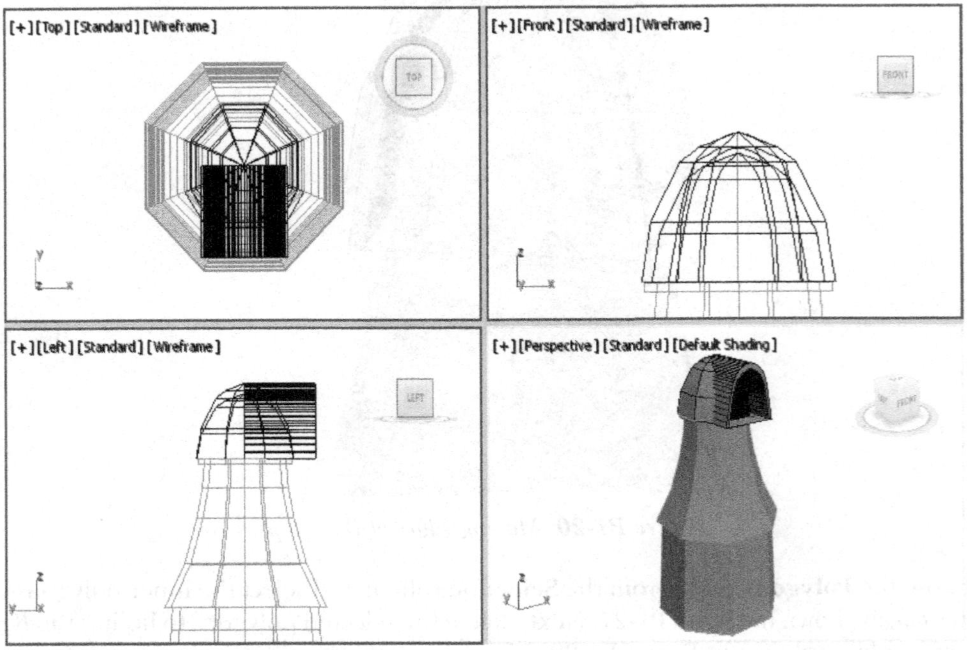

Figure P1-22 *Deleting the inner polygons of the Base_top*

Creating the Second Level Deck

In this section, you will create the deck and its multiple copies at the top of the windmill. This deck will be placed on the base structure and will serve as a stand for the middle structure of the windmill.

1. Select *Base_mid* and choose the **Display** tab in the **Command Panel**. In the **Hide** rollout, choose the **Hide Unselected** button; the middle section of the building remains visible.

2. Invoke the **Zoom Extents All** tool to zoom the viewports to their extents.

3. Choose **Create > Shapes** in the **Command Panel**. Invoke the **Line** tool from the **Object Type** rollout. Make sure the **Corner** radio button is selected in the **Initial Type** and **Drag Type** area from the **Creation Method** rollout.

4. Activate the Top viewport. Expand the **Keyboard Entry** rollout and draw an isosceles triangle starting at (**0,0,0**), through points (**50,-120,0**) and (**-50,-120 ,0**), and back to (**0,0,0**), and then choose the **Close** button to close the spline. In the **Name and Color** rollout, enter **Deck01**.

Note

If you are using the cursor to draw line, you may need to first zoom in to display the subgrid.

5. Make sure *Deck01* is selected and then choose the **Modify** tab in the **Command Panel**. Next, select **Extrude** in the **Modifier List** drop-down list under the **OBJECT-SPACE MODIFIERS** section. Set **–5.0** in the **Amount** spinner of the **Parameters** rollout for the **Extrude** modifier; the *Deck01* is extruded. Now, select **Cap Holes** in the **Modifier List** drop-down list under the **OBJECT-SPACE MODIFIERS** section.

6. In the **Modifier List** drop-down list, select **UVW Map** under the **OBJECT-SPACE MODIFIERS** section. Make sure the **Real-World Map Size** check box is cleared in the **Mapping** area of the **Parameters** rollout.

7. Activate the Front viewport and invoke the **Zoom Extents** tool so that the triangle is visible in the viewport.

8. Move *Deck01* up in the Front viewport until it is even with the bottom of *Base_mid*, as shown in Figure P1-23.

9. Make sure the **Box** radio button is selected and then set values as given next:

   Length: **50** Width: **100** Height: **100**

   This adds a **UVW Map** modifier to the object so that the material appears correct when it is added later. Since the object is going to be arrayed, it is more appropriate to apply the modifier one time to the original object, rather than to select and apply the modifier to each of the copies once the array is created.

10. Activate the Top viewport. Invoke the **Use Transform Coordinate Center** tool from the Pivot Point flyout. Next, select **World** in the **Reference Coordinate System** drop-down list. Select *Deck01*, if it is not already selected.

    These settings will cause all transformations to be centered on the origin.

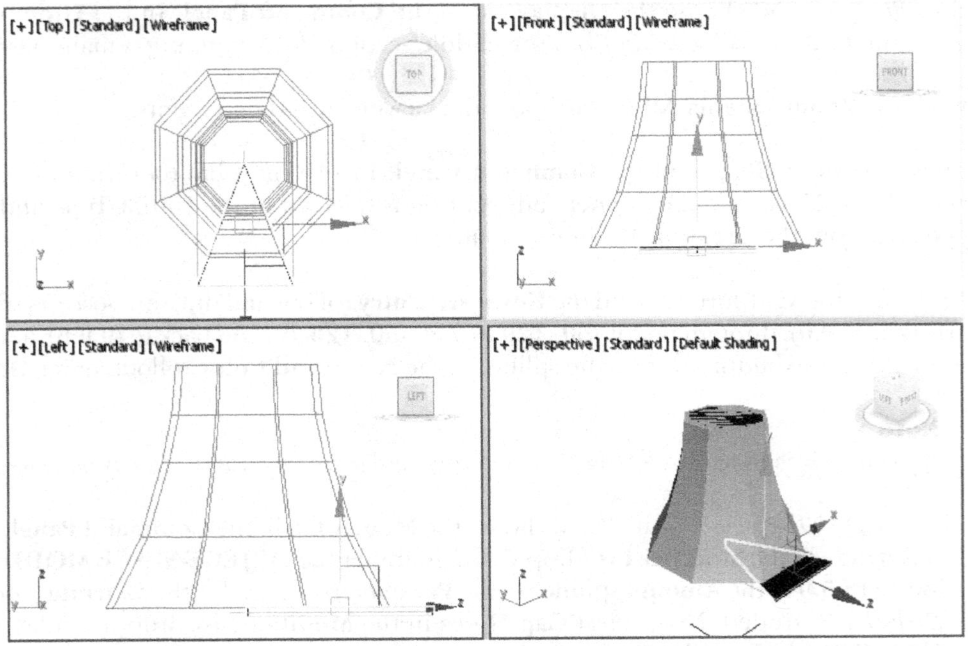

***Figure P1-23** Deck01 extruded in all viewports*

11. Choose **Tools > Array** from the menu bar; the **Array** dialog box is displayed, as shown in Figure P1-24.

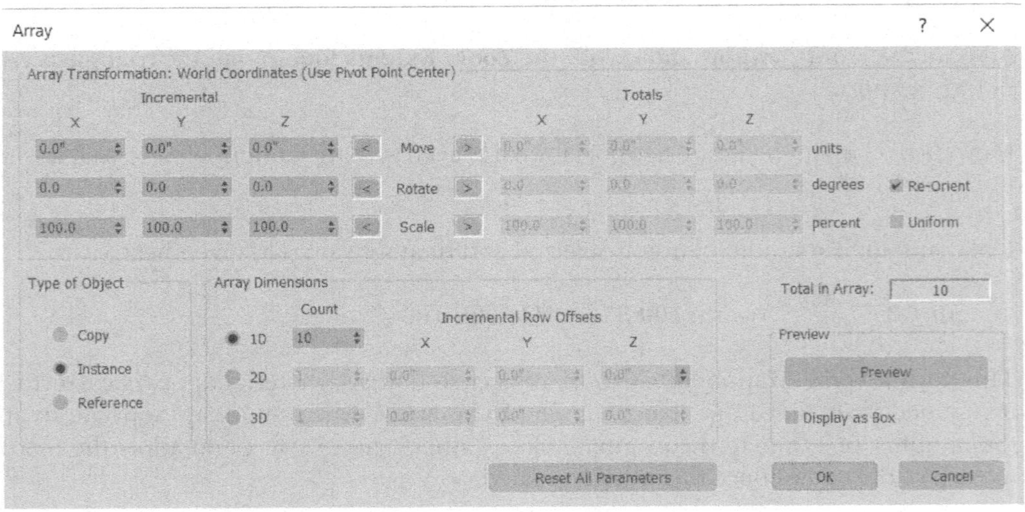

***Figure P1-24** The **Array** dialog box*

12. Choose the **Reset All Parameters** button in the **Array** dialog box. Set **45** in the **Incremental Z** spinner in the **Rotate** row. In the **Array Dimensions** area, make sure the **1D** radio button is selected and set **8** in the **Count** spinner. Select the **Copy** radio button in the **Type of Object**

area and choose the **OK** button to create the array, as shown in Figure P1-25. However, there may be a small overlap of the objects that need to be adjusted in order to get a perfect shape.

Note

If time permits, you may need to reposition the UVW map modifier gizmo for each of the Deck objects so that each deck segment has a slightly different material pattern. This will add to the realism of the model.

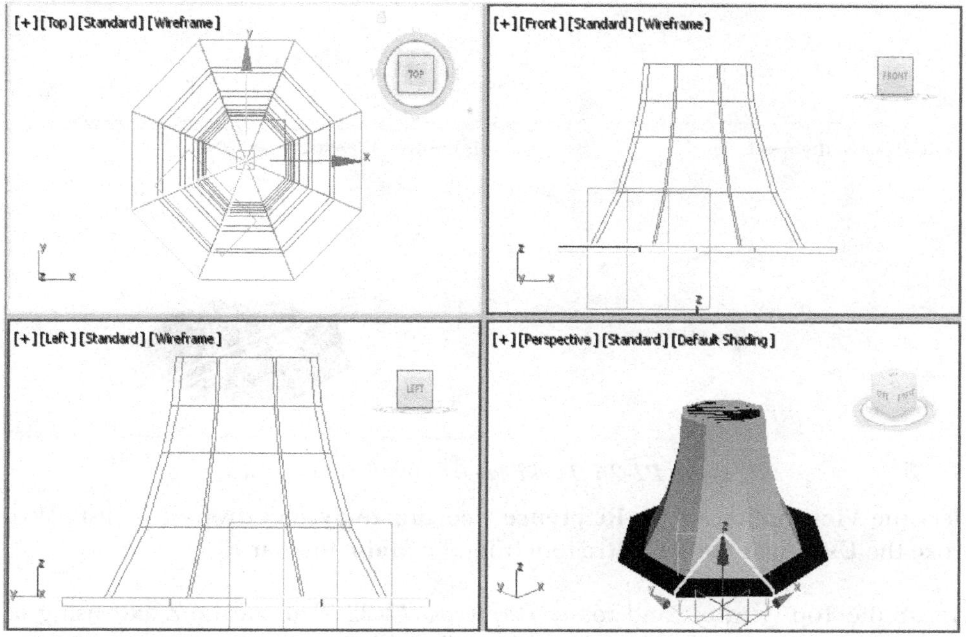

Figure P1-25 *The array created*

13. Select only the decks and choose the **Display** tab in the **Command Panel**. In the **Hide** rollout, choose the **Hide Unselected** button; only the decks are visible.

14. Activate the Top viewport. Choose **Create > Geometry** in the **Command Panel** and select **Standard Primitives** in the drop-down list.

15. Invoke the **Box** tool from the **Object Type** rollout. Expand the **Keyboard Entry** rollout and set values as given next:

    Length: **250** Width: **5** Height: **10**

16. In the **Parameters** rollout, make sure the **Real-World Map Size** check box is cleared.

17. Choose the **Create** button in the **Keyboard Entry** rollout; a box is created. Enter **DeckBeam01** in the **Name and Color** rollout. Next, exit the **3D Snap Toggle** tool and the **Percent Snap** tool from the **Main Toolbar**.

18. Activate the Left viewport and invoke the **Zoom Extents** tool. Move *DeckBeam01* up until its top is aligned with the bottom of the deck edge, as shown in Figure P1-26.

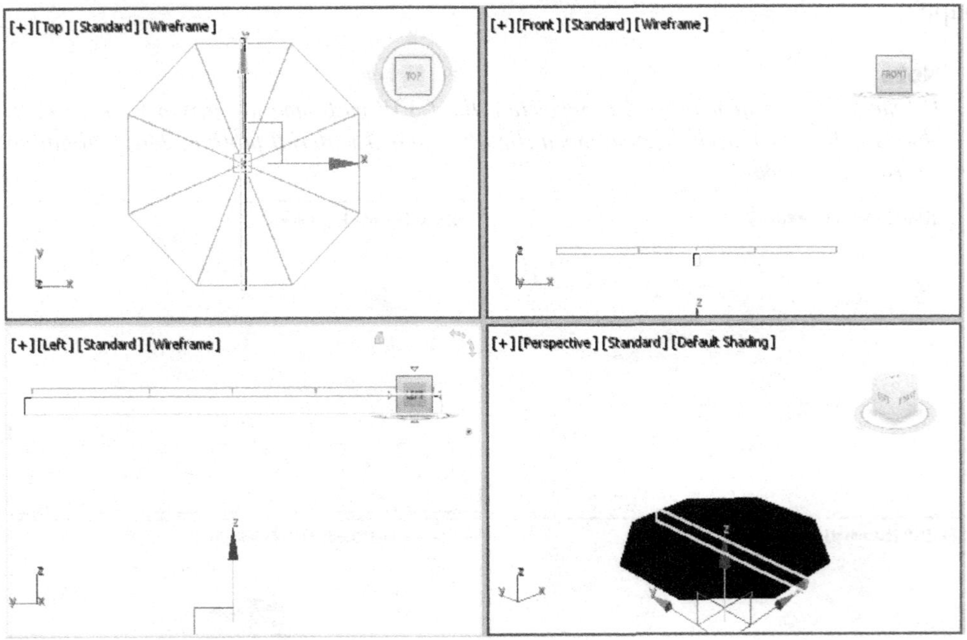

Figure P1-26 *DeckBeam01 created and aligned*

19. Select the **View** option in the **Reference Coordinate System** drop-down list. Also, invoke the **Use Pivot Point Centre** tool from the **Main Toolbar**.

20. Activate the Top viewport and rotate *DeckBeam01* –22.5° about the Z axis using its pivot point as the center of transformation. The default pivot point should be at 0,0,0.

21. Choose **Tools > Array** from the menu bar; the **Array** dialog box is displayed. Choose the **Reset All Parameters** button. Enter the value **45** in the **Incremental Z** spinner in the **Rotate** row. In the **Array Dimensions** area, select the **1D** radio button and enter the value **4** in the **Count** spinner. Select the **Copy** radio button in the **Type of Object** area and then choose the **OK** button; copies of *DeckBeam01*are created, as shown in Figure P1-27.

22. Make sure the Top viewport is active and then choose **Create > Geometry** in the **Command Panel**. Invoke the **Box** tool from the **Object Type** rollout.

23. Enter the following values in the **Keyboard Entry** rollout:

X:	**40**
Y:	**-120**
Z:	**144**
Length:	**40**
Width:	**3**
Height:	**6**

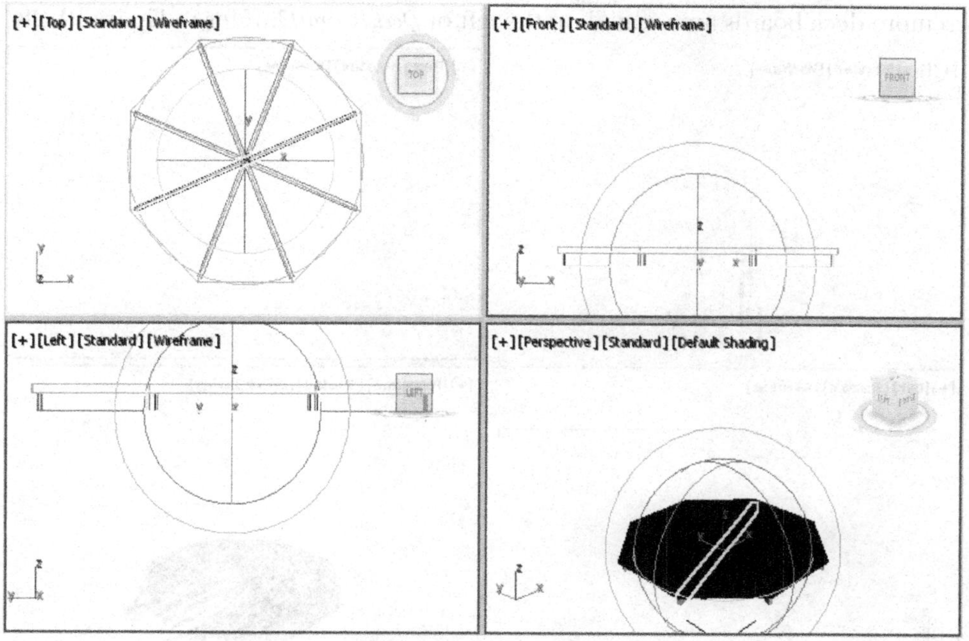

Figure P1-27 *The copies of DeckBeam01 created*

24. Make sure the **Real-World Map Size** check box in the **Parameters** rollout is cleared. Choose the **Create** button in the **Keyboard Entry** rollout and enter **DeckBoard01** in the **Name and Color** rollout.

25. Enter the values given next in the **Keyboard Entry** rollout to create two more boxes in the Top viewport. For each box, clear the **Real-World Map Size** check box in the **Parameters** rollout and then choose the **Create** button from the **Keyboard Entry** rollout. Next, rename each box in the **Name and Color** rollout.

Name and Color:	**DeckBoard02**	**DeckBoard03**
X:	**20**	**0**
Y:	**–95**	**–70**
Z:	**144**	**144**
Length:	**90**	**140**
Width:	**3**	**3**
Height:	**6**	**6**

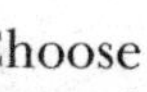

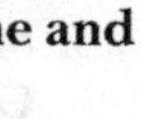

26. Select *DeckBoard01* and *DeckBoard02* in the Top viewport. Invoke the **Use Selection Center** tool from the Pivot Point flyout in the **Main Toolbar**. Select the **Screen** option from the **Reference Coordinate System** drop-down list. Next, invoke the **Mirror** tool; the **Mirror: Screen Coordinates** dialog box is displayed.

27. Make sure the **X** radio button in the **Mirror Axis** area is selected and set **– 60** in the **Offset** spinner. Select the **Copy** radio button in the **Clone Selection** area and choose the **OK** button to mirror the objects, refer to Figure P1-28.

Two more deck boards are created on the left of *DeckBoard03*, refer to Figure P1-28.

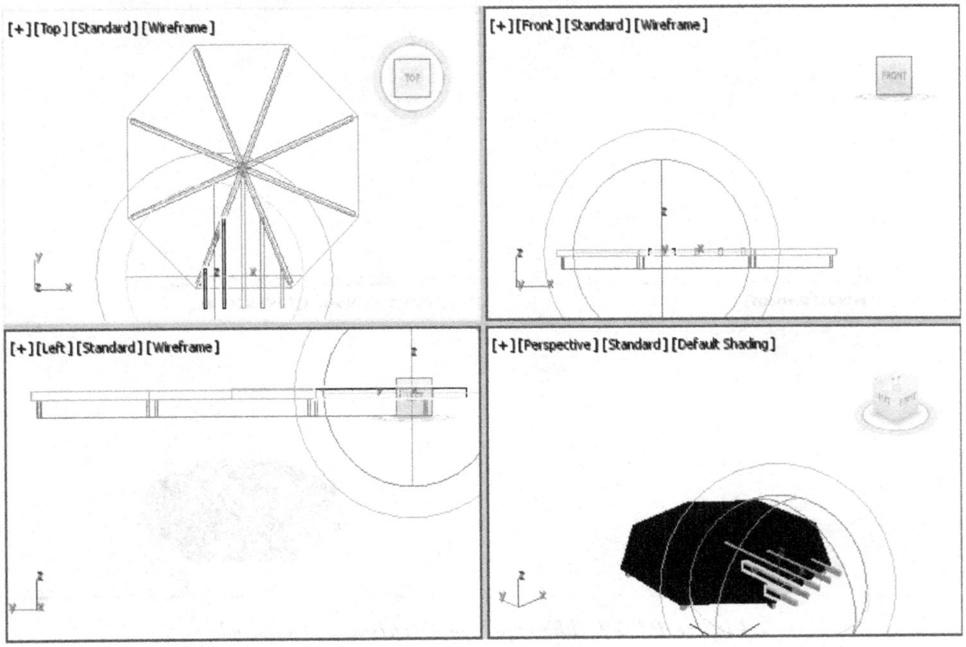

Figure P1-28 *Two more deck boards created*

28. Select all the five deck board objects. Choose **Group > Group** from the menu bar; the **Group** dialog box is displayed. Enter **DeckBoard_Group01** in the **Group name** text box and choose the **OK** button; the dialog box is closed and the five deckboards are grouped together.

29. Invoke the **Use Transform Coordinate Center** tool from the Pivot Point flyout. Also, select **World** from the **Reference Coordinate System** drop-down list.

Note

By default, the ***View*** *option is selected in the* ***Reference Coordinate System*** *drop-down list. If you want to rotate objects about the world origin, the* ***Reference Coordinate System*** *settings may need to be changed to* ***World*** *if the viewport is not centered on the origin. After rotating the objects, you may need to set the* ***Reference Coordinate System*** *back to* ***View*** *to avoid confusion.*

30. Choose **Tools > Array** from the menu bar; the **Array** dialog box is displayed. Choose the **Reset All Parameters** button and set **45** in the **Incremental Z** spinner in the **Rotate** row. In the **Array Dimensions** area, select the **1D** radio button, if it is not selected and set **8** in the **Count** spinner. Select the **Copy** radio button in the **Type of Object** area and then choose the **OK** button; the dialog box is closed and the array is created, as shown in Figure P1-29.

Creating the Deck Rails

In this section, you will create the deck rails by using the **Line** tool.

1. Activate the Front viewport and maximize it to zoom in on the right side of the objects.

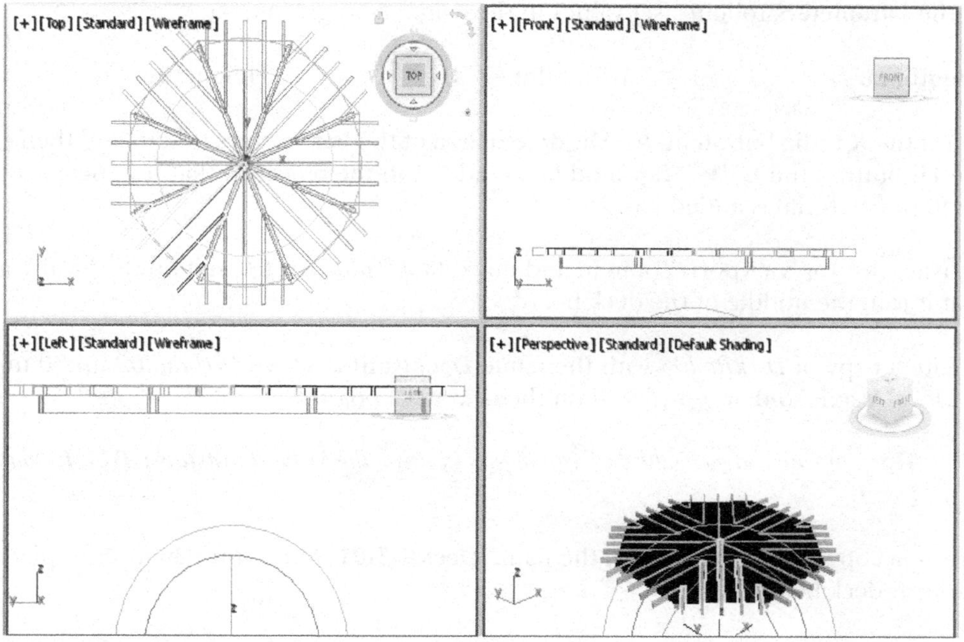

***Figure P1-29** The array of deck support boards created using the **Array** tool*

2. Choose **Create > Shapes** in the **Command Panel**. Invoke the **Line** tool from the **Object Type** rollout. Make sure the **Corner** radio button is selected in the **Initial Type** and **Drag Type** areas from the **Creation Method** rollout.

3. Expand the **Keyboard Entry** rollout and set values in the spinners as given next. Make sure you choose the **Add Point** button after entering each set of XYZ coordinates.

X: **120**	Y: **155**	Z: **0**
X: **130**	Y: **190**	Z: **0**
X: **135**	Y: **190**	Z: **0**
X: **130**	Y: **170**	Z: **0**
X: **140**	Y: **150**	Z: **0**
X: **135**	Y: **150**	Z: **0**
X: **130**	Y: **160**	Z: **0**
X: **125**	Y: **150**	Z: **0**
X: **120**	Y: **150**	Z: **0**

4. Close the shape and enter **DeckRail03** in the **Name and Color** rollout.

5. Make sure *DeckRail03* is selected and then choose the **Modify** tab in the **Command Panel**. Select **Extrude** in the **Modifier List** drop-down list under the **OBJECT-SPACE MODIFIERS** section. In the **Parameters** rollout for the extrude modifier, enter **2** in the **Amount** spinner.

6. Select **UVW Map** in the **Modifier List** drop-down list under the **OBJECT-SPACE MODIFIERS** section. Make sure the **Real-World Map Size** check box is cleared in the **Mapping** area of the **Parameters** rollout, and select the **Box** radio button.

7. In the **Parameters** rollouts, set values in the spinners as given next:

   Length: **20** Width: **20** Height: **20**

8. Select the **X** radio button in the **Alignment** area of the **Parameters** rollout and then choose the **Fit** button; the **UVW Map** modifier is added to the object so that it appears correctly when the material is added later.

9. Activate the Top viewport. Zoom in and move *DeckRail03* up 1.0 unit on the local Y axis so that it is in the middle of the deck board.

10. Create a copy of *DeckRail03* with the name **DeckRail02**. Move *DeckRail02* up 20 units on the local Y axis so that it is placed on the next deck board.

> **Tip:** *You can also move the deck rail objects by using the* ***Move Transform Type-In*** *dialog box.*

11. Create a copy of *DeckRail02* with the name **DeckRail01**. Move it up **20** units to place it on the next deck board.

12. Create a copy of *DeckRail01* with the name **DeckRail04** and move it down **–60** units on the local Y axis so that it is placed on the lower deck board.

13. Create a copy of *DeckRail04* with the name **DeckRail05** and move it down **–20** units so that it is placed on the next lower deck board.

14. Invoke the **Zoom Extents All** tool to view the objects properly in all viewports.

15. Select all the five deck rail objects and choose **Group > Group** from the menu bar; the **Group** dialog box is displayed. Enter **DeckRail_Group01** in the **Group name** text box and choose the **OK** button; the group of *DeckRails* is created, as shown in Figure P1-30.

16. Invoke the **Use Transform Coordinate Center** tool from the Pivot Point flyout. 

17. Choose **Tools > Array** from the menu bar; the **Array** dialog box is displayed. Choose the **Reset All Parameters** button in it. Set **45** in the **Incremental Z** spinner in the **Rotate** row. In the **Array Dimensions** area, select the **1D** radio button, if it is not selected and set **8** in the **Count** spinner. Select the **Copy** radio button in the **Type of Object** area. Next, choose the **OK** button; the dialog box is closed and the array of *DeckRails* is created, as shown in Figure P1-31.

Creating the Handrails and Deck Braces

In this section, you will create the handrails and deck braces by using the **NGon** tool.

1. Choose the **Display** tab in the **Command Panel**. In the **Hide** rollout, choose the **Unhide by Name** button; the **Unhide Objects** dialog box is displayed. Select *NGon001* from the list and choose the **Unhide** button; *NGon001* is displayed in the viewports; the other hidden objects remain hidden.

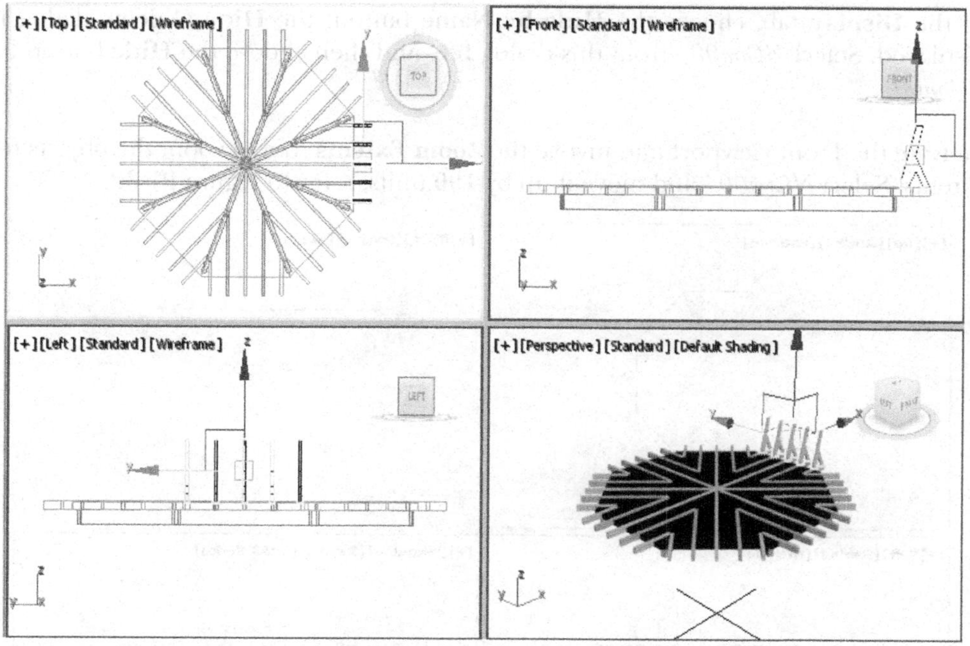

***Figure P1-30** DeckRail_Group01 created in all viewports*

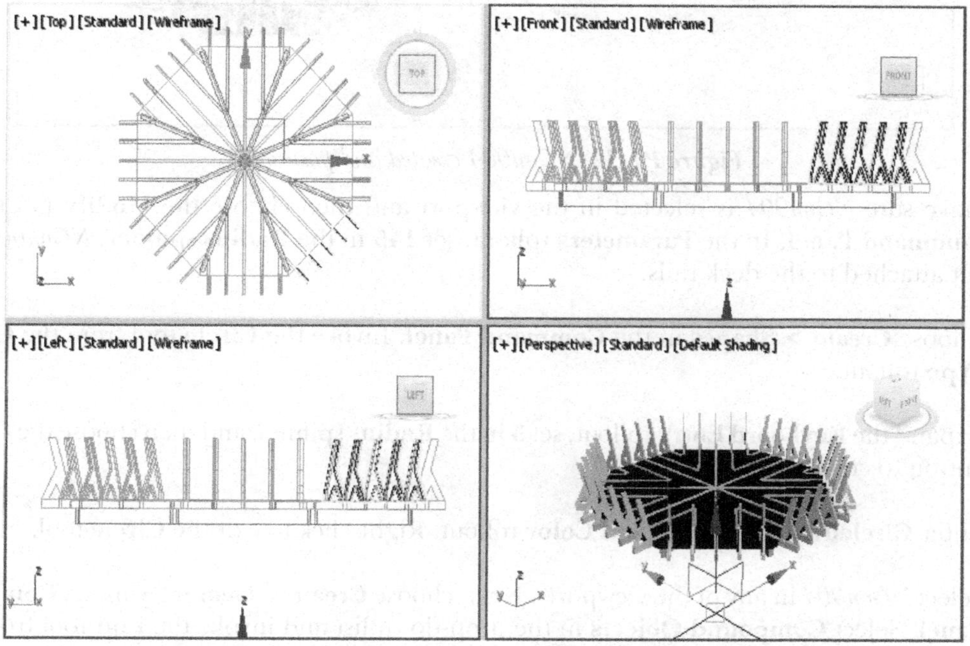

***Figure P1-31** The set of DeckRail_Group objects created*

2. Select *NGon001* and create its copy with the name **NGon004**.

3. In the **Display** tab, choose the **Hide by Name** button; the **Hide Objects** dialog box is displayed. Select *NGon001* from this dialog box and then choose the **Hide** button to hide *NGon001*.

4. Activate the Front viewport and invoke the **Zoom Extents** tool to zoom the objects to their extents. Select *NGon004* and move it up by **190** units, refer to Figure P1-32.

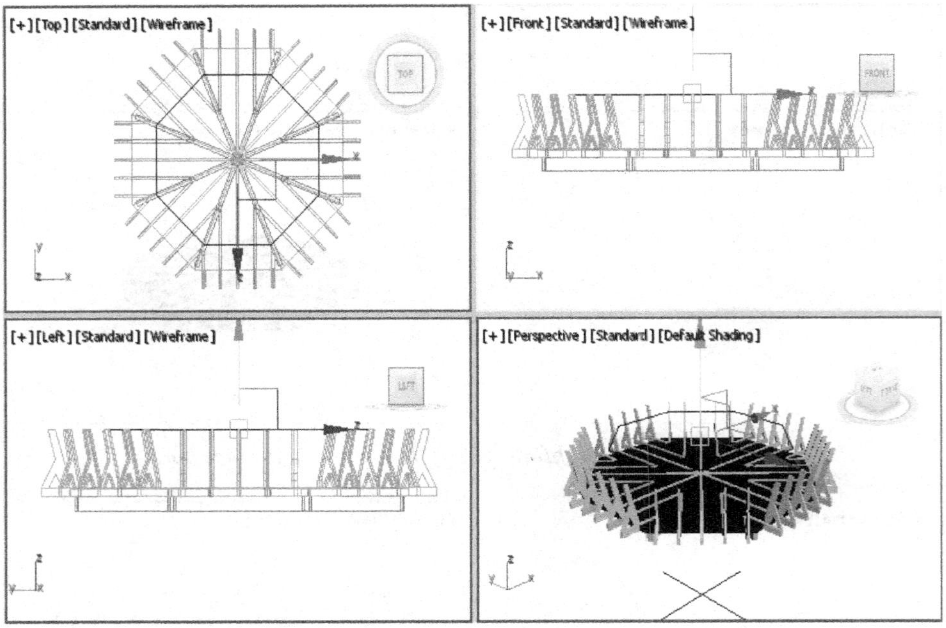

Figure P1-32 *NGon004 created and positioned*

5. Make sure *NGon004* is selected in the viewport and then choose the **Modify** tab in the **Command Panel**. In the **Parameters** rollout, set **145** in the **Radius** spinner; *NGon004* will get attached to the deck rails.

6. Choose **Create > Shapes** in the **Command Panel**. Invoke the **Circle** tool from the **Object Type** rollout.

7. Expand the **Keyboard Entry** rollout, set **3** in the **Radius** spinner, and then choose the **Create** button to create a circle.

8. Enter **Circle01** in the **Name and Color** rollout. Right-click to exit the **Circle** tool.

9. Select *NGon004* in any of the viewports. Next, choose **Create > Geometry** in the **Command Panel**. Select **Compound Objects** in the drop-down list and invoke the **Loft** tool from the **Object Type** rollout.

10. In the **Smoothing** area of the **Surface Parameters** rollout, make sure the **Smooth Length** and **Smooth Width** check boxes are selected. In the **Mapping** area, select the **Apply Mapping** check box. Set **8** in the **Length Repeat** spinner and **2** the **Width Repeat** spinner. Make sure the value **0** is entered in the **Shape Steps** spinner of the **Skin Parameters** rollout.

11. Choose the **Get Shape** button in the **Creation Method** rollout. Select *Circle01* in any of the viewports; *Circle01* is lofted along *NGon004* path to create a handrail, as shown in Figure P1-33. In the **Name and Color** rollout, assign the name **Handrail** to the loft object.

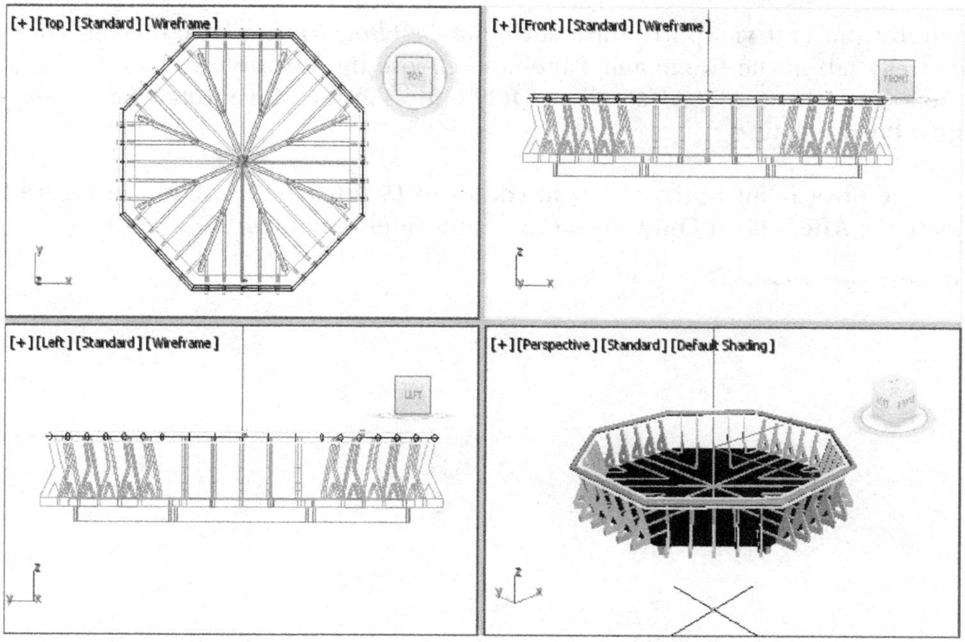

Figure P1-33 *Handrail created in viewports*

12. Choose the **Display** tab in the **Command Panel**. In the **Hide** rollout, choose the **Unhide by Name**; the **Unhide Objects** dialog box is displayed. Select *Base_lo* from the list and choose the **Unhide** button to unhide *Base_lo*.

13. Activate the Left viewport. Choose **Create > Geometry** in the **Command Panel**. Select **Standard Primitives** in the drop-down list. Next, invoke the **Box** tool from the **Object Type** rollout.

14. Expand the **Keyboard Entry** rollout, enter the values in the spinners as given next:

X:	**115**
Y:	**115**
Z:	**–2**
Length:	**70**
Width:	**6**
Height:	**4**

15. Choose the **Create** button to create the box and enter **DeckBrace01** in the **Name and Color** rollout.

16. Choose the **Modify** tab in the **Command Panel**. Select **UVW Map** in the **Modifier List** drop-down list under the **OBJECT-SPACE MODIFIERS** section. In the **Mapping** area of the **Parameters** rollout, make sure the **Real-World Map Size** check box is cleared.

17. Select the **Box** radio button, and enter the values in the spinners as given next:

    Length: **70** Width: **70** Height: **70**

18. Maximize the Left viewport. Make sure that *DeckBrace01* is selected. Next, choose the **Hierarchy** tab in the **Command Panel** and choose the **Pivot** button, if it is not already chosen. Now, choose the **Affect Pivot Only** button in the **Move/Rotate/Scale** area of the **Adjust Pivot** rollout.

19. Move the pivot point to the top right corner of *DeckBrace01*, as shown in Figure P1-34. Choose the **Affect Pivot Only** button to exit the adjustment mode.

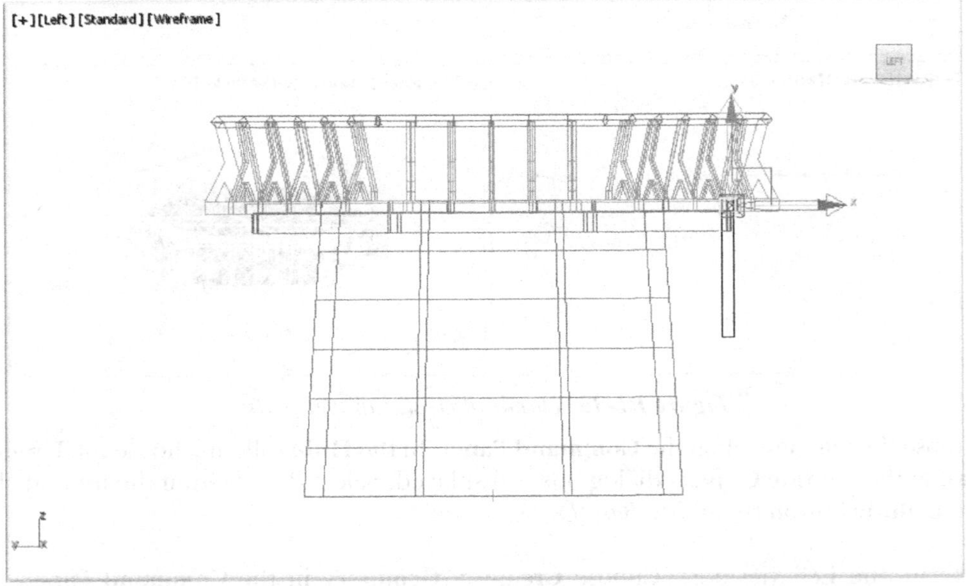

Figure P1-34 *The pivot point of DeckBrace01 moved to the top right corner*

20. Invoke the **Zoom Extents** tool to zoom the objects to their extents in the viewport.

21. Invoke the **Select and Rotate** tool and rotate *DeckBrace01* about the local X axis, as shown in Figure P1-35. Use the object's pivot point as the center of rotation and make sure that **View** is selected in the **Reference Coordinate System** drop-down list.

22. Invoke the **Maximize Viewport Toggle** tool to return to the four viewport configuration. Make sure the **Use Transform Coordinate Center** tool in the Pivot Point flyout is activated. Next, select **World** in the **Reference Coordinate System** drop-down list.

23. Activate the Top viewport. Make sure *DeckBrace01* is selected and choose **Tools > Array** from the menu bar; the **Array** dialog box is displayed. Choose the **Reset All Parameters** button. Set **45** in the **Incremental Z** spinner in the **Rotate** row. In the **Array Dimensions** area, select the **1D** radio button, if it is not selected and set **8** in the **Count** spinner. Now, select the **Copy** radio button in the **Type of Object** area and then choose the **OK** button; the copies of *DeckBrace01* are created, as shown in Figure P1-36.

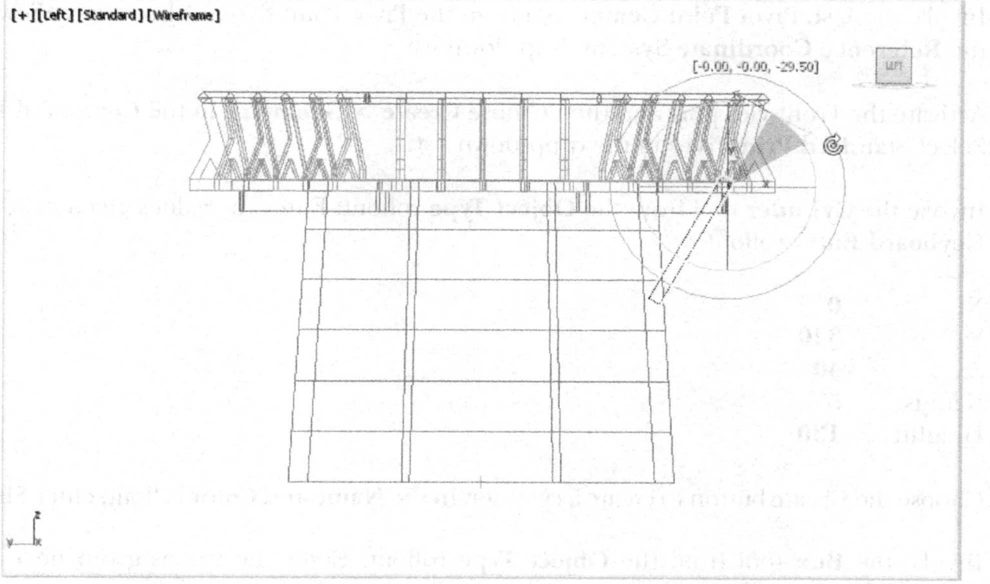

Figure P1-35 DeckBrace01 rotated in the Left viewport

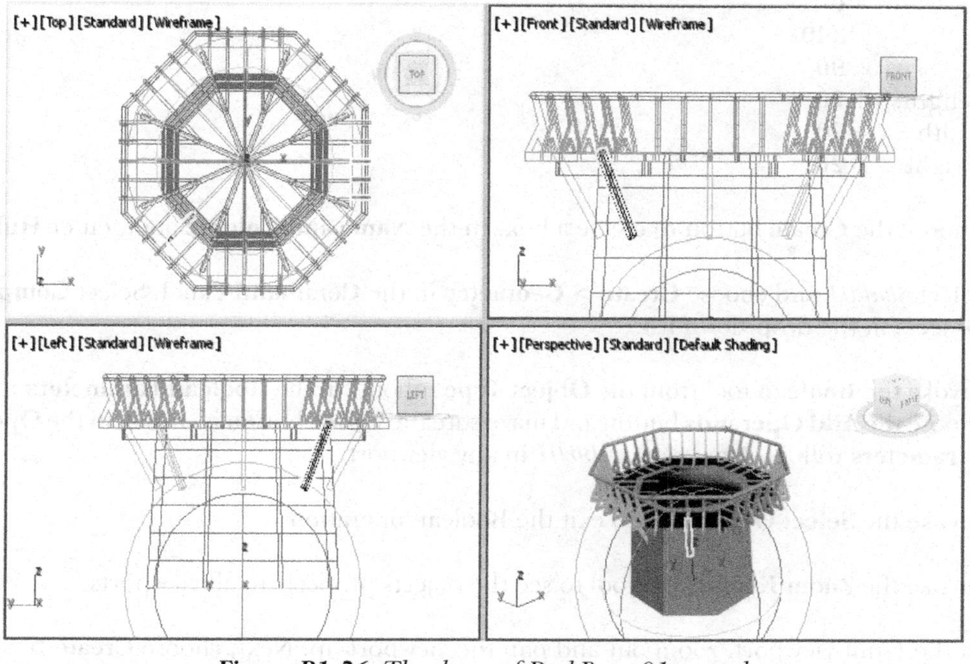

Figure P1-36 The clones of DeckBrace01 created

Creating the Windmill Blades

In this section, you will create the blades of windmill by using the **Cylinder** tool.

1. Select all objects in any of the viewports and right-click; a quad menu is displayed. Choose **Hide Selection** from the quad menu; all objects get hidden.

2. Invoke the **Use Pivot Point Center** tool from the Pivot Point flyout. Also, select **View** from the **Reference Coordinate System** drop-down list.

3. Activate the Front viewport and then choose **Create > Geometry** in the **Command Panel**. Select **Standard Primitives** in the drop-down list.

4. Invoke the **Cylinder** tool from the **Object Type** rollout. Enter the values given next in the **Keyboard Entry** rollout:

X:	**0**
Y:	**340**
Z:	**–40**
Radius:	**5**
Height:	**130**

5. Choose the **Create** button to create a cylinder. In the **Name and Color** rollout, enter **Shaft01**.

6. Invoke the **Box** tool from the **Object Type** rollout. Enter the values given next in the **Keyboard Entry** rollout:

X:	**0**
Y:	**340**
Z:	**90**
Length:	**20**
Width:	**20**
Height:	**20**

7. Choose the **Create** button to create a box. In the **Name and Color** rollout, enter **Hub01**.

8. Select *Shaft01* and choose **Create > Geometry** in the **Command Panel**. Select **Compound Objects** in the drop-down list.

9. Invoke the **Boolean** tool from the **Object Type** rollout. In the **Boolean Parameters** rollout, choose the **Add Operands** button and make sure the **Union** button is chosen in the **Operand Parameters** rollout. Now, select *Hub01* in any viewport.

10. Invoke the **Select Object** tool to exit the Boolean operation.

11. Invoke the **Zoom Extents All** tool to see the objects properly in all viewports.

12. In the Front viewport, zoom out and pan the viewport up. Next, choose **Create > Shapes** in the **Command Panel**. Invoke the **Line** tool from the **Object Type** rollout. Make sure the **Corner** radio button is selected in the **Initial Type** and **Drag Type** areas from the **Creation Method** rollout.

13. Draw a **130** unit line vertically down from the center of the hub's lower edge. Use snaps as needed.

14. Invoke the **Rectangle** tool from the **Object Type** rollout. Draw a 10-unit square anywhere in the Front viewport, as shown in Figure P1-37.

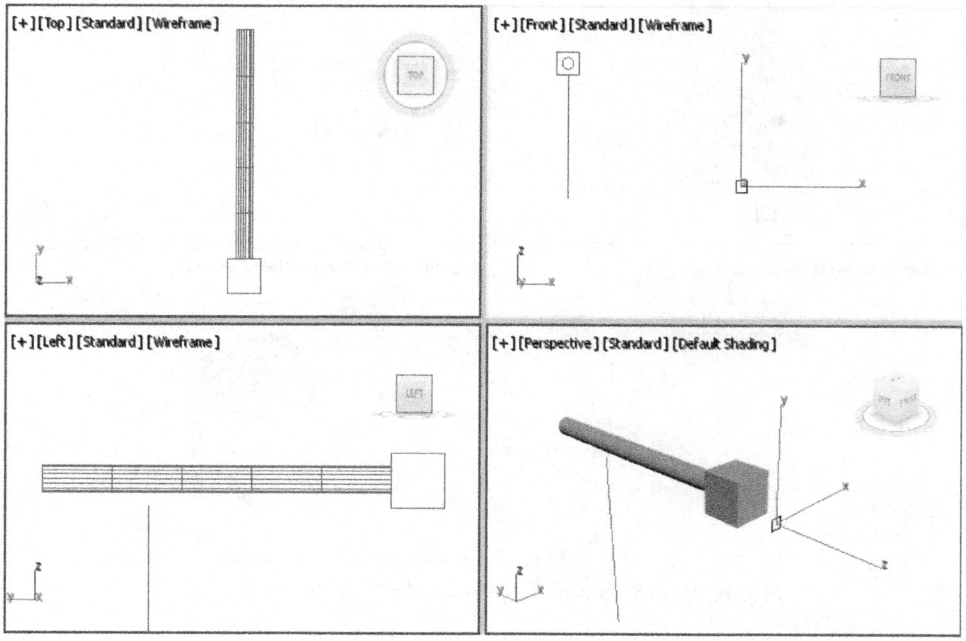

Figure P1-37 *A line and a square created*

15. Select the line and choose **Create > Geometry** in the **Command Panel**. Select **Compound Objects** from the drop-down list. Invoke the **Loft** tool from the **Object Type** rollout.

16. In the **Creation Method** rollout, choose the **Get Shape** button. Next, select the square in any viewport; the square is lofted along the line. In the **Name and Color** rollout, enter **MainRib01**.

17. Make sure the loft object is selected in the viewport and then choose the **Modify** tab in the **Command Panel**. Expand the **Deformations** rollout and choose the **Scale** button; the **Scale Deformation(X)** dialog box is displayed. Now, move the right hand vertex down to **50**. Close the dialog box.

18. Activate the Left viewport and invoke the **Zoom Extents** tool. Move *MainRib01* to the right so that it is centered on the hub. Choose the **Display** tab of the **Command Panel**. In the **Hide** rollout, choose the **Unhide by Name** button; the **Unhide Objects** dialog box is displayed. Select *Base_top* from this dialog box and then choose the **Unhide** button; the hidden objects are displayed, as shown in Figure P1-38.

19. Invoke the **Zoom Extents All** tool to zoom the objects to their extents.

20. Activate the Front viewport. Choose **Create > Geometry** in the **Command Panel**. Select **Standard Primitives** in the drop-down list.

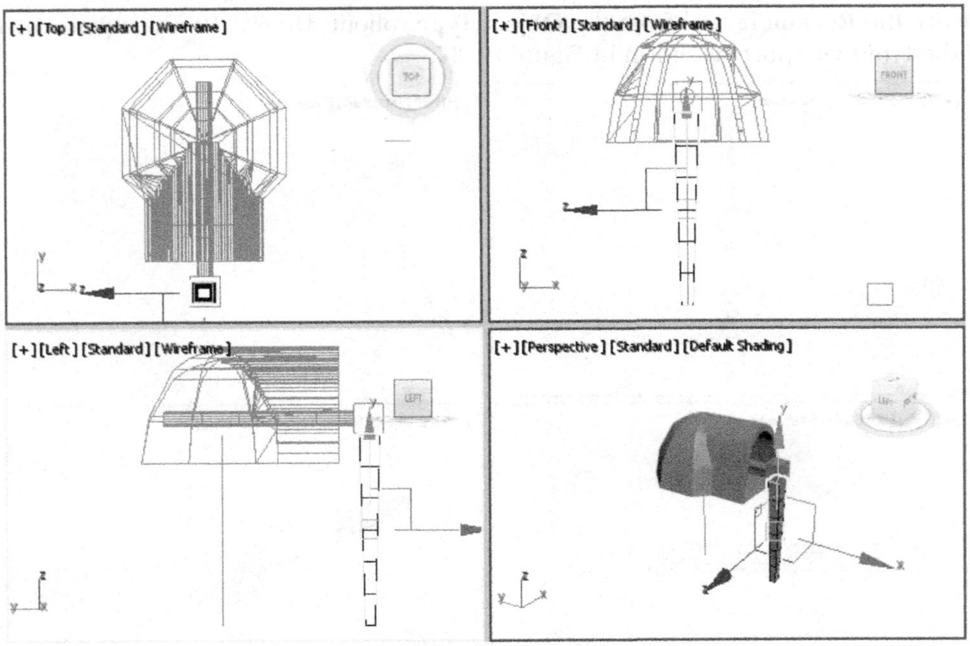

***Figure P1-38** Hidden objects displayed in the viewports*

21. Invoke the **Box** tool from the **Object Type** rollout. Set values as given next in the **Keyboard Entry** rollout:

X:	**10**
Y:	**260**
Z:	**99.5**
Length:	**118**
Width:	**2**
Height:	**1**

22. Choose the **Create** button; a box is created. Enter **Rib01** in the **Name and Color** rollout.

23. Create a copy of *Rib01* and name it as **Rib02**. In the Front viewport, move *Rib02* to the right by **10** units; two vertical ribs are created on which you will place the horizontal ribs, as shown in Figure P1-39.

24. Invoke the **Box** tool from the **Object Type** rollout. Make sure the Front viewport is active. In the **Keyboard Entry** rollout, set values as given next:

X:	**15**
Y:	**310**
Z:	**99.5**
Length:	**2**
Width:	**28**
Height:	**1**

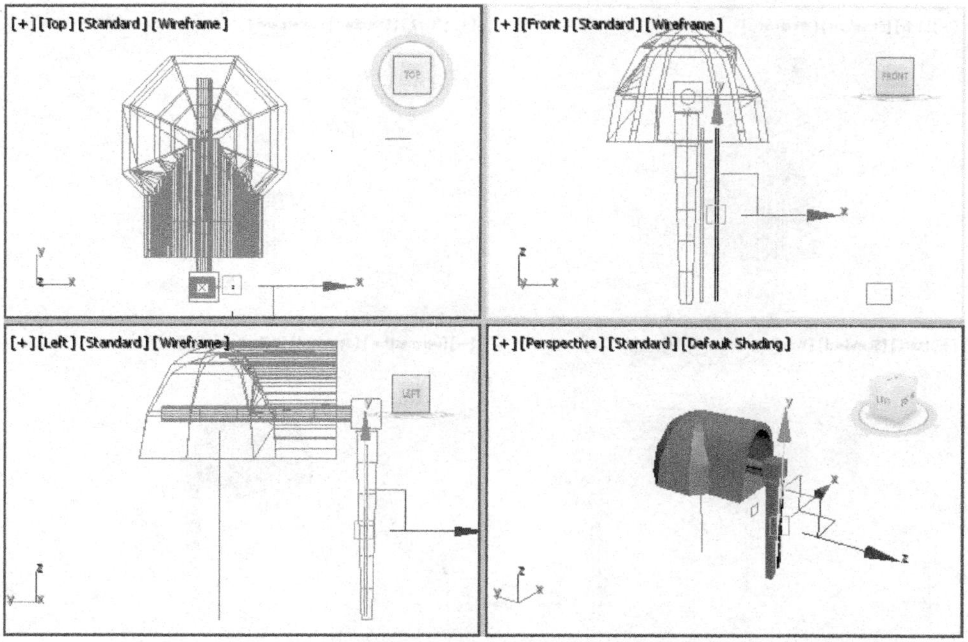

***Figure P1-39** RibO1 and Rib02 created in viewports*

25. Choose the **Create** button; a box is created. In the **Name and Color** rollout, enter **Rib03**; a horizontal rib is created on the vertical ribs.

26. Right-click in the Front viewport to exit the **Box** tool.

27. With *Rib03* selected, choose **Tools > Array** from the menu bar; the **Array** dialog box is displayed.

28. Choose the **Reset All Parameters** button. Set **–10** in the **Incremental Y** spinner in the **Move** row. In the **Array Dimensions** area, select the **1D** radio button, if it is not selected and set **11** in the **Count** spinner. Select the **Copy** radio button from the **Type of Object** area. Choose the **OK** button; the array is created, as shown in Figure P1-40.

29. Select *Rib01* and choose **Create > Geometry** in the **Command Panel**. Select **Compound Objects** from the drop-down list.

30. Invoke the **Boolean** tool from the **Object Type** rollout. In the **Operand Parameters** area of the **Boolean Parameters** rollout, make sure the **Union** button is chosen. In the **Boolean Parameters** rollout, choose the **Add Operands** button and then select *Rib02* in any of the viewports.

31. Invoke the **Select Object** tool to exit the **Boolean** operation.

32. Invoke the **Boolean** tool from the **Object Type** rollout again. Choose the **Add Operands** button in the **Boolean Parameters** rollout and select *Rib03* in any of the viewports. Similarly, combine all the ribs together using the **Boolean** tool.

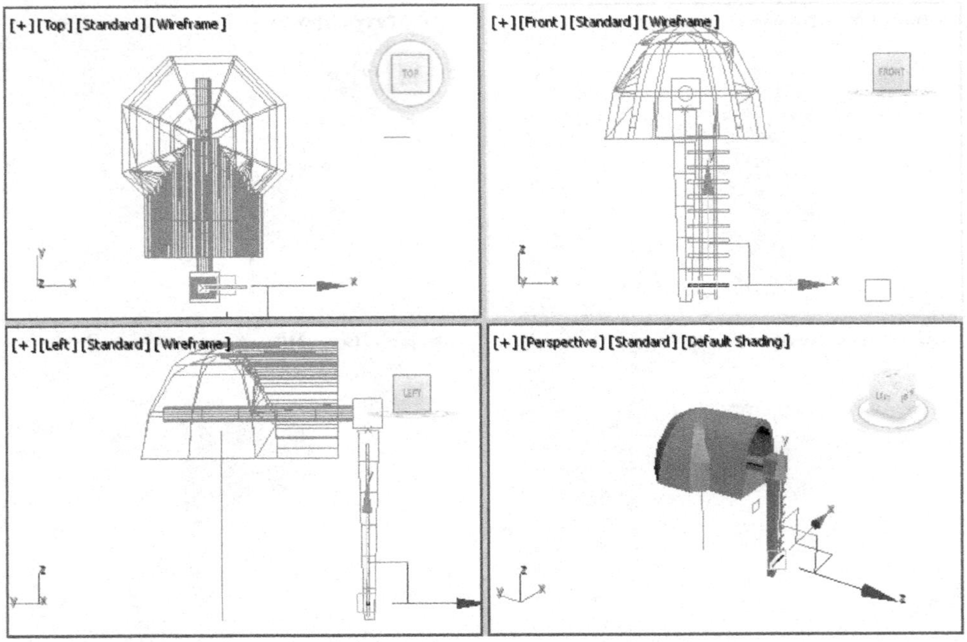

Figure P1-40 *The array of horizontal ribs created*

33. Invoke the **Select Object** tool to exit the Boolean tool.

34. Activate the Top viewport and select *MainRib01*. Rotate it through an angle of –45 degrees about the local Z axis using the object's pivot point as the center.

35. With *MainRib01* selected, invoke the **Boolean** tool from the **Object Type** rollout. Make sure the **Union** button is chosen in the **Operand Parameters** rollout and then choose the **Add Operands** button in the **Boolean Parameters** rollout and select *Rib01* in any of the viewports. The first blade of the windmill is created.

36. Invoke the **Select Object** tool to exit the Boolean command.

37. Make sure *MainRib01* is selected in the viewport and then choose the **Hierarchy** tab in the **Command Panel**. Choose the **Affect Pivot Only** button and move the pivot point to the center of *Shaft01* in the Front viewport. Choose the **Affect Pivot Only** button to exit the adjustment mode.

38. Make sure the Front Viewport is active. With *MainRib01* selected, choose **Tools > Array** from the menu bar; the **Array** dialog box is displayed. Choose the **Reset All Parameters** button.

39. Set **90** in the **Incremental Z** spinner in the **Rotate** row. In the **Array Dimensions** area, select the **1D** radio button, if it is not selected and set **4** in the **Count** spinner.

40. Select the **Copy** radio button in the **Type of Object** area and choose the **OK** button; the blades of the windmill are created, as shown in Figure P1-41.

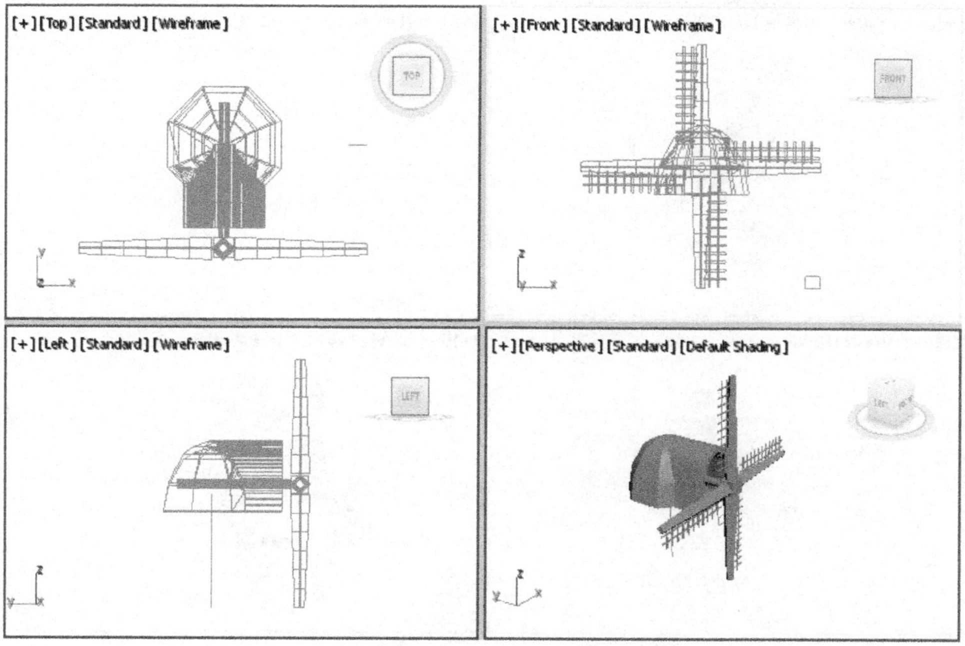

Figure P1-41 *The blades of the windmill created*

41. Invoke the **Zoom Extents All** tool to zoom the objects to their extents.

42. Choose the **Display** tab in the **Command Panel**. In the **Hide** rollout, choose the **Unhide All** button; all hidden objects are displayed, as shown in Figure P1-42. You need to hide *Top_drill* and *Drill02* if they are displayed along with other objects.

43. Invoke the **Zoom Extents All** tool to zoom all objects to their extents, refer to Figure P1-42.

Applying Materials

In this section, you will assign materials to all objects you have created.

1. Select *MainRib01, MainRib002, MainRib003, MainRib004*, and *Shaft01*.

2. Choose **Rendering > Material Editor > Compact Material Editor** from the **Main Toolbar**; the **Material Editor** dialog box is displayed. Choose the **Physical Material** button; the **Material Map/Browser** dialog box is displayed. Choose **Materials > Scanline > Standard (Legacy)** from the **Material Map/Browser** dialog box.

3. Select **Blinn** from the drop-down list in the **Shader Basic Parameters** rollout, if it is not already selected.

4. Expand the **Maps** rollout. Next, choose the **Diffuse Color** map button which is currently labeled as **No Map**; the **Material/Map Browser** dialog box is displayed showing the types of maps that can be added.

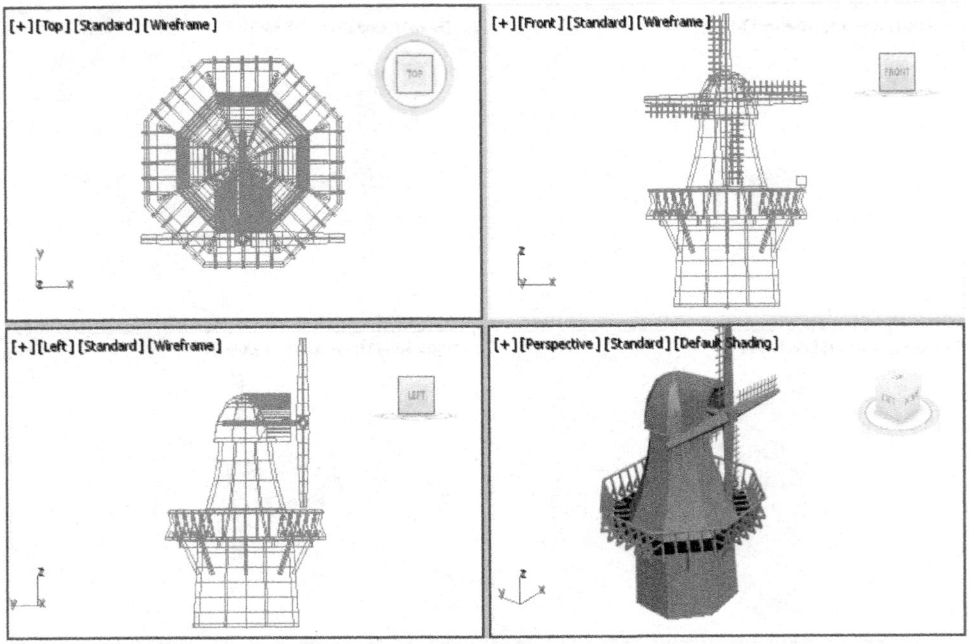

Figure P1-42 All objects of the windmill displayed

5. Select the **Bitmap** map from the **Maps > General** rollouts and choose the **OK** button; the **Material/Map Browser** dialog box is closed and the **Select Bitmap Image File** dialog box is displayed. As the project folder is already set, the *images* folder of this project is displayed in the **Look in** drop-down list of this dialog box.

6. Select the **WoodMaterial.jpg** file and then choose the **Open** button; the image file is loaded in the first material sample slot and the parameters are displayed for the map in the **Material Editor** dialog box.

7. In the **Material Editor** dialog box, choose the **Assign Material to Selection** button. Next, choose the **Show Shaded Material in Viewport** button; the material is applied to the selected objects.

8. Choose the **Modify** tab in the **Command Panel**. Select the **UVW Map** modifier from the **Modifier List** drop-down list. In the **Mapping** area of the **Parameters** rollout, make sure the **Real-World Map Size** check box is cleared.

9. Select the **Box** radio button and set values in the spinners as given next:

   Length: **220** Width: **220** Height: **220**

   The **UVW Map** modifier is applied to all selected objects.

10. Enter the name **WoodMaterial** in the material name drop-down list to rename the first material sample slot in the **Material Editor** dialog box.

Note

*By grouping the objects together and applying single **UVW Map** modifier, the map is spread out across those objects. This is an easy way to apply materials to cloned objects so that each object has a slightly different appearance.*

11. Similarly, apply map and materials to the objects in the scene as explained next:

- All Decks(*Deck01, Deck002* through *Deck008*) — Create a new standard material and name it **Old_Floor**. Use a **Mix** procedural map for the diffuse color map. For **Color #1** of the mix, use **CEDFENCE.jpg** that you have downloaded from *www.cadcim.com*. For **Color #2** of the mix, use **GRYPLANK.JPG** that you have downloaded from *www.cadcim.com*. Set in the **Mix Amount** spinner to **25** to give the material a weathered appearance. Assign the material to the *Deck* objects.

- *Floor_top* — Apply the material **Old_Floor**. Also, apply a **UVW Map** modifier. Make sure the **Real-World Map Size** check box is cleared.

  Select the **Box** radio button and set values in the spinners as given next:

Length: **90**	Width: **90**	Height: **90**
U Tile: **1**	V Tile: **1**	W Tile: **1**

  In the Top viewport, rotate the **UVW Map** gizmo **-22.5°** around the local Z axis.

- *Base_lo* — Apply the map **StoneMaterial.jpg** that you have downloaded from *www.cadcim.com*. Apply the **UVW Map** modifier and make sure the **Real-World Map Size** check box is cleared. Select the **Box** radio button. Next, select the **X** radio button in the **Alignment** area of the **Parameters** rollout. Also, choose the **Fit** button.

- *Base_mid* — Apply the map **DarkBrownStoneMaterial.jpg** that you have downloaded from *www.cadcim.com*. Add the **UVW Map** modifier and select the **Box** radio button in the **Mapping** area of the **Parameters** rollout. Make sure the **Real-World Map Size** check box is cleared.

- *Base_top* — Apply the map **DarkBrownStoneMaterial.jpg** that you have downloaded from *www.cadcim.com*. Add **UVW Map** modifier and select the **Box** radio button in the **Mapping** area of the **Parameters** rollout. Make sure the **Real-World Map Size** check box is cleared.

  In the **Alignment** section of the **Parameters** rollout, select the **X** radio button.

- On all *DeckBeam*, *DeckBoard*, *DeckRail*, and *Handrail* objects, apply the map **Pine.jpg** that you have downloaded from *www.cadcim.com*.

- On all *DeckBrace* objects, apply the map **Wood.Red Oak.jpg** that you have downloaded from *www.cadcim.com*.

- *Floor_lo* — Apply the map **StoneMaterial.jpg** that you have downloaded from *www.cadcim.com*.

Apply the **UVW Map** modifier and select the **Box** radio button in the **Mapping** area of the **Parameters** rollout. Make sure the **Real-World Map Size** check box is cleared. Next, select the **Y** radio button in the **Alignment** area of the **Parameters** rollout. Next, close the **Material Editor** dialog box.

Adding Lights and Camera

In this section, you will add lights and camera in the scene.

1. Invoke the **Zoom Extents All** tool to zoom the objects to their extents.

2. Choose **Create > Lights** in the **Command Panel**. Select the **Standard** option in the drop-down list. Invoke the **Omni** tool from the **Object Type** rollout. 

3. In the Left viewport, click at any point on the right side of the viewport midway up *Base_mid* to place the first omni light; an omni light (*Omni001*) is placed at the selected point.

4. Click at any point on the left side of *Base_mid*; an omni light (*Omni002*) is placed at the selected point. Zoom out of the Left viewport to get more space.

5. Click at any point on the far right, approximately at twice the height of the windmill; an omni light (*Omni003*) is placed at the selected point, refer to Figure P1-43.

6. Choose **Create > Cameras** in the **Command Panel**. Invoke the **Target** tool from the **Object Type** rollout. 

7. Create the camera and position it, as shown in Figure P1-43.

8. Activate the Perspective viewport and press the C key; the Perspective viewport becomes the Camera viewport.

9. With the Camera viewport active, use the various camera tools to place the objects at the center of the viewport. Right-click to exit each command. Also, move the camera in the other viewports so that you can see an angled view of the windmill.

Rendering a Still Image

In this section, you will render the still image for the final output. Before rendering, you need to change the background color of the scene to white, refer to Tutorial 1 of Chapter 2.

1. Make sure the Camera viewport is active. If possible, rotate the image to render it in an appropriate direction.

2. Invoke the **Render Production** tool from the **Main Toolbar**; the rendered image is displayed in the render window, as shown in Figure P1-44.

3. Close the Render window.

4. Move the lights and adjust their intensity as required.

5. Save the scene.

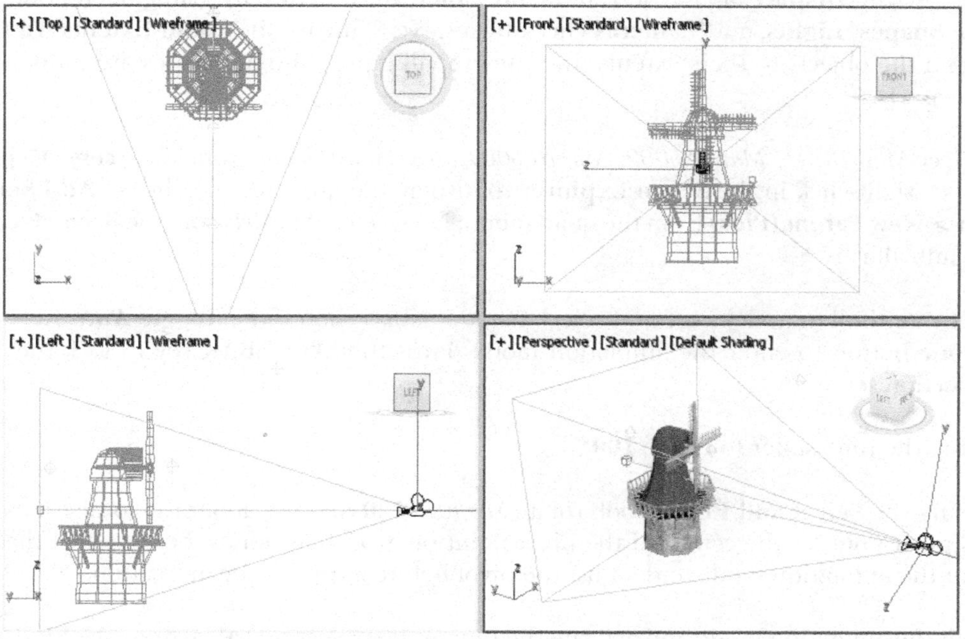

Figure P1-43 *Lights and Camera created and positioned in the scene*

Figure P1-44 *The scene displayed after rendering at frame 1*

Adding Animation in the Windmill

In this section, you will add animation in the windmill.

1. Choose the **Display** tab of the **Command Panel**. In the **Hide by Category** rollout, select the **Shapes**, **Lights**, and **Cameras** check boxes. Next, invoke the **Zoom Extents All** tool to zoom the objects to their extents; the camera, all lights, and all shapes are hidden in all viewports.

2. Select *MainRib01, MainRib002, MainRib003, and MainRib004* from the **Scene Explorer**. Next, right-click in the **Scene Explorer** to display the quad menu. Choose **Add Selected To > New Parent (Pick)** from the quad menu; Now, select *Shaft01* from the **Scene Explorer** to link all ribs.

3. Activate the Front viewport and select *Shaft01*. Next, choose the **Toggle Auto Key Mode** button to enter the animation mode. Press the SPACEBAR key to lock the selection set.

4. Drag the time slider to frame **100**.

5. Using the **Select and Rotate** tool, rotate *Shaft01* **-720** degrees about the local Z axis using its pivot point as the center. If the pivot point on *Shaft01* is not at the center of the shaft, exit the animation mode and adjust the pivot before applying the animation.

6. Exit the animation mode and activate the Camera001 viewport. Also, press the SPACEBAR key to unlock the selection set.

7. Choose the **Play Animation** button; the animation is played in the Camera001 viewport. The windmill blades and shaft rotate two complete turns over the length of the animation.

8. Choose the **Stop Animation** button to stop playback.

9. Save the scene.

Rendering the Animation

In this section, you will play the animation in the Windows Media Player.

1. With the Camera viewport active, invoke the **Render Setup** tool; the **Render Setup** dialog box is displayed.

2. In the **Time Output** area of the **Common Parameters** rollout, select the **Active Time Segment** radio button. In the **Render Output** area, choose the **Files** button; the **Render Output File** dialog box is displayed.

   As the project folder is already set, the *renderoutput* folder of this project is displayed in the **Save in** drop-down list of this dialog box.

3. In the **Render Output File** dialog box, select **AVI File (*.avi)** from the **Save as type** drop-down list. In the **File name** text box, enter the name **Project01**. Next, choose the **Save** button; the **AVI File Compression Setup** dialog box is displayed.

4. Choose the **OK** button in the **AVI File Compression Setup** dialog box.

5. Make sure the **Save File** check box is selected in the **Render Output** area of the **Render Setup** dialog box and then choose the **Render** button in the dialog box.

   It may take several minutes to render the animation. After rendering is complete, close the **Rendered Frame** window and the **Render Setup** dialog box.

6. Navigate to the *renderouput* folder of the project folder *Prj_1* and double-click on *Project01.AVI*; the AVI file is played in Movies and TV Player, refer to Figure *P1-45*.

Note

*By default, the **Movies** & **TV** is displayed as a media player in Windows 10. You can change it to Windows Media Player, refer to Figure P1-46.*

Figure P1-45 *The animation played in Movies & TV Player*

Figure P1-46 The animation played in Windows Media Player

Project 2

Creating a Diner

PROJECT DESCRIPTION

In this project, you will create a scene of the diner, as shown in Figure P2-1. The scene consists of floor, walls, booths, tables with pedestals, and hanging lamps above each table. You will create a window in the wall, a set of blinds in front of the window, and add animation to the blinds to give a scene more realistic look. To add more realistic effect to the scene, you will set up and adjust lights and camera, create glasses and plates on the tables, create text on the window, and assign materials to various objects. You will also set the window background to give more natural effect to the scene. Lastly, you will make necessary adjustments and render the scene for the final output.

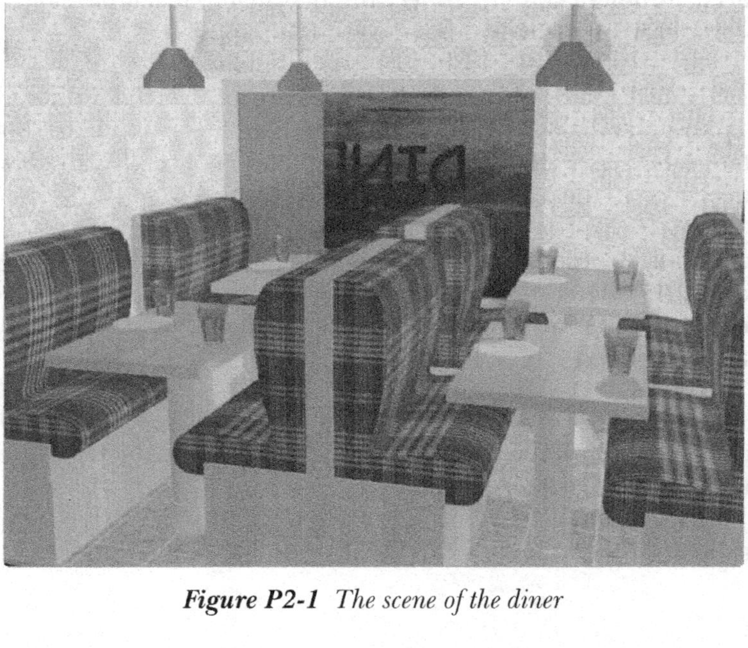

Figure P2-1 *The scene of the diner*

Creating the Project Folder

Create a new project folder with the name *Prj_2* at *\Documents\3dsmax2022* and then save the file with the name, as discussed in Tutorial 1 of Chapter 2.

Downloading the Files

Before starting the project, you need to download the *prj2_3dsmax_2022.zip* file from *www.cadcim.com*. The path of the file is as follows: *Textbooks > Animation and Visual Effects > Autodesk 3ds Max > Autodesk 3ds Max 2022 for Beginners: A Tutorial Approach.*

Extract the contents of the zip file to *Documents*. Open Windows Explorer and then browse to *\Documents\prj1_3dsmax_2022*. Next, copy all files to *\Documents\3dsmax 2022\Prj2\sceneassets\images.*

Creating the Floor

In this section, you will create the floor by using the **Box** tool and then you will apply material to the floor to give it a realistic look.

1. Activate the Top viewport and choose **Create > Geometry** in the **Command Panel**. Then, invoke the **Box** tool from the **Object Type** rollout.

2. Expand the **Keyboard Entry** rollout and set the values in the spinners as given next:

   Length: **100** Width: **130** Height: **1**

3. Choose the **Create** button; the floor is created. In the **Name and Color** rollout, enter **Floor**.

4. Choose **Rendering > Material Editor > Compact Material Editor** from the menu bar; the **Material Editor** dialog box is displayed. Select the first material sample slot.

5. In the **Material Editor** dialog box, select the first material sample slot. In the **Material Name** drop-down list, enter **floor**.

6. Choose the **Physical Material** button available in the **Material Editor** dialog box; the **Material/Map Browser** dialog box is displayed. In this dialog box, choose **Standard Surface** from the **Materials > Arnold > Surface** rollout and choose **OK**; the **Standard Surface** material is displayed in the **Material Editor** dialog box.

7. In the **Base** rollout, choose the map next to color swatch in the **Base Color** area of the **Material Editor**; the **Material/Map Browser** dialog box is displayed. In this dialog box, choose **Bitmap** from the **Maps > General** rollout and then choose the **OK** button; the **Select Bitmap Image File** dialog box is displayed.

8. As the project folder is already set, the *images* folder is displayed in the **Look in** drop-down list of this dialog box. Select the file **floor.jpg** and then choose the **Open** button; the image is displayed in the selected sample slot.

9. In the **Co-ordinates** rollout, set **5** in the **U Tiling** and **V Tiling** spinners.

10. Choose the **Go to Parent** button to return back to the parent level and then choose the **Assign Material to Selection** tool; the tile material is assigned to *floor*.

11. Choose the **Show Shaded Material in Viewport** tool; the applied material is displayed in the Perspective viewport. Next, rotate the Perspective viewport to set the angle of view, as shown in Figure P2-2.

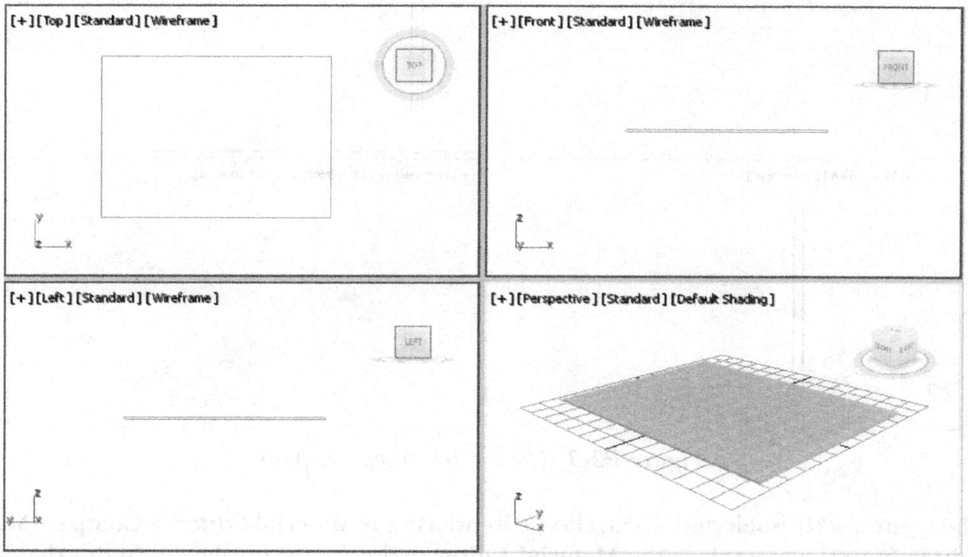

***Figure P2-2** Floor created and a material assigned to it*

12. Close the **Material Editor** dialog box and choose the **Zoom Extents All** tool to display *floor* to its extent.

Creating Walls

In this section, you will create the walls by using the **Box** tool.

1. Choose the **2D Snap** tool from the **Snaps Toggle** flyout. Next, right-click on it; the **Grid and Snap settings** dialog box is displayed. In this dialog box, select the **Vertex** and the **Edge/Segment** check boxes in the **Snaps** tab and close the dialog box.

2. Activate the Top viewport and then choose **Create > Geometry** in the **Command Panel**. Next, select **AEC Extended** from the drop-down list.

3. Invoke the **Wall** tool from the **Object Type** rollout.

4. Click on the lower left corner of *floor*. Next, drag the cursor to the upper left corner of *floor* and click. Again, drag the cursor to the upper right corner of *floor* and click. Now, right-click in the viewport to exit the tool.

5. In the **Parameters** rollout, make sure 5 is displayed in the **Width** spinner and enter **100** in the **Height** spinner.

6. Enter **wall01** in the **Name and Color** rollout. Figure P2-3 shows *wall01* in the viewport. Next, invoke the **2D Snap** tool again to deactivate it.

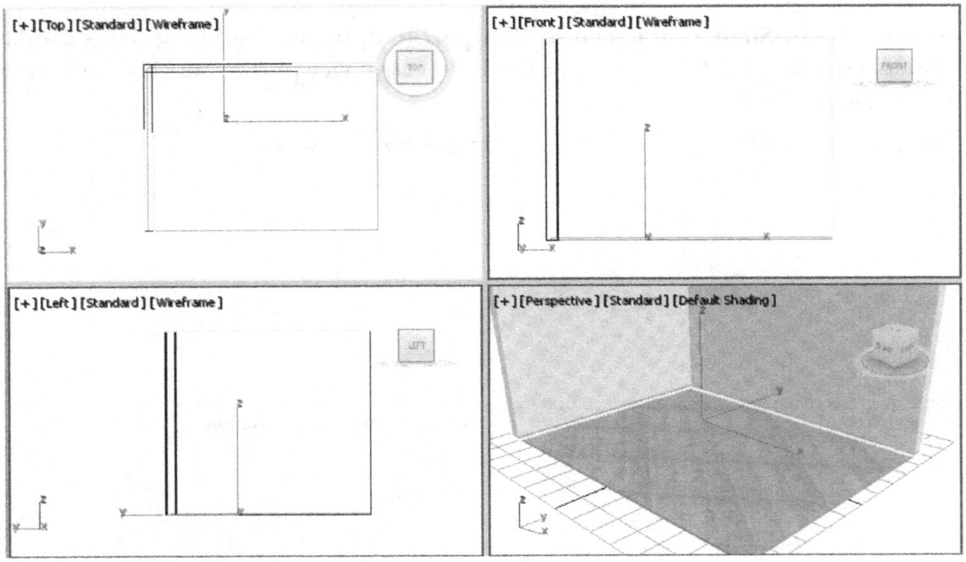

Figure P2-3 *The wall01 in the viewports*

7. Make sure *wall01* is selected. Next, choose **Rendering > Material Editor > Compact Material Editor** from the menu bar; the **Material Editor** dialog box is displayed. In this dialog box, select the second material sample slot.

8. Choose the **Physical Material** button available in the **Material Editor** dialog box; the **Material/Map Browser** dialog box is displayed. In this dialog box, choose **Standard Surface** from the **Materials > Arnold > Surface** rollout and choose **OK**; the **Standard Surface** material is displayed in the **Material Editor** dialog box.

9. In the **Base** rollout, choose the map next to the color swatch in the **Base Color** area of the **Material Editor**; the **Material/Map Browser** dialog box is displayed. In this dialog box, choose **Bitmap** from the **Maps > General** rollout and then choose the **OK** button; the **Select Bitmap Image File** dialog box is displayed.

10. As the project folder is already set, the *images* folder is displayed in the **Look in** drop-down list of this dialog box. Select the **wall.jpg** file and then choose the **Open** button; the image is displayed in the selected sample slot.

11. In the **Coordinates** rollout, make sure that the **Use Real-World Scale** check box is cleared. Next, set **8** in the **U-Tiling** and **V-Tiling** spinners.

12. In the **Material Editor** dialog box, choose the **Go To Parent** button to return back to the parent level. Now, choose the **Assign Material to Selection** tool to assign the material to the object.

13. Choose the **Show Shaded Material in Viewport** tool to display the wall with the assigned material in the viewport, as shown in Figure P2-4. Close the **Material Editor** dialog box.

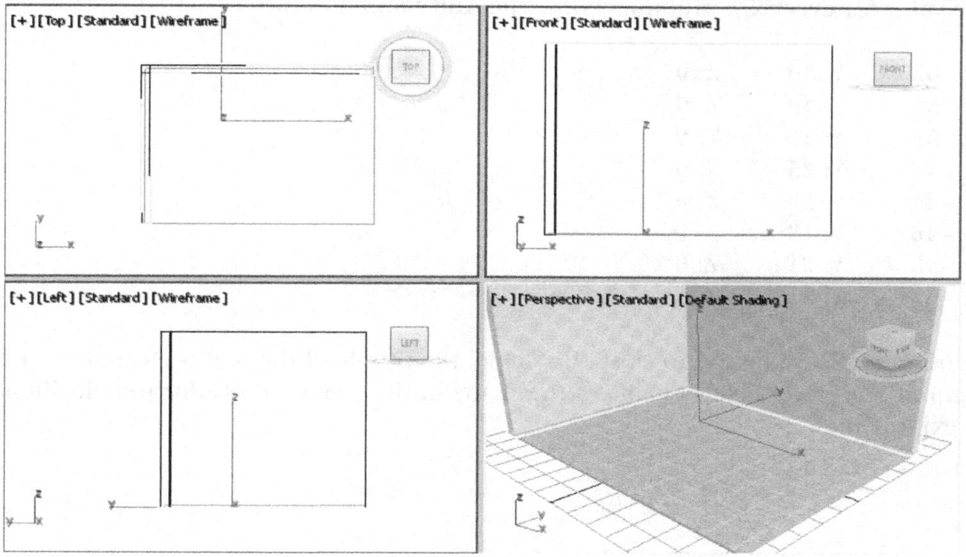

Figure P2-4 *The material assigned to wall01 displayed in the viewport*

Creating the Booth

In this section, you will create the profile of the booth by using the **Line** tool.

1. Right-click on the **Snaps Toggle** tool from the **Main Toolbar**; the **Grid and Snap Settings** dialog box is displayed. In this dialog box, choose the **Home Grid** tab and set the value **2** in the **Grid Spacing** spinner. Next, clear the **Inhibit Grid Subdivision Below Grid Spacing** check box and close the dialog box.

2. Activate the Front viewport and invoke the **Maximize Viewport Toggle** tool to maximize it. Zoom in on the lower-left portion of the viewport.

3. Choose **Create > Shapes** from the **Command Panel**. Choose the **Line** tool from the **Object Type** rollout. Expand the **Keyboard Entry** rollout and set the following values in the **X**, **Y**, and **Z** spinners. After entering the value for every x, y, z point as given below, choose the **Add Point** button:

X: **–64**	Y: **1**	Z: **0**
X: **–64**	Y: **49**	Z: **0**
X: **–62**	Y: **49**	Z: **0**
X: **–62**	Y: **21**	Z: **0**
X: **–46**	Y: **21**	Z: **0**
X: **–46**	Y: **1**	Z: **0**

4. Choose the **Close** button; the shape is created. Enter **Booth** as name of the shape; the profile of *Booth* is created. Right-click to exit the tool. Zoom out if you need to see the entire shape. Now, you need to create the seat for *Booth*.

5. Make sure the Front viewport is activated. Choose the **Line** tool again. Expand the **Keyboard Entry** rollout and set the values in the **X**, **Y**, and **Z** spinners as given next. After entering the value for every x, y, z point, choose the **Add Point** button:

X: **–62**	Y: **49**	Z: **0**
X: **–58**	Y: **49**	Z: **0**
X: **–58**	Y: **25**	Z: **0**
X: **–42**	Y: **25**	Z: **0**
X: **–42**	Y: **19**	Z: **0**
X: **–46**	Y: **19**	Z: **0**
X: **–46**	Y: **21**	Z: **0**
X: **–62**	Y: **21**	Z: **0**

6. Choose the **Close** button to close the shape; the profile of the seat is created, as shown in Figure P2-5. Enter **Seat** as name of the shape in the **Name and Color** rollout. Right-click to exit the tool.

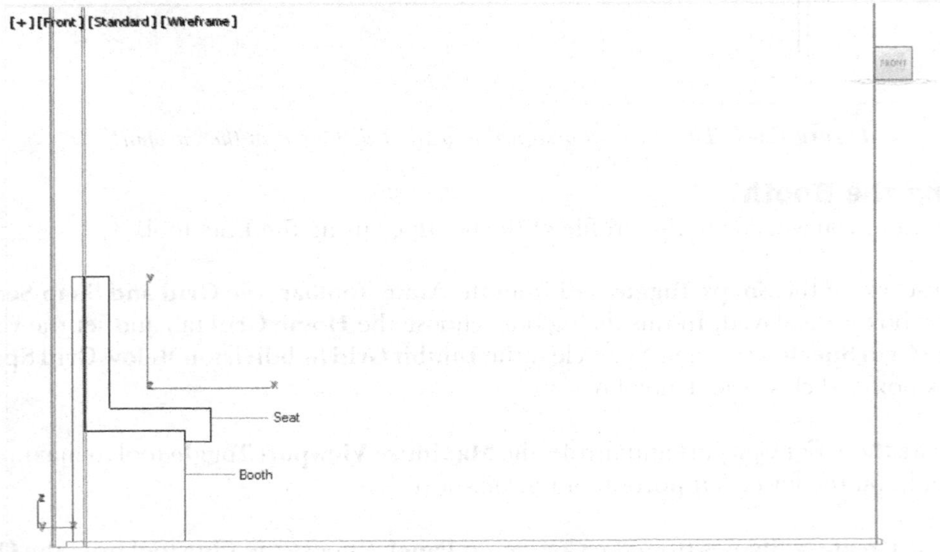

***Figure P2-5** The profile of seat created*

7. Make sure *Seat* is selected. Next, choose the **Modify** tab in the **Command Panel**. In the **Selection** rollout, choose the **Vertex** button; the **Vertex** sub-object mode is activated.

8. Right-click on the top-right vertex of *Seat*; the quad menu is displayed. Select **Bezier** from the upper-left quadrant; the bezier handles are displayed on the vertex and one segment adjoining the vertex is curved.

9. Using the **Select and Move** tool, move the bezier handles to form a curve for the back. Similarly, modify the middle-right vertex of *Seat*, as shown in Figure P2-6. Choose the **Vertex** button in the **Selection** rollout to exit the **Vertex** sub-object mode.

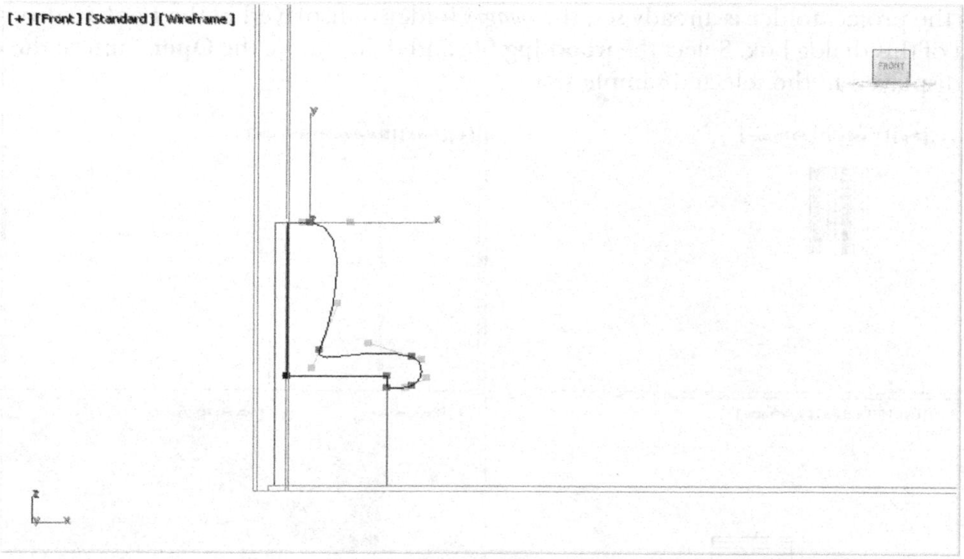

Figure P2-6 The vertices of Seat modified using the bezier handles

10. Invoke the **Maximize Viewport Toggle** tool to shift the display back to the four viewport display.

11. Select both *Booth* and *Seat* in any viewport and make sure the **Modify** tab of the **Command Panel** is chosen.

12. In the **Modifier List** drop-down list, select **Extrude** in the **OBJECT-SPACE MODIFIERS** section; the **Extrude** modifier is applied to both the shapes.

13. In the **Parameters** rollout, set the value **40** in the **Amount** spinner; both line profiles are extruded.

14. Activate the Top viewport and move both the extruded objects to the top-left corner of the walls. Next, activate the Perspective viewport and rotate and zoom the *Booth* and *Seat* for better visualization. On doing so, *Booth* and *Seat* are positioned, as shown in Figure P2-7.

15. Select *Booth* and open the **Material Editor** dialog box by pressing the M key. Select the third material sample slot and enter **wood** in the **Material Name** drop-down list.

16. Choose the **Physical Material** button available in the **Material Editor** dialog box; the **Material/Map Browser** dialog box is displayed. In this dialog box, choose **Standard Surface** from the **Materials > Arnold > Surface** rollout and choose **OK**; the **Standard Surface** material is displayed in the **Material Editor** dialog box.

17. In the **Base** rollout, choose the map next to the color swatch in the **Base Color** area of the **Material Editor**; the **Material/Map Browser** dialog box is displayed. In this dialog box, choose **Bitmap** from the **Maps > General** rollout and then choose the **OK** button; the **Select Bitmap Image File** dialog box is displayed.

18. As the project folder is already set, the *images* folder is displayed in the **Look in** drop-down list of this dialog box. Select the **wood.jpg** file and then choose the **Open** button; the image is displayed in the selected sample slot.

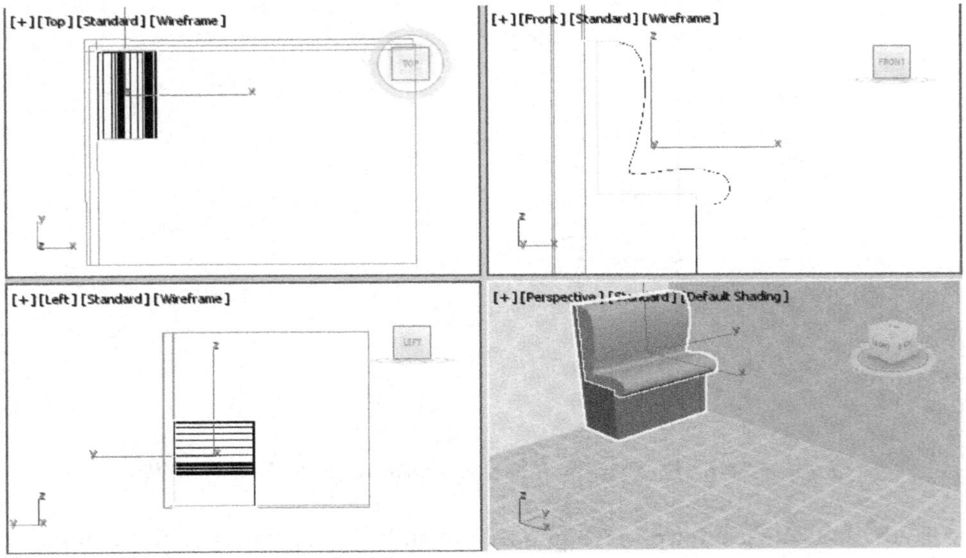

Figure P2-7 *Booth and Seat created in the viewport*

19. Choose the **Go to Parent** button to return to the parent level. Next, make sure *Booth* is selected and then choose the **Assign Material to Selection** and **Show Shaded Material in Viewport** buttons. Close the **Material Editor** dialog box.

20. Choose the **Modify** tab from the **Command Panel**. Next, click on the **Modifier List** drop-down list and select **UVW Map** modifier from it; the **UVW Map** modifier is applied to *Booth*.

21. In the **Parameters** rollout, select the **Box** radio button from the **Mapping** area. Next, select the **Y** radio button and choose the **Fit** button from the **Alignment** area.

22. Select *Seat* and invoke the **Material Editor** dialog box. Next, select the fourth sample slot in the **Material Editor** dialog box. Now, load the map **FabricPlaid.jpg**, as discussed earlier; the material is displayed in the selected sample slot.

23. In the **Coordinates** rollout, make sure the **Use Real-World Scale** check box is cleared. Next, choose the **Modify** tab in the **Command Panel** and select the **Generate Mapping Coords** check box from the **Parameters** rollout.

24. Make sure *Seat* is selected and then choose the **Assign Material to Selection** and **Show Shaded Material in Viewport** tools. Next, close the **Material Editor** dialog box. Choose the **Modify** tab from the **Command Panel**. Now, click on the **Modifier List** drop-down list and select **UVW Map** modifier from it; the **UVW Map** modifier is applied to *Seat*. Now, select the **Box** radio button in the **Mapping** area. Next, make sure the **Z** radio button is selected in the **Alignment** area and then choose the **Fit** button.

25. Activate the Top viewport and select *Booth* and *Seat*. Choose **Group > Group** from the menu bar; the **Group** dialog box is displayed. In this dialog box, enter **Booth01** in the **Group name** textbox and then choose the **OK** button.

    Now, the group can be transformed or modified as a single object.

26. Activate the Front viewport. With the group selected, make sure the **Use Selection Center** tool is invoked in the **Pivot Point** flyout of the **Main Toolbar**.

27. Invoke the **Mirror** tool from the **Main Toolbar**; the **Mirror: Screen Coordinates** dialog box is displayed. In the **Clone Selection** area of this dialog box, select the **Copy** radio button. In the **Mirror Axis** area, select the **X** radio button and set the value **40** in the **Offset** spinner. Next, choose the **OK** button; a mirror image is created at an offset of 40 units to the right of *Booth01*, as shown in Figure P2-8.

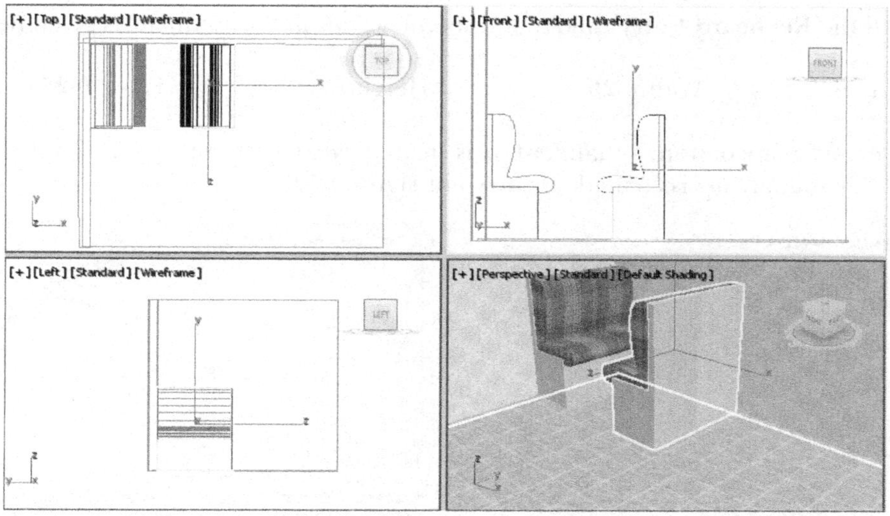

Figure P2-8 *Mirror image created 40 units to the right of Booth01*

28. Enter the name of the mirror image as **Booth02** in the **Name and Color** rollout.

29. Invoke the **Zoom Extents All** tool to zoom the objects to their extents in the viewports.

Creating the Table

In this section, you will create the table. You will also create the base for the table by using the **Cone** tool. Next, you will create the pedestal for the table by using the **Cylinder** tool. The pedestal will be placed on the base of the table. Finally, you will create the table top which will be placed on the top of the pedestal.

1. Activate the Top viewport and then choose **Create > Geometry** in the **Command Panel**. Choose the **Cone** tool from the **Object Type** rollout. Expand the **Keyboard Entry** rollout, and set the values of the parameters as given next.

   Radius 1: **6** Radius 2: **3** Height: **4**

2. Choose the **Create** button. Next, enter the name **Base** in the **Name and Color** rollout; a cone is created, which will become the base of the table.

3. In the **Object Type** rollout, invoke the **Cylinder** tool. In the **Keyboard Entry** rollout, set the value **3** in the **Radius** spinner and **28** in the **Height** spinner.

4. Choose the **Create** button; a cylinder is created. Next, enter the name **Pedestal** in the **Name and Color** rollout.

   Pedestal is created on the base. Now, you will create a tabletop which will be placed on the top of *Pedestal*.

5. Select **Extended Primitives** from the drop-down list below the **Geometry** button. In the **Object Type** rollout, invoke the **ChamferBox** tool.

6. Expand the **Keyboard Entry** rollout and set the values in the spinners as given next:

   Length: **40** Width: **26** Height: **2** Fillet: **0.25**

7. Choose the **Create** button; a chamfer box is created. Next, enter the name **Top** in the **Name and Color** rollout; *Top* is created, as shown in Figure P2-9.

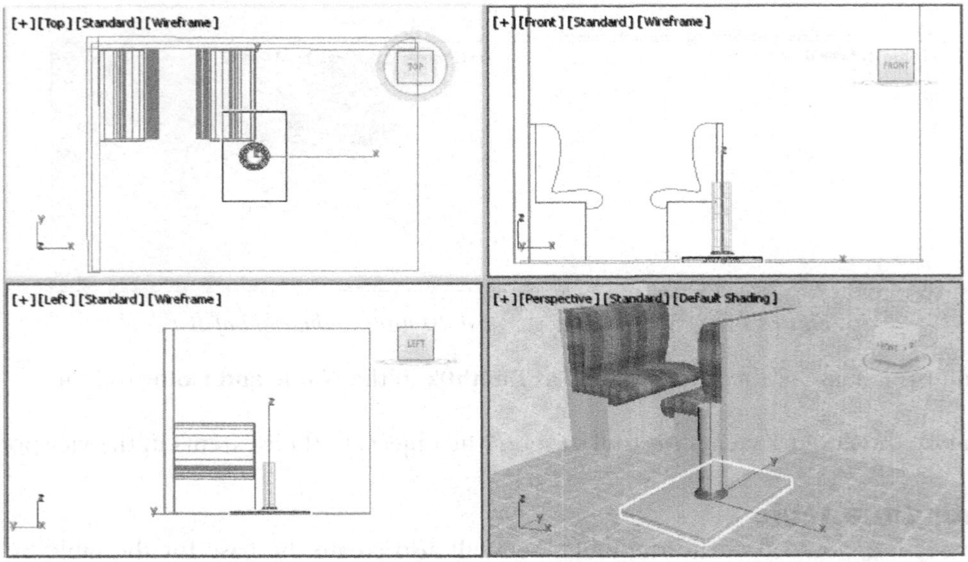

Figure P2-9 *Top created*

8. Maximize the Front viewport and zoom in on the three table objects until the subgrid is displayed. Use the **Select and Move** tool or the **Align** tool to move *Base* so that its lower edge is even with the top of *floor*.

9. Select *Pedestal* and use the **Select and Move** tool or the **Align** tool to move it upward so that its lower edge is even with the top edge of *Base*.

10. Select *Top* and use the **Select and Move** tool or the **Align** tool to move it upward so that its lower edge is even with the top edge of *Pedestal*.

11. Restore the four-viewport display. Maximize the Top viewport and zoom in to display the subgrid.

12. Select *Base*, *Pedestal*, and *Top* and move them such that they are centered between the two booths, as shown in Figure P2-10.

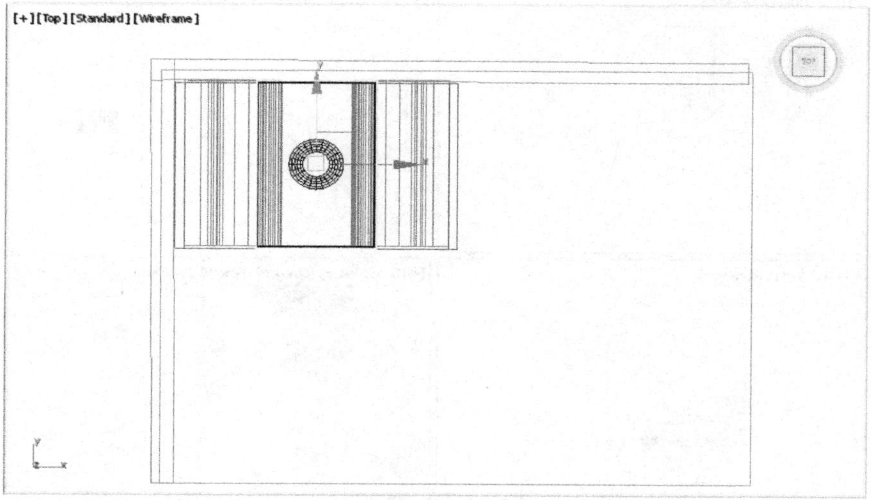

***Figure P2-10** The table components aligned*

13. Select *Top, Base* and *Pedestal* in any viewport and apply the *wood* material to them.

14. Render the Perspective viewport.

Creating the Lamp, Tube, and Target Light

In this section, you will create lights by using various light tools to illuminate the diner.

1. Activate the Top viewport and then choose **Create > Shapes** in the **Command Panel**. Invoke the **Star** tool from the **Object Type** rollout. In the **Parameters** rollout, set the value **30** in the **Points** spinner.

2. In the **Keyboard Entry** rollout, set the value **4** in the **Radius 1** spinner and **4.8** in the **Radius 2** spinner. Next, choose the **Create** button; a star shape is created in the Top viewport with the name *Star001*. Right-click to exit the tool.

3. Maximize the Front viewport and zoom to display the subgrid. Choose the **Line** tool from the **Object Type** rollout.

4. Draw a vertical line of about 6 units anywhere in the viewport. Right-click to exit the **Line** tool.

5. Make sure the line is selected. Next, choose **Create > Geometry** in the **Command Panel**. Select **Compound Objects** from the drop-down list. Invoke the **Loft** tool from the **Object Type** rollout.

6. In the **Skin Parameters** rollout, set the value **0** in both the **Shape Steps** and the **Path Steps** spinners. In the **Creation Method** rollout, choose the **Get Shape** button and select *Star001*; the star shape is lofted along the line, refer to Figure P2-11.

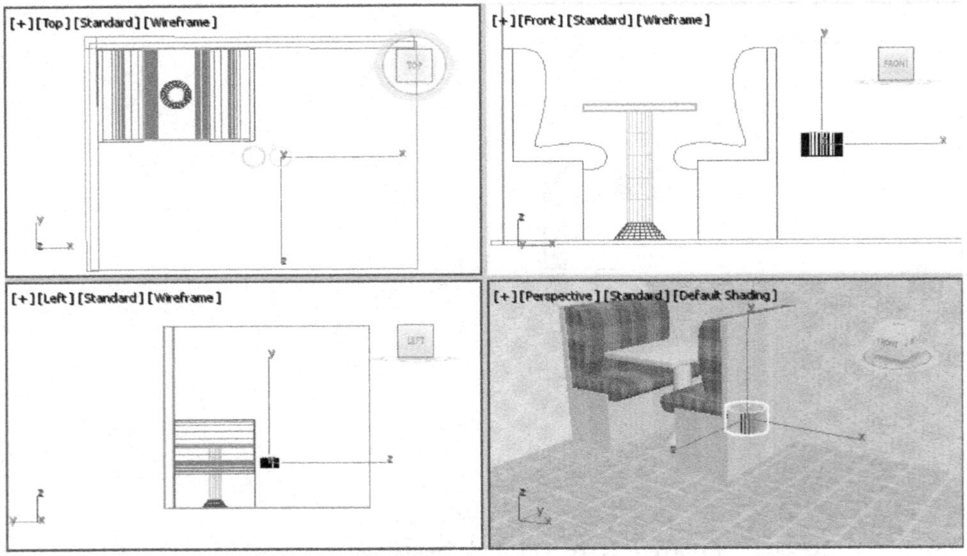

Figure P2-11 *The lofted star shape*

7. Enter the name **Shade** in the **Name and Color** rollout.

8. Make sure the loft object is selected and then choose the **Modify** tab in the **Command Panel**. Expand the **Deformations** rollout and choose the **Scale** button; the **Scale Deformation(X)** dialog box is displayed.

9. In the **Scale Deformation(X)** dialog box, choose the **Insert Corner Point** button and then insert vertices at points 10 and 90.

   After the vertices are inserted, they can be precisely positioned by entering a value in the text box at the bottom right of the dialog box.

10. Choose the **Move Control Point** button and move the right-hand vertex at point 100 and the vertex at point 90 to 30, as shown in Figure P2-12; the lamp shade is created. Close the **Scale Deformation(X)** dialog box.

11. Choose **Edit > Clone** from the menu bar; the **Clone Options** dialog box is displayed. In this dialog box, select the **Copy** radio button in the **Object** area and enter the name **Shade01** in the **Name** text box. Next, choose the **OK** button.

12. Make sure *Shade01* is selected. Choose the **Hierarchy** tab in the **Command Panel**. Make sure the **Pivot** button is chosen. Next, choose the **Affect Pivot Only** button in the **Move/Rotate/Scale:** area of the **Adjust Pivot** rollout. Move the pivot point in the Front viewport so that it is aligned with the bottom-middle of *Shade*. Choose the **Affect Pivot Only** button again to exit the pivot adjustment mode.

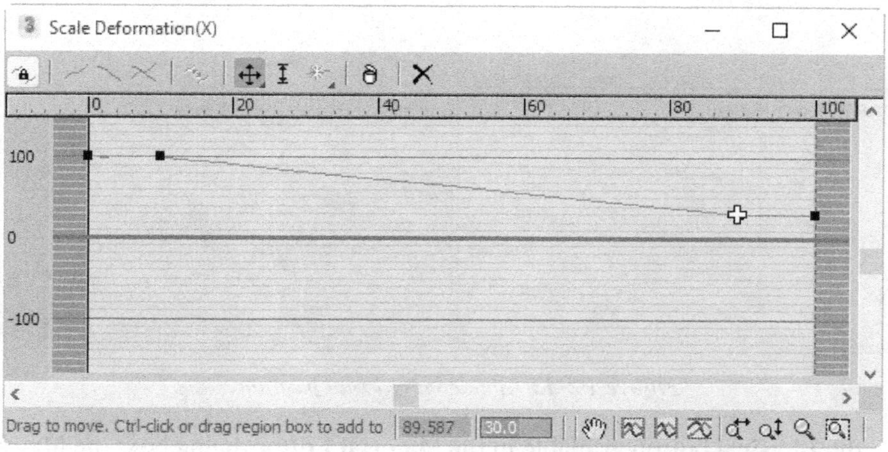

Figure P2-12 *The vertices created and moved in the* ***Scale Deformation****(X) dialog box*

13. Make sure *Shade01* is selected. Next, invoke the **Select and Uniform Scale** and **Use Pivot Point Center** tools. Scale *Shade01* in the Front viewport to 90%, as indicated in the **X**, **Y**, and **Z** text boxes. Alternatively, you can use the **Scale Transformation Type-In** dialog box to scale *Shade01*, which is displayed by right-clicking on the **Select and Uniform Scale** tool.

14. Move *Shade01* slightly upward. Next, select *Shade* from the **Scene Explorer** and then choose **Create > Geometry** in the **Command Panel**. Make sure the **Compound Objects** is selected in the drop-down list.

15. Invoke the **Boolean** tool from the **Object Type** rollout. Choose the **Add Operands** button in the **Boolean Parameters** rollout. Select *Shade01* in any viewport and then choose the **Subtract** button in the **Operand Parameters** rollout; *Shade* is hollowed out.

16. Delete the star and line shapes that were used to create the lamp.

17. Invoke the **Zoom Extents All** tool to view the objects to their extents. In the Top viewport, move *Shade* so that it is centered above the table. In the Front viewport, move and rotate *Shade* to the center of the table, as shown in Figure P2-13.

18. Choose **Rendering > Material Editor > Compact Material Editor** from the menu bar; the **Material Editor** dialog box is displayed. Select the next unused sample slot in it.

19. In the **Material Name** drop-down list, enter **shade**.

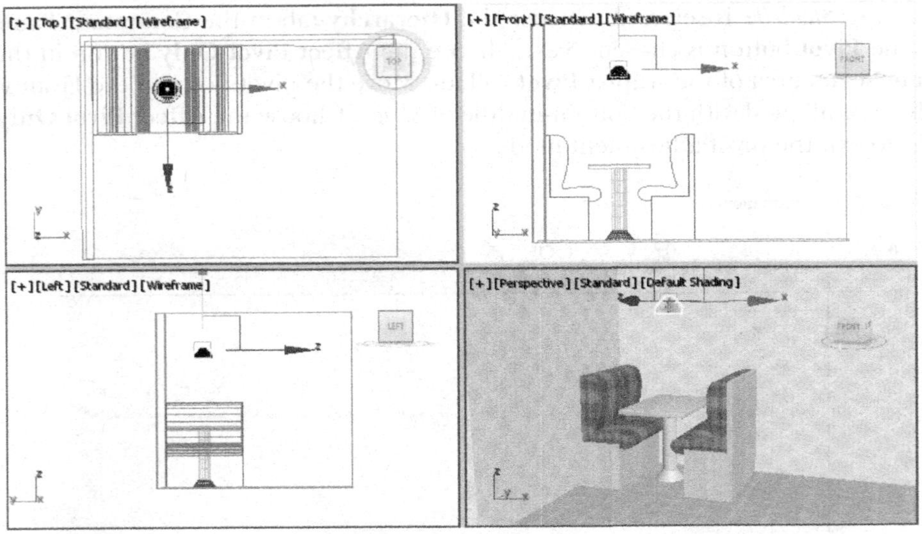

Figure P2-13 Shade moved into position

20. Choose the **Physical** button available in the **Material Editor** dialog box; the **Material/Map Browser** dialog box is displayed. In this dialog box, choose **Standard Surface** from the **Materials > Arnold > Surface** rollout and choose **OK**; the **Standard Surface** material is displayed in the **Material Editor** dialog box.

21. In the **General** area of the **Transmission** rollout, set **1** in the spinner located at the left side of the color swatch and then click on the color swatch; the **Color Selector: transmission_color** dialog box is displayed. In this dialog box, set the values as given next.

Red: **0.722** Green: **0.059** Blue: **0.231**

Choose **OK** to close the dialog box.

22. In the **Advanced** area of the **Transmission** rollout, set **0.6** in the **Extra Roughness** spinner. Notice the grains in the selected sample slot. Figure P2-13(a) shows the parameters set in the **Transmission** rollout.

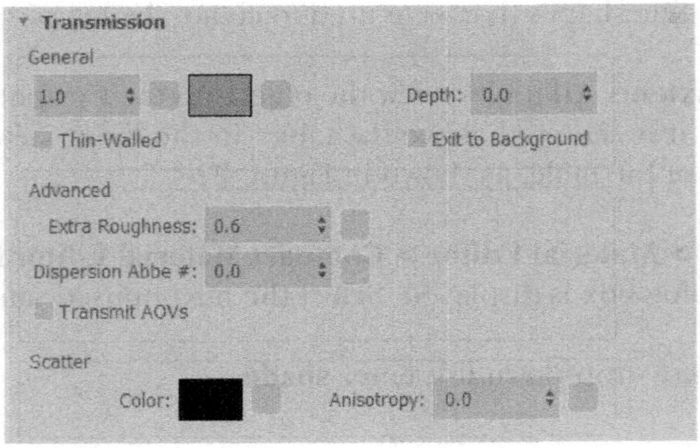

***Figure P2-13 (a)** Parameters set in the **Transmission** rollout*

23. Select *shade* in the viewport. Next, choose the **Assign Material to Selection** and the **Show Shaded Material in Viewport** buttons from the **Material Editor** dialog box; the **Shade** material is assigned to *shade*.

24. Activate the Top viewport and choose **Create > Geometry** in the **Command Panel**. Select **Standard Primitives** in the drop-down list. Invoke the **Cylinder** tool from the **Object Type** rollout.

25. In the **Keyboard Entry** rollout, enter **0.5** in the **Radius** spinner and **15** in the **Height** spinner. Next, choose the **Create** button; a cylinder is created.

26. In the Front viewport, move the cylinder so that its lower edge is aligned with the upper edge of *Shade*. In the Top viewport, place the cylinder at the center of *Shade,* as shown in Figure P2-14.

27. Assign *wood* material to *Cylinder001,* as discussed earlier. Close the **Material Editor** dialog box.

28. Invoke the **Zoom Extents All** tool to zoom the objects to their extents.

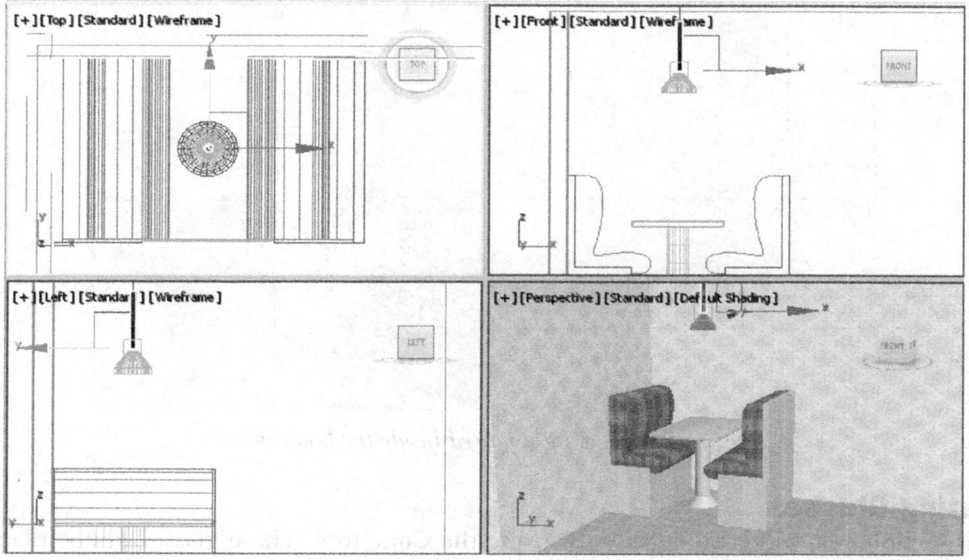

Figure P2-14 *A suspension cylinder created*

29. Activate the Top viewport. Choose **Create > Lights** in the **Command Panel** and select the **Arnold** option from the drop-down list. Next, choose **Arnold Light** from the **Object Type** rollout and then click in the viewport; the Arnold light is displayed in the viewport and it is automatically named as *ArnoldLight01*. Rename this light as *toplight1*. In the **Shapes** rollout, select **Quad** from the **Type** drop-down list and then set **127** and **92** in the **Quad X** and **Quad Y** spinners, respectively. Now, align the *toplight1* in all the viewports, as shown in Figure P2-15.

30. In the **Color/Intensity** rollout, set the value **30** in the **Intensity** spinner and then choose the color swatch located next to the **Color** radio button; the **Color Selector: Color** dialog box is displayed. In this dialog box, set the values as follows and close the dialog box.

    Red: **253** Green: **255** Blue: **111**

31. Activate the Left viewport. Choose the **Arnold Light** tool from the **Object Type** rollout. Next, select **Spot** from the **Type** drop-down list in the **Shapes** rollout and then click at a point near *Shade* and then drag the cursor towards *Table*; the Arnold light and its target are displayed in the viewport and they are automatically named as *ArnoldLight002*. Rename the light as *spot01*.

    You may need to zoom in on *Shade* and *Table*. Make sure the light is centered inside *Shade*.

32. In the **General** rollout, set **35** in the **Target Dist.** spinner. In the **Color/Intensity** rollout, set the value **20** in the **Intensity** spinner. Now, align *spot01* to the center of *Shade*, as shown in Figure P2-15.

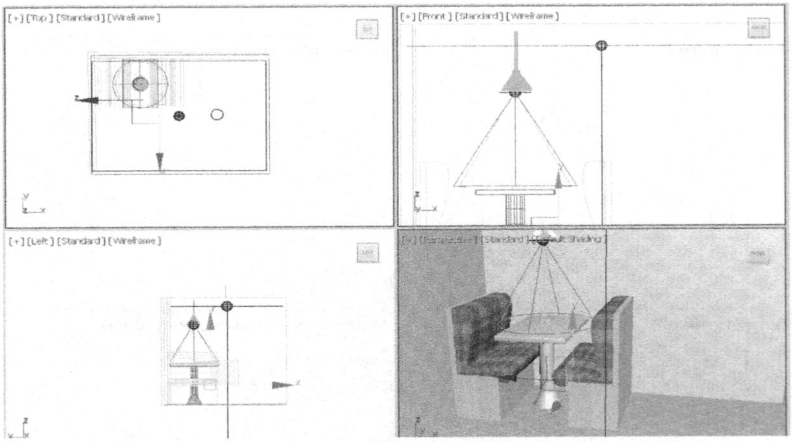

Figure P2-15 *A light placed inside the lamp shade*

Creating Glasses

In this section, you will create glasses by using the **Cone** tool. These glasses will be placed on the table top.

1. Make sure the Top viewport is active. Next, choose **Create > Geometry** in the **Command Panel**. In the **Object Type** rollout, invoke the **Cone** tool. In the **Parameters** rollout, set the value **1** in the **Height Segments** spinner.

2. Expand the **Keyboard Entry** rollout and enter the values in the spinners as given next:

    Radius 1: **1.5** Radius 2: **2.5** Height: **6**

3. Choose the **Create** button; a cone shape for the glass is created. Enter the name of the shape as **Glass01** in the **Name and Color** rollout and right-click on the viewport to exit the tool.

4. Zoom in on the glass in the Front viewport. Next, choose **Edit > Clone** from the menu bar; the **Clone Options** dialog box is displayed. In this dialog box, make sure the **Copy** radio button is selected in the **Object** area. Enter the name **Glass02** in the **Name** text box. Next, choose the **OK** button.

5. Make sure *Glass02* is selected and then choose the **Hierarchy** tab in the **Command Panel**. Make sure the **Pivot** button is chosen and then choose the **Affect Pivot Only** button. Move the pivot point to the top-center of the cone(*Glass02)* in the Front viewport. Choose the **Affect Pivot Only** button again to exit the pivot adjustment mode.

6. Invoke the **Select and Uniform Scale** tool and scale *Glass02* to 95%, as shown in the **X**, **Y**, and **Z** spinners on the status bar. Next, move *Glass02* slightly upward.

7. Select *Glass01* and then choose **Create > Geometry** in the **Command Panel**. Select **Compound Objects** from the drop-down list.

8. Invoke the **Boolean** tool from the **Object Type** rollout. Choose the **Add Operands** button in the **Boolean Parameters** rollout. Select *Glass02* in any viewport and then choose the **Subtract** button in the **Operand Parameters** rollout; *Glass01* is hollowed out. Now, exit the Boolean operation.

9. Activate the Top viewport. Next, choose **Create > Geometry** in the **Command Panel** and select **Standard Primitives** from the drop-down list. In the **Object Type** rollout, invoke the **Cone** tool.

10. Expand the **Keyboard Entry** rollout and set the values in the spinners as given next:

    Radius 1: **1.4** Radius 2: **2** Height: **4**

11. Choose the **Create** button; the water level effect is created in *Glass01*, as shown in Figure P2-16. Next, enter the name **Water** in the **Name and Color** rollout and right-click on the viewport to exit the tool.

12. Open the **Material Editor** dialog box and then choose **Options > Cycle 3X2, 5X3, 6X4 Sample Slots** from the Material Editor menu bar twice; 24 sample slots are displayed in the **Material Editor** dialog box. Next, select the 19th sample slot named **19 - Glass (Thin wall)**. Enter **water** in the **Material Name** drop-down list.

13. In the **Basic Parameter and Reflection** area, set **1.33** in the **IOR** spinner of the **Basic Parameters** rollout.

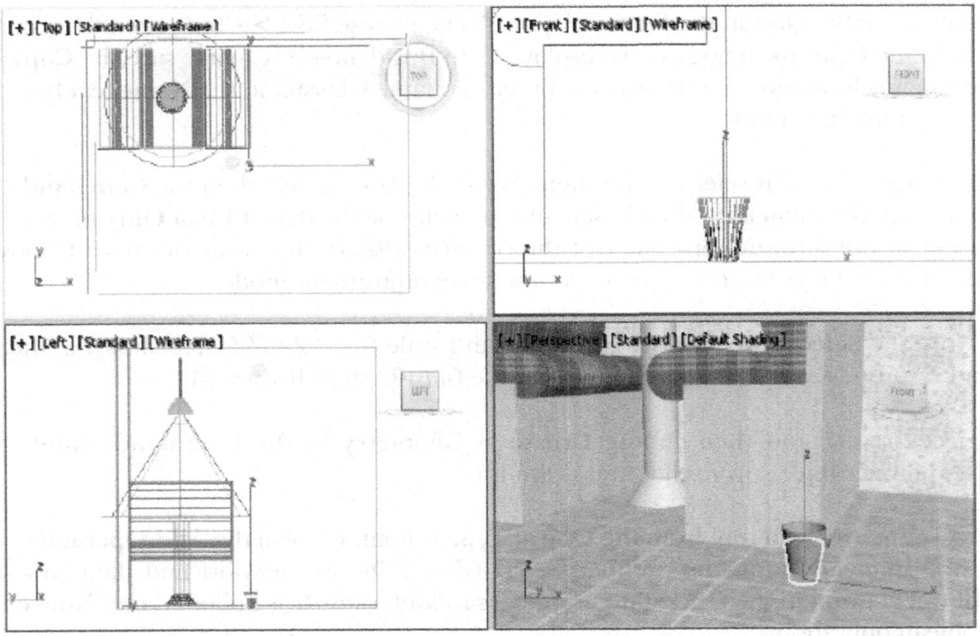

Figure P2-16 A glass with water created

14. Invoke the **Go to Parent** tool. Assign the *water* material to *Water*. Next, align *Water* with *Glass01*.

15. Select *Glass01* in any viewport. In the **Material Editor** dialog box, select the next unused material sample slot.

16. Choose the **Physical Material** button available in the **Material Editor** dialog box; the **Material/Map Browser** dialog box is displayed. In this dialog box, choose **Standard Surface** from the **Materials > Arnold > Surface** rollout and choose **OK**; the **Standard Surface** material is displayed in the **Material Editor** dialog box.

17. In the **General** area of the **Transmission** rollout, set **1** in the spinner located at the left and then click on the color swatch; the **Color Selector: transmission_color** dialog box is displayed. In this dialog box, set the values as given next.

    Red: **0.294** Green: **0.467** Blue: **0.729**

    Choose **OK** to close the dialog box.

18. In the **Advanced** area of the **Transmission** rollout, set **0.6** in the **Extra Roughness** spinner. Notice the grains in the selected sample slot.

19. Enter the name of the material as **Glass** in the **Material Name** drop-down list and assign it to *Glass01*. Close the **Material Editor** dialog box.

20. Select *Glass01* and *Water*. Choose **Group > Group** from the menu bar; the **Group** dialog box is displayed. Enter the name **Drink01** in the **Group** dialog box. Choose the **OK** button; the glass and water are grouped together.

21. Activate the Front viewport and choose the **Zoom Extents** tool. Move *Drink01* to the top of the table so that its bottom edge is aligned with the upper edge of *Top*.

22. In the Top viewport, move *Drink01* upward and toward the left, so that it is placed just to the left of the centerline and 3/4 of the length of the table.

23. Activate the Front viewport and then make sure the **Select and Move** tool is invoked. Press and hold the left mouse button along with the SHIFT key and move *Drink01* toward the right of the table. Now, release the SHIFT key and the left mouse button; the **Clone Options** dialog box is displayed.

24. Make sure the **Copy** radio button is selected in this dialog box and enter **Drink02** in the **Name** textbox. Choose the **OK** button; the dialog box is closed and a copy of *Drink01* is created on the right side of the table. Activate the Top viewport and align *Drink02* so that it is placed at the right of the center line and about ¼th of the length of the table, as shown in Figure P2-17.

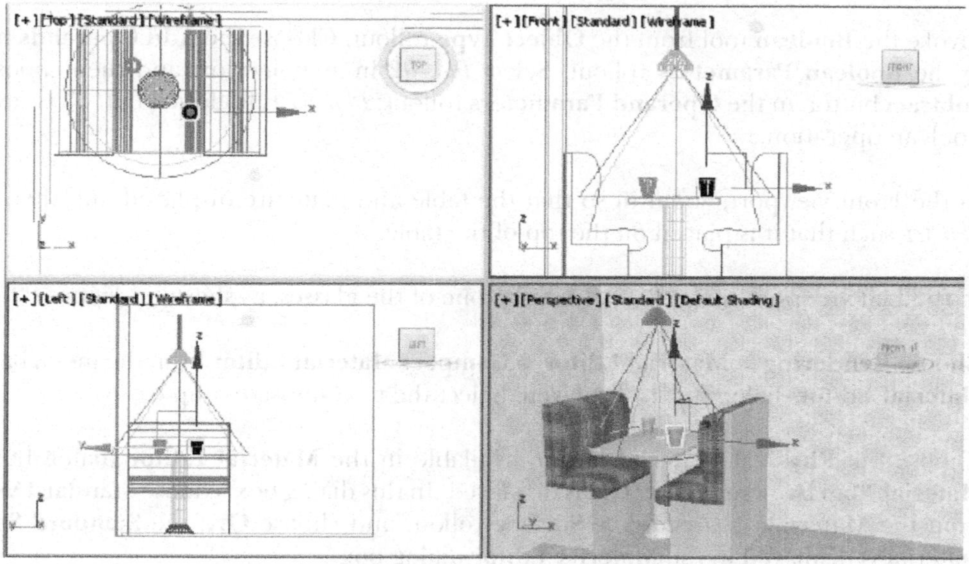

Figure P2-17 *Drink02 created and positioned*

Creating the Plates

In this section, you will create plates by using the **Cone** tool and then place them on the table top along with glasses.

1. Make sure the Top viewport is activated and then choose **Create > Geometry** in the **Command Panel**. Select **Standard Primitives** in the drop-down list below the **Geometry** button, if it is not selected. Invoke the **Cone** tool from the **Object Type** rollout.

2. In the **Parameters** rollout, make sure the value is set to **1** in the **Height Segments** spinner.

3. Expand the **Keyboard Entry** rollout and set the values in the spinners as given next:

   Radius 1: **2.5** Radius 2: **5.0** Height: **0.75**

4. Choose the **Create** button; a truncated cone is created. Enter the name **Plate01** in the **Name and Color** rollout and right-click on the viewport to exit the tool. Next, *Plate01* will be hollowed out to create a plate.

5. Create a copy of *Plate01,* as discussed earlier and name it as **Plate02**. Zoom in on the plates in the Front viewport.

6. Choose the **Hierarchy** tab in the **Command Panel**. Make sure the **Pivot** button is chosen and then choose the **Affect Pivot Only** button. Move the pivot point to the top-center of *Plate02* and scale *Plate02* from its pivot point to 95% of its original size, as discussed earlier. Next, move *Plate02* slightly upward.

7. Select *Plate01* and choose **Create > Geometry** in the **Command Panel**. Select **Compound Objects** in the drop-down list below the **Geometry** button.

8. Invoke the **Boolean** tool from the **Object Type** rollout. Choose the **Add Operands** button in the **Boolean Parameters** rollout. Select *Plate02* in any viewport and then choose the **Subtract** button in the **Operand Parameters** rollout; *Plate01* is hollowed out. Next, exit the Boolean operation.

9. In the Front viewport, zoom in so that the table and plate are displayed and then move *Plate01* such that it is placed on the top of the table.

10. In the Left viewport, move *Plate01* next to one of the glasses, as shown in Figure P2-18.

11. Choose **Rendering > Material Editor > Compact Material Editor** from the menu bar; the **Material Editor** dialog box is displayed. Select the next unused sample slot.

12. Choose the **Physical Material** button available in the **Material Editor** dialog box; the **Material/Map Browser** dialog box is displayed. In this dialog box, choose **Standard Surface** from the **Materials > Arnold > Surface** rollout and choose **OK**; the **Standard Surface** material is displayed in the **Material Editor** dialog box.

13. In the **Base** rollout, choose the **Base Color** color swatch; the **Color Selector: Base Color** dialog box is displayed. In this dialog box, set the values as given next:

    Red: **0.9** Green: **0.9** Blue: **0.9**

14. Choose the **OK** button to close the **Color Selector: Base Color** dialog box. Name the material as **Plate** and assign it to *Plate01*. Close the **Material Editor** dialog box.

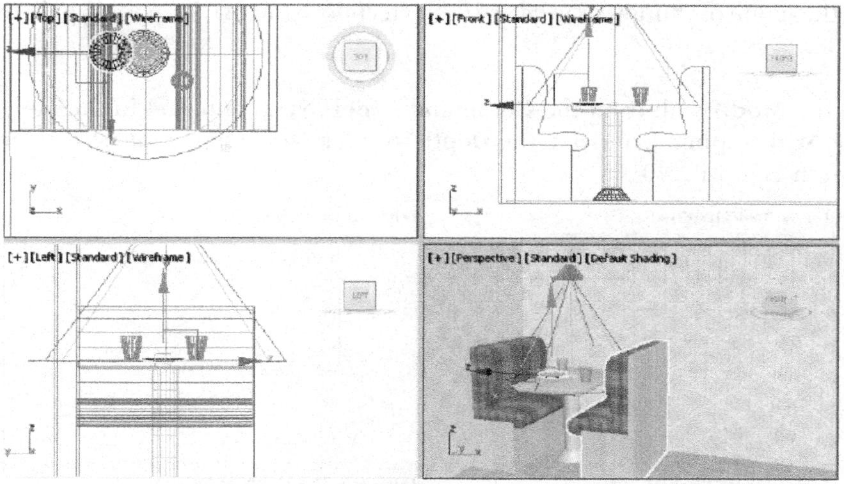

Figure P2-18 Plate01 created and positioned

Creating the Window

In this section, you will create a window by using the **Fixed** tool. This window will be positioned on the wall next to the diner.

1. Choose the **2D Snap** tool from the **Snaps Toggle** flyout. Next, right-click on it; the **Grid and Snap settings** dialog box is displayed. In this dialog box, make sure the **Edge/Segment** check box is selected in the **Snaps** tab. Close the **Grid and Snap settings** dialog box.

2. Activate the Top viewport and then choose **Create > Geometry** in the **Command Panel**. Next, select **Windows** from the drop-down list.

3. Invoke the **Fixed** tool from the **Object Type** rollout. Next, click on the upper edge of *wall01*, as shown in Figure P2-19, and drag the cursor horizontally to specify the width of the window. Next, release the left mouse button and move the cursor down to specify the depth of the window and click on the lower edge of *wall01*. Now, move the cursor up to define the height of the window and click on the screen; the fixed window is created.

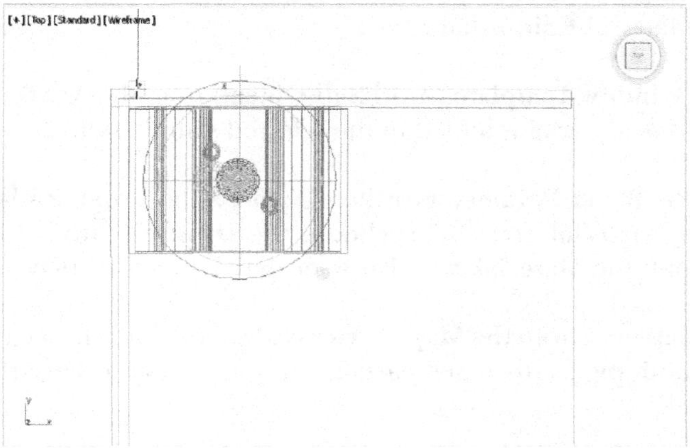

Figure P2-19 Clicking on the upper edge of wall01

4. Modify the name of window as *window01*. Next, choose the **2D Snap** tool again to deactivate it.

5. Choose the **Modify** tab from the **Command Panel**. Next, enter **54** in the **Height** spinner, **80** in the **Width** spinner, and **5** in the **Depth** spinner. Now, align *window01* in the viewports, as shown in FigureP2-20.

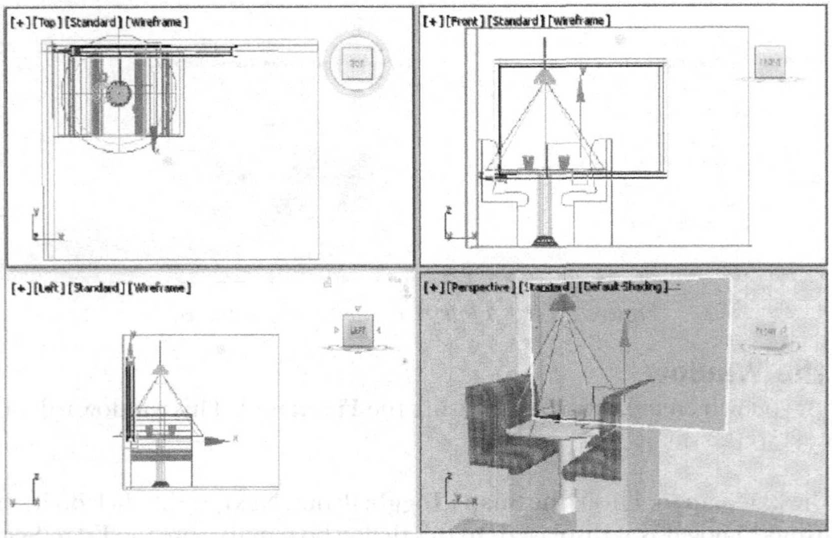

Figure P2-20 *window01 aligned in the viewports*

6. Choose **Rendering > Material Editor > Compact Material Editor** from the menu bar; the **Material Editor** dialog box is displayed. Select the next unused sample slot.

7. Choose the **Get Material** button; the **Material/Map Browser** dialog box is displayed. In this dialog box, choose the **Material/Map Browser Options** button; a flyout is displayed.

8. Choose **Open material Library** from the flyout; the **Import material Library** dialog box is displayed. Navigate to the location *C:\Program Files\Autodesk\3ds Max 2022\materiallibraries* and select **AecTemplates** from this dialog box and choose **Open**; the **AecTemplates** library is added in the **Material Editor** dialog box.

9. Double-click on **Window-Template(Multi/Sub-Object)** from the **AecTemplates.mat** rollout; the *Window-Template* material is added to the selected sample slot.

10. In the **Multi/Sub-Object Parameters** rollout, choose the **Front Rails (Standard)** button corresponding to **material ID1**. Next, choose the **Diffuse** button from the **Blinn Basic Parameters** rollout; the **Material/Map Browser** dialog box is displayed.

11. Choose the **Bitmap** map from the **Maps > General** rollout and choose the **OK** button. Now, load the map **wood.jpg**, as discussed earlier. Choose the **Go to Parent** button to return to the previous level.

12. Repeat the process given in steps 10 and 11 for material IDs **2**, **4**, and **5**.

13. Choose the **Panels (Standard)** button for the **material ID 3**. Next, choose the **Standard(Legacy)** button on the right of the **Material Name** drop-down list; the **Material/Map Browser** dialog box is displayed. In this dialog box, choose **Standard Surface** from the **Materials > Arnold > Surface** rollout and choose **OK**; the **Standard Surface** material is displayed in the **Material Editor** dialog box.

14. In the **General** area of the **Transmission** rollout, set **1** in the spinner located at the left of the color swatch and then click on the color swatch; the **Color Selector: transmission_color** dialog box is displayed. In this dialog box, set the values as given next.

    Red: **1** Green: **1** Blue: **1**

    Choose **OK** to close the dialog box.

15. In the **Advanced** area of the **Transmission** rollout, set **0.6** in the **Extra Roughness** spinner. Notice the grains in the selected sample slot.

16. Choose the **Go to Parent** tool. Next, assign the *Window-Template* material to *window01*.

17. Close the **Material Editor** and the **Material/Map Browser** dialog boxes.

Creating the Text DINER

In this section, you will create the text DINER by using the **Text** tool. This text will be positioned on the outer face of the wall.

1. Activate the Front viewport. Next, choose **Create > Shapes** in the **Command Panel**. Invoke the **Text** tool from the **Object Type** rollout.

2. In the **Parameters** rollout, select **Comic Sans MS Bold** from the **Font** drop-down list. Set the value **20** in the **Size** spinner and in the **Text** area, highlight the current text and type the word **DINER**.

3. Click at a point at the center of *window01* in the Front viewport to place the text, as shown in Figure P2-21.

   You may need to slightly reposition the text later in the modeling process.

4. Activate the Top viewport and move the DINER text to the outer (top) face of *wall01*.

5. Make sure the text object is selected and then choose the **Modify** tab in the **Command Panel**. In the **Modifier List** drop-down list, select **Extrude** in the **OBJECT-SPACE MODIFIERS** section. In the **Parameters** rollout for the modifier, set the value **0.2** in the **Amount** spinner.

6. Make sure the Top viewport is activated. With the text object selected, invoke the **Mirror** tool; the **Mirror: Screen Coordinates** dialog box is displayed. Next, make sure the **No Clone** and **X** radio buttons are selected and then choose the **OK** button; the text appears reversed so that it can be read from outside the building.

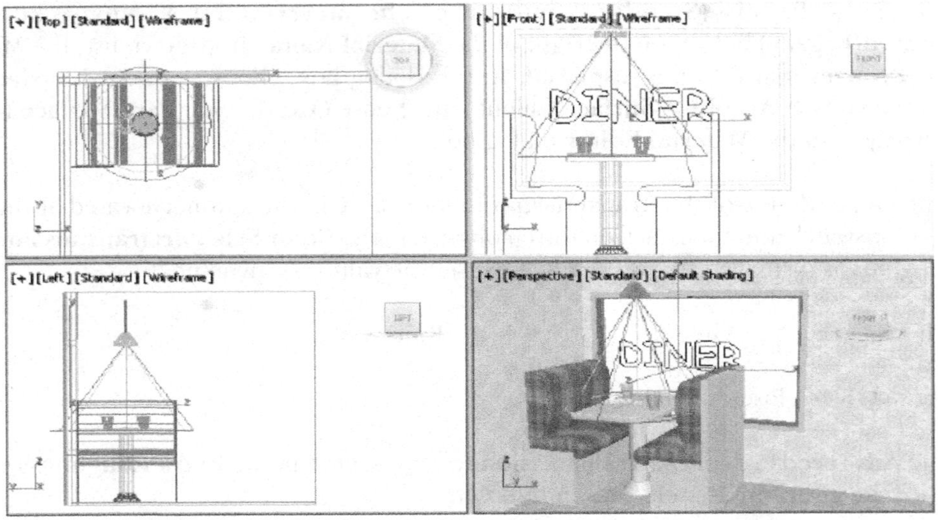

Figure P2-21 The text created

7. Choose **Rendering > Material Editor > Compact Material Editor** from the menu bar; the **Material Editor** dialog box is displayed. Select the 20th sample slot named **20 - Polished Aluminium**. Next, assign the selected material to the text.

8. Choose **Create > Lights** in the **Command Panel**. Make sure the **Standard** option is selected in the drop-down list. Next, invoke the **Arnold Light** tool from the **Object Type** rollout. In the **Shapes** rollout, make sure **Quad** is selected in the **type** drop-down list. Set the value **10** in the **Intensity** spinner of the **Color/Intensity** rollout.

9. In the Front viewport, click at a point; arnold light is created. Align it in the viewports, as shown in Figure P2-22.

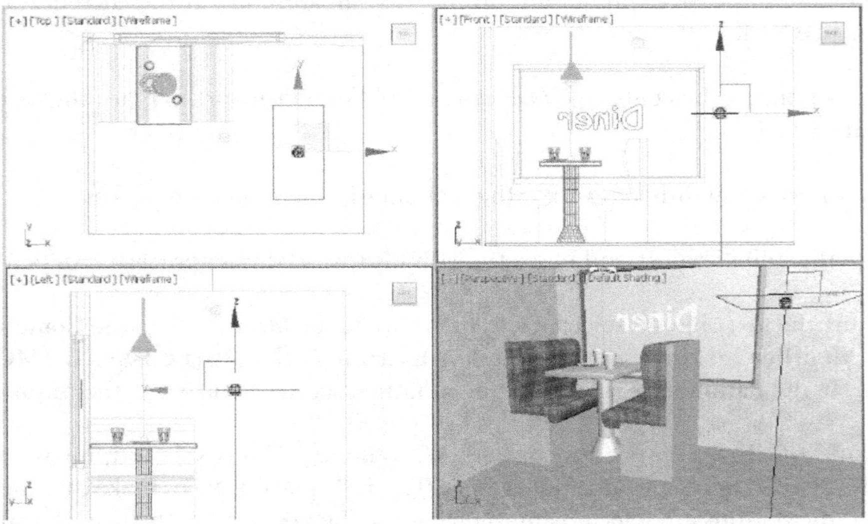

Figure P2-22 Arnold light created and positioned

Creating Blinds

In this section, you will create blinds for the window by using the **Box** tool.

1. Activate the Front viewport and then choose **Create > Geometry** in the **Command Panel** and select **Standard Primitives** from the drop-down list below the **Geometry** button. In the **Object Type** rollout, invoke the **Box** tool.

2. Expand the **Keyboard Entry** rollout and set the values in the spinners as given next:

   Length: **53** Width: **4** Height: **0.125**

3. Choose the **Create** button; a box is created. Enter the name **Blind01** in the **Name and Color** rollout.

4. In the Front viewport, move *Blind01* near the top-left corner of *Window01*. To do so, you may need to zoom into the area around the top-left corner of *Window01*.

5. Activate the Top viewport. Invoke the **Select and Rotate** and **Transform Gizmo Z Constraint** tools. Next, rotate *Blind01* at -20 degrees about the Z axis in the Top viewport.

6. Align *Blind01* in the viewports, as shown in Figure P2-23.

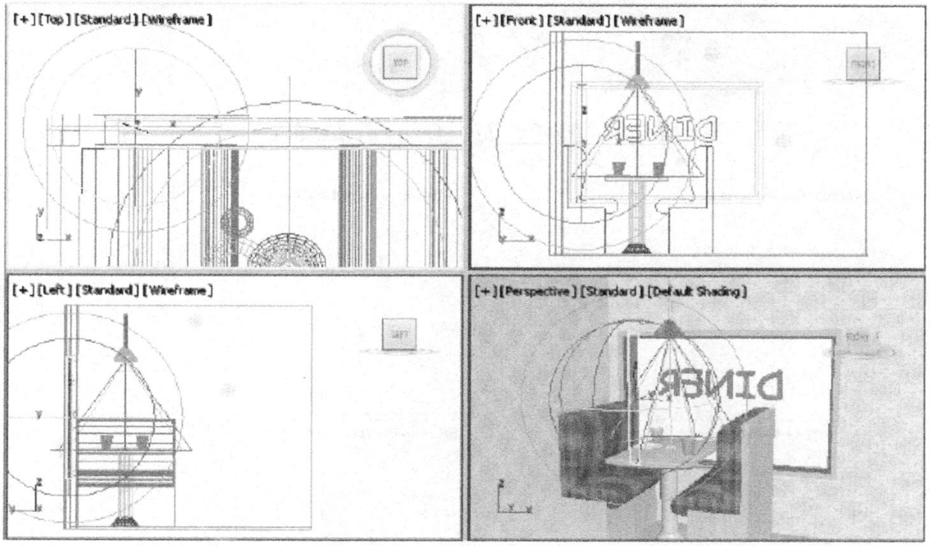

Figure P2-23 *Blind01 created and positioned*

7. Select *Blind01* and then choose **Rendering > Material Editor > Compact Material Editor** from the menu bar; the **Material Editor** dialog box is displayed. Select the next unused sample slot.

8. Choose the **Diffuse** color swatch; the **Color Selector:Diffuse Color** dialog box is displayed. Set the values as given next:

   Red: **255** Green: **244** Blue: **198**

9. Choose the **OK** button to close the **Color Selector:Diffuse Color** dialog box. Enter the name of the material as **Blinds** and assign it to *Blind01*. Next, close the **Material Editor** dialog box.

10. Make sure *Blind01* is selected and choose **Tools > Array** from the menu bar; the **Array** dialog box is displayed, as shown in Figure P2-24. Choose the **Reset All Parameters** button in this dialog box. Set the value **3.6** in the **Incremental X** spinner in the **Move** row. In the **Array Dimensions** area, make sure the **1D** radio button is selected and set the value **22** in the **Count** spinner. Next, select the **Copy** radio button in the **Type of Object** area and choose the **OK** button; 22 blinds are created extending across the window, as shown in Figure P2-25.

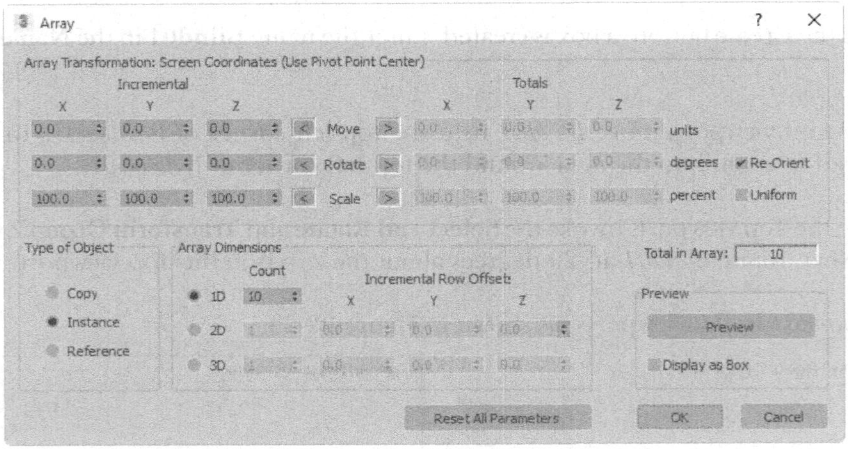

Figure P2-24 *The **Array** dialog box*

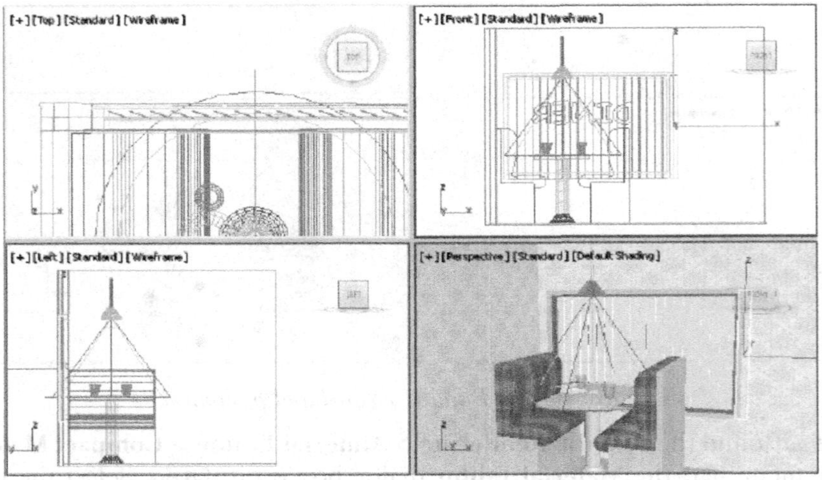

Figure P2-25 *The blinds created using the **Array** tool*

11. Invoke the **Zoom Extents All** tool to view the objects properly in all viewports and then save the scene.

Next, you will create multiple copies of the objects by using the **Clone Options** dialog box.

12. Activate the Front viewport and select *Drink01, Drink02, Booth01, Booth02, Base, Pedestal, Plate01, Shade, ArnoldLight001, Top,* and *Cylinder001* from the **Scene Explorer**. Group them with the name *Dinerset.*

13. Make sure the **Select and Move** tool is invoked. Then, choose the **Transform Gizmo X Constraint** tool. Press and hold the left mouse button along with the SHIFT key and move the selected objects to the right of the original objects. Release the SHIFT key and the left mouse button when the left edge of the copied booth lines up with the right edge of the original position; the **Clone Options** dialog box is displayed.

14. In this dialog box, enter the name **DinerSet01** in the **Name** text box and select the **Copy** radio button. Now, choose the **OK** button; another set of selected objects is created, as shown in Figure P2-26.

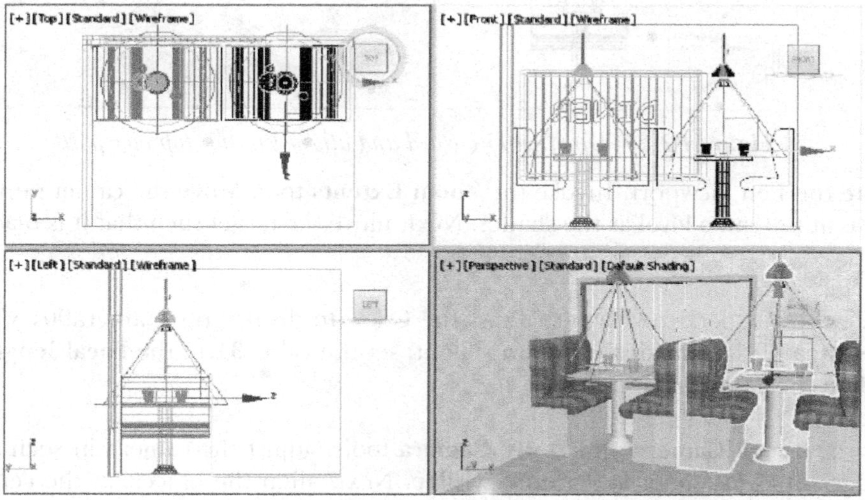

Figure P2-26 *The selected objects copied and positioned*

15. Select *Dinerset* and *DinerSet01* from the **Scene Explorer** and group them with the name **DinerSet02**.

16. Activate the Top viewport and create a copy of *DinerSet02* with the name **DinerSet03**. Next, align *DinerSet03* in the Top viewport, as shown in Figure P2-27.

Adding a Camera

In this section, you will add a camera by using the **Physical** tool.

1. Activate the Top viewport. Zoom out so that the objects are displayed in almost one-third of their original size.

2. Choose **Create > Cameras** in the **Command Panel**. Choose the **Physical** tool from the **Object Type** rollout. Click at a point at the bottom-middle of the viewport and a little toward the right of center. Press and hold the left mouse button, drag the cursor, and place the target on right side on top.

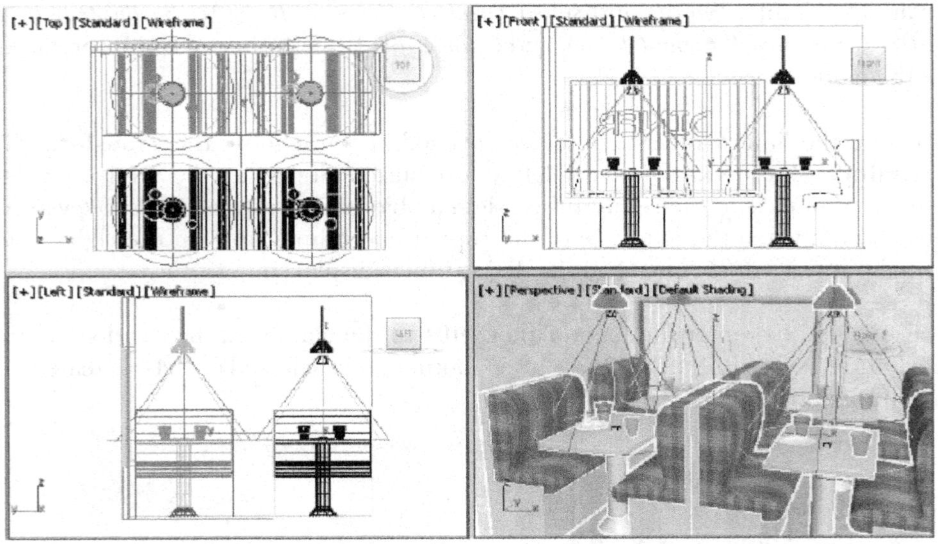

Figure P2-27 *DinerSet03 created and aligned in the Top viewport*

3. Activate the Left viewport. Invoke the **Zoom Extents** tool. Move the camera upward such that it is at the same level as the shades. Next, move the target such that it is placed on the table.

4. Activate the Perspective viewport. Press the C key to display the Camera001 view. In the **Lenses** area of the **Physical Camera** rollout, set the value **31** in the **Focal length** spinner and **16** in the **Aperture** spinner.

5. Using the **Truck Camera** and **Dolly Camera** tools, adjust the camera in such a way that the edges of *wall01* and *floor* are not visible. Next, align the objects at the center of the PhysCamera001 viewport, refer to Figure P2-28.

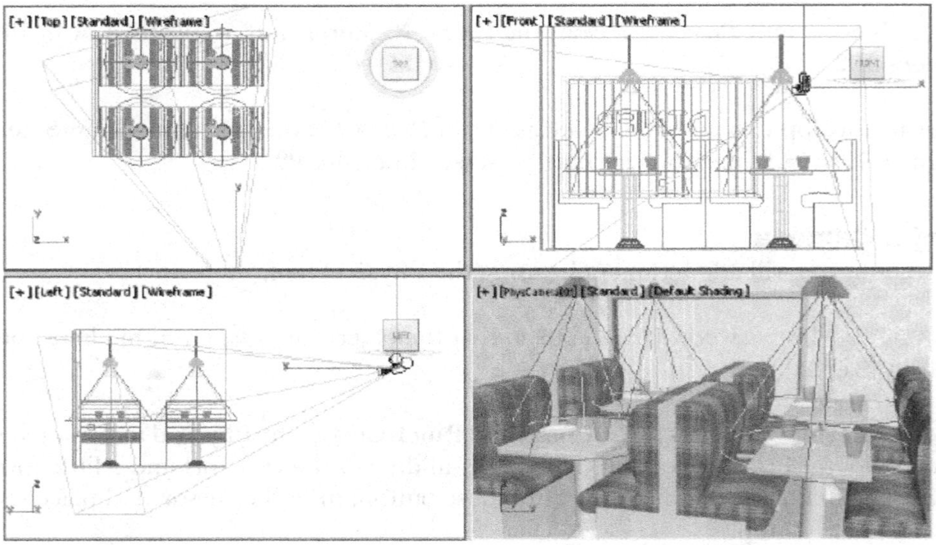

Figure P2-28 *Aligning objects at the center of the PhysCamera001 viewport*

Adding More Lights

In this section, you will add lights to illuminate the scene.

1. Choose **Create > Light** in the **Command Panel**. Next, select the **Arnold** option from the drop-down list displayed.

2. Activate the Top viewport and invoke the **Arnold Light** tool from the **Object Type** rollout. In the Top viewport, click at a point near the lower-right corner of the viewport and then drag the cursor to place the target on the window. In the **Shape** rollout, select **Spot** from the **Type** drop-down list.

3. In the Left viewport, invoke the **Zoom Extents** tool. Next, move the Arnold spotlight upward so that it is in level with the tabletop or slightly above it. Finally, move the spotlight target and place it in the middle of the window, refer to Figure P2-29.

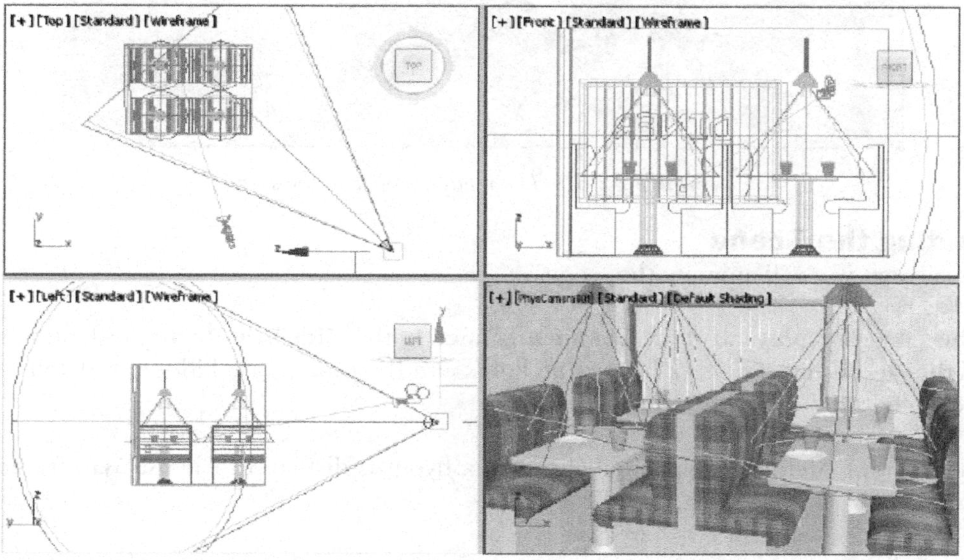

Figure P2-29 *The spotlight target moved and placed in the middle of the window*

4. Set the value **5** in the **Intesity** spinner of the **Color/Intensity** rollout. Also, set the required values in the **Cone Angle** spinner and the **Radius** spinner such that the cone created covers the whole room. Next, set **10** in the **Samples** spinner of the **Rendering** rollout.

Rendering the Still Image

In this section, you will render the still image.

1. Activate the Camera001 viewport. Choose the **Render Production** tool from the **Main Toolbar**; a rendered view of the scene is displayed in the **Rendered Frame** window, as shown in Figure P2-30.

2. Close the **Rendered Frame** window.

3. Make the necessary adjustments in the lights of the scene, if required.

Figure P2-30 The rendered view of the scene

Animating the Scene

In this section, you will animate the scene.

1. Choose the **Display** tab of the **Command Panel**. In the **Hide by Category** rollout, select the **Lights** and **Cameras** check boxes; the lights and the camera are hidden in the viewports. Next, choose the **Zoom Extents All** tool.

2. Click on the Shaded viewport label menu; a flyout is displayed. Choose **Wireframe** from the flyout.

   This will improve the system performance for previewing the animation in the viewport.

3. Activate and maximize the Top viewport. Zoom in on the blinds.

4. Choose the **Time Configuration** button to display the **Time Configuration** dialog box. In the **Animation** area, set the value **42** in the **Length** spinner. Choose the **OK** button; the number of frames in the active time segment changes to 42.

5. Select **World** from the **Reference Coordinate System** drop-down list from the **Main Toolbar**.

6. Drag the time slider to frame 0, if it is not at frame one and then choose the **Toggle Auto Key Mode** button to turn it on (red).

7. Select *Blind022* (the farthest-right blind) and right-click on the **Select and Move** tool; the **Move Transform-Type In** dialog box is displayed. In this dialog box, set the value **–2.7** in the **Offset World X** spinner; *Blind022* is moved to the left by 2.7 units. Then, close the **Move Transform Type-In** dialog box.

8. Move the time slider to Frame 1 and select *Blind021* along with *Blind022*. Using the **Move Transform Type-In** dialog box, move these two blinds left by -2.7 units.

9. Move the time slider to Frame 2 and select *Blind020* along with *Blind021* and *Blind022*. Using the **Move Transform Type-In** dialog box, move these three blinds left by -2.7 units.

10. Continue with this pattern until all the blinds are selected and moved except the last one (*Blind01*), as shown in Figure P2-31. The whole process will be completed in 20 frames.

11. Move the time slider to Frame 21 and start reversing the above process frame by frame. When you are done, choose the **Toggle Auto Key Mode** button to turn off the animation mode.

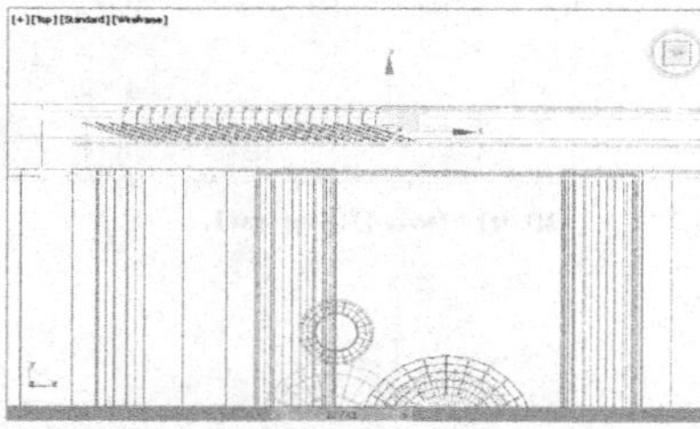

Figure P2-31 *The blinds at the midpoint of the animation*

12. Display the four-viewport configuration and then activate the Camera viewport. Next, choose the **Play Animation** button; the animation is played in the Camera001 viewport.

13. Choose the **Stop Animation** button to stop the playback.

Setting the Background

Now, you need to set the background for the scene.

1. Choose **Create > Geometry** in the **Command Panel**. Invoke the **Plane** tool from the **Object Type** rollout.

2. In the Front viewport, create a plane and align it on the back of *Wall01* in the Left viewport, refer to Figure P2-32.

3. In the **Parameters** rollout, enter **120** in the **Length** spinner and **170** in the **Width** spinner.

4. Choose the **Material Editor** dialog box and select the next unused material type sample slot.

5. Choose the **Physical Material** button available in the **Material Editor** dialog box; the **Material/Map Browser** dialog box is displayed. In this dialog box, choose **Standard Surface** from the **Materials > Arnold > Surface** rollout and choose **OK**; the **Standard Surface** material is displayed in the **Material Editor** dialog box.

6. In the **Base** rollout, choose the map next to the color swatch in the **Base Color** area of the **Material Editor**; the **Material/Map Browser** dialog box is displayed. In this dialog box, choose the **Bitmap** map button from the **Maps > General** rollout and choose the **OK** button; the **Select Bitmap Image File** dialog box is displayed.

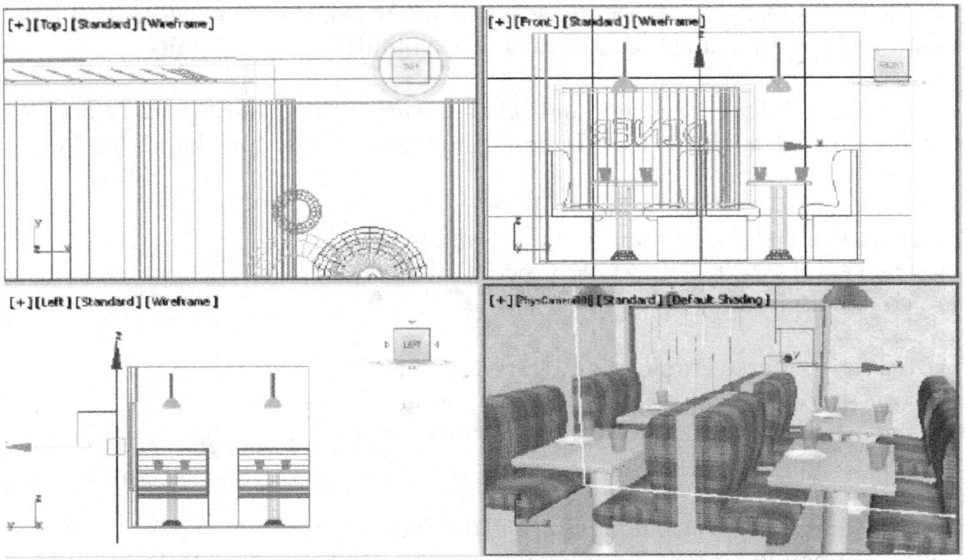

Figure P2-32 The Plane created and aligned on back of Wall01

7. In the **Select Bitmap Image File** dialog box, select the **Sunset.jpg** file and choose the **Open** button to return to the **Material Editor** dialog box.

8. In the **Coordinates** rollout, make sure the **User Real-World Scale** check box is cleared and **1.0** is set in the **U: Tiling** and **V: Tiling** spinners.

9. Invoke the **Go To Parent** button to go one level up. Make sure *Plane001* is selected in the viewport and assign this material to *Plane001*.

Rendering the Animation

In this section, you will render the final animation.

1. Activate the Camera001 viewport. Next, choose the **Render Setup** tool; the **Render Setup: Arnold** dialog box is displayed.

2. In the **Time Output** area of the **Common Parameters** rollout, select the **Active Time Segment** radio button. In the **Render Output** area, choose the **Files** button; the **Render Output File** dialog box is displayed.

3. In the **Render Output File** dialog box, select **AVI File (*.avi)** from the **Save as type** drop-down list. In the **File name** text box, enter the name **Project01** and then choose the **Save** button; the **AVI File Compression Setup** dialog box is displayed. Choose the **OK** button in this dialog box.

4. Make sure the **Save File** check box is selected in the **Render Output** area of the **Render Setup** dialog box and then choose the **Render** button in the dialog box.

It may take several minutes to render the animation frame by frame. After rendering is complete, you need to set some parameters available in the render window to improve the quality of render. Figure P2-33 shows the rendered image at frame 20.

Figure P2-33 *The final animation rendered at frame 20*

This page is intentionally left blank

Project 3

Architectural Project

PROJECT DESCRIPTION

This project guides you through the creation of the structure of a residential garage, as shown in Figure P3-1. In this project, you will first create a concrete floor slab and roof. You will then create overhead door, pivot door, and sliding windows in the wall. Next, you will add animation to doors and windows to give the scene a realistic look. To add more natural effect to the scene, you will set up and adjust lights and camera. Finally, you will make necessary adjustments and render the scene for the final output. The procedures used in this project can be applied to any architectural project.

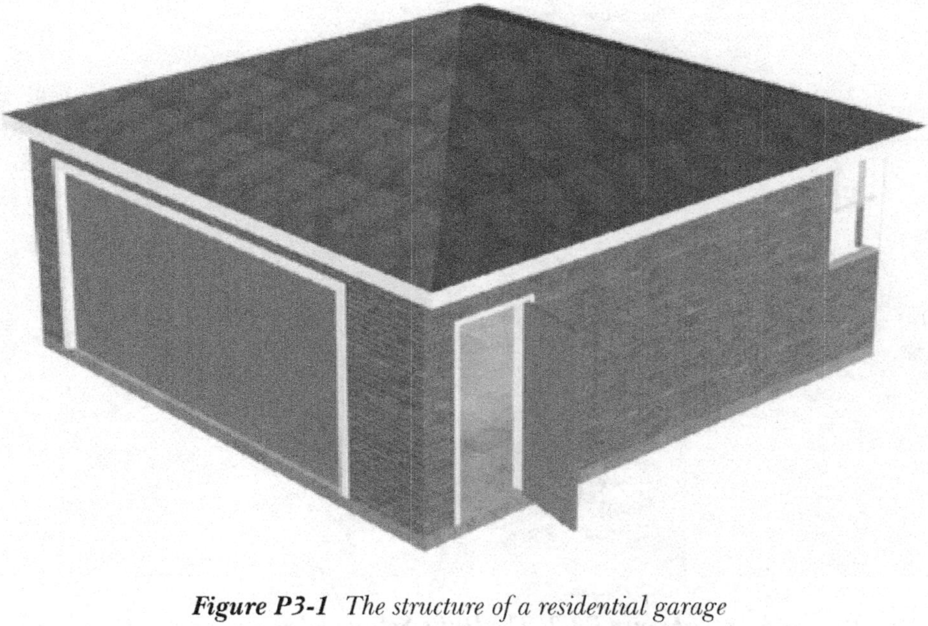

Figure P3-1 *The structure of a residential garage*

Creating the Project Folder

Create a new project folder with the name *Prj_3 at \Documents\3dsmax 2022* and then save the file with the name *Prj3* as discussed in Tutorial 1 of Chapter 2.

Downloading the Files

Before starting the project, you need to download the *prj3_3dsmax_2022.zip* file from *www.cadcim.com*. The path of the file is as follows: *Textbooks > Animation and Visual Effects > Autodesk 3ds Max > Autodesk 3ds Max 2022 for Beginners: A Tutorial Approach*

Extract the content of the zip file to *Documents*. Open Windows Explorer and browse to *\Documents\prj3_3dsmax_2022* and then copy all files to *\Documents\3dsmax 2022\Prj3\sceneassets\images*.

Changing the Units Setting

To start the project, you will change the units with which you are working. Since the units used in this project are in inches, you will set the units to decimal inches.

1. Choose **Customize > Units Setup** from the menu bar; the **Units Setup** dialog box is displayed.

2. In this dialog box, select the **US Standard** radio button in the **Display Unit Scale** area. Next, make sure the **Feet w/Fractional Inches** option is selected in the drop-down list below this radio button. In the right drop-down list, select 1/2, as shown in Figure P3-2. Choose the **OK** button to apply the settings and close the dialog box. If you want to work in the metric unit system, select the **Metric** radio button from the **Display Unit Scale** area of the **Units Setup** dialog box. Also, make sure the **Meters** option is selected in the drop-down list below the **Metric** radio button.

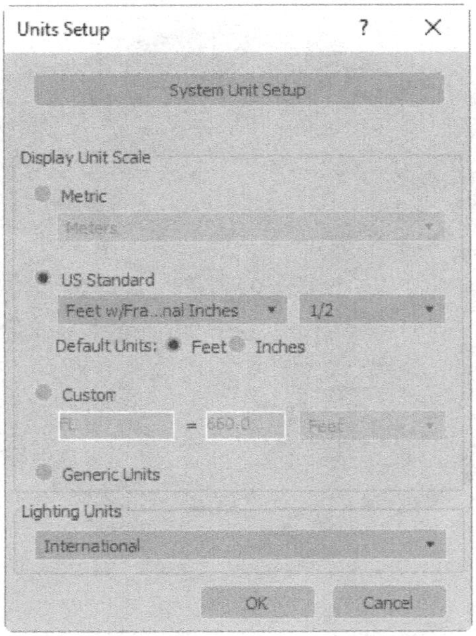

***Figure P3-2** The **Units Setup** dialog box*

3. Right-click on the **Snaps Toggle** tool from the **Main Toolbar**; the **Grid and Snap Settings** dialog box is displayed. Choose the **Home Grid** tab in this dialog box.

4. Set the value **1'0"(0.305m)** in the **Grid Spacing** spinner of the **Home Grid** tab. Also, set the value **2** in the **Major Lines every Nth Grid Line** spinner. Next, select the **All Viewports** radio button. Close the **Grid and Snap Settings** dialog box to apply these changes.

Creating the Walls

In architectural projects, you can use 2D shapes as paths for lofting other shapes. The 2D shapes can also describe the outlines or elevation views of the walls. The outlines are then extruded to the required thickness or height.

1. Activate the Top viewport. Choose **Create > Shapes** in the **Command Panel**. Invoke the **Line** tool from the **Object Type** rollout. Make sure the **Corner** radio button is selected in the **Initial Type** and **Drag Type** areas from the **Creation Method** rollout.

2. In the **Creation Method** rollout, make sure the **Corner** radio button is selected in the **Initial Type** area.

3. Expand the **Keyboard Entry** rollout and enter the values in the **X**, **Y**, and **Z** spinners as given next. Choose the **Add Point** button after every XYZ coordinate entry:

X: **0'**	Y: **0'**	Z: **0'**
X: **0'**	Y: **22'(6.706m)**	Z: **0'**
X: **22'(6.706m)**	Y: **22'(6.706m)**	Z: **0'**
X: **22'(6.706m)**	Y: **0'**	Z: **0'**

4. After adding a point for the last entry, choose the **Finish** button. In the **Name and Color** rollout, enter the name **Wall Path**.

5. Invoke the **Zoom Extents All** tool to zoom the object to its extent. The exterior *Wall Path* is now visible.

Creating 3D Wall Objects

3ds Max allows you to automatically create 3D walls. The wall objects have height, width, and justification that need to be altered according to the project work. Door and window objects can then be added to the wall objects and the necessary openings in the wall are automatically created.

1. Choose **Create > Geometry** in the **Command Panel**. Next, select **AEC Extended** in the drop-down list. Invoke the **Wall** tool from the **Object Type** rollout.

2. In the **Parameters** rollout, set the value **0'6"(0.152m)** in the **Width** spinner and **8'0"(2.438m)** in the **Height** spinner. Select the **Left** radio button in the **Justification** area.

3. Expand the **Keyboard Entry** rollout. In this rollout, choose the **Pick Spline** button and then select *Wall Path* line in any viewport; a 3D wall object (Wall001) is created on top of the selected path, as shown in Figure P3-3.

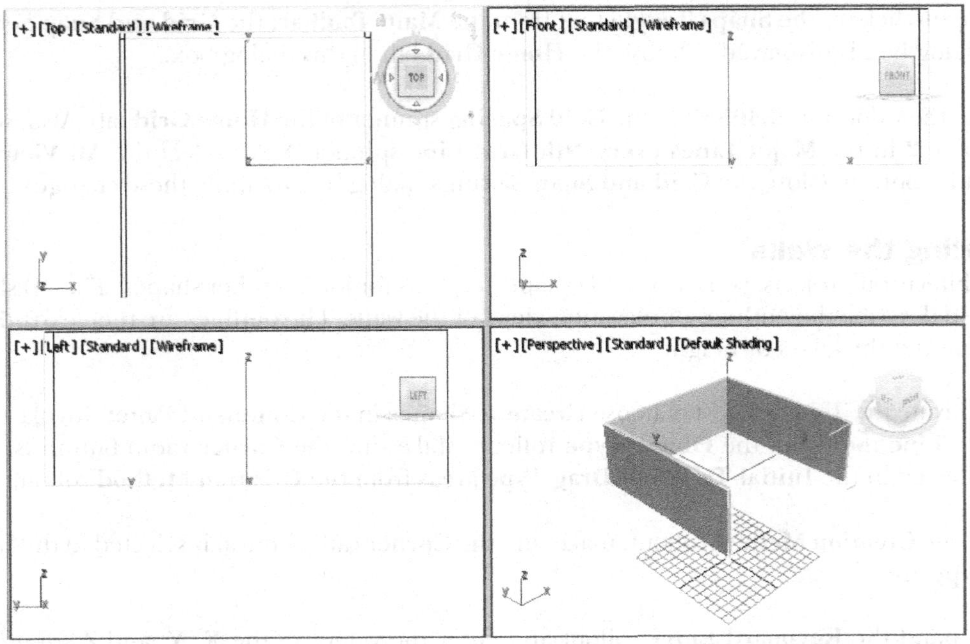

Figure P3-3 The 3D Wall001 created on top of the selected path

Creating an Extruded Wall

In this section, you will create the front wall of the garage as a 2D outline or elevation view of the wall. The outline will be extruded to proper thickness matching with the walls, door, and windows.

1. Activate the Front viewport and then choose **Create > Shapes** in the **Command Panel**. Invoke the **Line** tool from the **Object Type** rollout. Make sure the **Corner** radio button is selected in the **Initial Type** and **Drag Type** areas from the **Creation Method** rollout.

2. Expand the **Keyboard Entry** rollout and enter the values in the **X**, **Y**, and **Z** spinners, as given next. Choose the **Add Point** button after every XYZ coordinate entry:

X: **0'**	Y: **0'**	Z: **0'**
X: **0'**	Y: **8'(2.438m)**	Z: **0'**
X: **22'(6.706m)**	Y: **8'(2.438m)**	Z: **0'**
X: **22'(6.706m)**	Y: **0'**	Z: **0'**
X: **18'4"(5.588m)**	Y: **0'**	Z: **0'**
X: **18'4"(5.588m)**	Y: **7'4"(2.235m)**	Z: **0'**
X: **1'8"(0.508m)**	Y: **7'4"(2.235m)**	Z: **0'**
X: **1'8"(0.508m)**	Y: **0'**	Z: **0'**

3. After the last entry, choose the **Close** button to add the last segment; the front wall is created, as shown in Figure P3-4. In the **Name and Color** rollout, enter **Wall Front**.

4. Make sure *Wall Front* is selected and then choose the **Modify** tab in the **Command Panel**. In the **Modifier List** drop-down list, select **Extrude**. Next, in the **Parameters** rollout of

the modifier, set the value **0'6"(0.152m)** in the **Amount** spinner. Also, select the **Generate Mapping Coords** check box at the bottom of the rollout.

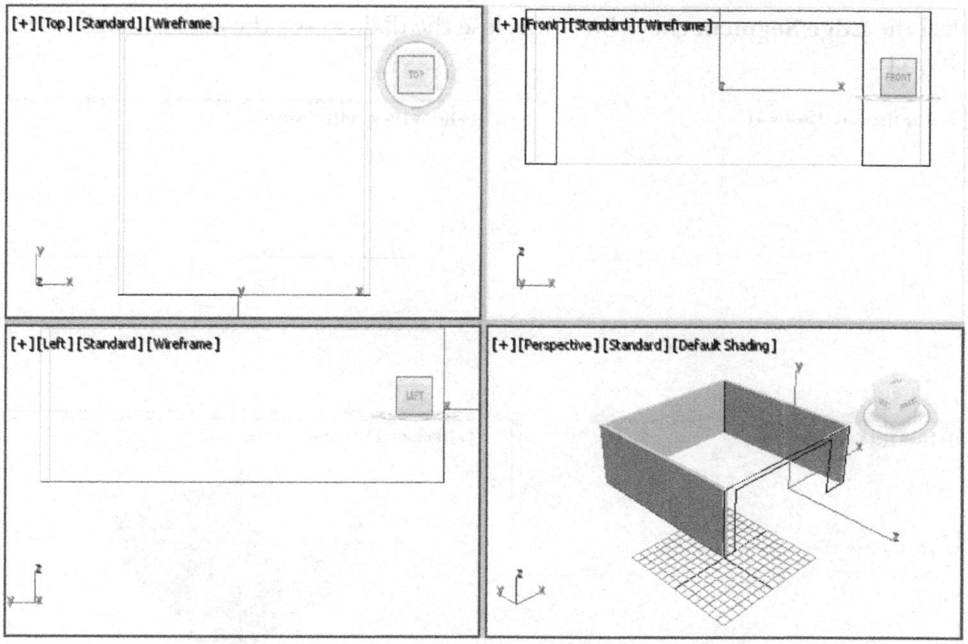

***Figure P3-4** The front wall created*

Creating the Floor Slab

The floor of the garage is created as a standard 3D box primitive.

1. Activate the Top viewport. Choose **Create > Geometry** in the **Command Panel**. Select **Standard Primitives** from the drop-down list. Next, invoke the **Box** tool. In the **Parameters** rollout, make sure the **Generate Mapping Coords** check box is selected at the bottom of the rollout.

2. Expand the **Keyboard Entry** rollout and enter the values as given next:

   X: **11'(3.353m)** Y: **10'9"(3.277m)** Z: **0'**
   Length: **22'6"(6.858m)** Width: **22'(6.706m)** Height: **0'5"(0.127m)**

3. Choose the **Create** button; the floor is created, as shown in Figure P3-5. In the **Name and Color** rollout, enter **Floor Slab**.

Creating the Doors and Windows

3ds Max can automatically create 3D doors and windows. The door and window objects must be placed perfectly in the wall object for the required opening to be automatically created. You will create the side door that will be placed in the wall object with required settings.

1. Right-click on the **Snaps Toggle** tool in the **Main Toolbar**; the **Grid and Snap Settings** dialog box is displayed.

2. Select the **Edge/Segment** check box and close the dialog box; the cursor snaps to the edges of objects.

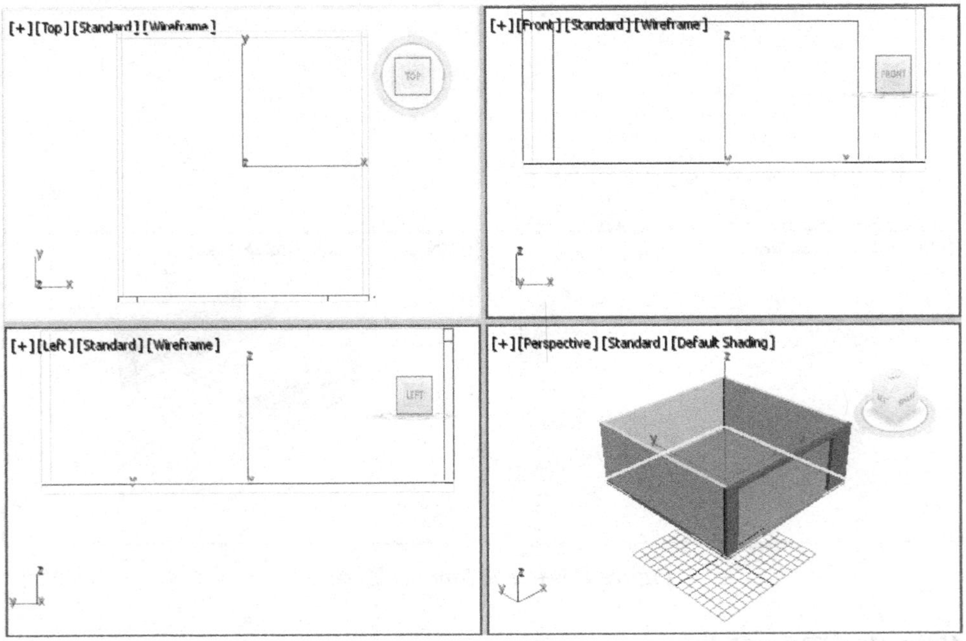

***Figure P3-5** The floor and walls created*

3. Invoke the **2D Snap** tool from the **Snap Toggle** flyout to make 2D snaps active. Also, make sure the Top viewport is activated. 

4. Choose **Create > Geometry** in the **Command Panel**. Next, select **Doors** from the drop-down list and then invoke the **Pivot** tool from the **Object Type** rollout.

5. In the **Creation Method** rollout, make sure the **Width/Depth/Height** radio button is selected.

6. In the **Frame** area of the **Parameters** rollout, make sure the **Create Frame** check box is selected. Also, select the **Generate Mapping Coords** check box at the bottom of the rollout.

7. Click and hold a point on the interior side of the right wall near the bottom. Make sure the snap helper highlights the edge. With the mouse button held down, drag the door to any distance up along the inside of the wall to set the door width and then release the mouse button; an approximate door width is set. The width will be modified to the exact required dimension later.

8. Highlight and select the outside edge of the right wall; the door depth is set.

9. Highlight and select the outside top edge of the right wall at any point; an approximate door height is set. The door parameters can now be set to their exact required dimensions.

10. Make sure the door is selected and then choose the **Modify** tab in the **Command Panel**. In the **Parameters** rollout, set the values in the spinners as given next:

    Height: **6'8"(2.032m)** Width: **3'0"(0.914m)** Depth: **0'6"(0.152m)**

11. In the **Parameters** rollout, enter the value **90** in the **Open** spinner.

    Verify that the door is now open at 90 degrees and there is an opening in the wall. Rotate the view in the Perspective viewport as needed to see the door. The door should open inside the garage toward the main door. If there is no door opening, then delete the door, check if the 2D snaps are active, and create the door again.

12. In the **Parameters** rollout, select the **Flip Swing** and **Flip Hinge** check boxes, if they are not already selected. Rotate the Perspective viewport to set the angle of view. The door is created, as shown in Figure P3-6.

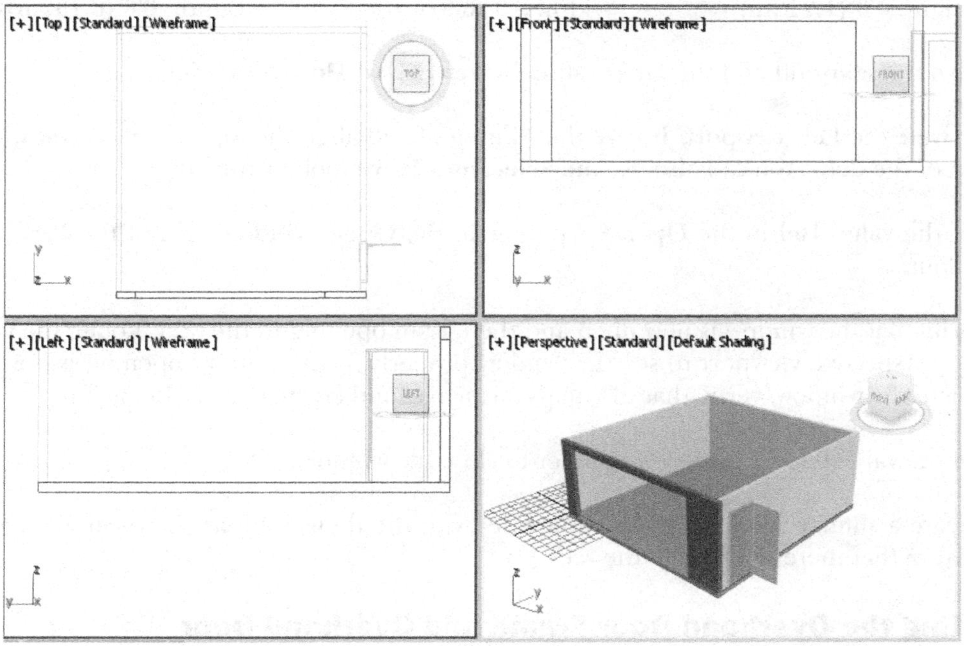

Figure P3-6 *The door created*

13. In the **Parameters** rollout, enter the value **0** in the **Open** spinner to close the door.

    Next, you will create a window object that will be placed in the wall object with the required openings automatically created in the wall object.

14. Activate the Top viewport and then choose **Create > Geometry** in the **Command Panel**. Select **Windows** from the drop-down list. Then, invoke the **Sliding** tool from the **Object Type** rollout.

15. In the **Creation Method** rollout, make sure the **Width/Depth/Height** radio button is selected.

16. In the **Parameters** rollout, select the **Generate Mapping Coords** check box and make sure the **2D Snap** tool is invoked.

17. Click and hold a point on the inside of the left wall near the top. Make sure the snap helper highlights the edge. With the mouse button held down, drag the window to any distance down along the inside of the wall to set the width of the window and release the button; an approximate window width is set. The width will be later modified to the required dimension.

18. Highlight and select the outside edge of the left wall; the window depth has been set.

19. Highlight and select the top edge of the left wall at any point; an approximate window height is set. The window parameters can now be set to their exact required dimensions.

20. Make sure the window is selected and then choose the **Modify** tab in the **Command Panel**. In the **Parameters** rollout, enter the values in the spinners as given below:

    Height: **4'0"(1.219m)**　　Width: **3'0"(0.914m)**　　Depth: **0'6"(0.152m)**

21. Turn the snaps off and then make sure the **Select and Move** tool is invoked.

22. Activate the Left viewport. Invoke the **Align** tool and align the top of the window with the top of the door. You can also use the **Select and Move** tool for this step.

23. Set the value **100** in the **Open %** spinner in the **Open Window** area of the **Parameters** rollout.

    Verify that the window is now open and there is an opening in the wall. Rotate the view in the Perspective viewport to see the window properly. If the window opening is not there, delete the window, verify that 2D snaps are active, and create the window again.

24. Set the value **0** in the **Open %** spinner to close the window.

25. Create a similar window on the right side (with the door) wall, as shown in Figure P3-7. Ensure that there is a hole in the wall.

Creating the Overhead Door Frame and Overhead Door

In this section, you will create the overhead door frame and the overhead door as 2D outlines (shapes). The outlines will then be extruded to the proper thickness.

1. Activate the Front viewport and then choose **Create > Shapes** in the **Command Panel**. Next, invoke the **Line** tool from the **Object Type** rollout. Make sure the **Corner** radio button is selected in the **Initial Type** and **Drag Type** areas of the **Creation Method** rollout.

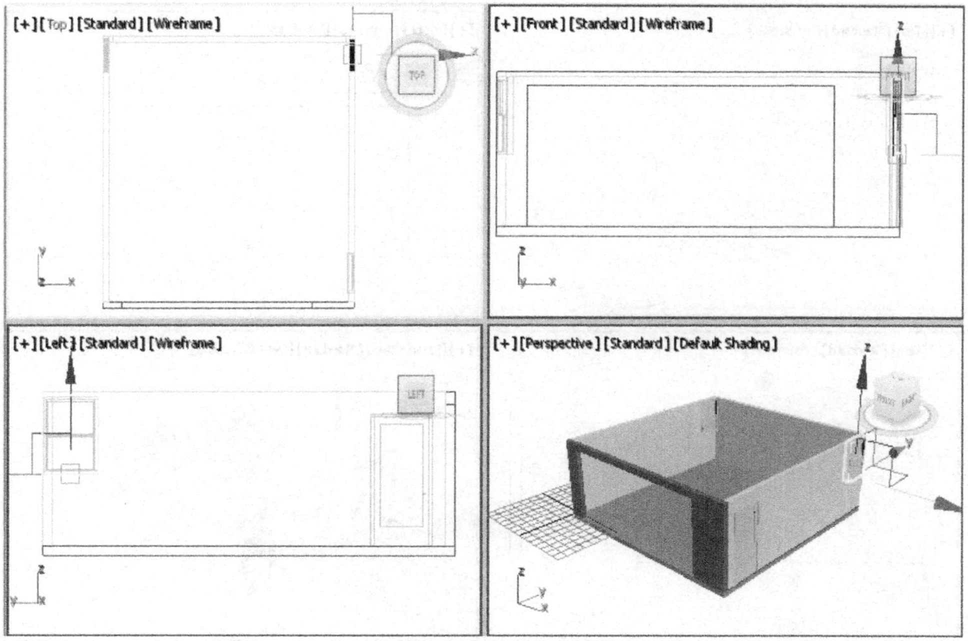

Figure P3-7 *The sliding windows created*

2. In the **Keyboard Entry** rollout, enter the following values. Choose the **Add Point** button after entering each set of coordinates:

X: **1'8"(0.508m)**	Y: **0'**	**Z: 0'**
X: **1'8"(0.508m)**	Y: **7'4"(2.235m)**	**Z: 0'**
X: **18'4"(5.588m)**	Y: **7'4"(2.235m)**	**Z: 0'**
X: **18'4"(5.588m)**	Y: **0'**	**Z: 0'**
X: **18'0"(5.486m)**	Y: **0'**	**Z: 0'**
X: **18'0"(5.486m)**	Y: **7'0"(2.134m)**	**Z: 0'**
X: **2'0"(0.61m)**	Y: **7'0"(2.134m)**	**Z: 0'**
X: **2'0"(0.61m)**	Y: **0'**	**Z: 0'**

3. Choose the **Close** button and enter the name **Overhead Door Frame** in the **Name and Color** rollout.

4. Make sure *Overhead Door Frame* is selected and then choose the **Modify** tab in the **Command Panel**. Select **Extrude** in the **Modifier List** drop-down list under the **OBJECT-SPACE MODIFIERS** section. In the **Parameters** rollout for the modifier, set the value **0'8"(0.203m)** in the **Amount** spinner. Also, make sure the **Generate Mapping Coords** check box is selected at the bottom of the rollout.

5. Activate the Top viewport and enter **0'1"(0.025)** in the **Y**-coordinate axis to move the frame up so that it is centered on the wall; *Overhead Door Frame* is created and positioned, as shown in Figure P3-8.

Next, you will create the overhead door by using the **Line** tool.

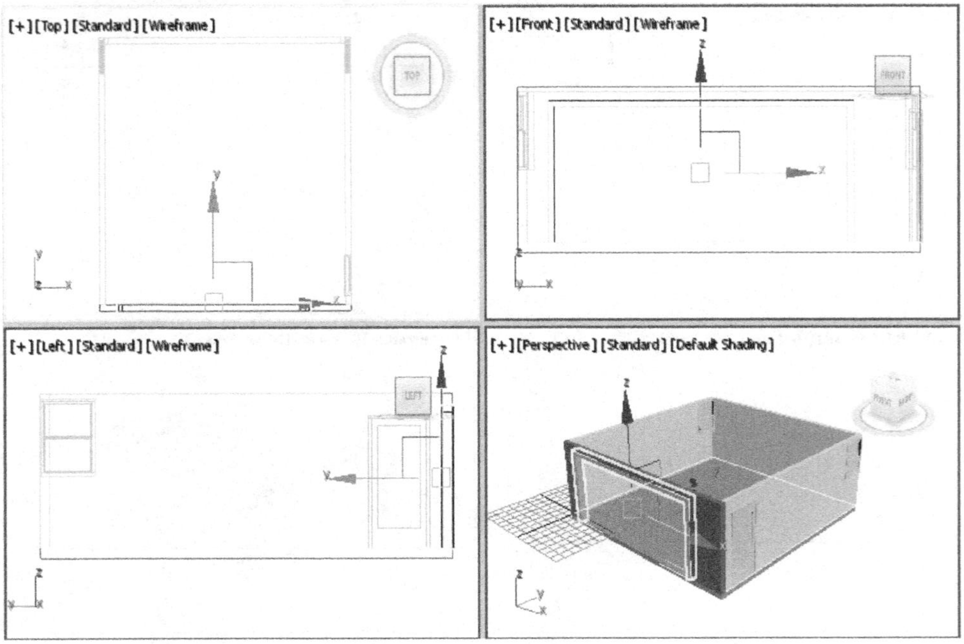

Figure P3-8 *Overhead Door Frame created and positioned*

6. Activate the Front viewport and then choose **Create > Shapes** in the **Command Panel**. Next, invoke the **Line** tool from the **Object Type** rollout. Make sure the **Corner** radio button is selected in the **Initial Type** and **Drag Type** areas from the **Creation Method** rollout.

7. In the **Keyboard Entry** rollout, enter the following values. Choose the **Add Point** button after entering each set of coordinates:

X: **2'0"(0.61m)**	Y: **0'**	Z: **0'**
X: **2'0"(0.61m)**	Y: **7'0"(2.134m)**	Z: **0'**
X: **18'0"(5.486m)**	Y: **7'0"(2.134m)**	Z: **0'**
X: **18'0"(5.486m)**	Y: **0'**	Z: **0'**

8. Choose the **Close** button. In the **Name and Color** rollout, enter **Overhead Door**.

9. Make sure *Overhead Door* is selected and then choose the **Modify** tab in the **Command Panel**. Select **Extrude** in the **Modifier List** drop-down list under the **OBJECT-SPACE MODIFIERS** section. In the **Parameters** rollout for the modifier, set the value **0'4"(0.102m)** in the **Amount** spinner. Also, select the **Generate Mapping Coords** check box at the bottom of the rollout.

10. Activate the Top viewport and move *Overhead Door* down 0'1"(0.025m) so that it is centered on the wall. Alternatively, you can enter the same value in the **Y**-coordinate axis.

    Overhead Door is created and positioned, as shown in Figure P3-9.

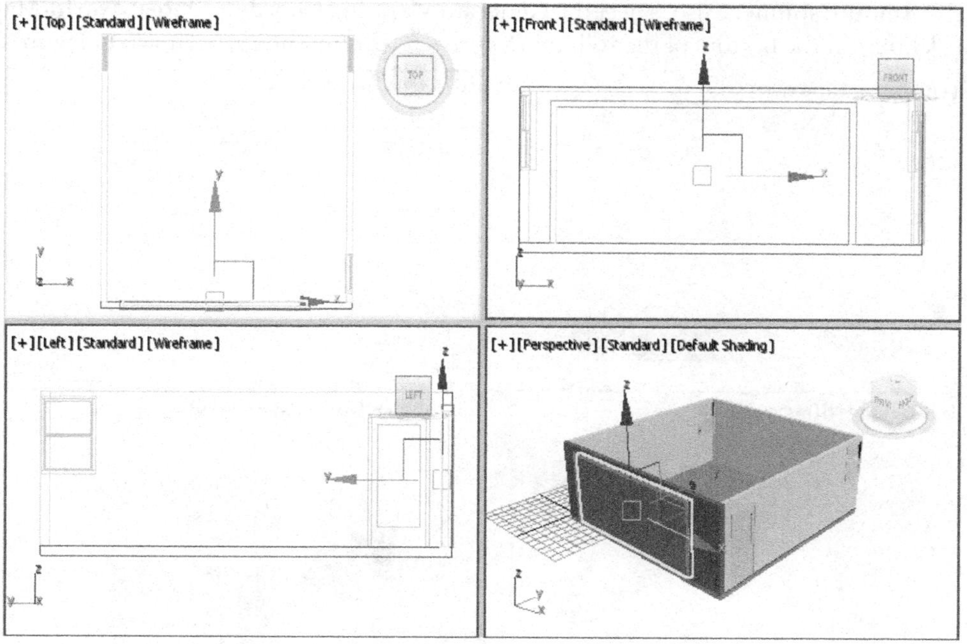

Figure P3-9 Overhead Door created and positioned

Creating the Roof

The roof overhangs the garage walls by 1'0"(0.304m) on each side. The length of the flat edge of the roof is 24'0"(7.3152m) and the angled lines is 12'0"(3.658m) in both the X and Y axis. Since the garage is a square, one-quarter of the roof can be created and then copied by using the **Array** command.

1. Make sure the Top viewport is active and then invoke the **Zoom Extents** tool to zoom in the objects.

2. Choose **Create > Shapes** in the **Command Panel**. Invoke the **Line** tool from the **Object Type** rollout. Make sure the **Corner** radio button is selected in the **Initial Type** and **Drag Type** areas from the **Creation Method** rollout.

3. Expand the **Keyboard Entry** rollout and enter the following values. Choose the **Add Point** button after entering each set of coordinates:

   X: **–1'0"(–0.305m)** Y: **–1'0"(–0.305m)** Z: **8'0"(2.438m)**
   X: **11'0"(3.353m)** Y: **11'0"(3.353m)** Z: **8'0"(2.438m)**
   X: **23'0"(7.01m)** Y: **–1'0"(–0.305m)** Z: **8'0"(2.438m)**

4. Choose the **Close** button. In the **Name and Color** rollout, enter **Roof Section01**.

5. Make sure *Roof Section01* is selected and then choose the **Modify** tab in the **Command Panel**. Select **Extrude** in the **Modifier List** drop-down list under the **OBJECT-SPACE MODIFIERS** section. In the **Parameters** rollout for the modifier, enter the value **0'6"(0.152m)**

in the **Amount** spinner. Also, select the **Generate Mapping Coords** and **Real-World-Map Size** check boxes at the bottom of the rollout; *Roof Section01* is modified, as shown in Figure P3-10.

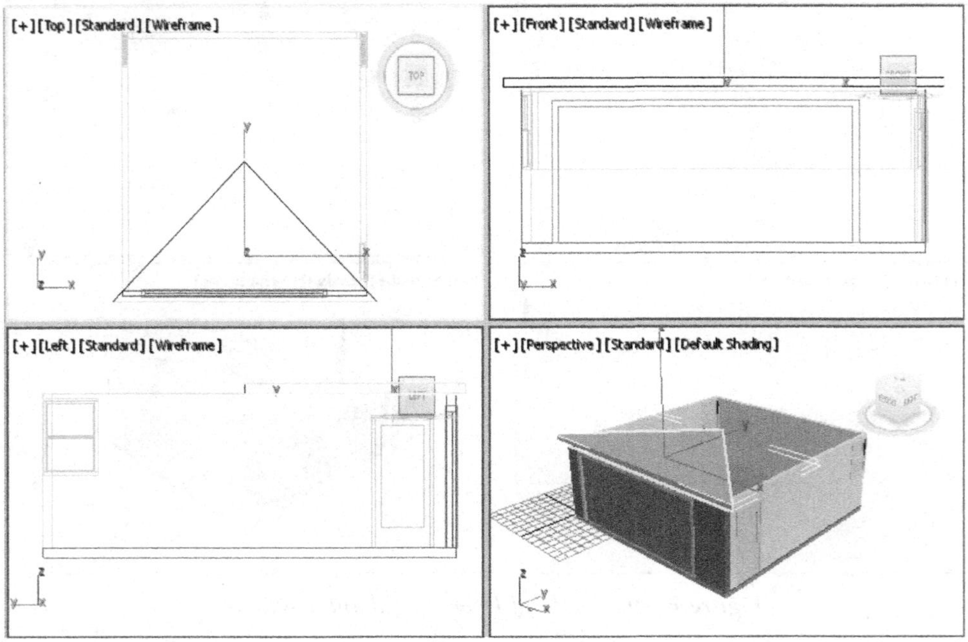

Figure P3-10 *Roof Section01 modified*

Next, you will add slope to the roof to modify the vertices at the center of the roof. By moving these vertices straight up, the peak is added to the roof.

6. Make sure *Roof Section01* is selected and the **Modify** tab in the **Command Panel** is chosen. Select **Edit Mesh** in the **Modifier List** drop-down list under the **OBJECT-SPACE MODIFIERS** section.

7. In the **Selection** rollout for the modifier, choose the **Vertex** button; a **Vertex** sub-object mode is activated. 

8. In the Top viewport, select the vertices at the center of the roof (top of the triangle) using the marque selection.

9. Right-click on the **Select and Move** tool; the **Move Transform Type-In** dialog box is displayed.

10. Set the value **4'0"(1.21m)** in the **Offset:Screen Z** text box in the **Move Transform Type-In** dialog box and press the ENTER key; the peak of the roof is moved, as shown in Figure 3-11.

11. In the **Selection** rollout, choose the **Vertex** button to exit the sub-object mode. Now, close the **Move Transform Type-In** dialog box.

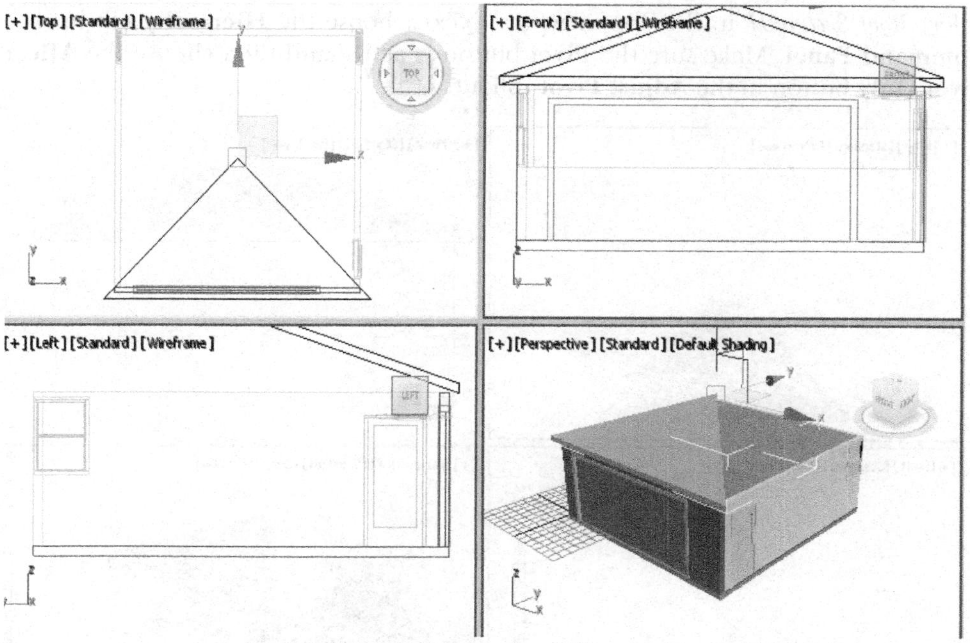

Figure P3-11 *The peak of Roof Section01 moved*

The roof consists of a trim board or fascia running around the edge and it shingles on the top. You can apply two materials to create this effect in one of the two ways; either you can use a multi/sub-object material and the material modifier or you can detach a face of the *Roof Section01* object so that different materials can be directly assigned to the roof and the detached face. You will use the second method in this section.

12. Make sure *Roof Section01 is* selected and the **Modify** tab is chosen in the **Command Panel**.

13. In the **Selection** rollout, choose the **Polygon** button; the **Polygon** sub-object mode is activated. In this sub-object mode, you can detach faces.

14. Select the front face of *Roof Section01* in the Front viewport; it is displayed in red and the front face of *Roof Section01* is selected, as shown in Figure P3-12.

15. In the **Edit Geometry** rollout, choose the **Detach** button; the **Detach** dialog box is displayed, as shown in Figure P3-13. Enter the name **Fascia01** and choose the **OK** button.

    The selected face is now a separate object in the scene.

16. In the **Selection** rollout, choose the **Polygon** button again to exit the sub-object mode.

Next, you will use the **Array** command to complete the roof. Before creating the array, you need to adjust the pivot point of the roof and fascia.

17. Select *Roof Section01* in the Top viewport. Next, choose the **Hierarchy** tab in the **Command Panel**. Make sure the **Pivot** button is active and then choose the **Affect Pivot Only** button in the **Adjust Pivot** rollout.

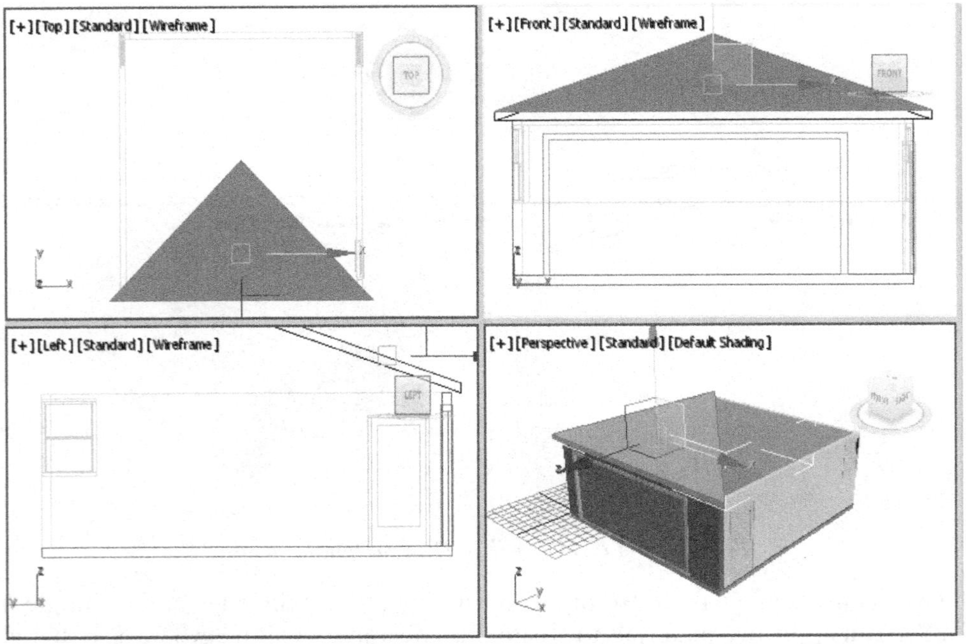

Figure P3-12 *The front face of the Roof Section01 selected*

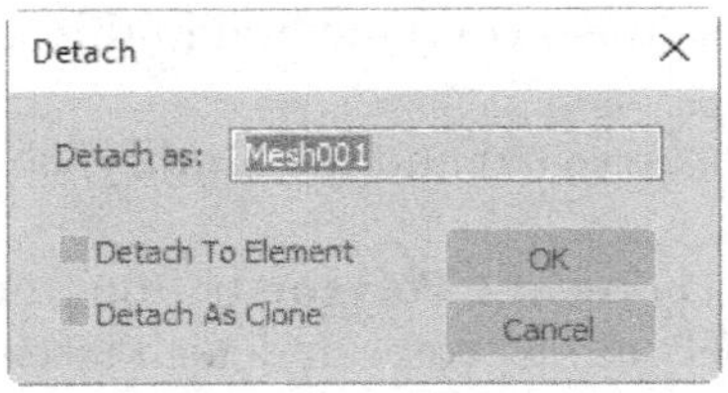

Figure P3-13 *The* ***Detach*** *dialog box*

18. Invoke the **Align** tool and then select *Fascia01*; the **Align Selection (Fascia01)** dialog box is displayed.

19. In this dialog box, make sure the **Pivot Point** radio button is selected in the **Current Object** area. Also, in the **Target Object** area, select the **Maximum** radio button. At the top of the dialog box, select the **Y Position** and **Z Position** check boxes and clear the **X Position** check box. Next, choose the **OK** button; the pivot point of each object is now at the topmost point of the roof.

20. Choose the **Affect Pivot Only** button again to exit the pivot mode.

21. Make sure the **Use Pivot Point Center** tool is activated in the Pivot Point flyout of the **Main Toolbar**.

22. Select *Roof Section01* and *Fascia01*. Next, choose **Tools > Array** from the menu bar; the **Array** dialog box is displayed, as shown in Figure P3-14.

23. Choose the **Reset All Parameters** button to restore the default values in this dialog box.

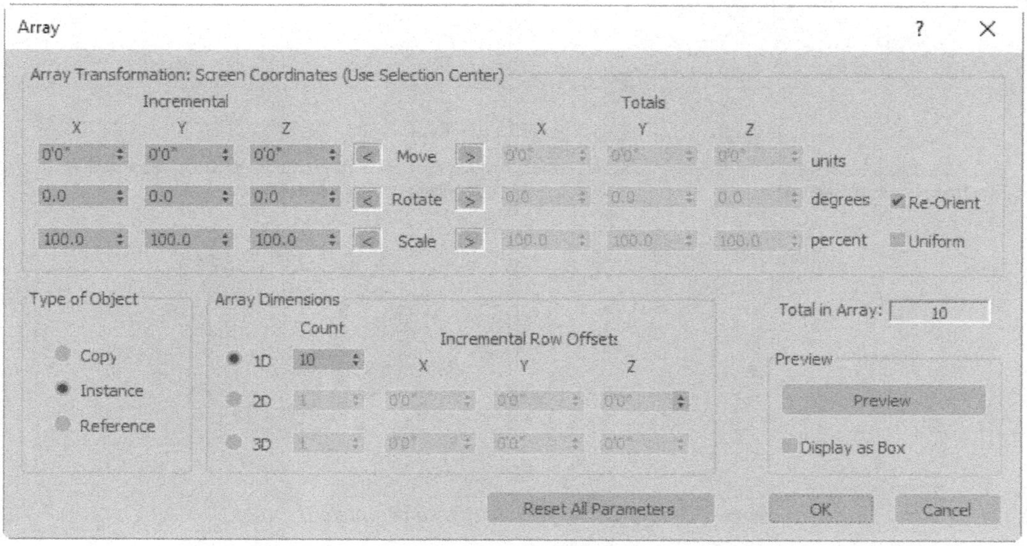

Figure P3-14 The ***Array*** *dialog box*

24. In the **Incremental** area, set the value **90** in the **Z** column corresponding to the **Rotate** row. Make sure the **Re-Orient** check box located on the right of the **Rotate** row is selected.

25. In the **Array Dimensions** area, make sure the **1D** radio button is selected. Next, set the value **4** in the **Count** column. Also, make sure the **Instance** radio button in the **Type of Object** area is selected and then choose the **OK** button; the array is created to complete the roof. You may need to properly align and adjust the array by using the **Select and Move** tool in Top, Front, and Left viewports, refer to Figure P3-15.

Adding Lights and Camera

In this section, two spotlights and one omni light will be used to illuminate the scene. In addition to it, a camera will be used to control the movement of the objects and set an angle of view in the Perspective viewport.

1. Activate the Top viewport and zoom out to make more space available for placing the lights and camera.

2. Choose **Create > Lights** in the **Command Panel**. Select **Standard** from the drop-down list and then invoke the **Omni** tool from the **Object Type** rollout.

3. Select a point above the outside wall of the garage, refer to Figure P3-16.

4. Invoke the **Target Spot** tool from the **Object Type** rollout. Select a point near the lower-right corner of the Top viewport and drag the target to the center of the garage.

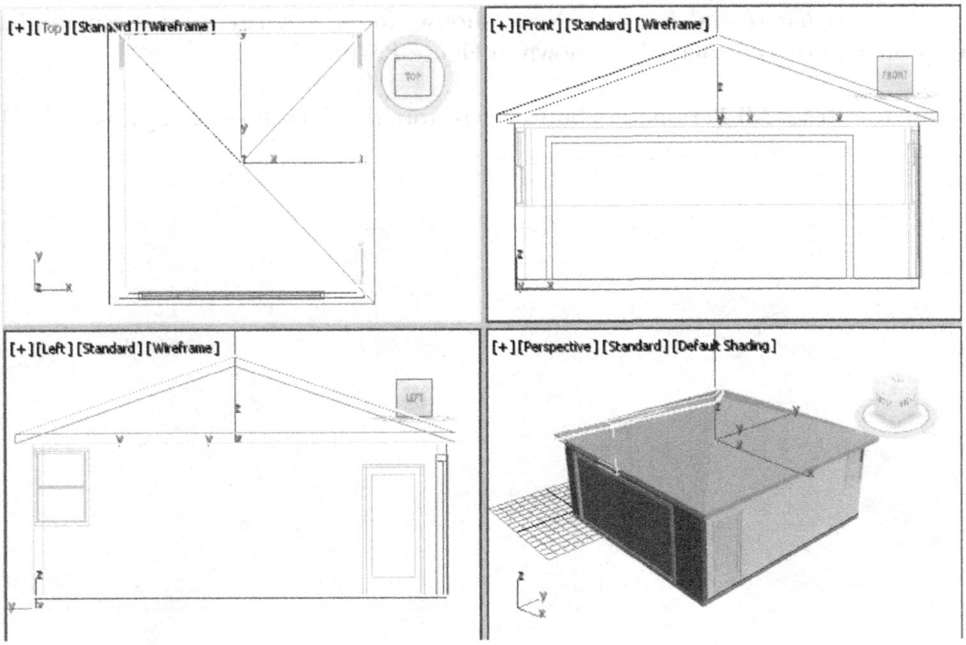

Figure P3-15 The array created to complete the roof

5. Place another spotlight by selecting a point in the lower-left corner of the Top viewport and dragging the target to the center of the garage.

6. Activate the Front Viewport. Next, invoke the **Select and Move** tool and move the lights to their approximate positions, as shown in Figure P3-16.

7. Choose **Create > Cameras** in the **Command Panel**. Next, invoke the **Target** tool from the **Object Type** rollout.

8. In the Top viewport, click at some point near the lower-right corner to place the camera and drag the target to the center of the garage.

9. Activate the Perspective viewport and press the C key; the Perspective viewport switches to the Camera001 viewport.

10. Invoke the **Zoom Extents All** tool to zoom the objects to their extents.

11. Use the **Truck Camera**, **Orbit Camera**, and **Field-of-View** tools and move the camera until an appropriate view is displayed in the Camera001 viewport, as shown in Figure P3-17.

Creating the Material Library

In this section, you will create a new material library. All materials that you create will be stored in the new material library.

1. Choose **Rendering > Material Editor > Compact Material Editor** from the **Main Toolbar**; the **Material Editor** dialog box is displayed.

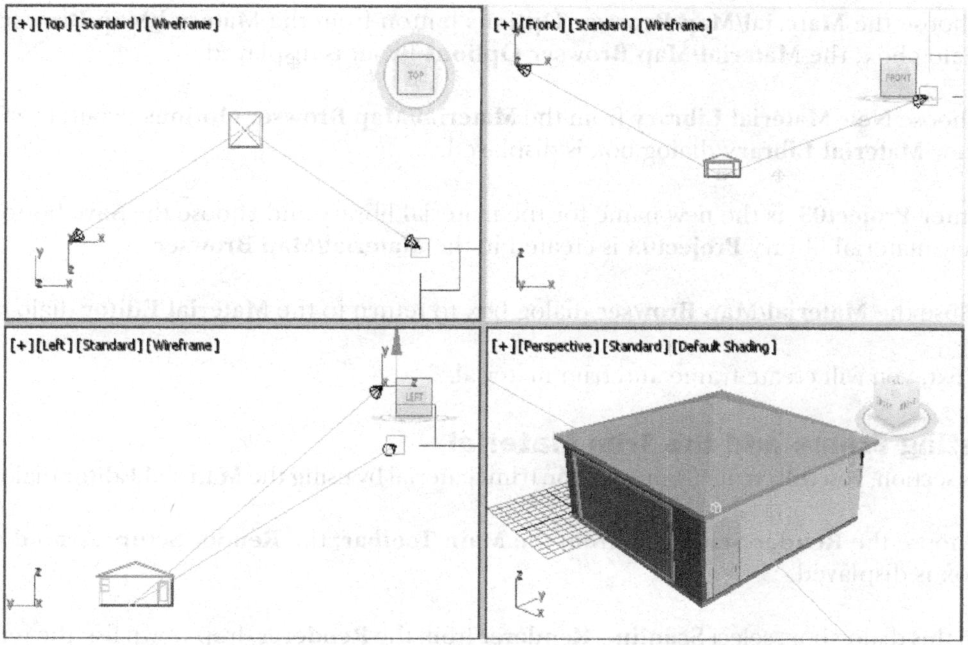

***Figure P3-16** The lights displayed and positioned in all viewports*

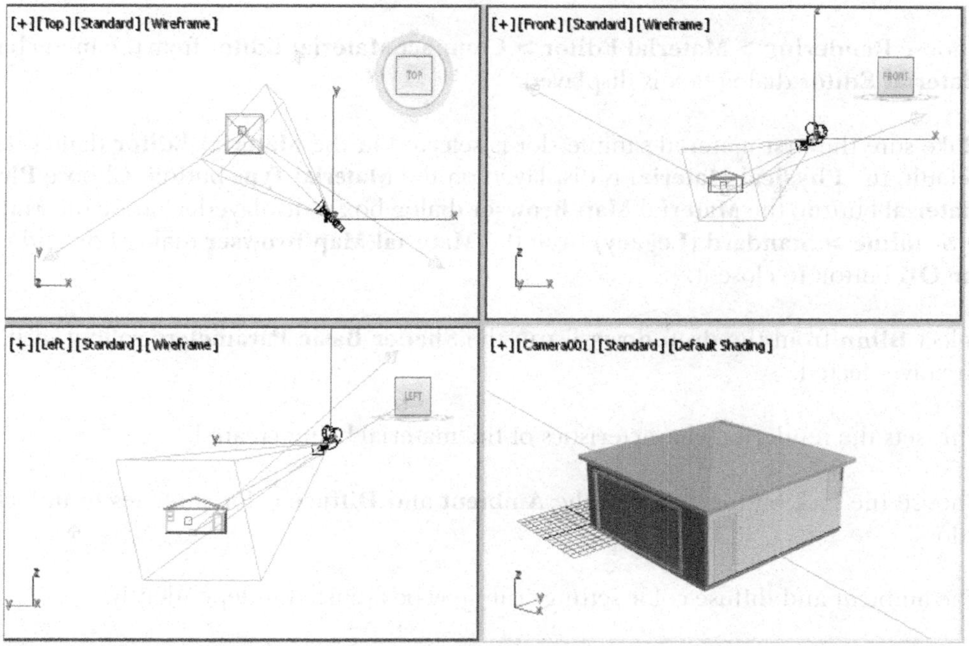

***Figure P3-17** The view displayed in the Camera001 viewport*

2. Choose the **Get Material** button in the **Material Editor** dialog box; the **Material/Map Browser** dialog box is displayed. 

3. Choose the **Material/Map Browser Options** button from the **Material/Map Browser** dialog box; the **Material/Map Browser Options** flyout is displayed. 

4. Choose **New Material Library** from the **Material/Map Browser Options** flyout; the **Create New Material Library** dialog box is displayed.

5. Enter **Project03** as the new name for the material library and choose the **Save** button; the new material library **Project03** is created in the **Material/Map Browser**.

6. Close the **Material/Map Browser** dialog box to return to the **Material Editor** dialog box.

   Next, you will create frame and trim material.

Creating Frame and the Trim Material

In this section, you will create frame and the trim material by using the **Material Editor** dialog box.

1. Choose the **Render Setup** tool from the **Main Toolbar**; the **Render Setup: Arnold** dialog box is displayed.

2. In this dialog box, select **Scanline Renderer** from the **Renderer** drop-down list; the renderer is changed to Scanline and the name of the dialog box is modified to the **Render Setup: Scanline Renderer** dialog box. Next, close this dialog box.

3. Choose **Rendering > Material Editor > Compact Material Editor** from the menu bar; the **Material Editor** dialog box is displayed.

4. Make sure the first material sample slot is selected in the **Material Editor** dialog box. By default, the **Physical Material** is displayed on the **Material Type** button. Choose **Physical Material** button; the **Material Map/Browser** dialog box is displayed. Choose the **Materials > Scanline > Standard (Legacy)** from the **Material Map/Browser** dialog box and choose the **OK** button to close it.

5. Select **Blinn** from the drop-down list in the **Shader Basic Parameters** rollout, if it is not already selected.

   This sets the rendering characteristics of the material being created.

6. Choose the lock button between the **Ambient** and **Diffuse** color swatches to unlock colors. 

   The ambient and diffuse color settings can now be adjusted independently.

7. Choose the **Ambient** color swatch in the **Material Editor** dialog box; the **Color Selector: Ambient Color** dialog box is displayed. Change the color settings as given next and choose the **OK** button to close this dialog box.

   Red: **50** Green: **50** Blue: **50**

8.. Choose the **Diffuse** color swatch in the **Material Editor** dialog box; the **Color Selector: Diffuse Color** dialog box is displayed. Change the color settings as given next and choose the **OK** button to close this dialog box.

Red: **230** Green: **230** Blue: **230**

9. Choose the **Specular** color swatch in the **Material Editor** dialog box; the **Color Selector: Specular Color** dialog box is displayed. Change the color settings as given next and choose the **OK** button to close this dialog box.

Red: **255** Green: **255** Blue: **255**

10. In the **Specular Highlights** area of the **Blinn Basic Parameters** rollout, set the value **25** in the **Specular Level** spinner and **5** in the **Glossiness** spinner.

11. Enter the name **Frame and Trim** in the material name drop-down list located below the material sample slots and then press the ENTER key.

12. Choose the **Put to Library** button in the **Material Editor** dialog box; a flyout is displayed.

13. Choose **Project03.mat** from the flyout; the **Put To Library** dialog box is displayed and the name **Frame and Trim** appears in the text box of the **Put To Library** dialog box.

14. Choose the **OK** button; the current material is assigned the name *Frame and Trim* and added to the material library.

Next, you will create window glass material.

Creating Window Glass Material

In this section, you will create the window glass material for the window by using the **Material Editor** dialog box.

1. Select the next unused sample slot in the **Material Editor** dialog box and choose **Physical Material** button; the **Material Map/Browser** dialog box is displayed. Choose **Materials > Scanline > Standard (Legacy)** and choose the **OK** button to close it.

   Now, enter the name **Glass** in the material name drop-down list.

2. Make sure **Blinn** is selected in the drop-down list in the **Shader Basic Parameters** rollout, if it is not already selected.

3. Unlock the **Ambient** and **Diffuse** color swatches, if required.

4. Choose the **Ambient** color swatch; the **Color Selector: Ambient Color** dialog box is displayed. Change the color settings as given below:

   Red: **15** Green: **15** Blue: **15**

5. Choose the **Diffuse** color swatch; the **Color Selector: Diffuse Color** dialog box is displayed. Change the color settings as given below:

   Red: **146** Green: **146** Blue: **133**

6. Choose the **Specular** color swatch; the **Color Selector: Specular Color** dialog box is displayed. Change the color settings as given below:

   Red: **255** Green: **255** Blue: **255**

7. In the **Specular Highlights** area, set the value **75** in the **Specular Level** spinner and **85** in the **Glossiness** spinner.

8. In the **Blinn Basic Parameters** rollout, set the value **50** in the **Opacity** spinner; a very shiny dark glass material is created.

9. Choose the **Put to Library** button in the **Material Editor** dialog box; a flyout is displayed.

10. Choose **Project03.mat** from the flyout; the **Put To Library** dialog box is displayed.

11. Accept the name **Glass** and choose the **OK** button.

Creating Texture Map Material

In this section, you will create the texture map material that can turn flat walls into brick walls with mortar joints.

1. Select *Wall001* in any viewport. Choose the **Modify** tab in the **Command Panel** and select the **Real-Word-Map Size** check box in the **Edit Object** rollout.

2. Select *Wall Front* in any viewport. Choose the **Modify** tab in the **Command Panel** and select the **Real-Word-Map Size** check box in the **Parameters** rollout.

3. Select the third material sample slot in the **Material Editor** dialog box and choose **Physical Material** button; the **Material Map/Browser** dialog box is displayed. Choose the **Materials > Scanline > Standard (Legacy)** from the **Material Map/Browser** dialog box and choose the **OK** button to close it.

4. In the material name drop-down list, enter the name **Brick**. Expand the **Maps** rollout below the **Blinn Basic Parameters** rollout.

5. Choose the **Diffuse Color** map button which is currently labeled as **No Map**; the **Material/ Map Browser** dialog box is displayed showing the types of maps that can be added.

6. Select the **Bitmap** map from the **Maps > General** rollout and choose the **OK** button; the **Material/Map Browser** dialog box is closed and the **Select Bitmap Image File** dialog box is displayed. As the project folder is already set, the *images* folder of this project is displayed in the **Look in** drop-down list.

7. Select the **Brkwea.jpg** file and then choose the **Open** button.

8. In the **Coordinates** rollout for the map, select the **Use Real-World Scale** check box. Set the value **5'8"(1.727m)** in the **Width Size** spinner and **5'0"(1.524m)** in the **Height Size** spinner. Since the **Use Real-World Scale** option is being used, the map should be sized according to actual measurements. Standard bricks are 0'2"(0.0508m) high and 0'8"(0.203m) long with 0'1/2"(0.0127m) of mortar between bricks. The map depicts a patch of wall, 8 bricks wide and 24 bricks tall.

9. Choose the **Go To Parent** button in the **Material Editor** dialog box; you will return to the root level in the material definition.

10. In the **Maps** rollout, choose the **Bump** map button which is currently labeled as **No Map**; the **Material/Map Browser** dialog box is displayed. Select **Bitmap** in the **Material/Map Browser** dialog box and then choose the **OK** button; the **Select Bitmap Image File** dialog box is displayed. As the project folder is already set, the *images* folder of this project is displayed in the **Look in** drop-down list.

11. Select the file **Brkwea_bump** and then choose the **Open** button.

12. Select the **Use Real-World Scale** check box. In the **Coordinates** rollout, set the value **5'8"(1.727m)** in the **Width Size** spinner and **5'0"(1.524m)** in the **Height Size** spinner.

    The map chosen for the bump map is a black and white version of the diffuse map, and therefore, must have the same scaling.

13. Choose the **Go to Parent** button to go back to the parent level. In the **Maps** rollout, set the value **150** in the **Bump** spinner.

    This sets the intensity of the bump map.

14. Choose the **Show Shaded Material in Viewport** button in the **Material Editor** dialog box to display the material in the viewport.

15. Choose the **Put to Library** button in the **Material Editor** dialog box; a flyout is displayed. Choose **Project03.mat** from the flyout; the **Put To Library** dialog box is displayed. Accept the name **Brick** and choose the **OK** button in the **Put To Library** dialog box.

16. Repeat this procedure to create three more texture-mapped materials using the following information for the materials. Remember to name each material and put it in the material library when completed. Select the **Use Real-World Scale** check box in the **Coordinates** rollout and the **Real-World-Map Size** check box from the **Object Type** rollout in the **Modify** tab for the diffuse map of each material. Use an instance of the diffuse map for the bump map. After defining each material, choose the **Put to Library** button to add it to the material library:

Name	Diffuse Color and Bump Maps	Bump Amount	Width/Height Size
Concrete	*GRYCON.jpg* and *GRYCON_bump.jpg*	**30**	**22'6"/22'(6.858m/6.706m)**
Wood	*TUTASH.jpg* and *TUTASH_bump.jpg*	**60**	**0'4"/0'4"(0.101m/0.101m)**
Pro Materials	*Thermal&Moisture.jpg* and *Thermal&Moisture_bump.jpg*	**150**	**2'0"/3'0"(0.609m/0.914m)**

Creating Multi/Sub-Object Materials for the Walls

The multi/sub-object material type allows you to apply different types of materials to a single object. Material IDs are used to control the kind of material to be applied to different faces of the object. This is very useful for architectural projects where you want the inner and outer sides of the walls to be of different materials.

The first multi/sub-object material you will create is for the walls. Each wall is divided into five elements: vertical end caps, outside wall, inside wall, top of the wall, and bottom of the wall. Therefore, this material needs five different sub-materials.

1. In the **Material Editor** dialog box, select the next unused material sample slot.

   To display more unused material samples, you can use the scroll bar. However, you can also right-click on the active material sample and choose **6 × 4 Sample Windows** from the shortcut menu. This will display more samples (24) at one time and make it easier to create the multi/sub-object material.

2. Choose the **Get Material** button in the **Material Editor** dialog box; the **Material/Map Browser** dialog box is displayed. In this dialog box, choose the **Material/map Browser options** button; a flyout is displayed. Next, choose **Open Material Library** from the flyout; the **Import Material Library** dialog box is displayed.

   Navigate to *C:\Program Files\Autodesk\3ds Max 2022\materiallibraries.*

3. Select the **AecTemplates** material library and then choose the **Open** button.

4. In the **Material/Map Browser** dialog box, double-click on the **Wall-Template (Multi/Sub-Object)** material type to display the material type in the selected sample slot. Rename the material as **Walls**.

   The **AecTemplates** material library contains five previously created multi/sub-object materials that can be assigned to objects in the scene. The material IDs and sub-material names are preset, as shown in Figure P3-18.

5. Make sure the material sample *Walls* is active and then press and hold the mouse button on material sample *Brick*, drag to the first (top) sub-material button in the **Multi/Sub-Object Basic Parameters** rollout, and then drop it; the **Instance (Copy) Material** dialog box is displayed. Next, make sure the **Instance** radio button is selected in this dialog box and then choose the **OK** button.

By selecting **Instance**, any change made in the original *Brick* material is also applied to the multi/sub-object material. *Brick* material is now part of the five elements of the new material and has 1 as material ID. All parts of the object having 1 as material ID and to which material is applied will be rendered as bricks.

Figure P3-18 *The* ***Multi/Sub-Object*** *material created*

6. Similarly, drag the *Brick* material sample and drop it on the second, fourth, and fifth sub-material buttons. These sub-materials have material IDs as 2, 4, and 5, respectively. Make sure that after each drop, you select the **Instance** radio button in the **Instance (Copy) Material** dialog box and choose the **OK** button.

7. Drag *Frame and Trim* material sample and drop it on the third sub-material button corresponding to the material ID **3**. Select the **Instance** radio button in the **Instance (Copy) Material** dialog box and then choose the **OK** button.

   Notice how each material sample in the **Multi/Sub-Object Basic Parameters** rollout reflects the sub-material. Also, notice how the main material sample reflects the sub-material.

8. Choose the **Put to Library** button in the **Material Editor** dialog box; a flyout is displayed. Choose **Project03.mat** from the flyout; the **Put To Library** dialog box is displayed. Accept the name *Walls* and choose the **OK** button in the **Put To Library** dialog box.

Creating Multi/Sub-Object Materials for the Windows

Each window is divided into five elements: front rails, back rails, panels, front frame, and back frame. Therefore, a multi/sub-object material is needed that contains five different sub-materials.

1. In the **Material Editor** dialog box, select the next unused material sample slot.

2. Choose the **Get Material** button; the **Material/Map Browser** dialog box is displayed.

3. In this dialog box, double-click on the **Window-Template (Multi/Sub-Object)** material type in the **AecTemplates** material library to be displayed in the selected material sample slot.

4. Rename the material as **Windows**. Make sure the *Windows* material sample is active. Next, drag *Frame and Trim* material sample and drop it on the first sub-material button in the **Multi/Sub-Object Basic Parameters** rollout. The **Instance (Copy) Material** dialog box is displayed. Select the **Instance** radio button in this dialog box and then choose the **OK** button.

5. Similarly, drag *Frame and Trim* material sample and drop it on the second, fourth, and fifth sub-material buttons. These sub-materials have material IDs as 2, 4, and 5, respectively. After each drop, select the **Instance** radio button in the **Instance (Copy) Material** dialog box and then choose the **OK** button.

6. Drag the material sample *Glass* and drop it on the third sub-material button corresponding to the material ID **3**. Select the **Instance** radio button in the **Instance (Copy) Material** dialog box and choose the **OK** button.

7. Choose the **Put to Library** button in the **Material Editor** dialog box; a flyout is displayed.

8. Choose **Project03** from the flyout; the **Put To Library** dialog box is displayed.

9. Accept the **Windows** name and choose the **OK** button in the **Put To Library** dialog box.

Creating the Multi/Sub-Object Material for the Side Door

The side door is divided into five elements: front, back, inner bevel (panels), frame, and inner door. Therefore, a multi/sub-object material with five sub-materials will be created.

1. In the **Material Editor** dialog box, select the next unused material sample slot.

2. Choose the **Get Material** button; the **Material/Map Browser** dialog box is displayed. In this dialog box, double-click on **Door-Template (Multi/Sub-Object)** in the **AecTemplates** material library to be displayed in the selected material sample slot.

3. Rename the material as **Side Door**. Make sure *Side Door* material sample is active, press and hold the mouse button on *Wood* material sample, drag it to the first sub-material button in the **Multi/Sub-Object Basic Parameters** rollout, and drop it. The **Instance (Copy) Material** dialog box is displayed. In this dialog box, select the **Instance** radio button and then choose the **OK** button.

4. Similarly, drag *Wood* material sample and drop it on the second, third, and fifth sub-material buttons. Make sure that you select the **Instance** radio button after each drop in the **Instance (Copy) Material** dialog box and then choose the **OK** button.

5. Drag *Frame* and *Trim* material sample and drop it on the fourth sub-material button. Select the **Instance** radio button in the **Instance (Copy) Material** dialog box and choose the **OK** button.

6. Choose the **Put to Library** button in the **Material Editor**; a flyout is displayed. Choose **Project03.mat** from the flyout; the **Put To Library** dialog box is displayed.

7. Accept the **Side Door** name and choose the **OK** button in the **Put To Library** dialog box.

Saving the Material Library

The changes made in the material library must be saved or they will be lost when you exit 3ds Max.

1. Choose the **Get Material** button in the **Material Editor** dialog box; the **Material/Map Browser** dialog box is displayed.

2. In this dialog box, right-click on the **Project03.mat** title bar; a flyout is displayed. Choose **Save** from the flyout, as shown in Figure P3-19. The material library is saved with the name **Project03**.mat.

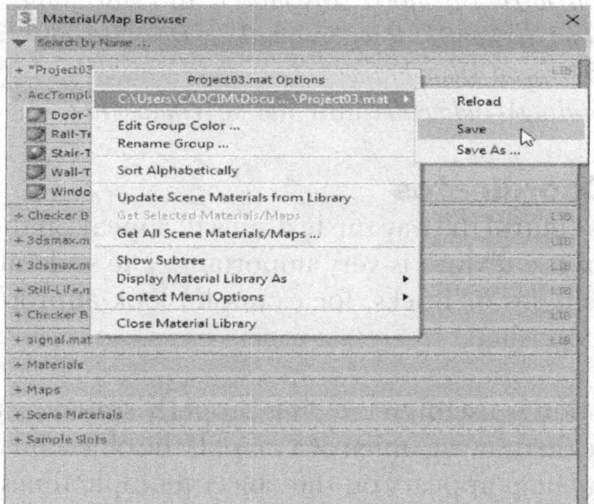

Figure P3-19 Saving the material library

3. Close the **Material/Map Browser** dialog box.

Applying Materials

The materials have been created and the material library is saved. Now, the materials can be assigned to objects.

1. Select *Wall001* in any viewport and invoke the **Material Editor** dialog box from the **Main Toolbar**. Make sure the **Real-World-Map Size** check box is selected in the **Edit Object** rollout of the **Modify** tab.

2. Select the *Walls* material from the **Material Editor** dialog box and then choose the **Assign Material to Selection** button.

3. Similarly, select *Wall Front* in any viewport. Select *Brick* material sample from the **Material Editor** dialog box and then choose the **Assign Material to Selection** button to assign the materials on the objects in the viewport. Make sure the **Real-World-Map Size** check box is selected in the **Parameters** rollout of the **Modify** tab.

4. Similarly, assign the following materials and then close the **Material Editor** dialog box:

Material	Object(s)
Concrete	*Floor Slab*
Wood	*Overhead Door*
Side Door	*PivotDoor001*
Pro Materials	*Fascia01, Fascia002, Fascia003,* and *Fascia004*
Frame and Trim	*Roof Section01, Roof Section002, Roof Section003,* and *Roof Section004* *Overhead Door Frame*
Glass	*SlidingWindow001* and *SlidingWindow002*

Applying Mapping Coordinates

The mapping coordinates control the way the texture will appear on the object. In architectural projects, correct scaling of the texture is very important. The scale of the texture determines whether the materials (shingles or bricks, for example) look appropriate and are oversized, undersized, stretched, or squashed.

The real-world scaling option is useful in creating properly scaled materials. With this option, you can determine the size (in drawing units) of a map. If the object on which the map is applied is larger than this size, the map appears on the object multiple times. If the object is smaller than the size specified for the map, only a portion of the map will appear on the object. This greatly reduces the problems of poorly scaled materials. Since the materials in this model use real world scaling, they should appear realistic in the rendering.

1. Change the background color to white. Make sure the Camera01 viewport is active and then invoke the **Render Production** tool from the **Main Toolbar**; the Camera01 viewport is rendered. Since no warning dialog box appears, all objects have default mapping coordinates. However, the brick pattern on *Wall Front* needs to be properly aligned to the brick pattern on *Wall001*.

2. Close the render window. Zoom in the area where *Wall001* meets *Wall Front* in the Left viewport, as shown in Figure P3-20. Right-click on the Left viewport label and choose **Default Shading** from the shortcut menu.

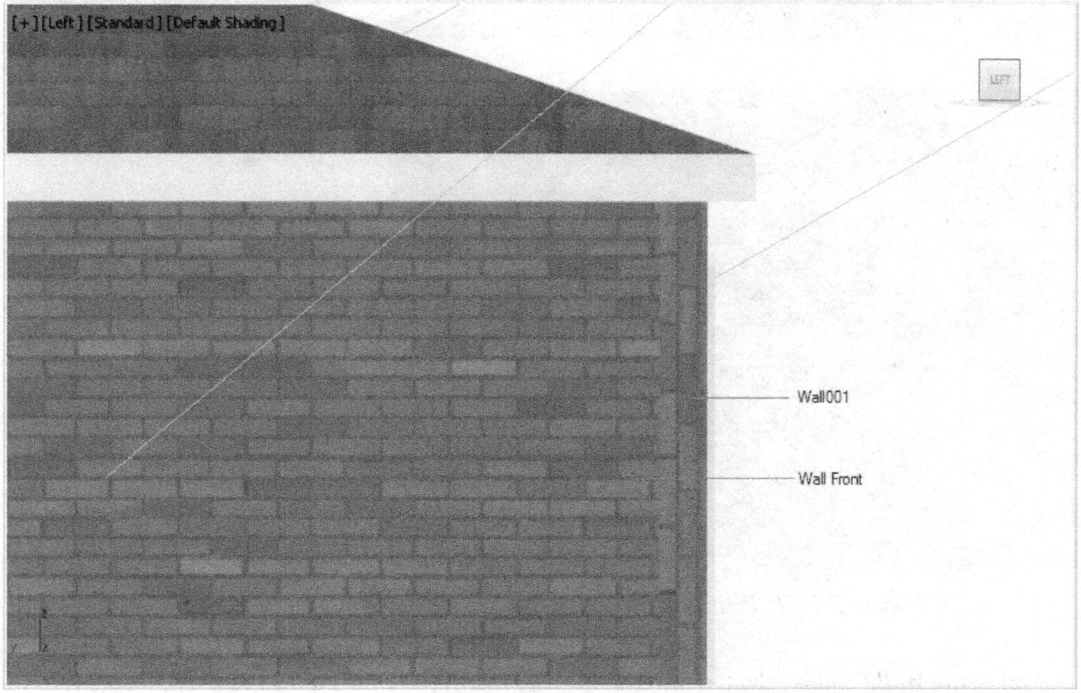

Figure P3-20 *Wall001 and Wall Front zoomed in the Left viewport*

3. Select *Wall Front* in the Front viewport and choose the **Modify** tab in the **Command Panel**. Select **UVW Map** in the **Modifier List** drop-down list under the **OBJECT-SPACE MODIFIERS** section.

4. In the **Parameters** rollout, make sure the **Real-World-Map Size** check box is selected.

5. In the **Alignment** area of the **Parameters** rollout, select the **Y** radio button. In the **Mapping** area, select the **Box** radio button. Also, select the **U Tile: Flip** and **V: Tile Flip** check boxes. Render the Left viewport again.

   The brick patterns between *Wall001* and *Wall Front* are not lined up correctly, as shown in Figure P3-21. To fix this, adjust the position of the **UVW Map Gizmo** by adding the **UVW Map** modifier. The **UVW Map** modifier will fix the problem of the bricks at the edge of the wall.

6 Choose **Views > Set Active Viewport > Right** from the menu bar. Now, maximize the viewport.

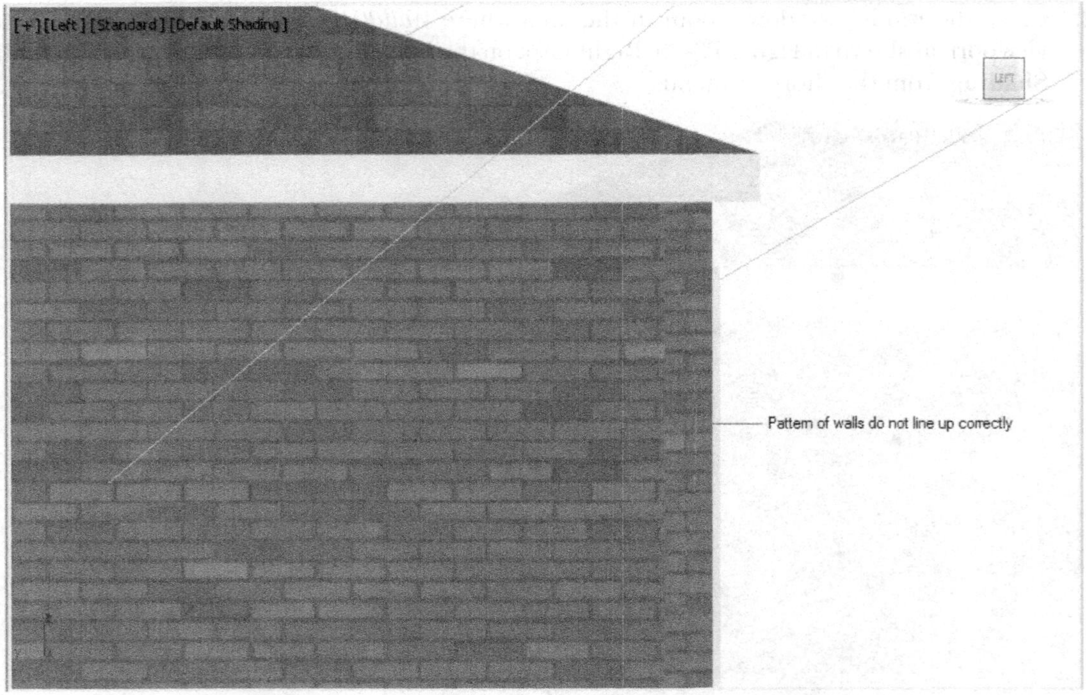

Figure P3-21 *Pattern of Wall001 and Wall Front not lined up correctly*

7. Make sure *Wall Front* is selected and the **Modify** tab in the **Command Panel** is chosen. Expand **UVW Map** in the modifier stack by clicking on the arrow on left side. Select **Gizmo** in the modifier stack.

8. Invoke the **Select and Move** tool from the **Main Toolbar**. Next, look at the modifier stack. If the highlighted selection has jumped from **Gizmo** to **UVW Map**, select the **Gizmo** again. In the Right viewport, move the gizmo until the brick patterns on the two walls match up correctly, as shown in Figure P3-22. Choose **Left** from the **Views** cascading menu to switch back to the Left view.

   The brick patterns displayed in the shaded viewport may get shifted slightly from the position displayed in rendering. For this reason, use the shaded viewport to get the gizmo close to the required location, and then render the Left viewport as required to precisely position the gizmo.

9. Select **UVW Map** in the modifier stack to deselect the gizmo. Right-click on the viewport label and choose **Wireframe Override** from the shortcut menu.

10. Select the Camera001 viewport and then invoke the **Render Production** tool from the **Main Toolbar**; the **Rendered Frame** window is displayed. In this window, you will notice that all mapping coordinates in the scene now appear correct.

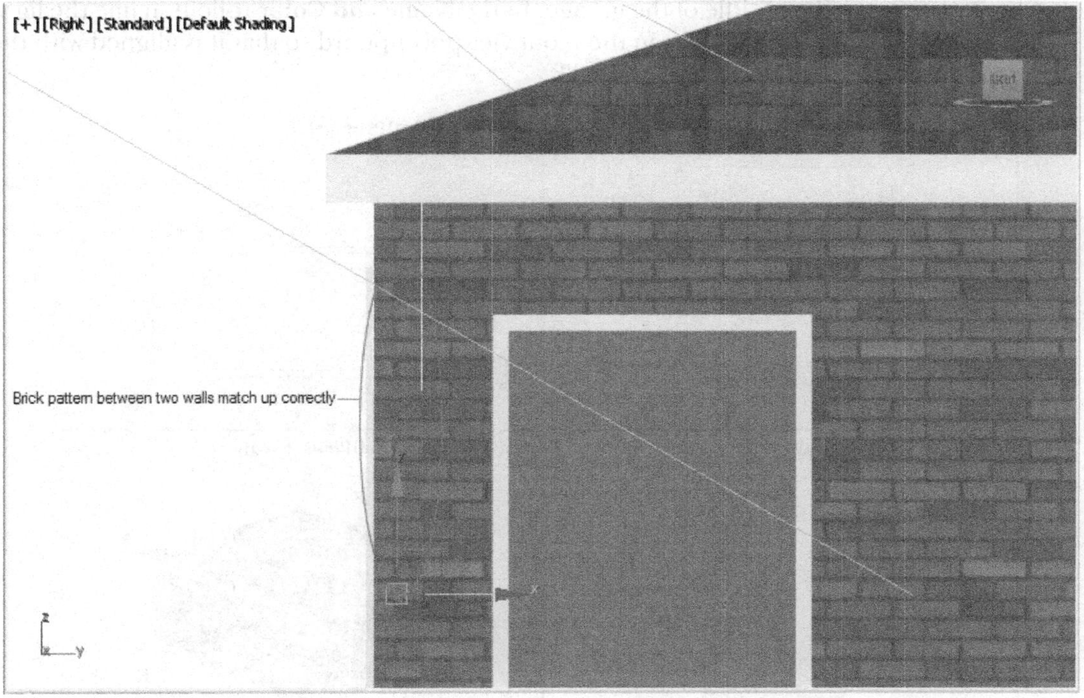

Figure P3-22 *The brick pattern between two walls matched up correctly*

Setting Frames for Animation

In this section, animation will be added to the scene by opening and closing the doors and one window. This will simulate the action of someone parking the car, opening the window, and leaving the garage through the side door. The first thing to be set in the animation is the number of frames.

1. Choose the **Time Configuration** button in the animation controls area; the **Time Configuration** dialog box is displayed.

2. In the **Animation** area, set the value **300** in the **Length** spinner and choose the **OK** button. This sets the total number of frames in the animation to 300 which is about ten seconds of animation.

Adding an Inside Light

You will now add a light inside the garage. This light will simulate the light on an automatic garage door opener which turns on and off with the opening and closing of the door.

1. Activate the Top viewport and then choose **Create > Lights** in the **Command Panel**. Make sure the **Standard** option is selected in the drop-down list. Next, invoke the **Omni** tool from the **Object Type** rollout.

2. Choose a point in the middle of the garage. In the **Name and Color** rollout, name the light as **Opener Light**. Move the light in the Front viewport upward so that it is aligned with the top of the walls, as shown in Figure P3-23.

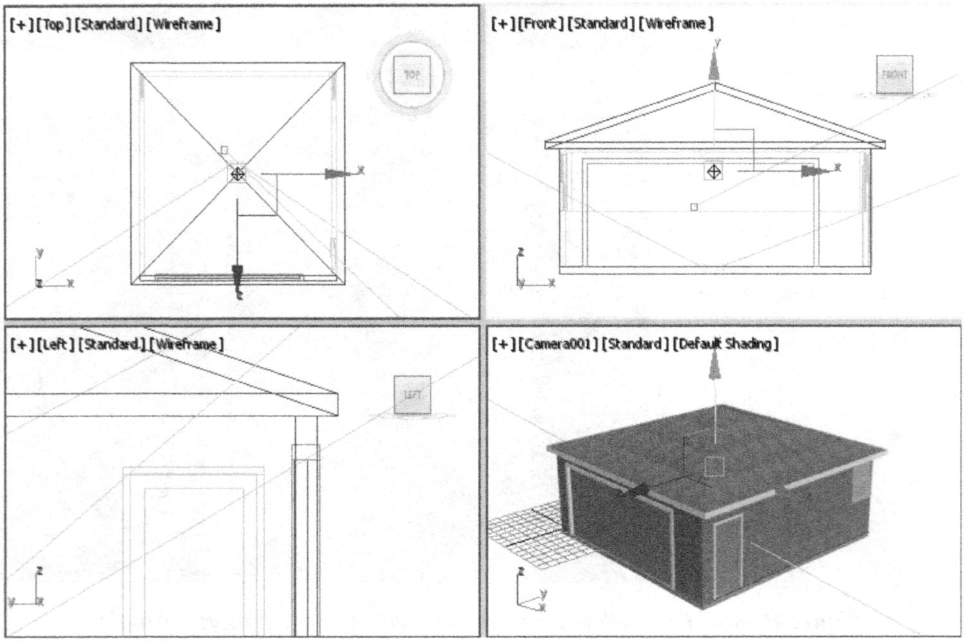

Figure P3-23 *The light aligned on the top of walls*

3. Make sure *Opener Light* is selected and then choose the **Modify** tab in the **Command Panel**. Set the value **2** in the **Multiplier** spinner in the **Intensity/Color/Attenuation** rollout. Choose the color swatch next to the **Multiplier** spinner and set the values as given below and then choose the **OK** button:

   Red: **255** Green: **226** Blue: **180**

   The light is adjusted to give it a slight orange hue. This simulates incandescent lighting.

4. Choose **Graph Editors > Track View - Dope Sheet** from the menu bar; the **Track View-Dope Sheet** window is displayed.

5. Expand the tree for *Opener Light*. Choose the plus sign **(+)** on left of *Opener Light*. Next, expand the tree for the **Transform** and then **Object (Omni Light)**; you will notice that there are no animation keys in the **Track View - Dope Sheet** window.

6. Choose the **Add/Remove Key** button on the top in the **Track View-Dope Sheet** window. 

7. Choose the **Multiplier** track and insert keys at frames **0**, **65**, and **200**, refer to Figure P3-24.

8. Right-click on the first key (on frame 0); the **Opener Light\Multiplier** dialog box is displayed.

9. Set the value **0** in the **Value** spinner and select the square mechanical transition for the **In** and **Out** image tiles, as shown in Figure P3-25.

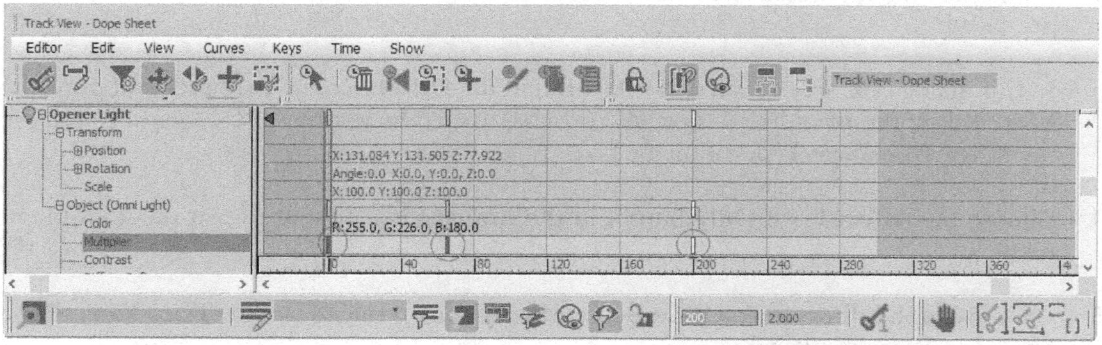

Figure P3-24 *Keys inserted at frames* ***0, 65,*** *and* ***200*** *in the* ***Multiplier*** *track*

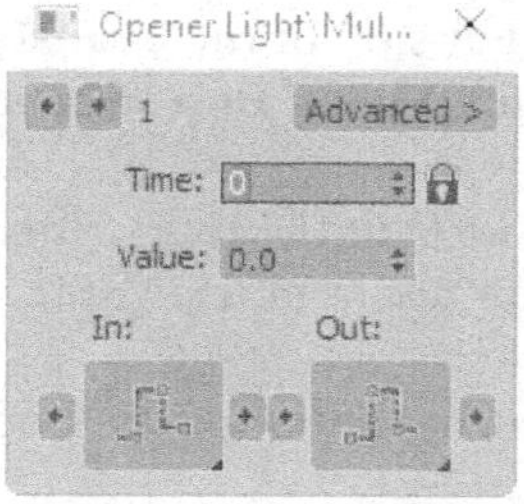

Figure P3-25 *The* ***Opener Light\Multiplier*** *dialog box*

10. Choose the right arrow button at the top in the **Opener Light\Multiplier** dialog box to move to the next key.

11. Make sure the **Value** spinner displays 2.0 and select the square mechanical transition for the **In** and **Out** image tiles.

12. Move to the next key (on frame **200**). Set the value **0** in the **Value** spinner and select the square mechanical transition for the **In** and **Out** image tiles.

13. Close the **Opener Light\Multiplier** dialog box and the **Track View - Dope Sheet** window.

Animating the Overhead Door

The animation of the overhead door must work in conjunction with the light. The door should start opening at frame 60 which is five frames before the light turns on. In addition, the door must open, pause, fully open, and then close at frame 150. In this case, the animation of the door is a combination of motion and rotation. Before animation can be added, the overhead door's pivot point must be relocated.

1. Select *Overhead Door* in the Left viewport. Next, choose the **Hierarchy** tab in the **Command Panel**. Make sure the **Pivot** button is chosen. In the **Move/Rotate/Scale** area of the **Adjust Pivot** rollout, choose the **Affect Pivot Only** button.

The pivot point is currently located in the middle of the door on the back (inside) surface.

2. Right-click on the **Select and Move** tool; the **Move Transform Type-In** dialog box is displayed.

3. In this dialog box, set the value **–3(–0.914m)** in the **Offset:Screen Y** spinner and press the ENTER key; the pivot axis is now *0'6"(0.153m)* above the bottom of *Overhead Door*. Close the **Move Transform Type-In** dialog box.

4. Choose the **Affect Pivot Only** button in the **Adjust Pivot** rollout to exit the adjustment mode.

5. Choose **Graph Editors > Track View - Dope Sheet** from the menu bar; the **Track View - Dope Sheet** window is displayed.

6. In this window, choose the plus sign **(+)** on the left side of *Overhead Door* and then again choose the plus sign **(+)** on the left side of the **Transform** track; you will notice that there are no animation keys in the **Track View - Dope Sheet** window.

7. Choose the **Add/Remove Key** button at the top of the **Track View** dialog box. Insert a key in both the **Position** and **Rotation** tracks at frame 60.

   These keys hold the same transform values for the door as on frame 0 or closed.

8. Close the **Track View - Dope Sheet** window.

9. Drag the time slider to frame 90 and choose the **Toggle Auto Key Mode** button to turn the animation mode on (red).

10. Make sure *Overhead Door* is selected and invoke the **Select and Rotate** tool. In the Left viewport, rotate the door 90 degrees about the local Z axis, as shown in Figure P3-26.

11. Invoke the **Select and Move** tool. In the Left viewport, move the door up by 6'0"(1.82m) on the Y axis, as shown in Figure P3-27.

12. Drag the time slider to frame 150.

13. In the Left viewport, rotate the door –90° about the local Z axis. Also, move the door down by –6'0"(–1.82m) on the Y axis.

14. Choose the **Toggle Auto Key Mode** button to exit the animation mode.

15. Make sure *Overhead Door* is selected. Now, select the key at frame 90 on the track bar below the time slider; the selected key turns white.

16. Hold down the SHIFT key and drag the key to frame 120. Next, release the key; the rotation key will be copied from frame 90 to frame 120.

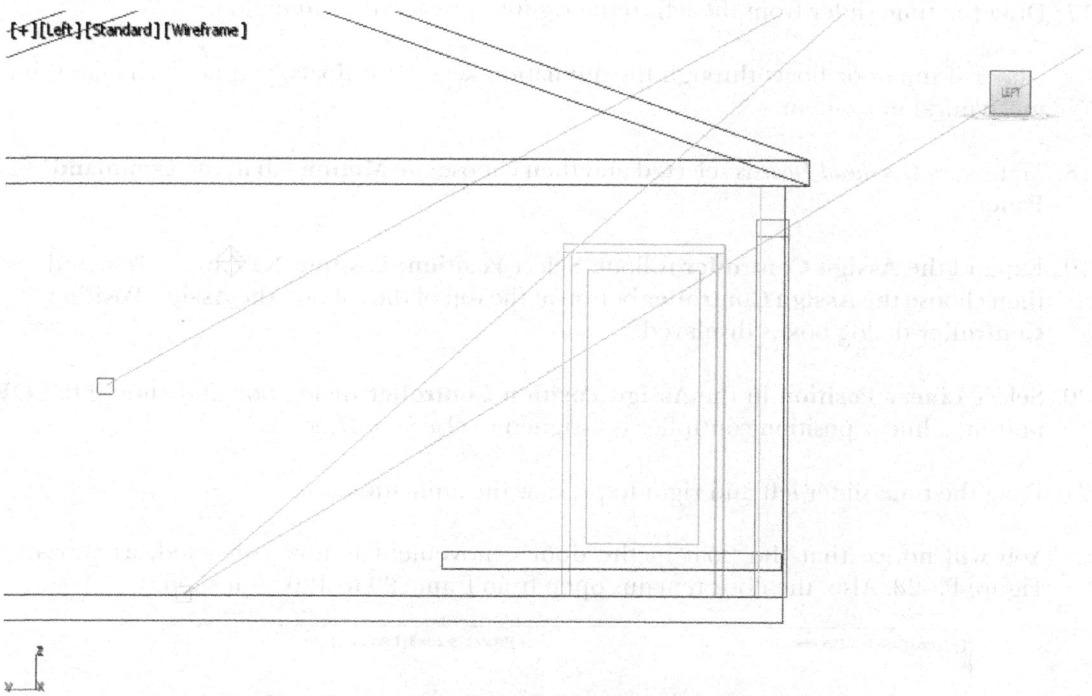

Figure P3-26 The door rotated 90 degrees about the local Z axis

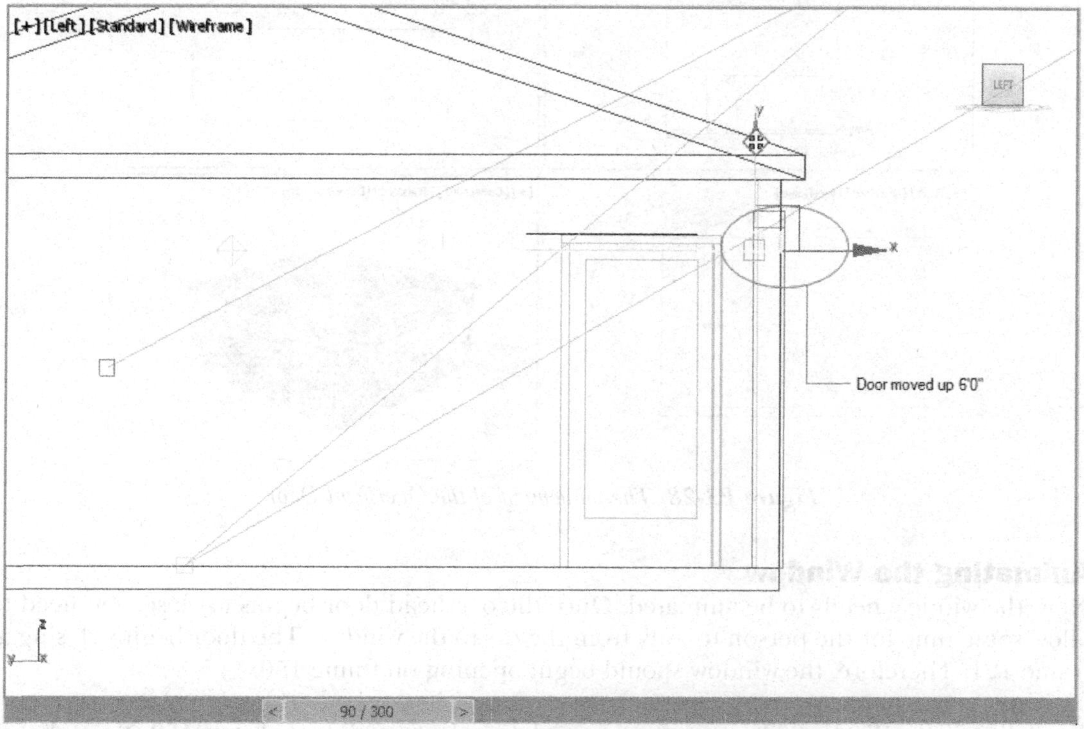

Figure P3-27 The door moved up 6'0"(1.82m) on the Y axis

17. Drag the time slider from the left to the right to preview the animation.

    Check if the door floats through the animation keys. If it floats, you need change it into mechanical movement.

18. Make sure *Overhead Door* is selected and then choose the **Motion** tab in the **Command Panel**.

19. Expand the **Assign Controller** rollout. Select **Position: Position XYZ** in the tree and then choose the **Assign Controller** button at the top of the rollout; the **Assign Position Controller** dialog box is displayed.

20. Select **Linear Position** in the **Assign Position Controller** dialog box and choose the **OK** button; a linear position controller is assigned to *Overhead Door*.

21. Drag the time slider left and right to preview the animation.

    You will notice that the float in the door's movement is now corrected, as shown in Figure P3-28. Also, the door remains open from frame 90 to 120, as it should.

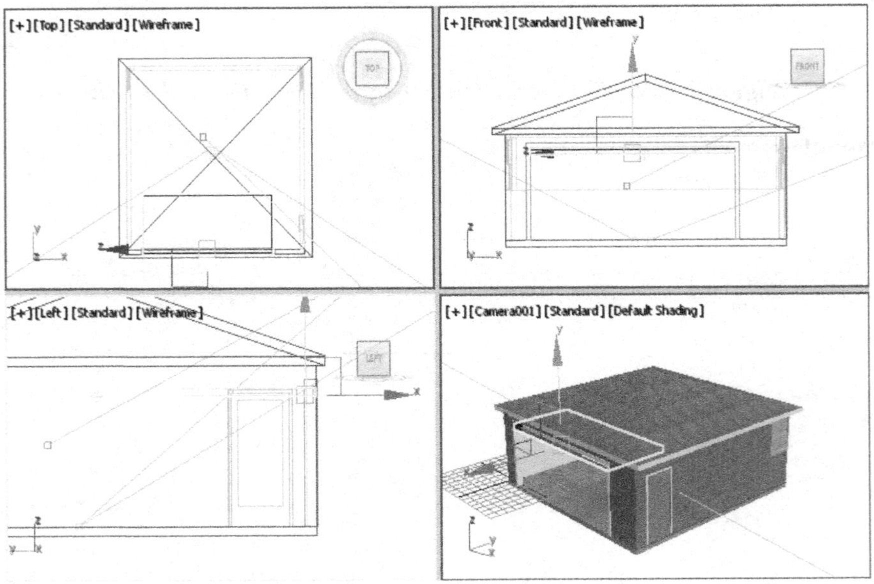

Figure P3-28 *The movement of the Overhead Door*

Animating the Window

Now, the window needs to be animated. Once the overhead door begins to close, you need to allow some time for the person to walk from the car to the window. The door begins closing at frame 121. Therefore, the window should begin opening on frame 150.

1. Select *SlidingWindow002* in any viewport and drag the time slider to frame **160**. Next, choose the **Toggle Auto Key Mode** button to turn the animation mode on (red).

2. Choose the **Modify** tab in the **Command Panel**. In the **Open Window** area of the **Parameters** rollout, set the value **100** in the **Open** spinner.

3. Choose the **Toggle Auto Key Mode** button again to exit the animation mode.

4. In the track bar below the time slider, choose the key at frame 0 and drag the key to frame 150. Next, release the key.

   Moving the key from frame 0 to frame 150 means the window will not start opening until frame 150.

Animating the Side Door

You will now animate the side door. At this point, the person has parked the car, closed the overhead door, walked to the window, and opened it. Now, it will take some time to walk to the side door and begin opening it.

1. Select *PivotDoor001* in any viewport and drag the time slider to frame 180. Next, choose the **Toggle Auto Key Mode** button to turn the animation mode on (red).

   This is the frame on which the side door should open completely.

2. Choose the **Modify** tab in the **Command Panel**, if it is not chosen. In the **Parameters** rollout, enter the value **90** in the **Open** spinner; the side door opens by 90 degrees.

3. Drag the **Time Slider** to frame 220.

4. In the **Parameters** rollout, enter the value **0** in the **Open** spinner; the side door is now closed.

5. Choose the **Toggle Auto Key Mode** button to exit the animation mode.

6. In the track bar below the time slider, choose the key at frame 0 and drag the key to frame 175. Next, release the key.

   Moving the key indicates the door will not start opening till frame 175.

Rendering the Animation

In this section, you will render the final animation to see the finished project. The final animation will show all materials, highlights, shadows, and lighting.

1. Make the Camera01 viewport active and choose **Rendering > Render Setup** from the menu bar or invoke the **Render Setup** tool from the **Main Toolbar**; the **Render Setup: Scanline Renderer** dialog box is displayed.

2. In the **Time Output** area of the **Common Parameters** rollout, select the **Active Time Segment** radio button to make it active.

   This sets all the active frames in the animation to be rendered.

3. In the **Output Size** area of the dialog box, choose the **640×480** button.

   This sets the animation to be rendered at a resolution of 640 × 480 pixels.

4. In the **Render Output** area of the dialog box, choose the **Files** button; the **Render Output File** dialog box is displayed. As the project folder is already set, the *renderoutput* folder of this project is displayed in the **Save as type** drop-down list.

5. Enter **Project03** in the **File name** text box.

6. Select **AVI File (*.avi)** from the **Save as type** drop-down list and choose the **Save** button; the **AVI File Compression Setup** dialog box is displayed if this file type has not been saved during the current drawing session. This dialog box allows you to adjust the quality and keyframe rate of the animation.

7. Choose the **OK** button to accept the default AVI settings and then choose the **Render** button to start rendering.

8. When the rendering of animation is done, close the render window and the **Render Setup: Scanline Renderer** dialog box. Figure P3-29 shows the rendered image at frame number 180.

Figure P3-29 *The rendered image at frame 180*

9. Navigate to the *renderoutput* folder and select the **Project03.avi** and choose the **Open** button; the avi file is opened and played.

10. Close the avi file and save the final scene.

Project 4

Corporate Design Project

PROJECT DESCRIPTION

In this project, you will create a corporate logo along with columns, floor slab, and stairs. You will then apply materials to the objects to make them look realistic. In addition to this, you will create lights to illuminate the logo. Finally, you will animate the lights and render the animation, as shown in Figure P4-1. The procedures used can be applied to any corporate design project.

Creating the Project Folder

Create a new project folder with the name *Prj_4* at *\Documents\3dsmax 2022* and then save the file with the name *Prj4*, as discussed in Tutorial 1 of Chapter 2.

Downloading the Files

Before starting the project, you need to download the *prj4_3dsmax_2022.zip* file from *www.cadcim.com.* The path of the file is as follows: *Textbooks > Animation and Visual Effects > Autodesk 3ds Max > Autodesk 3ds Max 2022 for Beginners: A Tutorial Approach*

Extract the contents of the zip file to *Documents*. Open Windows Explorer and browse to *\Documents\prj4_3dsmax_2022* and then copy all files to *\Documents\3dsmax 2022\Prj4\sceneassets\images.*

Figure P4-1 *The model of the logo*

Choose the **Render Setup** tool from the **Main Toolbar** and make sure the **Render Setup: Scanline Renderer** dialog box is displayed. Next, close this dialog box.

Changing the Unit Settings

In this section, you will change the units with which you are working. Since the measurement in this project is in inches, you will set the units to decimal inches.

1. Choose **Customize > Units Setup** from the menu bar; the **Units Setup** dialog box is displayed.

2. In the **Display Unit Scale** area, select the **US Standard** radio button and then select the **Decimal Inches** from the drop-down list below this radio button, as shown in Figure P4-2. Choose the **OK** button to apply the settings and close the dialog box.

3. Right-click on the **Snaps Toggle** tool; the **Grid and Snap Settings** dialog box is displayed. Choose the **Home Grid** tab in the dialog box.

Note

*If you want to work in the metric unit system, select the **Metric** radio button from the **Display Unit Scale** area of the **Units Setup** dialog box. Also, make sure the **Meters** option is selected in the drop-down list below the **Metric** radio button.*

4. In the **Grid Dimensions** area, set **1(0.3m)** in the **Grid Spacing** spinner. Next, select the **All Viewports** radio button from the **Dynamic Update** area and close the dialog box.

Creating the Column

In many designs, 2D shapes are used as the building blocks for creating 3D objects in a scene. In this project, the column will be created as a shape to be lofted along a path to give it a proper height. You will first create the circle shape and then the line path.

1. Activate the Top viewport and then choose **Create > Shapes** in the **Command Panel**. Next, invoke the **Circle** tool from the **Object Type** rollout.

2. Expand the **Keyboard Entry** rollout and set **19(0.483m)** in the **Radius** spinner. Next, choose the **Create** button; the circle for the column is created. In the **Name and Color** rollout, enter **Column Shape**.

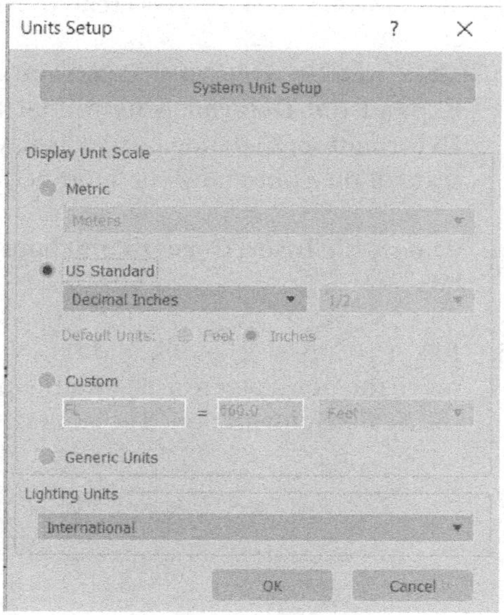

***Figure P4-2** The **Units Setup** dialog box*

3. Invoke the **Line** tool from the **Object Type** rollout. Make sure the **Corner** radio button is selected in the **Initial Type** and **Drag Type** areas of the **Creation Method** rollout.

4. Expand the **Keyboard Entry** rollout and set the values in the spinners as given next:

   X: **0** Y: **0** Z: **0**

5. Choose the **Add Point** button and set the values again in the spinners as given next:

   X: **0** Y: **0** Z: **140(3.556m)**

6. Choose the **Add Point** button and then choose the **Finish** button; the path for the column is created.

7. In the **Name and Color** rollout, enter **Column Path**.

Lofting the Column

The path and shape are created so now you need to loft the column. Once the loft object is created, a scale deformation will be added to control the shape of the column. To do so, the path needs more vertices. These extra vertices will be used to give desired shape to the column.

1. Make sure *Column Path* is selected and then choose **Create > Geometry** in the **Command Panel**. Next, select **Compound Objects** in the drop-down list displayed below the **Geometry** button.

2. Invoke the **Loft** tool from the **Object Type** rollout. Expand the **Skin Parameters** rollout and set **10** in the **Shape Steps** spinner.

3. In the **Creation Method** rollout, choose the **Get Shape** button and select *Column Shape* in any viewport; a column is created. Next, in the **Name and Color** rollout, enter **Column01**.

4. Make sure *Column01* is selected and choose the **Modify** tab in the **Command Panel**. Expand the **Deformations** rollout and then choose the **Scale** button; the **Scale Deformation(X)** dialog box is displayed. In this dialog box, the line representing the scale of the column will be modified by adding points at different positions. 

5. Choose the **Insert Corner Point** button at the top of the **Scale Deformation(X)** dialog box. 

6. Insert points at 2, 5, 95, and 98. Next, choose the **Scale Control Point** button and move the point at 5 to 70% and the point at 95 to 50%, as shown in Figure P4-3. Close the **Scale Deformation(X)** dialog box. 

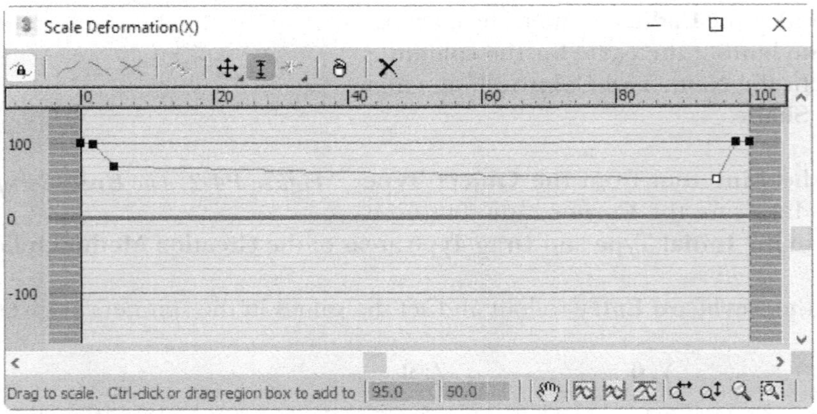

***Figure P4-3** The **Scale Deformation(X)** dialog box after scaling at different points*

The lofted object is scaled using the **Scale Deformation(X)** dialog box.

7. Invoke the **Zoom Extents All** tool to zoom the object to its extent. *Column01* is modified using the **Scale Deformation(X)** dialog box, as shown in Figure P4-4. The effects of scale deformation on the column are clearly visible.

Creating the Bottom and the Top for the Column

You will now create a box which will act as the base plate under the column. This box will be copied for the top plate of the column.

1. Make sure the Top viewport is activated and then choose **Create > Geometry** in the **Command Panel**. Select **Standard Primitives** in the drop-down list displayed below the **Geometry** button. Next, invoke the **Box** tool from the **Object Type** rollout.

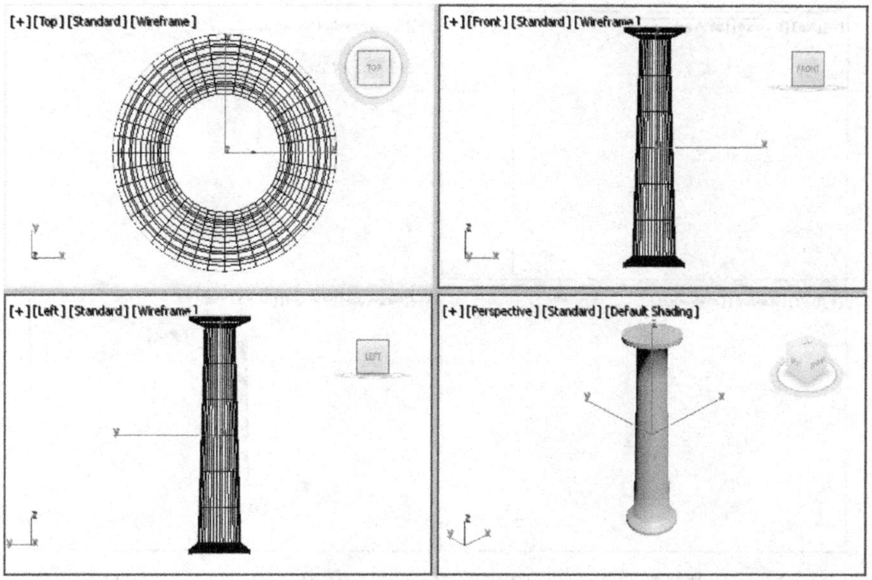

***Figure P4-4** Column01 modified using the **Scale Deformation(X)** dialog box*

2. Expand the **Keyboard Entry** rollout and set the values in the spinners as given next:

   Length: **45(1.143m)** Width: **45(1.143m)** Height: **–4(–0.102m)**

3. Choose the **Create** button; the base plate of the column is created. In the **Name and Color** rollout, enter **Column01Base**. Right-click to exit the tool.

4. Make sure *Column01Base* is selected and choose **Edit > Clone** from the menu bar; the **Clone Options** dialog box is displayed. In this dialog box, select the **Copy** radio button in the **Object** area. Enter **Column01Top** in the **Name** text box and choose the **OK** button; a copy of *Column01Base* is created with the name *Column01Top*.

5. Make sure *Column01Top* is selected. Next, invoke the **Align** tool from the **Main Toolbar**; the cursor gets modified. Now, select *Column01* from the Scene Explorer; the **Align Selection (Column 01)** dialog box is displayed. In this dialog box, select the **Z Position** check box only. Select the **Minimum** radio button in the **Current Object** area and the **Maximum** radio button in the **Target Object** area. Next, choose the **OK** button; the bottom of *Column01Top* is aligned with the top of *Column01*, as shown in Figure P4-5.

Copying the Column

In this section, you will create and move a copy of the column assembly. To make this easier, you will group the first column with its base and top.

1. Select *Column01*, *Column01Base,* and *Column01Top* in any viewport.

2. Choose **Group > Group** from the menu bar; the **Group** dialog box is displayed. Enter **Column01Assembly** in the **Group name** text box. Next, choose the **OK** button.

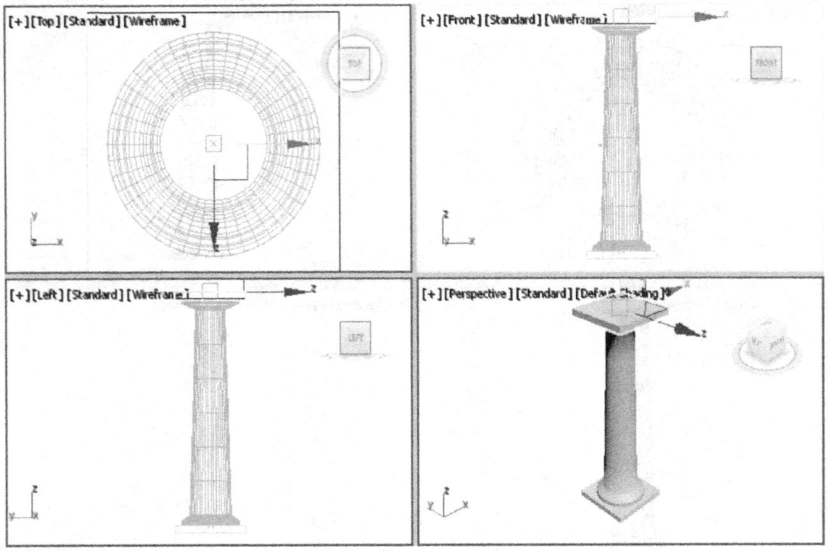

***Figure P4-5** Column01Top is aligned with the top of Column01*

3. Create a copy of *Column01Assembly* as discussed earlier. Rename it as **Column02Assembly**.

4. Activate the Top viewport. Next, right-click on the **Select and Move** tool; the **Move Transform Type-In** dialog box is displayed. 

5. In the **Offset: Screen** area of this dialog box, set **–45(–1.143m)** in the **X** spinner and press ENTER; the edges of the column bases are aligned. Next, close the dialog box; *Column02 Assembly* is displayed, as shown in Figure P4-6. Invoke the **Zoom Extents All** tool to zoom the objects to their extents.

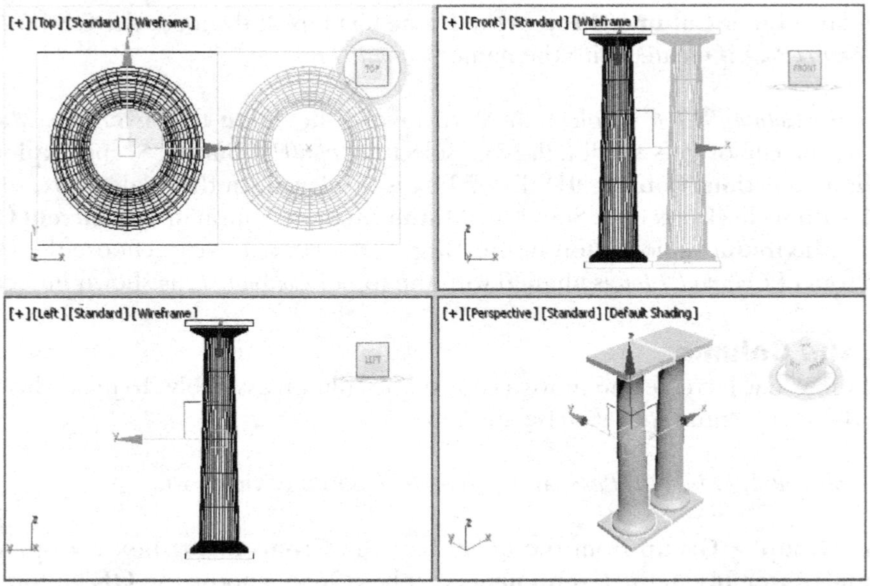

***Figure P4-6** Column02 Assembly created*

Creating the Horns

Now, you will create a horn by lofting a shape along a path. Next, you will mirror it to create the other horn.

1. Activate the Left viewport and choose **Create > Shapes** in the **Command Panel**. Next, invoke the **Circle** tool in the **Object Type** rollout.

2. Expand the **Keyboard Entry** rollout and set **8(0.203m)** in the **Radius** spinner. Choose the **Create** button; a circle is created, which is the cross-section for the horn. In the **Name and Color** rollout, enter **Horn Shape**.

3. Activate the Front viewport and then invoke the **Line** tool from the **Object Type** rollout.

4. Select the **Smooth** radio button in the **Initial Type** area of the **Creation Method** rollout.

5. Expand the **Keyboard Entry** rollout and set the values as given next. Choose the **Add Point** button after entering each set of XYZ coordinates.

X: **95(2.413m)**	Y: **15(0.381m)**	Z: **35(0.889m)**
X: **65(1.651m)**	Y: **15(0.381m)**	Z: **35(0.889m)**
X: **45(1.143m)**	Y: **15(0.381m)**	Z: **40(1.016m)**
X: **25(0.635m)**	Y: **25(0.635m)**	Z: **50(1.27m)**
X: **15(0.381m)**	Y: **45(1.143m)**	Z: **52(1.321m)**

6. Choose the **Finish** button; the path for the horn is created, as shown in Figure P4-7.

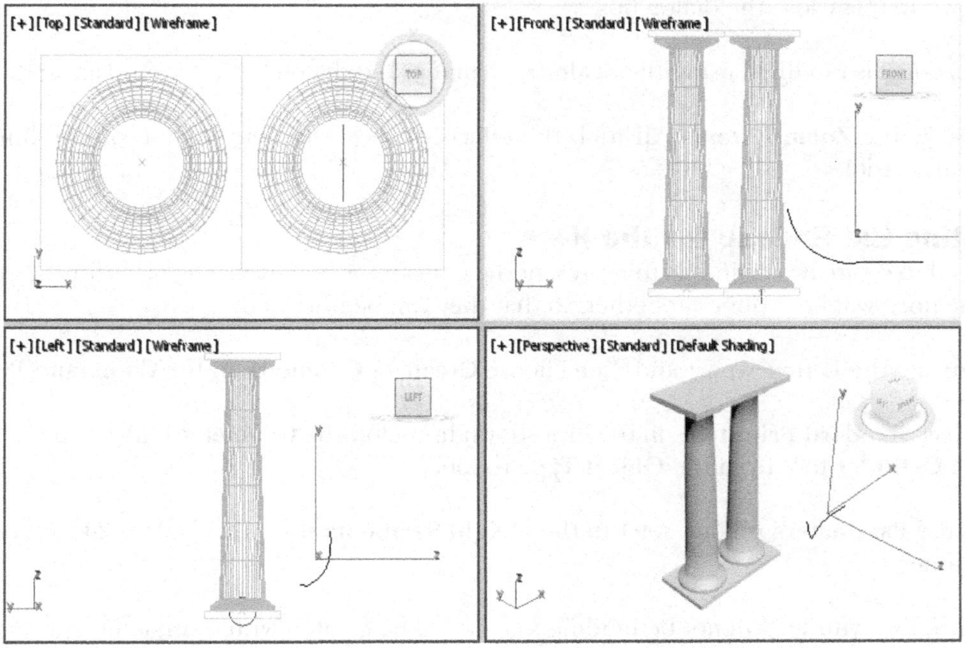

***Figure P4-7** The path created for the left-hand horn*

7. In the **Name and Color** rollout, enter **Horn Path**. Invoke the **Zoom Extents All Selected** tool to zoom the objects to their extents.

Lofting the Horn

After the path and shape are created, the first horn can be lofted. Once the loft object is created, its shape will be modified.

1. Make sure *Horn Path* is selected and choose **Create > Geometry** in the **Command Panel**. Next, select **Compound Objects** from the drop-down list. Invoke the **Loft** tool from the **Object Type** rollout.

2. Expand the **Skin Parameters** rollout and enter the value **10** in the **Shape Steps** spinner.

3. In the **Creation Method** rollout, choose the **Get Shape** button. Next, select *Horn Shape* in any viewport; the horn is lofted. In the **Name and Color** rollout, enter **Horn-Left**.

4. Make sure *Horn-Left* is selected and then choose the **Modify** tab in the **Command Panel**. In the **Deformations** rollout, choose the **Scale** button; the **Scale Deformation(X)** dialog box is displayed.

5. Make sure the **Move Control Point** button is chosen at the top of the **Scale Deformation(X)** dialog box.

6. Select (highlight) the control point at **100** (the right-hand control point). Enter **0** in the text box at the bottom right of the **Scale Deformation(X)** dialog box to move the control point down to 0%. Close the dialog box.

   Horn-Left is modified using the **Scale Deformation(X)** dialog box, refer to Figure P4-8.

7. Invoke the **Zoom Extents All** tool; the effects of the scale deformation on the horn are clearly visible.

Creating the Endcap for the Horn

You need to create two cylinders to represent the endcap for the horn. The parts for the endcap and the horn will be grouped together, so that they can be mirrored.

1. Activate the Left viewport and then choose **Create > Geometry** in the **Command Panel**.

2. Select **Standard Primitives** in the drop-down list below the **Geometry** button. Next, invoke the **Cylinder** tool from the **Object Type** rollout.

3. In the **Parameters** rollout, set **1** in the **Height Segments** spinner. Also, set **24** in the **Sides** spinner.

   Since the cylinder will not be modified, it should be created with a single height segment. This reduces the overall face count of the model. However, if modifiers are to be applied to the object, it may be necessary to increase the height segments to achieve the desired results.

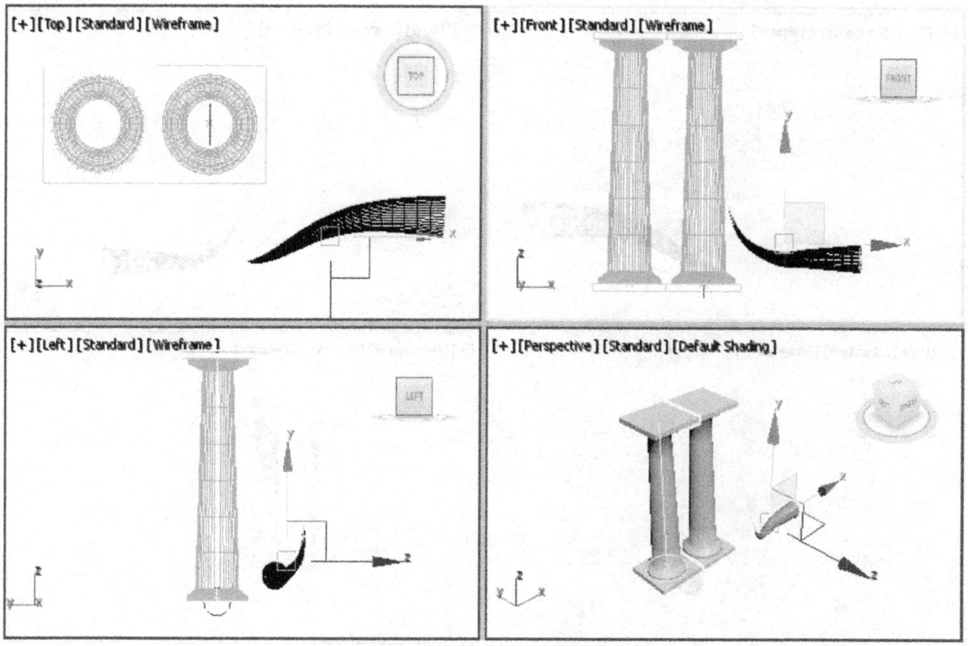

*Figure P4-8 Horn-Left modified using the **Scale Deformation(X)** dialog box*

4. Expand the **Keyboard Entry** rollout and set the values as given next:

X: **35(0.889m)** Y: **15(0.381m)** Z: **–95(–2.413m)**
Radius: **9(0.229m)** Height: **–1(–0.025m)**

5. Choose the **Create** button; the cylinder is created. In the **Name and Color** rollout, enter **Cap-Left01**.

6. Create the second cylinder by setting the following values in the **Keyboard Entry** rollout:

X: **35(0.889m)** Y: **15(0.381m)** Z: **–95(–2.413m)**
Radius: **8.5(0.216m)** Height: **3(0.076m)**

7. Choose the **Create** button; another cylinder is created. In the **Name and Color** rollout, enter **Cap-Left02**. Next, right-click to exit the tool.

8. Select *Horn-Left*, *Cap-Left01*, and *Cap-Left02*. Next, choose **Group > Group** from the menu bar; the **Group** dialog box is displayed. Enter **HornAssemblyLeft** in the **Group Name** text box. Then, choose the **OK** button; *HornAssemblyLeft* is created, as shown in Figure P4-9.

Mirroring the Horn Assembly

Now, you will mirror the left horn assembly to create the right horn assembly.

1. Activate the Front viewport.

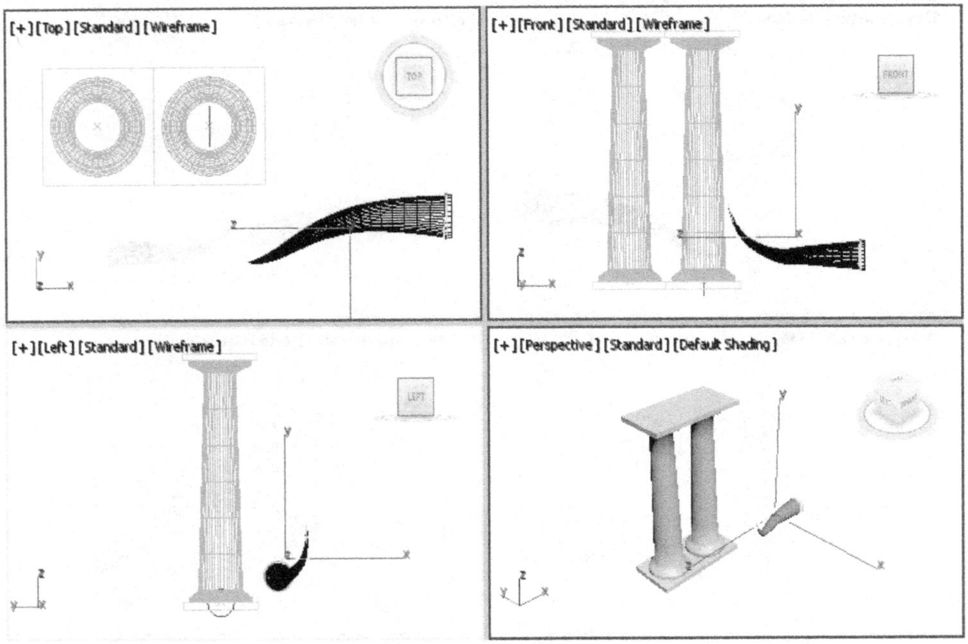

***Figure P4-9** HornAssemblyLeft created*

2. Make sure *HornAssemblyLeft* is selected and then invoke the **Mirror** tool; the **Mirror: Screen Coordinates** dialog box is displayed.

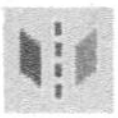

3. In the **Mirror: Screen Coordinates** dialog box, select the **Copy** radio button in the **Clone Selection** area. Set **100(2.54m)** in the **Offset** spinner in the **Mirror Axis** area. Also, make sure the **X** radio button is selected in the **Mirror Axis** area. Next, choose the **OK** button to create the mirrored copy.

4. In the **Name and Color** rollout, enter **HornAssemblyRight**.

5. Invoke the **Zoom Extents All** tool to zoom the objects to their extents.

Creating the Text

The text is created as a 2D shape and it will be lofted along a straight line path. The text will be lofted so that its shape can be refined by deforming it.

1. Make sure the Front viewport is activated and then choose **Create > Shapes** in the **Command Panel**. Next, invoke the **Text** tool from the **Object Type** rollout.

2. In the **Parameters** rollout, select **Bookman Old Style Bold** from the list of available fonts in the drop-down list at the top of the rollout. Next, set **30(0.762m)** in the **Size** spinner.

3. Enter **BULLISH** in the **Text** text box. Press the ENTER key and then enter **FINANCIAL** in the second line.

4. Click in the middle of the horn shapes in Front viewport to place the text; the text is displayed as a single object, refer to Figure P4-10.

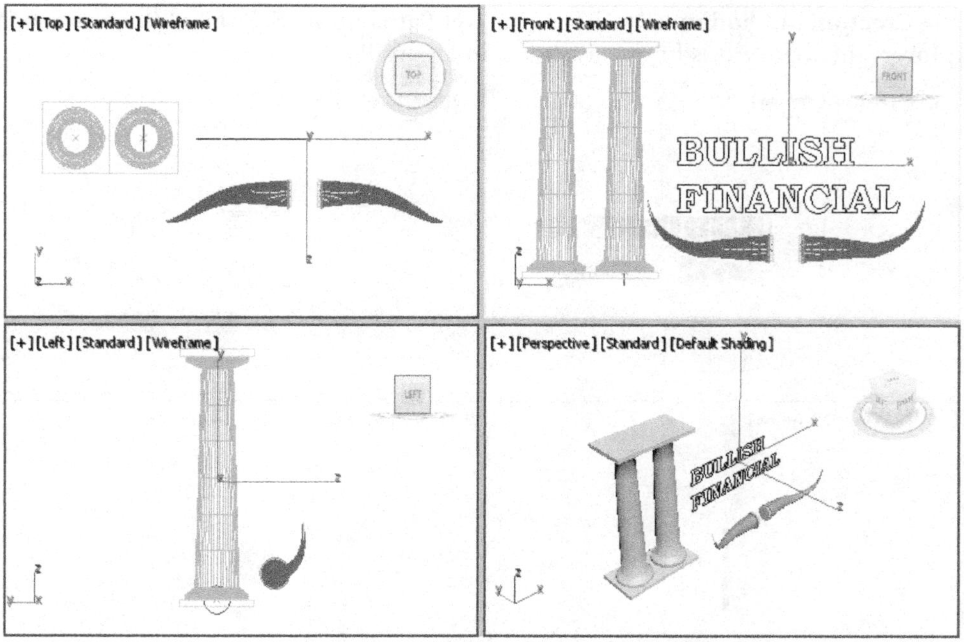

Figure P4-10 *The text displayed as a single object*

5. In the **Name and Color** rollout, enter **Text Shape**.

6. Activate the Top viewport and invoke the **Line** tool from the **Object Type** rollout.

7. In the **Initial Type** area of the **Creation Method** rollout, make sure the **Corner** radio button is selected.

8. Expand the **Keyboard Entry** rollout and set the values in the spinners as given next:

   X: **0** Y: **0** Z: **0**

9. Choose the **Add Point** button and set the values in the spinners as given next:

   X: **0** Y: **16(0.406m)** Z: **0**

10. Choose the **Add Point** button and then the **Finish** button; a line is created. In the **Name and Color** rollout, enter **Text Path**. Next, right-click in the viewport to exit the tool.

Lofting the Text

In this section, you will loft the text by using the lofting technique.

1. Select *Text Shape* in any viewport and then choose **Create > Geometry** in the **Command Panel**. Next, select **Compound Objects** in the drop-down list.

2. Invoke the **Loft** tool from the **Object Type** rollout. Next, expand the **Skin Parameters** rollout and set **0** in the **Shape Steps** spinner.

3. In the **Creation Method** rollout, choose the **Get Path** button. Select *Text Path* from the **Scene Explorer**; the logo text is lofted, as shown in Figure P4-11.

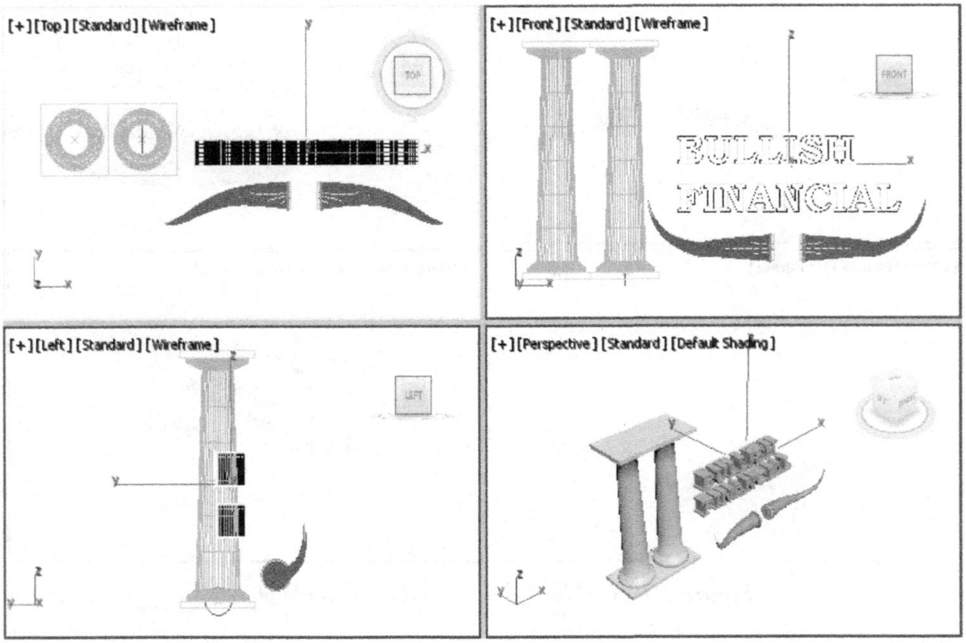

Figure P4-11 *The logo text lofted*

4. In the **Name and Color** rollout, enter **Logo Text**.

Modifying the Shape of the Text

In this section, you will add a bevel deformation to control the shape of the text.

1. Make sure *Logo Text* is selected and then choose the **Modify** tab in the **Command Panel**. Expand the **Deformations** rollout and choose the **Bevel** button; the **Bevel Deformation** dialog box is displayed.

2. Choose the **Insert Corner Point** button at the top of the **Bevel Deformation** dialog box and insert control points at 20 and 80.

3. Choose the **Move Control Point** button. Next, press and hold the CTRL key and select the newly inserted control points. Next, enter **–2.0** in the text box at the bottom right of the dialog box; the text is beveled, as shown in Figure P4-12.

4. Close the dialog box and invoke the **Zoom Extents All** tool to zoom the objects to their extents.

Note

*You can use the **TextPlus** tool in place of the **Text** tool to create the text for the logo. To invoke this tool, choose **Create > Geometry** in the **Command Panel** and then select **Standard Primitives** from the drop-down list below the **Geometry** button. Textplus object is a text object and it has the provision to extrude and bevel a text without lofting or applying any modifier.*

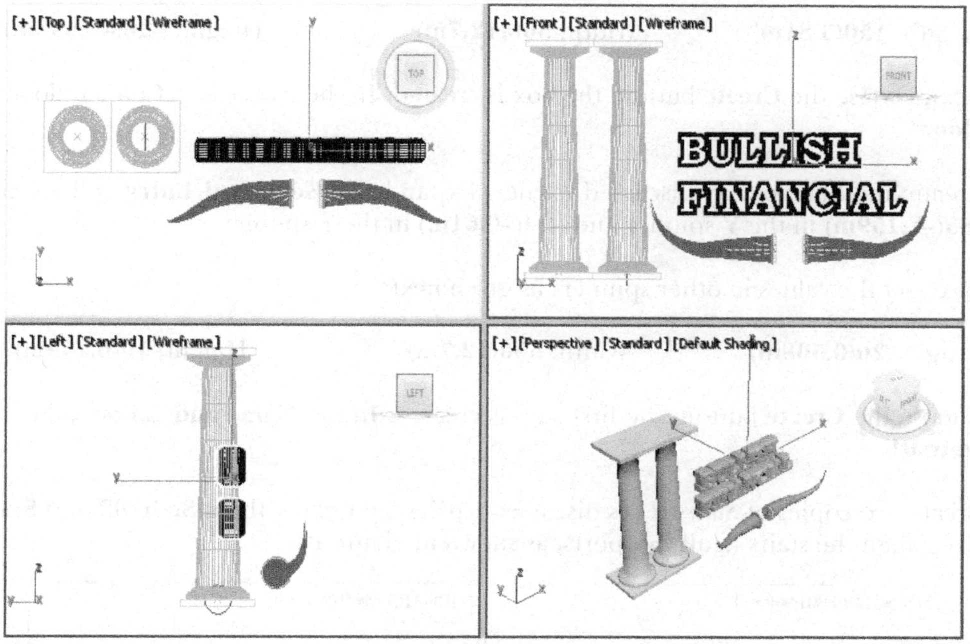

Figure P4-12 *The bevel deformation applied to Logo Text and the text centered over the horns*

Positioning the Text

In this section, you will move the text to a proper position.

1. Make sure the Front viewport is active. Invoke the **Select and Move** tool and then move the text so that it is placed approximately at the center in the space between the horns.

2. In the Top viewport, move the text so that it is centered above the portion of the horns that has the largest diameter.

3. Choose the **Display** tab in the **Command Panel**. Next, in the **Hide by Category** rollout, select the **Shapes** check box; the shapes are now hidden. Invoke the **Zoom Extents All** tool; the text is centered over the horns, refer to Figure P4-12. 

Creating the Floor and Steps

In this section, you will create boxes to represent the floor and steps. A box will be created for the first step and then it will be copied to add the rest of the steps.

1. Choose **Create > Geometry** in the **Command Panel**.

2. Select **Standard Primitives** in the drop-down list below the **Geometry** button if not selected. Next, invoke the **Box** tool from the **Object Type** rollout.

3. Expand the **Keyboard Entry** rollout and set **–4(–0.102m)** in the **Z** spinner. Next, set the values in other spinners as given next:

   Length: **150(3.81m)** Width: **500(12.7m)** Height: **–20(–0.508m)**

4. Next, choose the **Create** button; the box is created. In the **Name and Color** rollout, enter **Floor**.

5. Create another box, as discussed earlier. Expand the **Keyboard Entry** rollout and set **–85(–2.159m)** in the **Y** spinner and **–24(–0.61m)** in the **Z** spinner.

6. Next, set the values in other spinners as given next:

   Length: **20(0.508m)** Width: **500(12.7m)** Height: **10(0.254m)**

7. Choose the **Create** button; the first step is created. In the **Name and Color** rollout, enter **Stairs01**.

8. Create two copies of *Stairs01,* as discussed earlier, and name them **Stairs02** and **Stairs03**. Now, align the stairs in all viewports, as shown in Figure P4-13.

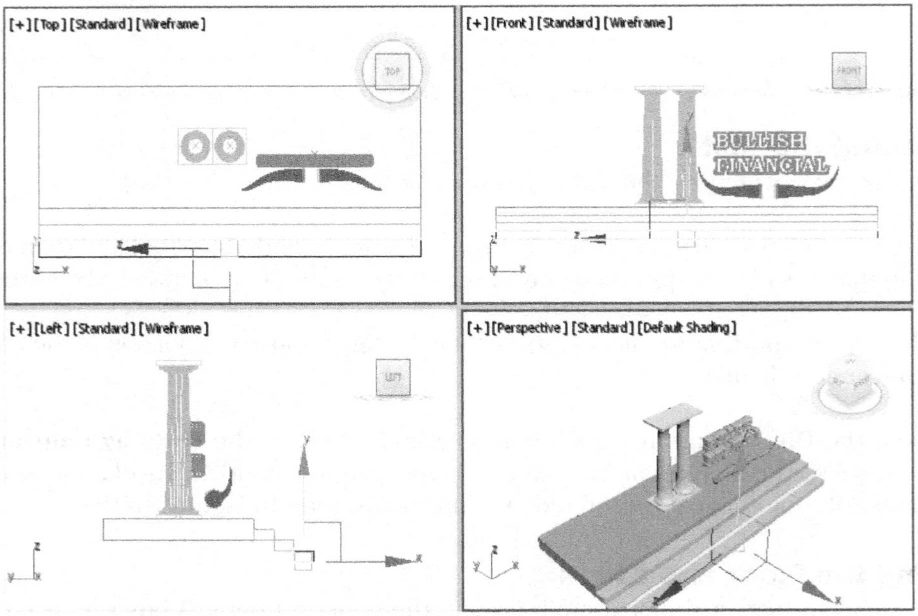

Figure P4-13 *The floor and three stairs created and aligned*

Creating a New Material Library

The creation of materials and texture maps will give this scene its needed detail. Texture mapping will turn the steps, base, and column into marble and the text into shiny gold. This project needs four materials, which will be stored in a new material library.

1. Choose **Rendering > Material Editor > Compact Material Editor** from the menu bar; the **Material Editor** dialog box is displayed.

2. In this dialog box, choose the **Get Material** button; the **Material/Map Browser** dialog box is displayed.

3. Choose the **Material/Map Browser Options** button from the **Material/Map Browser** dialog box; the **Material/Map Browser Options** flyout is displayed. Select **New Material Library** from the flyout; the **Create New Material Library** dialog box is displayed.

4. Enter **Project04** as the new name for the material library. Choose the **Save** button; the new material library *Project04.mat* is created and displayed in the **Material/Map Browser** dialog box.

5. Close the **Material/Map Browser** dialog box to return to the **Material Editor** dialog box.

Creating the Shiny Gold Material

In this section, you will create shiny gold material for the objects by using the **Material Editor** dialog box.

1. Make sure the first material sample slot is active in the **Material Editor** dialog box. By default, the **Physical Material** is displayed on the **Material Type** button. Choose **Physical Material** button; the **Material Map/Browser** dialog box is displayed. Choose **Materials > Scanline > Standard (Legacy)** from the **Material Map/Browser** dialog box and choose the **OK** button to close it.

2. In the **Shader Basic Parameters** rollout, select **Metal** from the drop-down list.

   This sets the rendering characteristics of the material being created.

3. In the **Metal Basic Parameters** rollout, choose the Lock button located between the **Ambient** and **Diffuse** color swatches to unlock the colors; the ambient and diffuse color settings can now be adjusted independently.

4. Choose the **Ambient** color swatch; the **Color Selector: Ambient Color** dialog box is displayed. In this dialog box, set the values as given next and then choose the **OK** button.

   Red: **25** Green: **15** Blue: **10**

5. Choose the **Diffuse** color swatch in the **Material Editor** dialog box; the **Color Selector: Diffuse Color** dialog box is displayed. In this dialog box, set the color values as given next and then choose the **OK** button.

   Red: **160** Green: **135** Blue: **30**

6. In the **Specular Highlights** area of the **Metal Basic Parameters** rollout, set the color value **80** in the **Specular Level** spinner. Set **75** in the **Glossiness** spinner.

7. In the **Material Name** drop-down list, enter **Gold** and press the ENTER key.

8. Choose the **Put to Library** button in the **Material Editor** dialog box; a flyout is displayed.

9. Choose **Project04.mat** from the flyout; the **Put to Library** dialog box is displayed and the name **Gold** appears in the **Name** text box. Choose the **OK** button; the current material is assigned the name *Gold* and added to the **Project04** material library.

Creating the Ivory Material

In this section, you will create the ivory material for the objects.

1. Select the next unused sample slot in the **Material Editor** dialog box. Choose the **Physical Material** button; the **Material Map/Browser** dialog box is displayed. Choose **Materials > Scanline > Standard (Legacy)** from the **Material Map/Browser** dialog box and choose the **OK** button to close it.

2. In the **Material Name** drop-down list, enter **Ivory** and press the ENTER key.

3. In the **Shader Basic Parameters** rollout, select **Blinn** in the drop-down list if not already selected.

4. In the **Blinn Basic Parameters** rollout, unlock the **Ambient** and **Diffuse** color swatches.

5. Choose the **Ambient** color swatch; the **Color Selector: Ambient Color** dialog box is displayed. In this dialog box, set the values as given next:

   Red: **15** Green: **16** Blue: **0**

6. Choose the **Diffuse** color swatch; the **Color Selector: Diffuse color** dialog box is displayed. In this dialog box, set the values as given next:

   Red: **240** Green: **240** Blue: **220**

7. Choose the **Specular** color swatch; the **Color Selector: Specular color** dialog box is displayed. In this dialog box, set the values as given next:

   Red: **255** Green: **255** Blue: **235**

   Close the **Color Selector: Specular Color** dialog box.

8. In the **Specular Highlights** area, set **80** in the **Specular Level** spinner and **50** in the **Glossiness** spinner.

9. Choose the **Put to Library** button in the **Material Editor** dialog box; a flyout is displayed.

10. Choose **Project04.mat** from the flyout; the **Put to Library** dialog box is displayed; the name **Ivory** appears in the **Name** text box. Choose the **OK** button in this dialog box.

Creating the Bump Map Marble Material

In this section, you will create the bump map marble material.

1. Activate the third material sample slot in the **Material Editor** dialog box. Choose the **Physical Material** button; the **Material Map/Browser** dialog box is displayed. Choose **Materials > Scanline > Standard (Legacy)** from the **Material Map/Browser** dialog box and choose the **OK** button to close it.

2. Expand the **Maps** rollout and then choose the **Diffuse Color** map button, currently labeled as **No Map**; the **Material/Map Browser** dialog box is displayed. This dialog box shows the types of maps that can be added.

3. Select the **Bitmap** map from **Maps > General** rollout and choose the **OK** button; the **Select Bitmap Image File** dialog box is displayed. As the project folder is already set, the *images* folder of this project is displayed in the **Look in** drop-down list of this dialog box.

4. Select the **BENEDETI.JPG** file and choose the **Open** button; the material is displayed in the sample slot.

5. In the **Coordinates** rollout, make sure the **Use Real-World Scale** check box is cleared. Also, make sure the **U: Tiling** and **V: Tiling** spinners are set to **1**.

6. Choose the **Go To Parent** button in the **Material Editor** dialog box to return to the root level. 

7. In the **Maps** rollout, select the **Bump** check box and set **200** in the **Bump Amount** spinner to set the intensity of the bump map.

8. Choose the **Bump Map** button, currently labeled as **No Map**; the **Material/Map Browser** dialog box is displayed.

9. Select the **Bitmap** map from **Maps > General** rollout and then choose the **OK** button; the **Select Bitmap Image File** dialog box is displayed. As the project folder is already set, the *images* folder of this project is displayed in the **Look in** drop-down list of this dialog box.

10. Select the **ConcreteMaterial.bump.jpg** file and then choose the **Open** button; the material is displayed in the sample slot.

11. In the **Coordinates** rollout, make sure the **Use Real-World Scale** check box is cleared. Set **0.75** in the **U: Tiling** spinner and **1** in the **V: Tiling** spinner.

12. Choose the **Go To Parent** button in the **Material Editor** dialog box to return to the parent level.

13. Choose the **Put to Library** button in the **Material Editor** dialog box; a flyout is displayed. Choose **Project04.mat** from the flyout; the **Put to Library** dialog box is displayed.

14. In this dialog box, enter **Column Marble** in the **Name** text box and then choose the **OK** button.

Creating the Shiny Marble Material

Marble floors are shiny and reflect the other objects around them. Since the material defined for the columns is not reflective, a second material needs to be created for the floor and steps. This material will be quite similar to the one you just created, but will appear highly polished.

1. In the **Material Editor** dialog box, click and hold the **Column Marble** material sample slot. Drag the sample to the next unused material sample slot. Then, rename the material as **Floor Marble**; a duplicate copy of the material is created.

2. Make sure the **Maps** rollout is expanded. Next, click and hold on any button labeled as **None**. Drag and drop the **No Map** button on top of the **Bump** map button; the bump map is cleared.

3. In the **Maps** rollout, choose the **Reflection** map button, currently labeled as **No Map**; the **Material/Map Browser** dialog box is displayed.

4. Select the **Reflect/Refract** map from the **Maps > Scanline** rollout and choose the **OK** button in the **Material/Map Browser** dialog box.

5. Choose the **Go To Parent** button in the **Material Editor** dialog box. In the **Maps** rollout, set **30** in the **Reflection Amount** spinner.

6. Make sure the **Show Shaded Material in Viewport** button is chosen. Next, choose the **Put to Library** button in the **Material Editor** dialog box; a flyout is displayed.

7. Choose **Project04.mat** from the flyout; the **Put to Library** dialog box is displayed.

8. In the **Put To Library** dialog box, the name **Floor Marble** appears in the **Name** text box. Choose the **OK** button.

Saving the Material Library

The changes to the material library must be saved, else they will be lost when you exit 3ds Max. By saving the material library, the newly created materials will be available the next time when you work in 3ds Max.

1. Choose the **Get Material** button in the **Material Editor** dialog box; the **Material/Map Browser** dialog box is displayed.

2. In this dialog box, right-click on the **Project04.mat** rollout; a cascading menu displayed. Choose **Save** from the cascading menu, as shown in Figure P4-14.

3. Close the **Material/Map Browser** dialog box.

The materials have been created and then saved in the material library.

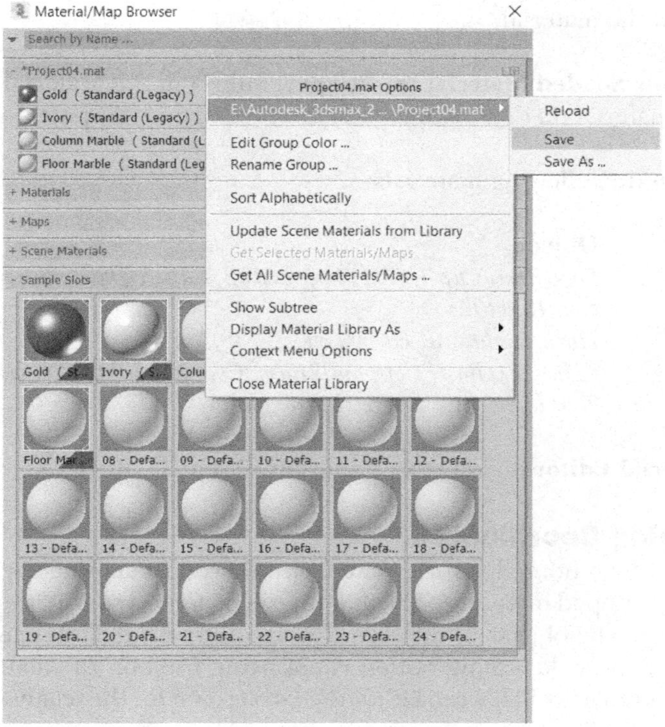

Figure P4-14 *Saving the material library*

Applying the Materials

In this section, the materials created will be assigned to the objects. You need to render the scene in the Perspective viewport to view the materials assigned to objects.

1. Before assigning materials, you need to ungroup all grouped objects. Press the CTRL+A keys to select all objects in the scene. Next, choose **Group > Ungroup** from the menu bar.

   Next, you need to rename some objects since they are created as clones of the original groups and automatically named.

2. Select and rename the following objects:

Current Name	**New Name**
Column002	*Column02*
Column01Top001	*Column02Top*
Column01Base001	*Column02Base*
Horn-Left001	*Horn-Right*
Cap-Left003	*Cap-Right01*
Cap-Left004	*Cap-Right02*

3. Select *Column01* and *Column02* in any viewport. In the **Material Editor** dialog box, select the material sample slot named *Column Marble*. Next, choose the **Assign Material to Selection** button to assign the material.

4. Choose the **Show Shaded Material in Viewport** button in the **Material Editor** dialog box to display the material on *Column Marble* in the Perspective viewport.

5. Similarly, assign the following materials:

Material	**Object**
Gold	*Logo Text, Cap-Left01, Cap-Left02, Cap-Right01,* and *Cap-Right02*
Ivory	*Horn-Right* and *Horn-Left*
Floor Marble	*Column01Base, Column02Base, Column01Top, Column02 Top, Floor*, and *Stairs01 through 03*.

6. Close the **Material Editor** dialog box and save the scene.

Applying Mapping Coordinates

Before the scene can be rendered, appropriate mapping coordinates need to be applied to the objects with texture-mapped materials. The mapping coordinates control the appearance of the texture on the object. The objects in the scene are created with default mapping coordinates. However, since the real-world scaling option is not being used on the mapped materials, the default mapping coordinates will have to be adjusted to account for the relative sizes of the objects.

1. Select *Floor* in the Top viewport. Choose the **Modify** tab in the **Command Panel**; the **Modifier List** drop-down list is displayed.

2. Select the **UVW Map** modifier from the **OBJECT-SPACE MODIFIERS** section in the **Modifier List** drop-down list; the **UVW Map** modifier is displayed in the modifier stack and the **Parameters** rollout is displayed in the **Modify** panel.

3. In the **Mapping** area of the **Parameters** rollout for the modifier, make sure the **Real-World Map Size** check box is cleared and then select the **Box** radio button.

4. Set the values in the spinners as given next:

   Length: **140 (3.556m)** Width: **140 (3.556m)** Height: **140(3.556m)**

   By setting the value in the spinners to **140(3.556m)**, the mapping effect on objects repeats every 140 units in all directions. The columns are **140(3.556m)** units tall, which will be used as the basis for all mappings.

5. Select *Stairs01* in the viewport and then choose the **Modify** tab in the **Command Panel**.

6. Select **UVW Map** from the **Modifier List** drop-down list.

7. In the **Mapping** area of the **Parameters** rollout, make sure the **Real-World Map Size** check box is cleared and then select the **Box** radio button.

8. Set the values in the spinners as given next:

   Length: **140(3.556m)** Width: **140(3.556m)** Height: **140(3.556m)**

9. Click on the arrow on left side to the **UVW Map** modifier in the modifier stack. Select **Gizmo** in the modifier stack. Invoke the **Select and Move** tool from the **Main Toolbar**.

10. In the Top viewport, move the gizmo along the X and Y axes to a random location. Invoke the **Select and Rotate** tool from the **Main Toolbar**. Rotate the gizmo to some degrees about the Z axis. Next, choose **Gizmo** in the modifier stack again to exit the **Gizmo** sub-object mode.

    As you move and rotate the gizmo in the Top viewport, you can observe the changes in map in the Perspective viewport. In this way, you can minimize repeating patterns on the object. This technique is also useful if you want to precisely align map patterns on different objects, such as brick patterns on adjacent walls.

11. Select *Stairs02* and apply a **UVW Map** modifier, as discussed earlier.

12. In the **Mapping** area of the **Parameters** rollout, make sure the **Real-World Map Size** check box is cleared and select the **Box** radio button.

13. Set the values in the spinners as given next:

    Length: **140(3.556m)** Width: **140(3.556m)** Height: **140(3.556m)**

14. Adjust the gizmo, as described in step 10.

15. Select *Stairs03* and apply the **UVW Map** modifier to it.

16. In the **Mapping** area of the **Parameters** rollout, make sure the **Real-World Map Size** check box is cleared and select the **Box** radio button.

17. Set the values in the spinners as given next:

    Length: **140 (3.556m)** Width: **140 (3.556m)** Height: **140 (3.556m)**

18. Adjust the gizmo, as described in step 10.

    The floor and each of the three steps will now have a slightly different appearance when rendered.

19. Select *Column01* and choose the **Modify** tab in the **Command Panel**; the **Modifier List** drop-down list is displayed.

20. Select **UVW Map** in the **Modifier List** drop-down list.

21. In the **Mapping** area of the **Parameters** rollout, make sure the **Real-World Map Size** check box is cleared and select the **Cylindrical** radio button. In the **Alignment** area, make sure the **Z** radio button is selected, and then choose the **Fit** button.

22. Click on the arrow on left side of the **UVW Map** modifier in the modifier stack. Select **Gizmo** in the modifier stack. Invoke the **Select and Rotate** tool from the **Main Toolbar**. Rotate the gizmo to some degrees about the Z axis. Choose **Gizmo** in the modifier stack again to exit the **Gizmo** sub-object mode.

23. Repeat the procedure followed in steps 19 through 22 for *Column02*.

    After rotating the **UVW Map** modifier gizmo, a different portion of the map is displayed on the front of the column. The two columns no longer look identical.

24. Select *Column01Base, Column02Base, Column01Top,* and *Column02Top* objects. With all the four objects selected, select **UVW Map** from the **OBJECT-SPACE MODIFIERS** section of the **Modifier List**. In the **Mapping** area of the **Parameters** rollout, make sure the **Real-World Map Size** check box is cleared. Next, select the **Box** radio button.

25. Set the values in the spinners as given next:

    Length: **140 (3.556m)** Width: **140 (3.556m)** Height: **140(3.556m)**

    The **UVW Map** modifiers are applied to all four objects at the same time. Each object has a different location in relation to the gizmo, and will therefore display a different section of the map. Since the two top objects are adjacent to each other, they will display adjacent areas of the map giving an illusion of a single object. The same is true for the bases.

Note

When a modifier is applied to multiple objects, as described here, changing the position of the modifier gizmo of one object will automatically change the position of the gizmo for the other objects as well.

Adding Lights

Lighting plays a very important role in a scene. Excessive light may spoil the visibility of a scene, whereas insufficient light may hide the details that you want to make visible. Spotlights can be used to cast shadows that add realism to the scene, but it slows down the rendering time. Therefore, in this scene, you will limit the lights to two spotlights and an Omni light.

1. Activate the Top viewport and zoom out to make the space available for creating the lights.

2. Choose **Create > Lights** from the **Command Panel**. Next, select the **Standard** option in the drop-down list and then invoke the **Target Spot** tool from the **Object Type** rollout.

3. Click on a point at the lower-left corner of the Top viewport and drag it to the center of *floor* and then release it to place the target.

4. Invoke the **Select and Move** tool and move the spotlight in the Left and Front viewports, as shown in Figure P4-15.

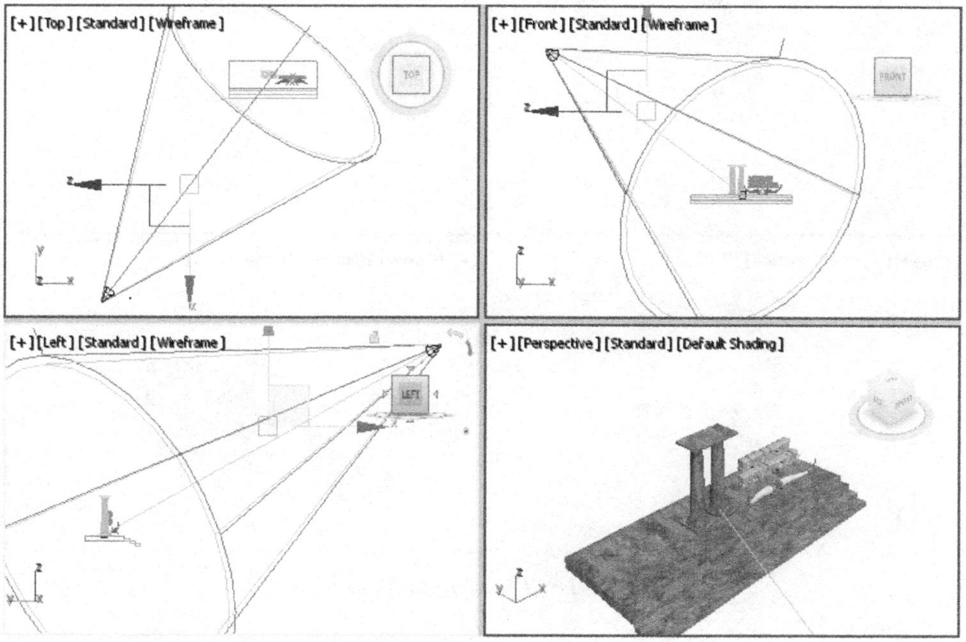

Figure P4-15 *The spotlights created and positioned*

5. Invoke the **Target Spot** tool from the **Object Type** rollout. Place another spotlight by clicking at a point in the lower-center of the Top viewport. Drag the spotlight target to the center of the horns and release it to place the target.

6. Move the spotlight in the Left and Front viewports so that it points directly to the center of the horns, refer to Figure P4-16.

   The spotlight is used to highlight the text and horns so that they can stand out in the scene. The other objects will be excluded from the effect of light.

7. Make sure that *Spot002* light is selected. Next, choose the **Modify** tab in the **Command Panel**. Choose the **Exclude** button at the bottom of the **General Parameters** rollout; the **Exclude/Include** dialog box is displayed.

8. Select the names of all objects in the scene except the two horns, text, and endcaps.

9. Choose the **>>** button to add selected objects to the exclusion list. Choose the **OK** button to close the **Exclude/Include** dialog box.

10. Select the **On** and **Use Global Settings** check boxes in the **Shadows** area of the **General Parameters** rollout if not already selected. Also, select **Ray Traced Shadows** from the drop-down list located below the rollout.

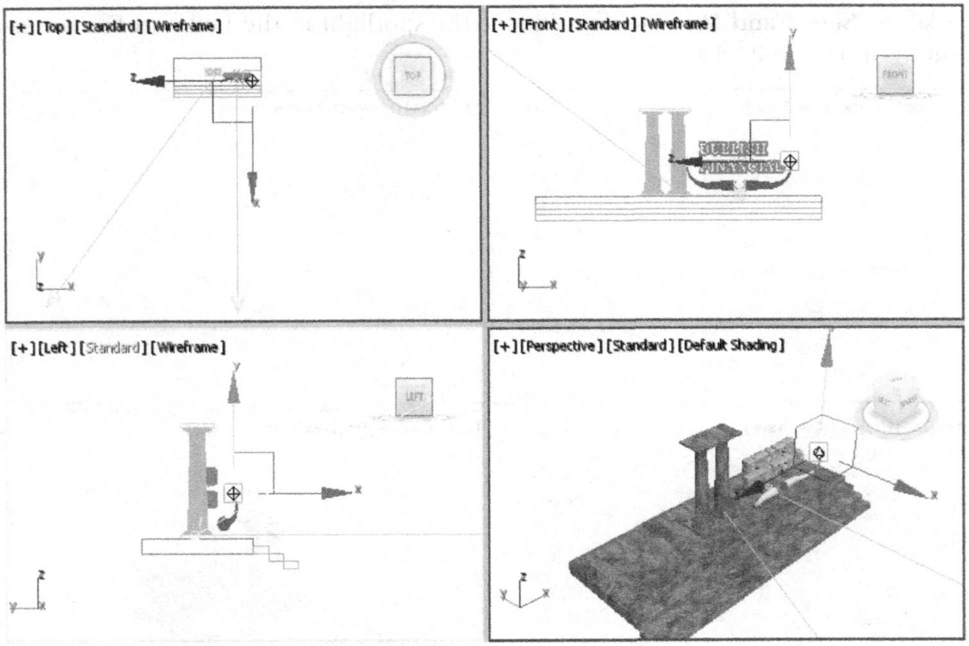

Figure P4-16 *The omni light positioned*

11. Repeat the procedure followed in steps 7 to 10 for *Spot001* light and then save the scene.

12. Activate the Front viewport and then choose **Create > Lights** in the **Command Panel**.

13. Invoke the **Omni** tool from the **Object Type** rollout. Click on a point in front and to the right of the text to create the light. Align the light in the Front and Left viewports, as shown in Figure P4-16.

14. Make sure the **Omni** light is selected. Next, choose the **Modify** tab in the **Command Panel**. In the **Intensity/Color/Attenuation** rollout, set **1.5** in the **Multiplier** spinner. Next, choose the **Exclude** button at the bottom of the **General Parameters** rollout; the **Exclude/Include** dialog box is displayed.

15. Select the **Include** radio button and then select **Horn-Left**, **Horn-Right**, and **Logo Text** from the **Scene Objects** area.

16. Choose the **>>** button in the **Exclude/Include** dialog box to add the selected objects to the inclusion list, as shown in Figure P4-17. Choose the **OK** button to close the **Exclude/Include** dialog box.

17. Make sure the **On** and **Use Global Settings** check boxes are selected in the **Shadows** area of the **General Parameters** rollout. Also, make sure **Ray Traced Shadows** is selected in the drop-down list located below it.

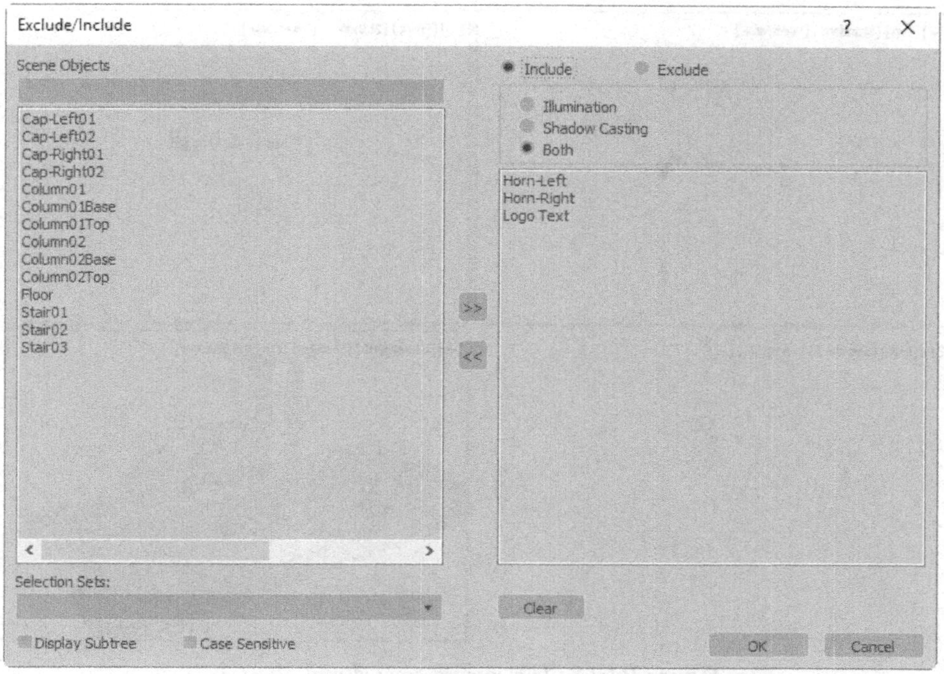

*Figure P4-17 The **Exclude/Include** dialog box*

Creating the Camera

You will now create a camera in the scene. Creating a camera will help you to view the scene properly.

1. Activate the Top viewport and then choose **Create > Cameras** in the **Command Panel**.

2. Invoke the **Target** tool from the **Object Type** rollout.

3. Click at a point in the lower-right corner of the Top viewport. Next, drag it to the center of the horns to place the target. Right-click to exit the tool.

4. Activate the Perspective viewport and press the C key; the Perspective viewport switches to Camera001 viewport.

5. Invoke the **Zoom Extents All** tool to zoom the objects to their extents.

6. Invoke the **Select and Move** tool and move the camera in all viewports until an appropriate view is displayed in the Camera001 viewport, as shown in Figure P4-18.

   When the Camera001 viewport is active, the **Field-of-View** and **Truck Camera** tools provide additional control of the view.

7. Choose **Rendering > Environment** from the menu bar or press 8; the **Environment and Effects** dialog box is displayed.

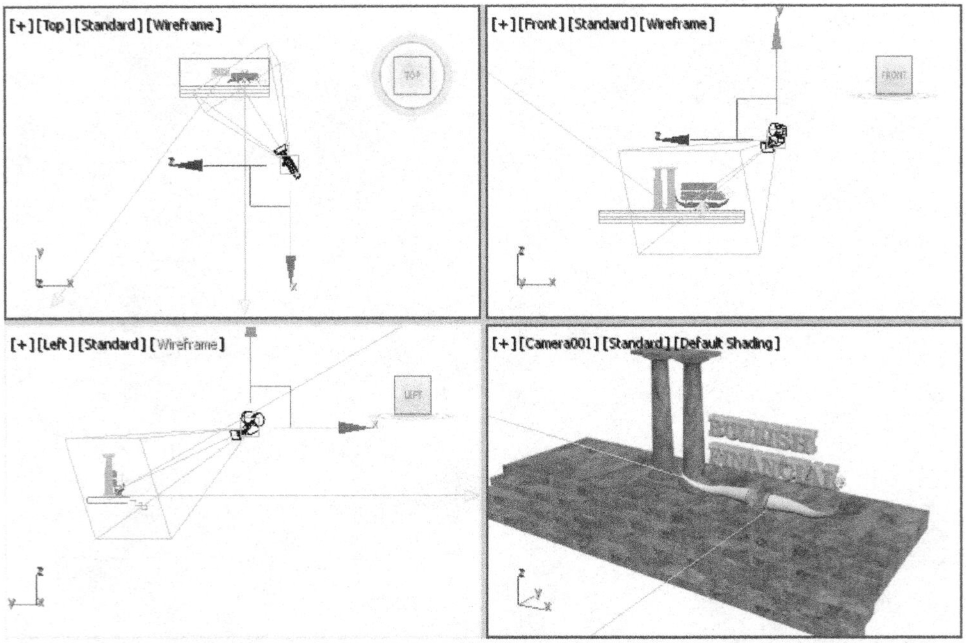

Figure P4-18 The camera created and aligned

8. In the **Background** area of the **Environment and Effects** dialog box, choose the **Color** swatch; the **Color Selector: Background Color** dialog box is displayed. In this dialog box, set the values as given next:

   Red: **255** Green: **255** Blue: **255**

   This sets the background color to white. Since the marble material on the floor and steps is reflective, therefore changing the background color from black to white will lighten the scene and bring out detailing in the marble material.

9. In the **Global Lighting** area, choose the **Ambient** color swatch; the **Color Selector: Ambient Light** dialog box is displayed. In this dialog box, set the values as given next:

   Red: **100** Green: **100** Blue: **100**

10. Close the **Environment and Effects** dialog box. Make required adjustments to the lights and ambient lighting. Next, invoke the **Render Production** tool; the final rendered image is displayed, as shown in Figure P4-19.

11. Save the scene.

    Next, you will increase the animation range.

12. Choose the **Time Configuration** button in the animation playback controls; the **Time Configuration** dialog box is displayed. 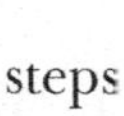

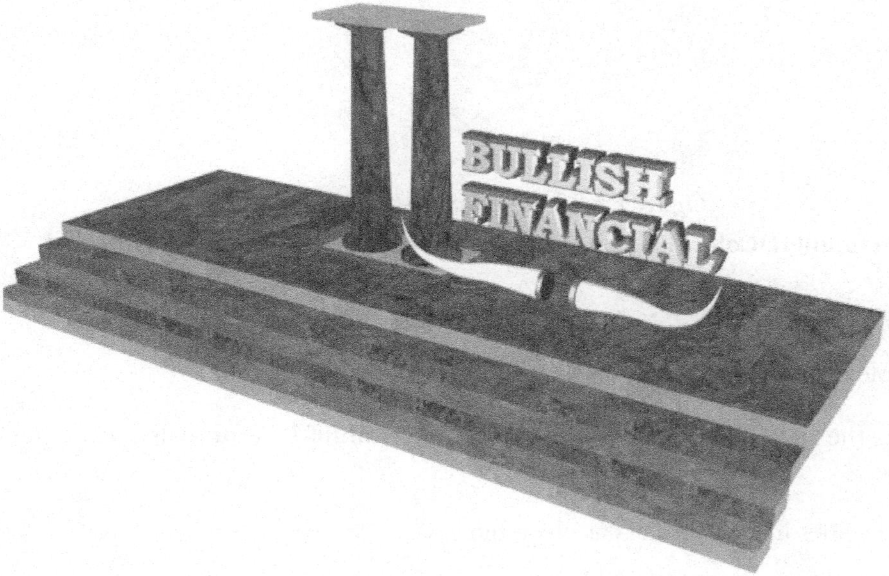

Figure P4-19 *The scene rendered after creating the lights and camera*

13. In the **Animation** area, set **120** in the **Length** spinner and choose the **OK** button.

    This sets the total number of frames in the animation at 120, which is about four seconds of animation.

Animating the Light Color

The color of a light source can be changed during an animation. By changing the color from black to white, an animated highlight can be created.

1. Select the **Omni** light in any of the viewports.
2. Choose **Graph Editors > Track View - Dope Sheet** from the menu bar; the **Track View - Dope Sheet** window is displayed.
3. Expand the tree for *Omni001* by clicking on the plus sign **(+)** on the left side of the *Omni001*. Also, expand the tree for **Transform** and then **Object (Omni Light)**. Notice that there is currently no animation key.
4. Choose the **Add Remove Key** button at the top of the **Track View-Dope Sheet** window.
5. Click on the **Color** track and add keys at frames 0, 5, 15, 110, and 120; five animation keys are created for the **Omni** light.
6. Right-click on the first key (on frame 0); the **Omni001\Color** dialog box is displayed, as shown in Figure P4-20.

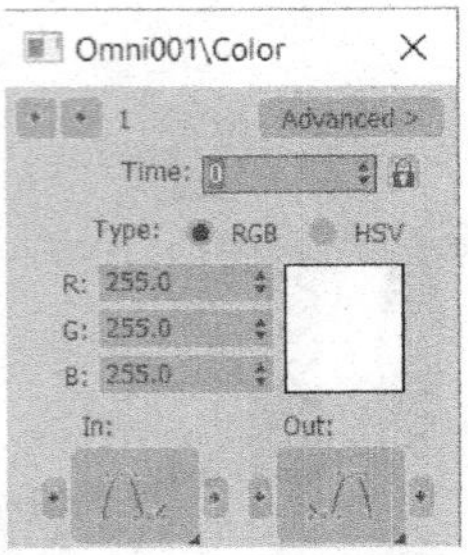

Figure P4-20 The ***Omni001\Color*** *dialog box*

7. In the **Omni001\Color** dialog box, set the values in the spinners as given next:

   R: **0** G: **0** B: **0**

   The color of the light is now black at frame 0.

8. Choose the right arrow button at the top of the **Omni001\Color** dialog box to move to the next key.

9. Set the values in the spinners as given next:

   R: **0** G: **0** B: **0**

   The color of the light is set to black from frame 0 to frame 5.

10. Move to the next key. Set the value of the **R**, **G**, and **B** spinners to **255** if not already set to this value.

    From frame 5 to frame 15, the color of the light changes to white.

11. Move to the next key. Set the values in the spinners as given next:

    R: **255** G: **255** B: **255**

12. Move to the next key. Set the values in the spinners as given next:

    R: **0** G: **0** B: **0**

    From frame 110 to frame 120, the color of the light changes from white to black.

13. Close the **Omni001\Color** dialog box and save the scene.

Animating the Light

In this section, you will animate the light in the scene.

1. In the **Track View-Dope Sheet** window, expand the tree for the **Position** track and add a keyframe in the Position track at frame 15 for *Omni001*. Then, close the **Track View-Dope Sheet** window.

This keyframe prevents the transformation of light till frame 15.

2. Drag the time slider to frame 110 and then choose the **Toggle Auto Key Mode** button to turn the animation mode on.

3. Make sure *Omni001* is selected and then right-click on the **Select and Move** tool; the **Move Transform Type-In** dialog box is displayed.

4. Activate the Front viewport. Set **–300(–7.62m)** in the **X** spinner in the **Offset: Screen** area of the **Move Transform Type-In** dialog box and press the ENTER key.

5. Exit the animation mode by choosing the **Toggle Auto Key Mode** button, close the **Move Transform Type-In** dialog box, and save the scene.

Rendering the Animation

Render the final animation to see the completed project. The final animation will show all materials, highlights, shadows, and lighting.

1. Activate the Camera001 viewport. Choose **Rendering > Render Setup** from the menu bar or invoke the **Render Setup** tool from the **Main Toolbar**; the **Render Setup: Scanline Renderer** dialog box is displayed.

2. In the **Time Output** area, select the **Active Time Segment** radio button to make it active; the active frames are set for rendering.

3. In the **Output Size** area of the dialog box, choose the **640×480** button; the animation to be rendered is set at a resolution of 640 × 480 pixels.

4. In the **Render Output** area of the dialog box, choose the **Files** button; the **Render Output File** dialog box is displayed.

   As the project folder is already set, the *renderoutput* folder of this project is displayed in the **Save in** drop-down list of this project. Enter **Project04** in the **File name** text box.

5. Select **AVI File (*.avi)** from the **Save as type** drop-down list.

6. Choose the **Save** button; the **AVI File Compression Setup** dialog box is displayed.

7. Choose the **OK** button to close the **AVI File Compression Setup** dialog box.

8. Make sure the Camera001 viewport is activated and then choose the **Render** button in the **Render Setup** dialog box; the rendering begins.

   After the rendering is done, close the render window and the **Render Setup** dialog box.

9. Navigate to the *renderoutput* folder of your project folder and play the **Project04.avi** file, refer to Figure P4-21.

Figure P4-21 A rendered image of the final animation at frame 0

Project 5

Creating a Computer Center

PROJECT DESCRIPTION

In this center, you will create a room with walls, floor, and also add roof at the top. Additionally, you will create an entrance door and window to be placed on the front and right side walls. Inside the center, you will create a set of computer station consisting of computer table, monitors, chairs, and tabletop. Next, you will create ground and a bed for placing the plants. At the end, you will create a pathway leading to the center. To make the scene more realistic, you will create lights and illuminate the center. To give a natural look, the exterior scene will be created with trees surrounding the computer center. Finally, you will apply materials to these objects and create a walkthrough, as shown in Figure P5-1.

Figure P5-1 *Walkthrough towards the computer center*

Creating the Project Folder

Create a new project folder with the name *Prj_5* at *\Documents\3dsmax 2022* and then save the file with the name *Prj5*, as discussed in Tutorial 1 of Chapter 2.

Downloading the Files

Before starting the project, you need to download the *prj5_3dsmax_2022* from *www.cadcim.com*. The path of the file is as follows: *Textbooks > Animation and Visual Effects > Autodesk 3ds Max > Autodesk 3ds Max 2022 for Beginners: A Tutorial Approach*

Extract the contents of the zip file to *Documents*. Open Windows Explorer and browse to *\Documents\prj5_3dsmax_2022* and then copy all the files to *\Documents\3dsmax 2022\Prj_5\sceneassets\images*.

Changing the Unit Settings

In this section, you need to change the current units you are working on. Since the units in this project are in inches, you will set the units to decimal inches.

1. Choose **Customize > Units Setup** from the menu bar; the **Units Setup** dialog box is displayed.

2. Select the **US Standard** radio button in the **Display Unit Scale** area. Next, select **Feet w/Fractional Inches** from the drop-down list below this radio button and make sure **1/2** is selected in the drop-down list on the right. Also, make sure the **Feet** radio button is selected, as shown in Figure P5-2. Choose the **OK** button to apply the settings.

   If you want to work in the metric unit system, select the **Metric** radio button from the **Display Unit Scale** area of the **Units Setup** dialog box. Also, make sure the **Meters** option is selected in the drop-down list below the **Metric** radio button.

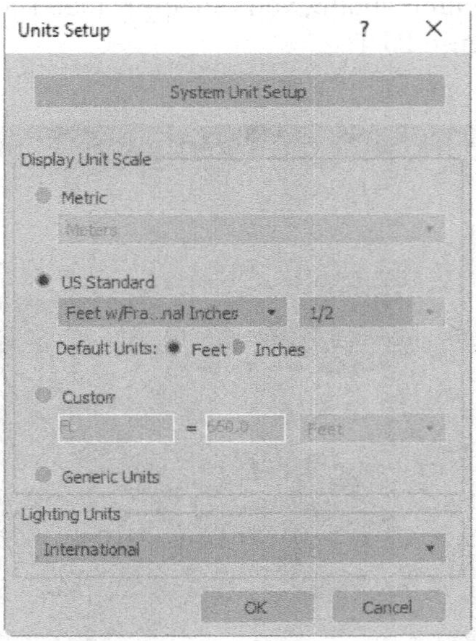

***Figure P5-2** The **Units Setup** dialog box*

Creating the Walls and Floor for the Room

In this section, you will create walls and floor for the computer center by using the **Wall** tool. Walls belonging to a group of 3ds Max objects are called AEC extended objects.

1. Switch to four viewport configuration. Activate the Top viewport and then choose **Create > Geometry** in the **Command Panel**. Select **AEC Extended** in the drop-down list. Next, invoke the **Wall** tool from the **Object Type** rollout.

2. In the **Parameters** rollout, set **9"(0.229m)** in the **Width** spinner and **10'(3.048m)** in the **Height** spinner.

3. In the **Keyboard Entry** rollout, set the following values in the spinners. Notice that the **Z** spinner is grayed out. Choose the **Add Point** button after specifying each set of coordinates:

   X: **0'** Y: **0'**
   X: **13'(3.962m)** Y: **0'**
   X: **13'(3.962m)** Y: **23'6"(7.163m)**
   X: **0'** Y: **23'6"(7.163m)**

4. Choose the **Close** button; the walls of the computer center are created, as shown in Figure P5-3. Right-click to exit the tool.

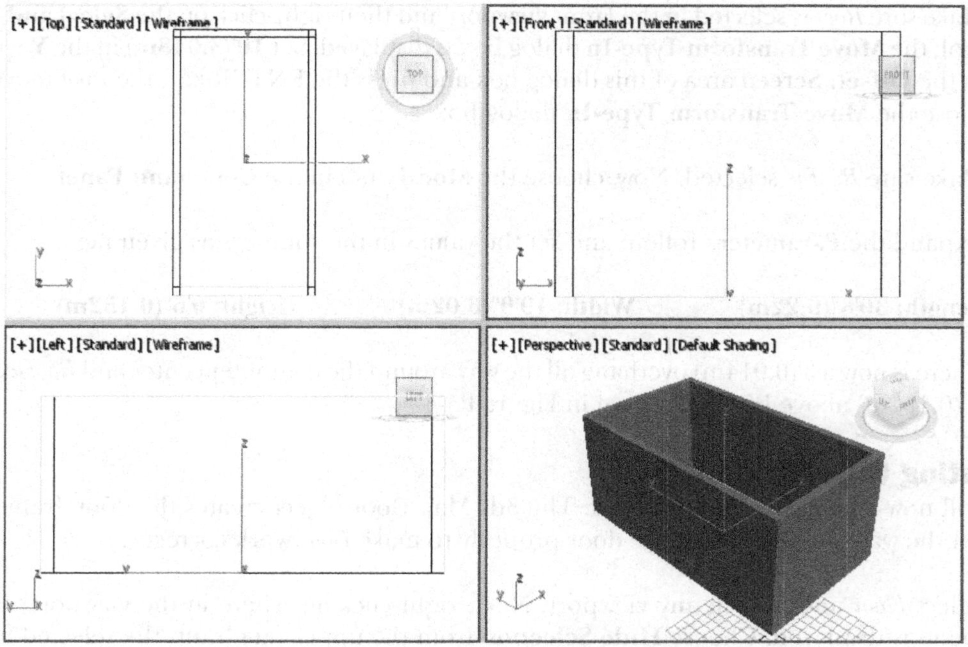

Figure P5-3 *Walls of the computer center*

5. In the **Name and Color** rollout, enter **Walls** and then invoke the **Zoom Extents All** tool to zoom the object to its extent.

6. Make sure the Top viewport is activated and then choose **Create > Geometry** in the **Command Panel**. Select **Standard Primitives** in the drop-down list. Next, invoke the **Box** tool from the **Object Type** rollout.

7. Expand the **Keyboard Entry** rollout and set the values as given next:

   X: **6'6"(1.981m)** Y: **11'9"(3.581m)** Z: **0'**
   Length: **24'3"(7.391m)** Width: **13'9"(4.191m)** Height: **–6"(–0.152m)**

8. Choose the **Create** button; the floor is created below *Walls*. In the **Name and Color** rollout, enter **Floor**. Right-click to exit the tool.

Adding Roof to the Building

The flat roof of the computer center will be created by copying *Floor* and then modifying the parameters of *Floor*.

1. Select *Floor* in the Front viewport and then choose **Edit > Clone** from the menu bar; the **Clone Options** dialog box is displayed. Next, select the **Copy** radio button and enter the name as **Roof** in the **Name** text box. Now, choose the **OK** button; a copy of *Floor* is created in the viewport.

2. Make sure *Roof* is selected in the Front viewport and then right-click on the **Select and Move** tool; the **Move Transform Type-In** dialog box is displayed. Set **10'(3.048m)** in the **Y** spinner in the **Offset: Screen** area of this dialog box and press the ENTER key; the *Roof* moves up. Close the **Move Transform Type-In** dialog box.

3. Make sure *Roof* is selected. Now, choose the **Modify** tab in the **Command Panel**.

4. Expand the **Parameters** rollout and set the values in the spinners as given next:

   Length: 30'3"(9.22m) **Width: 19'9"(6.02m)** **Height: 0'6"(0.152m)**

   There is now a 3'(0.914m) overhang all the way around the computer center and *Roof* extends 6"(0.152m) above *Walls*, as shown in Figure P5-4.

Creating the Entrance Door

You will now create the entrance door. The 3ds Max Door object creates the door, frame, and hole in the wall. You need to place door properly to make *Door* work correctly.

1. Select *Floor* and *Roof* in any viewport. Next, right-click anywhere in the viewport; a quad menu is displayed. Choose **Hide Selection** from the upper quadrant; the selected objects are hidden.

2. Invoke the **2D Snap Toggle** tool from the **Snaps Toggle** flyout to turn the snaps on. Next, right-click on the **2D Snap Toggle** tool; the **Grid and Snap Settings** dialog box is displayed.

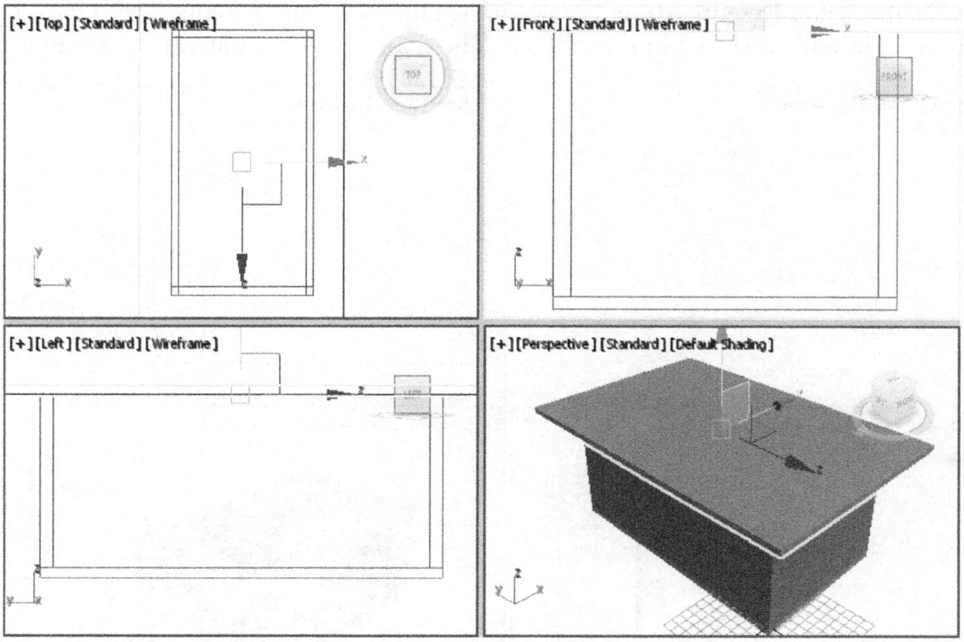

Figure P5-4 *Roof created and placed above Walls*

3. In the **Snaps** tab, select the **Edge/Segment** check box if not already selected. Now, close the dialog box.

4. Choose **Create > Geometry** in the **Command Panel**. Next, select **Doors** in the drop-down list and choose the **Pivot** button from the **Object Type** rollout.

5. In the Top viewport, click on the inside of the right-hand wall near the bottom. Make sure the inner wall is highlighted. Hold down the mouse button and drag the cursor straight to a short distance and release. Make sure the the inner wall is highlighted. Next, move the cursor to the outside wall and click making sure the outside wall is highlighted. Finally, move the cursor a short distance straight up and pick to set the height of the door. Right-click to exit the tool.

   The parameters will be set to the exact values in the next step.

6. Make sure the *Door* is selected and then choose the **Modify** tab in the **Command Panel**. In the **Parameters** rollout, set **7'(2.134m)** in the **Height** spinner and **4'(0.919m)** in the **Width** spinner.

7. Set **90** in the **Open** spinner to open the door. Then, select the **Flip Swing** and **Flip Hinge** check boxes; the door opens out toward the right. Also, check that the opening in the wall has been created.

   The *Door* should be approximately **2'(0.609m)** from the edge of the building. If it is not, move it as needed in the Top viewport.

8. Invoke the **Zoom Extents All** tool to zoom the objects to their extents. The door is created on the right wall, as shown in Figure P5-5. The door is placed at an approximate location.

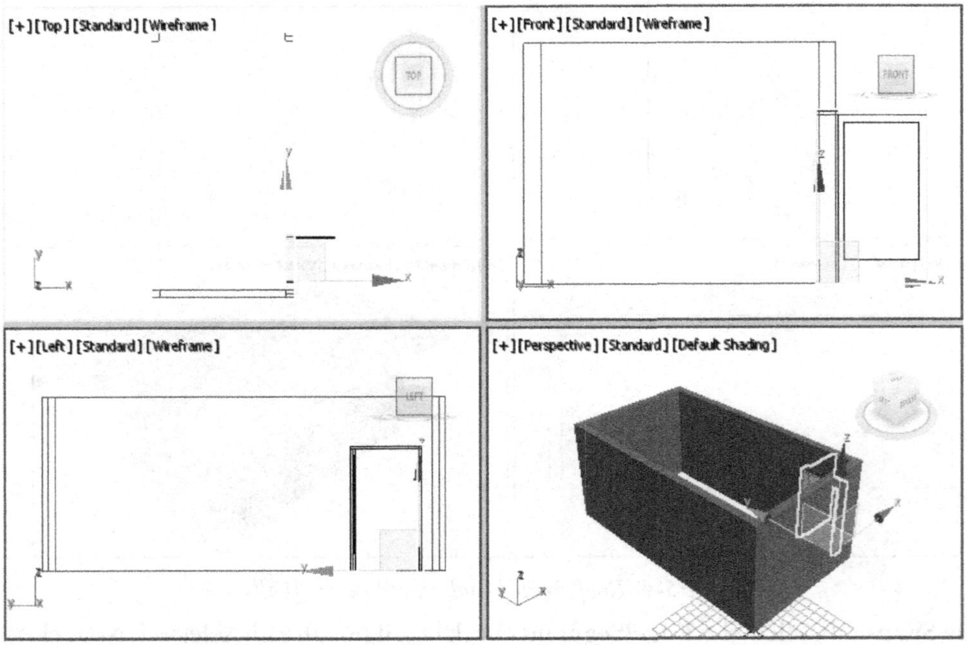

Figure P5-5 *The door created on the right wall*

9. Set **0** in the **Open** spinner of the **Parameters** rollout; the *Door* is closed.

Creating the Window

A window will now be placed at the end of the wall. Like the door, the window is created as a special 3ds Max object. The window object automates the creation and animation of windows. However, the window to be created will be fixed and will not be animated.

1. Choose **Create > Geometry** in the **Command Panel**. Select **Windows** from the drop-down list. Then, invoke the **Fixed** tool from the **Object Type** rollout.

2. In the Top viewport, click on the inside of the lower wall about 1/3 of the way inside from the left-hand side. Make sure the inner wall is highlighted. Then, drag straight and right to about 1/3 of the way inside from the right-hand side and click. Make sure the inner wall is highlighted. Next, move the cursor outside the wall and click to make sure the outside wall is highlighted. Finally, move the cursor a short distance to the left and click to set the height of the window.

   The parameters will be set to exact values in the next step.

3. Make sure the window is selected and then choose the **Modify** tab in the **Command Panel**. In the **Parameters** rollout, set **8'(2.438m)** in the **Height** and **Width** spinners.

4. Make the Top viewport active. Invoke the **Align** tool from the **Main Toolbar**. Next, select *Walls*; the **Align Selection (Walls)** dialog box is displayed. Select the **X Position** and **Z Position** check boxes and clear the **Y Position** check box. Then, make sure the **Center** radio button is selected in the **Current Object** and **Target Object** areas of this dialog box. Choose the **OK** button; the window is precisely centered in the front wall, as shown in Figure P5-6.

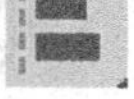

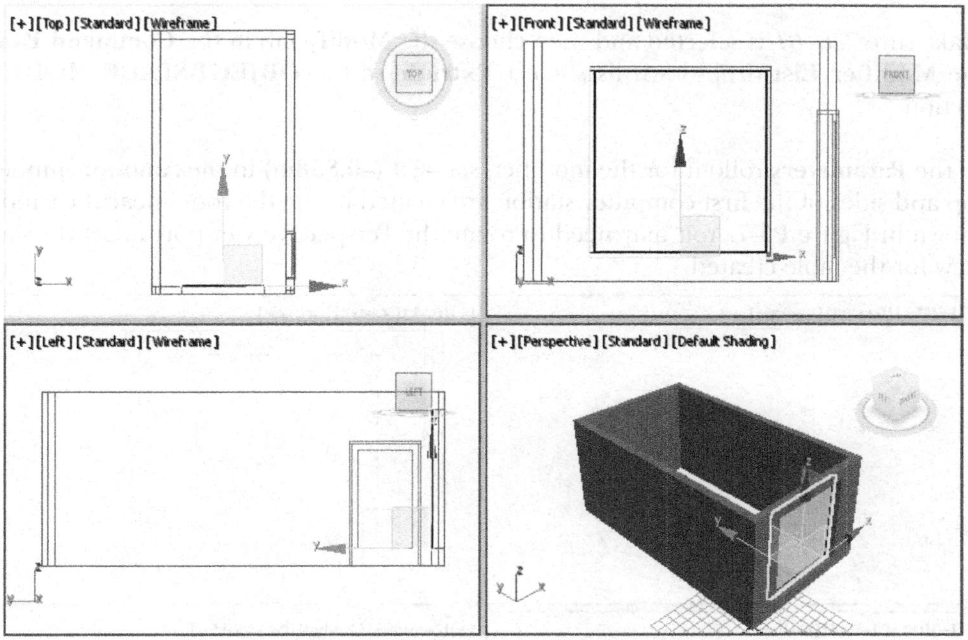

Figure P5-6 The alignment of the window on the front wall

Creating a Computer Station

You will create one set of computer station which consists of a computer table, monitors, and the chairs. Each computer station consists of a tabletop, two sides, and a partition. The tabletop and two sides are created as a single spline that is extruded. The partition is also a spline that is extruded. In this section, you will first start with a computer station.

1. Activate the Left viewport and then choose **Create > Shapes** in the **Command Panel**. Next, invoke the **Line** tool from the **Object Type** rollout. Make sure the **Corner** radio button is selected in the **Initial Type** and **Drag Type** areas in the **Creation Method** rollout.

2. In the **Keyboard Entry** rollout, set the values as given next. Choose the **Add Point** button after specifying each set of XYZ coordinates:

X: **–0'4 1/2"(–0.114m)**	Y: **0'0"**	Z: **–0'4 1/2"(–0.114m)**
X: **–0'4 1/2"(–0.114m)**	Y: **2'6"(0.762m)**	Z: **–0'4 1/2"(–0.114m)**
X: **–4'11 1/2"(–1.511m)**	Y: **2'6"(0.762m)**	Z: **–0'4 1/2"(–0.114m)**
X: **–4'11 1/2"(–1.511m)**	Y: **0'0"**	Z: **–0'4 1/2"(–0.114m)**
X: **–4'10 1/2"(–1.486m)**	Y: **0'0"**	Z: **–0'4 1/2"(–0.114m)**

X: **–4'10 1/2"(–1.486m)**	Y: **2'5"(0.737m)**	Z: **–0'4 1/2"(–0.114m)**
X: **–0'5 1/2"(–0.14m)**	Y: **2'5"(0.737m)**	Z: **–0'4 1/2"(–0.114m)**
X: **–0'5 1/2"(–0.14m)**	Y: **0'0"**	Z: **–0'4 1/2"(–0.114m)**

3. After the last set of coordinates is entered, choose the **Close** button. In the **Name and Color** rollout, enter **Table01**.

4. Make sure *Table01* is selected and then choose the **Modify** tab in the **Command Panel**. In the **Modifier List** drop-down list, select **Extrude** in the **OBJECT-SPACE MODIFIERS** section.

5. In the **Parameters** rollout for the modifier, set **–2'9"(–0.838m)** in the **Amount** spinner; the top and sides of the first computer station are created inside the room near the window, as shown in Figure P5-7. You may need to rotate the Perspective viewport to set the angle of view for the table created.

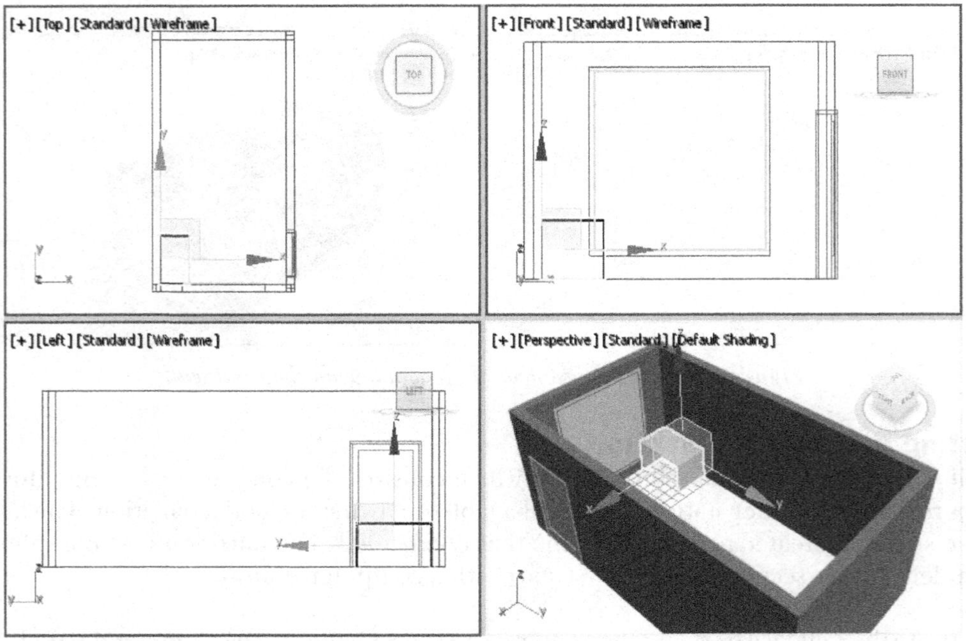

***Figure P5-7** The top and sides of the first computer station*

Now, you need to create partitions.

6. Activate the Perspective viewport and zoom in on *Table01*. Align *Table01* in the viewport. You may maximize the viewport.

7. Activate the Front viewport and choose **Create > Shapes** in the **Command Panel**. Next, invoke the **Line** tool from the **Object Type** rollout. Draw a triangle by entering the following coordinates in the **Keyboard Entry** rollout. Choose the **Add Point** button after entering the value of each set of the coordinates.

X: **0'4 1/2"(0.114m)** Y: **2'6"(0.762m)** Z: **–4'11 1/2"(–1.511m)**
X: **0'4 1/2"(0.114m)** Y: **4'0"(1.219m)** Z: **–4'11 1/2"(–1.511m)**
X: **3'1 1/2"(0.952m)** Y: **2'6"(0.762m)** Z: **–4'11 1/2"(–1.511m)**

8. After the last set of coordinates is entered, choose the **Close** button. In the **Name and Color** rollout, enter **Side01**.

9. Make sure *Side01 is* selected in the Front viewport and then choose the **Modify** tab in the **Command Panel**. In the **Interpolation** rollout, select the **Adaptive** check box. In the **Selection** rollout, choose the **Vertex** button. Right-click on the top vertex; a quad menu is displayed and choose **Bezier Corner** from the quad menu. Similarly, change the right-hand vertex to the bezier corner type.

10. Invoke the **Select and Move** tool. Now, move the handles on the top and right-hand vertices to produce a shape similar to that in the Front viewport, as shown in Figure P5-8. Next, choose the **Vertex** button to exit the sub-object mode.

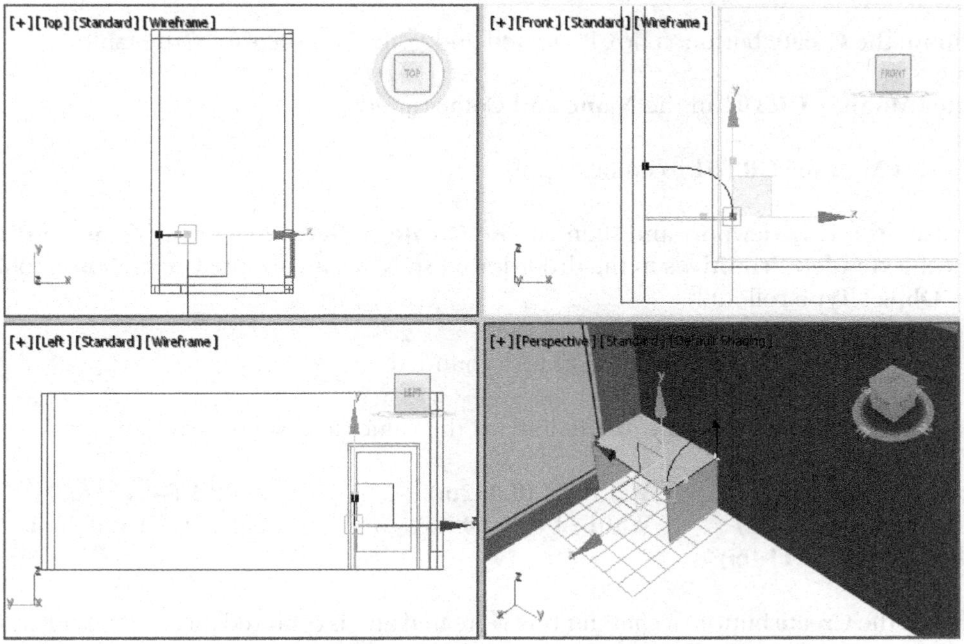

Figure P5-8 *Side01 spline modified in the Front viewport*

11. Make sure *Side01* is selected. Next, in the **Modifier List** drop-down list, select **Extrude** in the **OBJECT-SPACE MODIFIERS** section. In the **Parameters** rollout for the modifier, set **1"(0.025m)** in the **Amount** spinner.

A side partition is created on the right-hand side of the computer station. Each station has a side partition only on the right side which serves as the left-hand side partition for the next station.

Creating the Monitor

Now, a computer monitor will be created and placed on top of the computer station table. The monitor is created using three primitives: box, sphere, and chamfer box. These objects are modified to obtain the final shape.

1. Zoom in the computer station tabletop in the Top, Left, and Front viewports. Then, activate the Top viewport. Choose **Create > Geometry** in the **Command Panel**. Select **Standard Primitives** in the drop-down list. Invoke the **Box** tool from the **Object Type** rollout.

   It may be helpful to turn off the display of walls, door, and window.

2. In the **Parameters** rollout, set **2** in the **Width Segs** spinner. Next, in the **Keyboard Entry** rollout, set the values as given next:

   X: **1'9"(0.533m)** Y: **2'8"(0.813m)** Z: **2'6"(0.762m)**
   Length: **1'3"(0.381m)** Width: **1'1"(0.33m)** Height: **1'3"(0.381m)**

3. Choose the **Create** button; a box is created and placed on the top of the table.

4. Enter **Monitor CRT01** in the **Name and Color** rollout.

5. Convert **Monitor CRT01** to editable poly.

6. Activate the Left viewport and then choose **Create > Geometry** in the **Command Panel**. Select **Extended Primitives** in the drop-down list. Next, invoke the **ChamferBox** tool from the **Object Type** rollout.

   A chamfer box is similar to a box but with rounded edges.

7. Expand the **Keyboard Entry** rollout and set the values as given next:

   X: **–2'8"(–0.813m)** Y: **3'1 1/2"(0.952m)** Z: **–2'4"(–0.711m)**
   Length: **1'0"(0.305m)** Width: **1'0"(0.305m)** Height: **0'1"(0.025m)**
   Fillet: **0'4 1/2"(0.114m)**

8. Choose the **Create** button; a chamfer box is created and is centered on the monitor, as shown in Figure P5-9. The chamfer box is also centered on the front of the monitor.

9. Enter **CRT Tube01** in the **Name and Color** rollout.

10. Activate the Top viewport and then choose **Create > Geometry** in the **Command Panel**. Select **Standard Primitives** in the drop-down list. Next, invoke the **Sphere** tool from the **Object Type** rollout.

11. In the **Parameters** rollout, set **0.75** in the **Hemisphere** spinner.

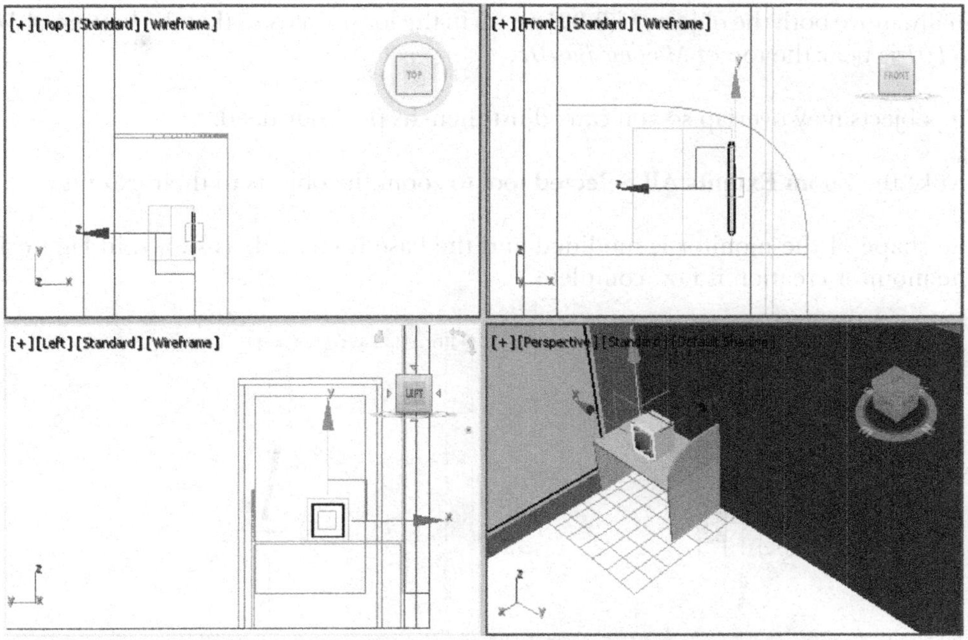

Figure P5-9 The chamfer box created and centered on the monitor

12. Expand the **Keyboard Entry** rollout and set the values as given next:

    X: **1'9"(0.533m)** Y: **2'8"(0.813m)** Z: **0'0"(0m)** Radius: **0'8"(0.203m)**

13. Choose the **Create** button; the object is created.

14. Enter **Monitor Base01** in the **Name and Color** rollout.

15. Invoke the **Select and Move** tool from the **Main Toolbar**. Next, select **Monitor Base01** and move it such that the base will rest on the top of the computer station, refer to Figure P5-10.

16. Select *Monitor CRT01* and *CRT Tube01* in the Top viewport and then right-click on the **Select and Move** tool; the **Move Transform Type-In** dialog box is displayed. Set **4"(0.101m)** in the **X** spinner in the **Offset: Screen** area; the objects are moved close to the front of the monitor.

17. Select the left side vertices in the Top viewport. Invoke the **Percent Snap Toggle** tool. Also, invoke the **Select and Uniform Scale** tool. Scale the vertices to 50%; the back of the monitor gets tapered.

18. Choose the **Vertex** button to exit the sub-object mode.

19. Activate the Front viewport and select *Monitor CRT01* and *CRT Tube01*. Next, invoke the **Select and Rotate** tool. Make sure the **Use Selection Center** tool is invoked in the transform center flyout. Rotate both objects 10 degrees about the local Z axis.

20. Finally, move both the objects **1"(0.025m)** up in the local Y axis so that the bottom of *Monitor CRT01* is near the top of *Monitor Base01*.

    The objects may overlap so you can adjust them as per your need.

21. Invoke the **Zoom Extents All Selected** tool to zoom the objects to their extents.

    The shape of the monitor is modified and the base is created, as shown in Figure P5-10. The monitor creation is now complete.

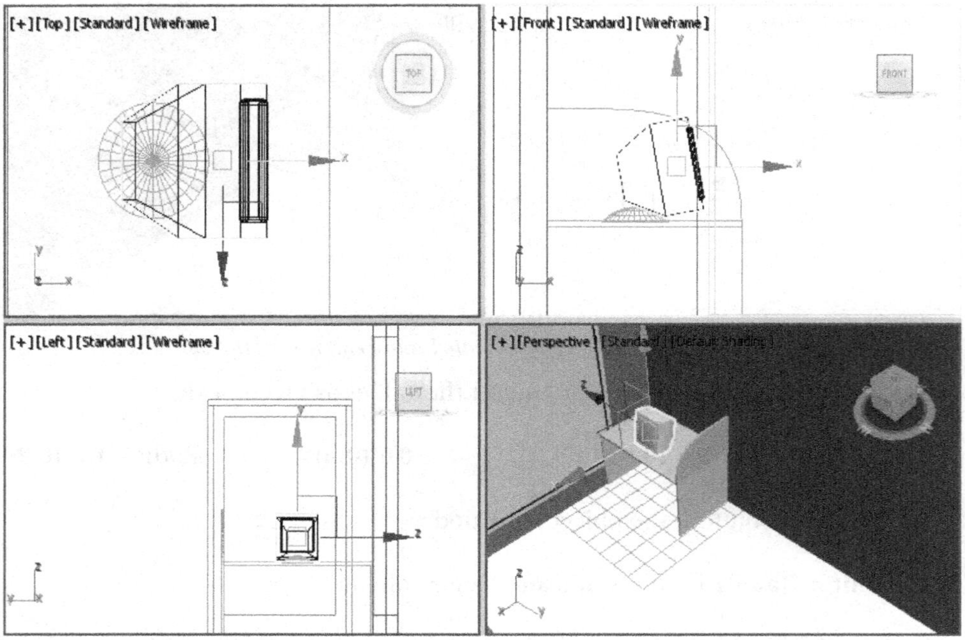

Figure P5-10 The shape of the monitor modified and the base created

Creating the Chair

In this section, you will create a chair. A loft object is used as a frame and two chamfer boxes are used as seat and back of the chair. The loft object is created by lofting two shapes, star and circle along a path.

1. Activate the Front viewport and zoom in the computer station. Choose **Create > Shapes** in the **Command Panel**. Invoke the **Line** tool from the **Object Type** rollout. You may need to maximize the viewport. Make sure the **Corner** radio button is selected in the **Initial Type** and **Drag Type** areas in the **Creation Method** rollout.

2. Expand the **Keyboard Entry** rollout and set the values as given next:

X: **4'0"(1.219m)**	Y: **0'0"(0m)**	Z: **–2'8"(–0.813m)**
X: **4'0"(1.219m)**	Y: **1'6"(0.457m)**	Z: **–2'8"(–0.813m)**
X: **5'0"(1.524m)**	Y: **1'6"(0.457m)**	Z: **–2'8"(–0.813m)**

X: **5'0"(1.524m)** Y: **3'0"(0.914m)** Z: **–2'8"(–0.813m)**

3. Choose the **Add Point** button after setting the values in each set of the XYZ coordinates.

4. Choose the **Finish** button; the path for the chair frame is created, as shown in Figure P5-11.

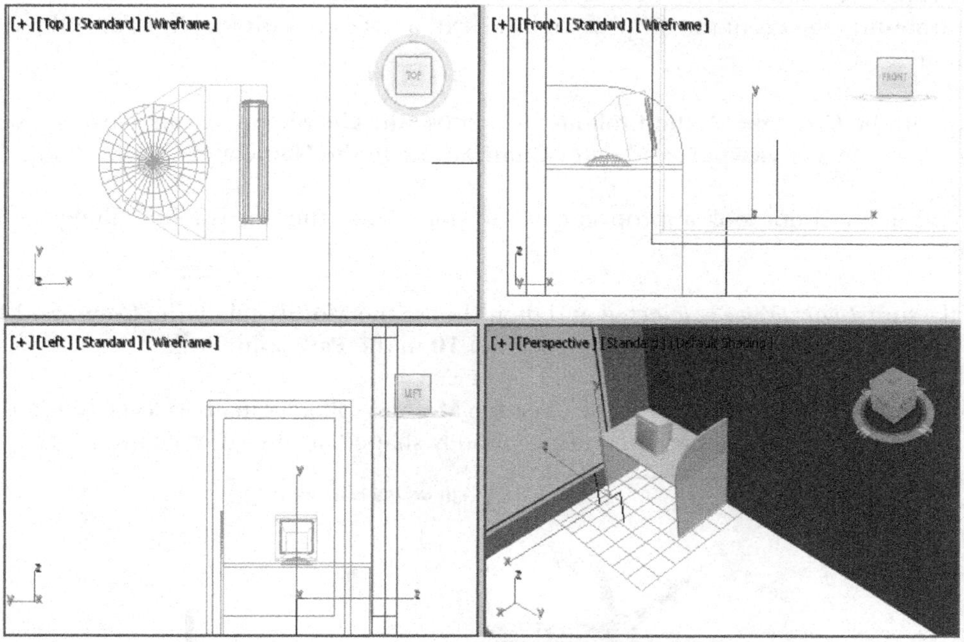

Figure P5-11 *The path created for the chair frame*

5. Enter **Chair Frame Path** in the **Name and Color** rollout.

6. Choose **Create > Shapes** in the **Command Panel**. Invoke the **Star** tool from the **Object Type** rollout.

7. In the **Parameters** rollout, set **5** in the **Points** spinner.

8. Expand the **Keyboard Entry** rollout and set the values as given next:

   Radius 1: **1'0"(0.305m)** Radius 2: **0'1 1/2"(0.038m)** Fillet Radius 1: **0'1"(0.025m)**
   Fillet Radius 2: **0'1"(0.025m)**

9. Choose the **Create** button; the shape is created.

10. Enter **Chair Base Shape** in the **Name and Color** rollout.

11. Choose **Create > Shapes** in the **Command Panel**. Invoke the **Circle** tool from the **Object Type** rollout.

12. Expand the **Keyboard Entry** rollout and set **1"(0.025m)** in the **Radius** spinner. Choose the **Create** button; the shape is created.

13. Enter **Chair Rail Shape** in the **Name and Color** rollout.

14. Select *Chair Frame Path* and then choose **Create > Geometry** in the **Command Panel**. Select **Compound Objects** in the drop-down list. Next, invoke the **Loft** tool from the **Object Type** rollout.

15. Expand the **Creation Method** rollout and choose the **Get Shape** button. Next, select *Chair Base Shape* in any viewport and enter **Chair Frame** in the **Name and Color** rollout.

    The frame will not look appropriate at this point since the star is lofted along the entire path.

16. Make sure *Chair Frame* is selected and then choose the **Modify** tab in the **Command Panel**. Expand the **Path Parameters** rollout and set **10** in the **Path** spinner.

17. Choose the **Get Shape** button in the **Creation Method** rollout. Finally, select *Chair Rail Shape* in any viewport; the frame is now appropriately shaped, as shown in Figure P5-12.

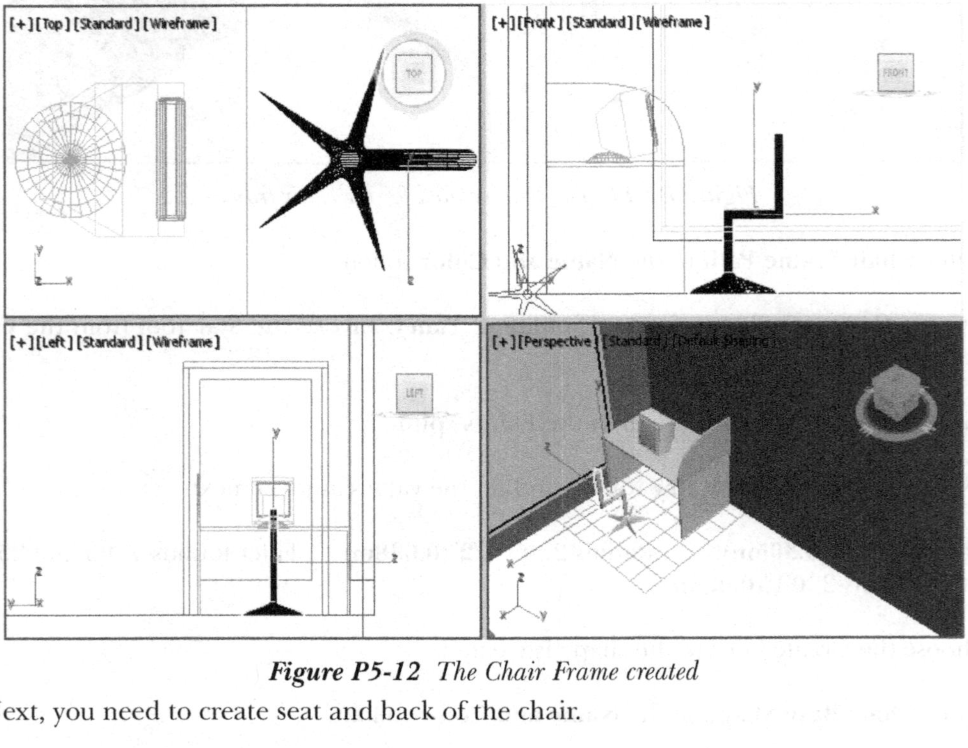

Figure P5-12 *The Chair Frame created*

Next, you need to create seat and back of the chair.

18. Activate the Top viewport. Choose **Create > Geometry** in the **Command Panel**. Next, select **Extended Primitives** in the drop-down list. Then, invoke the **Chamfer Box** tool from the **Object Type** rollout.

19. Expand the **Keyboard Entry** rollout and set the values as given next:

    X: **3'11"(1.194m)** Y: **2'8"(0.813m)** Z: **1'7"(0.483m)** Length: **1'6"(0.457m)**
    Width: **1'6"(0.457m)** Height: **0'3"(0.076m)** Fillet: **0'1"(0.025m)**

20. Choose the **Create** button; the object is created. Enter **Seat01** in the **Name and Color** rollout.

21. Make sure the **Chamfer Box** tool is activated. Next, activate the Left viewport and then set the following values in the **Keyboard Entry** rollout:

    X: **–2'8"(–0.813m)** Y: **3'0"(0.914m)** Z: **–4'11"(–1.499m)** Length: **0'8"(0.203m)**
    Width: **1'3"(0.381m)** Height: **0'3"(0.076m)** Fillet: **0'3"(0.076m)**

22. Choose the **Create** button; the chair is created, as shown in Figure P5-13.

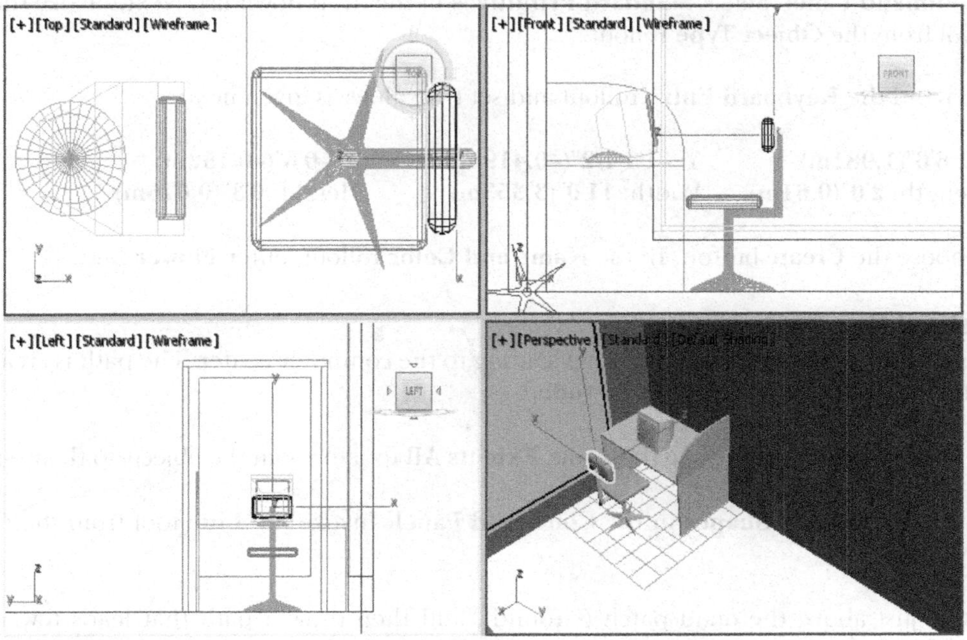

Figure P5-13 *The chair created*

23. Enter **Back01** in the **Name and Color** rollout.

Creating the Ground

In this section, you will create the ground and bed for plants. You will first create a quad patch for the ground in the scene and then you will create a box that will be used as a bed for plants in front of the window.

1. Activate the Top viewport. Choose **Create > Geometry** in the **Command Panel**. Select **Patch Grids** in the drop-down list. Choose the **Quad Patch** button from the **Object Type** rollout.

2. Expand the **Keyboard Entry** rollout and set the values as given next:

   X: **6'6"(1.981m)** Y: **11'9"(3.581m)** Z: **–0'6"(–0.152m)**
   Length: **150'(45.72m)** Width: **225'(68.58m)**

3. Choose the **Create** button; the floor is placed on top of the ground inside the computer center.

   Note that by setting the value **–0'6"(–0.152m)** for the Z coordinate, the floor is placed on top of the ground, thus creating a step into the computer center.

4. Enter **Ground** in the **Name and Color** rollout.

5. Make sure the Top viewport is activated and then choose **Create > Geometry** in the **Command Panel**. Select **Standard Primitives** in the drop-down list. Next, invoke the **Box** tool from the **Object Type** rollout.

6. Expand the **Keyboard Entry** rollout and set the values as given next:

   X: **6'6"(1.981m)** Y: **–1'4 1/2"(–0.419m)** Z: **–0'6"(–0.152m)**
   Length: **2'0"(0.61m)** Width: **11'0"(3.353m)** Height: **0'3"(0.076m)**

7. Choose the **Create** button. In the **Name and Color** rollout, enter **Flower Bed**.

Creating the Pathway

In this section, you will create a pathway leading to the computer center. The path is created as a spline that is modified and then extruded.

1. In the Top viewport, invoke the **Zoom Extents All** tool to zoom the objects to their extents.

2. Choose **Create > Shapes** in the **Command Panel**. Invoke the **Line** tool from the **Object Type** rollout.

3. Start just above the quad patch (Ground), and then draw a path that leads toward the window and curves around the room to the door, as shown in Figure P5-14. Next, close the spline. Enter the name of the spline as **Path**. To get the shape shown in Figure P5-14, you need to smoothen some of the vertices of *Path*.

4. Make sure the vertex sub-object mode is activated and then move and curve the vertices as needed. The two vertices of the door should meet the floor. After modifying the spline, exit the sub-object mode.

5. Make sure *Path* is selected and then choose the **Modify** tab in the **Command Panel**. Select **Extrude** in the **Modifier List** drop-down list under the **OBJECT-SELECTION MODIFIERS** section. In the **Parameters** rollout for the modifier, set **1'(0.025m)** in the **Amount** spinner and press the ENTER key; *Path* is extruded, as shown in Figure P5-15.

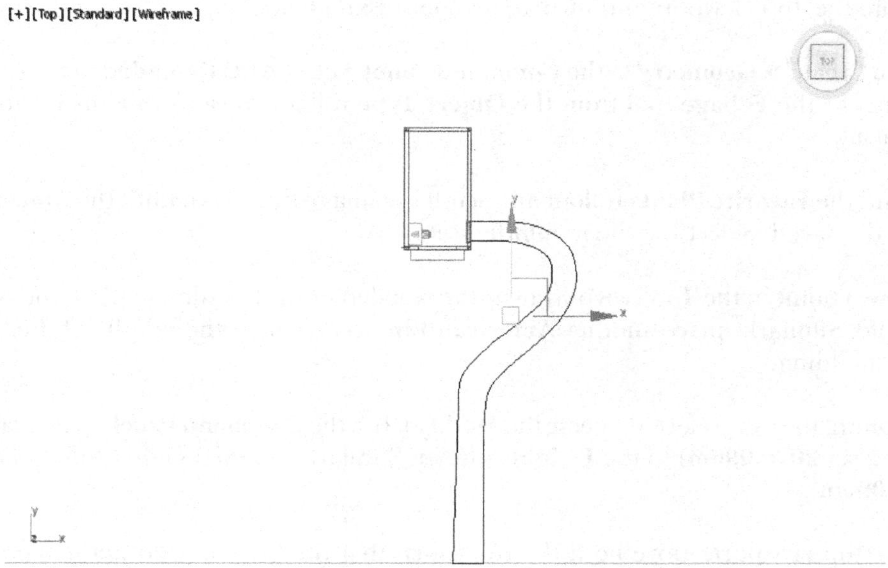

Figure P5-14 Path leading to computer center

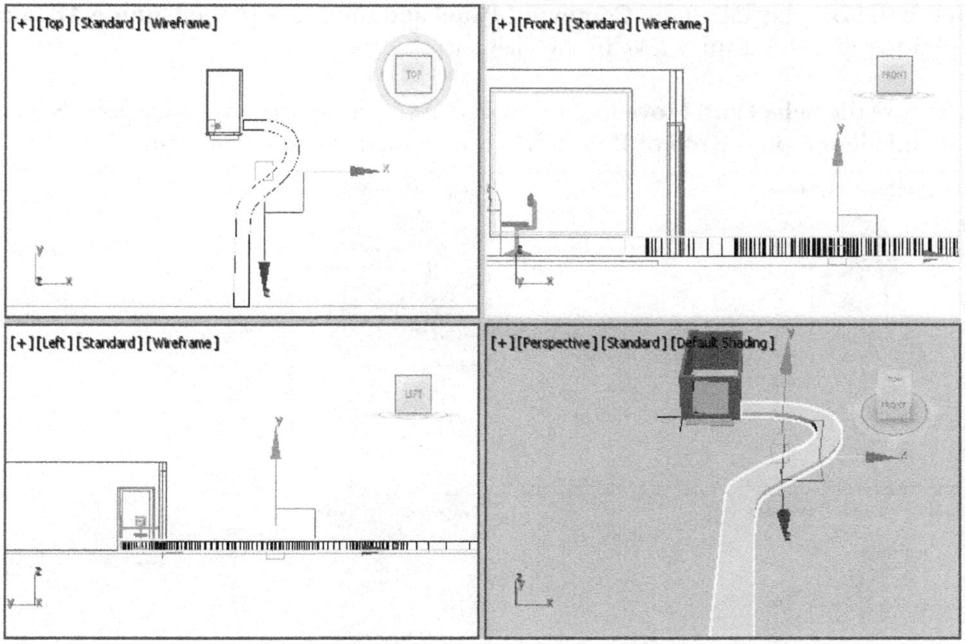

Figure P5-15 Path extruded

6. In the Front viewport, move *Path* –6"(–0.152m) on the Y axis so that it sets on the top of the ground.

Creating the Foliage

In this section, you will add foliage to the scene. 3ds Max has an automated foliage feature with several plants available in it. You will add two large trees and several small trees around the path.

1. Activate the Top viewport and invoke the **Zoom Extents** tool.

2. Choose **Create > Geometry** in the **Command Panel**. Select **AEC Extended** in the drop-down list. Invoke the **Foliage** tool from the **Object Type** rollout; various rollouts for foliage are displayed.

3. Expand the **Favorite Plants** rollout and scroll the image tiles down until the **American Elm** tile is displayed. Select the tile to highlight it.

4. Choose a point in the Top viewport near the rear-left of the building to place the **American Elm** tile. Similarly, place another **American Elm** foliage near the middle of the right side of the building.

5. With one of the trees selected, choose the **Modify** tab in the **Command Panel**. In the **Parameters** rollout, set **20'(6.096m)** in the **Height** spinner. Similarly, set the height of the other tree to **20'(6.096m)**.

6. In the Top viewport, move both the trees such that the branches do not extend into the building.

7. Choose the **Display** tab in the **Command Panel** and then choose the **Unhide All** button in the **Hide** rollout to display *Roof* in the viewport.

8. Make sure the **Select and Move** tool is invoked and activate the Front viewport. Next, select *Roof* and align it on the top of *Walls* if it is not aligned, as shown in Figure P5-16.

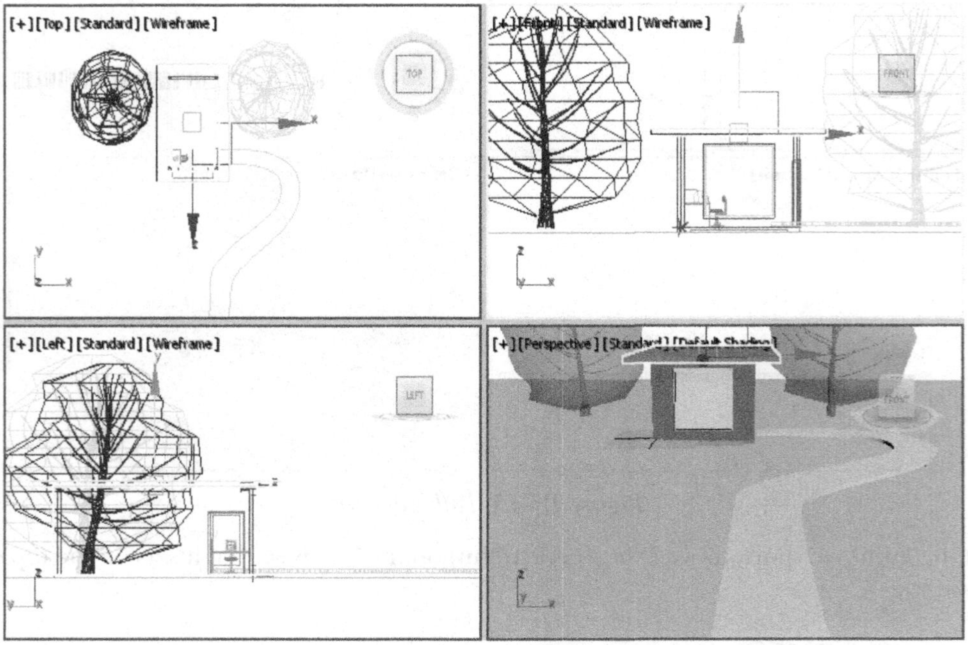

Figure P5-16 *The roof aligned on the top of walls*

9. In the Top viewport, zoom in on *Flower Bed* in front of the window.

10. Choose **Create > Geometry** in the **Command Panel**. Select **AEC Extended** in the drop-down list. Invoke the **Foliage** tool from the **Object Type** rollout.

11. Make sure the **Favorite Plants** rollout is expanded and scroll the image tiles down until the **Society Garlic** image tile is displayed. Select the tile to highlight it.

12. Choose a point near the center of *Flower Bed* to place a plant. Place four more plants, two on each side of the center plant.

13. With one of the plants selected, choose the **Modify** tab in the **Command Panel**. In the **Parameters** rollout, set **1'5"(0.305m)** in the **Height** rollout. Similarly, set the height of the other plants to **1'5"(0.305m)**, or as required.
    You may change the plant heights slightly to create a more realistic appearance.

14. In the Left viewport, move the plants so that they do not overlap with the wall or window. Move *Flower Bed* so that it is centered underneath the plants. Finally, move the plants down so that the bottom of the plant objects is at the same level as the top of *Flower Bed*.

    Next, you need to add more trees around the path to make the scene beautiful and realistic.

15. Expand the **Favorite Plants** rollout if not expanded and scroll the image tiles down until the **Scotch Pine** tile is displayed. Select the tile to highlight it.

16. Activate the Top viewport and place 8 Scotch Pine tiles around *Path*, 4 on the left side and 4 on the right side.

17. With the trees selected, choose the **Modify** tab in the **Command Panel**. Expand the **Parameters** rollout and set **12'0"(3.658m)** in the **Height** spinner. Similarly, set the height of the remaining trees to **12'0"(3.658m)**.

18. In the Top viewport, align the trees around *Path*, as shown in Figure 5-17.

Creating and Applying Multi/Sub-Object Materials to the Walls

The material applied to interior and exterior walls should be different. You will create multi/sub-object materials for the walls. You will then make sub-object selections on the walls and change the sub-object material IDs to achieve the desired result.

1. Choose the **Render Setup** tool from the **Main Toolbar** and make sure the **Render Setup: Scanline Renderer** dialog box is displayed. Next, close this dialog box.

2. Choose **Rendering > Material Editor > Compact Material Editor** from the menu bar; the **Material Editor** dialog box is displayed. Select one of the sample slots from the **Material Editor** dialog box. By default, the **Physical Material** is displayed on the **Material Type** button. Choose the **Physical Material** button; the **Material Map/Browser** dialog box is displayed. Choose **Materials > Scanline > Standard (Legacy)** from the **Material Map/Browser** dialog box and choose the **OK** button to close it.

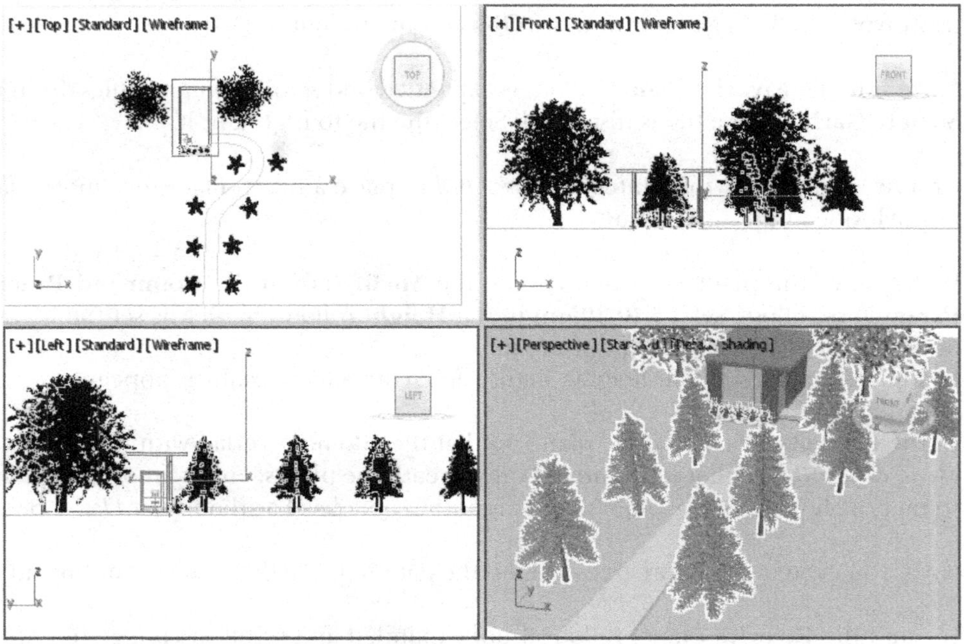

Figure P5-17 The trees aligned around Path

3. With the first material sample active, choose the **Diffuse** button from the **Blinn Basic Parameters** rollout; the **Material Map/Browser** dialog box is displayed.

4. Select **Bitmap** map from the **Maps > General** rollout. Next, choose the **OK** button; the **Select Bitmap Image File** dialog box is displayed. As the project folder is already set, the *images* folder is displayed in the **Look in** drop-down list.

5. Select the **Stucco.Brown.Jpg** file and then choose the **Open** button; the **Stucco.Brown** material is displayed in the first sample slot of the **Material Editor** dialog box.

6. In the **Coordinates** rollout, select the **Use Real-World Scale** check box, if it is not already selected. Set **5'(1.524m)** in the **Width: Size** and **Height: Size** spinners. Next, press the ENTER key.

7. Select the next unused material sample slot in the **Material Editor** dialog box. Choose the **Physical Material** button; the **Material Map/Browser** dialog box is displayed. Choose **Materials > Scanline > Standard (Legacy)** from the **Material Map/Browser** dialog box and choose the **OK** button to close it.

8. Choose the **Diffuse** button from the **Blinn Basic Parameters** rollout; the **Material/Browser** dialog box is displayed.

9. Select **Bitmap** map from the **Maps > General** rollout. Next, choose the **OK** button; the **Select Bitmap Image File** dialog box is displayed. As the project folder is already set, the *images* folder of this project is displayed in the **Look in** drop-down list.

10. Select the **Finishes.Painting.Paint.jpg** file and then choose the **Open** button; the **Finishes. Painting.Paint material** is displayed in the second sample slot of the **Material Editor** dialog box.

11. Choose the **Go to Parent** button. Next, choose the **Diffuse Color** swatch in the **Blinn Basic Parameters** rollout; the **Color Selector: Diffuse Color** dialog box is displayed.

12. Set the color values in the spinners as given next:

    Red: **192** Green: **220** Blue: **255**

13. Choose the **OK** button to close this dialog box.

14. Expand the **Maps** rollout and set **43** in the **Diffuse Color** spinner.

    This defines a light blue paint material that will be used on the interior walls of the building.

15. Select the next unused material sample slot in the **Material Editor** dialog box and choose the **Get Material** button; the **Material/Map Browser** dialog box is displayed. Choose the **Material/Map Browser Options** button from this dialog box; a flyout is displayed. Choose **Open Material Library** from the **Material/Map Browser Options** flyout; the **Import Material Library** dialog box is displayed.

16. Navigate to *C:\Program Files\Autodesk\3ds Max 2022\materiallibraries.*

17. Select the **AecTemplates.mat** material library and choose the **Open** button; the **Material/ Map Browser** dialog box is displayed.

18. In the **Material/Map Browser** window, double-click on the **Wall-Template (Multi/SubObject)** material.

19. With the **Wall-Template** material selected in the **Material Editor** dialog box, click and drag the **Stucco.Brown.Jpg** material to the first sub-material button in the **Walls-Template** material definition. Release the mouse button; the **Instance (Copy)** dialog box is displayed. Make sure the **Instance** radio button is selected and choose the **OK** button; an instance of the **Stucco.Brown.Jpg** material is assigned as the first sub-material in the **Walls-Template** material.

20. Replace the **Inside Wall**, **Top**, and **Bottom** sub-materials with instances of the **Stucco.Brown.Jpg** material using the drag-and-drop method described in the previous step. Also, replace the **Outside Wall** sub-material with an instance of the **Finishes.Painting.Paint** material.

21. In any viewport, select *walls* and then choose the **Assign Material to Selection** button in the **Material Editor** dialog box; the **Walls-Template** material is assigned to *Walls*.

Creating and Applying Multi/Sub-Object Materials to the Roof

The material applied to the interior and exterior of the roof should be different. You will create and apply multi/sub-object materials for the roof.

1. Select *Roof* and then select the next available sample slot in the **Material Editor** dialog box. Choose the **Physical Material** button; the **Material Map/Browser** dialog box is displayed. Choose **Materials > Scanline > Standard (Legacy)** from the **Material Map/Browser** dialog box and choose the **OK** button to close it.

2. Make sure that the **Blinn** shader is selected in the **Shader Basic Parameters** rollout.

3. Expand the **Maps** rollout and then choose the **Diffuse Color** button which is currently labeled as **No Map**; the **Material Map/Browser** dialog box is displayed.

4. Select **Bitmap** from the **Maps > General** rollout and then choose the **OK** button; the **Select Bitmap Image File** dialog box is displayed. As the project folder is already set, the *images* folder of this project is displayed in the **Look in** drop-down list.

5. Select the **RoofTiles.Jpg** file and choose the **Open** button to load it into the **Material Editor** dialog box. Next, choose the **Assign Material to Selection** button and then choose the **Show Shaded Material to Viewport** button; the material is assigned to *Roof*.

6. Make sure *Roof* is selected. Next, choose the **Modify** tab in the **Command Panel** and select **UVW Map** from the **Modifier List** drop-down list.

7. Select the **Box** radio button in the **Parameters** rollout and set **30' 3"** in the **Width** and **Height** spinners. The concrete material is applied to *Roof*.

Creating and Applying Multi/Sub-Object Materials to the Floor

In this section, you will create and apply multi/sub-object materials to the floor.

1. Select the next available material sample slot in the **Material Editor** dialog box. Choose the **Physical Material** button; the **Material Map/Browser** dialog box is displayed. Choose **Materials > Scanline > Standard (Legacy)** from the **Material Map/Browser** dialog box and choose the **OK** button to close it.

2. Make sure that the **Blinn** shader is selected in the **Shader Basic Parameters** rollout. Choose the **Diffuse** button in the **Blinn Basic Parameters** rollout; the **Material Map/Browser** dialog box is displayed.

3. Select **Bitmap** from the **Maps > General** rollout. Next, choose the **OK** button; the **Select Bitmap Image File** dialog box is displayed. As the project folder is already set, the *images* folder of this project is displayed in the **Look in** drop-down list.

4. Select the **Flooring.Tile.Square.Jpg** file and then choose the **Open** button to load it into the **Material Editor** dialog box. Next, choose the **Show Shaded Material in Viewport** button to activate it.

5. Make sure the **Coordinates** rollout is expanded. Also, make sure the **Use Real-World Scale** check box is cleared. Set **0.01** in the **U: Tiling** and **0.01** in the **V: Tiling** spinners in the **Coordinates** area.

6. Rename the material to **Carpet** and then select the next available material sample slot. Choose the **Physical Material** button; the **Material Map/Browser** dialog box is displayed Choose **Materials > Scanline > Standard (Legacy)** from the **Material Map/Browser** dialog box and choose the **OK** button to close it. Choose the **Diffuse Color** button currently labeled as **No Map** in the **Maps** rollout; the **Material Map/Browser** dialog box is displayed.

7. Select **Bitmap** from the **Maps > General** rollout. Next, choose the **OK** button; the **Select Bitmap Image File** dialog box is displayed. As the project folder is already set, the *images* folder of this project is displayed in the **Look in** drop-down list.

8. Select the **ConcreteFlatGrey.Jpg** file and then choose the **Open** button to load it into the **Material Editor** dialog box. Next, choose the **Show Shaded Material in Viewport** button to activate it. Name the material as **Concrete**.

9. Make sure the **Coordinates** rollout is expanded. Also, make sure the **Use Real-World Scale** check box is cleared. Set **1.5** in the **U: Tilling** and **6** in the **V: Tilling** spinners in the **Coordinates** area.

10. Select the *Carpet* material from the **Material Editor** dialog box. Choose the **Bitmap** button; the **Material Map/Browser** dialog box is displayed. Next, choose the **Multi/Sub-Object** material from the **Materials > General** rollout; the **Replace Material** dialog box is displayed.

11. In the **Replace Material** dialog box, select the **Keep old material as sub-material?** radio button and choose the **OK** button; the **Carpet** material is changed into the **Multi/Sub-Object** material. Note that the original material is loaded into the first sub-material slot.

12. In the **Multi/Sub-Object Basic Parameters** rollout, choose the **Set Number** button; the **Set Number of Materials** dialog box is displayed. Set **2** in the **Number of Materials** spinner and choose the **OK** button to exit the dialog box.

13. Drag and drop an instance of *Concrete* material onto the second sub-material sample slot; the **Instance(Copy) Material** dialog box is displayed. Select the **Copy** radio button and then choose **OK**.

14. In the Front viewport, select *Floor*. In the **Material Editor** dialog box, choose the **Assign Material to Selection** button to apply the material to *Floor*. Close the **Material Editor** dialog box.

15. Make sure *Floor* is selected and then choose the **Modify** tab in the **Command Panel**. Select the **Mesh Select** modifier from the **Selection Modifiers** section of the **Modifier List**.

16. In the **Mesh Select Parameters** rollout, choose the **Face** button. Next, invoke the **Select Object** tool from the **Main Toolbar**. In the Front viewport, pick a point below and to the right of *Floor*. Drag the cursor to a point to the left of and about halfway up *Floor*.
The faces on the sides and bottom of *Floor* are selected, but none of the faces are selected on the top.

17. Select **Material** from the **OBJECT-SPACE MODIFIERS** area in the **Modifier List**. Expand the **Parameters** rollout and set the value in the **Material ID** spinner to **2**.

    Next, you will change the names of the material objects.

Creating and Applying Remaining Materials

In this section, you will create and apply the remaining materials to the objects.

1. Select *Monitor Base01* and *Monitor CRT01* from the **Scene Explorer**.

2. Invoke the **Material Editor** dialog box by pressing the M key and select the next available sample slot. Choose the **Physical Material** button; the **Material Map/Browser** dialog box is displayed. Choose **Materials > Scanline > Standard (Legacy)** from the **Material Map/Browser** dialog box and choose the **OK** button to close it. Select **Blinn** from the drop-down list in the **Shader Basic Parameters** rollout if not already selected.

3. Choose the **Diffuse** color swatch in the **Blinn Basic Parameters** rollout; the **Color Selector:Diffuse Color** dialog box is displayed.

4. Set the color values in the spinners of this dialog box as given next and close the **Color Selector:Diffuse Color** dialog box:

    Red: **234** Green: **226** Blue: **211**

5. In the **Material Editor** dialog box, rename the material as **Computer_Case** and choose the **Assign Material to Selection** button to activate it.

6. In any of the viewport, select *CRT Tube01* and invoke the **Material Editor** dialog box if not already invoked.

7. Select the next available material sample slot. Choose the **Physical Material** button; the **Material Map/Browser** dialog box is displayed. Choose **Materials > Scanline > Standard (Legacy)** from the **Material Map/Browser** dialog box and choose the **OK** button to close it. Select **Blinn** from the drop-down list in the **Shader Basic Parameters** rollout, if it is not already selected.

8. Make sure the **Blinn Basic Parameters** rollout is expanded and set **75** in the **Opacity** spinner and then choose the **Diffuse** color swatch; the **Color Selector:Diffuse Color** dialog box is displayed.

9. Set **0** in the **Red**, **Green**, and **Blue** spinners and close the **Color Selector Diffuse Color** dialog box.

10. In the **Specular Highlights** area of the **Blinn Basic Parameters** rollout, set **95** in the **Specular Level** spinner and **45** in the **Glossiness** spinner.

11. Rename the material as **Computer_Screen** and then choose the **Assign Material to Selection**

button to assign the material to the object. Make sure the **Show Shaded Material in Viewport** button is chosen.

12. Load and assign materials to the remaining objects as given in the following table:

Object(s)	**Maps**
Ground	*Sitework.Planting.Grass.jpg*
Flower Bed	*Sitework.Planting.Soil.jpg*
Path	*Sitework.Paving & Surfacing.Pavers*
Table and Side01	*Woods & Plastics.Finish Carpentry.Wood.WhiteAsh.jpg*
Seat01and Back01	*Furnishings.Fabrics.Plaid.jpg*
Chair Frame	*Metals.Ornamental Metals.Chrome.Satin.jpg*
PivotDoor001	*Door_Template (Multi/Sub-Object)*
FixedWindow001	*Window_Template (Multi/Sub-Object)*

Mapping Coordinates

Now, all the materials are assigned. The objects, however, need their mapping coordinates adjusted to provide desired look to the materials.

1. Select *Ground* in the Top viewport and then choose the **Modify** tab in the **Command Panel**. In the **Modifier List** drop-down list, select **UVW Map** in the **OBJECT-SPACE MODIFIERS** section.

2. Make sure the **Planar** radio button is selected. Set **5'0"** in **Length** and **Width** spinners. If needed, set the mapping coordinates in the **U Tile** and **V Tile** spinners as per your requirements.

3. Select *Path* and apply the **UVW Map** modifier, refer to step1.

4. Similarly, set the required mapping coordinates for all other materials to get the desired output.

Copying the Furniture

Now with the **Array** tool, you will copy the station created to create remaining stations in the computer center.

1. Make sure the Top viewport is active and select *Back01, Chair Frame, CRT Tube01, Monitor Base01, Monitor CRT01, Seat01, Side01*, and *Table01* from the **Scene Explorer**.

2. Choose **Group > Group** from the menu bar; the **Group** dialog box is displayed. Enter **CompAssy01** in the **Group name** text box.

3. Choose **Tools > Array** from the menu bar; the **Array** dialog box is displayed, as shown in Figure P5-18.

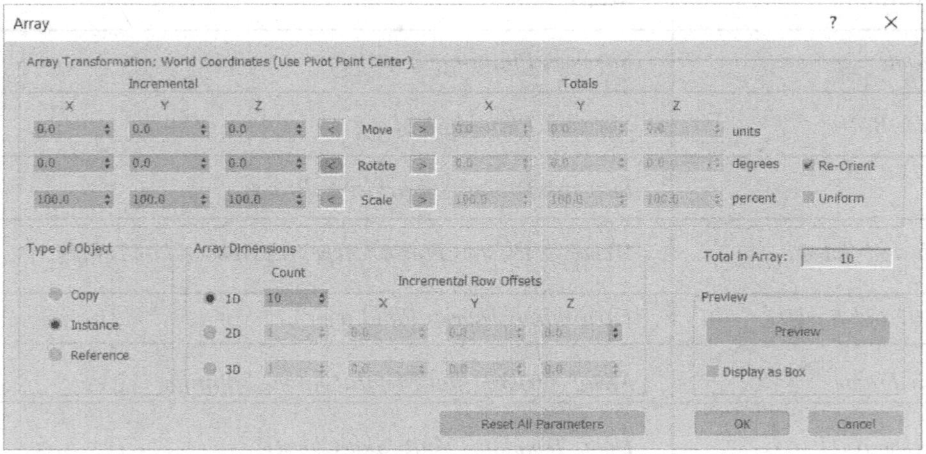

*Figure P5-18 The **Array** dialog box*

4. Set **4'7"(1.397m)** in the **Incremental Y** spinner in the **Move** row. In the **Type of Object** area, select the **Copy** radio button. In the **Array Dimensions** area, make sure the **1D** radio button is selected and then set **4** in the **Count** spinner. Choose the **OK** button; the array is created.

   There are now four computer stations inside the computer center on the left wall, as shown in Figure P5-19.

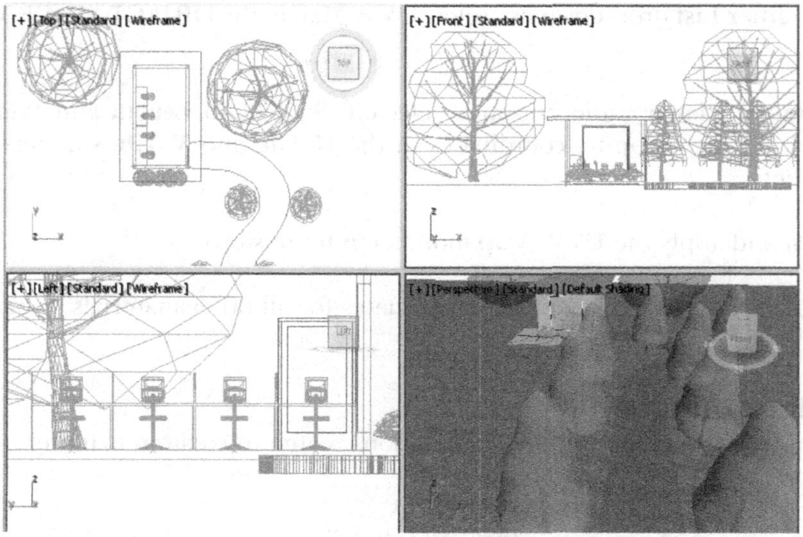

***Figure P5-19** The first row of computer stations created*

5. Hide *Chair Rail Shape* and *Chair Base Shape*.

6. Select all the four computer stations in the Top viewport. Invoke the **Use Selection Center** tool from the **Main Toolbar**. Next, invoke the **Mirror** tool; the **Mirror: Screen Coordinates** dialog box is displayed.

7. In the **Mirror Axis** area, select the **X** radio button. Also, set **7'4 1/2"(2.26m)** in the **Offset** spinner. Select the **Copy** radio button in the **Clone Selection** area if not selected.

8. Choose the **OK** button; the objects are mirrored and you will notice that four more computer stations are created on the right wall.

9. With four mirrored computer stations selected in the Top viewport, move them up **4'5"(1.378m)** on the local Y axis.

   The mirrored computer stations are aligned with the wall opposite the window, as shown in Figure P5-20. The inside of the computer center is completed.

   Make a note of the location of the door. If needed, move the door down in the Top viewport so that it does not overlap the first computer station.

Setting a Background

Now, you will set the background scene. You will begin by loading the map into the **Material Editor** dialog box. Then, you will adjust its position so that the desired portion of the map is visible above the horizon. Finally, you will load the map as background.

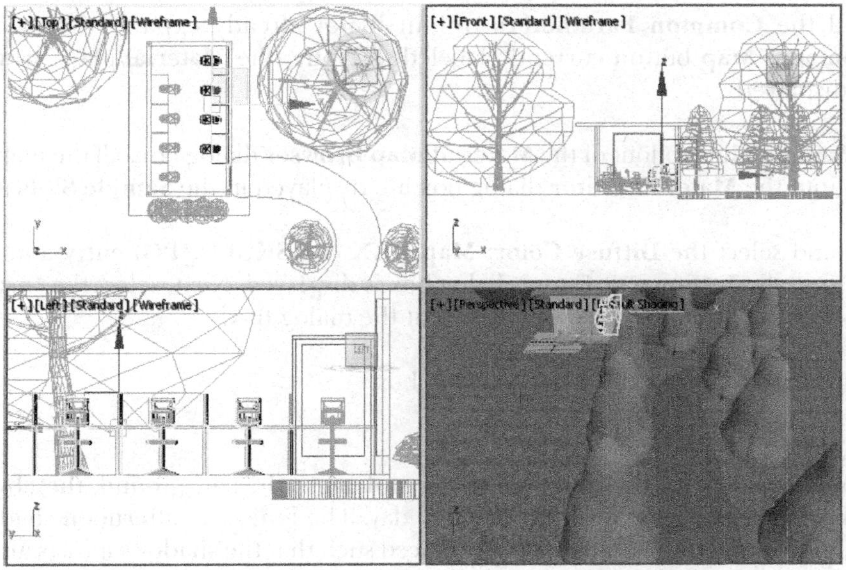

***Figure P5-20** The second row of computer stations created and aligned*

1. Invoke the **Material Editor** dialog box and select the next available material sample slot. Choose the **Physical Material** button; the **Material Map/Browser** dialog box is displayed.

Choose **Materials > Scanline > Standard (Legacy)** from the **Material Map/Browser** dialog box and choose the **OK** button to close it.

2. Select **Blinn** from the drop-down list in the **Shader Basic Parameters** rollout if not already selected.

3. Expand the **Maps** rollout and then choose the **Diffuse Color** map button currently labeled as **No Map**; the **Material/Map Browser** dialog box is displayed.

4. Select **Bitmap** from the **Maps > General** rollout and then choose the **OK** button; the **Select Bitmap Image File** dialog box is displayed. As the project folder is already set, the *images* folder of this project is displayed in the **Look in** drop-down list.

5. Select **DUSKCLD.JPG** and then choose the **Open** button to load it into the **Material Editor** dialog box. Next, choose the **Show Shaded Material in Viewport** button to activate it.

6. In the **Coordinates** rollout, select the **Environ** radio button and set **0.35** in the **V: Offset** spinner. Also, set **1.0** in the **U: Tiling** and **V: Tiling** spinners and close the **Material Editor** dialog box.
   This moves the map up so that the horizon in the map will roughly correspond with the horizon in the scene. You may need to fine tune the **V: Offset** spinner setting after the camera is added.

7. Choose **Rendering > Environment** from the menu bar; the **Environment and Effects** dialog box is displayed.

8. Expand the **Common Parameters** rollout if not already expanded and choose the **Environment Map** button currently labeled as **None**; the **Material/Map Browser** dialog box is displayed.

   In the **Sample Slots** rollout of the **Material/Map Browser** dialog box, all the maps currently loaded into the **Material Editor** dialog box are displayed in the **Sample Slots** area.

9. Locate and select the **Diffuse Color: Map #XX (DUSKCLD.JPG)** entry and choose the **OK** button; the **Instance or Copy?** dialog box is displayed. Next, select the **Instance** radio button if not selected and choose **OK** to exit the dialog box.

10. Close the **Environment and Effects** dialog box.

Adding Lights

You will now add lights to the scene that will complement the background, thereby creating a lighting effect simulating the dusk on a cloudy day. The long, late afternoon shadows will be cast by a single omni light. This light will be placed such that the shadows it casts are consistent with the sun's position in the background image. Two additional omni lights are added inside the building to create the appearance of fluorescent lighting.

1. Activate the Top viewport and choose **Create > Lights** in the **Command Panel**. Select the **Standard** option in the drop-down list. Invoke the **Omni** tool from the **Object Type** rollout; various rollouts are displayed.

2. Expand the **General Parameters** rollout and make sure the **On** check box in the **Shadows** area is selected. Next, select **Ray Traced Shadows** in the drop-down list under the **Shadows** area if not selected.

   Ray-traced shadows are required to cast illumination through transparent materials, such as the window glass.

3. Expand the **Intensity/Color/Attenuation** rollout and make sure **1** is set as the value in the **Multiplier** spinner. Select the color swatch next to the **Multiplier** spinner; the **Color Selector:Light Color** dialog box is displayed.

4. Set the color values in the spinners given next:

   Red: **161** Green: **152** Blue: **139**

5. Choose the **OK** button in the **Color Selector:Light Color** dialog box; the color of the omni light is adjusted to simulate the color of light produced by the setting sun.
6. In the Top viewport, select a point on immediate right of the building and just above *Ground*. In the Front viewport, move the omni light so that it is at the top of the canopy of the tree.

7. In the **Name and Color** rollout, rename the light as **Sun**.

   This location approximates the location of the *Sun* in the background map. From this location, the omni light will cast shadows and create highlights that are consistent with the background.

8. Expand the **Shadow Parameters** rollout and set **0.7** in the **Dens** spinner under the **Object Shadows** area. Also, use the **Color** swatch to set the color gray.

9. Choose **Create > Lights** in the **Command Panel** and then invoke the **Omni** tool from the **Object Type** rollout.

10. Expand the **Intensity/Color/Attenuation** rollout and choose the color swatch next to the **Multiplier** spinner; the **Color Selector:Light Color** dialog box is displayed.

11. Set the color values in the spinners as given next:

    Red: **226** Green: **237** Blue: **250**

12. Close the **Color Selector:Light Color** dialog box.

13. In the Top viewport, select a point centered left and right and about a third of the way up from the bottom of the computer center and then move the light in the Left viewport so

that it is near the ceiling. In the **Name and Color** rollout, enter the name of the omni light as **Fluorescent01**.

14. In the Top viewport, select the omni light you just created. Next, hold the SHIFT key and drag the omni light slightly down from the top of the computer center; a copy of the light is created, as shown in Figure P5-21.

15. Choose **Rendering > Environment** from the menu bar or press 8; the **Environment and Effects** dialog box is displayed. In the **Common Parameters** rollout, choose the **Ambient** color swatch; the **Color Selector:Ambient Color** dialog box is displayed.

16. Set the ambient light values in the spinners as given next:

    Red: **35** Green: **35** Blue: **35**

17. Close the **Color Selector:Ambient Color** and **Environment and Effects** dialog boxes.

    A small amount of ambient light is added to the scene to simulate the ambient light of an overcast late afternoon/early evening. The ambient light brings out a little bit of color and detail in the areas that are not directly illuminated by the lights in the scene.

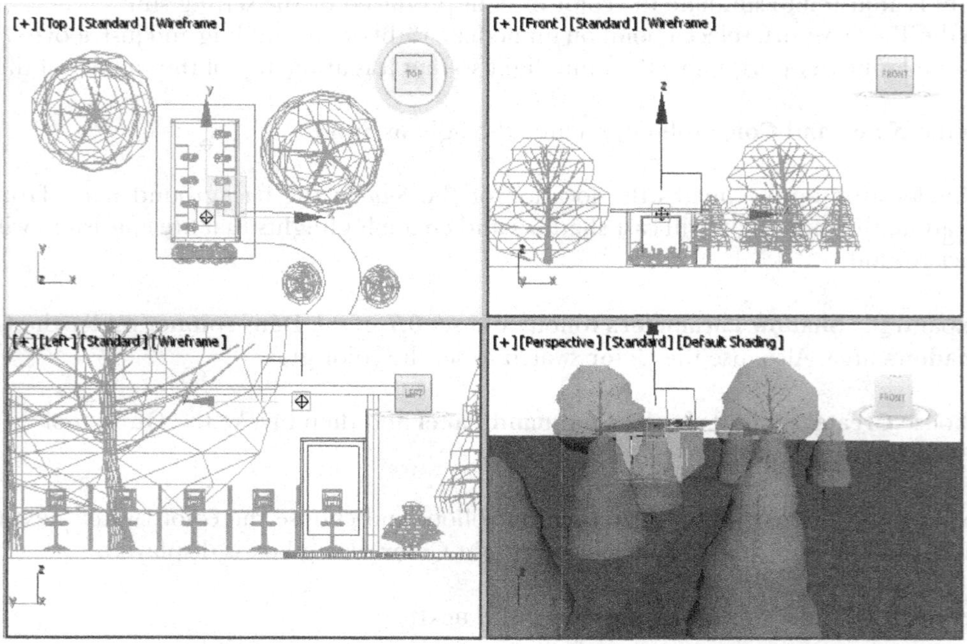

***Figure P5-21** The omni lights created and aligned in the viewports*

18. Activate the Perspective viewport and then invoke the **Render Production** tool from the **Main Toolbar** to see the effects of the lights. Make adjustments in the view and lights as needed.

Setting the Animation Length

In this section, you will create a path for a walkthrough animation. You will also animate the entrance door. First, set the total number of frames in the animation.

1. Choose the **Time Configuration** button; the **Time Configuration** dialog box is displayed.

2. In the **Animation** area, set **1000** in the **Length** spinner and then choose the **OK** button.

   The number of frames is set to 1000, which is about 33 seconds of playback.

Animating the Camera along the Path

1. Activate the Top viewport and choose **Create > Shapes** in the **Command Panel**. Next, invoke the **Line** tool from the **Object Type** rollout.

2. Choose a series of points in the Top viewport to create a line that roughly follows the path, as shown in Figure P5-22.

   The line should also pass through the door and enter the computer center. In the **Name and Color** rollout, enter **Camera Path**.

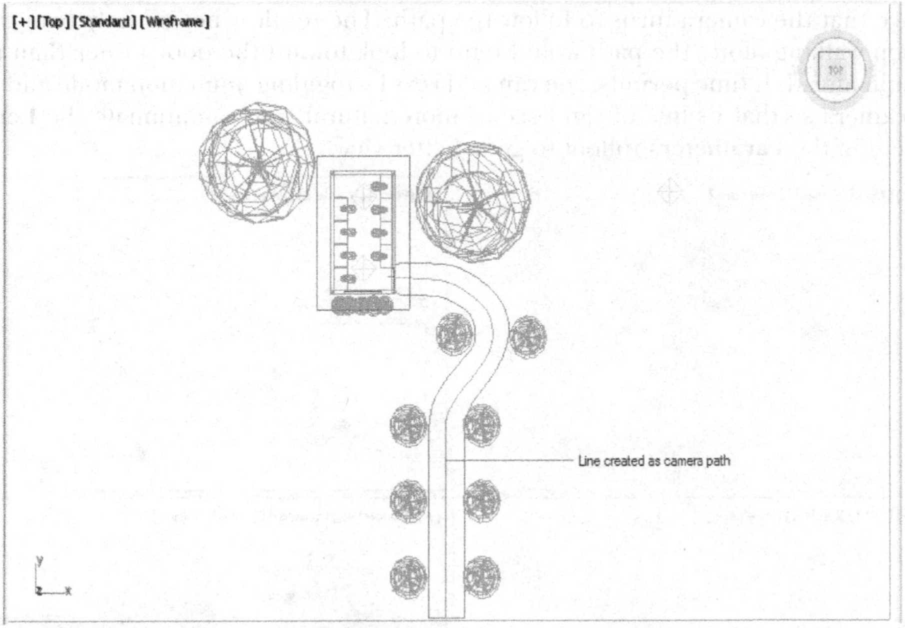

***Figure P5-22** Line created as Camera Path*

3. In the Left viewport, move the *Camera Path* up by **5'6"(1.676m)**.

4. With *Camera Path* selected, choose the **Modify** tab in the **Command Panel**. Choose the **Vertex** button in the **Selection** rollout to enter the vertex sub-object mode. Modify the path as needed to create a smooth curvature. Then choose the **Vertex** button again to exit the sub-object mode.

5. Choose **Create > Cameras** in the **Command Panel** and then invoke the **Free** tool from the **Object Type** rollout. In the **Parameters** rollout, choose the **35mm** button in the **Stock Lenses** area.

6. Place the camera anywhere in the Front viewport.

7. Choose **Animation > Constraints > Path Constraint** from the menu bar.

8. In the Top viewport, select *Camera Path* created for the camera; the camera is placed at the start point of *Camera Path*, as shown in Figure P5-23.

9. Choose the **Parameters** button in the **Selection Level** area, if it is not chosen. In the **Path Parameters** rollout, select the **Follow** check box in the **Path Options** area and the **Y** radio button in the **Axis:** area.

10. Activate the Perspective viewport and press the C key to switch to the Camera01 viewport.

11. Choose the **Play Animation** button to view the animation.

    Notice that the camera turns to follow the path. The result is not entirely realistic. A real person walking along the path would tend to look toward the door rather than perfectly straight ahead. If time permits, you can add keys by toggling animation mode and rotating the camera so that its line of sight seems more natural. You can animate the **Lens** of the camera in the **Parameters** rollout to get a better view.

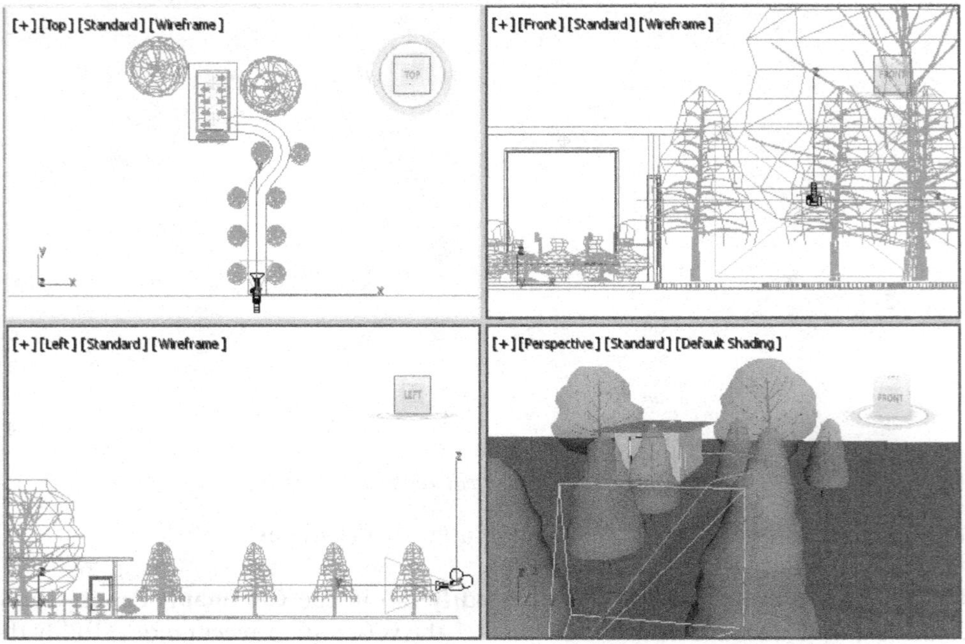

Figure P5-23 *The camera placed at the start point of Camera Path*

12. Move the **Time Slider** right and left to preview the animation. Determine the frame on which the camera is ready to enter the door.

Animating the Door

When you preview the animation, you may notice that the camera is passing through the door. Therefore, the door needs to be animated so that the camera can enter the computer center without passing through the door.

1. Move to the frame at which the camera is about to enter the computer center.

2. Choose the **Toggle Auto Key Mode** button to enter the animation mode.

3. Activate the Top viewport and select the door. Next, choose the **Modify** tab in the **Command Panel**.

4. In the **Parameters** rollout, set **90** in the **Open:** spinner.

5. Choose the **Toggle Auto Key Mode** button to exit the animation mode.
   The door now begins to open at frame 0. The animation needs to be adjusted so that the door remains closed until the camera approaches it.
6. With the door selected, highlight the key at frame 0 in the track bar below the time slider. Move the key to about 100 frames before the keyframe where the door is fully open.

Rendering the Final Animation

In this section, you will render the final animation.

1. Make sure the Camera viewport is activated and then invoke the **Render Setup** tool from the **Main Toolbar**; the **Render Setup: Scanline Renderer** dialog box is displayed.

2. In the **Time Output** area, select the **Active Time Segment** radio button. In the **Output Size** area, choose the **320×240** button.

3. In the **Render Output** area, choose the **Files** button; the **Render Output File** dialog box is displayed. As the project folder is already set, the *renderoutput* folder of this project is displayed in the **Save in** drop-down list.

4. Select **AVI File (*.avi)** from the **Save as type** drop-down list. In the **File name:** text box, enter **Project05** and then choose the **Save** button.

   If the **AVI File Compression Setup** dialog box appears, choose the **OK** button to accept the default settings.

5. In the **Render Setup** dialog box, choose the **Render** button.

   It will take some time to render the animation. After the rendering is complete, close the **Rendered Frame** window and the **Render Setup** dialog box.

6. Navigate to the *renderoutput* folder of the *P5_1* project and double-click on the **Project05.avi** file; the AVI file starts playing, as shown in Figure P5-24.

***Figure P5-24** The final animation being played leading the path to the computer center*

Student Project

PROJECT DESCRIPTION

Create a walkthrough animation in a house that consists of a drawing room, bedroom, kitchen, dining room, drawing room, bathroom, and lawn; refer to Figures SP-3 to SP-19.

To create the walkthrough animation, you need to download the AutoCAD drawing file *student_proj_3dsmax_2022.zip* from the CADCIM website, refer to Figure SP-1. This file will give you an idea about the dimensions of the objects in the project. The path of the file is as follows: *Textbooks > Animation and Visual Effects > Autodesk 3ds Max > Autodesk 3ds Max 2022 for Beginners: A Tutorial Approach*

Hint:
To measure the dimensions of the rooms in the house, you need to import the AutoCAD drawing file in 3ds Max 2022, refer to Figure SP-2. To do so, follow the steps given next.

1. Choose **File > Import** from the menu bar; the **Select File to Import** dialog box is displayed. In the **Files of type** drop-down list of this dialog box, select the **AutoCAD Drawing (*.DWG,*.DXF)** option. Browse to the *student_project_drawing.dwg* file that you have downloaded from the CADCIM website. Choose the **Open** button; the **AutoCAD DWG/DXF Import Options** dialog box is displayed. Use the default settings and choose the **OK** button; the AutoCAD drawing file, *student_project_drawing.dwg,* is displayed in all viewports.

2. Click anywhere in the viewport and press CTRL+A keys; all parts of the AutoCAD drawing file are selected. Now, group them as *reference drawing*.

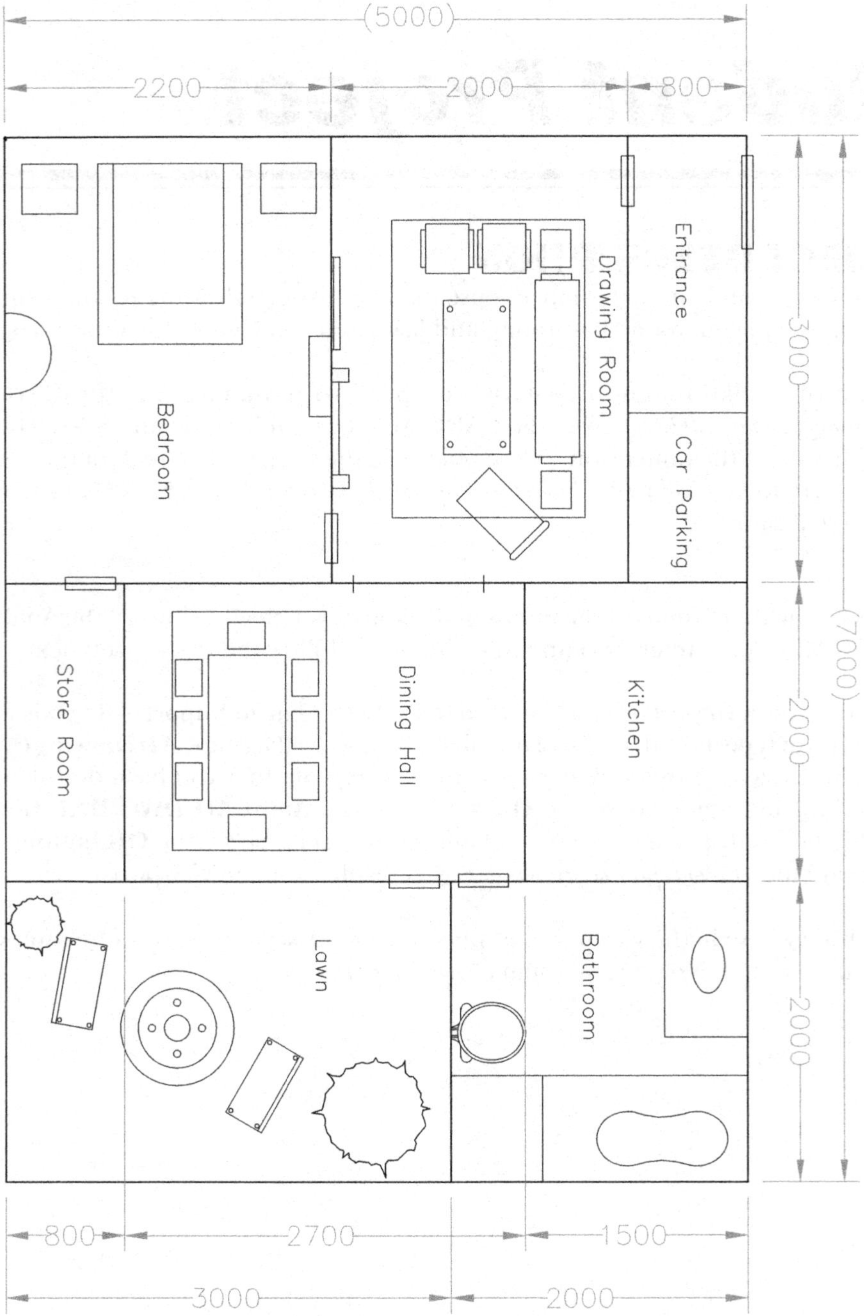

Figure SP-1 The AutoCAD drawing for the student project (house)

3. Select the *reference drawing* in the Left viewport and right-click on the **Select and Rotate** tool in the **Main Toolbar**; the **Rotate Transform Type-In** dialog box is displayed. In the **Offset: Screen** area, set the value **90** in the X spinner; the *reference drawing* gets rotated.

4. Now, create the scene in the Top viewport using the dimensions given in the *reference drawing* file.

You can view the rendered images of the scene by downloading them from *www.cadcim.com*. The path of the files is as follows: *Textbooks > Animation and Visual Effects > Autodesk 3ds Max > Autodesk 3ds Max 2022 for Beginners: A Tutorial Approach*

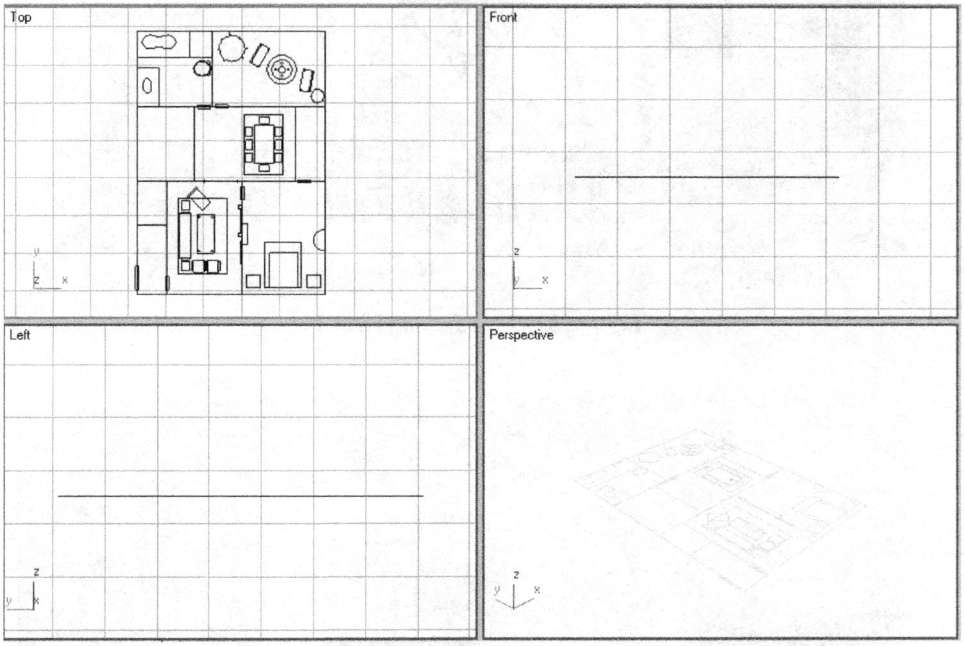

Figure SP-2 *The drawing file imported to Autodesk 3ds Max 2022*

Figure SP-3 *Bedroom (View 1)*

Figure SP-4 *Bedroom (View 2)*

Figure SP-5 *Bedroom (View 3)*

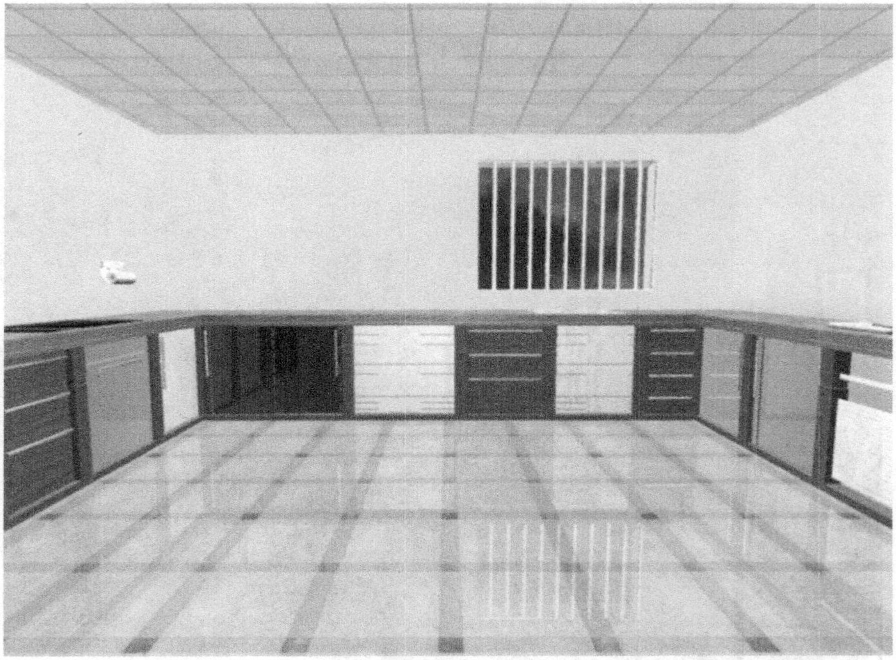

Figure SP-6 *Kitchen*

Figure SP-7 *Dining room (View 1)*

Figure SP-8 *Dining room (View 2)*

Figure SP-9 *Dining room (View 3)*

Figure SP-10 *Drawing room (View 1)*

Figure SP-11 *Drawing room (View 2)*

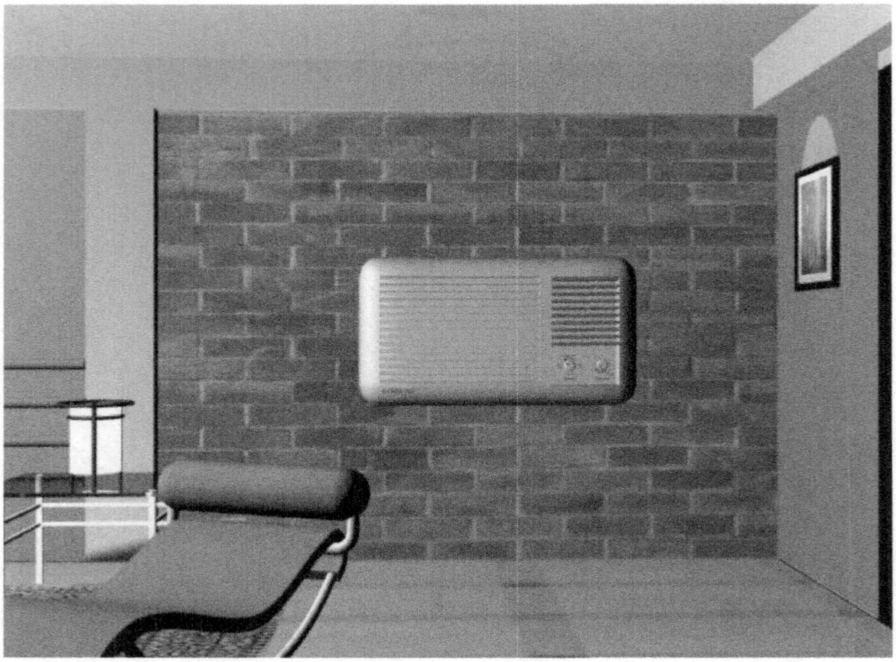

Figure SP-12 *Drawing room (View 3)*

Figure SP-13 Bathroom (View 1)

Figure SP-14 Bathroom (View 2)

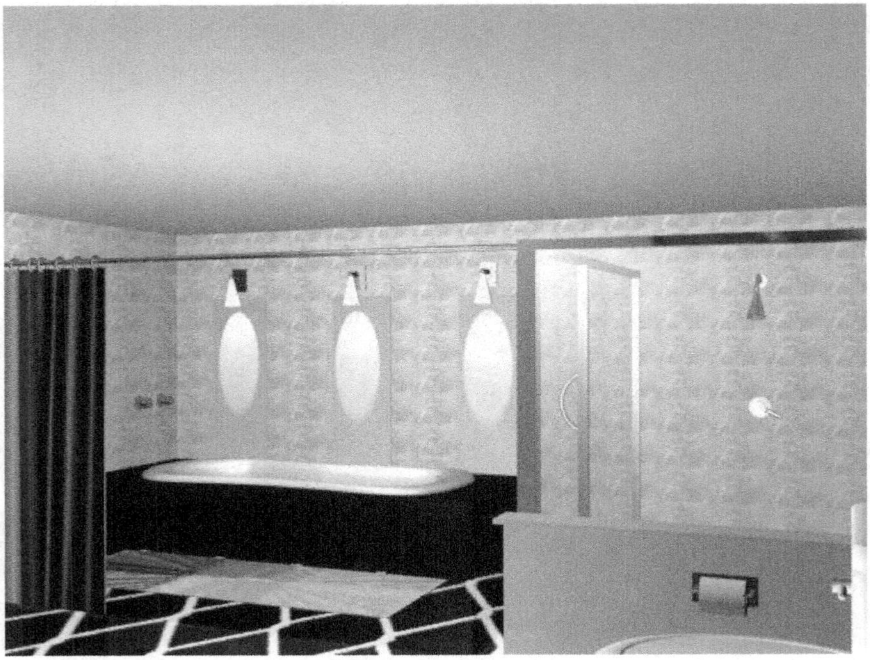

Figure SP-15 *Bathroom (View 3)*

Figure SP-16 *Bathroom (View 4)*

Figure SP-17 *Bathroom (View 5)*

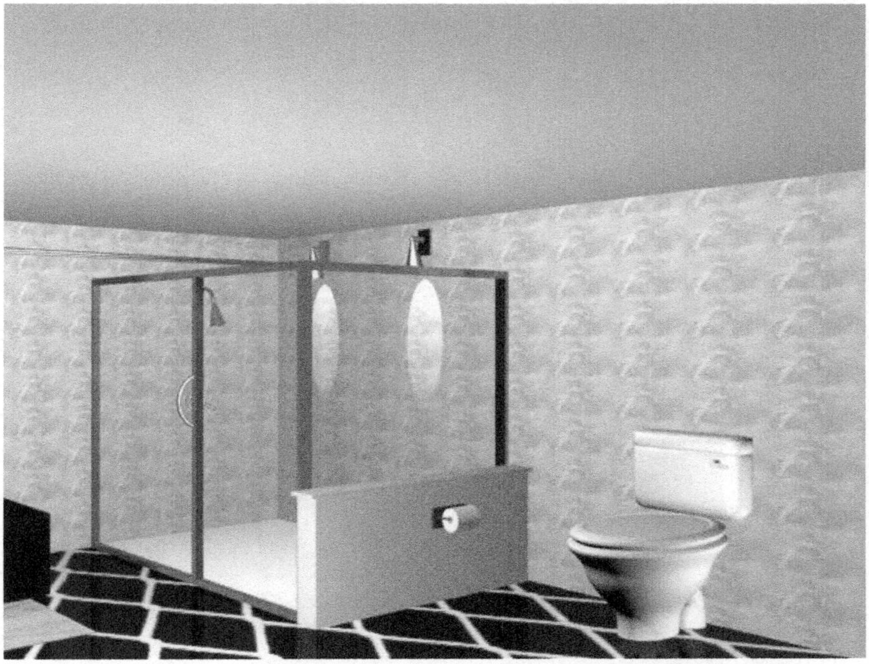

Figure SP-18 *Bathroom (View 6)*

Figure SP-19 *The lawn*

Index

F

G

H

I

K

L

M

N

O

P

Q

R

S

This page is intentionally left blank

Other Publications by CADCIM Technologies

The following is the list of some of the publications by CADCIM Technologies. Please visit *www.cadcim.com* for the complete listing.

Autodesk 3ds Max Textbooks

- Autodesk 3ds Max 2021 for Beginners: A Tutorial Approach, 21st Edition
- Autodesk 3ds Max 2020 for Beginners: A Tutorial Approach, 20th Edition
- Autodesk 3ds Max 2019 for Beginners: A Tutorial Approach, 19th Edition
- Autodesk 3ds Max 2018 for Beginners: A Tutorial Approach, 18th Edition
- Autodesk 3ds Max 2017 for Beginners: A Tutorial Approach, 17th Edition
- Autodesk 3ds Max 2016 for Beginners: A Tutorial Approach, 16th Edition
- Autodesk 3ds Max 2022: A Comprehensive Guide, 22nd Edition
- Autodesk 3ds Max 2021: A Comprehensive Guide, 21st Edition
- Autodesk 3ds Max 2020: A Comprehensive Guide, 20th Edition
- Autodesk 3ds Max 2019: A Comprehensive Guide, 19th Edition
- Autodesk 3ds Max 2018: A Comprehensive Guide, 18th Edition
- Autodesk 3ds Max 2017: A Comprehensive Guide, 17th Edition

Autodesk 3ds Max Design Textbooks

- Autodesk 3ds Max Design 2015: A Tutorial Approach, 15th Edition
- Autodesk 3ds Max Design 2014: A Tutorial Approach
- Autodesk 3ds Max Design 2013: A Tutorial Approach
- Autodesk 3ds Max Design 2012: A Tutorial Approach
- Autodesk 3ds Max Design 2011: A Tutorial Approach

Autodesk Maya Textbooks

- Autodesk Maya 2022: A Comprehensive Guide, 13th Edition
- Autodesk Maya 2020: A Comprehensive Guide, 12th Edition
- Autodesk Maya 2019: A Comprehensive Guide, 11th Edition
- Autodesk Maya 2018: A Comprehensive Guide, 10th Edition
- Autodesk Maya 2017: A Comprehensive Guide, 9th Edition
- Autodesk Maya 2016: A Comprehensive Guide, 8th Edition
- Autodesk Maya 2015: A Comprehensive Guide, 7th Edition

CINEMA 4D Textbooks

- MAXON CINEMA 4D S24: A Tutorial Approach, 8th Edition
- MAXON CINEMA 4D Studio R20: A Tutorial Approach, 7th Edition
- MAXON CINEMA 4D Studio R19: A Tutorial Approach, 6th Edition
- MAXON CINEMA 4D Studio R18: A Tutorial Approach, 5th Edition

Digital Modeling Textbook

- Exploring Digital Modeling Using 3ds Max and Maya 2015

Blender Textbook
• Blender 2.79 for Digital Artists

Fusion Textbooks
• Blackmagic Design Fusion 7 Studio: A Tutorial Approach, 3rd Edition
• The eyeon Fusion 6.3: A Tutorial Approach

Flash Textbooks
• Adobe Flash Professional CC 2015: A Tutorial Approach, 3rd Edition
• Adobe Flash Professional CC: A Tutorial Approach
• Adobe Flash Professional CS6: A Tutorial Approach

ZBrush Textbooks
• Pixologic ZBrush 2021: A Comprehensive Guide, 7th Edition
• Pixologic ZBrush 2020: A Comprehensive Guide, 6th Edition
• Pixologic ZBrush 2018: A Comprehensive Guide, 5th Edition
• Pixologic ZBrush 4R8: A Comprehensive Guide, 4th Edition

Premiere Textbooks
• Adobe Premiere Pro CC: A Tutorial Approach
• Adobe Premiere Pro CS6: A Tutorial Approach

Nuke Textbook
• The Foundry NukeX 7 for Compositors

Computer Programming Textbooks
• Introducing PHP7/MySQL
• Introduction to C++ Programming, 2nd Edition
• Learning Oracle 12c: A PL/SQL Approach
• Learning ASP.NET AJAX
• Introduction to Java Programming, 2nd Edition
• Learning Visual Basic.NET 2008

Coming Soon from CADCIM Technologies
• Flow Simulation using SOLIDWORKS 2021
• Finite Element Analysis Using ANSYS Workbench 2021

Online Training Program Offered by CADCIM Technologies
CADCIM Technologies provides effective and affordable virtual online training on various software packages including Computer Aided Design, Manufacturing, and Engineering (CAD/CAM/CAE), Computer Programming and Applications, GIS, Civil, Animation and Visual Effects, and related technologies. The training will be delivered 'live' via Internet at any time, any place, and at any pace to individuals as well as the students of colleges, universities, and CAD/CAM/CAE training centers. For more information, please visit the following link: ***www.cadcim.com***. CADCIM Technologies also provides consulting services in these domains.

Made in the USA
Monee, IL
08 November 2026

68028678R00367